Introduction to Soil Mechanics

About the companion website

This book's companion website is at www.wiley.com/go/bodo/soilmechanics and offers invaluable resources for students and lecturers:

- Supplementary problems
- Solutions to supplementary problems

www.wiley.com/go/bodo/soilmechanics

Introduction to Soil Mechanics

Béla Bodó and Colin Jones

WILEY Blackwell

Registered Office
John Wiley & Sons, Ltd, The Atrium, Southern Gate, Chichester, West Sussex, PO19 8SQ, United Kingdom.

Editorial Offices
9600 Garsington Road, Oxford, OX4 2DQ, United Kingdom.
The Atrium, Southern Gate, Chichester, West Sussex, PO19 8SQ, United Kingdom.

For details of our global editorial offices, for customer services and for information about how to apply for permission to reuse the copyright material in this book please see our website at www.wiley.com/wiley-blackwell.

Library of Congress Cataloging-in-Publication Data

Bodo, Bela, (Engineer)
Introduction to soil mechanics / Bela Bodo, Colin Jones.
pages cm
Includes bibliographical references and index.
ISBN 978-0-470-65943-4 (pbk. : alk. paper) – ISBN 978-1-118-55387-9 (emobi) – ISBN 978-1-118-55388-6 (epub) – ISBN 978-1-118-55389-3 (epdf) 1. Soil mechanics. I. Title.
TA710.B617 2013
624.1′5136–dc23

2012040913

A catalogue record for this book is available from the British Library.

Wiley also publishes its books in a variety of electronic formats. Some content that appears in print may not be available in electronic books.

Cover image courtesy of Shuttlestock.com
Cover design by Steve Thompson

1 2013

Contents

Appendices

About the companion website

This book's companion website is at www.wiley.com/go/bodo/soilmechanics and offers invaluable resources for students and lecturers:

- Supplementary problems
- Solutions to supplementary problems

www.wiley.com/go/bodo/soilmechanics

Preface

This book is intended to introduce the subject to students studying for BTEC Higher National Certificate/Diploma in Civil Engineering and Building Studies or for a Degree in Civil Engineering. It should also be practical reference to Architects, Geologists, Structural and Geotechnical Technicians.

The primary aim is to provide a clear understanding of the basic concepts of Soil Mechanics. We endeavoured to avoid the temptation of over-elaboration by providing excessively detailed text, unnecessary at this early stage of technical studies.

The purpose of this publication is threefold:

1. To introduce the student to the basics of soil mechanics.
2. To facilitate further advanced study.
3. To provide reference Information.

In order to satisfy the above requirements, the concepts of the subject are defined concisely, aided by diagrams, charts, graphs, tables and worked examples as necessary.

The text may appear to be excessively analytical at first sight, but all formulas are derived in terms of basic mathematics, except for a few requiring complicated theory, for those interested in working from first principles. They can be applied however, without reference to the derivation. The expressions are numbered and referred to throughout the text.

There are numerous worked examples on each topic as well as supplementary problems. All examples and problems are solved, many of them interrelated so that solutions can be compared and verified by means of several methods.

Some soil testing procedures are outlined only, as there are a number of excellent, detailed, specialized books and laboratory manuals available to cover this part of the subject.

There is some emphasis on the units employed and on the difference between mass and weight. This subject is discussed in Appendix A.

Béla Bodó and Colin Jones

Dedication

"I dedicate this book to my late wife Dorie."
Béla Bodó

Acknowledgments

We wish to express our appreciation to Mr. Norman Seward, Senior Lecturer in Civil Engineering at the University of Wales College, Newport for his technical advice as to the presentation of the subject.
We are also grateful to Mr. Gregory Williams for his help in the production of this book.
We would like to thank ELE International for their support in providing product images.

List of Symbols

Chapter 1

CBR California bearing ratio
C_r Relative compaction
D_r Relative Density
e Voids ratio
G_s Specific gravity
k CBR Load-ring factor
M Total Mass of sample
m Moisture (water) content
m_o Optimum moisture content
M_s Mass of solids
M_w Mass of water
n Porosity
P CBR applied force
P_a Percentage of air voids
Q CBR Load gauge reading
S_r Degree of saturation
V Total volume of sample
V_a Volume of air
V_c Volume of calibrating cylinder
V_s Volume of solids
V_v Volume of voids
V_w Volume of water
W Total weight of sample
W_s Weight of solids
W_w Weight of water
δ CBR Penetration distance (delta)
γ Bulk weight density (Gamma)
γ' Submerged weight density
γ_d Dry Weight density
γ_d Dry Unit weight to be achieved by compaction
γ_s Weight density of solids
γ_{sat} Saturated weight density
ρ Bulk mass density
ρ_d Dry mass density
ρ_{sat} Saturated mass density
ρ' Submerged mass density
ρ_s Mass density of solids

Chapter 2

C_d Correction for dispersing agent
C_m Meniscus correction
D Equivalent particle diameter
D_{10} Effective size of a particle
f Specific Volume change
H Height from the top of the bulb to surface
h_b Length of bulb
H_R Height of centre of bulb to surface
LI Liquidity index
LL Liquid limit
M_p Mass passing the nth sieve
M_r Mass retained on the nth sieve
m_T Temperature correction
N Number of blows
PI Plasticity index
PL Plastic limit
P_n Percentage of soil passing the nth sieve
R Mixing ratio
R'_h Recorded hydrometer reading
R_h Corrected hydrometer reading
RI Relative consistence index
SL Shrinkage limit
T Temperature
t Time
U Uniformity coefficient
u Velocity of sedimentation
V_b Volume of hydrometer bulb
V_o Volume of over-dried specimen
$\approx$ Volume at SL
x Magnitude of linear shrinkage or swelling
Z Saturation limit
η Dynamic viscosity <eta>

Chapter 3

A Cross-sectional area of specimen
a Cross-sectional area of standpipe
A_s Cross-sectional area of solids in specimen
A_v Cross-sectional area of voids in specimen
EPL Equipotential line
FL Flow Line
F_s Factor of safety
GL Ground level
GWL Groundwater level (Water Table)

h Head loss
H_T Total head at x
H_x Head loss to point x
h_x Pressure head at x
i Hydraulic gradient
i_{av} Average hydraulic gradient
i_c Critical hydraulic gradient
i_e Exit gradient
k Coefficient of permeability
L Length of flow path
N_e Number of squares (head drops)
N_f Number of flow channels
N_x Number of head drops to point x
P Hydrostatic force
Q Flowrate
q Quantity of flow in time (t)
R Radius of influence
r Radius to observation well
r_o Radius of central well
S Seepage force
u_x Seepage pore pressure at x
Δh Head Loss between equipotential line
v Discharge velocity
v_s Seepage velocity

Chapter 4

I Influence factor
n Number of elements on the Newmark chart
Q Concentrated point load
q Uniformly distributed load (UDL)
r Radius
z Depth
σ Horizontal pressure
σ_v Vertical pressure
τ Shear stress

Chapter 5

dh Total deformation of specimen of thickness h
h_A Artesian pressure head
h_c Capillary head
h_s Seepage pressure head
i_c Critical hydraulic gradient
m_E Equilibrium moisture content
m_o Optimum moisture content
pF Soil suction index
PI Plasticity index
S_r Degree of saturation

S_s	Soil suction
T	Surface tension
u	Pore pressure
u_{cs}	Pore pressure in the capillary fringe
u_h	Static pore pressure at depth h
u_s	Seepage pore pressure
z_c	Critical depth for piping
Δu	Small change in u
$\Delta\gamma$	Change in unit weight
$\Delta\sigma$	Small change in σ
$\Delta\sigma'$	Small change in σ'
δ	Deformation of specimen at time t
σ	Total pressure
σ'	Effective pressure
σ_A	Artesian pressure

Chapter 6

A	Pore pressure coefficient
$\bar{A}$	Pore pressure coefficient
B	Pore pressure coefficient
c	Cohesion
c_u	Undrained shear strength
CD	Consolidated-drained test
CU	Consolidated-undrained test
ESP	Effective stress path
NCC	Normally consolidated clay
n	Proving ring constant
OCC	Over consolidated clay
$p\&q$	Stress path coordinates
$p_f\&q_f$	Stress path coordinates at failure
QU	Quick-undrained test
r_x	Force dial reading at x
TSP	Total stress path
UU	Unconsolidated-undrained test
x	Strain gauge reading
Δu_d	Change in pore pressure due to $\Delta\sigma_d$
Δu_c	Change in pore pressure due to $\Delta\sigma_c$
$\Delta\sigma_c$	Change in cell pressure
$\Delta\sigma_d$	Change in the deviator stress
ε	Strain at x
ϕ	Angle of friction
σ_n	Normal pressure
σ_x	Deviator stress at x
σ_u	Unconfined compression strength
τ	Shear stress
τ_f	Shear stress at failure
τ_p	Shear stress on a plain
τ_m	Maximum shear stress

Chapter 7

A_c Area indicating completed consolidation
A_t Area under an isochrone
a_v Coefficient of compressibility
C_α Coefficient of Secondary settlement () to consolidation
C_c Compression index
C_v Coefficient of consolidation
D_x Dial reading at stage x
dH_i Initial settlement
E Modulus of elasticity
e_0 Initial voids ratio
e_f Final voids ratio
e_s Voids ratio after swelling
e_x Voids ratio at stage x
H Layer thickness
H_0 Flow path
h_x Height of specimen at stage x
I_p Influence factor
k Coefficient of permeability
m_v Coefficient of volume change
OCR Overconsolidation ratio
q Bearing pressure
T_v Time factor
t Time
U Average degree of consolidation
U_z Degree of consolidation
u Pore pressure at time t
u_0 Initial pore pressure
ΔH Long-term consolidation settlement
$\Delta\sigma'$ Effective consolidating pressure
δ Depth factor (Delta)
$\propto$ Poisson's ratio (My)
σ'_x Effective pressure at stage x

Chapter 8

c_u Unconfined compression strength
c_W Adhesion between soil and wall
e Eccentricity
F_ϕ Factor of safety in terms of friction angle
f_{max} Maximum compressive stress
f_{min} Minimum compressive stress
F_s Factor of safety
H Height of wall
H_0 Height of unsupported clay
K Coefficient of lateral pressure

K_0 Coefficient of earth pressure at rest
K_a Coefficient of active earth pressure
K_f Coefficient of earth pressure at failure
K_p Coefficient of passive earth pressure
L Length of slip surface
M_{max} Maximum bending moment
M_0 Overturning moment
M_R Resisting moment
P_a Active force
P_p Passive force
P_W Force of water in tension crack
R Force on wedge
T Tension force in tie rod
z_c Pile penetration
z_0 Depth of tension crack
δ Angle of wall friction
ϕ'_m Mobilised friction
μ Coefficient of friction
σ_a Active earth pressure
σ_c Cell pressure in triaxial test
σ_d Deviator stress in triaxial test
σ_p Passive earth pressure
σ'_a Effective active earth pressure
σ'_p Effective passive earth pressure
$\bar{\sigma}$ Average pressure
τ_f Shear stress at failure

Chapter 9

$\bar{c}_u$ Average undrained shear strength
A_e End bearing area
A_s Surface area of pile
B Width of footing
c Cohesion
F_0 Overall factor of safety
F_s Factor of safety
K_s Average coefficient of earth pressure
l Length of pile
N Number of SPT blows
n Number of piles
N' Corrected value of N
N_c, N_q, N_γ Bearing capacity factors
P Failure load on pile
Q Design working load
Q_a Allowable carrying capacity of pile
Q_{ag} Allowable carrying capacity of pile group

Q_e End bearing resistance
Q_f Negative skin friction
Q_S Shaft resistance
Q_u Ultimate carrying capacity of pile
Q_{ug} Ultimate carrying capacity of pile group
q_n Net ultimate bearing capacity
q_s Safe bearing capacity
q_{sn} Safe net bearing capacity
q_u Ultimate bearing capacity
SPT Standard penetration test
W_P Weight of pile
α Adhesion factor (Alpha)
δ Angle of friction between soil and pile (Delta)
η Efficiency of pile group (Eta)
ϕ Angle of friction
σ Safe bearing pressure of footing
σ_n Net bearing pressure of footing
$\bar{\sigma}'_o$ Average effective overburden pressure
σ'_o Effective overburden pressure

Chapter 10

c_u Shear strength
F Friction force
F_C Factor of safety with respect to cohesion
F_S Factor of safety
F_ϕ Factor of safety with respect to friction
L Length of slip surface
M_D Disturbing moment
M_R Resisting moment
N Normal (or radial) component of W
N_C Stability number
R Radius of slip circle
r_u Pore pressure ratio
S Shear force
T Tangential component of W
W Weight

Chapter 11

The comprehensive list of symbols for EC7 is given in *Eurocode 7. Geotechnical design Part 1: General rule*. Only some of the symbols, applied in this book, are reproduced here:

E_d Design value of the effect of actions
$E_{dst;d}$ Design value of the effect of destabilizing action
$E_{stb;d}$ Design value of the effect of stabilizing action
F_d Design value of an action
F_{rep} Representative value of an action
F_s Factor of safety

$G_{dst;d}$ Design value of destabilising permanent action
$G_{stb;d}$ Design value of stabilising permanent action
$Q_{dst;d}$ Design value of destabilising variable action
R_d Design value of resistance action
$S_{dst;d}$ Design value of destabilising seepage force
T_d Design value of total shear resistance
$U_{dst;d}$ Design value of destabilising pore water pressure
$V_{dst;d}$ Design value of destabilising vertical action
X_d Design value of a material property
X_k Characteristics value of a material property
γ_G Partial factor for a permanent action
$\gamma_{G;dist}$ Partial factor for a destabilising action
$\gamma_{G;stb}$ Partial factor for a stabilising action
γ_m Partial factor for soil parameters (material property)
γ_Q Partial factor for a variable action
$\gamma_{R;h}$ Partial factor for sliding resistance

Chapter 1
Soil Structure

Soils consist of solid particles, enclosing voids or pores. The voids may be filled with air or water or both. These three soil states (or phases) can be visualized by the enlargement of three small samples of soil.

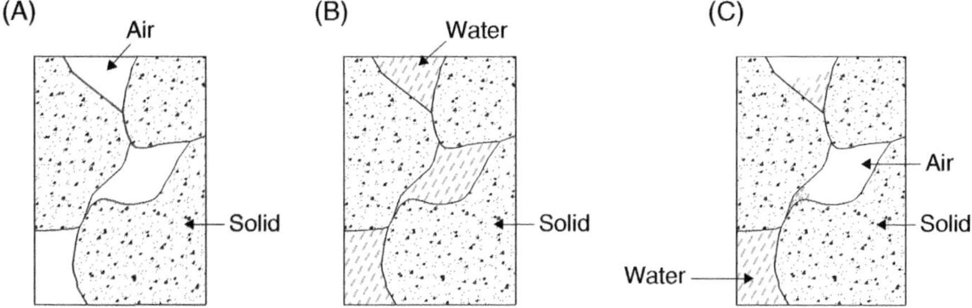

Figure 1.1

Sample A: The soil is oven-dry, that is there is only air in the voids.
Sample B: The soil is saturated, that is the voids are full of water.
Sample C: The soil is partially saturated, that is the voids are partially filled with water.

The above three soil states can be described mathematically by considering:

1. Volume occupied by each constituent.
2. Mass (or weight) of the constituents.

1.1 Volume relationships

The expressions derived in this section will answer two questions:

1. How much voids and solids are contained in the soil sample?
2. How much water is contained in the voids?

In order to obtain these answers, the partially saturated sample (C) is examined. It is assumed, for the purpose of analysis, that the soil particles are lumped together into a homogeneous mass. Similarly, the voids are combined into a single volume, which is

Introduction to Soil Mechanics, First Edition. Béla Bodó and Colin Jones.

partly occupied by a volume of water. The idealisation of the sample, indicating the volumes occupied by the constituents, is shown diagrammatically in Figure 1.2b.

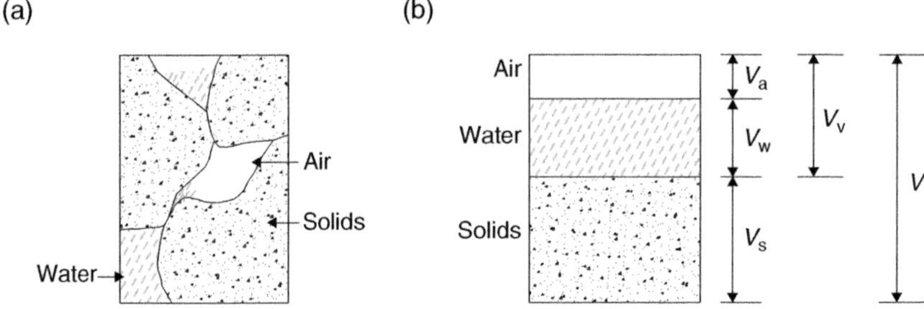

Figure 1.2

Idealized representation of sample C.
Where: V = Total volume of the sample
V_v = Volume of voids in the sample
V_s = Volume of soil in the sample
V_w = Volume of water in the sample
V_a = Volume of air in the sample

The basic relationships between the volumes can be seen in the diagram.

Total volume: $$V = V_s + V_v \quad (1.1)$$

Volume of voids: $$V_v = V_w + V_a \quad (1.2)$$

Hence: $$V = V_s + V_w + V_a \quad (1.3)$$

Three important relationships are derived from the basic ones. These are:

e = voids ratio (or void ratio)
n = porosity
S_r = degree of saturation

1.1.1 Voids ratio (*e*)

This shows the percentage of voids present in the sample, compared to the volume of solids. Thus, if V_s is considered to be 100%, then V_v is e%.

Hence: $$\boxed{e = 100\frac{V_v}{V_s}\,\%} \quad (1.4)$$

For example: if $V_s = 60\,\text{cm}^3$
and $V_v = 15\,\text{cm}^3$
then $e = 100\frac{15}{60} = 25\%$

That is, the volume of voids is 25% of the volume of solids, in this particular sample. Alternatively, the voids ratio maybe expressed as a decimal e.g. $e = 0.25$.

Formula (1.4) now becomes: $$\boxed{e = \frac{V_v}{V_s}} \quad (1.5)$$

The ratio of voids to solids in a sample is represented by Figure 1.3.

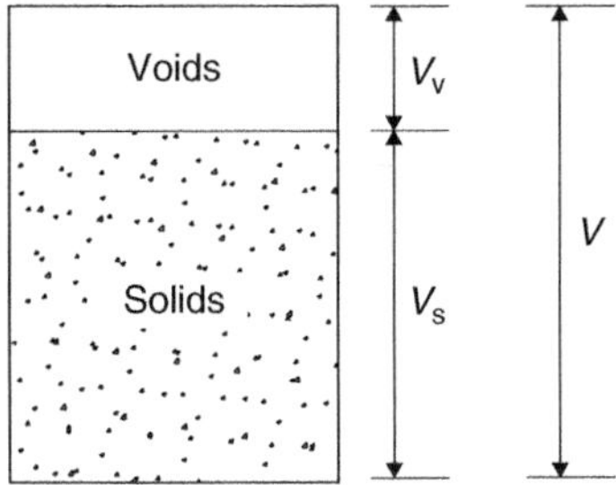

Figure 1.3

1.1.2 Porosity (n)

This shows how many percent of voids are present in the sample, compared to the total volume V. Thus, if V is considered to be 100%, then V_v is n%.

$$\boxed{n = 100\frac{V_v}{V} \ \%} \tag{1.6}$$

For example: if $V = 75\,\text{cm}^3$
and $V_v = 15\,\text{cm}^3$
then $n = 100\dfrac{15}{75} = 20\%$

That is, the volume of voids is 20% of the total volume of the sample of soil.

Again, n maybe expressed as a decimal number $n = 0.2$.

Formula (1.6) now becomes:

$$n = \frac{V_v}{V} \tag{1.7}$$

The diagrammatic representation of porosity is:

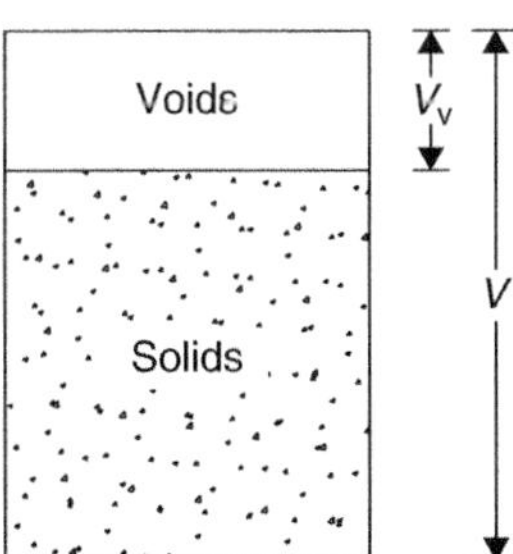

Figure 1.4

1.1.3 Degree of saturation (S_r)

This shows the percentage of voids filled with water. Thus, if V_v is considered to be 100%, then V_w is S_r%.

$$\boxed{S_r = 100\frac{V_w}{V_v} \ \%} \tag{1.8}$$

For example, if $V_w = 6\,\text{cm}^3$
and $V_v = 15\,\text{cm}^3$
then $S_r = 100\dfrac{6}{15} = 40\%$

That is, water fills 40% of the volume of voids. In decimal form $S_r = 0.4$ and formula (1.8) becomes:

$$\boxed{S_r = \frac{V_w}{V_v}} \tag{1.9}$$

Diagrammatically,

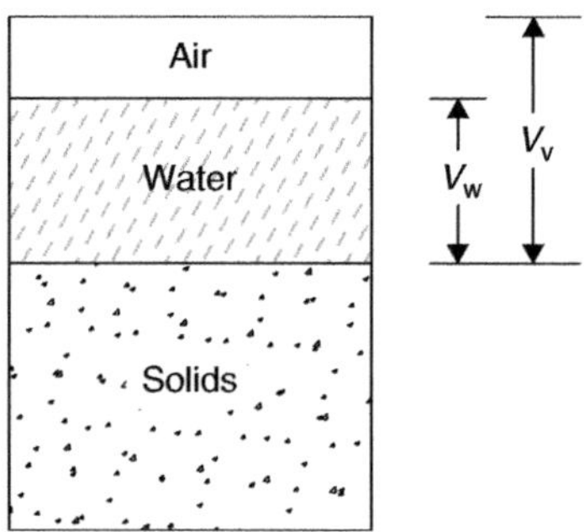

Figure 1.5

Note: For oven-dry soil (Sample A, Figure 1.1):

$V_w = 0$, hence $S_r = 0$

For fully saturated soil (Sample B, Figure 1.1):

$V_w = V_v$, hence $S_r = 1$

For partially saturated soil therefore: $0 < S_r < 1$

Combined formulae

The quantities defined by formulae (1.1) to (1.9) can be interrelated:

From (1.1): $V = V_s + V_v$
From (1.5): $V_v = eV_s$

either $V = V_s + eV_s$ $\therefore \boxed{V = (1+e)V_s}$ (1.10)

or $V = \dfrac{V_v}{e} + V_v$

$= \left(\dfrac{1}{e} + 1\right)V_v$ $\therefore \boxed{V_v = \left(\dfrac{1+e}{e}\right)V_v}$ (1.11)

From (1.7): $n = \dfrac{V_v}{V}$
From (1.10): $V = (1+e)V_s$

$n = \dfrac{eV_s}{(1+e)V_s}$ $\therefore \boxed{n = \dfrac{e}{1+e}}$ (1.12)

From (1.12): $n = \dfrac{e}{1+e}$

$n + ne = e$

$n = e(1-n)$ $\boxed{e = \dfrac{n}{1-n}}$ (1.13)

From (1.9): $S_r = \dfrac{V_w}{V_v}$

From (1.11): $V_v = \dfrac{eV}{1+e}$

$$S_r = \frac{V_w}{\dfrac{eV}{1+e}} \quad \therefore \quad \boxed{S_r = \left(\frac{1+e}{e}\right)\frac{V_w}{V}} \tag{1.14}$$

From (1.12): $n = \dfrac{e}{1+e}$

$$\text{or} \quad \boxed{S_r = \frac{V_w}{nV}} \tag{1.15}$$

Example 1.1

Given: $V = 946\,\text{cm}^3$ Calculate: V_v, V_a, e, n and S_r

$V_s = 533\,\text{cm}^3$

$V_w = 303\,\text{cm}^3$

From (1.1): $V_v = V - V_s = 946 - 533 = 413\,\text{cm}^3$

From (1.2): $V_a = V_v - V_w = 413 - 303 = 110\,\text{cm}^3$

From (1.5): $e = \dfrac{V_v}{V_s} = \dfrac{413}{533} = 0.775$, that is the volume of voids is 77.5% that of solids.

From (1.7): $n = \dfrac{V_v}{V} = \dfrac{413}{946} = 0.437$

or From (1.12): $n = \dfrac{e}{1+e} = \dfrac{0.775}{1.775} = 0.437$

That is, the volume of voids is 43.7% of the sample.

From (1.9): $S_r = \dfrac{V_w}{V_v} = \dfrac{303}{413} = 0.73$

or From (1.15): $S_r = \dfrac{V_w}{nV} = \dfrac{303}{(0.437 \times 946)} = 0.73$

That is, water fills 73% of voids. The sample is partially saturated.

Example 1.2

A sample of sand was taken from below the ground water table. The volumes measured were:

$V = 1000\,\text{cm}^3$ Calculate: V_v, V_a, V_s, e and n

$V_w = 400\,\text{cm}^3$

Note: Assume sand samples taken from above the water table as partially saturated ($S_r < 1$) and saturated ($S_r = 1$) if taken from below.

In this example, therefore, $S_r = 1 \quad \therefore \quad V_a = 0$.

From (1.8) $$S_r = \frac{V_w}{V_v} = 1 \quad \therefore \quad \boxed{V_w = V_v} \tag{1.16}$$

$$V_v = 400\,\text{cm}^3$$

From (1.2): $V_a = V_v - V_w = 400 - 400 = 0$ The voids are full of water

From (1.1): $V_s = V - V_v = 1000 - 400 = 600\,\text{cm}^3$

From (1.5): $e = \dfrac{V_v}{V_s} = \dfrac{400}{600} = 0.67$ | V_v is 67% of V_s

From (1.7): $n = \dfrac{V_v}{V} = \dfrac{400}{1000} = 0.4$ | V_v is 40% of V

1.2 Weight–volume relations

As the title implies, the formulae derived in this section take into account the weights of V_s and V_w. It is assumed that air is weightless. The weight volume relations are shown diagrammatically:

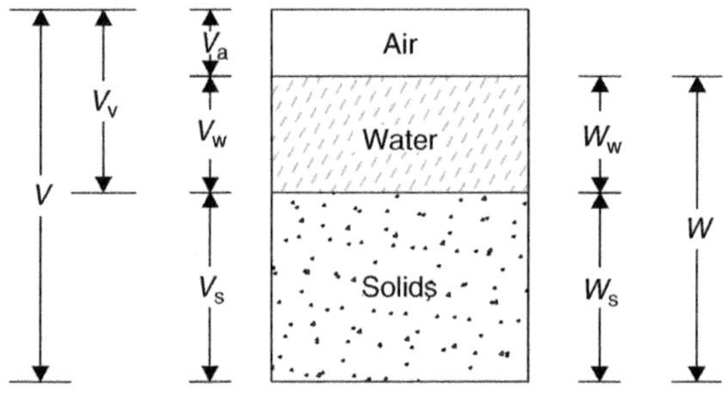

Figure 1.6

Where: W_s = Weight of solids
W_w = Weight of water | From Figure 1.6 $\boxed{W = W_s + W_w}$ (1.17)
W = Total weight

Note: The concepts of mass and weight are defined in Appendix A. Suffice to say here, that if mass (M) is given in kilograms, then weight (W) is calculated from:

$$W = 9.81 \times \text{mass } (M)\, N \qquad \therefore \boxed{W = 9.81 \times 10^{-3} \times M\ \text{kN}} \tag{1.18}$$

Several important relationships are derived below in terms of mass, weight and volume. These are:

ρ = bulk mass density
γ = bulk weight density (unit weight)
ρ_d = dry mass density
γ_d = dry weight density
ρ_{sat} = saturated mass density
γ_{sat} = saturated weight density

ρ' = submerged mass density
γ' = submerged weight density
ρ_s = mass density of solids
γ_s = weight density of solids.

Note: Normally, the mass density of materials is expressed in kg/m^3. For instance, the average mass of reinforced concrete is quoted in tables as ρ=2400 kg/m^3. Sometimes, especially in laboratory work, it is more convenient to use gram as the unit of mass. Possibly for this reason ρ is often expressed in g/cm^3 or Mg/m^3 (Mg/m^3=g/cm^3).

For a reason, justified in the Appendix, the unit adopted in this book is kg/m^3, unless otherwise stated.

1.2.1 Bulk densities

These are the densities of a partially saturated soil sample, taken from above ground water level.

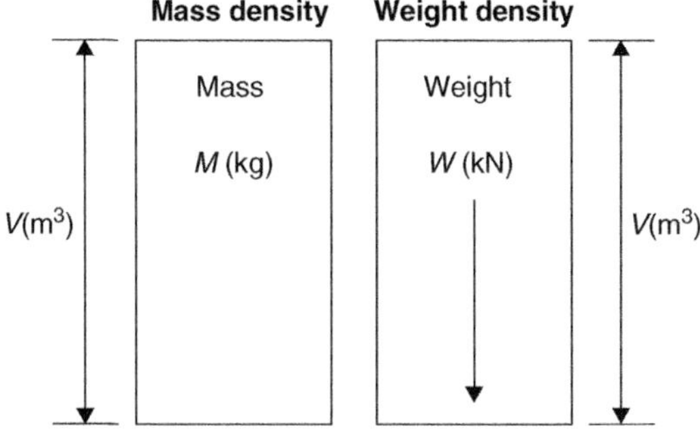

Figure 1.7

$$\boxed{\rho = \frac{M}{V} \text{ kg/m}^3} \quad (1.20) \qquad \boxed{\gamma = \frac{W}{V} \text{ kN/m}^3} \quad (1.19)$$

$$\gamma = \frac{9.81 \times 10^{-3} M}{V} = 9.81 \times 10^{-3}\left(\frac{M}{V}\right) \qquad \therefore \boxed{\gamma = 9.81 \times 10^{-3}\rho \text{ kN/m}^3} \quad (1.21)$$

For water: $$\boxed{\rho_w = \frac{M_w}{V_w} = 1000 \text{ kg/m}^3} \quad (1.22) \qquad \boxed{\gamma_w = \frac{W_w}{V_w} = 9.81 \text{ kN/m}^3} \quad (1.23)$$

All practical problems in soil mechanics are concerned with forces acting in one way or another. As the weight density (or unit weight) itself is a force, its application is a matter of necessity. For this reason the formulae derived in the rest of this section are mostly in terms of weight. Remember, however, that 1 kg=1000 g and 1 g/cm^3=10^3 kg/m^3. Therefore, if the mass density is given in gram and centimeter units as:

$$\boxed{\rho = \frac{M}{V} \text{ g/cm}^3} \quad \text{then} \quad \boxed{\gamma = 9.81\rho \text{ kN/m}^3} \quad (1.24)$$

Example 1.3

Partially saturated sand was tested in a laboratory. Its volume was measured to be 75.4 cm³ and weighed 136.2 g. Calculate the unit weight in kN/m³.

Mass: $M = 136.2\,\text{g} = 136.2 \times 10^{-3}\,\text{kg}$

Volume: $V = 75.4\,\text{cm}^3 = 75.4 \times 10^{-6}\,\text{m}^3$

Weight: $W = 9.81 \times 10^{-3} \times 136.2 \times 10^{-3} = 1336 \times 10^{-6}\,\text{kN}$

Mass density: $\rho = \frac{M}{V} = \frac{136.2 \times 10^{-3}}{75.4 \times 10^{-6}} = 1806\,\text{kg/m}^3$

Weight density: $\gamma = \frac{M}{V} = \frac{1336 \times 10^{-6}}{75.4 \times 10^{-6}} = 17.72\,\text{kg/m}^3$

or by (1.21): $\gamma = 9.81 \times 10^{-3}\,\rho = 9.81 \times 10^{-3} \times 1806 = 17.72\,\text{kg/m}^3$

1.2.2 Dry densities

These are the densities of oven-dry soil, after the excavated sample has completely dried out.

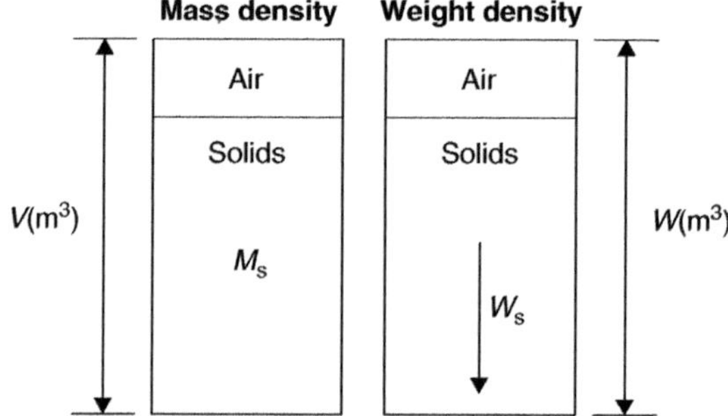

Figure 1.8

$$\boxed{\rho_d = \frac{M_s}{V}\;\text{kg/m}^3} \quad (1.25) \qquad \boxed{\gamma_d = \frac{M_s}{V}\;\text{kg/m}^3} \quad (1.26)$$

Therefore,

$$\boxed{\gamma_d = 9.81 \times 10^{-3}\rho_d\;\;\text{kN/m}^3} \quad (1.27)$$

1.2.3 Saturated densities

These are the bulk densities of a sample, taken from below the ground water level (GWL), hence the sample is fully saturated and the degree of saturation is unity.

From formula (1.9): $S_r = \frac{V_w}{V_v} = 1 \quad \therefore \quad V_w = V_v$

and from (1.1): $V = V_s + V_v = V_s + V_w$

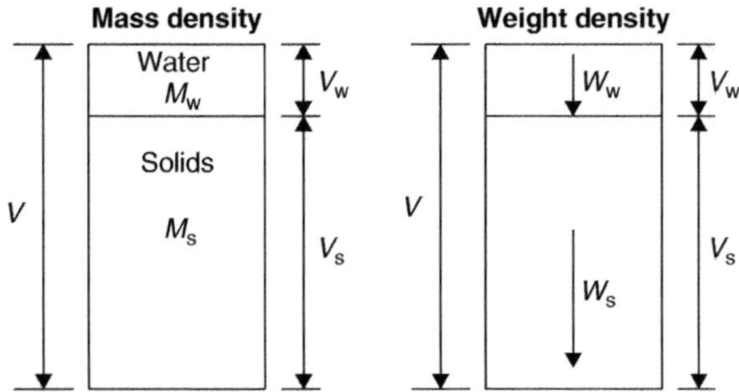

Figure 1.9

For water $\rho_w = 1\,\text{g/cm}^3$

Or $\rho_w = 1000\,\text{kg/m}^3$ and $\gamma_w = 9.81\,\text{kN/m}^3$

Total mass: $M_{sat} = M_s + \rho_w \times V_v$ and $W_{sat} = W_s + \gamma_w \times V_v$

Either $\rho_{sat} = \dfrac{M_{sat}}{V}$ Or $\rho_{sat} = \dfrac{M_s + \rho_w V_v}{V}$ kN/m³ (1.28)

Either $\gamma_{sat} = \dfrac{W_{sat}}{V}$ Or $\gamma_{sat} = \dfrac{W_s + \gamma_w V_v}{V}$ kN/m³ (1.29)

Therefore, $\gamma_{sat} = 9.81 \times 10^{-3} \rho_{sat}$ kN/m³ (1.30)

Example 1.4

The partially saturated sand in Example 1.3 was saturated by the addition of water and then dried out completely. The quantities measured were:

Dry mass: $M_s = 122.9\,\text{g} = 122.9 \times 10^{-3}\,\text{kg}$

Mass of water lost: $M_w = 29.0\,\text{g} = 29 \times 10^{-3}\,\text{kg}$

Total volume: $V = 75.4\,\text{cm}^3 - 75.4 \times 10^{-6}\,\text{m}^3$

Calculate the saturated unit weight of the sample:

$$M_{sat} = 122.9 \times 10^{-3} + 29 \times 10^{-3} = 151.9 \times 10^{-3}\,\text{kg}$$

$$\rho_{sat} = \frac{M_{sat}}{V} = \frac{151.9 \times 10^{-3}}{75.4 \times 10^{-6}} = 2015\,\text{kg/m}^3$$

$$\therefore \gamma_{sat} = 9.81 \times 10^{-3} \times 2015 = 19.76\,\text{kN/m}^3$$

1.2.4 Submerged density (γ')

It is the saturated density of soil, taking its buoyancy into account. In other words, as long as the saturated sample remains under water, an uplift force is exerted on it in accordance with Archimedes' Principle.

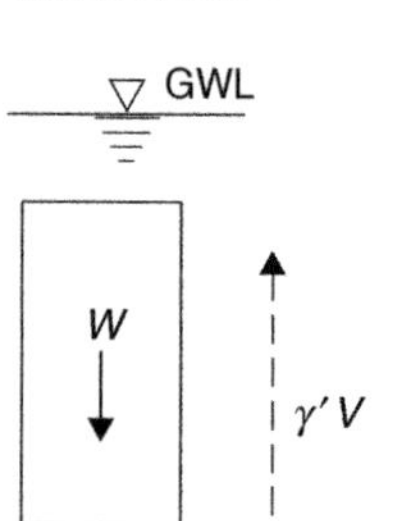

Weight of sample: $W = \gamma_{sat} V \downarrow$

Buoyant force: $B = \gamma_w V \uparrow$

Resultant force on sample $= \gamma' V$

Now: $\gamma' V = W - B$

$= \gamma_{sat} \times V - \gamma_w \times V$

Figure 1.10

Cancelling volume V: $$\boxed{\gamma' = \gamma_{sat} - \gamma_w} \tag{1.31}$$

Note: The submerged density is to be used, when assessing the stresses induced in the soil below GWL by surface loading. This type of problem includes the determination of:

a) Effective pressure
b) Load bearing capacity of a soil.

1.2.5 Density of solids (γ_s)

It is the unit weight of the soil particles, occupying the volume V_s. Particle mass density is denoted by ρ_s.

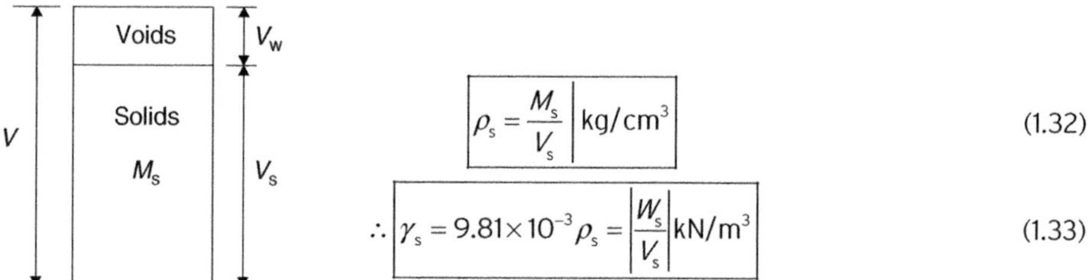

$$\boxed{\rho_s = \frac{M_s}{V_s} \text{ kg/cm}^3} \tag{1.32}$$

$$\therefore \boxed{\gamma_s = 9.81 \times 10^{-3} \rho_s = \frac{W_s}{V_s} \text{ kN/m}^3} \tag{1.33}$$

Figure 1.11

1.2.6 Specific gravity (G_s)

It is also called "Particle specific gravity", as it shows how heavy the solids are compared to water. In other words, the weight of V_s volume of solids is compared with the same volume of water.

Weight of solids: $W_s = \gamma_s \times V_s$

Weight of water: $W_w = \gamma_w \times V_s$

and $G_s = \frac{W_s}{W_w} = \frac{\gamma_s V_s}{\gamma_w V_s}$

Cancelling V_s to obtain $$\boxed{G_s = \frac{\gamma_s}{\gamma_w} = \frac{\rho_s}{\rho_w}} \tag{1.34}$$

Table 1.1 Average values of G_s

Soil	G_s
Clay	2.75
Silt	2.68
Sand	2.65
Gravel	2.65

1.2.7 Moisture content (*m*)

This expresses the mass or weight of water as a percentage of the mass or weight of solids.

$$\boxed{m = 100\frac{M_w}{M_s} = 100\frac{W_w}{W_s}\%} \tag{1.35}$$

Or in decimal form:

$$\boxed{m = \frac{M_w}{M_s} = \frac{W_w}{W_s}} \tag{1.35a}$$

Note: The quantities given in formulae (1.1) to (1.35) can be calculated from these four laboratory results:

1. Total mass of the sample M (g).
2. Mass of solids M_s (g).
3. Total volume of the sample V (cm^3).
4. Specific gravity of the soil particles G_s.

Example 1.5

Using the laboratory results of Examples 1.3 and 1.4, tabulate the calculations for all soil characteristics introduced this far, in Table 1.2. Assume G_s=2.65.

Table 1.2

Formula Number	Soil characteristic	Calculations and Results	Unit
	Total mass	M=136.2×10^{-6}	mg
	Mass of solids	M_s=122.9×10^{-6}	mg
	Total volume	V=75.4×10^{-6}	m^3
	Specific gravity	G_s=2.65	-
1.18	Total weight	W=9.81×10^{-3}× 136.2×10^{-3}=1336×10^{-6}	kN
1.18	Weight of solids	W_s=9.81×10^{-3}×122.9×10^{-3}=1206×10^{-6}	kN
1.17	Weight of water	W_w=W – W_s=(1336–1206)×10^{-6}=130×10^{-6}	kN
1.35	Water content	$m = 100 \times \frac{W_w}{W_s} = 100\frac{130 \times 10^{-6}}{1206 \times 10^{-6}} = 10.8$	%
1.20	Bulk mass density	$\rho = \frac{M}{V} = \frac{136.2 \times 10^{-3}}{75.4 \times 10^{-6}} = 1806$	kg/m^3

Table 1.2 *(continued)*

	Soil characteristic	Calculations and Results	Unit
1.21	Bulk unit weight	$\gamma = 9.81 \times 10^{-3} \rho = 9.81 \times 10^{-3} \times 1806 = 17.7$	kN/m³
1.25	Dry mass density	$\rho_d = \frac{M_s}{V} = \frac{122.9 \times 10^{-3}}{75.4 \times 10^{-6}} = 1630$	kg/m³
1.27	Dry unit weight	$\gamma_d = 9.81 \times 10^{-3} \rho_d = 9.81 \times 10^{-3} \times 1630 = 16$	kN/m³
1.34	Unit weight of solids	$\gamma_s = G_s \times \gamma_w = 2.65 \times 9.81 = 26$	kN/m³
1.33	Volume of solids	$V_s = \frac{W_s}{\gamma_s} = \frac{1206 \times 10^{-6}}{26} = 46.4 \times 10^{-6}$	m³
1.32	Mass density of solids	$\rho_s = \frac{M_s}{V_s} = \frac{122.9 \times 10^{-3}}{46.4 \times 10^{-6}} = 2649$	kg/m³
1.1	Volume of voids	$V_v = V - V_s = (75.4 - 46.4) \times 10^{-6} = 29 \times 10^{-6}$	m³
1.5	Voids ratio	$e = 100 \times \frac{V_v}{V_s} = 100 \times \frac{29 \times 10^{-6}}{46.4 \times 10^{-6}} = 62.5$	%
1.6	Porosity	$n = 100 \times \frac{V_v}{V} = 100 \times \frac{29 \times 10^{-6}}{75.4 \times 10^{-6}} = 38.5$	%
1.29	Saturated unit weight	$\gamma_{sat} = \frac{W_s + \gamma_w \times V_v}{V} = \frac{(1206 + 9.81 \times 29) \times 10^{-6}}{75.4 \times 10^{-6}} = 19.8$	kN/m³
1.28	Saturated mass density	$\rho_{sat} = \frac{\gamma_{sat}}{9.81 \times 10^{-3}} = \frac{19.8 \times 10^{-3}}{9.81} = 2018$	kg/m³
1.23	Volume of water	$V_w = \frac{W_w}{\gamma_w} = \frac{130 \times 10^{-6}}{9.81} = 13.3 \times 10^{-6}$	m³
1.2	Volume of air	$V_a = V_v - V_w = (29 - 13.3) \times 10^{-6} = 15.7 \times 10^{-6}$	m³
1.8	Degree of saturation	$S_r = 100 \times \frac{V_w}{V_v} = 100 \times \frac{13.3 \times 10^{-6}}{29 \times 10^{-6}} = 45.9$	%
1.31	Submerged unit weight	$\gamma' = \gamma_{sat} - \gamma_w = 19.8 - 9.81 = 10$	kN/m³

Further useful relationships can be derived by the combination of the above formulae.

1.2.8 Partially saturated soil

It has already been mentioned, that soil is normally partially saturated above ground water level that is the degree of saturation is less than unity. In fine-grained soil capillary action may saturate the soil somewhat above GWL. In any case, always assume partial saturation, unless proven otherwise.

From (1.9): $V_v = \dfrac{V_w}{S_r}$

From (1.23): $V_w = \dfrac{W_w}{\gamma_w}$

From (1.33): $V_s = \dfrac{W_s}{\gamma_s}$

From (1.34): $G_s = \dfrac{\gamma_s}{\gamma_w}$

From (1.35) $m = \dfrac{W_w}{W_s}$

$W_w = m \times W_s$

From (1.5): $e = \dfrac{V_v}{V_s} = \dfrac{\dfrac{V_w}{S_r}}{\dfrac{W_s}{\gamma_s}}$

$$= \frac{\dfrac{W_w}{S_r\gamma_w}}{\dfrac{W_s}{\gamma_s}} = \frac{1}{S_r} \times \frac{W_w}{W_s} \times \frac{\gamma_s}{\gamma_w} = \frac{1}{S_r} \times m \times G_s$$

$$\therefore \boxed{e = \frac{mG_s}{S_r}} \tag{1.36}$$

From (1.17) $W = W_s + W_w = W_s + mW_s$

$$\boxed{W = (1+m)W_s} \tag{1.37}$$

From (1.10): $V = (1+e) \times V_s$

From (1.36): $m = \dfrac{S_r e}{G_s}$

From (1.33): $\gamma_s = \dfrac{W_s}{V_s}$

From (1.34): $\gamma_w = \dfrac{\gamma_s}{G_s}$

From (1.19): $\gamma = \dfrac{W}{V} = \dfrac{(1+m)W_s}{(1+e)V_s}$

$$\gamma = \frac{1 + \dfrac{S_r e}{G_s}}{1+e} \times \gamma_s = \frac{G_s + S_r e}{1+e} \times \frac{\gamma_s}{G_s}$$

$$\boxed{\gamma = \left(\frac{G_s + S_r e}{1+e}\right)\gamma_w} \tag{1.38}$$

$$\gamma = \left(\frac{G_s + m \times G_s}{1+e}\right)\gamma_w$$

$$\boxed{\gamma = \left(\frac{1+m}{1+e}\right)G_s\gamma_w} \tag{1.39}$$

From (1.37): $W_s = \dfrac{W}{1+m}$

From (1.19) $\gamma = \dfrac{W}{V}$

From (1.26): $\gamma_d = \dfrac{W_s}{V} = \dfrac{\dfrac{W}{1+m}}{V} = \dfrac{\dfrac{W}{V}}{1+m}$

$$\boxed{\gamma_d = \frac{\gamma}{1+m}} \tag{1.40}$$

From (1.38): $\gamma = \left(\dfrac{G_s + S_r e}{1+e}\right)\gamma_w$

For dry soil $S_r = 0$ and $\gamma = \gamma_d$

$$\boxed{\gamma_d = \left(\frac{G_s}{1+e}\right)\gamma_w} \tag{1.41}$$

Note: Dry density is an important factor in the compaction of soils.

For fully saturated soil $S_r = 1$ and $\gamma = \gamma_{sat}$

From (1.38): $\gamma = \left(\dfrac{G_s + S_r e}{1+e}\right)\gamma_w$ hence $\boxed{\gamma_{sat} = \left(\dfrac{G_s + e}{1+e}\right)\gamma_w}$ (1.42)

From (1.31):
$$\gamma' = \gamma_{sat} - \gamma_w = \left(\frac{G_s + e}{1+e}\right)\gamma_w - \gamma_w$$
$$= \left(\frac{G_s + e - 1 - e}{1+e}\right)\gamma_w$$

Hence the submerged density:
$$\gamma' = \left(\frac{G_s - 1}{1+e}\right)\gamma_w \qquad (1.43)$$

Table 1.3 (Comparison of formulae)

Partially saturated soil	Saturated soil	Dry soil
$S_r < 1$	$S_r = 1$	$S_r = 0$
$V_w < V_v$	$V_w = V_v$	$V_w = 0$
$m = \frac{S_r e}{G_s}$	$m = \frac{e}{G_s}$	$m = 0$
$W = (1+m)\ W_s$	$W = (1+m)\ W_s$	$W = W_s$
$\gamma = \left(\frac{G_s + S_r e}{1+e}\right)\gamma_w$	$\gamma_{sat} = \left(\frac{G_s + e}{1+e}\right)\gamma_w$	
$\gamma_d = \frac{\gamma}{1+m}$	$\gamma_d = \frac{\gamma_{sat}}{1+m}$	$\gamma_d = \left(\frac{G_s}{1+e}\right)\gamma_w$

Example 1.6

Clay of $G_s = 2.8$ was compacted into six standard ASTM moulds at different water contents. The internal volume of each mould was 944 cm³. The total and dry masses of samples were found to be:

Table 1.4

	Sample					
Quantity	**1**	**2**	**3**	**4**	**5**	**6**
M (g)	1743	1827	1880	1890	1880	1834
M_s (g)	1449	1502	1533	1542	1510	1467

1. Calculate the quantities contained in Table 1.2 (Example 1.5) for sample No.1, in both mass and weight units. Show calculations in Table 1.5.
2. Complete Table 1.6 by evaluating for each sample the:
   a. Water content (m %)
   b. Bulk unit weight (γ kN/m³)
   c. Dry unit weight (γ_d kN/m³)
   d. Voids ratio (e %)
   e. Volume of air (V_a cm³)
3. Plot γ, γ_d, e and V_a against m on Graph 1.1, indicating their variation with increasing water content.

Table 1.5 For sample No. 1

In mass units		In weight units	
$M=1743\,g=1.743$	kg	$W=9.81\times10^{-3}\times1.743=0.0171$	kN
$M_s=1449\,g=1.449$	kg	$W_s=9.81\times10^{-3}\times1.449=0.0142$	kN
$V=944$	cm^3	$V=944\times10^{-6}$	m^3
$G_s=2.8$	-	$G_s=2.8$	-
$M_w=M-M_s=1.743-1.449=0.294$	Kg	$W_w=W-W_s$ $=0.0171-0.0142=0.0029$	kN
$m=100\times\frac{M_w}{M_s}=\frac{100\times0.294}{1.449}=20.3$	%	$m=100\times\frac{W_w}{W_s}=\frac{100\times0.29}{14.2}=20.4$	%
$\rho=\frac{M}{V}=\frac{1.743}{944\times10^{-6}}=1846$	kg/m^3	$\gamma=9.81\times10^{-3}\times1846=18.1$ $\gamma=\frac{W}{V}=\frac{0.0171}{944\times10^{-6}}=18.1$	kN/m^3 kN/m^3
$\rho_d=\frac{M_s}{V}=\frac{1.449}{944\times10^{-6}}=1535$	kg/m^3	$\gamma_d=9.81\times10^{-3}\times1535=15.1$	kN/m^3
$\rho_d=\frac{\rho}{1+\frac{m}{100}}=\frac{1846}{1.203}=1535$	kg/m^3	$\gamma_d=\frac{W_s}{V}=\frac{0.0142}{944\times10^{-6}}=15.1$	kN/m^3
$\rho_s=G_s\rho_w=2.8\times1000=2800$	kg/m^3	$\gamma_s=9.81\times10^{-3}\times2800=27.5$ $\gamma_s=G_s\gamma_w=2.8\times9.81=27.5$	kN/m^3 kN/m^3
$V_s=\frac{M_s}{\rho_s}=\frac{1.449}{2800}\times10^6=518$	cm^3	$V_s=\frac{W_s}{\gamma_s}=\frac{0.0142}{27.5}=516\times10^{-6}$	m^3
$V_v=V-V_s=944-518=426$	cm^3	$V_v=V-V_s=(944-516)\times10^{-6}$ $=428\times10^{-6}$	m^3
$e=100\times\frac{V_v}{V_s}=\frac{100\times426}{518}=82$	%	$e=82$	%
$n=100\times\frac{e}{1+e}=\frac{100\times0.82}{1.82}=45$	%	$n=45$	%
$\rho_{sat}=\frac{M_s+\rho_w V_v}{V}$ $=\frac{1.449+10^3\times428\times10^{-6}}{944\times10^{-6}}=1981$	kg/m^3	$\gamma_{sat}=9.81\times10^{-3}\times1981=19.4$ $\gamma_{sat}=\frac{W_s+\gamma_w V_v}{V}=$ $\frac{0.0142+9.81\times428\times10^{-6}}{944\times10^{-6}}=19.5$	kN/m^3 kN/m^3
$V_w=\frac{M_w}{\rho_w}=\frac{0.294}{1000}\times10^{-6}=294$	cm^3	$V_w=\frac{W_w}{\gamma_w}=\frac{0.0029}{9.81}=296\times10^{-6}$	m^3
$V_a=V_v-V_w=426-294=132$	cm^3	$V_a=V_v-V_w$ $=(428-296)\times10^{-6}=132\times10^{-6}$	m^3
$S_r=100\times\frac{V_w}{V_v}=\frac{100\times294}{426}=69$	%	$S_r=\frac{100\times296\times10^{-6}}{428\times10^{-6}}=69$	%
$\rho'=\rho_{sat}-\rho_w=1981-1000=981$	kg/m^3	$\gamma'=\gamma_{sat}-\gamma_w=19.5-9.81=9.69$	kN/m^3

Revision

In mass units		**In weight units**
$\rho_w = 1000\dfrac{kg}{m^3} = 1 g/cm^3 = 1 Mg/m^3$		$\gamma_w = 9.81 kN/m^3$
$\rho = \dfrac{10^3 \gamma}{9.81} kg/m^3$	(1.21)	$\gamma = 9.81 \times 10^{-3} \rho\, kN/m^3$
$M = \dfrac{10^3 W}{9.81} kg$	(1.18)	$W = 9.81 \times 10^{-3} M\ kN$

Table 1.6 could be completed by calculating all of the quantities in succession as in Table 1.5. Instead, formulae are derived, where necessary, for the determination of the five unknowns.

From (1.35) $$m = 100\frac{M_w}{M_s} = \frac{100 M - M_s}{M_s} = 100\left(\frac{M}{M_s} - 1\right)(\%)$$

From (1.21): $$\gamma = 9.81 \times 10^{-3} \rho = 9.81 \times 10^{-3} \times \frac{M}{V}\left(kN/m^3\right)$$

From (1.40): $$\gamma_d = \frac{\gamma}{1+m}(kN/m^3)$$

From (1.34): $\gamma_s = G_s \times \gamma_w$

From (1.33): $V_s = \dfrac{W_s}{\gamma_s}$

$$\boxed{V_s = \frac{W_s}{G_s \gamma_w}} \text{ or } \boxed{V_s = \frac{9.81 \times 10^{-3} M_s}{G_s \gamma_w}(m^3)} \qquad (1.44)$$

From (1.5): $$e = \frac{V_v}{V_s} = \frac{V - V_s}{V_s} = \boxed{\frac{V}{V_s} - 1} \text{ and } \boxed{V_s = \frac{V}{1+e}} \qquad (1.45)$$

From (1.2): $$V_a = V_v - V_w = V - V_s - V_w = V - \frac{V}{1+e} - \frac{W_w}{\gamma_w}$$

$$\boxed{V_a = \frac{eV}{1+e} - \frac{W_w}{\gamma_w}} \text{ or } \boxed{V_a = \frac{eV}{1+e} - \left(\frac{9.81 \times 10^{-3} M_w}{\gamma_w}\right) \ m^3} \qquad (1.46)$$

But $\gamma_w = 9.81 (kg/m^3)$, $V = 944\,(cm^3)$, M and M_s (kg)

$$\therefore \boxed{V_a = \frac{eV}{1+e} - (M - M_s) \times 10^{-3} \ cm^3}$$

$$\text{or } \boxed{V_a = \frac{eV}{1+e} - 10^3 \times M_w \ cm^3} \qquad (1.46a)$$

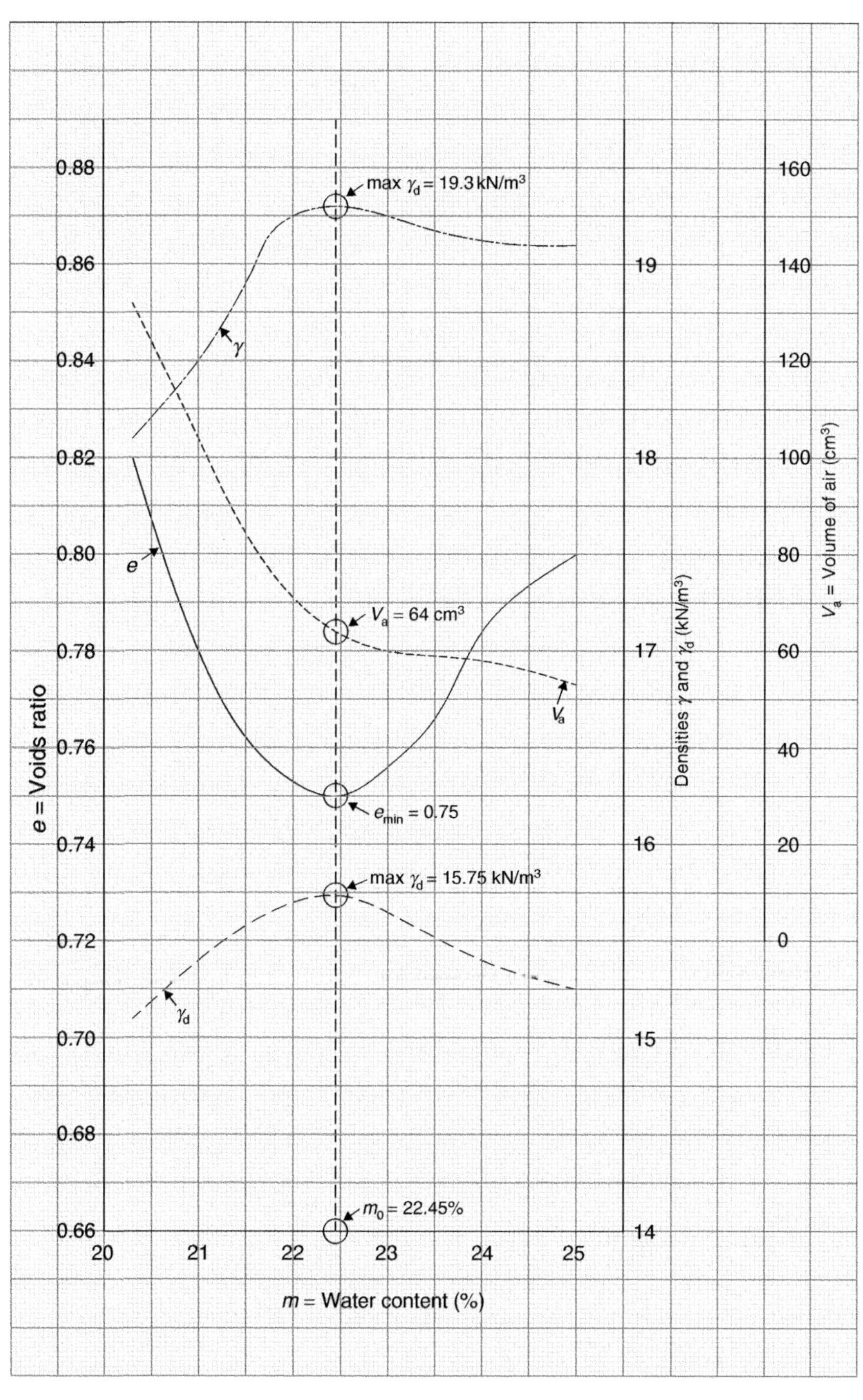
max γ_d = 19.3 kN/m³
γ
e
V_a = 64 cm³
V_a
e_{min} = 0.75
max γ_d = 15.75 kN/m³
γ_d
m_0 = 22.45%
e = Voids ratio
Densities γ and γ_d (kN/m³)
V_a = Volume of air (cm³)
m = Water content (%)
0.88
0.86
0.84
0.82
0.80
0.78
0.76
0.74
0.72
0.70
0.68
0.66
20
21
22
23
24
25
19
18
17
16
15
14
160
140
120
100
80
60
40
20
0

Graph 1.1

Table 1.6 (See Graph 1.1)

Quantity		Sample					
		1	2	3	4	5	6
M	kg	1.743	1.827	1.855	1.846	1.838	1.834
M_s	kg	1.449	1.502	1.514	1.496	1.479	1.467
m	%	20.3	21.6	22.5	23.4	24.3	25.0
γ	kN/m³	18.1	19.0	19.3	19.2	19.1	19.1
γ_d	kN/m³	15.1	15.6	15.73	15.55	15.4	15.25
e	-	0.82	0.76	0.75	0.77	0.79	0.80
V_a	cm³	132.5	82.6	62.3	59.7	56.8	53.1

Notes:

1. As m increases, the air voids hence e decrease due to compaction, resulting in higher densities.
2. At the "optimum moisture content" m_o, the densities reach their maximum values, whilst e attains its minimum. The volume of air is also reduced considerably. In this example, the changes are:

Table 1.7

	m=20%	m_o=22.45%	Change	
			+2.45	%
γ	18.1	19.3	+1.2	%
γ_d	15.1	15.75	+0.65	%
e	0.82	0.75	−0.07	–
V_a	132.5	64	−68.5	cm³

3. If the water content is increased beyond the optimum value, the soil becomes less compact. This is indicated by the decreasing values of γ and γ_d. The increase in the volume of water in the voids is reflected in the changed value of e.
4. It is not possible to compact partially saturated soil so, that all air is expelled (V_a=0). In this example, the minimum amount of air voids remaining beyond m=25% is about V_a=50 cm³.

1.2.9 Relative density (D_r)

Granular soil, sand in particular, is often described as either loose or dense. The relative density, alternatively called "density index" compares the voids ratio of sand, in its natural state, with those in its most loose and most dense states. It is formulated as:

$$D_r = 100\left(\frac{e_{max} - e}{e_{max} - e_{min}}\right)\% \qquad (1.47)$$

Where e = in-situ voids ratio
e_{min} = voids ratio in loosest state
e_{max} = voids ratio in densest state

The values of D_r tabulated below should be taken as indicative only, because of the uncertainties in obtaining minimum and maximum voids ratios or densities.

Table 1.8

Description of soil	D_r (%)
Very loose	0–15
Loose	15–35
Medium	35–65
Dense	65–85
Very dense	85–100

D_r can be expressed in terms of dry unit weight by means of formula (1.41) from which:

$$e = \frac{G_s \gamma_w}{\gamma_d} - 1$$

$$\text{and } e_{max} = \frac{G_s \gamma_w}{\min \gamma_d} - 1$$

$$\text{Also, } e_{min} = \frac{G_s \gamma_w}{\max \gamma_d} - 1$$

Substituting these into formula (1.47) we get:

$$D_r = 100 \times \frac{(\gamma_d - \min \gamma_d) \max \gamma_d}{(\max \gamma_d - \min \gamma_d) \gamma_d} \% \tag{1.48}$$

Determination of e_{max} and min γ_d

Dry sand is poured slowly into a cylinder through a funnel, keeping its end near the surface of the material to prevent compaction. When the cylinder is full, measure the weight of the contained sand.

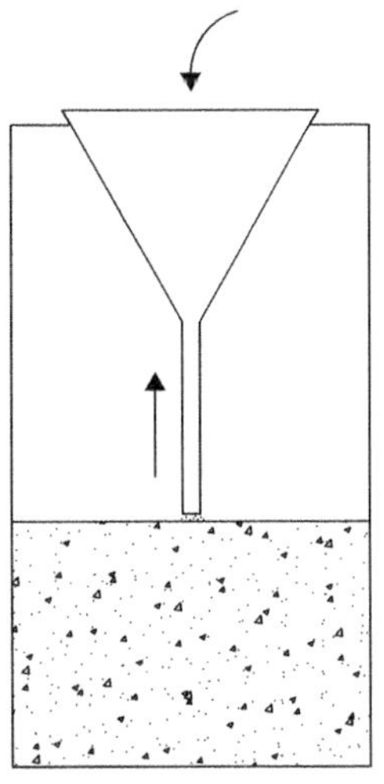

Figure 1.12

V = volume of the cylinder (m^3)

W_s = weight of soil (kN)

From formula (1.26): $\min \gamma_d = \frac{W_s}{V}$

Hence from (1.41): $e_{max} = \frac{V G_s \gamma_w}{W_s} - 1$

Determination of e_{min} and max γ_d

The sand is compacted into cylinders at different water contents. Plot the voids ratio and dry density against moisture content as in Example 1.6.

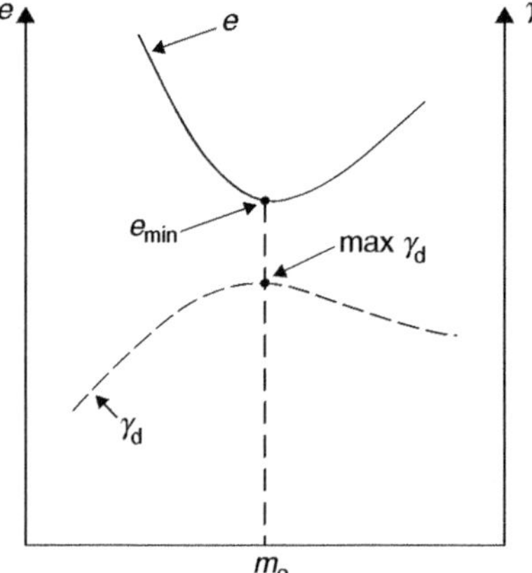

m_o = optimum moisture content, where:
Voids ratio = e_{min}
Dry density = max γ_d

Figure 1.13

Example 1.7

The results of density test conducted on sand were:

$e = 58.2\%$, $e_{max} = 62.4\%$ and $e_{min} = 41.5\%$

Calculate the relative density of the sand.

$$D_r = 100 \times \left(\frac{e_{max} - e}{e_{max} - e_{min}} \right) = 100 \times \left(\frac{62.4 - 58.2}{62.4 - 41.5} \right) = 20\%$$

The sand can be considered as loose, hence not load-bearing. It is not suitable for foundation construction.

1.3 Alteration of soil structure by compaction

It often occurs that soil has to be excavated at one place and deposited elsewhere for various reasons. Some of the reasons are:

1. Construction of embankments.
2. Construction of large horizontal areas for housing, roads, runways, etc.
3. Exchanging soil of unsuitable bearing strength with strong, compacted soil, prior to erection of structures.

The excavated soil is in loose condition; hence it has to be compacted during deposition. The purpose of compaction is to:

1. increase density by decreasing the air voids, hence the voids ratio;
2. decrease permeability by the reduction of voids;
3. increase shear strength by packing the soil particles closer.

The soil is partially saturated during compaction. The process, therefore, must not be confused with consolidation, where water is expelled from fully saturated soil, whereas in compaction, air is expelled from partially saturated one. In effect, the voids ratio in well compacted soil is low and the grains are packed so, that future consolidation settlement is minimized.

The efficiency or rather the degree of compaction is measured in terms of either dry mass density (ρ_d) or dry unit weight (γ_d) and moisture content (m%). Some amount of water added helps compaction by reducing surface tension. However, if m% is in excess of the so called "Optimum moisture content (m_o)", then the void ratio begins to increase and the soil becomes looser. The variation of γ_d, γ, e and V_a with m% is illustrated in Graph 1.1.

Soil stabilization is carried out in five stages:

1. Retrieval of soil samples from the area to be quarried.
2. The samples are compacted in a laboratory and the maximum value of γ_d at the m_o% is obtained.
3. The engineer or architect specifies these values in the earthworks contract.
4. The contractor should compact the imported soil as specified.
5. The engineer or architect initiates spot checks on site, in order to determine the in-situ dry density, hence the efficiency of the compaction.

1.3.1 Laboratory compaction tests (BS 1377-4: 1990)

There are three British standard and two American tests in use:

1. B.S. 'light' -2.5 kg rammer test
2. B.S. 'heavy' – 4.5 kg rammer test
3. B.S. vibrating hammer test
4. American (ASTM) light and heavy tests.

The British standard tests are outlined below. Figure 1.14 shows the equipment used.

B.S. 'light' test

It is carried out, using either a 1000 cm^3 or a 2305 cm^3 (CBR) mould and a 2.5 kg rammer. (See Figure 1.14). The procedure is as follows:

Step 1: Compact the soil in three layers, by dropping the rammer from a height of 300 mm. The number of drops (or blows) depends on the mould used.
1000 cm^3 mould requires 27 blows/layer
2305 cm^3 mould requires 62 blows/layer

Step 2: Obtain the mass of the soil.

Step 3: Measure its moisture content.

Step 4: Mix a little water to the soil.

Step 5: Repeat the procedure from step 1 at least five times.

Step 6: Calculate the dry unit weight and the volume of air voids for each moisture content and plot the compaction curve.

B.S. 'heavy' test

In this test a 4.5 kg rammer is applied.

Step 1: Compact the soil in five layers. The number of blows again depends on the mould used.
1000 cm^3 mould requires 27 blows/layer
2305 cm^3 mould requires 62 blows/layer
The rammer is dropped from a height of 450 mm.

Steps 2–6: As for the 'light'test.

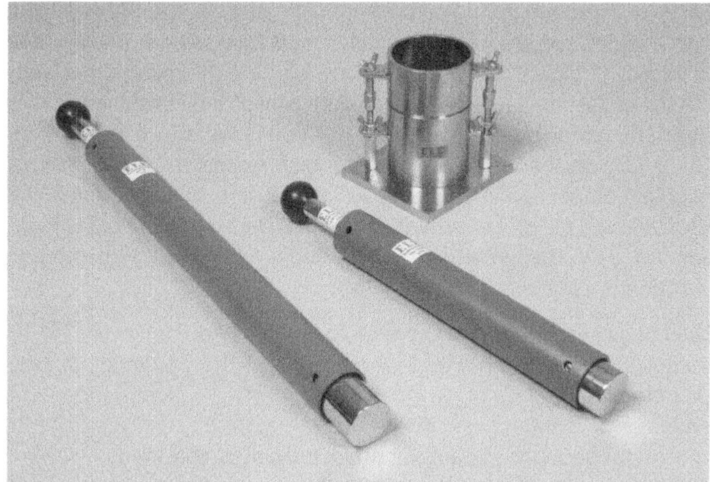

Figure 1.14 BS Compaction mould and rammers EL24-9002. Reproduced by permission of ELE International.

B.S. vibrating hammer test

The 2305 cm^3 CBR (California Bearing Ratio) mould is used in this test, which is applicable to granular soil only. The soil is compacted in three layers. Each layer is vibrated for one minute by a 32–41 kg vibrating hammer.

Notes:

a) The ASTM tests are similar in principle to the B.S. ones. The difference is in the size of the moulds, number of layers, mass of the rammer and number of blows per layer.
b) The CBR mould is normally used in the CBR test, which helps in the determination of the strength of soil layers under roads and pavements.

Presentation of results

The usual way to present the outcome of a compaction test is by plotting the dry unit weight against moisture content. Also, curves indicating 0, 5 and 10% air voids are drawn to determine the efficiency of the compaction.

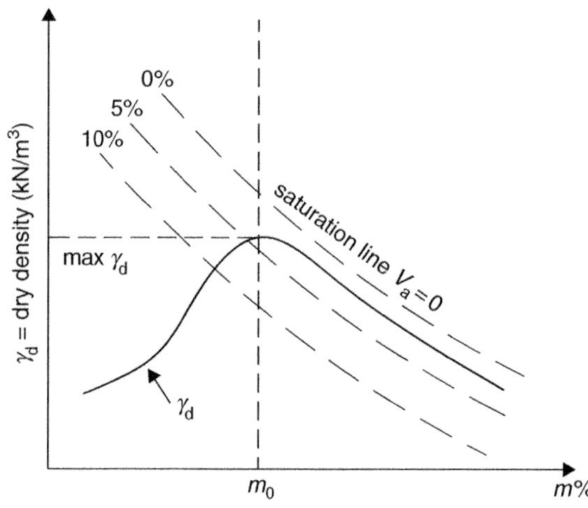

Where:
m_o = Optimum moisture content, corresponding to the maximum value of dry unit weight (max γ_d).

Figure 1.15

The dry unit weight/moisture graph is drawn by means of formula (1.40):

$$\gamma_d = \frac{\gamma}{1+m}$$

The dry unit weight has to be expressed in terms of m and percentage of air voids so, that the saturation lines may be drawn. The first step is to define the volume of air as a percentage of the total volume:

$$\boxed{P_a = 100 \times \frac{V_a}{V} \quad \%} \qquad (1.49)$$

Of course, P_a is expressed in decimals in formulae just like *m*, e, *n*, or s, that is $P_a = \frac{V_a}{V}$

From (1.3): $V = V_s + V_w + V_a$

From (1.33): $V_s = \frac{W_s}{\gamma_s}$

From (1.34): $\gamma_s = G_s \times \gamma_w$

$$V_s = \frac{W_s}{G_s\gamma_w}$$

From (1.23): $V_w = \frac{W_w}{\gamma_w}$

From (1.35a): $W_w = m \times W_s$

$$V = \frac{W_s}{G_s\gamma_w} + \frac{W_w}{\gamma_w} + V_a$$

$$I = \frac{W_s}{VG_s\gamma_w} + \frac{mW_s}{V\gamma_w} + \frac{V_a}{V}$$

$$I = \frac{\gamma_s}{G_s\gamma_w} + \frac{m\gamma_d}{\gamma_w} + \frac{V_a}{V}$$

$$I = \frac{\gamma_d}{\gamma_w} \times \left(\frac{1}{G_s} + m\right) + \frac{V_a}{V}$$

Expressing

$$\gamma_d = \frac{\left(1 - \frac{V_a}{V}\right)\gamma_w}{\frac{1}{G_s} + m}$$

Or

$$\boxed{\gamma_d = \frac{(1-P_a)G_s\gamma_w}{1+mG_s} \quad \text{kN/m}^3} \qquad (1.50)$$

For $P_a = 0\%$

$$\boxed{\gamma_d = \frac{G_s\gamma_w}{1+mG_s} \quad \text{kN/m}^3} \qquad (1.51)$$

Example 1.8

Table 1.9 contains the results of a compaction test carried out on soil to be placed in a 3 m thick layer under a heavy industrial building.

The dry unit weight (γ_d) is plotted against *m*% on Graph 1.2 and the results noted. In this example:

$$\max \gamma_d = 16.85 \text{kN/m}^3$$

$$m_o = 19.15\%$$

Table 1.9

Compaction							
Sample number		**1**	**2**	**3**	**4**	**5**	**6**
Mass of wet soil + mould	g	6141	6498	6602	6556	6441	7271
Mass of mould	g	1900	1900	1900	1900	1900	1900
M_c = mass of wet soil	g	4241	4598	4702	4656	4541	5371
V = volume of mould	cm³	2305	2305	2305	2305	2305	2305
Mass density: $\rho = M_c/V$	g/cm³	1.84	1.99	2.04	2.02	19.70	2.33
Unit weight: $\gamma = 9.81\,\rho$	kN/m³	18.1	19.5	20.0	19.8	19.3	22.9
Moisture content (G_s = 2.7)							
Mass of wet soil + container	g	173.1	140.7	121.2	129.3	142.7	153.6
Mass of container	g	8.71	10.28	7.95	8.92	9.51	8.53
M = mass of wet soil	g	164.39	130.42	113.25	120.38	133.19	145.07
M_s = mass of dry soil	g	143.01	106.30	95.14	98.02	106.20	113.16
Mass of water: $M_w = M - M_s$	g	21.38	24.12	18.11	22.36	26.99	31.91
Water content: $m = 100\,M_w/M_s$	%	14.95	17.40	19.03	22.81	25.41	28.20
Dry unit weight: $\gamma_d = \gamma/(1+m)$	kN/m³	15.7	16.7	16.9	16.2	15.4	17.8
Air voids lines							
$\gamma_d = \dfrac{(1-P_a)G_s\gamma_w}{1+mG_s}$ P_a = 0%	kN/m³	19.3	17.9	17.2	16.4	15.7	15.0
P_a = 5%		17.9	17.0	16.6	15.6	14.9	14.8
P_a = 10%		17.0	16.2	15.8	14.8	14.1	13.5

The volume of air in the soil at maximum dry density can be determined by interpolating between the 0% and 5% lines. From Graph 1.2:

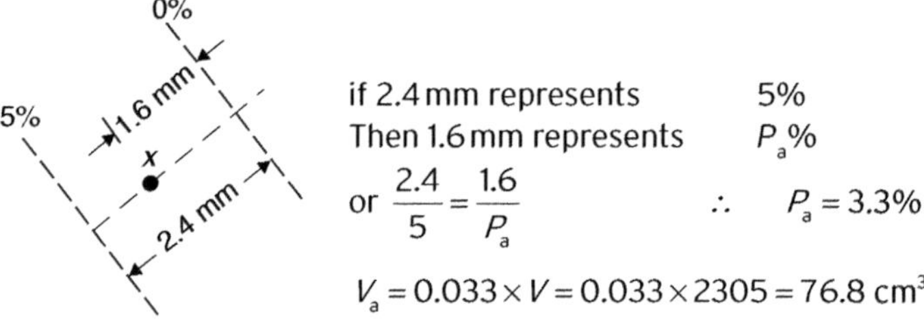

if 2.4 mm represents 5%
Then 1.6 mm represents P_a%

or $\dfrac{2.4}{5} = \dfrac{1.6}{P_a}$ $\therefore$ $P_a = 3.3\%$

$$V_a = 0.033 \times V = 0.033 \times 2305 = 76.8\ \text{cm}^3$$

Figure 1.16

Alternatively, by formula (1.50):

$$\max\gamma_d = \frac{(1-P_a)G_s\gamma_w}{1+m_o \times G_s} \quad \text{or} \quad 16.85 = \frac{(1-P_a)\times 2.7 \times 9.81}{1+0.1915\times 2.7}$$

$$1-P_a = \frac{16.85\times(1+0.1915\times 2.7)}{2.7\times 9.81} = 0.965$$

$$P_a = \frac{V_a}{V} = 1-0.965 = 0.0359\ (3.5\%)$$

$$\therefore \quad V_a = 0.035\times V = 0.035\times 2305 = 80.7\text{cm}^3$$

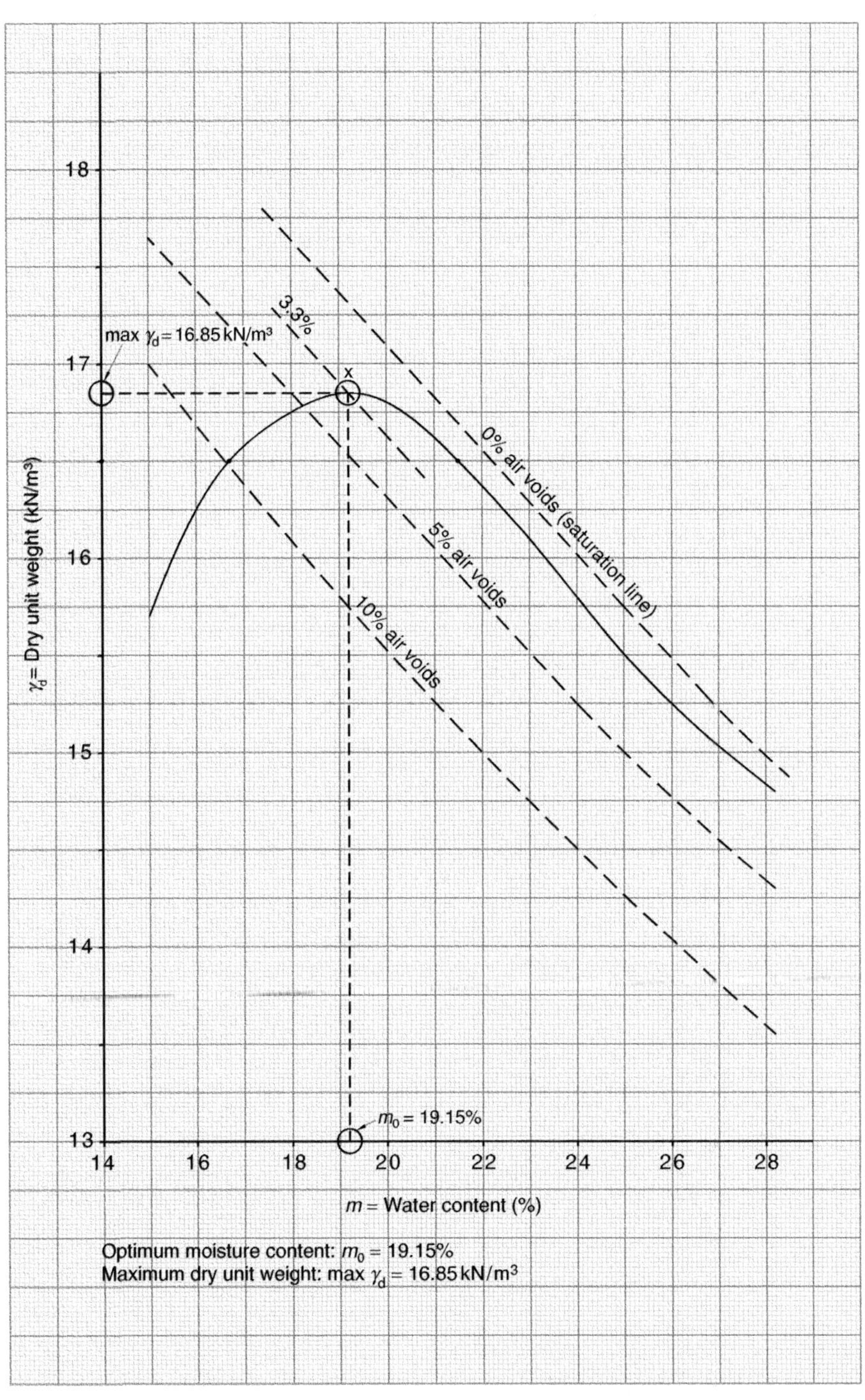

18
17
16
15
14
13
γ_d = Dry unit weight (kN/m³)
14
16
18
20
22
24
26
28
m = Water content (%)
max γ_d = 16.85 kN/m³
x
3.3%
0% air voids (saturation line)
5% air voids
10% air voids
m_0 = 19.15%
Optimum moisture content: m_0 = 19.15%
Maximum dry unit weight: max γ_d = 16.85 kN/m³

Graph 1.2

Alternatively, V_a can be determined from the basic formulae.

Known: $m_o = 19.15\%$ and $\gamma_d = 16.85\,\text{kN/m}^3$

From (1.40): $\gamma = (1+m_o)\,\gamma_d = 1.1915 \times 16.85 = 20.08\,\text{kN/m}^3$

From (1.19): $W = \gamma V = 20.08 \times \left(\dfrac{2305}{10^6}\right) = 46.3 \times 10^{-3}\,\text{kN}$

From (1.37): $W_s = \dfrac{W}{1+m_o} = \dfrac{46.3 \times 10^{-6}}{1.1915} = 38.9 \times 10^{-3}\,\text{kN}$

From (1.17): $W_w = W - W_s = (46.3 - 38.9) \times 10^{-3} = 7.4 \times 10^{-3}\,\text{kN}$

From (1.3): $V_a = V - V_w - V_s$

But, $V_w = \dfrac{W_w}{\gamma_w} = \dfrac{7.4 \times 10^{-3}}{9.81} = 0.754 \times 10^{-3}\,\text{m}^3$

$= (0.754 \times 10^{-3}) \times 10^6\,\text{cm}^3$

$= 754\,\text{cm}^3$

And $V_s = \dfrac{W_s}{\gamma_s} = \dfrac{W_s}{G_s \gamma_w} = \dfrac{38.9 \times 10^{-3}}{2.7 \times 9.81} = 1.4686 \times 10^{-3}\,\text{m}^3$

$= (1.4686 \times 10^{-3}) \times 10^6\,\text{cm}^3$

$= 1469\,\text{cm}^3$

Therefore, $V_a = 2305 - 754 - 1469 = 82\,\text{cm}^3$

In percentage terms: $P_a = 100 \times \dfrac{82}{2305} = 3.6\%$

The three results, therefore, are comparable.

1.3.2 Practical considerations

It is very much unlikely, or rather impossible to achieve the same compaction in the field as predicted by the laboratory results. It is somewhat difficult to maintain the soil at optimum water content because of rain or very dry weather. It is therefore unreasonable to expect a contractor to produce the exact dry density shown on a compaction curve. For

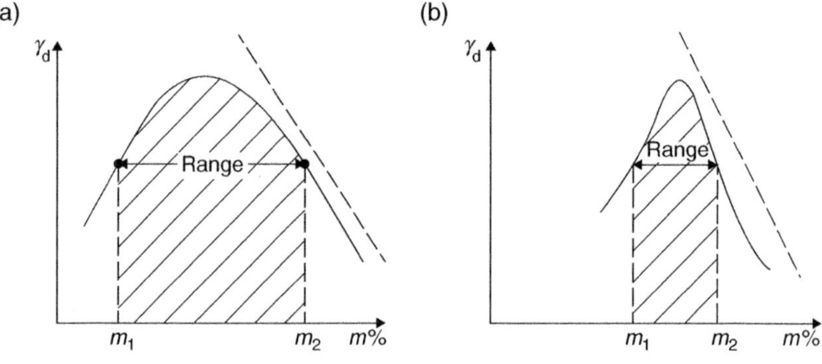

Figure 1.17

this reason, the specification should state an acceptable range of moisture content. This range is chosen by observing the variation of γ_d with m on the compaction curve. For example, the dry bulk density for $m=16.7\%$ and $m=21.6\%$ is $\gamma_d=16.5$ kN/m^3 on Graph 1.2. For this range of moisture content, the deviation from the maximum value is 16.85–16.5=0.35 kN/m^3. Should this small variation be acceptable, this range would be specified.

In general, the flatter is the compaction curve the less sensitive is the dry density to the variation in $m\%$.

Soils with flatter curves (Figure 1.17a) need less compactive effort than those with steeper ones (Figure 1.17b). On the other hand, however higher values of dry density can be achieved by soils sensitive to moisture content variations.

1.3.3 Relative compaction (C_r)

The allowed deviation described above may be specified by the relative compaction (C_r).

$$\boxed{C_r = 100 \times \frac{\overline{\gamma}_d}{\max \gamma_d} \Bigg| \%} \tag{1.52}$$

where $\overline{\gamma}_d$ = dry unit weight to be achieved in the field.

In example 1.7, for the range discussed above:

$$C_r = \frac{100 \times 16.5}{16.85} = 98\%$$

Therefore, for the latitude of moisture content variation only 2% of the density was lost. It is not desirable to depart too far from $m_o\%$ because:

1. If $m<m_o$ then more air voids can remain in the soil, after compaction, than intended.
2. If $m>m_o$, then the additional moisture could make the soil weaker than intended.

1.3.4 Compactive effort

There are various types of compactors used in the field, depending on the soil treated. The efficiency of their compacting effort is a function of the:

1. thickness of the compacted layer
2. number of passes over the layer
3. mass of the compactor
4. moisture content of the soil.

Types of compaction plant:

vibratory roller
smooth-wheeled roller
sheepfoot roller
pneumatic-tyred roller
grid roller
power rammer.

The thickness of each layer and the number of passes depend on the mass of the plant used. On the average, 4–5 passes are sufficient, as long as the roller is of large enough mass. Detailed information on the types and capability of compaction plants is available in the relevant literature. In general, vibrating rollers are applicable to cohesive as well as well graded granular soil, as long as their mass is over 1800 kg. The thickness of layers is within the 150–300 mm range.

1.3.5 Under- and overcompaction

Dry density increases, whilst the optimum moisture content decreases, and vice versa, with compactive effort.

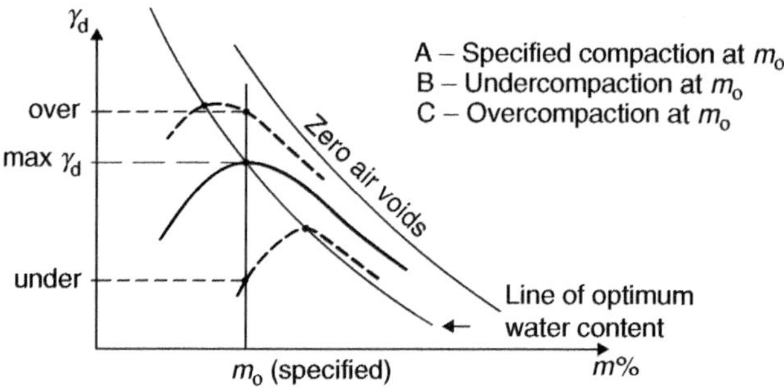

Figure 1.18

Undercompaction means that the compactive effort is less than necessary, when the soil is at the specified moisture content. It is now too dry (at point B) and not at its maximum dry density (point D) for the particular effort.

Conversely, overcompaction means unnecessary extra effort, when the soil is worked at the specified m_o%. Although the dry density is increased (point C), the soil becomes wetter than it would be at the new optimum (point E), hence it becomes weaker than intended.

1.3.6 Site tests of compaction

It is imperative to carry out daily checks on the dry density achieved by the compactor. There are five well known in-situ methods to do this:

1. core cutter method
2. sand replacement method
3. water displacement method
4. penetration needle measurement
5. nuclear radiation

Of these, only the first two will be outlined.

1.3.6.1 Core cutter method

It is applicable to cohesive soils. A cylindrical steel cutter of volume V is driven into the layer. The soil mass is measured and the dry density determined on site. The moisture content is obtained normally by the drying-out process. The cutter shown is pressed into the soil by a rammer-dolly assembly.

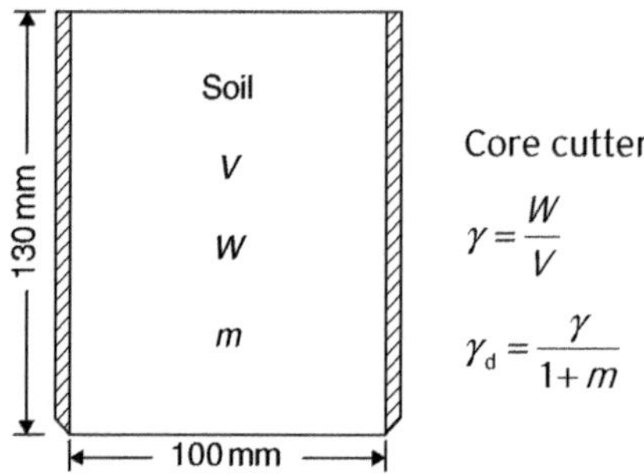

Figure 1.19

1.3.6.2 Sand replacement method

It is used mainly for granular soil, as the dimensions of the hole dug for a sample are irregular and cannot be measured normally.

Calibration of the apparatus (Figure 1.20):

a) Fill the pourer with sand.
b) Place the pourer on a flat surface and release sand, filling the cone. Weigh the sand released (M_1).

The bulk mass density of the sand (ρ_s) has to be determined.

c) Fill the pourer with sand and weigh it (M_2).
d) Place the pourer on the calibrating cylinder and release sand to fill it as well as the cone.
e) Weigh the pourer (M_3).
f) Calculate $\rho_s = \dfrac{M_2 - M_3 - M_1}{V_c}$ where V_c=volume of cylinder

Measurement of soil mass density on site:

g) Excavate a round hole, approximately 100 mm in diameter.
h) Weigh the excavated soil (M_4)
i) Completely fill the pourer with sand and place it over the hole.
j) Release sand until it fills the hole.
k) Weigh the pourer (M_5).
l) Calculate the mass of sand filling the hole: $M_s = M_2 - M_1 - M_5$
m) Calculate the volume of the hole:

$$V = \frac{M_s}{\rho_s}$$

Figure 1.20 Photograph of ELE29-4000.100 calibrating container, metal tray. Reproduced by permission of ELE International.

n) Calculate the mass density of the soil sample:

$$\rho = \frac{M_4}{V}$$

o) Determine the moisture content (m) and hence the dry density ρ_d by (1.40)

1.4 California bearing ratio (CBR) test

The test is entirely empirical and the CBR value depends on the degree of compacted that is on the dry density and the moisture content of the soil to be tested. The result is used in the design of pavements, roads and air-field runways.

Definition

The CBR value of a material is the ratio of the force required to penetrate the compacted soil to a standard force, causing the same penetration. In other words, if the standard force is 100%, then the measured force is CBR% i.e.

$$\boxed{\text{CBR} = 100 \times \left(\frac{\text{Measured force}}{\text{Standard force}}\right)\%} \qquad (1.53)$$

It can also be considered as an index of shear strength of a soil in known state of compaction.

Outline of the laboratory test

The sketch of the CBR apparatus is shown below:

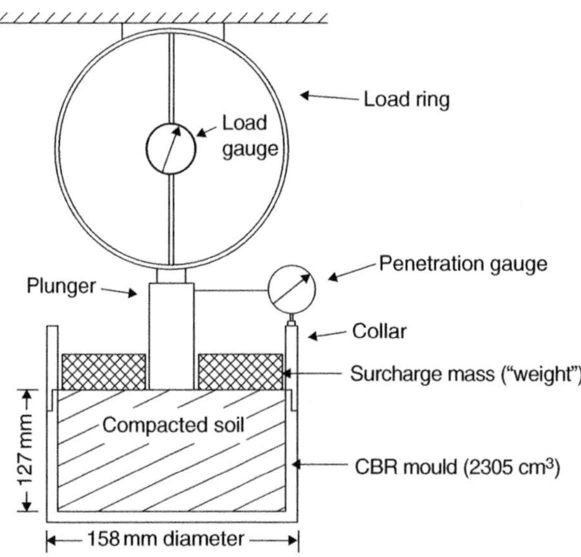

Figure 1.21

Step 1: Compact the soil in five layers into the mould by either of the 2.2 kg or 4.5 kg rammer.

Step 2: Place surcharge rings on the top of the soil, if necessary, to simulate possible overburden pressure.

Step 3: Seat the 49.6 mm diameter plunger on the surface of the soil and apply seating load according to the expected CBR value:

Table 1.10

CBR %	Seating load (N)
≤30	50
>30	250

Step 4: Start motor and read the load gauge (Q) at every 0.25 mm indicated on the penetration gauge up to 7.5 mm maximum penetration.

Step 5: Remove the soil from the mould. Obtain its moisture content and dry density.

Step 6: Calculate the applied force (P) from:

$$\boxed{P = \frac{Qk}{1000}\,\text{kN}} \tag{1.54}$$

where k = load ring factor (N / division).

Step 7: Plot the value of (P) against penetration (δ). The curve can have either of the two shapes shown:

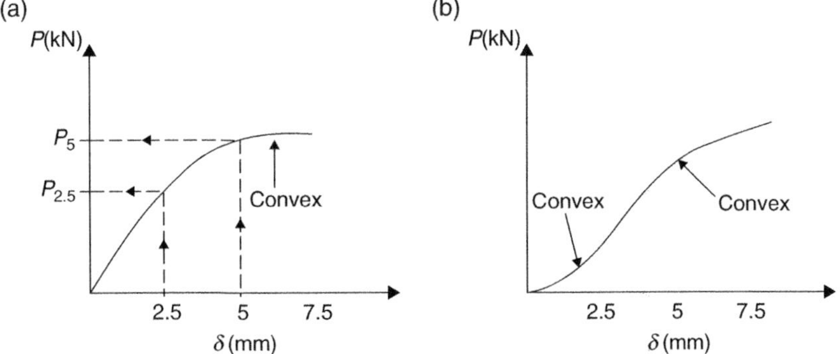

Figure 1.22

Step 8: Calculate the CBR values by comparing the loads at $\delta = 2.5$ mm and $\delta = 5$ mm to the loads on the standard 100% CBR curve at the same penetrations. The standard curve is given by:

Table 1.11

δ (mm)	2	2.5	4	5	6	8	10	12
Standard P_{100}	11.5	13.24	17.6	19.96	22.2	26.3	30.3	33.5

The results for a curve type Figure 1.22(a) can be calculated directly from formula (1.53).

$$\text{CBR1} = 100 \times \left(\frac{\text{Measured force at 2.5mm}}{\text{Standard force at 2.5mm}}\right) = \boxed{100 \times \frac{P_{2.5}}{13.24}\,\%}$$
$$\text{CBR2} = 100 \times \left(\frac{\text{Measured force at 5mm}}{\text{Standard force at 5mm}}\right) = \boxed{100 \times \frac{P_{5}}{19.96}\,\%} \tag{1.55}$$

Alternative graphical procedure: Plot the experimental curve on standard charts as shown in the following example.

Construction of the charts

In order to draw the curve for a particular CBR% (say CBR=60%), the standard force in Table 1.11 has to be multiplied by 0.6 at each penetration. Table 1.12 contains the figures for CBR=12% and 60%

Table 1.12

δ (mm)	2	2.5	4	5	6	8	10	12
Standard P_{100}	11.5	13.24	17.6	19.96	22.2	26.3	30.3	33.5
12% CBR=0.12 P_{100}	1.38	1.59	2.11	2.40	2.66	3.16	3.64	4.02
60% CBR=0.6 P_{100}	6.90	7.94	10.56	11.98	13.32	15.78	18.18	20.10

The table can be completed this way for any CBR value and either one or several charts drawn. In this case, Chart 1.1a has been drawn for easier interpretation under CBR=12%.

Example 1.9

Two soils A and B were tested in the CBR mould and the results tabulated. Determine the CBR value for each, analytically and graphically.

Table 1.13

δ (mm)	P_A(kN)	P_B(kN)	δ(mm)	P_A(kN)	P_B(kN)
0.00	0.00	0.00	4.00	0.98	7.14
0.25	0.01	1.10	4.25	1.07	7.28
0.5	0.03	2.00	4.50	1.16	7.42
0.75	0.05	2.82	4.75	1.26	7.58
1.00	0.07	3.51	**5.00**	1.34	**7.65**
1.25	0.09	3.95	5.25	1.39	7.74
1.50	0.13	4.22	5.50	1.47	7.80
1.75	0.17	4.80	5.75	1.54	7.81
2.00	0.23	5.21	6.00	1.10	7.85
2.25	0.26	5.53	6.25	1.64	7.88
2.50	0.34	**5.84**	6.50	1.72	7.92
2.75	0.43	6.17	6.75	1.76	7.95
3.00	0.54	6.35	7.00	1.79	7.98
3.25	0.66	6.57	7.25	1.83	7.99
3.50	0.77	6.76	7.50	1.87	8.00
3.75	0.87	6.98	-	-	-

Note: It is prudent to draw the load-penetration curves in order to ascertain their shapes. A correction has to be made if the shape is as indicated in Figure 1.22(b). In this example, P_A is plotted on Chart 1.1a. Its curve is convex downwards near the origin, hence it has to be corrected as shown.

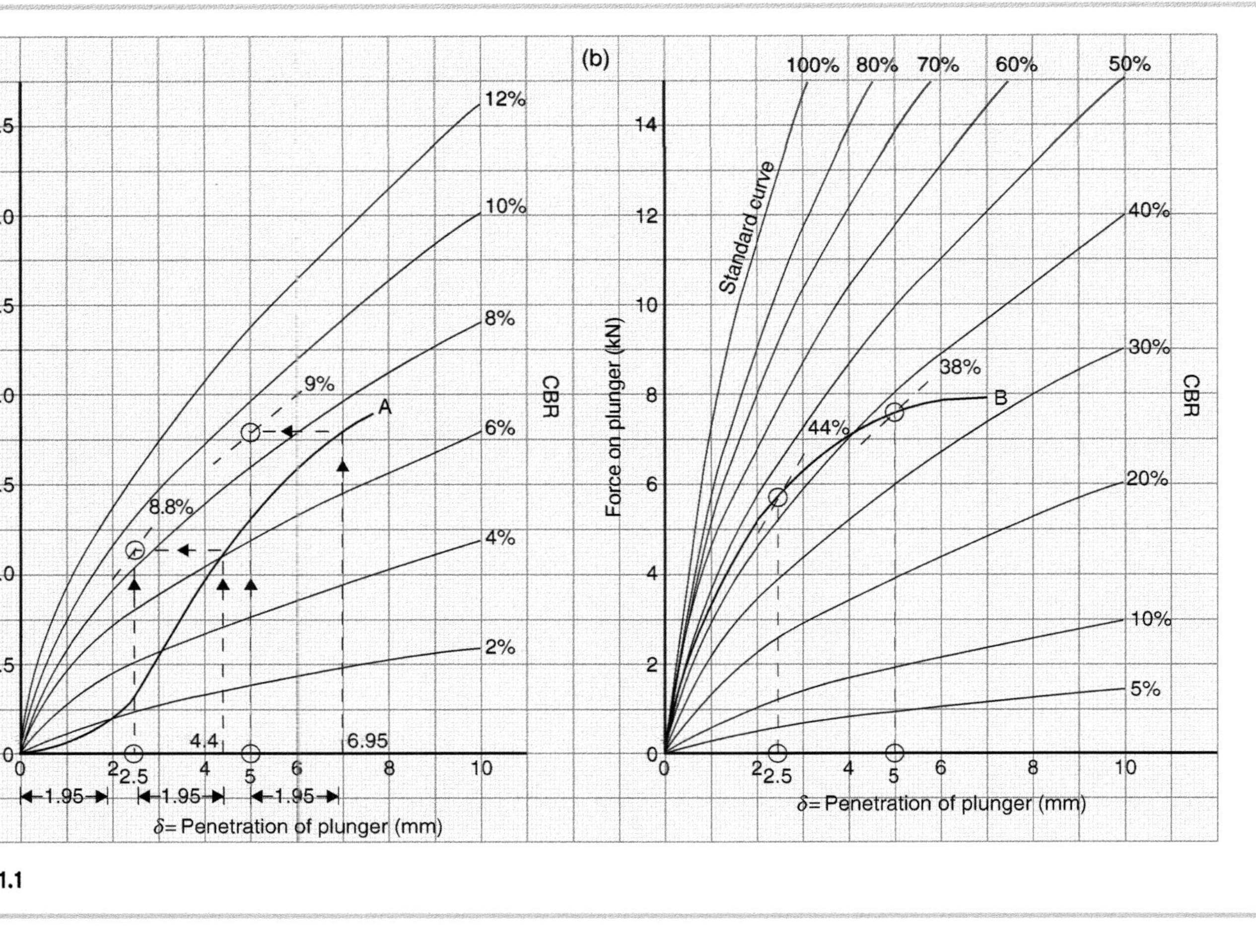

(a)
Force on plunger (kN)
3.5
3.0
2.5
2.0
1.5
1.0
0.5
0
12%
10%
8%
6%
4%
2%
CBR
9%
8.8%
A
4.4
6.95
1.95
1.95
1.95
δ= Penetration of plunger (mm)
0
2
2.5
4
5
6
8
10
(b)
Force on plunger (kN)
14
12
10
8
6
4
2
0
100%
80%
70%
60%
50%
40%
30%
20%
10%
5%
CBR
Standard curve
38%
44%
B
δ= Penetration of plunger (mm)

Chart 1.1

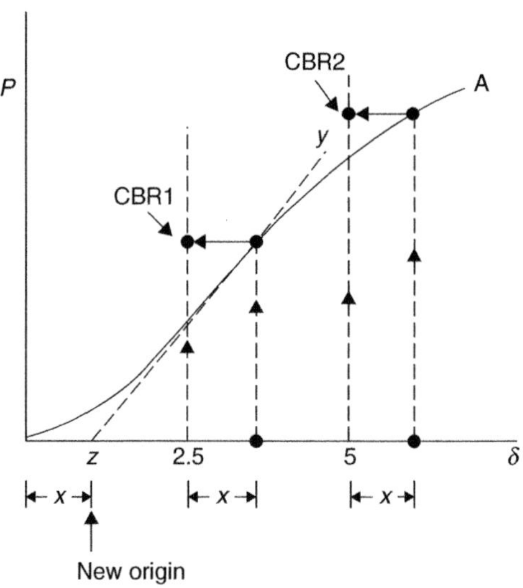

Figure 1.23

Draw line y–z. Point z is the new origin for δ. Interpolate for the two CBR values in Chart 1.1a.

From Chart 1.1a: $x = 1.95 \quad \therefore \quad 2.5 + x = 4.45\,\text{mm}$

And $5 + x = 6.95\,\text{mm}$

Interpolating between curves 8% and 10%.

$$\text{CBR}\,1 = 8.8\%$$
$$\text{CBR}\,2 = 9\%$$

Accept the larger figure as the CBR value for material A, that is 9%.

Curve B for P_B is convex upwards along its entire length, hence no correction is necessary. In this case, it is easier to calculate the CBR values by formula (1.55), then by interpolation on Chart 1.1b.

From Table 1.13: For $\delta = 2.5$ mm, $P_B = 5.84$ kN $\quad \therefore \text{CBR1} = 100 \times \dfrac{5.84}{13.24} = 44.1\%$

For $\delta = 5$ mm, $P_B = 7.65$ kN $\quad \therefore \text{CBR2} = 100 \times \dfrac{7.65}{19.96} = 38.3\%$

The larger answer is taken as the CBR value rounded to 44%. The graphical solution is given on Chart 1.1b.

Note: The shape of curve A near the origin is assumed to be due to inadequate compaction of the surface layer compared to the rest of the material.

Comparative CBR values of various soils are tabulated below.

Table 1.14

Soil type	Plasticity index (PI, %)	CBR (%)	Strength
Heavy clay	70	1–2	Weak
	60	1.5–2.0	
	50	2.0–2.5	
	40	2–3	
Silty clay	30	3–5	Normal
Sandy clay	20	4–6	
	10	5–7	
Sand	Non-plastic		Stable
Poorly graded	Non-plastic	10–20	
Well graded	Non-plastic	15–40	
Sandy gravel	Non-plastic	20–60	

Guidance is given in Road Note 29 of the Road Research Laboratory as to the design of flexible and concrete roads and pavements in terms of CBR values and estimated traffic intensities.

1.5 The pycnometer

It is a glass jar, fitted with a conical screw-top with a 6m circular orifice at the apex, as shown schematically below. The pycnometer is used to determine:

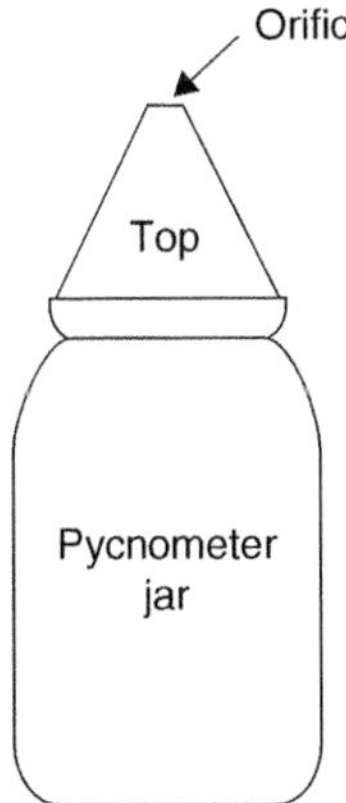

Figure 1.24

1. Specific gravity (G_s)
2. Moisture content (m)

The test is based on formula (1.34).

$$G_s = \frac{\rho_s}{\rho_w} = \frac{\dfrac{M_s}{V_s}}{\dfrac{M_w}{\gamma_s}} = \frac{M_s}{M_w} = \frac{\text{Mass of soil}}{\text{Mass of equal volume of water}}$$

It is necessary to find that mass of water, which is displaced by the soil.

Notes: The units used are:

a) volume in cm^3
b) mass in grams.

Therefore:

$$\rho_w = 1 g/cm^3$$

$$G_s = \frac{\rho_s}{1} = \rho_s$$

$$M_w = \rho_w \times V_w = V_w \text{ cm}^3$$

Outline of the test

Step 1: Weigh the empty jar + top (M_p). Fill the pycnometer to the orifice with water. Weigh the pycnometer + water (M_1)

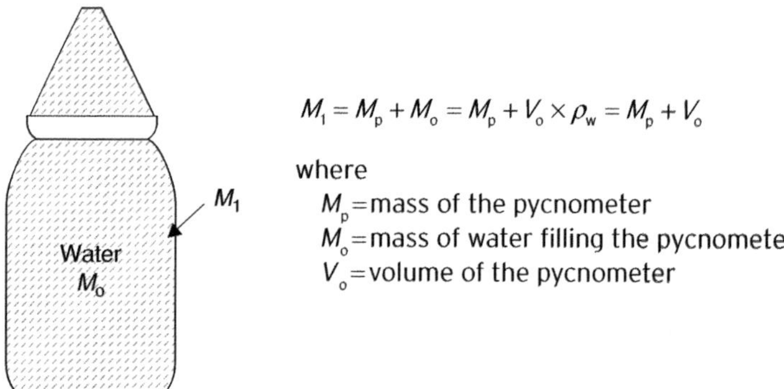

$$M_1 = M_p + M_o = M_p + V_o \times \rho_w = M_p + V_o$$

where
M_p = mass of the pycnometer
M_o = mass of water filling the pycnometer
V_o = volume of the pycnometer

Figure 1.25

Step 2: Place about 200 g oven-dried fine-grained or 400 g coarse-grained soil (M_s) into the dry, empty pycnometer. Add water at room temperature. Stir the mixture to remove air bubbles.

Step 3: Screw on the conical top and fill the pycnometer completely. Cover the orifice with a finger and shake the jar to remove any air from the soil and water.

Step 4: Weigh the pycnometer + soil + water (M_2).

Step 5: Apply the formulae derived below.

1. **Derivation of G_s** (in terms of dry mass M_s)
   In order to determine the mass (M_A) of the displaced water, imagine that **wet** soil of mass M is dropped into the full pycnometer and the discharge is collected.

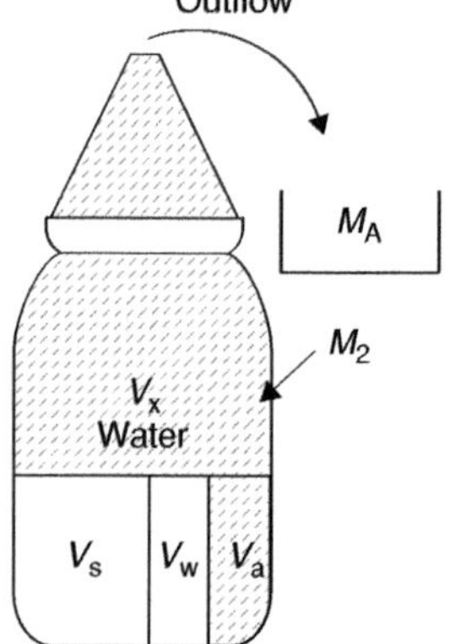

Figure 1.26

Volume of partially saturated soil:

$$V = V_s + V_w + V_a$$

Mass of partially saturated soil:

$$M = M_s + M_w$$
$$= V_s \times \rho_s + V_w \times \rho_w = V_s \times G_s + V_w$$

Volume of water displaced from the jar by the volume of solids and water in the soil:

$$V_A = V_x + V_a = V_o - (V_s + V_w)$$

The mass of the displaced water is given by:

$$M_A = V_A \times \rho_w = V_A = V_o - V_s - V_w = M_o - \frac{M_s}{G_s} - M_w$$

The formula in terms of the dry mass (M_s) and the measurements made (M_1 and M_2) during the test:

From step 1: $M_1 = M_p + M_o$

From step 2:
$$M_2 = M_p + M + (V_x + V_a)\times\rho_w$$
$$= M_p + M + M_A$$

Change in the mass of pycnometer in steps 1 and 2.

$$\Delta M = M_2 - M_1 = M_p + M + M_A - M_p - M_o$$
$$= M_p + (M_s + M_w) + \left(M_o - \frac{M_s}{G_s} - M_w\right) - M_p - M_o$$
$$= M_s + M_w + M_o - \frac{M_s}{G_s} - M_w - M_o$$

$$M_2 - M_1 = M_s - \frac{M_s}{G_s} = \left(1 - \frac{1}{G_s}\right)\times M_s = \left(\frac{G_s - 1}{G_s}\right)\times M_s$$

Expressing dry mass:
$$\boxed{M_s = (M_2 - M_1)\left(\frac{G_s}{G_s - 1}\right)} \qquad (1.56)$$

Therefore, should G_s of a partially or fully saturated soil be known from another source, then its dry mass can be determined by a pycnometer.

Expressing G_s from:
$$M_2 - M_1 = \left(1 - \frac{1}{G_s}\right)\times M_s$$

$$\frac{1}{G_s} = 1 - \frac{M_2 - M_1}{M_s} = \frac{M_s + M_1 - M_2}{M_s}$$

Hence,
$$\boxed{G_s = \frac{M_s}{M_s + M_1 - M_2}} \qquad (1.57)$$

2. **Derivation of *m*** (in terms of total mass *M*)

From (1.35a):
$$m = \frac{M_w}{M_s} = \frac{M - M_s}{M_s} = \frac{M}{M_s} - 1$$
$$= \frac{M}{(M_2 - M_1)\times\left(\frac{G_s}{G_s - 1}\right)} - 1$$

Hence,
$$m = \left(\frac{M}{M_2 - M_1}\right)\times\left(\frac{G_s - 1}{G_s}\right) - 1 \qquad (1.58)$$

Therefore, the moisture content can be found by a pycnometer if G_s is known.

Example 1.10

A partially saturated soil specimen, weighing 1743 g was tested by placing its oven-dried mass of 1449 g in a pycnometer. The following results were obtained:

Step 1: $M_p = 610\,g$

$M_1 = 1923\,g$

Step 4: $M_2 = 2854\,g$

Calculate:

a) volume of the pycnometer
b) specific gravity
c) moisture content of the soil
d) volume of water
e) volume of solids

Check: Total and dry mass of the specimen

a) Volume of pyconmeter = volume of water to fill it

$$M_o = M_1 - M_p = 1923 - 610 = 1313\,g$$

b) From (1.57): $G_s = \dfrac{M_s}{M_s + M_1 - M_2} = \dfrac{1449}{1449 + 1923 - 2854} = 2.8$

c) From (1.58):

$$m = \left(\frac{M}{M_2 - M_1}\right) \times \left(\frac{G_s - 1}{G_s}\right) - 1 = \left(\frac{1743}{2854 - 1923}\right) \times \left(\frac{2.8 - 1}{2.8}\right) - 1 = 0.203$$

d) From (1.35): $m = \dfrac{M_w}{M_s}$ But, $M_w = V_w$ $\quad \therefore \quad V_w = m \times M_s = 0.203 \times 1449 = 294\text{ cm}^3$

e) From (1.32): $V_s = \dfrac{M_s}{\rho_s}$ But, $\rho_s = G_s$ $\quad \therefore \quad V_s = \dfrac{M_s}{G_s} = \dfrac{1449}{2.8} = 518\,\text{cm}^3$

Check: (1.56): $M_s = (M_2 - M_1) \times \left(\dfrac{G_s}{G_s - 1}\right) = (2854 - 1923) \times \left(\dfrac{2.8}{1.8}\right) = 1448\text{ g}$

And $M = M_s + M_w = V_s \times G_s + V_w = 518 \times 2.8 + 294 = 1744\,g$

Note: See also Supplementary problem 1.11.

Problem 1.1

A site test was carried out in order to check the compacting efficiency of a contractor. 1500 cm³ soil was removed and tested. The available results are:

Volume of sample:	$V=1500\,cm^3$
Dry Density:	$\gamma_d=17\,kN/m^3$
Degree of saturation:	$S_r=53\%$
Specific gravity:	$G_s=2.7$

Calculate:

V_a = Volume of air in the sample
W_w = weight of water in the sample
M_w = mass of water in the sample

Problem 1.2

A compacted, partially saturated sand sample has to be fully saturated by the addition of water. Calculate, in the light of the following information, the weight of water to be added.

Volume of sample:	$V=5260\,cm^3$
Water content:	$m_1=15\%$
Porosity:	$n=35\%$
Specific gravity:	$G_s=2.67$

Problem 1.3

The following information is known about a sample of soil:

Volume:	$V=3000\,cm^3$
Water content:	$m=15\%$
Specific gravity:	$G_s=2.65$
Submerged density:	$\gamma'=8.69\,kN/m^3$

Calculate:

a) How many percent of voids are filled with water?
b) Weight of the pore water.
c) Mass of the pore water.

Problem 1.4

Starting from formula (1.38), expressing the bulk unit weight of partially saturated soil, derive the formulae:

1. $\gamma = \left(\frac{1+m}{m}\right)\left(\frac{e}{1+e}\right) S_r \gamma_w$ 2. $\gamma = (1+m)(1-n) G_s \gamma_w$	Partially saturated soil
3. $\gamma_{sat} = [(1-n) G_s - n] \gamma_w$ 4. $\gamma_{sat} = \left(\frac{1+m}{m}\right)\left(\frac{e}{1+e}\right)\gamma_w$	Saturated soil
5. $\gamma_d = (1-n) G_s \gamma_w$ 6. $\gamma_d = \gamma_{sat} - n\gamma_w$	Dry soil
7. $\gamma' = \frac{(e-m)\gamma_w}{(1+e)m}$ 8. $\gamma' = \gamma_d - (1-n)\gamma_w$	Submerged soil

Problem 1.5

Test of site compaction was carried out by means of sand pourer equipment. The apparatus was calibrated prior to its application.

The results were:

Calibration stage

Mass of pourer and sand =4.991 kg
Mass of sand released into the cone =0.58 kg

Final mass of pourer after filling cylinder and cone =1.19 kg
Volume of cylinder=2000 cm^3

Testing stage

Mass of excavated soil =2.574 kg
Water content of excavated soil =19%
Mass of pourer after filling the hole =2.321 kg

Estimate the dry density of the compacted soil.

Problem 1.6

Dry sand weighing 100 kg, is compacted to a voids ratio of 52%. The available information on the sand is:

Specific gravity = 2.66
Minimum voids ratio = 31%
Density Index = 40%

Estimate:

a) the Volume of sand in its most loose, most dense as well as compacted state.
b) the moisture content in the above three states, given that the degree of saturation is 80% in each case.
c) The saturated density in all three states in kN/m^3.

Problem 1.7

The weight of 2.5 m^3 saturated soil is 48.5 kN. Given that the specific gravity of the material is 2.7, calculate the volume of water in the voids.

Problem 1.8

An embankment of 12 m^2 cross-sectional area is to be constructed. Site survey indicates that 40,000 m^3 suitable soil of $G_s = 2.66$ can be excavated near the site. Tests carried out on 1 m^3 of the material yielded the following average values:

$$W = 18.1\,\text{kN}$$
$$W_s = 16\,\text{kN}$$

If the soil is compacted at in-situ moisture content to dry density = 18.2 kN/m^3, then:

a) compare the voids ratios as well as the percentages of air in the excavated and compacted soil.
b) determine the length embankment, that can be built with the available material, in kilometres.

Problem 1.9

Partially saturated clay was tested and its characteristics calculated. Most of the results were lost however, except the following four:

Volume:	$V=0.15\,m^3$
Moisture content:	$m=12\%$
Degree of saturation:	$S_r=49\%$
Dry unit weight:	$\gamma_d=16\,kN/m^3$

Find:

a) Bulk unit weight
b) Voids ratio
c) Specific gravity
d) Saturated density

Problem 1.10

Suppose the available results in Problem 1.9 are:

$m=12\%$
$e=66\%$
$\gamma=17.9\,kN/m^3$
$V=0.15\,m^3$

Calculate:

1. Total weight
2. Weights of solids and water
3. Volumes of solids, water and voids.

Problem 1.11

The results of pycnometer test, carried out on a saturated specimen are:

$M=519\,g$
$M_s=412\,g$
$M_1=1923\,g$
$M_2=2185\,g$

Show that, for a saturated soil, the entire list of soil characteristics (see Table 1.2 and 1.5) can be determined by the pycnometer test.

Chapter 2
Classification of Cohesive Soils

The engineering properties of fine-grained soils depend largely on their moisture content. For the same $m\%$, the consistency of two soils could be markedly different. For example, at $m=29\%$ a sample of silt could be very soft, whilst clay would be somewhat stiff and unpliable. This fact can, therefore, be conveniently used to differentiate between different types of soil.

2.1 Atterberg Limits

The accepted empirical method for the determination of consistency was devised by A. Atterberg (1911), based on the fact that, if sufficient water is added to a soil, its state changes from solid to liquid. Considering it the other way round, if the soil in liquid state is dried, it solidifies. The transformation during the drying process occurs in three stages as defined by the three Atterberg Limits:

1. Liquid Limit (LL)
2. Plastic Limit (PL)
3. Shrinkage Limit (SL)

2.1.1 Liquid Limit (LL)

When soil paste, wet enough to flow under its own weight, is dried, it changes into a plastic mass. The moisture content at which the change occurs is the Liquid Limit (LL). The volume of soil has decreased and it is fully saturated ($S_r=1$) at this stage.

In general, the value of LL increases as the size of soil particles gets smaller. Average values for:

sandy loam:	15%<LL<20%
silty soil:	20%<LL<50%
clay:	40%<LL<80%

If the natural moisture content of a soil is near to its LL, then it is susceptible to large deformation and shear failure when loaded.

Usually, the Liquid Limit of a soil is defined as that moisture content at which the soil passes from a plastic to a liquid state.

Introduction to Soil Mechanics, First Edition. Béla Bodó and Colin Jones.

2.1.1.1 Tests to find LL

The standard test was introduced by Professor Casagrande (1932). The apparatus employed is illustrated in Figure 2.1. This test is now less favoured than the Cone Penetration test.

Figure 2.1 ELE Liquid Limit device with assessories EL24-0410. Reproduced by permission of ELE international.

The test is described fully in BS1377-4:1990 as well as in laboratory manuals, hence only its outline is given below:

Step 1: Break up about 200 g dry soil and remove particles not passing the 0.425 mm sieve.
Step 2: Place about 100–130 g of the material on the glass plate, mixing it thoroughly with a small amount of distilled water, until the mass becomes a thick paste.
Step 3: Making sure that the cup of the apparatus rests on the rubber block, place some of the paste into the cup and level it off parallel with the base to a maximum depth of 10 mm.
Step 4: Cut a groove in the sample, at right angles to the crank, by drawing the grooving tool across it.
Step 5: Turn the crank of the apparatus at a rate of two rotations per second, until the two parts of the soil come into contact, along the bottom of the groove, for a distance of 13 mm. Record the number of rotations, that is, the number of blows *N*.
Step 6: Transfer a small quantity of the tested soil into a container for the determination of its moisture content.
Step 7: Repeat operations 2–6 at least four times, using the same sample. Each time add a small amount of water, thus making the soil progressively wetter. The moisture contents have to be so chosen, that the number of blows should progressively increase, within the range 10–50.
Step 8: Obtain the moisture content, corresponding to each test.
Step 9: Plot $m\%$ against $\log N$ and draw a straight line (the flow curve) though the points.
Step 10: Draw a vertical line from $N=25$ to intersect the flow curve and read off the corresponding moisture content. This value of m is the Liquid Limit.

Example 2.1

The results of five Liquid Limit tests on a clay sample are tabulated below (Table 1). Specific gravity was found to be 2.75.

Table 2.1

Test number	**1**	**2**	**3**	**4**	**5**
Moisture content (m%)	72	68	62	59	57
Number of blows (N)	70	16	44	60	80
Log N	1	1.2	1.64	1.78	1.9

Estimate:

1. the Liquid Limit by plotting the figures on Graph 2.1
2. the voids ratio at LL.

1. From the graph: $m = \text{LL} = 65.4\%$
2. As the sample is saturated at LL, its voids ratio is given by $e = M\,G_s = 0.654 \times 2.75 = 1.8$. Note that e can be larger than unity. This means, that the volume of voids is larger than the volume of solids.

2.1.1.2 Liquid Limit from a single test

According to the results of investigations by the US Waterways Experiment Station, Vicksburg in 1949, the Liquid Limit may be found from:

$$\boxed{\text{LL} = m\left(\frac{N}{25}\right)^{0.121}} \tag{2.1}$$

where N is the number of blows corresponding to water content m in a single test.

Example 2.2

Calculate the LL in Example 2.1 from three individual tests.

From $m = 72\%, N = 10$ $\quad \text{LL} = 0.72 \times \left(\frac{10}{25}\right)^{0.121} = 0.644 \quad (64.4\%)$

From $m = 62\%, N = 44$ $\quad \text{LL} = 0.62 \times \left(\frac{44}{25}\right)^{0.121} = 0.664 \quad (66.4\%)$

From $m = 57\%, N = 80$ $\quad \text{LL} = 0.57 \times \left(\frac{80}{25}\right)^{0.121} = 0.656 \quad (65.6\%)$

These figures compare well with $m = 65.4\%$. Theoretically, therefore LL can be found from a single test. It is, however, prudent to prevent major blunders, by repeating operations 2-6 at least three times and taking the average value.

Explanation: It was found that the flow curves of different soils were parallel lines, when the graph was drawn on log-log paper (log m% against log N), instead of semi-log one (m% against log N) as on Graph 2.1. Formula (2.1) was derived on this basis.

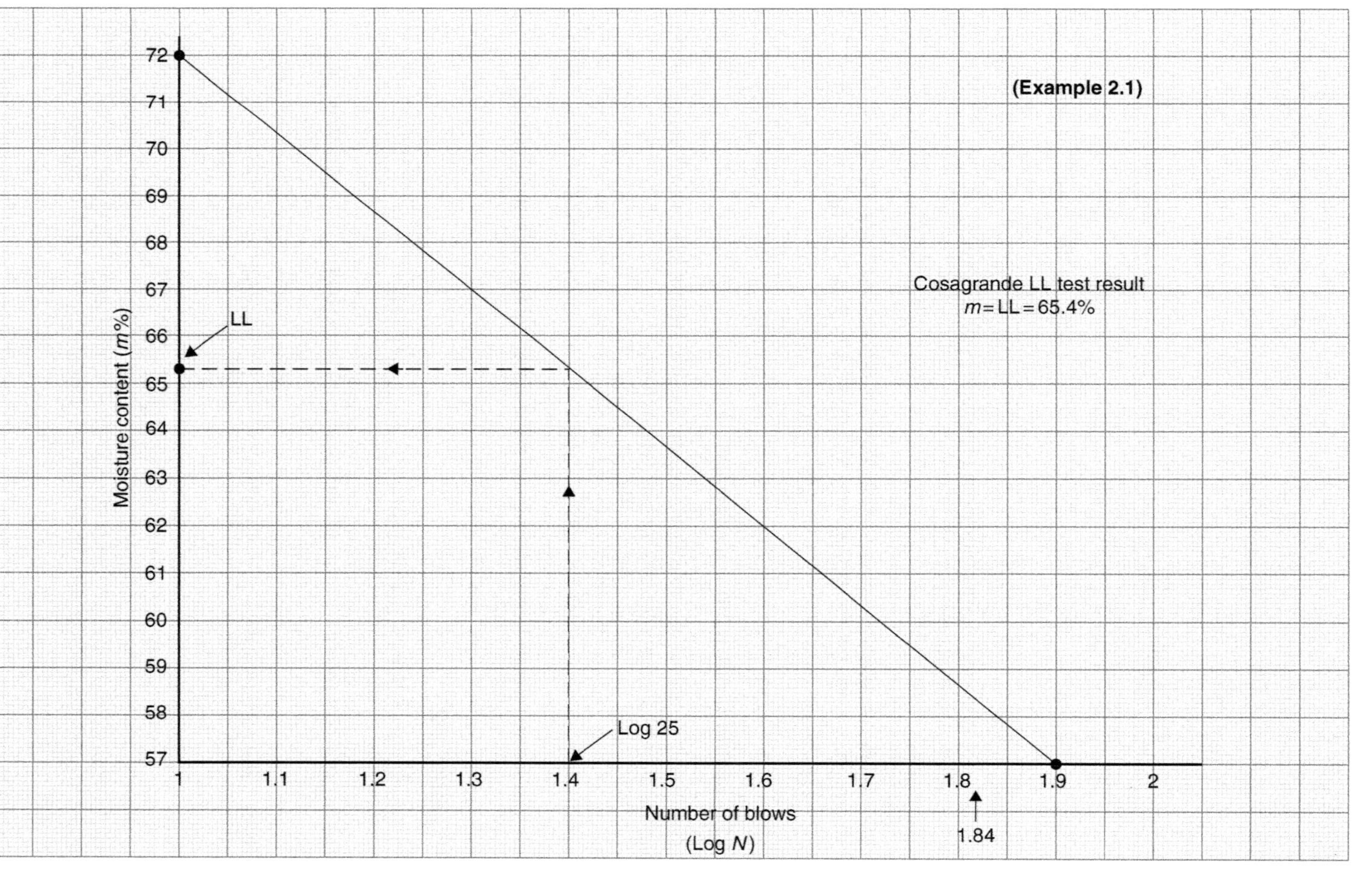
(Example 2.1)
Cosagrande LL test result
m = LL = 65.4%
Moisture content (m%)
72
71
70
69
68
67
66
65
64
63
62
61
60
59
58
57
LL
Log 25
1
1.1
1.2
1.3
1.4
1.5
1.6
1.7
1.8
1.9
2
1.84
Number of blows
(Log N)

Graph 2.1

2.1.1.3 Cone penetrometer test

This is an alternative method for the measurement of Liquid Limit. It has some advantages over the standard Casagrande procedures. For one, the result is largely independent of human and instrumental errors.

The main parts of the penetrometer are outlined below:

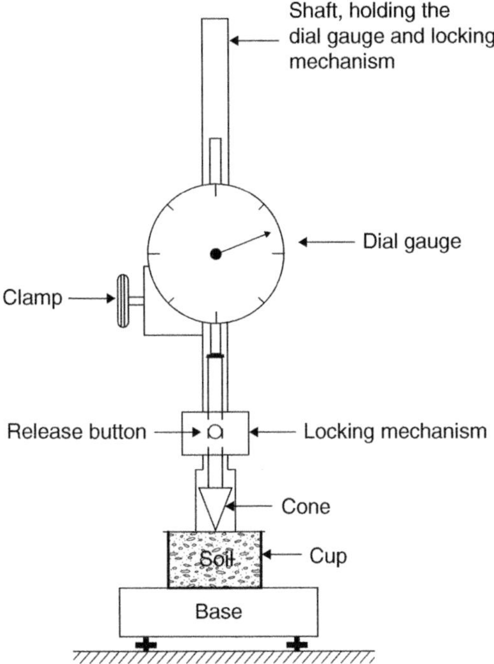

Figure 2.2

Test procedure

Step 1: As for the Casagrande test.
Step 2: As for the Casagrande test.
Step 3: Fill the cup with the paste and level its surface.
Step 4: Lower the cone so that its point just touches the surface of the specimen. Note the reading on the gauge.
Step 5: Release the cone for 5 seconds and read the gauge at the end of this period. The cone penetration is the difference between the two readings.
Step 6: Make the surface of the sample good, by adding soil at the same water content and repeat steps 4-5. Take the average value of the two results, if the difference between them is less than 0.5 mm. If the difference is more, then carry out a third test at the same value of $m\%$.
Step 7: Repeat steps 3-6 at slightly different water content at least four times.
Step 8: Plot the values of penetrations (in mm) against $m\%$. In this test the Liquid Limit is the water content corresponding to 20 mm penetration.

Example 2.3

The results of cone penetrometer tests on clay are tabulated below (Table 2.2). The natural moisture content was $m = 38\%$.

Table 2.2

Sample number	1	2	3	4	5
Moisture content (%)	19	31	40	57	71
Penetration (mm)	15.5	16	17.5	19	21.5

Estimate:

1. the Liquid Limit
2. the penetration when $m = 38\%$

1. From Graph 2.2: LL = 61%
2. Also Graph 2.2: Penetration = 17.3 mm

2.1.2 Plastic Limit

As the soil gets drier, it reaches a water content below which the material becomes friable. This particular value of $m\%$ is the Plastic Limit (PL) of the soil.

Usually, the Plastic Limit is defined as the lowest moisture content at which the soil is plastic. At PL the soil crumbles and falls apart easily, under pressure. Average values for:

Sandy loam: $17\% < PL < 20\%$
Silt: $20\% < PL < 25\%$
Clay: $25\% < PL < 35\%$

If the natural moisture content of a soil is near its PL then, from an engineering point of view, it is easy to:

1. excavate;
2. compact to its smallest volume.

2.1.2.1 Test to find PL (outline)

Step 1: Take the soil set aside during the LL test and add just enough water to make it plastic, after mixing it thoroughly.

Step 2: Roll the plastic soil on a glass plate, under the palm of your hand with just enough pressure to form a 3 mm diameter thread.

Step 3: If the thread crumbles, then measure its moisture content. This value of $m\%$ is the PL of the soil. If however, the thread does not crumble, then knead the soil into a ball and repeat step 2. Repeat step 3 as many times as required to reduce the amount of water in the specimen to PL.

Note: During the drying-out process, the soil remains saturated. At PL, therefore $S_r = 1$, although the accuracy is subject to the skill of the technician.

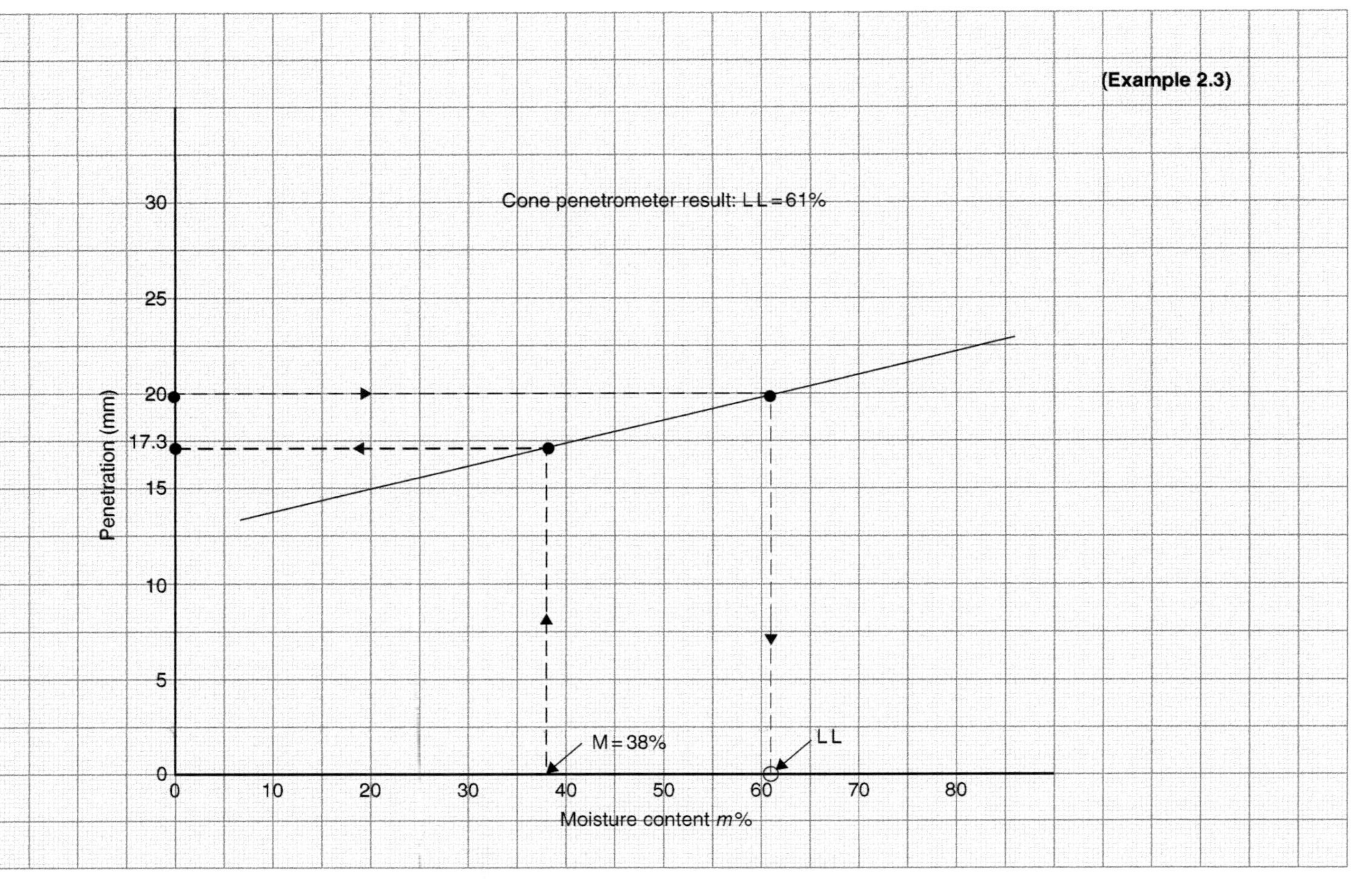
(Example 2.3)
Cone penetrometer result: L L = 61%
Penetration (mm)
30
25
20
17.3
15
10
5
0
M = 38%
L L
0
10
20
30
40
50
60
70
80
Moisture content m%

Graph 2.2

2.1.3 Shrinkage Limit

As fine-grained soil loses water it shrinks, and its volume decreases. This applies particularly to clayey soils. The determination of Shrinkage Limit is very important, therefore, when this type of soil can lose moisture for various reasons, for example:

1. Roots of trees can extract large amounts of water from clays supporting the foundations of buildings, and may induce large enough settlement to damage the structures.

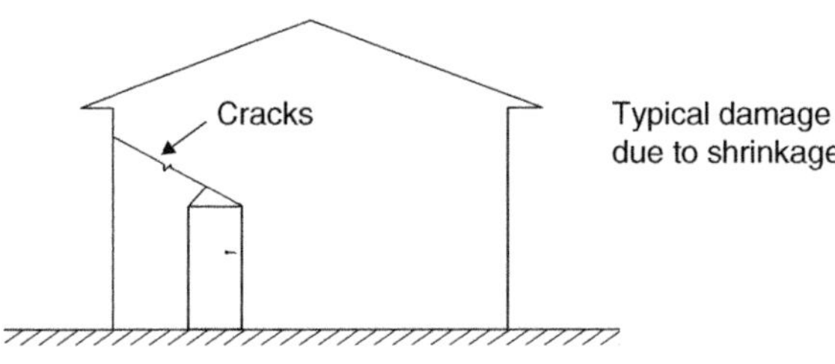

Figure 2.3

2. Foundations of boilers, brick kilns and furnaces can transmit heat to the soil, causing considerable distortion, unless insulated.
3. Climatic conditions, such as droughts, could dry the clay below footings. The effect of shrinkage usually extends to a depth of 1 metre. This is one reason for placing footings at this depth below the surface.
4. Road and pavement surfaces become undulating, because of differential volume changes in the underlying clay.
5. Pipelines, laid in the shrinkage zone could deform and split apart.

Definition

Shrinkage Limit (SL) is that moisture content below which the volume of the drying soil does not decrease, even after further loss of water. The volume at the Shrinkage Limit is denoted by V_0. This means that the volume remains the same during subsequent heating, in an oven, to oven-dry state.

Notes:

1. Shrinkage is caused by the capillary forces developing in the voids as the cohesive soil loses its moisture.
2. The soil is still fully saturated (S_r=1) at its Shrinkage Limit.
3. The drying-out process causing shrinkage is slow, e.g. drying by the ambient temperature.

2.1.3.1 Test to find SL (outline)

Step 1: Measure the mass and volume of a sample of soil to be tested.
Step 2: Air-dry the sample at constant temperature.
Step 3: Measure its mass (*M*) and volume (*V*) at intervals.

Step 4: Repeat steps 2 and 3, until there is no perceptible change in volume.
Step 5: Dry specimen completely in an oven and measure its mass (M_s) and volume (V_0) at this oven-dry state.
Step 6: Plot the results as shown in Figure 2.4.

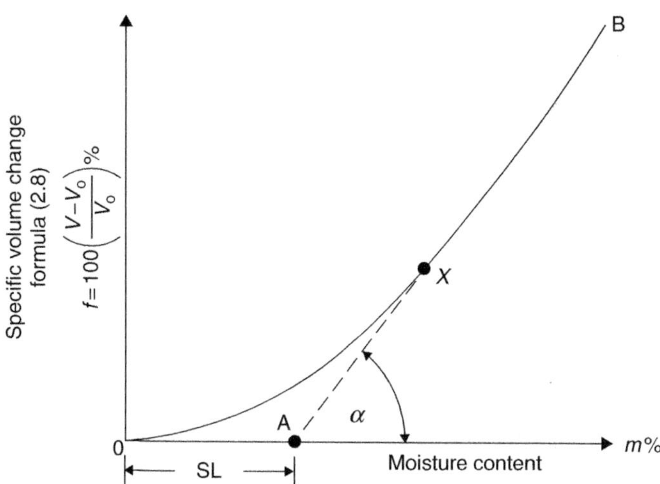

Figure 2.4

The Shrinkage Limit is given by the intersection of the extended straight line BX and the horizontal axis at point A.

Example 2.4

The results of a shrinkage test, carried out on a clay sample, are tabulated below. The initial volume of the specimen was $V = 50.27\ \text{cm}^3$, weighing $M = 94.11\ \text{g}$. Its natural moisture content was found to be 29.7%.

Preliminary tests on the clay yielded:

Specific gravity: $G_s = 2.75$
Liquid Limit: LL = 65%
Plastic Limit: PL = 21%

The air-dried soil was dried out completely in an oven and its dry mass and oven-dry volume were measured as:

$M_s = 72.55\ \text{g}$
$V_o = 36.19\ \text{cm}^3$

Plot Graph 2.3 and determine the Shrinkage Limit.

Table 2.3

Time (hours)	V (cm^3)	M (g)	$M_w = M - M_s$ (g)	$m = 100\frac{M_w}{M_s}$ (%)	$f = 100\frac{V - V_o}{V_o}$ (%)	$e = \frac{VG_s - \rho_w}{M_s} - 1$	$S_r = \frac{mG_s}{e}$
0	50.27	94.11	21.56	29.72	38.91	0.905	0.903
3	45.69	90.16	17.61	24.27	26.25	0.732	0.912
8	41.98	86.92	14.37	19.81	16.00	0.591	0.922
20	39.15	84.09	11.54	15.65	8.13	0.484	0.889
25	37.60	80.76	8.21	11.14	3.90	0.425	0.721
28	37.00	78.35	5.80	7.87	2.24	0.402	0.538
44	36.52	74.41	1.86	2.52	0.91	0.384	0.180
48	36.46	73.94	1.39	1.89	0.75	0.382	0.136
51	36.30	73.12	0.57	0.77	0.30	0.376	0.056
60	36.28	73.09	0.54	0.73	0.25	0.375	0.054
63	36.27	73.08	0.53	0.72	0.22	0.375	0.053
68	36.27	73.08	0.53	0.72	0.22	0.375	0.053

After plotting the above results on Graph 2.3 (see Table 2.3), the Shrinkage Limit was found to be 13%. The derivation of the formula for the voids ratio is as follows.

From (1.5): $e = \frac{V_v}{V_s}$

From (1.1): $V_v = V - V_s$

From (1.34): $\rho_s = G_s \times \rho_w$

From (1.32): $V_s = \frac{M_s}{\rho_s}$

$$e = \frac{V - V_s}{V_s} = \frac{V}{V_s} - 1$$

$$= \frac{V\rho_s}{M_s} - 1$$

$$= \frac{VG_s\,\rho_w}{M_s} - 1$$

$$\therefore \quad e = \frac{VG_s\,\rho_w}{M_s} - 1 \qquad (2.2)$$

Calculate the voids ratio, volume of voids, volume of solids and the total volume at each limit.

Solution:

As the specimen is saturated at each limit, the voids ratio can be obtained from formula (1.36), taking S_r=1. Therefore, $e = m \times G_s$

From (1.32) & (1.34):

$$V_s = \frac{M_s}{G_s\,\rho_w} \qquad (2.3)$$

From (1.5)

$$V_v = e \times V_s$$

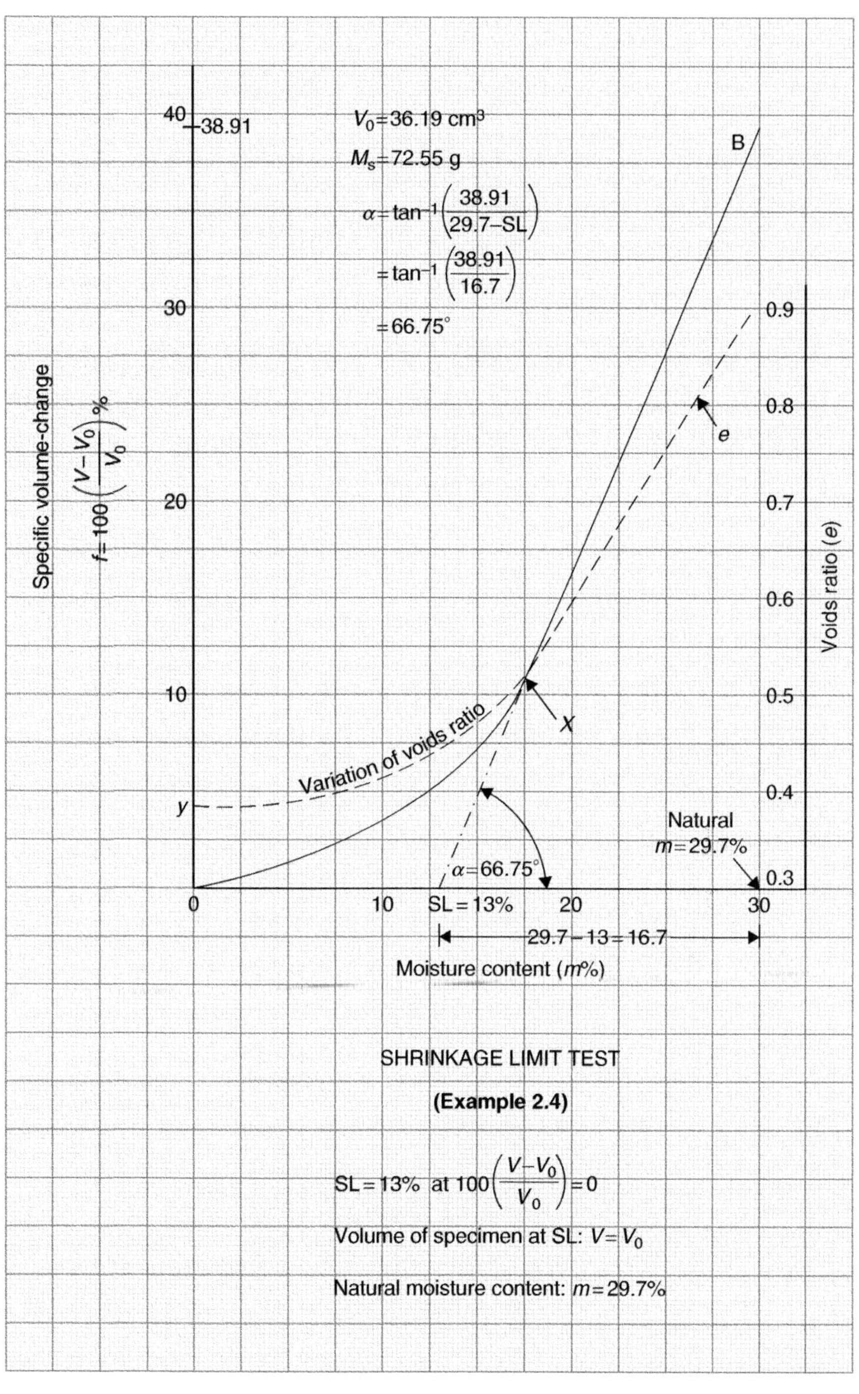

40
38.91
30
20
10
y
0
10
SL = 13%
20
30
Specific volume-change
$f = 100\left(\frac{V - V_0}{V_0}\right)\%$
$V_0 = 36.19\ cm^3$
$M_s = 72.55$ g
$\alpha = \tan^{-1}\left(\frac{38.91}{29.7 - SL}\right)$
$= \tan^{-1}\left(\frac{38.91}{16.7}\right)$
$= 66.75°$
B
0.9
0.8
e
0.7
0.6
0.5
0.4
0.3
Voids ratio (e)
X
Variation of voids ratio
Natural
$m = 29.7\%$
$\alpha = 66.75°$
29.7 − 13 = 16.7
Moisture content (m%)
SHRINKAGE LIMIT TEST
(Example 2.4)
SL = 13% at $100\left(\frac{V - V_0}{V_0}\right) = 0$
Volume of specimen at SL: $V = V_0$
Natural moisture content: $m = 29.7\%$

Graph 2.3

Tabulating the calculations (see Table 2.4): the values for SL are approximate ones, owing to its graphical determination (Graph 2.3).

Table 2.4

LL = 65%	PL = 21%	SL = 13%
$e = (\text{LL})G_s = 0.65 \times 2.75 = 1.788$	$e = (\text{PL})G_s = 0.21 \times 2.75 = 0.578$	$e = (\text{SL})G_s = 0.13 \times 2.75 = 0.358$ (approx)
$V_s = \dfrac{72.55}{2.75 \times 1} = 26.38\,\text{cm}^3$	$V_s = 26.38\,\text{cm}^3$	$V_s = 26.38\,\text{cm}^3$
$V_v = 1.788 \times 26.38 = 47.17\,\text{cm}^3$	$V_v = 0.578 \times 26.38 = 15.24\,\text{cm}^3$	$V_v = 0.358 \times 26.38 = 9.44\,\text{cm}^3$
$V = 26.38 + 47.17 = 73.55\,\text{cm}^3$	$V = 26.38 + 15.24 = 41.62\,\text{cm}^3$	$V = 26.38 + 9.44 = 35.82\,\text{cm}^3$
$V_w = V_v$: 47.17 cm³; V_s: 26.38 cm³; V	$V_w = V_v$: 15.24 cm³; V_s: 26.38 cm³; V	$V_w = V_v$: 9.44 cm³; V_s: 26.38 cm³; V

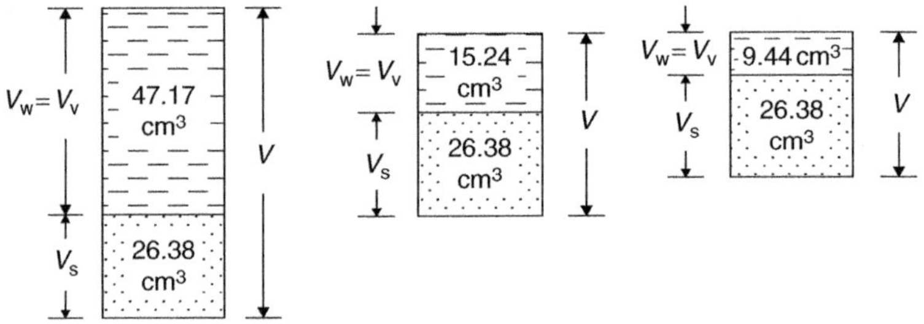

Note that:

1. The voids ratio can be larger than unity. This is the case at LL, where the volume of water is nearly twice the volume of solids.
2. The calculated $V = 35.82\,\text{cm}^3$ at the Shrinkage Limit is practically equal to $V_0 = 36.19\,\text{cm}^3$.

2.1.3.2 Approximate determination of Shrinkage Limit

Theoretically, the Shrinkage Limit (SL) may be estimated without carrying out the entire test. The formula derived for this is based on the, by now well known, facts:

1. The soil is saturated at SL, hence $V_v = V_w$.
2. The volume (V) of the specimen at SL equals to its volume (V_0) at oven-dry state.
3. M_s is the mass of oven-dry specimen.
4. The specific gravity of solids in V_0 is G_s.
5. $V_0 > V_s$

In the knowledge of the above quantities, the saturated moisture content, that is the Shrinkage Limit, can be estimated.

From (1.35a): $m = \dfrac{M_w}{M_s}$

But (1.22): $M_w = V_v \times \rho_w$

From (1.1): $V_v = V_o - V_s$

From (2.3): $V_s = \dfrac{M_s}{G_s \rho_w}$

From (1.34): $\rho_s = G_s \times \rho_w$

$$m = \frac{V_v \rho_w}{M_s} = \frac{(V_o - V_s)\rho_w}{M_s}$$

$$= \frac{\left(V_o - \dfrac{M_s}{G_s \rho_w}\right)\rho_w}{M_s} = \frac{V_o \rho_w - \dfrac{M_s}{G_s}}{M_s}$$

$$\therefore m = \boxed{SL = \frac{V_o \rho_w - \dfrac{M_s}{G_s}}{M_s}} \qquad (2.4)$$

Expressed as a percentage

$$\boxed{SL = 100 \times \left(\frac{V_o \rho_w - \dfrac{M_s}{G_s}}{M_s}\right) \%}$$

If V_o is given in cm^3

M_s is given in grams

then

$$\boxed{SL = 100 \times \left(\frac{V_o - \dfrac{M_s}{G_s}}{M_s}\right) \%} \qquad (2.5)$$

and ρ_w=1 g/cm^3

Applying this method to Example 2.4, given that

$V_o = 36.19$ cm^3

$M_s = 72.55$ g

$G_s = 2.75$

$$SL = 100 \times \left(\frac{36.19 - \dfrac{72.55}{2.75}}{72.55}\right) = 13.52\%$$

which is a good approximation to 13%

It is, therefore, only necessary to dry out the sample completely in order to estimate the Shrinkage Limit.

Example 2.5

A clod of clay was dried out slowly and its dry mass measured as M_s=59.10 g. The oven-dry specimen was inserted in mercury safely in laboratory conditions and the displaced amount of Hg was noted. This gave the volume of the clod as V_0=27.67 cm^3.

Calculate the:

1. Shrinkage Limit of the clay
2. Volume of solids
3. Voids ratio at SL
4. Volume of voids at SL

Assume that $G_s = 2.75$

From (2.5): $$SL = 100 \times \left(\frac{27.67 - \frac{59.10}{2.75}}{59.10} \right) = 10.46\%$$

From (2.3): $$V_s = \frac{M_s}{G_s \rho_w} = \frac{59.10}{2.75} = 21.49\,\text{cm}^3$$

From (1.36): $e = (SL)\ G_s = 0.1046 \times 2.75 = 0.288$

From (1.5): $V_v = eV_s = 0.288 \times 21.49 = 6.19\,\text{cm}^3$.

2.1.4 Swelling of cohesive soils

In contrast to shrinkage, clays swell when they absorb water. The resulting increase in the volume of soil supporting buildings can cause substantial damage to the structures. The most frequently occurring problems are caused by:

1. Leaking water pipes or mains.
2. Heavy rainfall after a prolonged dry period.
3. The removal of old trees and shrubs. In this case, the clay is no longer dried by the roots, hence it absorbs water and swells. This process can take several years. It is imperative, therefore, to determine the history of tree or hedge clearance, before building on shrinkable clays.

As the foundations of structures are heavier than internal floor slabs, the swelling pressures could cause differential movement between the two structural elements, resulting in floor heave and general cracking of the walls. Also, horizontal swelling pressures could push shallow footings sideways.

Preventative foundation construction may include either short bored piles or shallow footings with adequate compressible backfill, to accommodate the swelling clay. Also, there should be movement joints between the foundations or the walls of the building and the ground floor.

2.1.5 Saturation Limit ($Z\%$)

As the soil absorbs water, its volume increases. The absorption ceases at a certain moisture content. This value of $m\%$ is the Saturation Limit.

Determination of *Z*%

This consistency limit, additional to the Atterberg ones, may be obtained in two ways:

1. Laboratory test (outline)

   Step 1: Remove particles larger than 2 mm from the sample.

   Step 2: Mix the soil with water to a paste so that its moisture content is well below its Liquid Limit.

   Step 3: Place the paste into a bowl, forming its surface flat and smooth.

   Step 4: Apply water to the surface in drops. When the drops are not absorbed any more, measure the moisture content, which is the Saturation Limit of the soil.

2. Casagrande's formula

   Professor Casagrande (1932) found that the relationship between Liquid Limit and Saturation Limit may be expressed by:

$$\boxed{Z = \sqrt{15.2\left(\mathrm{LL}\% - 16.3\right)} + 9\%} \qquad (2.6)$$

   Theoretically therefore, it is not necessary to carry out the laboratory test, because this formula yields a satisfactory average value for $Z\%$, provided $\mathrm{LL} > 16.3\%$.

Note: The absorption of water by fine-grained soils is due to capillary suction (see section 5.8.2).

2.1.6 Relationship between the limits

The Saturation Limit is the maximum moisture content of a clay specimen, whose volume cannot increase any further by swelling. The Shrinkage Limit is that moisture content at which the volume of the same clay sample cannot decrease any further.

It can be seen in Figure 2.4 that the variation of volume change relative to the moisture content is linear. In the knowledge of $Z\%$ and $\mathrm{SL}\%$ therefore, the equation of the line (Figure 2.5) extending between these two limits can be derived in terms of f, m, SL and α.

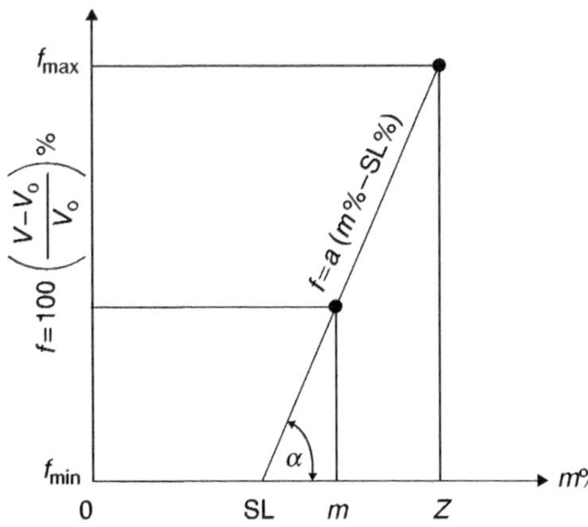

Figure 2.5

General equation of a straight line:

$$f = am + b$$

For: $f = 0$, $m = \mathrm{SL}$

$$0 = a \times \mathrm{SL} + b$$

$$\therefore b = -a \times \mathrm{SL}$$

For any value of f:

$$f = \mathrm{am} - \mathrm{a} \times \mathrm{SL}$$
$$= \mathrm{a}(\mathrm{m} - \mathrm{SL})$$

The gradient of the line $a = \tan\alpha$

Hence, the equation is: $$\boxed{f = \tan\alpha(m\% - SL\%)\ \%} \tag{2.7}$$

Where f is the specimen volume change, indicating that the volume change $V - V_0$ is f% of the minimum volume V_0. It can, therefore, be expressed as:

$$\boxed{f = 100\frac{V - V_0}{V_0}\ \%} \tag{2.8}$$

Equating (2.7) and (2.8):

$$f = 100\left(\frac{V - V_0}{V_0}\right) = \tan\alpha\left(m\% - SL\%\right)$$

The volume (V) of the specimen can be expressed in terms of moisture content m% within the range $SL\% \leq m\% \leq Z\%$; between point c and x on the extended experimental line (Graph 2.4).

$$V - V_0 = \frac{V_0 \tan\alpha}{100} \times \left(m\% - SL\%\right)$$

$$\boxed{V = \left[1 + \frac{\tan\alpha}{100}\left(m\% - SL\%\right)\right]V_0} \tag{2.9}$$

Example 2.6

Calculate the maximum and minimum volume of the clay in Example 2.4 caused by shrinkage and swelling respectively, using:

$LL = 65\%$ $\quad V_0 = 36.19\,\text{cm}^3$
$SL = 13\%$ $\quad \alpha = 66.75°$ (From Graph 2.3)
$m = 29.7\%$

Calculating the Saturation Limit from (2.6):

$$Z = \sqrt{15.2\left(65 - 16.3\right)} + 9 = 36.2\%$$

Substituting α and V_0 into (2.9):

$$V = \left[1 + \frac{\tan 66.75}{100}\left(m\% - SL\%\right)\right] \times 36.19$$

$$= 36.19 + \frac{2.327 \times 36.19}{100}\left(m\% - SL\%\right)$$

$$\therefore \quad \boxed{V = 36.19 + 0.842(m\% - SL\%)} \tag{2.10}$$

Formula (2.10) can be applied to estimate the maximum volume to which the experimental clay specimen could swell.

At $m = Z\% = 36.2\%$ $V_{max} = 36.19 + 0.842 \times (36.2 - 13)$
$= 36.19 + 19.53 = 55.72\,\text{cm}^3$

Alternatively, V_{max} may be estimated by means of Graph 2.4, which is a reproduction of Graph 2.3 with line AB extended to point C, corresponding to $Z = 36.2\%$.

Read off $f_{max} = 54\%$, corresponding to $Z\%$ from Graph 2.4:

$$\therefore f_{max} = 100 \times \frac{V_{max} - V_0}{V_0} = 54$$

From which, $V_{max} = \dfrac{54\,V_0}{100} + V_0 = \dfrac{54 \times 36.19}{100} + 36.19$
$= 19.54 + 36.19 = 55.73\,\text{cm}^3$

Similarly, the volume of the sample at natural moisture content $m = 29.7\%$ is estimated as:

$$V_1 = 36.19 + 0.842 \times (29.7 - 13) = 50.25\,\text{cm}^3$$

The minimum volume to which the specimen could shrink is $V_{min} = V_0 = 36.19\,\text{cm}^3$, hence the volume decrease relative to the natural state is approximately:

$$V_1 - V_0 = 50.25 - 36.19 = 14.1\,\text{cm}^3$$

The total estimated volume change between full swelling and shrinkage is:

$$V_{max} - V_{min} = 55.73 - 36.19 = 19.54\,\text{cm}^3.$$

2.1.7 Linear shrinkage and swelling

So far, volumetric change has been discussed. However, the engineer is more interested in the magnitude of horizontal and vertical shrinkage and swelling. To estimate these, one has to take into account the change in length of a cube of sides 'h', assuming that each side is shortened by a distance x during shrinkage.

Initial volume: $V_1 = h^3$
Final volume: $V_2 = (h - x)^3$
Change in volume: $V_1 - V_2 = h^3 - (h - x)^3$

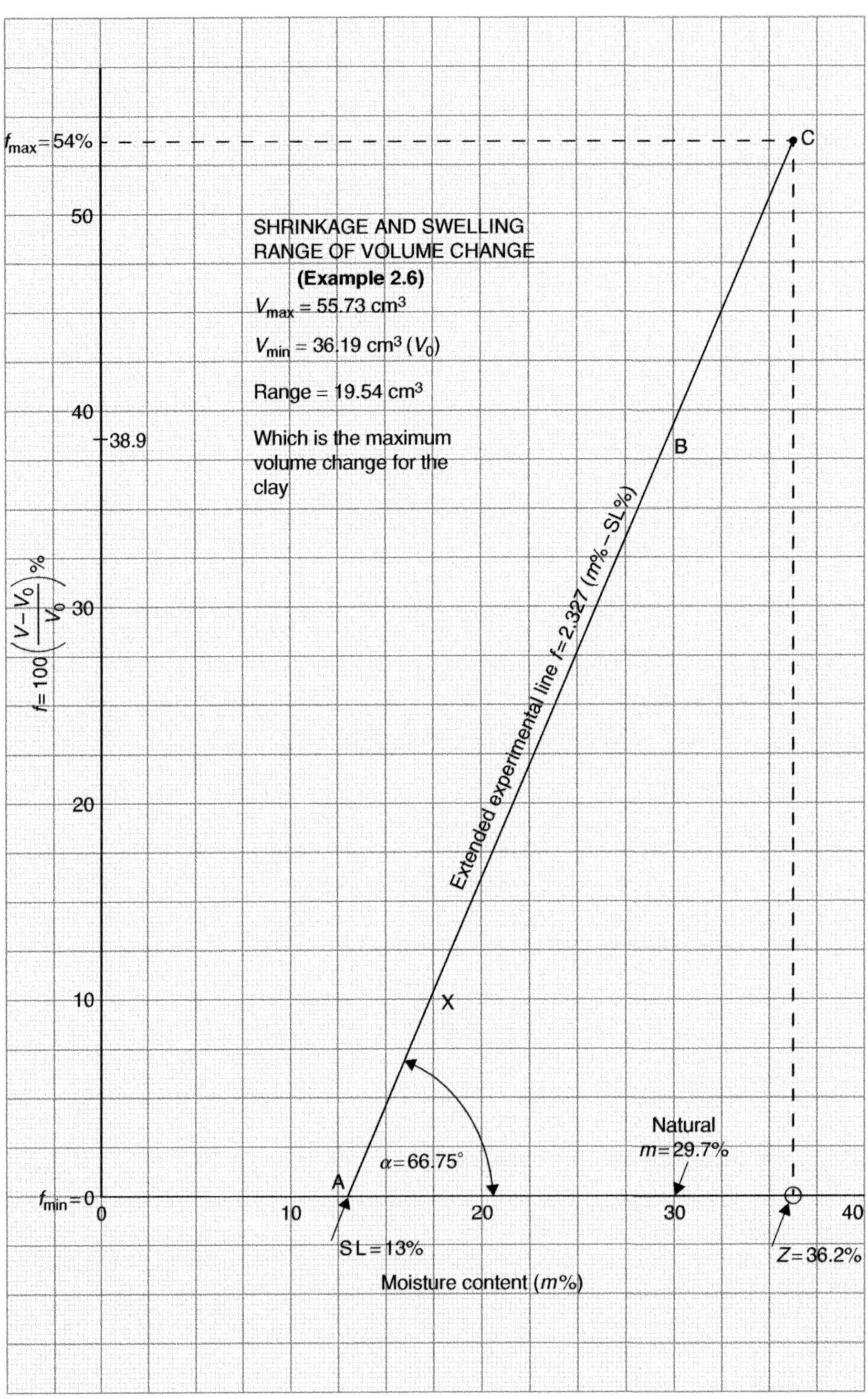

f_{max}=54%
C
50
SHRINKAGE AND SWELLING
RANGE OF VOLUME CHANGE
(Example 2.6)
V_{max} = 55.73 cm³
V_{min} = 36.19 cm³ (V_0)
Range = 19.54 cm³
Which is the maximum volume change for the clay
40
38.9
B
Extended experimental line f=2.327 (m% − SL%)
$f = 100\left(\frac{V - V_0}{V_0}\right)\%$
30
20
10
X
Natural
m=29.7%
α=66.75°
A
f_{min}=0
0
10
20
30
40
SL=13%
Z=36.2%
Moisture content (m%)

Graph 2.4

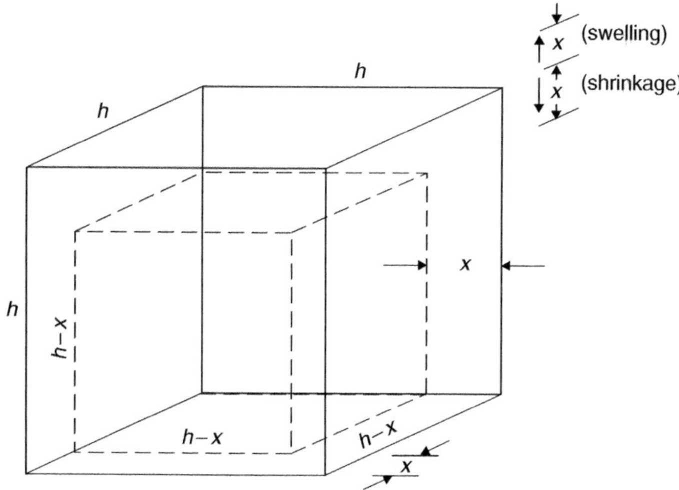

Figure 2.6

Specific volume change:

$$\Delta f\% = 100\frac{V_1 - V_2}{V_2} = 100\frac{h^3 - (h-x)^3}{(h-x)^3} = 100\frac{h^3}{(h-x)^3} - 100$$

So $(h-x)^3 = \dfrac{100\,h^3}{\Delta f\% + 100}$

$$h - x = h\left(\frac{100}{\Delta f\% + 100}\right)^{1/3} \quad \therefore \quad \boxed{x = h\left(1 - \sqrt[3]{\frac{100}{\Delta f\% + 100}}\right)} \qquad (2.11)$$

Note, that in (2.11) $\Delta f = f_1 - f_2 = 100\dfrac{V_1 - V_2}{V_0}$

Where f_1 and f_2 are the values of specific volume change at the initial and final moisture contents respectively. For swelling, the increase in length h is given by:

$$x = h\left(\sqrt[3]{\frac{\Delta f + 100}{100}} - 1\right) \qquad (2.11a)$$

It must be appreciated that the calculation values of shrinkage or swelling are not exact but only indicative of possible magnitudes.

Example 2.7

A bungalow was built so that its shallow footings and ground floor slab were placed directly on the top of shrinkable clay. The natural moisture content of the clay at the time of construction was 29.7%. Subsequently, two oak trees were planted at a distance of 6 m from the building. Estimate the maximum possible magnitude of shrinkage under the footings and the slab over the following years, due to the roots of the maturing trees. The shrink/swell characteristics of the clay are given in Graph 2.4. A schematic cross-section of the structure is shown below.

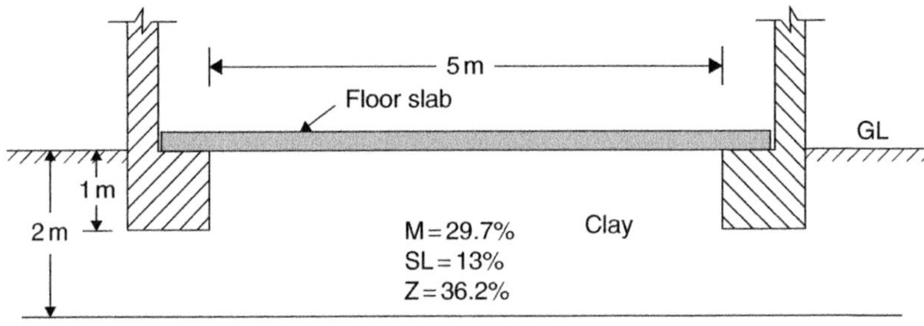

Figure 2.7

From Graph 2.4: For $m = 29.7\%$ $\quad f_1 = 38.91\%$

For maximum value, take SL = 13% $\quad f_2 = 0\%$

$\therefore \quad \Delta f = 38.91\%$

From (2.11): $$x = h\left(1 - \sqrt[3]{\frac{100}{38.91 + 100}}\right) = 0.1037h$$

For $h = 1\,\text{m}$ $$x = 100\left(1 - \sqrt[3]{\frac{100}{138.91}}\right) = 100 \times 0.1037$$

$$= 10.37\,\text{cm}$$

For $h = 2\,\text{m}$ $\quad x = 200 \times 0.1037 \quad = 20.75\,\text{cm}$

For $h = 5\,\text{m}$ $\quad x = 500 \times 0.1037 \quad = 51.85\,\text{cm}$

It appears from these figures that the centre of the slab would settle twice as much as the footings. Depending on the extent of the root system, the magnitude of shrinkage under each footing could be entirely different. This would be reflected by the mode of failure (cracking, bending and tilting) observed at various parts of the structure as seen below.

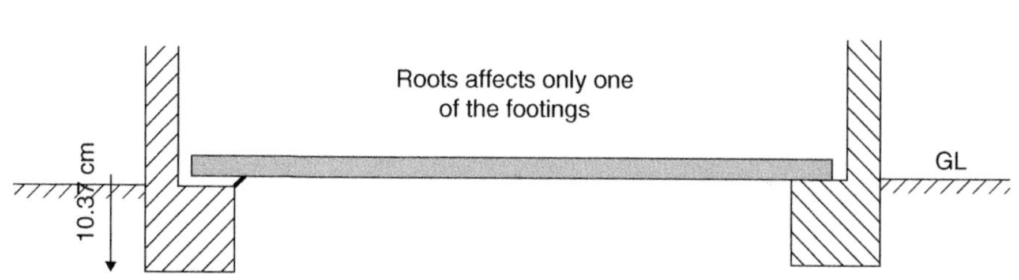

Figure 2.8

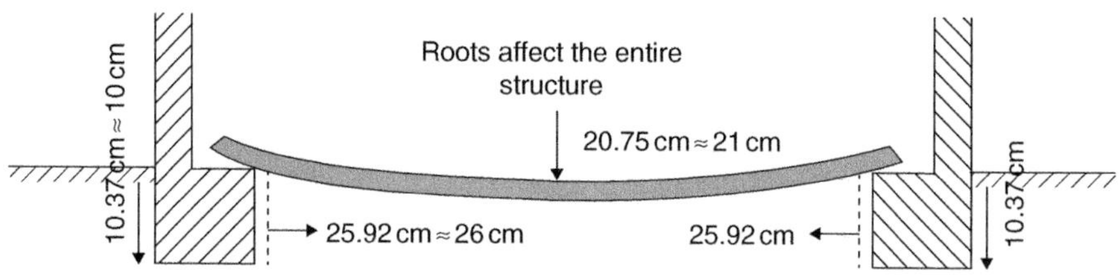

Figure 2.9

Shrinkage occurs along the length of the footings. This could pull a brick construction apart.

Note: In most calculations, the results may be rounded to the nearest whole number.

Suppose the bungalow is built on land recently cleared of trees, shrubs and hedges for this purpose. Calculate the magnitude of swelling under the structure, taking the natural moisture content as 29.7%.

From Graph 2.4: $Z = 36.2\,\%$ $\quad\therefore\quad$ $f_1 = 54\,\%$

$m = 29.7\,\%$ $\qquad f_2 = 38.91\%$

$\therefore\quad \Delta f = 15.09\%$

From (2.11a): $$x = h\left(\sqrt[3]{\frac{15.09 + 100}{100}} - 1\right) = 0.048\,h$$

For $h = 1\,\text{m}$ $\quad x = 0.048 \times 100 = 4.8\,\text{cm} \approx 5\,\text{cm}$

For $h = 2\,\text{m}$ $\quad x = 0.048 \times 200 = 9.6\,\text{cm} \approx 10\,\text{cm}$

For $h = 5\,\text{m}$ $\quad x = 0.048 \times 500 = 24.0\,\text{cm}$

Notes:

a) The upwards pressure caused by swelling is opposed by the downward pressure of the structural elements. As the weight of the ground slab is probably small compared to the upward pressure, its uplift is likely.
b) The horizontal pressure on the two footings could push them apart to some extent, depending on the passive resistance of soil.
c) The magnitude of pressure due to swelling of undisturbed clay may be obtained in an oedometer test.

2.2 Consistency indices

Having found the consistency limits of a particular soil, they can be used to describe it in its natural state, by means of the three consistency indices:

1. Plasticity index (PI)
2. Relative consistency index (RI)
3. Liquidity index (LI)

2.2.1 Plasticity index (PI)

This indicates the range of moisture content over which the soil can be considered plastic. It is given by the difference between LL and PL.

$$\boxed{\mathrm{PI} = \mathrm{LL} - \mathrm{PL}}\ \% \tag{2.12}$$

The larger the value of PI is, the more cohesive is the soil, assuming no organic contamination.

2.2.2 Relative consistency index (RI)

This shows the position of the natural moisture content (m), relative to the Liquid Limit within the plastic range. Thus, if (LL − PL) is considered to be 100% then (LL − m) is RI%.

Hence,

$$\boxed{\mathrm{RI} = 100\left(\frac{\mathrm{LL} - m}{\mathrm{LL} - \mathrm{PL}}\right) = 100\left(\frac{\mathrm{LL} - m}{\mathrm{PI}}\right)\%} \tag{2.13}$$

2.2.3 Liquidity index (LI)

This shows the position of the natural moisture content, relative to the Plastic Limit within the plastic range. Thus, if (LL–PL) is 100%, then (m – PL) is LI%.

Hence,

$$\boxed{\mathrm{LI} = 100\left(\frac{m - \mathrm{PL}}{\mathrm{LL} - \mathrm{PL}}\right) = 100\left(\frac{m - \mathrm{PL}}{\mathrm{PI}}\right)\%} \tag{2.14}$$

Graphical representation

The consistency limits and indices may be represented as in Figure 2.10.

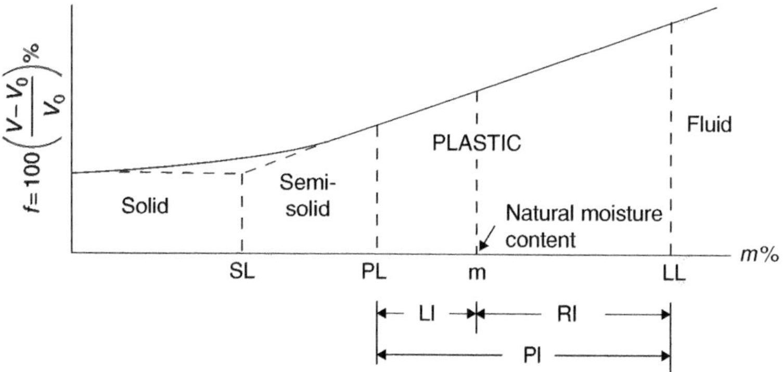

Figure 2.10

From Figure 2.10: LI% + RI% = PI%

But, PI = LL − PL = 100% $\therefore$ $\boxed{\text{LI} = 100 - \text{RI}\,\%}$ (2.15)

Application

The purpose of assessing the consistency parameters is to get an idea of the suitability of a cohesive soil for engineering purposes. It must be remembered that apart from the natural moisture content and the Shrinkage Limit, tests are carried out on reconstructed materials. Because of the empirical nature of the experiments, the achievement of standardized results depends largely on the skill of the person carrying them out. Despite the inherent errors a reasonable, notional, estimate can be made of the consistency and the type of soil at hand. The relationship between the Relative consistency index and water content of a particular soil may be drawn as in Figure 2.11.

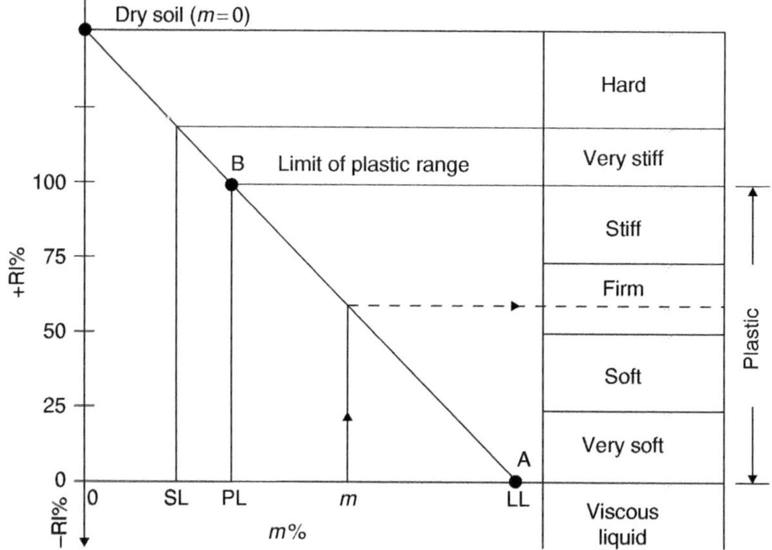

Figure 2.11

Note that the straight line is drawn by connecting points A and B. Point A is located at RI=0% and LL, whilst B at RI=100% and PL.

The position of the natural moisture content locates the consistency definition. The soil is normally classified by Chart 2.1 and Table 2.5, devised by Casagrande after tests on many types of soil.

Example 2.8

Taking, yet again, the results of Example 2.4 in order to classify the clay as discussed:

$$\text{LL} = 65\%$$
$$\text{PL} = 21\%$$
$$\text{SL} = 13\%$$
$$\text{M} = 29.7\% \approx 30\%$$

From (2.12) PI=LL−PL=65−21=44%

$$\left.\begin{aligned} &\text{From (2.13) RI} = 100\left(\frac{\text{LL}-m}{\text{LL}-\text{PL}}\right) = 100\times\frac{65-29.7}{44} &&= 80\% \\ &\text{From (2.15) LI} = 100-\text{RI} = 100-80 &&= 20\% \end{aligned}\right\} \quad 100\%$$

After examining Figure 2.11 at RI=80%, it may be concluded that the sample is of stiff consistency. Check, by drawing Figure 2.12 to scale.

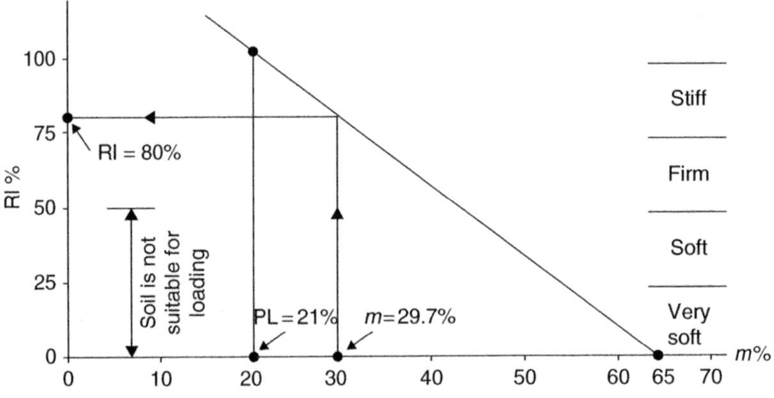

Figure 2.12

Also, by plotting LL=65% and PI=44% on Chart 2.1, the soil is depicted to be inorganic clay of high plasticity (CH).

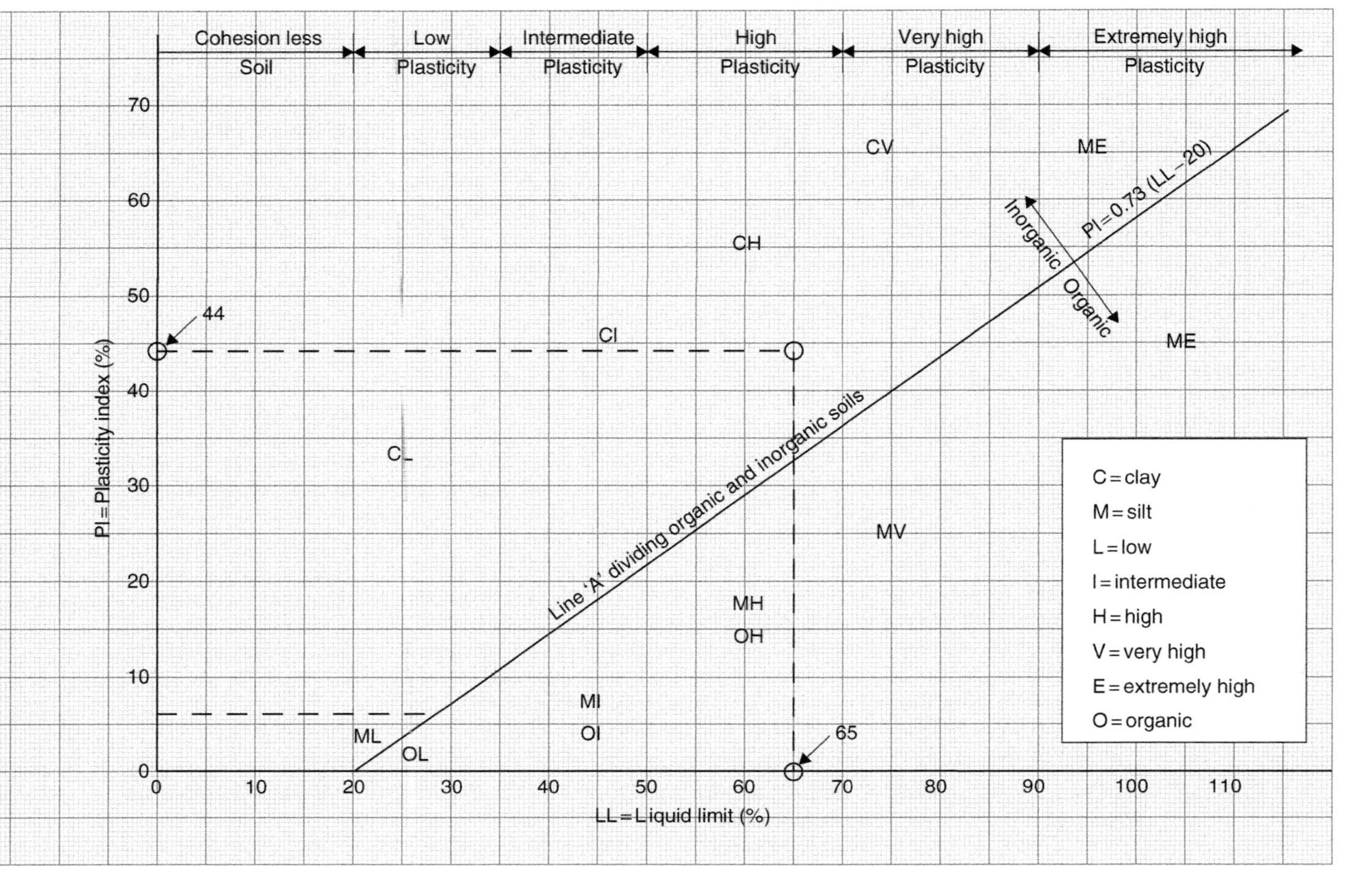
Cohesion less
Soil
Low
Plasticity
Intermediate
Plasticity
High
Plasticity
Very high
Plasticity
Extremely high
Plasticity
PI = Plasticity index (%)
LL = Liquid limit (%)
44
65
CV
ME
CH
CI
CL
MV
MH
OH
MI
OI
ML
OL
ME
PI = 0.73 (LL − 20)
Inorganic
Organic
Line 'A' dividing organic and inorganic soils
C = clay
M = silt
L = low
I = intermediate
H = high
V = very high
E = extremely high
O = organic

Chart 2.1

Table 2.5 Fine-grained soils

Major Divisions	Description and field identification	Sub-groups	Group symbol	Value as a road foundation when not subject to frost action	Potential frost action	Shrinkage or swelling properties	Drainage characteristics
Fine-grained soils having low plasticity (silts)	Soils with an appreciable fraction passing the 0.075 mm sieve, and with Liquid Limits less than 35. Not gritty between the fingers. Cannot be readily rolled into threads when moist. Exhibit dilatancy.	Silts (inorganic) rock flour, silty fine sands with slight plasticity	ML	Fair to poor	Medium to very high	Slight to medium	Fair to poor
		Clayey silts (inorganic)	CL	Fair to poor	Medium to high	medium	Practically impervious
		Organic clays of low plasticity	OL	Poor	Medium to high	Medium to high	Poor
Fine-grained soils having medium plasticity	Soils with Liquid Limits between 35 and 50. Can be readily rolled into threads when moist. Do not exhibit dilatancy. Show some shrinkage on drying.	Silty clays (inorganic) and sandy clays	MI	Fair to poor	Medium	Medium to high	Fair to poor
		Clays (inorganic) of medium plasticity	CI	Fair to poor	Slight	High	Fair to practically impervious
		Organic clays of medium plasticity	OI	Poor	Slight	High	Fair to practically impervious
Fine-grained soils having high plasticity	Soils with Liquid Limits greater than 50. Can be readily rolled into threads when moist. Greasy to the touch. Show considerable shrinkage on drying. All highly compressible soils.	Highly compressible micaceous or diatomaceous soils	MH	Poor	Medium to high	High	Poor
		Clays (inorganic) of high plasticity	CH	Poor to very poor	Very slight	High	Practically impervious
		Organic clays of high plasticity	OH	Very poor	Very slight	High	Practically impervious

2.3 Classification of soils by particle size

As soils are made up of particles of various sizes, it is convenient to classify them in terms of these characteristics. This is done by the observation of the particle-size distribution of a given weight of soil. This would show what percentage (by mass) of each size is present in the material. The resulting distribution is plotted on Chart 2.2. This chart shows that there are four main soil types, corresponding to four ranges of size. These are shown in Table 2.6:

Table 2.6

Soil type	Size-range (mm)
Clay *(Fine grained)*	0.0001–0.002
Silt *(Fine grained)*	0.002–0.06
Sand *(Coarse grained)*	0.06–2
Gravel *(Coarse grained)*	2–60
Cobbles	>60

The method of testing for particle size depends on whether the soil is coarse-grained or fine-grained. These are:

For coarse-grained: Sieve analysis
For fine-grained: Sedimentation tests

2.3.1 Sieve analysis

There are two methods of sieve analysis, namely wet and dry, of which the wet process is favoured by BS1377, unless dry sieving has been shown to be satisfactory for the type of material under test. The procedures are well described in laboratory manuals, hence the subject is restricted to the presentation of results and their application. There are nineteen B.S. sieves used normally, having the following apertures in millimetres: 75, 63, 50, 37.5, 28, 20, 14, 10, 6.3, 5, 3.35, 2, 1.18, 0.6, 0.425, 0.3, 0.212, 0.15 and 0.063. The apparatus is shown in Figure 2.13. The results of the sieve test may be evaluated by either of these two procedures:

1. Standard
2. Recursive

Example 2.9

A soil sample, weighing 300 grams, was passed through a set of ten sieves and the mass retained on each tabulated.

1. Standard procedure

Table 2.7 Standard procedure to calculate the percentage (of mass) passing through each sieve

n	Sieve (mm)	Mass retained M_n (g)	Mass passing sieve M_p (g)	Percent passing P_n (%)
1	20	0	300	100
2	10	6	300−6=294	294/3=98
3	5	15	294−15=279	279/3=93
4	3.35	24	279−24=255	255/3=85
5	2	30	255−30=225	225/3=75
6	1.18	51	225−51=174	174/3=58
7	0.6	51	174−51=123	123/3=41
8	0.3	48	123−48=75	75/3=25
9	0.15	42	75−42=33	33/3=11
10	0.063	27	33−27=6	6/3=2
	Σ	294		

n=number indicating the position of a sieve in the set
M_n=Mass retained on the nth sieve
M_p=Mass passed the nth sieve
M_T=total mass of the sample=300 g
P_n=percentage of M_T passing through the nth sieve

The calculation of P_n% is self-explanatory in Table 2.7. If M_T is 100%, then M_p at each sieve is P_n%, hence:

$$\frac{300}{100\%}=\frac{M_p}{P_n\%}, \quad \text{that is} \quad \boxed{P_n=\frac{M_p}{3}\%}$$

The particle-size distribution, that is P_n% against sieve size, can now be drawn on Chart 2.2.

2. Recursive procedure

This method is somewhat shorter than the standard one as there is no need to calculate the intermediate (M_p) values.

Table 2.8

n	Sieve	M_n	$P_n\%=P_{n-1}-\frac{M_n}{3}$
1	20	0	100
2	10	6	100−6/3=98
3	5	15	98−15/3=93
4	3.35	24	93−24/3=85
5	2	30	85−30/3=75
6	1.18	51	75−51/3=58
7	0.6	51	58−51/3=41
8	0.3	48	41−48/3=25
9	0.15	42	25−42/3=11
10	0.063	27	11−27/3=2

Figure 2.13 ELE Sieve Shaker EL80-0200/01.
Reproduced by permission of ELE international.

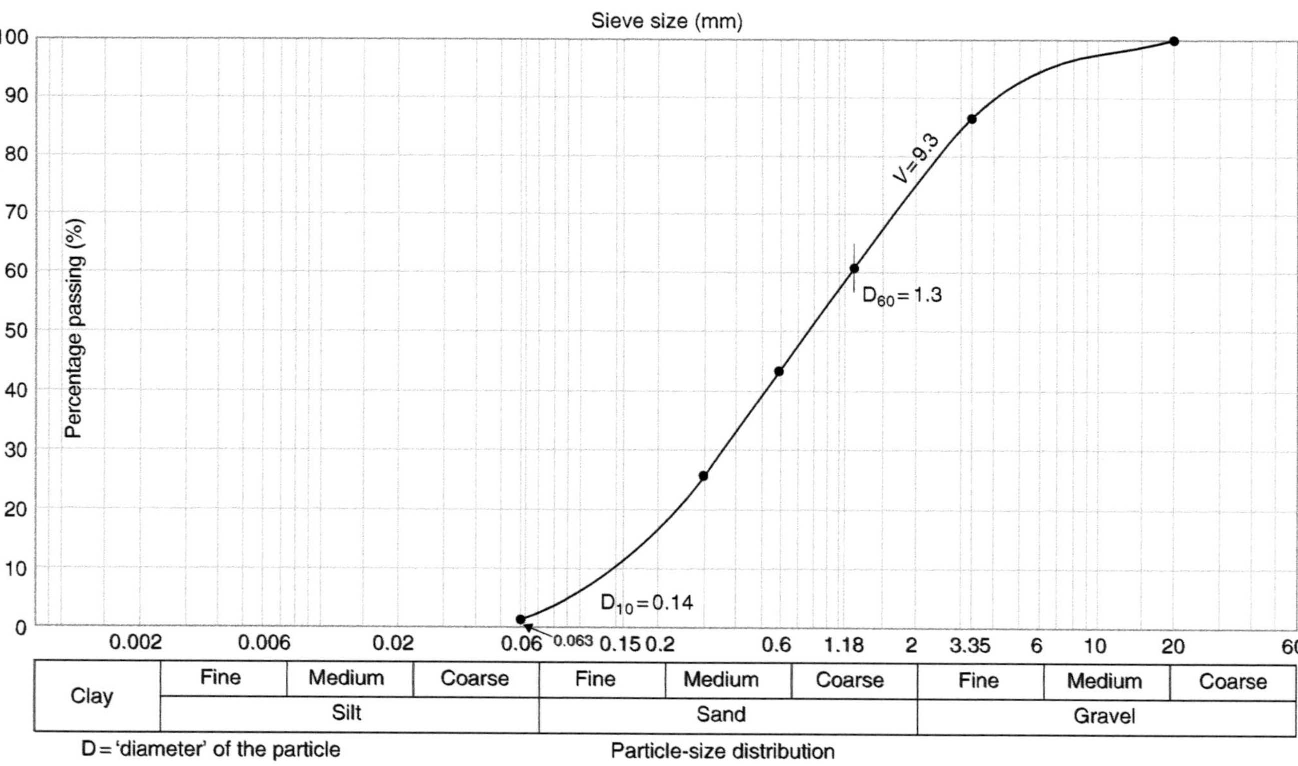
Sieve size (mm)
100
90
80
70
60
50
40
30
20
10
0
Percentage passing (%)
V=9.3
D60=1.3
D10=0.14
0.002
0.006
0.02
0.06
0.063
0.15
0.2
0.6
1.18
2
3.35
6
10
20
60
Clay
Fine
Medium
Coarse
Fine
Medium
Coarse
Fine
Medium
Coarse
Silt
Sand
Gravel
D='diameter' of the particle
Particle-size distribution
(Example 2.9)

Chart 2.2

The recursion formula used is: $$\boxed{P_n = P_{n-1} - \frac{100\,M_n}{M_T}\,\%}\tag{2.16}$$

In this example $M_T = 300\,\text{g}$, therefore (2.16) becomes:

$$\boxed{P_n = P_{n-1} - \frac{M_n}{3}\,\%}$$

If n = 1 $\quad M_1 = 0 \quad P_{1-1} = P_0 = 100\%$

$\therefore\ P_1$ is always taken as 100%

If n = 2 $\quad \left.\begin{matrix} M_2 = 6 \\ P_{2-1} = P_1 = 100 \end{matrix}\right\} \quad \therefore\ P_2 = 100 - \frac{6}{3} = 98\%$

If n = 3 $\quad \left.\begin{matrix} M_3 = 15 \\ P_{3-1} = P_2 = 98 \end{matrix}\right\} \quad \therefore\ P_3 = 98 - \frac{15}{3} = 93\%$

and so on....

Note that 2% of the soil passes through the 0.063 mm sieve, being either silt or clay.

Curve - characteristics

The particle-size distribution curve on Chart 2.2 enables the engineer to describe the soil according to its shape. The description is normally in terms of effective size and uniformity coefficient.

Effective size (D_{10})

It is the size of the particle at $P_n = 10\%$ indicating that 90% of the sample of soil is larger than D_{10} thus gives some comparative idea of the soil type. The effective size in this example is $D_{10} = 0.14\,\text{mm}$, which shows that 25% of the soil is gravel and 65% is sand so, it may be described as gravelly sand.

2.3.2 Uniformity coefficient (U)

Soils made up of a great range of particle sizes are more compact, hence stronger than uniformly graded ones. The degree of uniformity is expressed by:

$$\boxed{U = \frac{D_{60}}{D_{10}}}\tag{2.17}$$

where D_{60} is the size (loosely called 'diameter') of the particles at $P_n = 60\%$.

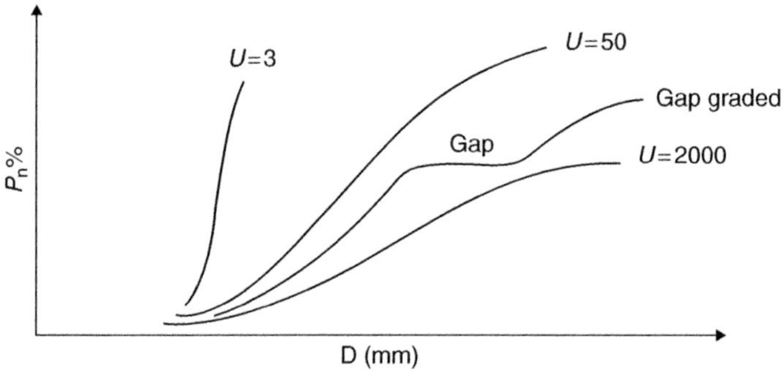

Figure 2.14

Typical values: $U<5$ - The soil is uniform
$5 \leq U \leq 15$ - It is non-uniform
$U>15$ - It is well graded
$U=1$ - All particles are of the same size.

If the value of U is very low, the soil is usually loose and 'floats', when water flows through it or subjected to vibration. Large values of U mean that the soil is well-graded and the smaller particles fill the voids between the large ones. In gap-graded soils, the horizontal line shows, that a range of particle size is missing.

The uniformity coefficient in this example is calculated from:

$$\left.\begin{array}{l} D_{60} = 1.3\text{mm} \\ D_{10} = 0.14\text{mm} \end{array}\right\} \quad \therefore U = \frac{1.3}{0.14} = 9.3$$

The soil falls within group GC in Table 2.9.

2.3.3 Filter design

This is an important application of the particle-size analysis. There are two cases to consider:

1. The movement of particles from one soil to another, caused by flowing water, has to be prevented. This is accomplished by placing a filter layer between the two soils in accordance with the following rule: The D_{15} of the filter material must be less than four times the D_{85} of the protected soil, that is:

$$\frac{D_{15\ (\text{filter})}}{D_{85\ (\text{soil})}} < 4 \tag{2.18}$$

2. The voids in the filters have to be large enough to allow unrestricted seepage through them, thus preventing the build up of hydrostatic pressures and seepage forces. This criterion is satisfied, if D_{15} of the filter material is at least four times the D_{15} of the protected soil, that is:

$$\frac{D_{15(\text{filter})}}{D_{15(\text{soil})}} > 4 \tag{2.19}$$

For more detailed discussion of this subject consult reference 3.

Table 2.9 The Extended Casagrande soil classification coarse-grained soils

Major Divisions	Description & Field Identification	Sub-groups	Group Symbol	Value as a road foundation when not subject to frost action	Potential frost action	Shrinkage or swelling properties	Drainage characteristics
Gravel and gravelly soils	Soils with an appreciable fraction between 60 mm and 2 mm sieves. Generally easily identifiable by visual inspection. A medium to high dry strength indicates that some clay is present. A negligible dry strength indicates the absence of clay.	Well graded gravel-sand mixtures, little or no fines	GW	Excellent	None to very slight	Almost none	Excellent
		Well graded gravel-sands with small clay content	GC	Excellent	Medium	Very slight	Practically impervious
		Uniform gravel with little or no fines	GU	Good	None	Almost none	Excellent
		Poorly graded gravel-sand mixtures, little or no fines	GP	Good to Excellent	None to very slight	Almost none	Excellent
		Gravel-sand mixtures with excess of fines	GF	Good to Excellent	Slight to medium	Almost none to slight	Fair to Practically impervious
Sands and sandy soils	Soils with appreciable fraction between the 2 mm and the 0.06 mm sieve. Majority of the particles can be distinguished by eye. Feel gritty when rubbed between the fingers. A medium to high dry strength indicates that some clay is present. A negligible dry strength indicates absence of clay.	Well graded sands and gravelly sands, little or no fines	SW	Excellent to good	None to very slight	Almost none	Excellent
		Well graded sands with small clay content	SC	Excellent to good	Medium	Very slight	Practically impervious
		Uniform sands, with little or no fines	SU	Fair	None to very slight	Almost none	Excellent
		Poorly graded sands, with little or no fines	SP	Fair to good	None to very slight	Almost none	Excellent
		Sands with excess of fines	SF	Fair to good	Slight to high	Almost none to medium	Fair to practically impervious

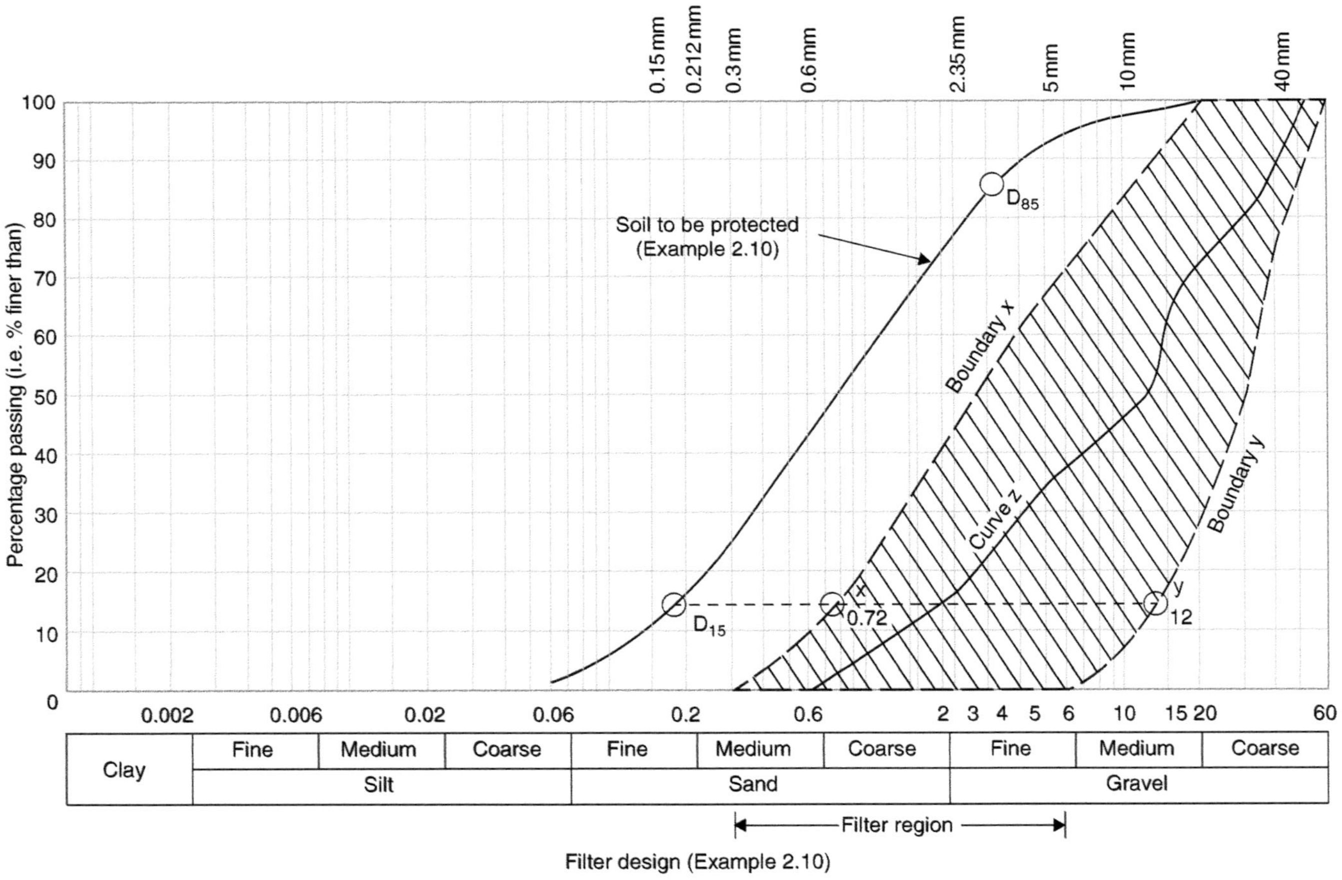

0.15 mm
0.212 mm
0.3 mm
0.6 mm
2.35 mm
5 mm
10 mm
40 mm
Percentage passing (i.e. % finer than)
100
90
80
70
60
50
40
30
20
10
0
Soil to be protected
(Example 2.10)
D_{85}
D_{15}
Boundary x
Boundary y
Curve z
x
0.72
y
12
0.002
0.006
0.02
0.06
0.2
0.6
2 3 4 5 6
10
15 20
60
Clay
Fine
Medium
Coarse
Silt
Fine
Medium
Coarse
Sand
Fine
Medium
Coarse
Gravel
Filter region
Filter design (Example 2.10)

Chart 2.3

Example 2.10

Design a suitable protective filter layer for the soil in Example 2.9 and draw the result on Chart 2.3.

From the chart: $D_{15\ (soil)} = 0.18$
$D_{85\ (soil)} = 3.00$

From (2.18): $D_{15\ (filter)} < 4 \times 3.00 = 12.00\,\text{mm}$ (Point y)

From (2.19): $D_{15\ (filter)} > 4 \times 0.18 = 0.72\,\text{mm}$ (Point x)

After plotting points x and y on Chart 2.3, the boundaries of the filter region are drawn, approximately parallel with the grading curve of the soil to be protected. The particle size distribution (z) of the filter material has to fall within the region and its shape should be roughly similar to that of the soil.

It must be remembered that (2.18) and (2.19) provide empirical solution to problems for well-graded materials and not for gap-graded or stratified soils. In some cases the construction of two or more filter layers may be necessary. In many problems, synthetic filter fabrics may be used efficiently, instead of filter layers.

2.3.4 Typical problems

Possible failures of structures, constructed in water-bearing erodible strata can be avoided by the use if filter layers designed so, that they do not clog up, thus permit water flow at all times. Figure 2.15 shows some practical problems.

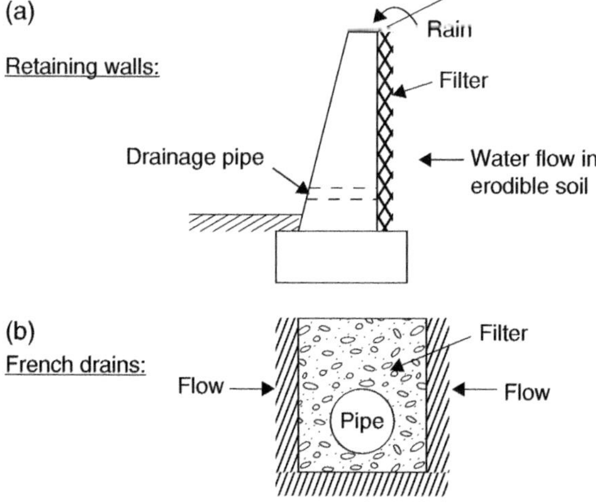

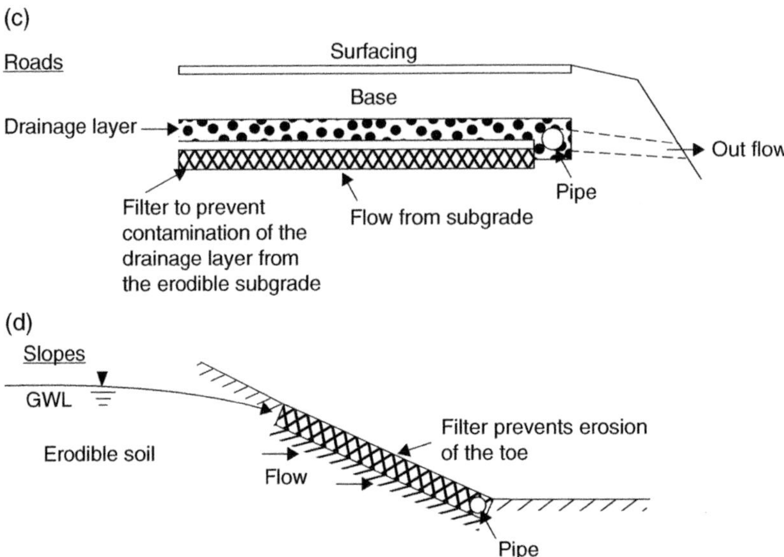

Figure 2.15

2.3.5 Combination of materials

See Barabás in Reference 10. Suppose there is no suitable soil available, having particle-size distribution situated within the boundaries of the filter zone. In that case, it is possible to mix two soils to satisfy this condition, provided their distribution curves encompass at least part of the filter region. The procedure is best explained by means of an example.

Example 2.11

Given the particle-size distribution of soils A and B to be mixed such, that the distribution curve of the resulting soil can be plotted within the boundaries *x* and *y*. The four distributions are tabulated in Table 2.10:

Table 2.10

	Percentage passing (P%)			
Sieve (mm)	**A (P_A)**	**B (P_B)**	**X (P_x)**	**Y (P_y)**
63	100	100	100	100
50	100	88	100	91
37.5	100	59	100	77
28	100	17	100	57
20	100	10	100	30
10	97.5	0	84	9
5	85	0	68	0
3.35	79	0	57	0
2	68	0	43	0
1.18	42	0	24	0
0.6	20	0	11	0
0.3	10	0	0	0
0.15	0	0	0	0

First, check whether soils A and B encompass the filter region, bounded by distributions x and y (See Chart 2.5). As they are outside the boundaries, proceed step-by-step.

Step 1

Draw two vertical lines A and B on Chart 2.4, at an arbitrary distance (d) from each other and scale each from 0% to 100%. In this case $d=70$ divisions.

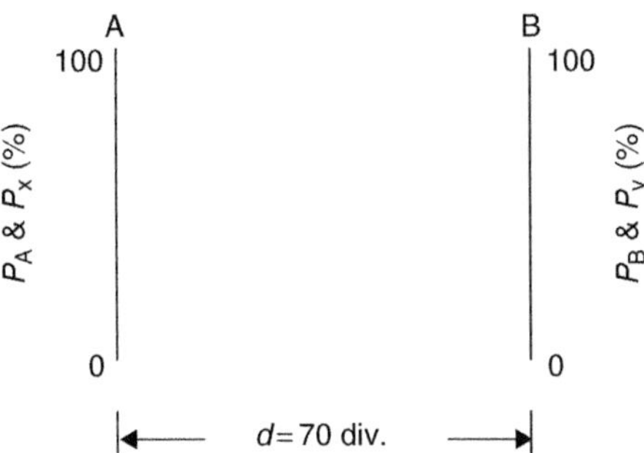

Figure 2.16

Step 2

Connect P_A and P_B, relating to a particular sieve by a diagonal line, e.g. for sieve = 10 mm

$P_A=97.5\%$ and $P_B=0\%$.

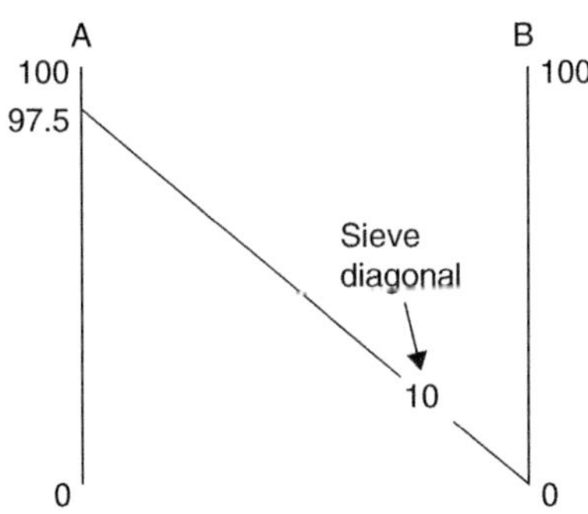

Figure 2.17

Step 3

Taking the P_x and P_y figures for the particular sieve-size, draw horizontal lines from P_x and P_y to this sieve-diagonal e.g.:

For sieve = 10 mm

$P_A = 97.5\%$ $P_X = 84\%$

$P_B = 0\%$ $P_Y = 9\%$

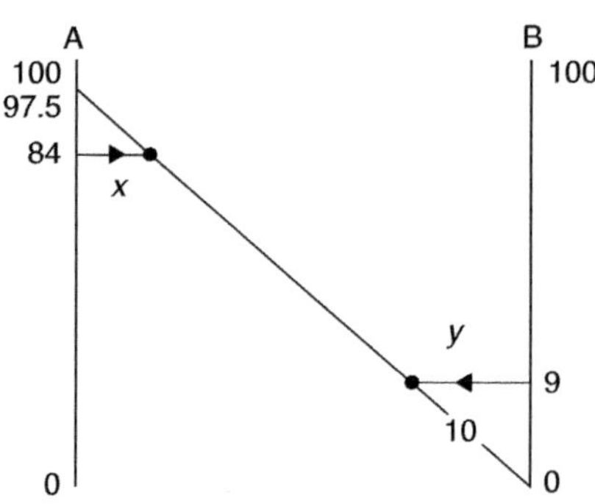

Figure 2.18

Step 4

Repeat steps 2 and 3 for each sieve and connect the resulting set of points on each side to represent the *x* and *y* boundaries.

Step 5

Draw a tangential vertical line to each curve.

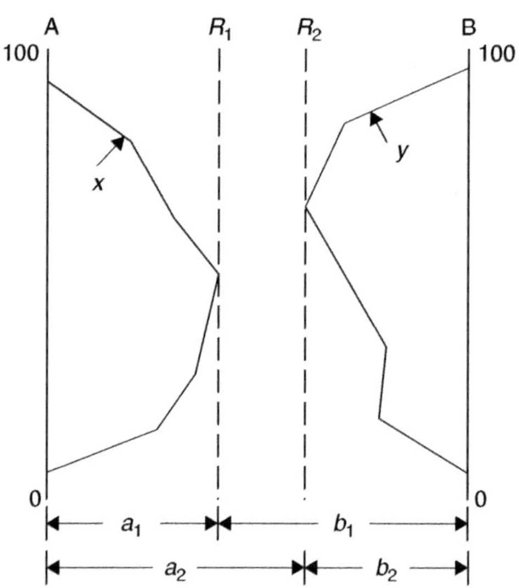

Figure 2.19

Note that the distance of a vertical line is denoted '*a*' and '*b*' from A and B respectively.

Also, if no vertical line can be drawn without intersecting the curves, than there is no solution to the problem. The given soils then cannot be mixed so, that the filter rules are satisfied.

The significance of a vertical line is, that its position indicates a mixing ratio (R), which is, in general, given by:

$$\boxed{R = \frac{b}{a}} \tag{2.20}$$

This may be written as $\frac{R}{1} = \frac{b}{a}$ implying, that the mass (M) of the mix is R-times the mass of A (M_A) plus the mass of B (M_B).

Therefore,
$$\boxed{M = RM_A + M_B \; \text{g}} \tag{2.21}$$

Similarly,
$$\boxed{P_n = RP_n + P_B \; \%} \tag{2.22}$$

There are two tangential verticals on Chart 2.4, hence there are two mixing ratios R_1 and R_2.

For R_1: $\left.\begin{matrix} a_1 = 32 \\ b_1 = 38 \end{matrix}\right\} \quad \therefore \; R_1 = \frac{38}{32} = 1.19$ (maximum)

For R_2: $\left.\begin{matrix} a_2 = 36 \\ b_2 = 34 \end{matrix}\right\} \quad \therefore \; R_2 = \frac{34}{36} = 0.94$ (minimum)

From practical point of view, it would be inconvenient to weigh 1.19 M_A or 0.94 M_A. Fortunately, the range of ratios include in this case R=1, with corresponding vertical between that of R_1 and R_2. The distance 'a' of this or of any vertical may be calculated from:

$$\boxed{a = \frac{d}{1+R}} \tag{2.23}$$

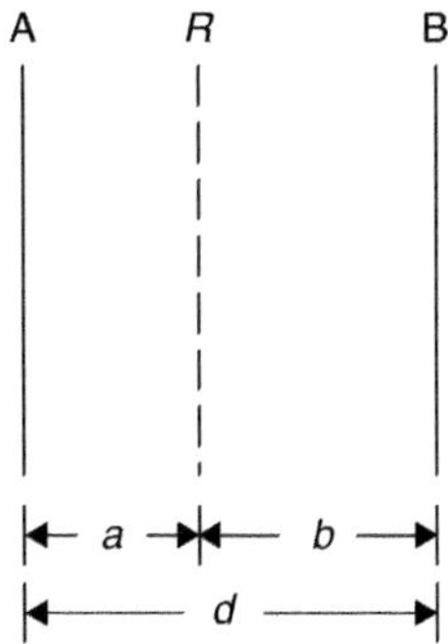

Derivation:

$$R = \frac{b}{a} = \frac{d-a}{a}$$

$$aR = d - a$$

$$a + aR = d$$

$$a(1+R) = d$$

$$\therefore a = \frac{d}{1+R}$$

Figure 2.20

It is easy to mix one unit mass of soil A with one unit of soil B. Therefore the task is now to obtain the expected particle-size distribution for the mix at R=1.

From Chart 2.4: $d = 70$ div. $\quad \therefore \quad a = \frac{70}{1+1} = 35$ div

Step 6

Having drawn the vertical for $R=1$ on Chart 2.4, its intersection with diagonals indicates the percentage passing at each sieve.

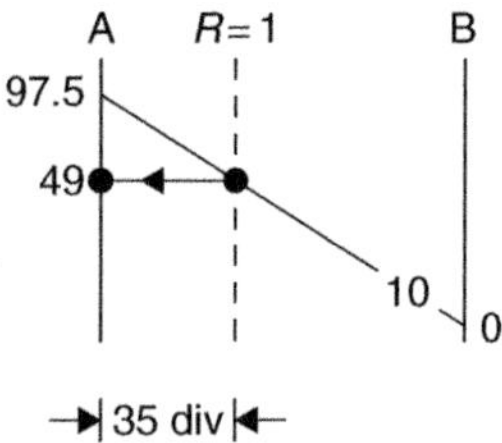

Figure 2.21

This applies to any ratio within the range $0.67 \le R \le 1.19$. The particle-size distributions P_2, P and P_1, corresponding to 0.94, 1.0 and 1.19 may now be tabulated (Table 2.11):

Table 2.11

Sieve (mm)	63	50	37.5	28	20	10	5	3.35	2	1.18	0.6	0.3	0.15
P_2 %	100	94	79	57	54	47	41	38	33	20	10	5	0
P %	100	94	80	58	55	49	43	39	34	21	10	5	0
P_1 %	100	94.5	81	62	59	53	46	43	37	23	11	5	0

Step 7

Plot the points in the above table on Chart 2.5. The shaded area between the boundaries, corresponding to R_1 and R_2 includes all possible mixtures of soils A and B such that they can be drawn within the filter zone. These mixtures are expected to be suitable as protective filters to the soil in Example 2.9.

Analytic alternative

The value of P% at each sieve can also be calculated, once R is known. The formula to do this is derived from similar triangles.

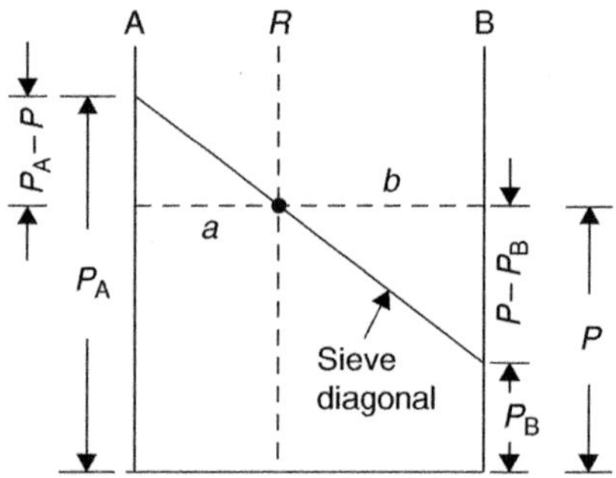

Figure 2.22

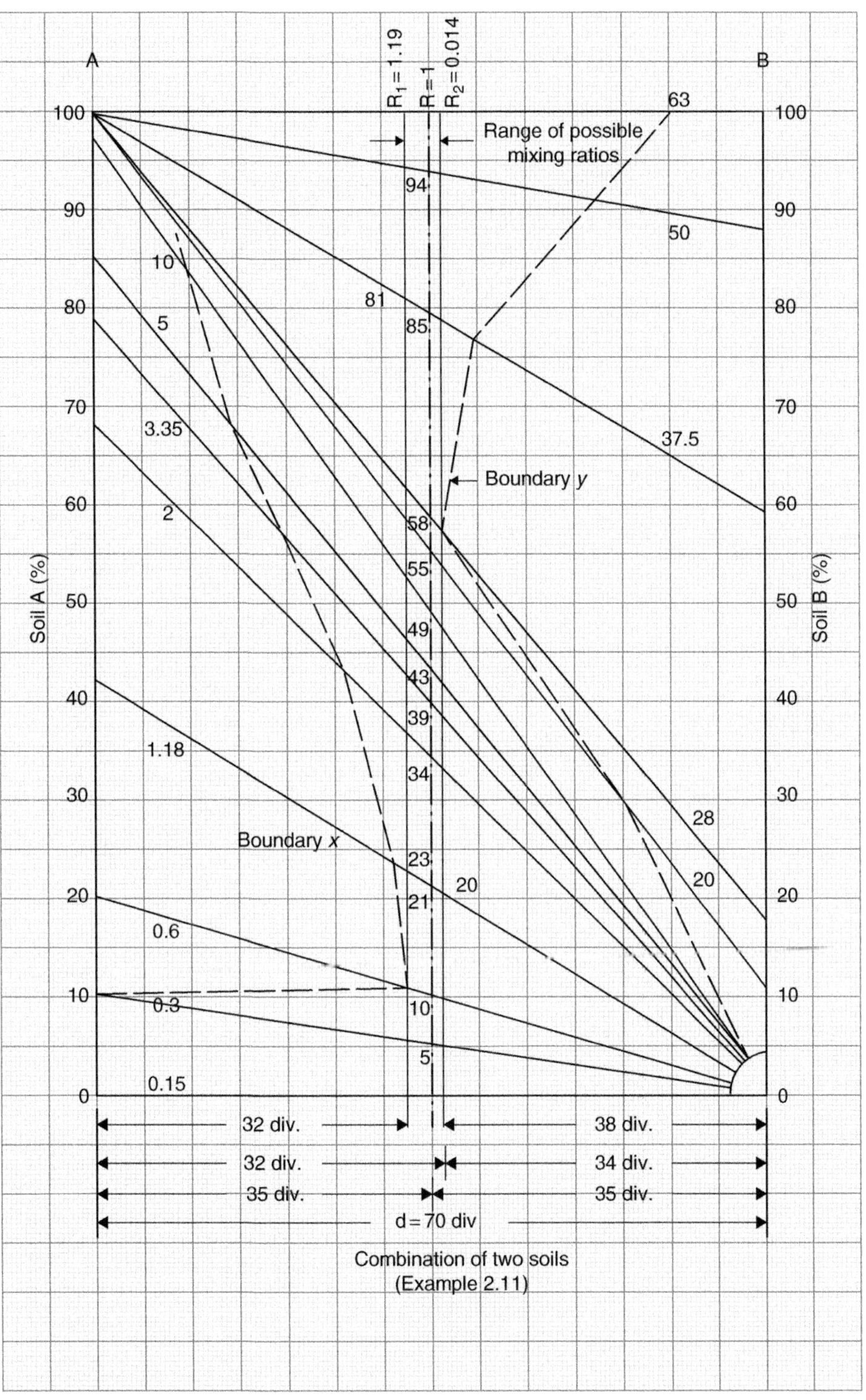

A
B
Soil A (%)
Soil B (%)
$R_1 = 1.19$
R = 1
$R_2 = 0.014$
Range of possible mixing ratios
Boundary y
Boundary x
10
5
3.35
2
1.18
0.6
0.3
0.15
81
94
85
58
55
49
43
39
34
23
21
10
5
63
50
37.5
28
20
20
32 div.
38 div.
32 div.
34 div.
35 div.
35 div.
d = 70 div
Combination of two soils
(Example 2.11)

Chart 2.4

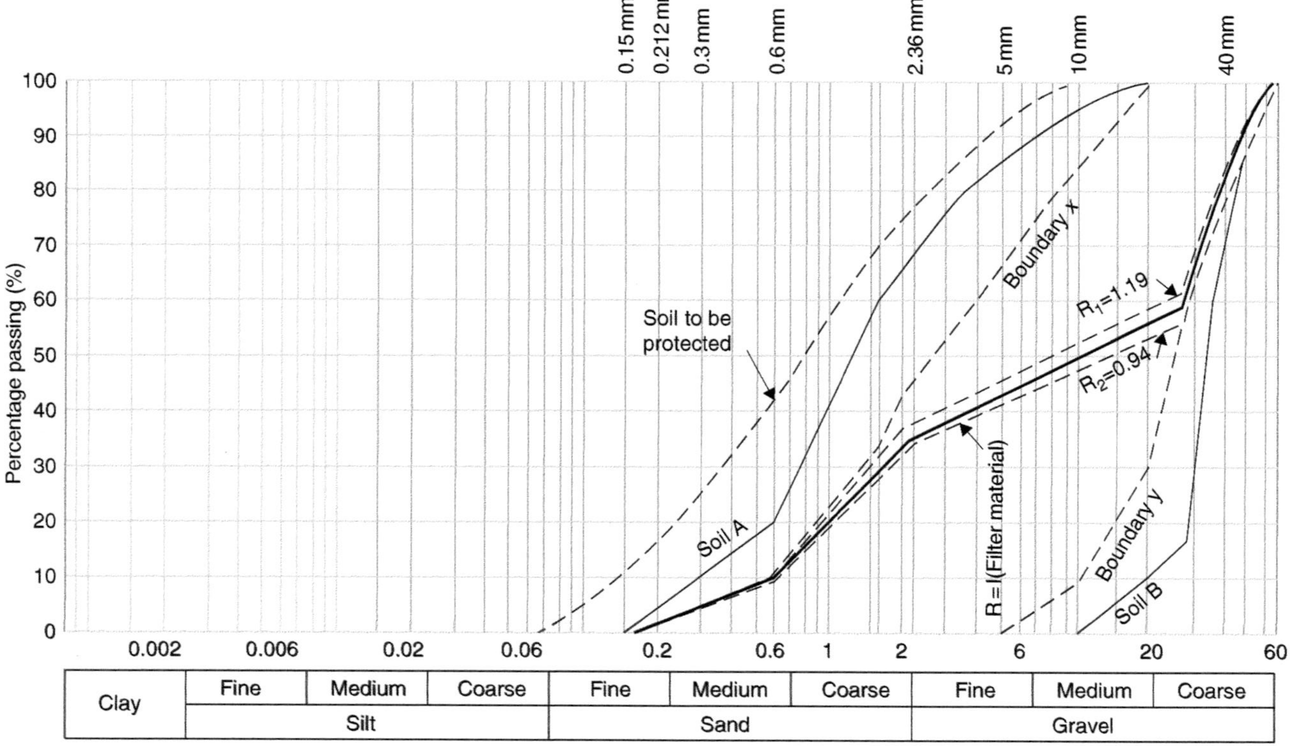

Filter layer: combination of two soils
(Example 2.11)

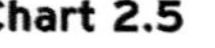

Chart 2.5

From (2.20): $R=\frac{b}{a}$

But, from similar triangles:

$$\frac{a}{P_A-P}=\frac{b}{P-P_B} \qquad \therefore \frac{b}{a}=\frac{P-P_B}{P_A-P}=R$$

From which: $RP_A-RP=P-P_B$

$$RP_A+R_B=P\,(1+R) \qquad \boxed{\therefore P=\frac{RP_A-P_B}{1+R}\%} \tag{2.24}$$

2.3.6 Sedimentation tests

The smallest B.S. sieve that can be used in coarse sieve analysis has a mesh size of 0.063 mm. In order to complete the particle size distribution curve of soil containing finer grains, one of the two sedimentation tests are carried out. These are:

1. Pipette analysis
2. Hydrometer analysis

The hydrometer test can be executed more easily in a site laboratory than the pipette test, with sufficient accuracy for engineering purposes; hence this analysis is outlined here. In this, the soil is mixed with distilled water and a suitable dispersing agent to form a uniform suspension.

The purpose of the hydrometer is to measure the variation in density of the suspension with time, at a particular height, within the measuring cylinder. The density of each elevation depends on the size of particles present. Knowing the density, the percentage of grains smaller than the largest size present at the level tested can be calculated.

The largest particle size at a particular height is computed from Stoke's Law, which deals with the fall of a single spherical object in large amount of water. As the soil particles are not normally of spherical shape, the calculated diameters are referred to as 'equivalent diameters'. Also, the testing cylinder should have large diameter compared to that of the hydrometer's bulb, in order to keep the soil grains some distance apart. Figure 2.23 depicts a suspension of three particle sizes d_1, d_2 and d_3 in separate compartments. By Stoke's Law each sinks at different velocity, say u_1, u_2 and u_3.

After time t, therefore, all particles above level x will be finer than d.

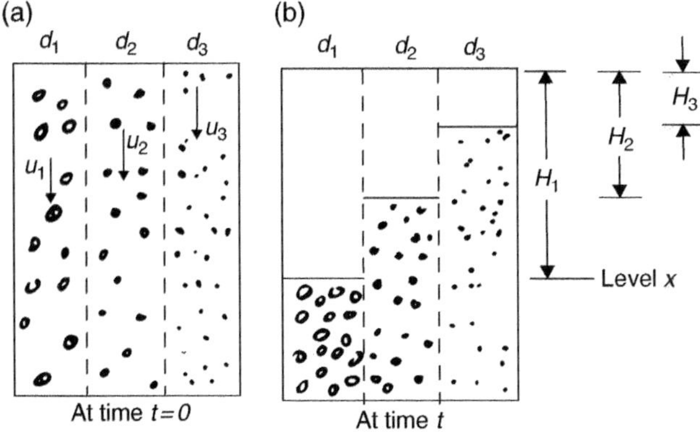

Figure 2.23

Reading a hydrometer

The hydrometer is read at the top of meniscus. The figure read is multiplied by 1000. For instance; if the reading is 1.015, then it is recorded as: $R_h' = 15$.

The usual range of hydrometer scale is 0.995–1.030.

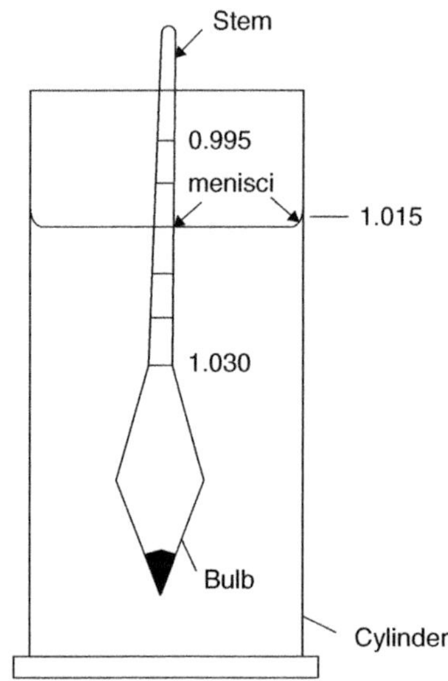

Figure 2.24

Meniscus correction

It is difficult or impossible to read the hydrometer at the level surface because of the non-transparent meniscus at the stem. That is why the reading is taken at the top of the meniscus and then a correction (C_m) is added to get the true reading (R_h). The meniscus correction is constant for a given hydrometer. It is determined by lowering the hydrometer into a cylinder containing distilled water and taking readings at the top and bottom of the meniscus.

Note: The hydrometer readings are increasing downwards $\therefore\ R_h = R_h' + C_m$

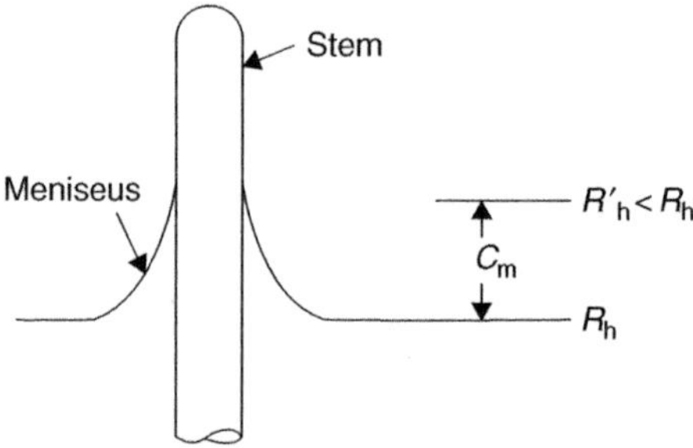

Figure 2.25

Correction for dispersing agent (C_d)

The agent is added to prevent flocculation of the grains. As a result, the specific gravity of the suspension is increased. The correction is the difference between the two readings:

1. In distilled water
2. In the same water after the addition of dispersing agent

Temperature correction (m_T)

The hydrometer is normally calibrated for use in suspension at 20 °C. If the suspension is at different temperature, then its specific gravity is altered. If therefore, the temperature during the test differs from the calibrated value, then the $\pm m_T$ correction from Chart 2.6 has to be added to the reading.

The corrected, true reading is given by:

$$\boxed{R_h = R_h' + C_m - C_d \pm m_T} \tag{2.25}$$

Effective depth (H_R)

The hydrometer measures the density of the suspension at the centre of the bulb. The effective depth is the true height from the surface to this centre. Figure 2.26 indicates the quantities required to calculate H_R.

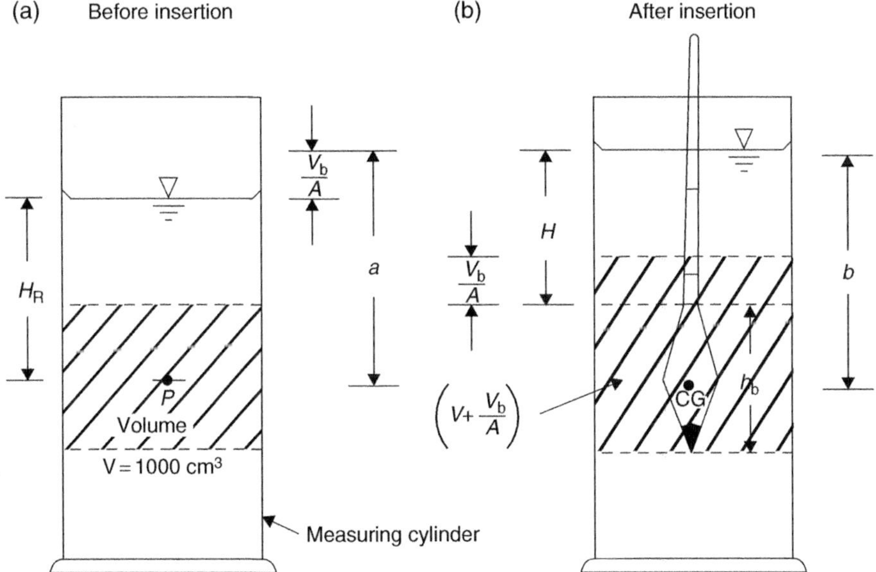

Figure 2.26

V_b = volume of the hydrometer bulb
h_b = length of the bulb
A = cross-sectional area of the 1000 ml cylinder
H = height from surface to the top of the bulb
V = volume of suspension = 1000 cm^3

When the bulb is inserted, its volume displaces equal volume of the suspension, thus increasing the surface level above point P at which the density is measured. The true height is found by equating $a = H_R + \frac{V_b}{A}$ and $b = H + \frac{h_b}{2} + \frac{1}{2}\left(\frac{V_b}{A}\right)$

$$\therefore \quad H_R + \frac{V_b}{A} = H + \frac{h_b}{2} + \frac{V_b}{2A} \qquad \therefore \quad \boxed{H_R = H + \frac{1}{2}\left(h_b - \frac{V_b}{A}\right)} \tag{2.26}$$

Calibration of the hydrometer

The purpose of calibration is to determine its effective depth in terms of hydrometer readings. Formula (2.26) indicates that H_R is dependent on the cross-sectional area of the cylinder used. For this reason the hydrometer has to be calibrated and used in the same cylinder. It is normally calibrated in distilled water at 20°C.

Step 1: Insert the hydrometer and note the change in the water level $\left(\frac{V_b}{A}\right)$.

Note: V_b in cm^3 is approximately equal to the weight of the hydrometer in grams.

Step 2: Measure:

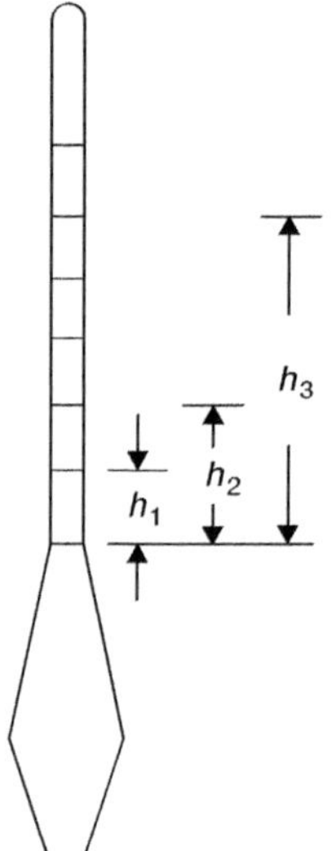

1. The cross-sectional area of cylinder (A)
2. The distance between each calibration mark and the lowest one (h_x)
3. Length of the stem
4. Length of the bulb (h_b)
5. Apply (2.26) in the form $H_R = h_x + \frac{1}{2}\left(h_b - \frac{V_b}{A}\right)$ to get a series of values for H_R.
6. Plot h_x against H_R as in Figure 2.28.

Figure 2.27

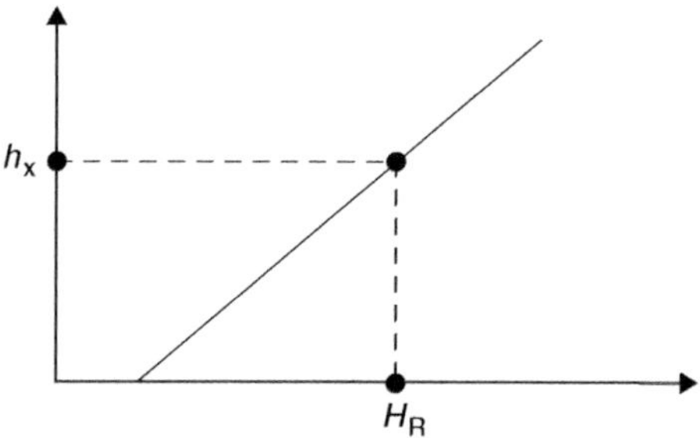

Figure 2.28

Equivalent particle diameter (*D*)

The formula is derived from Stoke's Law:

$$D = 0.005531\boxed{\sqrt{\frac{\eta H_R}{(G_S - 1)t}}}\text{ mm} \tag{2.27}$$

where η = dynamic viscosity ($Ns/m^2 = P_{as}$) at temperature T°C

G_s = specific gravity of particles

t = elapsed time (min)

Percentage finer than *D* (*P*%)

These results are plotted on the particle-size distribution chart. P_n% is given by:

$$P_n = \boxed{\frac{100\,G_s\,R_h}{M_s(G_s - 1)}}\% \tag{2.28}$$

where M_S = total dry mass of soil particles/1000 ml of suspension.

Example 2.12

Test temperature:	$T = 23$°C	$m_T = +0.56$
Total dry mass of soil:	$M_S = 54$ g	$\eta = 0.936$ m Pas
Reading:	$h_x = R'_h = 24$ at time $t = 130$ min	$h_b = 160$ mm
Specific gravity of particles:	$G_S = 2.7$	$V_b = 61\,cm^3$
Meniscus correction:	$C_m = 0.45$	$A = 78.5\,cm^2$
Correction for dispersing agent:	$C_d = 0.8$	$H = 60$ mm

From (2.25): $R_h = 24 + 0.45 - 0.8 + 0.56 = 24.21$ mm

From (2.26): $H_R = 60 + \frac{1}{2}\left(160 - \frac{10.61}{78.5}\right) = 136.1$ mm

From (2.27): $D = 0.005531\sqrt{\frac{0.936 \times 136.1}{(2.7 - 1) \times 130}} = 0.0042$ mm

From (2.28): $P_n = \frac{100 \times 2.7 \times 24.21}{54 \times (2.7 - 1)} = 71\%$

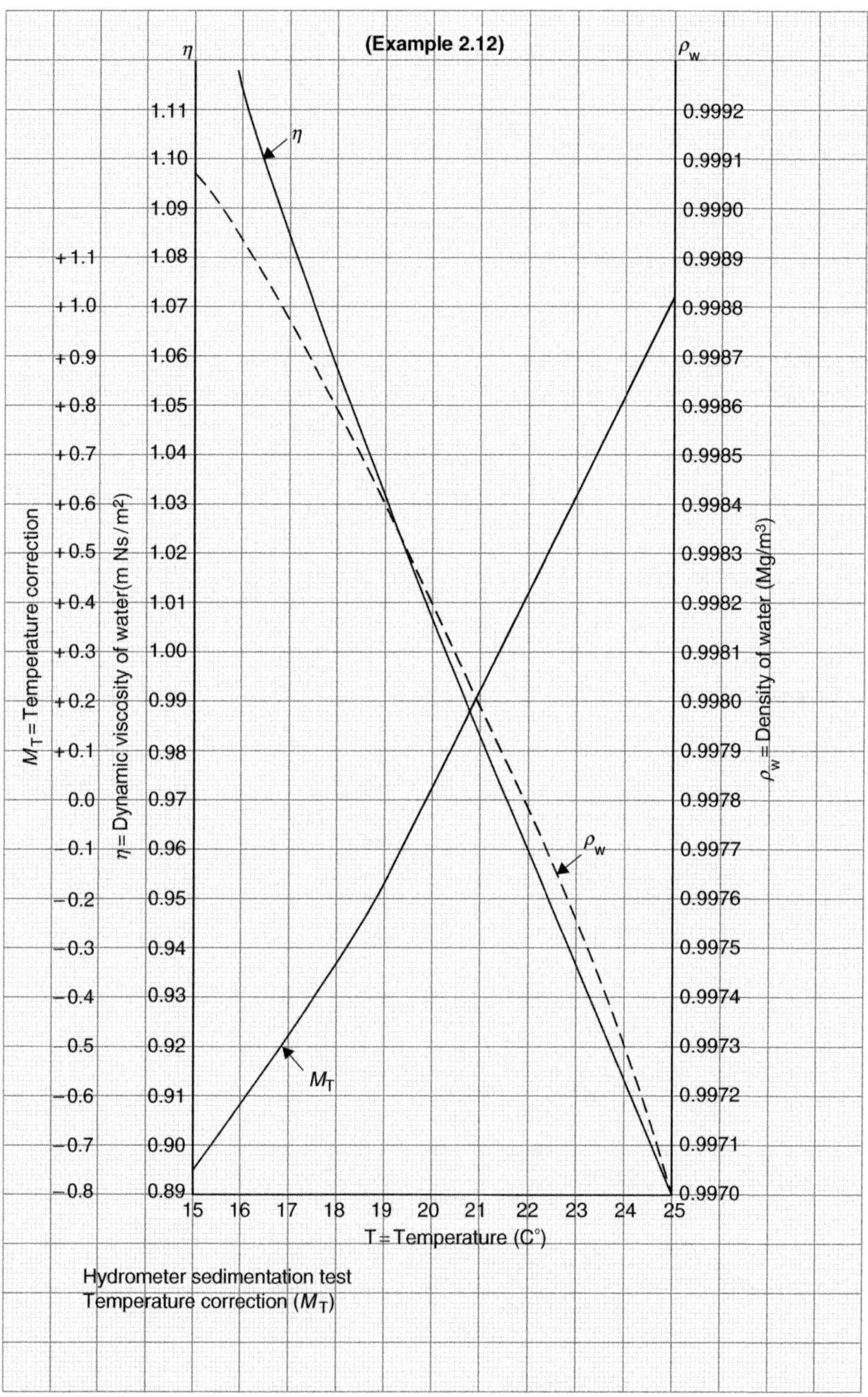
(Example 2.12)
η
ρw
1.11
1.10
1.09
1.08
1.07
1.06
1.05
1.04
1.03
1.02
1.01
1.00
0.99
0.98
0.97
0.96
0.95
0.94
0.93
0.92
0.91
0.90
0.89
0.9992
0.9991
0.9990
0.9989
0.9988
0.9987
0.9986
0.9985
0.9984
0.9983
0.9982
0.9981
0.9980
0.9979
0.9978
0.9977
0.9976
0.9975
0.9974
0.9973
0.9972
0.9971
0.9970
+1.1
+1.0
+0.9
+0.8
+0.7
+0.6
+0.5
+0.4
+0.3
+0.2
+0.1
0.0
−0.1
−0.2
−0.3
−0.4
−0.5
−0.6
−0.7
−0.8
MT = Temperature correction
η = Dynamic viscosity of water (m Ns/m2)
ρw = Density of water (Mg/m3)
MT
15 16 17 18 19 20 21 22 23 24 25
T = Temperature (C°)
Hydrometer sedimentation test
Temperature correction (MT)

Chart 2.6

Problem 2.1

A saturated sample of clay has the following properties:

Natural moisture content: $m=21\%$
Plastic Limit: $PL=16\%$
Volume in natural state: $V_1=2000\,cm^3$
Volume at Plastic Limit: $V_2=1832\,cm^3$
Volume at Shrinkage Limit: $V_0=1600\,cm^3$
Specific gravity: $G_s=2.65$

Estimate the approximate values of:

a) The Shrinkage Limit.
b) The voids ratios, saturated and dry unit weight in natural state, as well as at the Plastic and Shrinkage Limits.

Problem 2.2

Given three particle-size distributions A, B and C. Distributions A and B represent two granular materials available on site, whilst C indicates the material required for the construction of a filter layer.

Table 2.13

Sieve (mm)	0.1	0.15	0.3	0.6	1.18	2.36	5	10	20	40
P_A %	0	2	20	45	75	90	98	100	100	100
P_B %	0	0	0	0	0	0	5	40	97	100
P_C %	0	1	10	21	34	44	48	70	98	100

a) Determine the mixing ratio (R) graphically and check its value analytically.
b) Tabulate the resulting particle - size distribution (P%) and compare it with the given distribution (P_C%).

Chapter 3
Permeability and Seepage

It is evident, from the concepts introduced in Chapter 1, that soil characteristics are influenced to a great extent by its static water content. In many practical engineering problems, however, the pore water is not in a state of rest, but flows through the soil. The extent of this seepage depends largely on the porosity of the material as well as on the hydrostatic head inducing the flow. Figure 3.1 shows a typical example of this type.

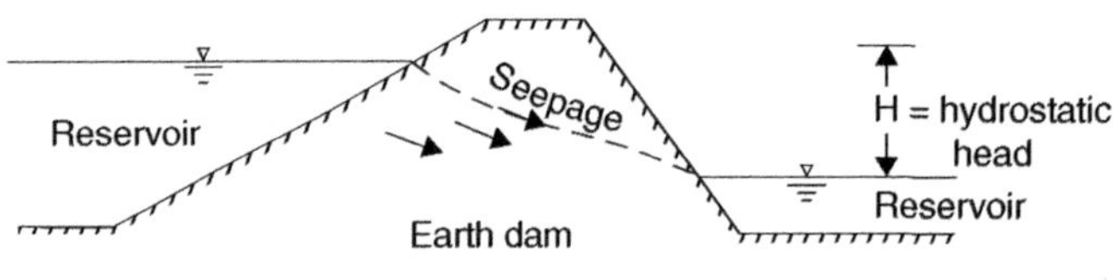

Figure 3.1

The characteristic of the soil which enables water to permeate it is called 'permeability'. Its measure is the coefficient of permeability, represented by the letter k, which varies significantly with:

1. Density $\begin{cases} \text{Voids ratio} \\ \text{Porosity} \end{cases}$

   High density means low porosity, hence low permeability and vice versa
2. Particle-size distribution:
   Large grain diameter means large voids ratio, hence high permeability and vice versa
3. Soil structure:
   Most soil layers were deposited by water. They are more permeable horizontally then vertically.
4. Discontinuities:
   Fissures, cracks in clay or joints in rock or intrusions of different soil types can increase their permeability.

Introduction to Soil Mechanics, First Edition. Béla Bodó and Colin Jones.

3.1 Coefficient of permeability (*k*)

The general expression for the coefficient is derived by means of Darcy's Law, introduced in 1856 after carrying out experiments in connection with laminar, that is steady streamline flow, in sand filters. Figure 3.2 depicts the experimental apparatus.

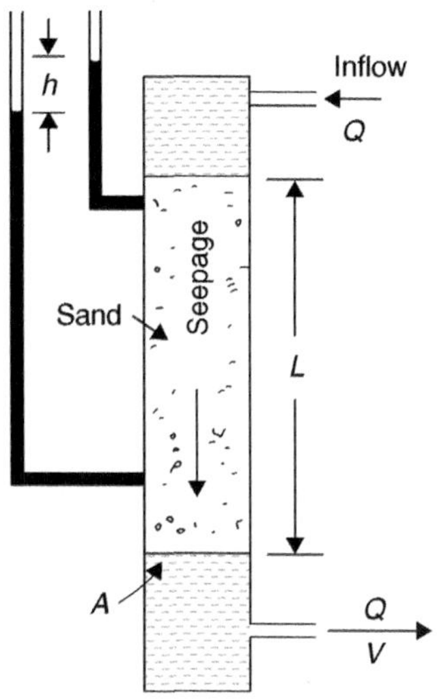

Figure 3.2

Q = flow rate (m^3/sec = cumecs)
L = Length of specimen = Length of flow path (m)
h = head loss (m)
A = cross-sectional area of the specimen (m^2)

Darcy's Law

It expresses the discharge velocity v:

$$v = ki \text{ m/s} \tag{3.1}$$

where i = hydraulic gradient, given by:

$$i = \frac{h}{L} \quad \text{(Dimensionless)} \tag{3.2}$$

Note: The dimensions of k are those of velocity since from 3.1:

$$k = \frac{v}{i} \text{ m/s} \tag{3.3}$$

The flow rate is given by:

$$Q = Av = Aki \text{ m}^3\text{/s} \tag{3.4}$$

From which

$$k = \frac{Q}{Ai} \text{ m/s} \tag{3.5}$$

3.2 Seepage velocity (v_s)

The actual cross-sectional area, through which water permeates, depends on the voids ratio or porosity of the soil. It is therefore smaller than area A of the specimen. For the same quantity of discharge flow rate (Q) therefore, the seepage velocity is larger than the discharge one. The relationship between the velocities can be derived in terms of voids ratio. From formula (3.4):

$$Q = Av = A_V v_S$$

where A_V=total area of voids at a cross-section of the sample.

But, $A = A_S + A_V$

where A_S=total area of solids at a cross-section of the sample.

Substituting: $Q=(A_S+A_V)v=A_V v_S$

From which, $v_S = \left(\frac{A_s + A_v}{A_v}\right) v$

Now, the volume of solids in the specimen is given by:

$$V_s = A_s L \quad \therefore \quad A_s = \frac{V_S}{L}$$

And the volume of voids is: $V_V = A_V L \quad \therefore \quad A_V = \frac{V_V}{L}$

Substituting:

$$v_s = \left(\frac{\frac{V_S}{L} + \frac{V_V}{L}}{\frac{V_V}{L}}\right) v = \left(\frac{V_S + V_V}{V_V}\right) v$$

$$= \left(\frac{V_S}{V_V} + 1\right) v$$

But, the voids ratio is given by: $e = \frac{V_V}{V_s}$

Hence, $v_s = \left(\frac{1}{e} + 1\right) v = \left(\frac{1+e}{e}\right) v$

But, porosity is: $n = \frac{e}{1+e}$

Therefore, the seepage velocity is given by:

Either $$\boxed{v_s = \left(\frac{1+e}{e}\right) v = \frac{v}{n} \text{ m/s}} \tag{3.6}$$

or $$\boxed{v_s = \frac{ki}{n} \text{ m/s}} \tag{3.7}$$

Example 3.1

Figure 3.3 shows a filter arrangement between two tanks. Calculate the coefficient of permeability and the seepage velocity.

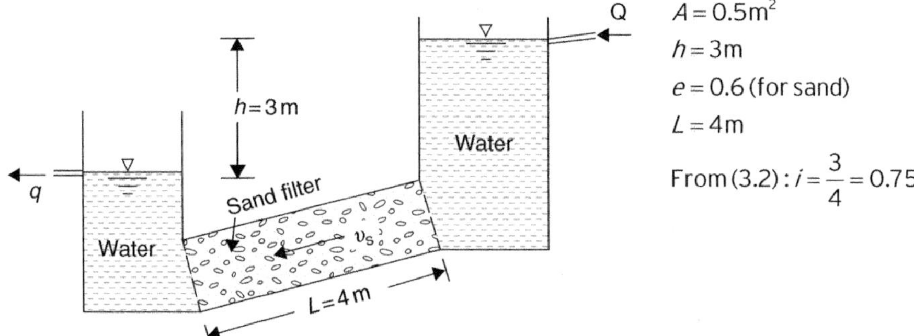

Figure 3.3

The amount of water collected at the discharge end in $t=60$ seconds was $q=0.0608\,\text{m}^3$. The flowrate Q in these terms is given by:

$$\boxed{Q=\frac{q}{t}\ \text{m}^3/\text{s}} \tag{3.8}$$

Therefore, $Q=\frac{0.0608}{60}=0.00101\,\text{m}^3/\text{s}$

From (3.5): $k=\frac{Q}{Ai}=\frac{0.00101}{0.5\times 0.75}=0.0027\,\text{m/s}$

Either from (3.1): $v=ki=0.0027\times 0.75=0.00202\,\text{m/s}$

Or from (3.4): $v=\frac{Q}{A}=\frac{0.00101}{0.5}=0.00202\,\text{m/s}$

From (3.6): $v_s=\left(\frac{1+e}{e}\right)v=\frac{1.6\times 0.00202}{0.6}=0.0054\,\text{m/s}$

Therefore, the seepage velocity is approximately two and a half times faster than the discharge velocity, in this example.

3.3 Determination of the value of *k*

The coefficient of permeability may be obtained by either laboratory or field tests of which the latter is more representative of the actual in-situ conditions. The standard tests are:

Laboratory:	(1) Constant head
	(2) Falling head
	(3) Pumping
In-situ:	(4) Borehole
Laboratory:	(5) Consolidation

Average values of *k*:

Table 3.1

Soil	k (m/s)
Gravel	10^{-2} to 1
Sand	10^{-2} to 10^{-5}
Silt	10^{-5} to 10^{-8}
Clay	10^{-8} to 10^{-12}

3.3.1 Constant head test

The test is suitable for coarse-grained soil, such as gravel or sand. The apparatus is drawn schematically in Figure 3.4. Because of the large voids ratio, the flowrate through the soil is fairly high. The water level in the tank is kept constant by maintaining a uniform rate of inflow at the same head *h*. Measurements are made only after steady seepage had been achieved.

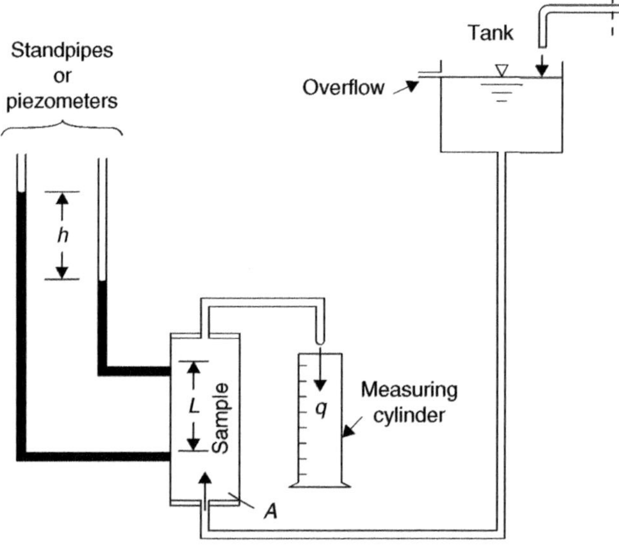

Figure 3.4

Procedure

Step 1: Eliminate all air bubbles from the system.
Step 2: Increase flow and note h.
Step 3: Collect q cm^3 of water in the cylinder over t seconds.
Step 4: Measure h and L in centimeters and obtain hydraulic gradient from (3.2): $i = \dfrac{h}{L}$
Step 5: Measure the cross-sectional area A (cm^2).
Step 6: Calculate the coefficient of permeability from:

(3.5): $k = \dfrac{Q}{Ai}$

(3.8): $Q = \dfrac{q}{t}$

(3.2): $i = \dfrac{h}{L}$

$$k = \frac{\frac{q}{t}}{A\frac{h}{L}}$$

$$\therefore \quad \boxed{k = \frac{qL}{Aht} \text{ cm/s}} \qquad (3.9)$$

Step 7: Calculate the velocity of flow from:

(3.4): $v = \dfrac{Q}{A}$

(3.8): $Q = \dfrac{q}{t}$

$$\boxed{v = \frac{q}{At} \text{ cm/s}} \qquad (3.10)$$

Step 8: Repeat steps 2-7 at least four times.
Step 9: Plot i against v, as shown in Figure 3.5.

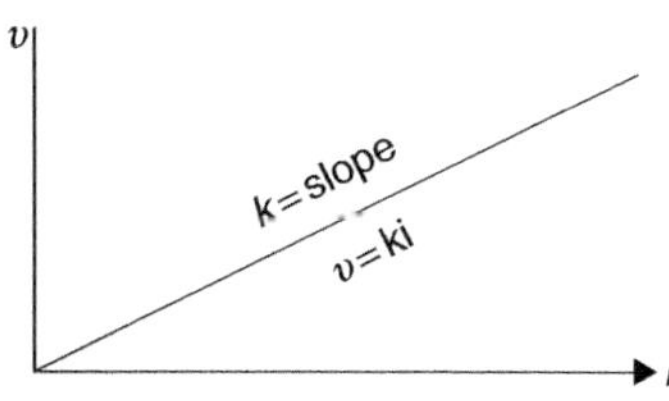

Figure 3.5

The slope of the line through the plotted points yields the average value of k.

Example 3.2

Tests were carried out on loose and compact sand and the results tabulated. Calculate the value of k for each flow rate as well as its average value for both materials. The apparatus was as shown in Figure 3.4, where $L = 20$ cm and $A = 45$ cm^2.

Table 3.2

	h (cm)	$i=\frac{h}{L}$	q (cm³/10 sec)	Q (cm³/s)	$v=\frac{Q}{A}$ (cm/s)	$k=\frac{v}{i}$ (cm/s)	Average k
Loose Sand	3.32	0.166	11.2	1.12	0.025	0.1500	0.1417 cm/s
	7.12	0.356	21.6	2.16	0.048	0.1348	
	9.28	0.464	29.2	2.92	0.065	0.1400	
	11.68	0.584	36.7	3.67	0.082	0.1396	
	15.26	0.763	49.5	4.95	0.110	0.1442	
Compact Sand	2.32	0.116	4.3	0.43	0.0096	0.0824	0.0784 cm/s
	6.56	0.328	11.7	1.17	0.026	0.0793	
	9.76	0.488	16.2	1.62	0.036	0.0738	
	13.30	0.665	22.9	2.29	0.051	0.0765	
	17.00	0.85	30.6	3.06	0.068	0.080	

From Graph 3.1, the coefficients of permeability for:

a) Loose sand: $k=0.141\,\text{cm/s}=1.41\times10^{-3}\,\text{m/s}$
b) Compact sand: $k=0.079\,\text{cm/s}=7.9\times10^{-4}\,\text{m/s}$

Therefore, the values of k obtained by means of Graph 3.1 verify the calculated average figures in Table 3.2.

3.3.2 Falling head test

The test is suitable for fine-grained, cohesive soils, such as clay. Because the permeability is very low, the piezometric water level falls very slowly. A sketch of the apparatus is shown in Figure 3.6.

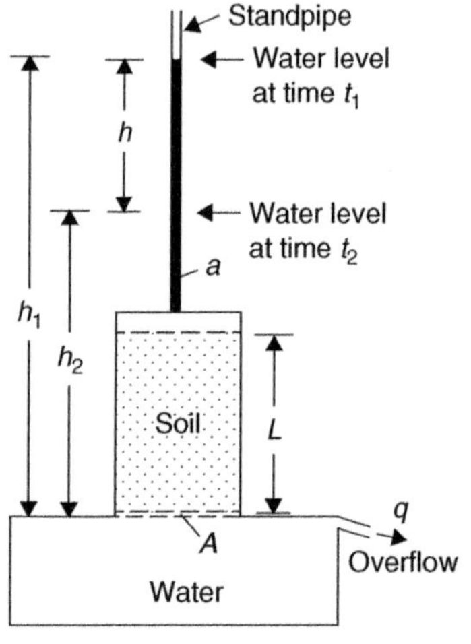

Figure 3.6

A = cross-sectional area of the specimen.
a = cross-sectional area of the standpipe.
L = length of the sample.
h_1 and h_2 are the water levels in the standpipe at time t_1 and t_2 respectively.
h = drop in water level.
$h=h_1-h_2$

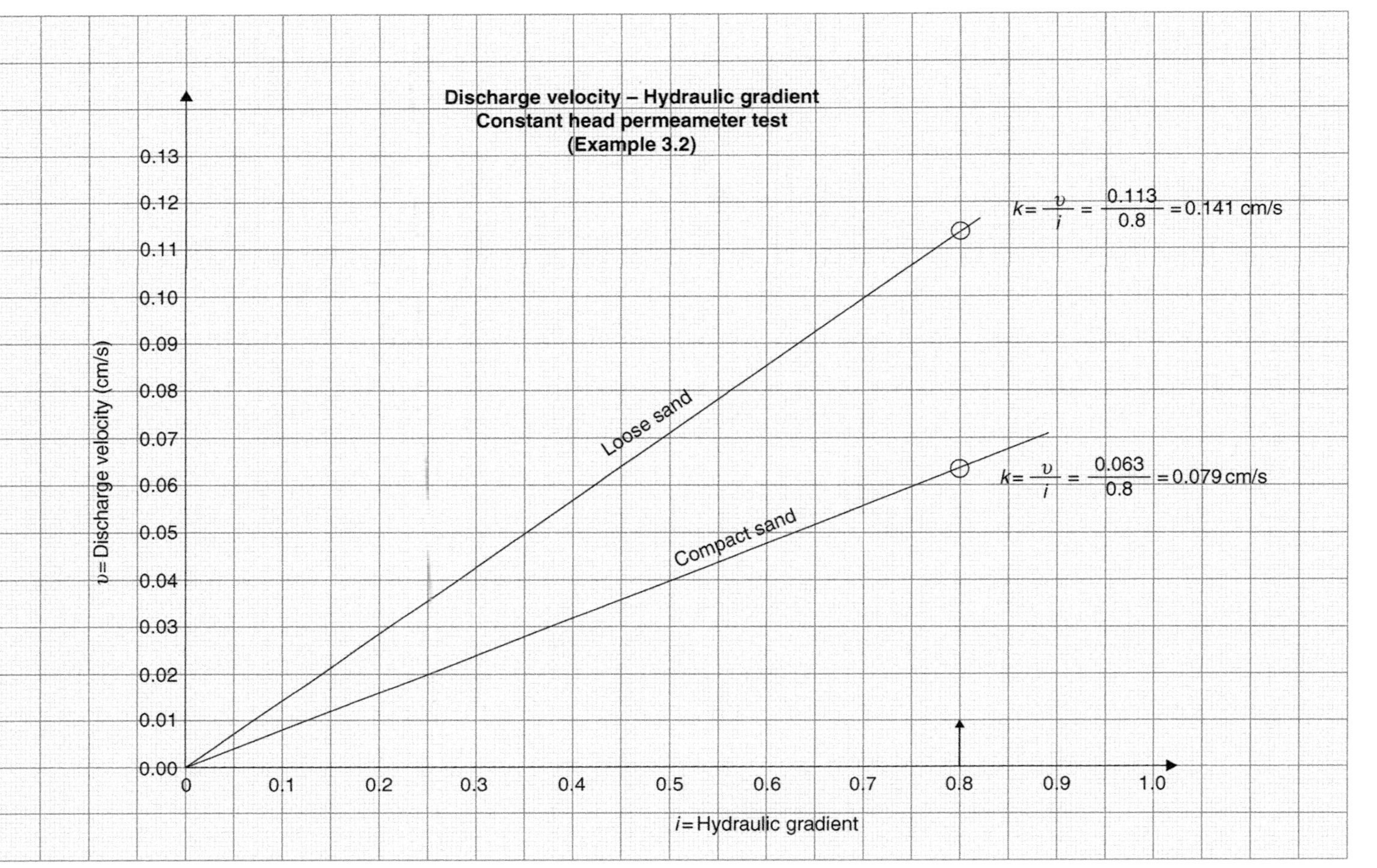

Discharge velocity – Hydraulic gradient
Constant head permeameter test
(Example 3.2)
υ = Discharge velocity (cm/s)
0.13
0.12
0.11
0.10
0.09
0.08
0.07
0.06
0.05
0.04
0.03
0.02
0.01
0.00
0
0.1
0.2
0.3
0.4
0.5
0.6
0.7
0.8
0.9
1.0
i = Hydraulic gradient
Loose sand
Compact sand
k = υ/i = 0.113/0.8 = 0.141 cm/s
k = υ/i = 0.063/0.8 = 0.079 cm/s

Graph 3.1

Step 1: Saturate the sample and eliminate any air bubbles from the system.
Step 2: Raise the water level in the piezometer standpipe and note the height (h_1).
Step 3: Allow the water level to drop to height h_2 and note the elapsed time $t = t_2 - t_1$.
Step 4: Calculate the coefficient of permeability from:

$$\boxed{k = \frac{aL}{At}\ln\left(\frac{h_1}{h_2}\right)} \tag{3.11}$$

Example 3.3

A sandy silt sample was tested at 20°C. The calculated results are tabulated below.

Table 3.3

Time t **(sec)**	h_2 **cm**	$\frac{h_1}{h_2}$	$\ln\left(\frac{h_1}{h_2}\right)$	k **cm/sec**
0	65.6	-	-	-
2880	62.2	1.055	0.054	4.7×10^{-6}
5190	59.4	1.104	0.099	4.7×10^{-6}
8100	55.8	1.176	0.162	5.0×10^{-6}
10800	53.1	1.235	0.211	4.8×10^{-6}
13545	50.7	1.294	0.258	4.7×10^{-6}
6752.5	←Averages→		0.13	4.78×10^{-6}

$A = 81\,\text{cm}^2$
$L = 25.4\,\text{cm}$
$h_1 = 65.6\,\text{cm}$
$a = 0.79\,\text{cm}^2$

$$k = \frac{0.79\times25.4}{81\times t}\times\ln\left(\frac{h_1}{h_2}\right) = \frac{0.248}{t}\times\ln\left(\frac{h_1}{h_2}\right)$$

Alternatively, using the average values:

Average $t = 6752.5$

Average $\ln\left(\frac{h_1}{h_2}\right) = 0.13$

$\therefore$ average $\frac{1}{t}\ln\left(\frac{h_1}{h_2}\right) = \frac{0.13}{6752.5} = 1.93\times10^{-5}$

and $k = \frac{0.79\times25.4\times1.93}{81}\times10^{-5}$

$= 0.478\times10^{-4}\,\text{mm/s}$

$= 4.78\times10^{-8}\,\text{m/s}$

Alternatively, k may be found from Graph 3.2 by the modified form of (3.11):

$$\boxed{k = \left(\frac{aL}{A}\right)\cot\theta\ \text{cm/s}} \tag{3.11a}$$

In this formula $\frac{1}{t}\times\ln\left(\frac{h_1}{h_2}\right) = \cot\theta$

$$\cot\theta = \frac{1}{\tan\theta} = \frac{0.13}{6752.5} = 1.93\times10^{-5}$$

$$\therefore\quad k = \frac{0.79\times25.4}{81}\times1.93\times10^{-5} = 0.478\times10^{-4}\,\text{mm/s}$$

$$= 4.78\times10^{-8}\,\text{m/s}$$

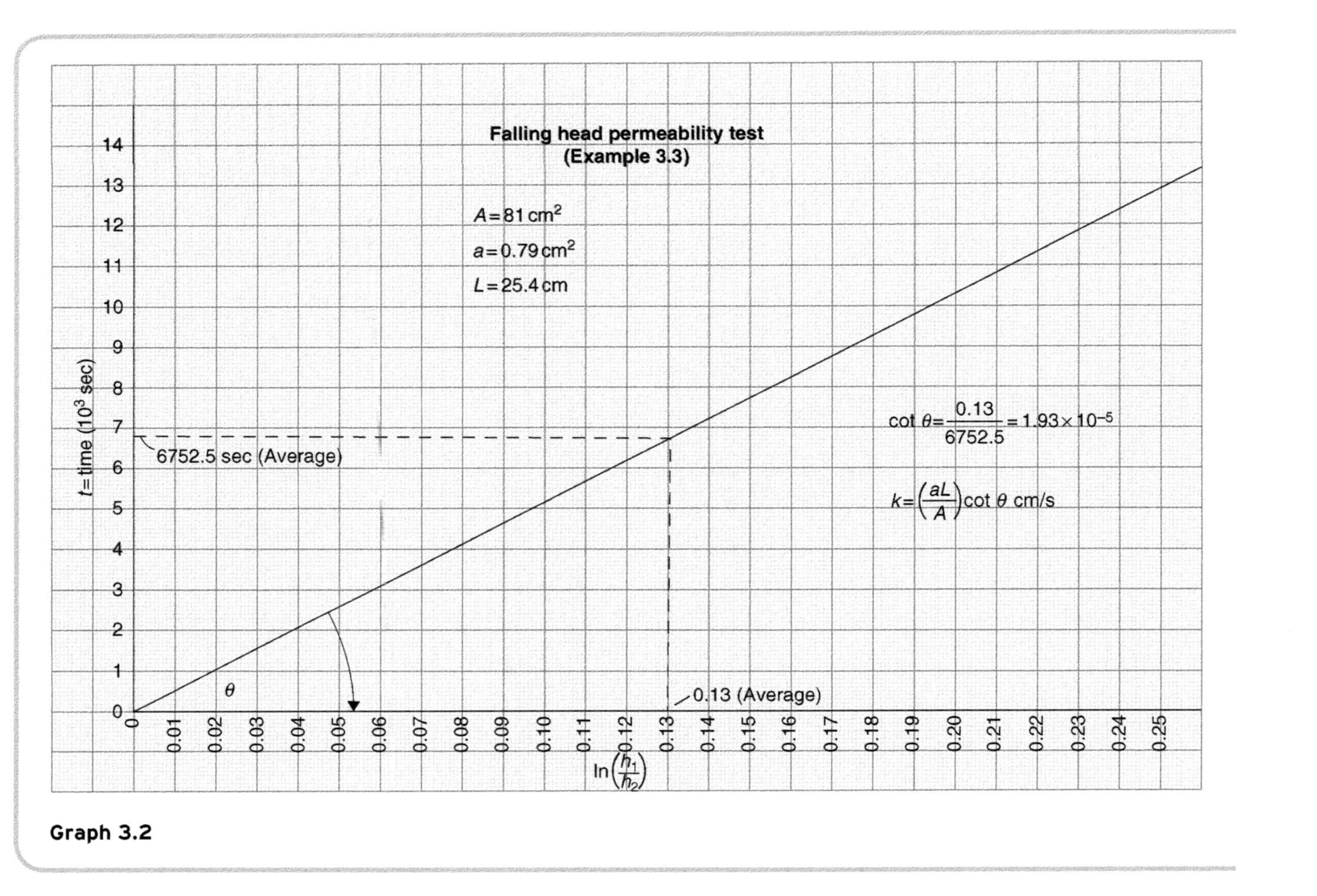

Falling head permeability test
(Example 3.3)
A = 81 cm²
a = 0.79 cm²
L = 25.4 cm
cot θ = 0.13 / 6752.5 = 1.93 × 10⁻⁵
k = (aL/A) cot θ cm/s
6752.5 sec (Average)
0.13 (Average)
θ
t = time (10³ sec)
ln(h1/h2)
0
1
2
3
4
5
6
7
8
9
10
11
12
13
14
0
0.01
0.02
0.03
0.04
0.05
0.06
0.07
0.08
0.09
0.10
0.11
0.12
0.13
0.14
0.15
0.16
0.17
0.18
0.19
0.20
0.21
0.22
0.23
0.24
0.25

Graph 3.2

3.4 Field pumping tests

These tests are more expensive than the laboratory ones, but the results are more realistic. They are applicable to non-cohesive, homogeneous soils. The applications of the coefficient of permeability, determined by the pumping test, are normally threefold:

1. Lowering the ground water table to create dry working environment.
2. To solve seepage problems in connection with structural stability.
3. For possible water supply.

There are two usual types of problems, depending on the position of the water-bearing layer:

a) Uniform coarse-grained soil extending from ground level to an impervious layer. Piping (boiling) failure could occur.
b) Uniform coarse-grained soil, between two impervious layers, containing water under artesian pressure. Shear failure (heaving) could occur.

Assumptions:

i. The ground water level is static in all directions.
ii. The permeable layer is homogeneous and horizontal.
iii. The pumping well penetrates the bottom impervious layer.
iv. The lining of the well is perforated up to the ground water table.
v. The coefficient of permeability of the layer to be pumped is larger than 10^{-4} m/s.
vi. The coefficient k of the soil is uniform and constant at every point.

3.4.1 Unconfined layer

Figure 3.7 shows the arrangement of the pumping scheme for a thick layer of permeable soil, underlain by impervious material. As a result of pumping from the central well, water seeps towards it, that is in a radial direction and the water surface falls, forming the so called 'drawdown curve'. The shape of this curve is found from observation wells placed as shown on the half-plan view below:

(a)

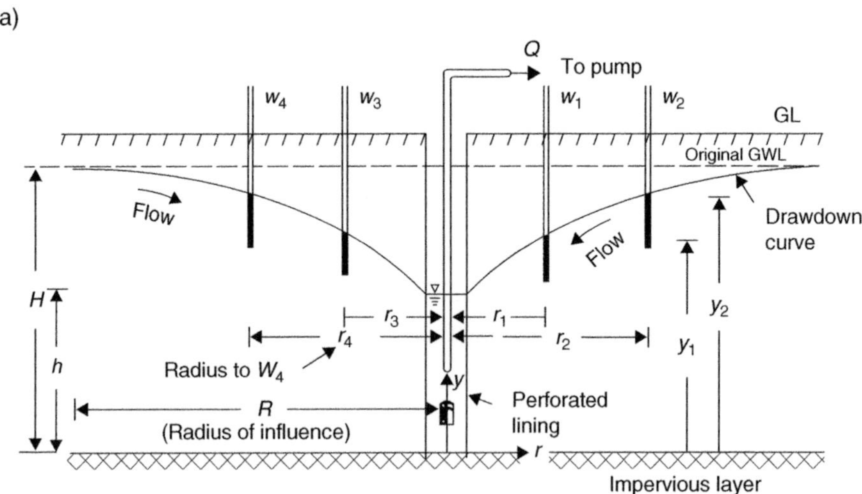

(b)

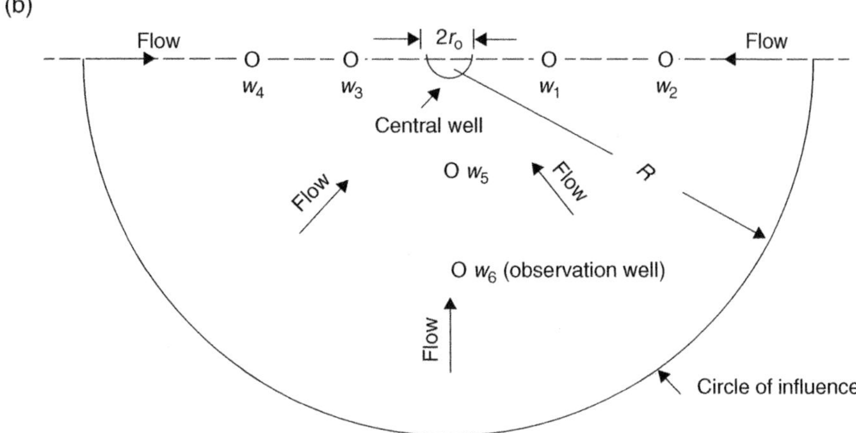

Figure 3.7

The formulae presented here are derived relative to the coordinate systems shown.

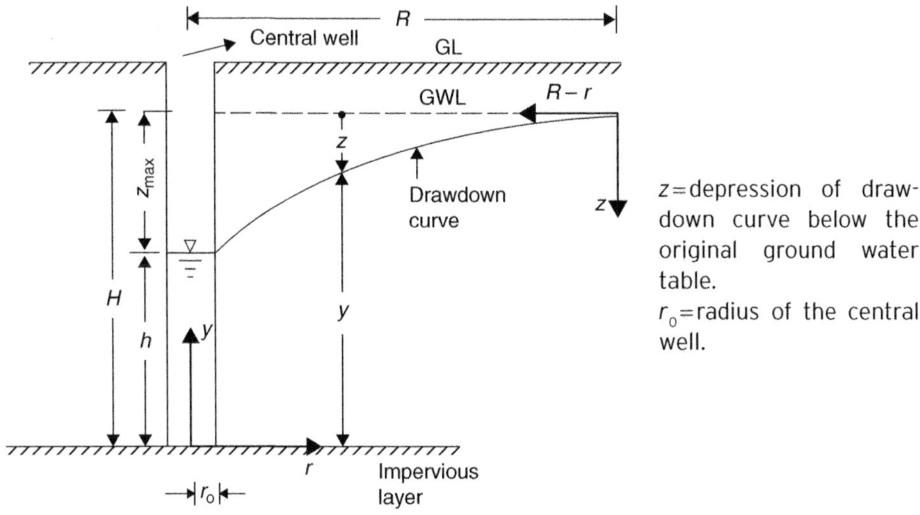

z=depression of drawdown curve below the original ground water table.

r_0=radius of the central well.

Figure 3.8

Equation of the drawdown curve:

$$y = \sqrt{H^2 - \frac{Q}{\pi k} \ln\left(\frac{R}{r}\right)} \tag{3.12}$$

Equation of depression:

$$z = H - y \tag{3.13}$$

When, $\left| \begin{matrix} r = r_0 \\ y = h \end{matrix} \right|$ then the maximum height of water in the central well is given by:

$$h = \sqrt{H^2 - \frac{Q}{\pi k} \ln\left(\frac{R}{r_0}\right)} \tag{3.14}$$

Hence, the maximum depression is: $$\boxed{z_{max} = H - h} \tag{3.15}$$

The flowrate pumped out of the central well is given by:

$$\boxed{Q = \frac{\pi k\left(H^2 - h^2\right)}{\ln\left(\frac{R}{r_0}\right)}} \tag{3.16}$$

Its maximum theoretical value occurs when h is practically zero.

Eliminating h from (3.16) gives: $$\boxed{Q_{max} = \frac{\pi k H^2}{\ln\left(\frac{R}{r_0}\right)}} \tag{3.17}$$

The coefficient of permeability may be expressed from (3.16) as:

$$\boxed{k = \frac{Q}{\pi\left(H^2 - h^2\right)} \times \ln\left(\frac{R}{r_0}\right)} \tag{3.18}$$

3.4.2 Radius of influence (*R*)

It is normally determined by placing observation wells at various distances from the central one. All the observations are made only after steady flow conditions had been attained. Its value can be high, depending on the particle size. Average figures given by Jumikis:

Table 3.4

Soil type	Particle size α (mm)	Radius (R) (m)
Coarse gravel	> 10	> 1500
Medium gravel	2–10	500–1500
Fine gravel	1–2	400–500
Coarse sand	0.5–1	200–400
Medium sand	0.25–0.5	100–200
Fine sand	0.1–0.25	50–100
Very fine sand	0.05–0.10	10–50
Silly sand	0.025–0.5	5–10

In view of these large figures and because the assumptions are least valid at the central well, formula (3.18) is modified for the calculation of k, using the data taken at two observation wells, say at r_1 and r_2 in Figure 3.7:

$$\boxed{k = \frac{Q}{\pi\left(y_2^2 - y_1^2\right)} \times \ln\left(\frac{r_2}{r_1}\right)} \tag{3.19}$$

Example 3.4

A pumping test was carried in a 4.7 m thick clean sand layer underlain by impervious soil. The original ground water table was 1.5 m below the surface. The steady-state discharge from the central well was 43 m³/hour. The water level observation at each well, at uniform flow is tabulated below. Calculate the coefficient of permeability of the sand.

Table 3.5

Well number	Central 0	1	2	3	4	5
Radius = r (m)	0.15	3	12	20	50	76
Depth to water (m)	1.96	1.77	1.65	1.61	1.53	1.5
Drawdown = z (m)	0.46	0.27	0.15	0.11	0.03	0

Central well diameter = 0.3 m $\therefore r_0 = 0.15$ m

H = impervious layer below GWL = 4.7 – 1.5 = 3.2 m

$h = H - 0.46 = 3.2 - z_0 = 2.74$ m

From Table 3.5: $R = 76$ m (At $z = 0$)

Discharge $$Q = 43\,\text{m}^3/\text{hour} = \frac{43}{3600}\,\text{m}^3/\text{s} = 1.194\times10^{-2}\,\text{m}^3/\text{s}$$

From (3.18): $$k = \frac{Q}{\pi\left(H^2 - h^2\right)} \times \ln\left(\frac{R}{r_0}\right) = \frac{1.194\times10^{-2}}{\pi\left(3.2^2 - 2.74^2\right)} \times \ln\left(\frac{76}{0.15}\right)$$
$$= 8.66\times10\,\text{mm/s}$$

Alternatively from (3.19), using the results from wells 3 and 4:

$$y_3 = H - z_3 = 3.2 - 0.11 = 3.09\,\text{m}$$
$$y_4 = H - z_4 = 3.2 - 0.03 = 3.17\,\text{m}$$
$$r_3 = 20\,\text{m}$$
$$r_4 = 50\,\text{m}$$

$$\therefore \quad k = \frac{Q}{\pi\left(y_4^2 - y_3^2\right)} \times \ln\left(\frac{r_4}{r_3}\right) = \frac{1.194\times10^{-2}}{\pi\left(3.17^2 - 3.09^2\right)} \times \ln\left(\frac{50}{20}\right)$$
$$= 6.95\times10^{-1}\,\text{mm/s}$$

This value of k is accepted as a reasonable approximation to the in-situ permeability.

3.4.3 Confined layer under artesian pressure (σ_A)

In this case, the permeable layer is confined and compressed by two impermeable ones. The drawdown curve should not be lowered below the bottom of the upper confining layer. The observation wells are arranged around the central one, as in the previous case.

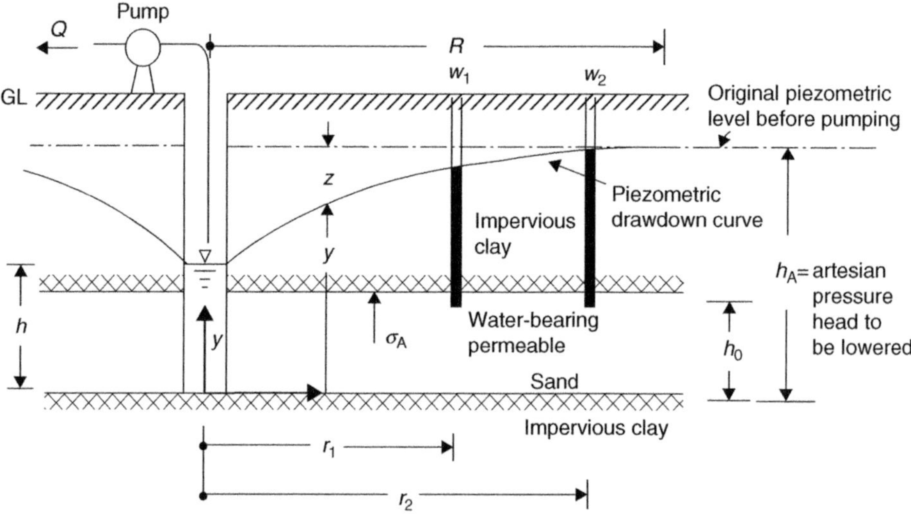

Figure 3.9

Note: In this case, the artesian pressure head is lowered by pumping so that excavation may be carried out, without base failure (heaving). (See Chapter 5.)

Equation of the drawdown curve:

$$y = h + \frac{Q}{2\pi k h_0} \times \ln\left(\frac{r}{r_0}\right) \tag{3.20}$$

where h_0 = thickness of the layer under pressure.

Equation of depression:

$$z = h_A - y \tag{3.21}$$

Coefficient of permeability of the permeable layer:

$$k = \frac{Q}{2\pi h_0 (h_A - h)} \times \ln\left(\frac{R}{r_0}\right) \tag{3.22}$$

Or, using data from two observation wells:

$$k = \frac{Q}{2\pi h_0 (y_2 - y_1)} \times \ln\left(\frac{r_2}{r_1}\right) \tag{3.23}$$

3.5 Permeability of stratified soil

A soil profile is normally made up of several layers, each having its own coefficient of vertical (k_v) and horizontal (k_H) permeability. Theoretically: $\boxed{k_H > k_v}$

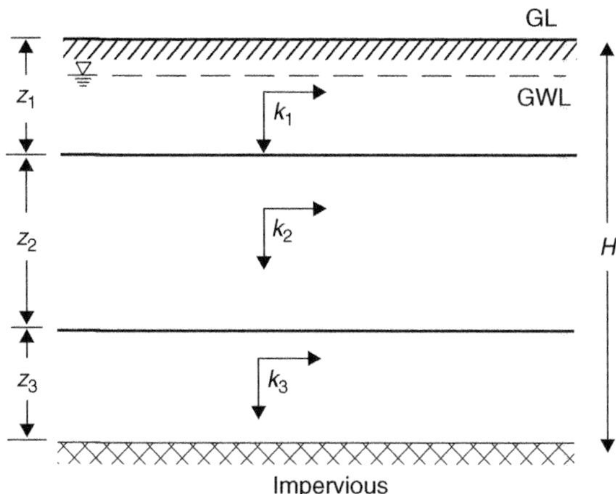

Figure 3.10

Where k_1, k_2 and k_3 indicate either horizontal or vertical coefficient of permeability, obtained in laboratory tests for each layer.

Equivalent horizontal coefficient

The average value of the horizontal coefficient of permeability for several layers is given by:

$$\boxed{k_H = \frac{1}{H}\sum(kz) = \frac{1}{H}\left(k_1 z_1 + k_2 z_2 + k_3 z_3 + \ldots\right)} \tag{3.24}$$

Equivalent vertical coefficient

The average value of vertical coefficient is given by:

$$\boxed{k_v = \frac{H}{\sum\left(\frac{z}{k}\right)} = \frac{H}{\frac{z_1}{k_1} + \frac{z_2}{k_2} + \frac{z_3}{k_3}}} \tag{3.25}$$

Example 3.5

Permeability tests were carried out on the three layers shown in Figure 3.11. The relevant horizontal and vertical coefficients of permeability is indicated for each layer. Calculate the average permeability in both directions.

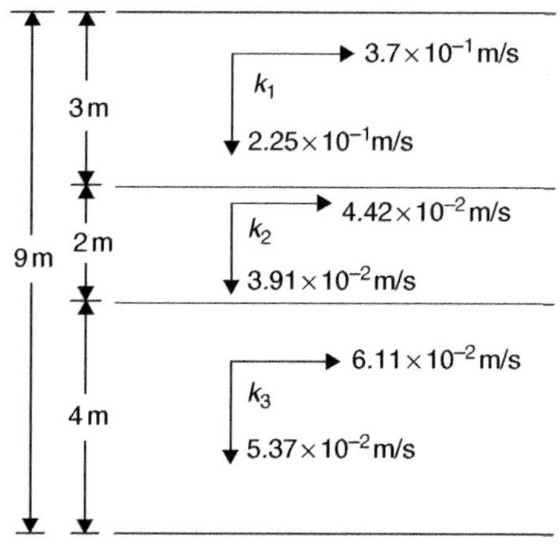

Figure 3.11

Note: The average vertical permeability is normally in the same order of magnitude as the smallest vertical coefficient.

From (3.24): $k_H = \frac{1}{9}(3\times3.7\times10^{-1} + 2\times4.42\times10^{-2} + 4\times6.11\times10^{-2})$

$= \frac{1}{9}(1.11 + 0.0884 + 0.244) = 16\text{ mm/s}$

From (3.25): $k_v = \dfrac{9}{\dfrac{3}{2.25\times10^{-1}} + \dfrac{2}{3.91\times10^{-2}} + \dfrac{4}{5.37\times10^{-2}}} = 6.48\text{ mm/s}$

Ratio: $\dfrac{k_H}{k_v} = \dfrac{1.6\times10^{-1}}{6.48\times10^{-2}} = 2.47 \quad \therefore\ k_H = 2.47k_v$

3.6 Flow nets

Flow nets are a graphical representation of the passage of water through a permeable material. They are made up of:

1. Flow lines
2. Equipotential lines

3.6.1 Flow lines (FL)

These represent the path of water through the soil. Two flow lines may be considered to form a sloping channel, carrying a quantity (q) of water in steady (laminar) flow per second. To be a channel, conducting laminar flow, the two flow lines should never cross, although they need not be exactly parallel curves. A typical seepage channel is drawn below.

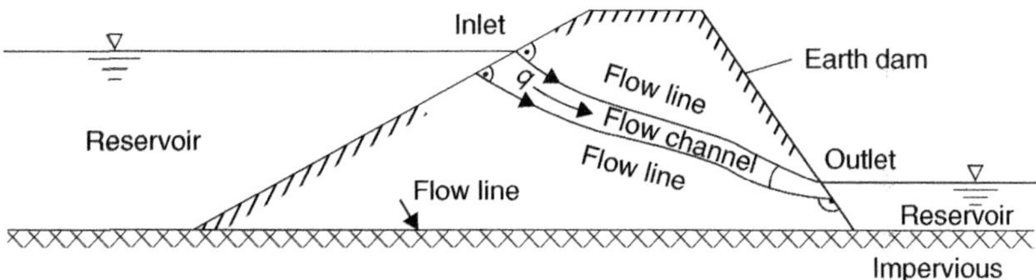

Figure 3.12

Rules for drawing flow lines

1. An impermeable surface is a flow line, as water has to flow along it.
2. Two flow lines do not cross.
3. A flow line always starts at the inlet end, at right angle to the soil boundary.

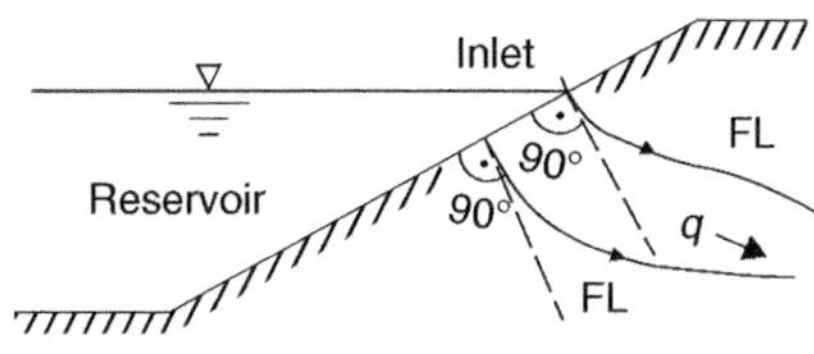

Figure 3.13

4. There are three cases at the outlet end:
   a) If the flow line emerges in water, then it is drawn at 90° to the soil boundary.

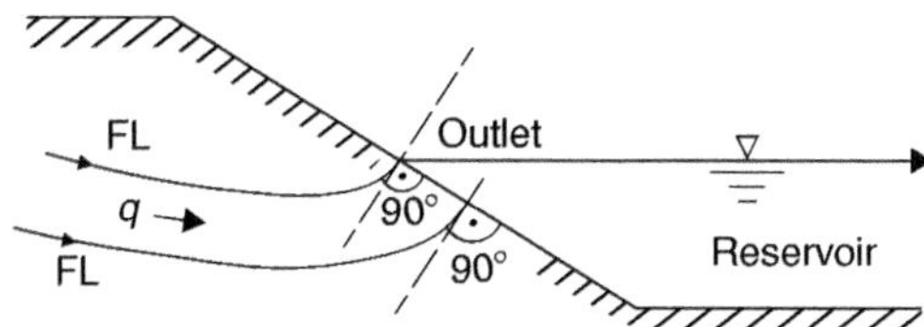

Figure 3.14

b) If the flow line ends in a filter drain, then it is drawn at 90° to the boundary of the filter.

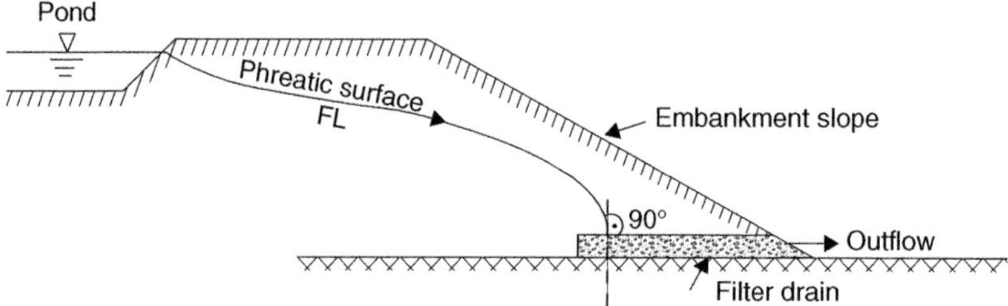

Figure 3.15

c) If the flow emerges from a slope, then the flow line is drawn tangentially to the surface.

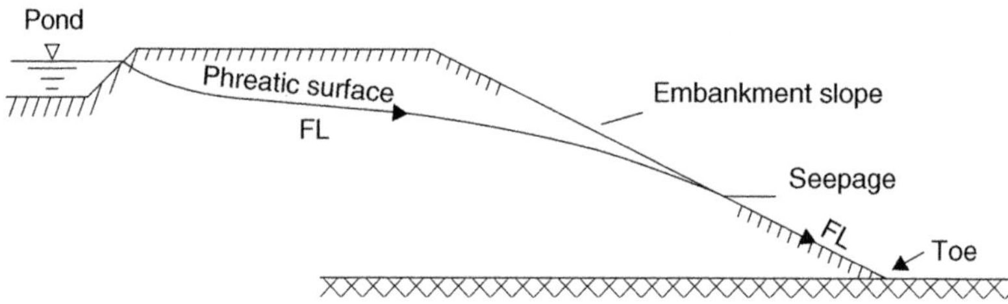

Figure 3.16

Note: Seepage from a slope is undesirable as it could cause washout or erosion of the toe. For this reason, new embankments should be provided with toe drains as shown in Figure 3.15.

In the case of natural slopes, counterfort drains or a surface toe filter should be constructed to prevent damage.

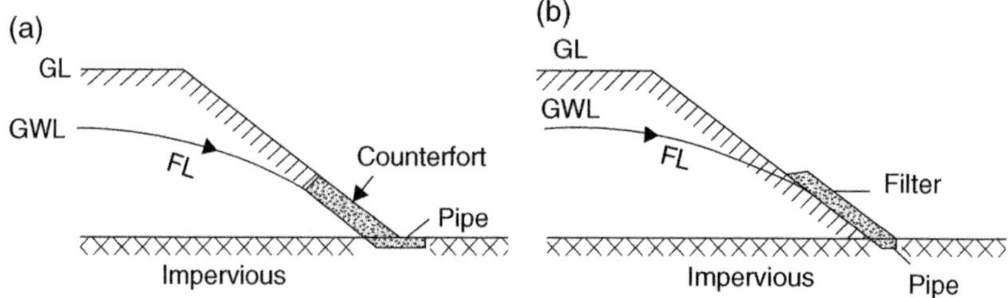

Figure 3.17

3.6.2 Head loss in a flow channel

Just like in pipe flow, some pressure head is lost as water flows between two flow lines. Piezometers placed at two points P_1 and P_2 along a flow line would indicate different water levels, that is a pressure head drop of Δh.

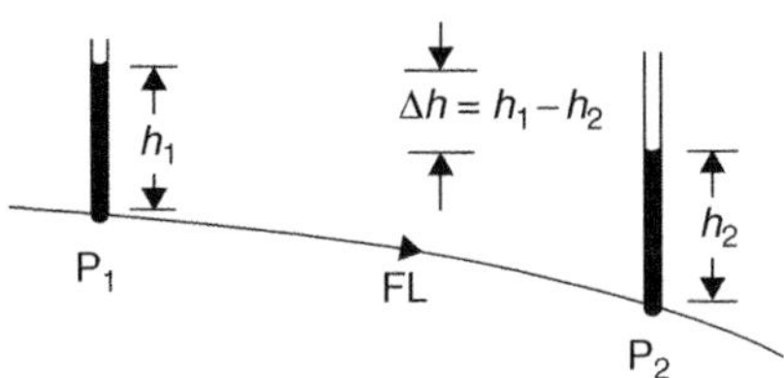

Figure 3.18

3.6.3 Equipotential lines (EPL)

These represent points of equal pressure heads within the soil mass, caused by steady seepage forces. Water surface in piezometers placed along an equipotential line would be at the same level.

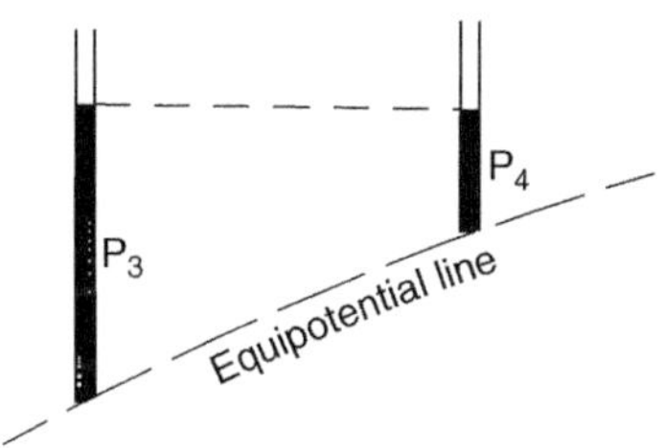

Figure 3.19

Rules for drawing EPL

1. It is drawn at 90° to a flow line. A piezometer placed at their junction indicates the pressure head common to both lines.

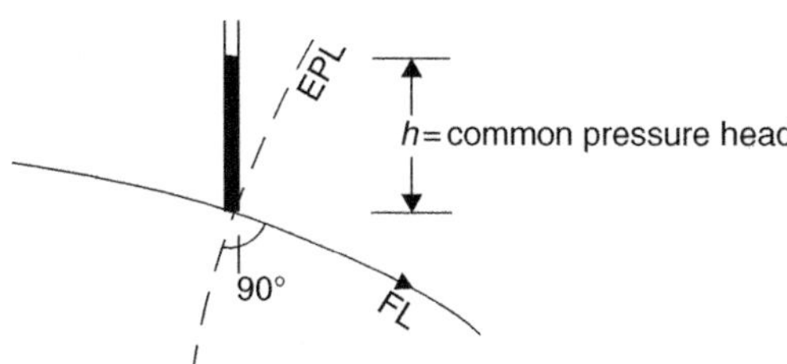

Figure 3.20

2. An EPL is drawn at 90° to impervious surfaces.

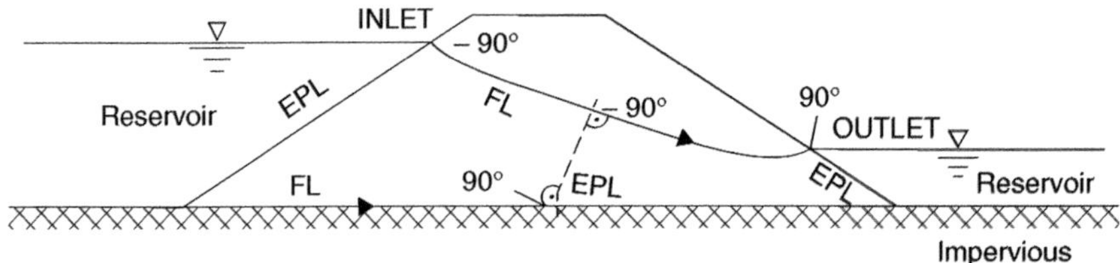

Figure 3.21

3. Ground surfaces at the inlet and outlet are equipotential lines. This is why the flow lines are drawn at 90° to them, as in Figure 3.13.
4. EPLs are drawn at such intervals that they form approximate, curved 'squares' with the flow lines.

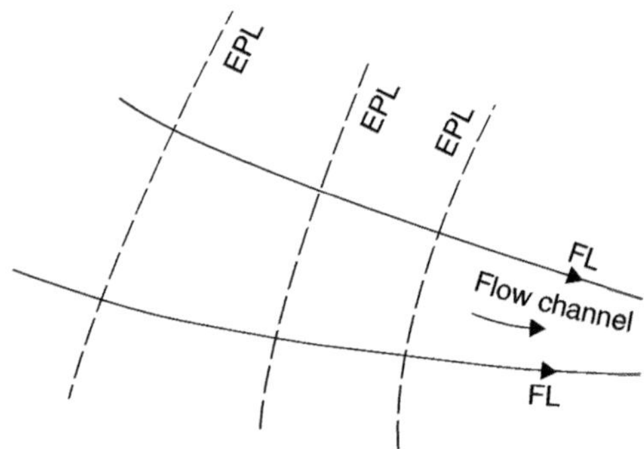

Figure 3.22

5. Permeable boundary, submerged under water is EPL.

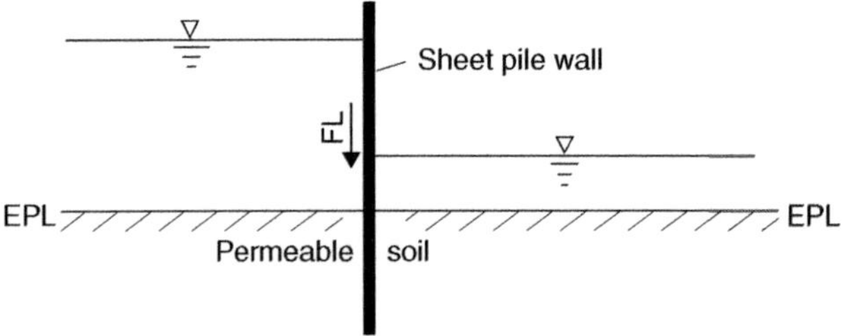

Figure 3.23

The ground is permeable, hence it is an EPL. The sheet pile wall is impermeable; therefore it is a flow line.

6. The top flow line in Figures 3.12 to 3.16 and 3.21 is at atmospheric pressure. It is often called 'phreatic' or 'free water' surface. The equipotential lines do not cross the phreatic surface.

3.6.4 Flow net construction

The trial and error procedure, attributed to Forcheimer, is best carried out on transparent paper. The structure is drawn to scale on one side and the flow net on the other. It is sufficient to draw at least four flow channels correctly. The final net may then be subdivided for more detail. Note that the lines must converge or diverge gradually. Sudden changes of direction may only occur at the boundaries. The method is best demonstrated by a simple example.

Example 3.6

Figure 3.24 shows a concrete dam constructed in permeable soil. The vertical and horizontal permeabilities are assumed to be equal. The dam itself is taken to be impervious. Sketch the flow net for the structure.

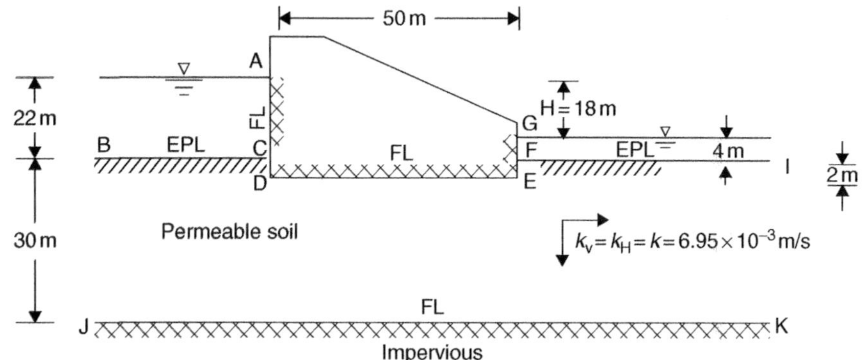

Figure 3.24

Step 1: Consideration of the boundary conditions:

a) Surface A-C-D-E-F-G and JK are impermeable, therefore flow lines, hence each EPL is drawn at 90° to them.

b) Surface BC and FI are submerged permeable ones, therefore equipotential lines, hence each FL is drawn at 90° to them.

Step 2: Locate the first flow channel by sketching the flow line 1 and draw the equipotential lines for this flow path.

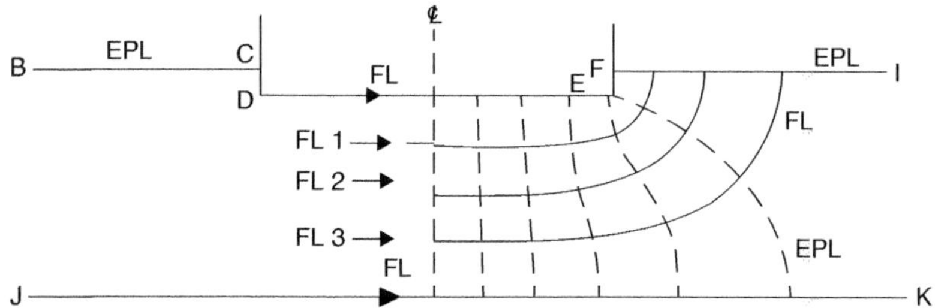

Figure 3.25

Note: As the base is symmetrical about its centre line, only half of the net need be drawn.

Step 3: Sketch flow line 2, forming approximate 'squares' with the equipotential ones.

Step 4: Repeat the process until a reasonably satisfactory flow net is the result. If not, then adjust the sketch.

Note that some elements of the mesh are not even 'squares'. Despite this and other assumptions, the flow net is useful tool in the assessment of seepage parameters, as long as at least four flow channels are drawn.

The completed flow net is given on Graph 3.3.

3.6.5 Application of flow nets

Once the flow net for a soil or soil-structure configuration has been constructed, it can be applied to the solution of three types of seepage problems:

1. To establish the seepage flowrate.
2. To calculate the seepage pressure acting on a structure.
3. To determine, whether piping, that is internal erosion of a particular soil, could be caused by seepage or not.

3.6.6 Seepage flowrate (*Q*)

It can be seen on Graph 3.3 that, in general, the level difference between the two reservoirs is *H*. This head is lost as water seeps through the flow channels. Figure 3.26 shows that there is a constant head loss across each square along a flow line, that is, between any pair of equipotential lines.

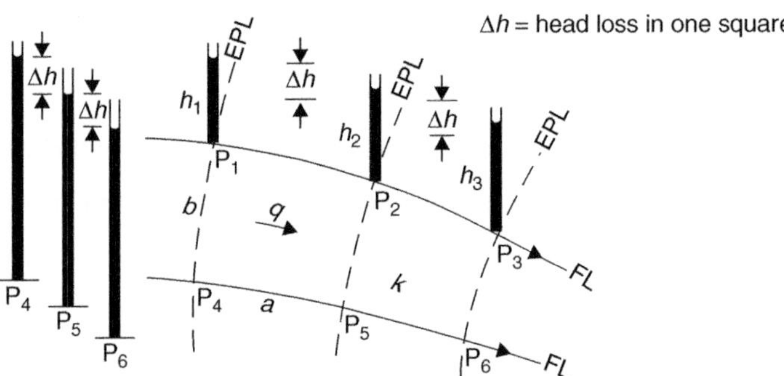

Figure 3.26

If N_e denotes the number of 'squares' drawn/channel in the flow net, then the total head loss is given by:

$$\boxed{H = N_e \times \Delta h}$$

By Darcy's Law the flowrate in a flow channel of unit width is: $\boxed{q = kbi}$

But hydrautic gradient in a square is: $i = \dfrac{\Delta h}{a}$

For an approximate 'square' $a \approx b$

Therefore, the flowrate is: $\boxed{q = k\Delta h}$ per channel

If N_f = number of flow channels in the flow net, then the total flowrate per unit width is:

$$\begin{aligned} Q &= qN_f \\ &= kN_f\,\Delta h \\ &= kN_f\frac{H}{N_e} \end{aligned} \qquad \boxed{Q = kH\left(\frac{N_f}{N_e}\right)} \tag{3.26}$$

For example 3.6 (Graph3.3): $H = 18\,\text{m}$, $k = 6.95\times10^{-3}\,\text{m/s}$, $N_f = 4$, $N_e = 12$

$$\boxed{\begin{aligned} Q &= \frac{6.95\times10^{-3}\times18\times4}{12} \\ &= 0.042\,\text{m}^3/\text{s per m} \end{aligned}}$$

3.6.7 Seepage pressure

Flow net is a very useful aid for the determination of seepage pore pressure at any point in the soil mass or under a structure. The relevant general formulae are:

Head loss up to point x:

$$\begin{aligned} H_x &= N_x\,\Delta h \\ &= N_x\frac{H}{N_e} \end{aligned} \qquad \boxed{H_x = H\left(\frac{N_x}{N_e}\right)} \tag{3.27}$$

Where, N_x = the number of squares between point x and the tailwater end. This is why it is convenient to number N_e from that end (see Graph 3.3).

Pressure head at point x is obtained by subtracting that the head loss from the total head (H_T) at x, as shown.

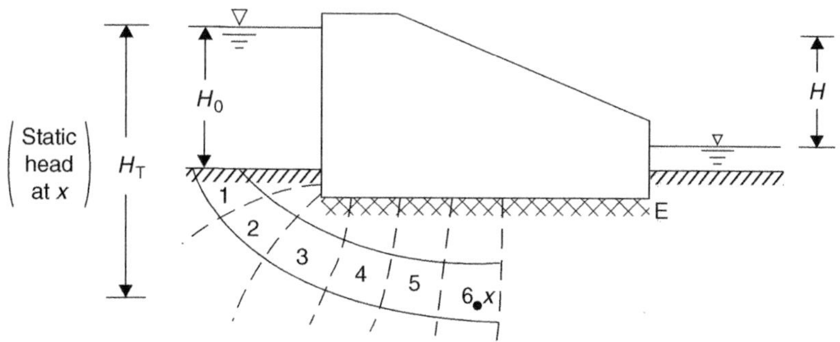

Figure 3.27

In this case $N_x = 5.5$ as x is in the middle of square 6.

The pressure head therefore is: $h_x = H_T - H_x$

or

$$\boxed{h_x = H_T - H\left(\frac{N_x}{N_e}\right)} \quad (3.28)$$

The seepage pore pressure at point x is given by:

$$\boxed{u_x = \gamma_w\, h_x = \gamma_w\left[H_T - H\left(\frac{N_x}{N_e}\right)\right]} \quad (3.29)$$

Continuing with Example 3.6, estimate the pressure head and seepage pressure at points P and E indicated in Graph 3.3.

At point P: $H_T = 22 + 12 = 34$ m
$N_x = 8.4$
$N_e = 12$
$H = 18$ m

Head loss up to P:

$$H_p = 18 \times \left(\frac{8.4}{12}\right) = 12.6\text{m}$$

Pressure head: $h_P = 34 - 12.6 = 21.4$ m
Seepage pressure: $u_P = 9.81 \times 21.4 = 210$ kN/m^2

At point E: $H_T = 22 + 2 = 24$ m
$N_x = 10.7$
$N_e = 12$
$H = 18$ m

Head loss up to E:

$$H_p = 18 \times \left(\frac{10.7}{12}\right) = 16.05\text{ m}$$

Pressure head: $h_E = 24 - 16.05 = 7.95$ m
Seepage pressure: $u_E = 9.81 \times 7.95 = 78$ kN/m^2

The 'uplift pressure', acting on the base of the dam at 10 m intervals, is tabulated below.

Table 3.6

Point x at	**D 0**	**10**	**20**	**30**	**40**	**E 50**
N_x (from Graph 3.3)	1	3.5	5.2	6.8	8.5	10.7
$H_x = 18\left(\frac{N_x}{12}\right)$(m)	1.5	5.25	7.8	10.2	12.75	16.05
H_T at x (m)	24	24	24	24	24	24
$h_x = H_T - H_x$ (m)	22.5	18.75	16.2	13.8	11.25	7.95
$u_x = 9.81\, h_x$ (kN/m^2)	221	184	159	135	110	78

Hydraulic gradient (i)

Formula (3.2) indicates that the hydraulic gradient may be defined by:

$$i = \frac{\text{Head loss in seepage}}{\text{Seepage distance}}$$

If can be obtained for each 'square' of a flow net as shown:

Note:
The average dimensions (a and b) of a square may be measured directly from the flow net.
a = seepage distance between two EPLs.

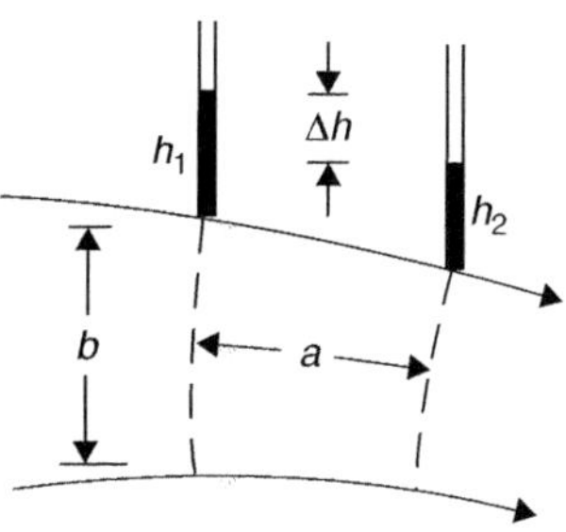

Figure 3.28

The hydraulic gradient is expressed in these terms:

$$\boxed{i = \frac{\Delta h}{a}} = \frac{h_1 - h_2}{a} \qquad (3.30)$$

Example 3.7

Calculate the average hydraulic gradient between the base of the dam and the impervious layer, for squares 8 to 10 on Graph 3.3.

Head loss per square:

$$\Delta h = \frac{H}{N_e} = \frac{18}{12} = 1.5\text{m}$$

The average value of the hydraulic gradient is calculated from the tabulated ones for the 12 squares measured.

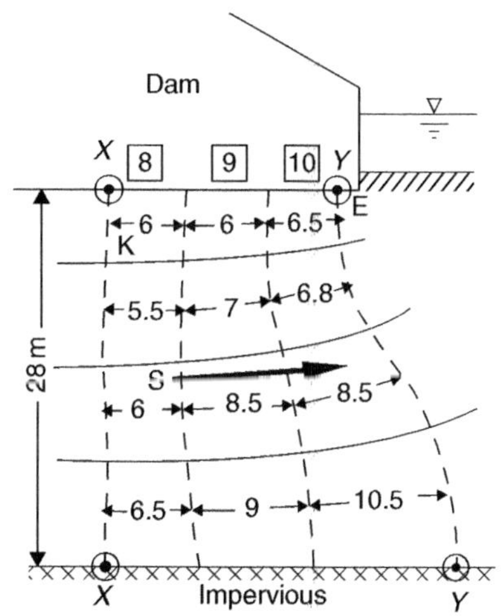

Figure 3.29

Table 3.7

a (m)	6	5.5	6	6.5	6	7	8.5	9	5.5	6.2	8.5	10.5	SUM
$i = \frac{1.5}{a}$	0.25	0.27	0.25	0.23	0.25	0.21	0.18	0.17	0.27	0.24	0.18	0.14	2.64

Therefore, the average value is: $i = \frac{2.64}{12} = 0.22$

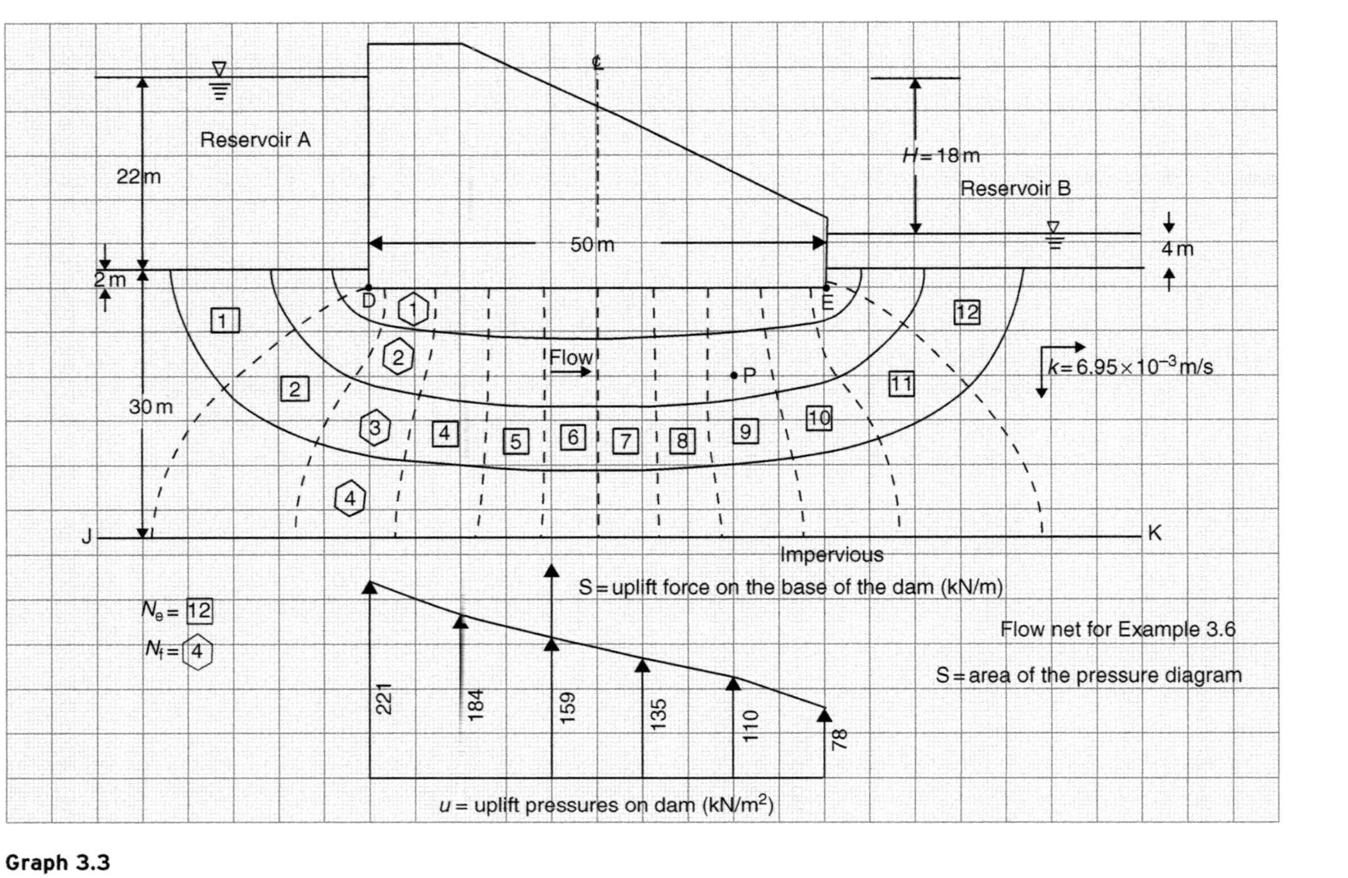

Reservoir A
22 m
2 m
30 m
50 m
H = 18 m
Reservoir B
4 m
D
E
P
Flow
k = 6.95 × 10⁻³ m/s
1
2
3
4
1
2
4
5
6
7
8
9
10
11
12
J
K
Impervious
S = uplift force on the base of the dam (kN/m)
Flow net for Example 3.6
S = area of the pressure diagram
Nₑ = 12
N_f = 4
221
184
159
135
110
78
u = uplift pressures on dam (kN/m²)

Graph 3.3

3.6.8 Seepage force (*S*)

The general expressions for the force, exerted on the soil by seeping water, is derived by considering the hydrostatic forces acting on a square of the flow net. It is then given by:

$$\boxed{S = i\gamma_w V} \tag{3.31}$$

where V = volume of soil through which the average hydraulic gradient is known.

Example 3.8

Continuing Example 3.7, estimate for the volume between sections *x-x* and *y-y* the following:

1. Seepage force
2. Flowrate
3. Flow velocity
4. Seepage velocity

Assume: Voids ratio for the sand: $e = 0.6$
Coefficient of permeability: $k = 6.95 \times 10^{-3}$ m/s
$i = 0.22$

Average length: $= \frac{1}{4}(17.5 + 18.7 + 23 + 26) = 21.3$ m

Volume: $V = 28 \times 21.3 \times 1 = 596.4$ m^3/m thickness of dam
Seepage force: $S = i\gamma_w V = 0.22 \times 9.81 \times 596.4 = 1287$ kN/m

From (3.4): Flowrate: $Q = Aki$
$A = 28.1 = 28$ m^2/m thickness of dam
$\therefore$ $Q = 28 \times 6.95 \times 10^{-3} \times 0.22 = 0.043$ m^3/s

From (3.3): Flow velocity: $v = ki = 6.95 \times 10^{-3} \times 0.22 = 1.53 \times 10^{-3}$ m/s

From (3.6): Seepage velocity: $v_s = \left(\frac{1+e}{e}\right)v$

$$= \frac{1.6}{0.6} \times 1.53 \times 10^{-3} = 4.1 \times 10^{-3} \text{ m/s}$$

Derivation of formula (3.31)

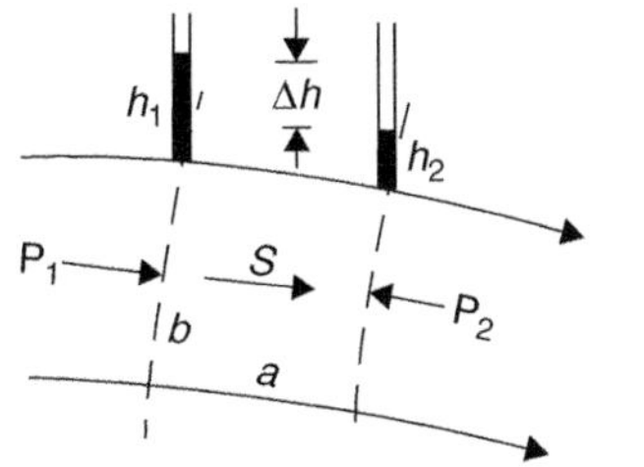

Figure 3.30

Hydrostatic forces:
$P_1 = bh_1\gamma_w$
$P_2 = bh_2\gamma_w$
b = cross-sectional area of 1m wide flow channel

The resultant force acting on soil particles is equal to the hydraulic pressure-force on the opposite faces of a square.

$$S = P_1 - P_2 = bh_1\,\gamma_w - bh_2\,\gamma_w$$

$$\boxed{S = b\gamma_w(h_1 - h_2)} \qquad (3.32)$$

Note: This formula may also be used to estimate the seepage force due to upward flow. See section 3.7 Erosion due to seepage.

The volume of the square $V = a\,b$ (m³/unit width)

Hence, $b = \dfrac{V}{a}$ $\qquad \therefore \quad S = \dfrac{\Delta h}{a}\gamma_w V$

Also, $\Delta h = h_1 - h_2$

and $i = \dfrac{\Delta h}{a}$ $\qquad \therefore \quad \boxed{S = i\gamma_w V} \qquad (3.31)$

Example 3.9

Calculate the magnitude of the seepage force, acting on the square indicated by the letter K, in Figure 3.29, using formulae (3.31) and (3.32).

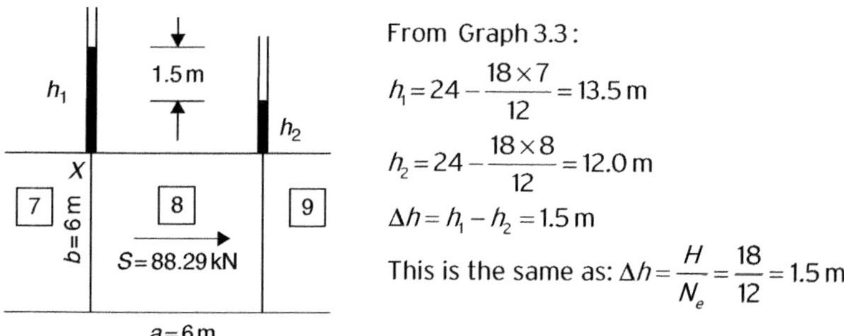

Figure 3.31

From Graph 3.3:

$$h_1 = 24 - \frac{18 \times 7}{12} = 13.5\text{ m}$$

$$h_2 = 24 - \frac{18 \times 8}{12} = 12.0\text{ m}$$

$$\Delta h = h_1 - h_2 = 1.5\text{ m}$$

This is the same as: $\Delta h = \dfrac{H}{N_e} = \dfrac{18}{12} = 1.5\text{ m}$

From (3.32): $S = b\gamma_w(h_1 - h_2)$

$= 6 \times 9.81 \times 1.5 = 88.29$ kN/m width of dam

From (3.30): $i = \dfrac{\Delta h}{a} = \dfrac{1.5}{6} = 0.25$

$V = ab \times 1 = 6 \times 6 = 36\text{ m}^3$

$\therefore S = i\gamma_w V$

$= 0.25 \times 9.81 \times 36$

$= 88.29$ kN (as before)

3.7 Erosion due to seepage

The usual names given to the phenomenon are 'piping', 'boiling' or 'quicksand'. It is caused by upward flow of water at critical velocity in fine sand. Soil in this state has no bearing capacity, hence cannot support structures. The classic example on piping is a sheet pile wall, separating water at different surface levels.

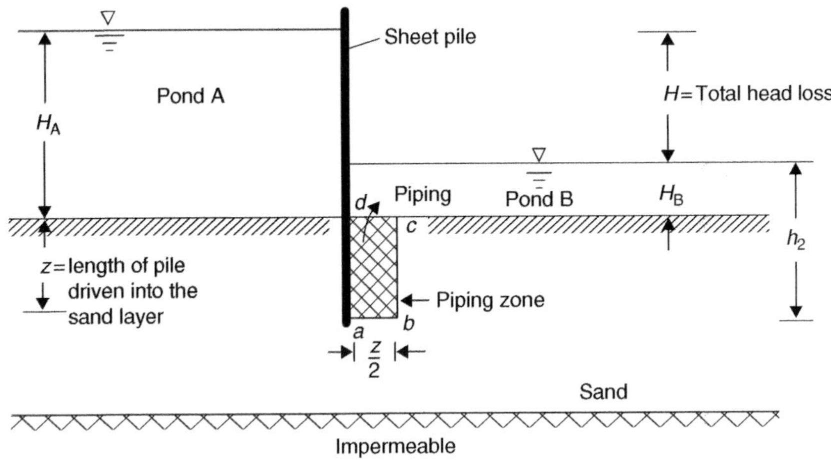

Figure 3.32

According to Terzaghi's experiments, the width of the piping zone is approximately half the length of the sheet pile below the surface. In order to prevent this type of failure, the weight (W) of the soil prism in the zone must be greater than upward seepage force (S).

The factor of safety against piping failure is given by:

$$\boxed{F_s = \frac{W}{S} > 3} \tag{3.33}$$

The problem is solved by means of flow net based on the fact that the head loss over one square is:

$$\boxed{\Delta h = \frac{H}{N_e}} \tag{3.34}$$

Step 1

Estimate the average seepage pressure-head (h_1) at point P along the base (a-b) of the soil prism of unit thickness. The variation of pressure over the base is considered to be parabolic, as shown. See its derivation in Supplementary problem 3.8.

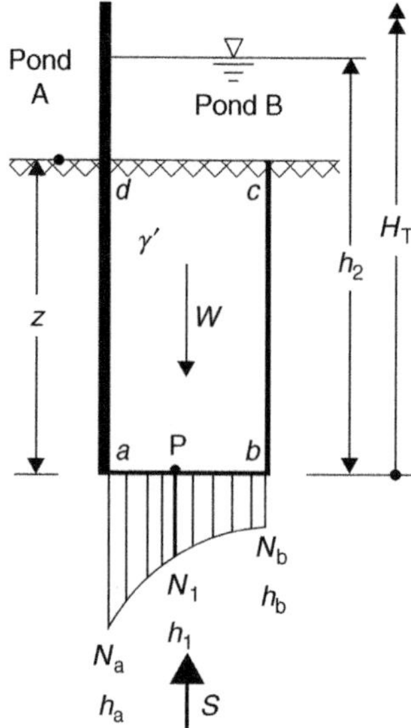

N_a = number of squares between pond A and the base of pile at point *a*

N_b = Ditto at point *b*

Figure 3.33

It can be shown that for a parabolic variation, the average value of N_x is given by:

$$N_1 = \frac{N_a + 2N_b}{3} \tag{3.35}$$

Also, the average pressure head is:

$$h_1 = \frac{h_a + 2h_b}{3} \tag{3.36}$$

The magnitude of this head may be estimated in two ways:

1. By using N_a and N_b in formula (3.28) to calculate h_a and h_b:
   $h_a = H_T - \Delta h\, N_a$
   $h_b = H_T - \Delta h\, N_b$
   and then applying (3.26)
2. By using N_1 to get $h_1 = H_T - \Delta h N_1$

Step 2

Determine the actual seepage pressure-head (h_s) to be dissipated through the soil prism. This may be done in two ways:

1. By subtracting the existing static hydraulic head (h_2) above the base from h_1.

$$\boxed{h_s = h_1 - h_2} \tag{3.37}$$

2. By equating h_s to the total head loss through the prism.
   $N_s = N_e - N_1$ = number of squares between base and ground surface within the prism.
   Total head loss between the base and the surface:

$$\boxed{h_s = (N_e - N_1)\Delta h} \tag{3.38}$$

Step 3
Estimate the upward seepage force (S) by (3.32):

$$S = A\gamma_w(h_1 - h_2)$$

where $A = \frac{Z}{2}$ is the surface area of the base of unit thickness.

$$\boxed{S = \frac{Z}{2}\gamma_w h_s} \tag{3.39}$$

Alternatively, $u_s = \gamma_w h_s$: The seepage pressure acting on the base.
$i_{av} = \frac{h_s}{Z}$: The average hydraulic gradient across the prism.

Therefore, either

$$\boxed{S = \frac{Z}{2}u_s} \tag{3.40}$$

or

$$\boxed{S = \frac{Z}{2}\gamma_w(Zi_{av}) = i_{av}\,\gamma_w\,V} \tag{3.40a}$$

where $V = \frac{Z^2}{2}$ is the volume of the prism.

Step 4
Calculate the submerged weight (W') of the prism and hence the factor of safety.

$$W' = \gamma'\frac{Z^2}{2} = \gamma' V$$

Therefore,

$$F_s = \frac{W'}{S} = \frac{\gamma' V}{i_{av}\gamma_w V} = \boxed{\frac{\gamma'}{i_{av}\gamma_w}} = \boxed{\frac{Z\gamma'}{h_s\gamma_w}} \tag{3.41}$$

Note, that the hydraulic gradient at failure is called the 'critical hydraulic gradient' (see Chapter 5, problem 4), given by:

Either

$$i_c = \frac{\gamma'}{\gamma_w} \tag{3.42}$$

Or in terms of specific gravity and voids ratio:

$$\boxed{i_c = \frac{G_s - 1}{1 + e}}$$

Hence,

$$\boxed{F_s = i_c \frac{z}{h_s} = \frac{i_c}{i_{av}}} \tag{3.43}$$

Alternatively, divide the pressure of the submerged weight (see also Chapter 5, Effective Pressure), acting downwards on the base, by the seepage pressure.

Pressure of W'

$$\sigma' = \frac{W'}{A} = \frac{\gamma' \frac{z^2}{2}}{\frac{z}{2}} = z\gamma'$$

Therefore,

$$\boxed{F_s = \frac{\sigma'}{u_s}} \tag{3.44}$$

Harza's method

In this, the exit gradient (i_e) is estimated by measuring the length (x) of the exit square, adjacent to the pile at location d.

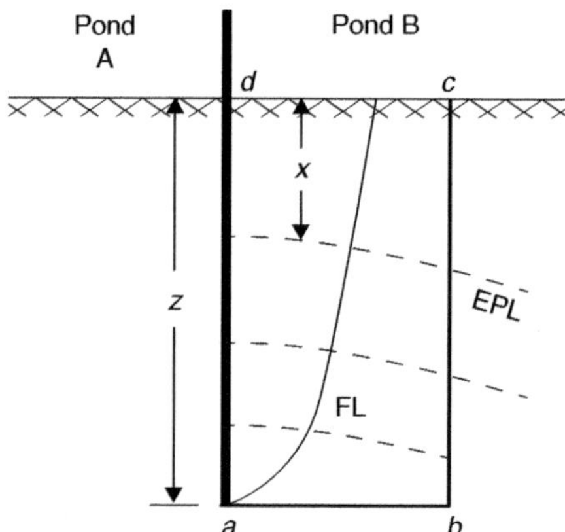

Figure 3.34

Exit gradient:

$$\boxed{i_e = \frac{\Delta h}{x}} \approx i_{av} \tag{3.45}$$

and

$$\boxed{F_s = \frac{i_c}{i_e}} \tag{3.46}$$

Example 3.10

The sheet pile arrangement depicted in Figure 3.35 is to be examined for adequacy.

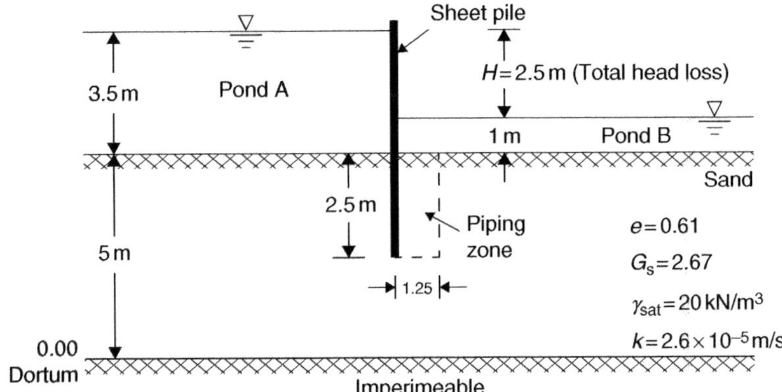

Figure 3.35

Determine from the flow net drawn on Graph 3.4a:

1. The water pressure distribution on both sides of the sheet pile wall, tabulating the calculations.
2. The factor of safety against piping failure.

(1) Pressure distribution

Table 3.8

Elevation (m)	8.5	5	4	3	2.5	3	4	5	6
H_T (m)	0	3.5	4.5	5.5	6	5.5	4.5	3.5	2.5
N_x	0	0	1	2.4	4	5.5	7	8	0
H_x (m)	0	0	0.313	0.75	1.25	1.72	2.19	2.50	2.50
h_x (m)	0	3.5	4.19	4.75	4.75	3.78	2.31	1	0
u_x (kN/m²)	0	34.3	41.1	46.6	46.6	37	22.7	9.81	0

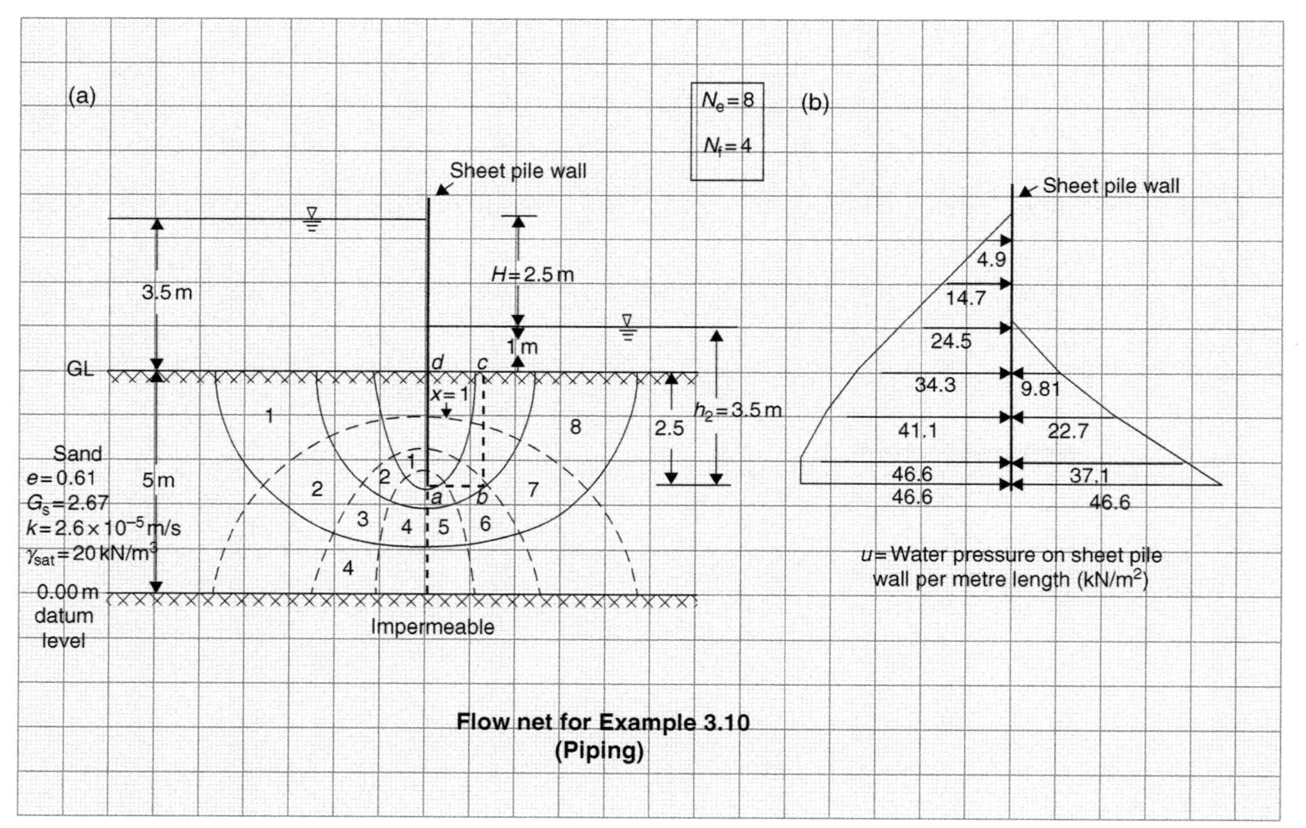
(a)
Sheet pile wall
3.5 m
H = 2.5 m
1 m
GL
d
c
x = 1
a
b
Sand
e = 0.61
Gs = 2.67
k = 2.6 × 10^-5 m/s
γsat = 20 kN/m3
5 m
2.5
h2 = 3.5 m
0.00 m
datum
level
Impermeable
1
2
3
4
5
6
7
8
Ne = 8
Nf = 4
(b)
Sheet pile wall
4.9
14.7
24.5
34.3
9.81
41.1
22.7
46.6
37.1
46.6
46.6
u = Water pressure on sheet pile
wall per metre length (kN/m2)
Flow net for Example 3.10
(Piping)

Graph 3.4

2. Factor of safety

Head loss in each square: $\Delta h = \frac{H}{N_e} = \frac{2.5}{8} = 0.313$ m

Submerged density: $\gamma' = 20 - 9.81 = 10.19$ kN/m^3

Critical hydraulic gradient: $i_c = \frac{\gamma'}{\gamma_w} = \frac{10.19}{9.81} = 1.04$

or $i_c = \frac{G_s - 1}{1 + e} = \frac{2.67 - 1}{1 + 0.61} = 1.04$

Weight of the prism: $W' = \gamma' \frac{z^2}{2} = 10.19 \times \frac{2.5^2}{2} = 31.8$ kN

Step 1

From Graph 3.4: $\left. \begin{array}{l} N_a = 4 \\ N_b = 5.8 \end{array} \right| \quad \therefore N_1 = \frac{N_a + 2N_b}{3} = \frac{4 + 2 \times 5.8}{3} = 5.2$

$h_a = H_T - \Delta h \times N_a = (3.5 + 2.5) - 0.313 \times 4 = 4.75$ m

$h_b = H_T - \Delta h \times N_b = 6 - 0.313 \times 5.8 = 4.18$ m

From (3.36): $h_1 = \frac{h_a + 2h_b}{3} = \frac{4.75 + 2 \times 4.18}{3} = 4.37$ m

Alternatively, $h_1 = H_T - \Delta h N_1 = 6 - 0.313 \times 5.2 = 4.37$ m

Step 2

$h_2 = 2.5 + 1 = 3.5$ m

From (3.37): $h_s = h_1 - h_2 = 4.37 - 3.5 = 0.87$ m

Or from (3.38): $h_s = (N_e - N_1)\Delta h = (8 - 5.2) \times 0.313 = 0.87$ m

Step 3

Seepage pressure: $u_s = \gamma_w h_s = 9.81 \times 0.87 = 8.53$ kN/m^2

Average gradient: $i_{av} = \frac{h_s}{z} = \frac{0.87}{2.5} = 0.348 < i_c \quad \therefore$ safe

From (3.39): $S = \frac{z}{2} \gamma_w h_s = \frac{2.5}{2} \times 9.81 \times 0.87 = 10.7$ kN

From (3.40): $S = \frac{z}{2} u_s = \frac{2.5}{2} \times 8.53 = 10.7$ kN

From (3.31): $S = i_{av} \gamma_w V = 0.348 \times 9.81 \times \frac{2.5^2}{2} = 10.7$ kN

Step 4

From (3.33): $F_s = \frac{W'}{S} = \frac{31.8}{10.7} = 2.97$

From (3.41): $F_s = \frac{z\gamma'}{h_s\gamma_w} = \frac{2.5 \times 10.19}{8.53} = 2.98$

From (3.43): $F_s = \frac{i_c}{i_{av}} = \frac{1.04}{0.348} = 2.99$

From (3.44): $F_s = \frac{\sigma'}{u_s} = \frac{z\gamma'}{8.53} = \frac{2.5 \times 10.19}{8.53} = 2.99$

By Harza's method: From the flow net $x = 1\,\text{m}$

From (3.45): $i_e = \frac{\Delta h}{x} = \frac{0.313}{1} = 0.313$

From (3.46): $F_s = \frac{i_c}{i_e} = \frac{1.04}{0.313} = 3.32 > 2.99$

Note: The differences in the results are negligible considering the approximate nature of the flow net construction. Despite the inherent inaccuracies, flow nets provide valuable insight into the problem of seepage and its consequence as to the stability of structure.

3.8 Prevention of piping

In order to increase the factor of safety against failure, four methods are normally adopted:

1. Placing filter material over the danger zone.

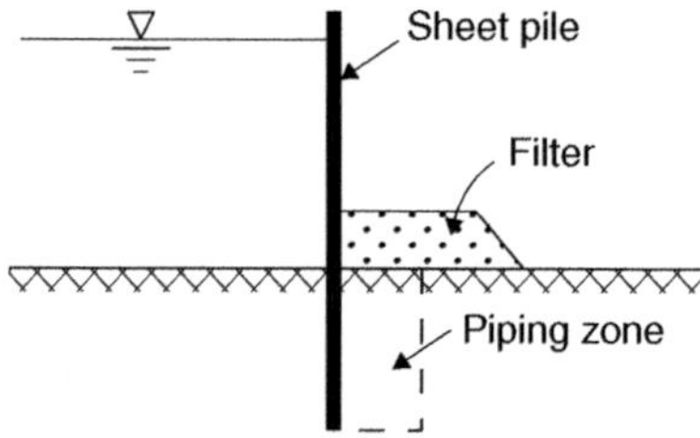

Figure 3.36

2. Lengthening the flow lines, by driving the sheet pile deeper as in Figure 3.37, or by installing sheet piles at one or both ends of a concrete dam.

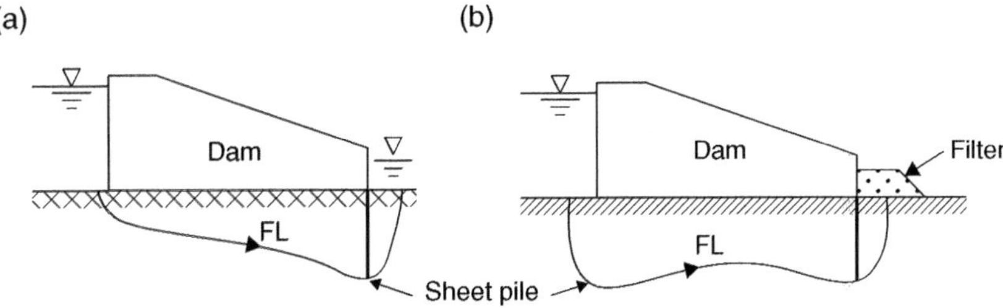

Figure 3.37

3. Shortening the flow lines by installing toe drainage for earth or concrete dams.

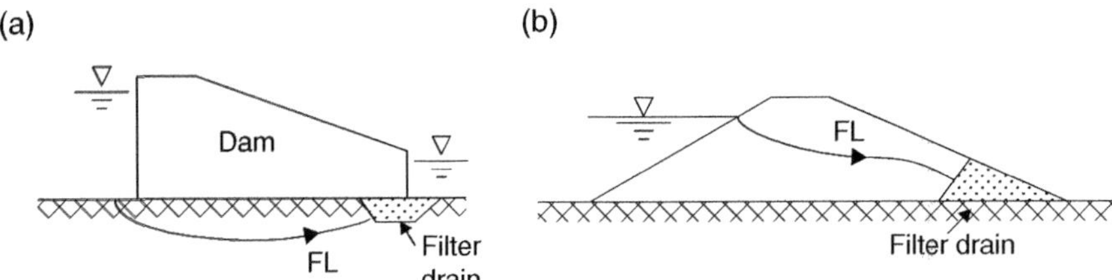

Figure 3.38

4. Lengthening the flow lines at concrete dams by constructing upstream or downstream concrete aprons.

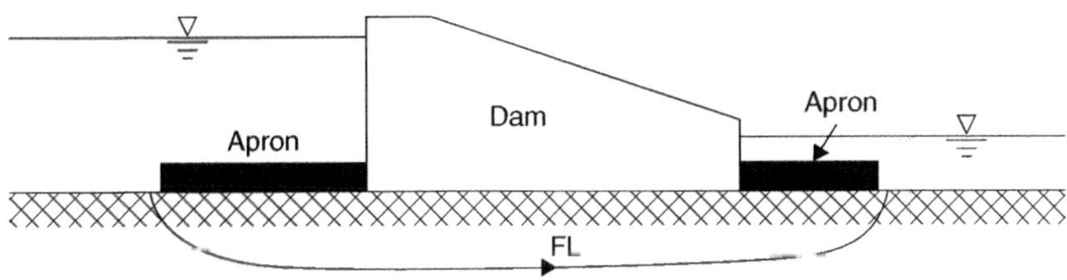

Figure 3.39

The advantage of lengthening the flow lines is to increase the number of equipotential lines. This means larger number of pressure drops, hence quicker dissipation of pressure head (H) and smaller hydraulic gradient (i_e) at the exit.

3.9 Flow net for earth dams

In most problems, such as for sheet piles and concrete dams, the boundary conditions, that is the positions of flow lines, along the impermeable surface are known. In the case of a lagoon embankment or dam, constructed from porous material, the phreatic surface, that is the position of the uppermost flow line, is unknown. There are several methods for the location of the phreatic line. Only the parabolic solution, evolved by Kozeny/Casagrande, shall be introduced here.

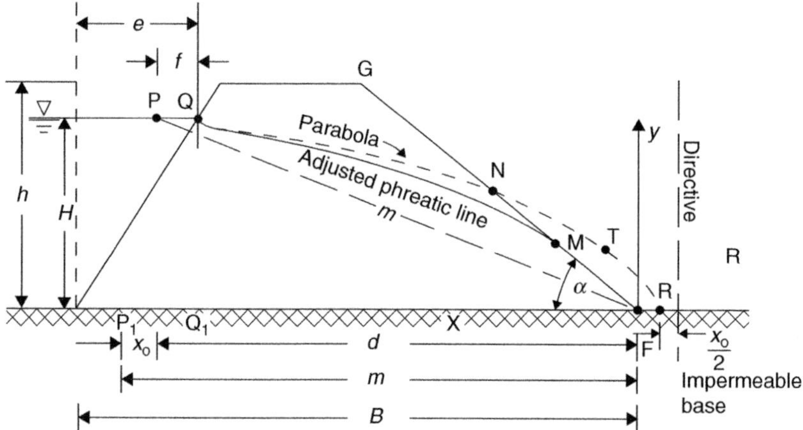

Figure 3.40

Step 1

Experiments indicate that the parabola intersects the water surface at P. The distance PQ is given by $f=0.3e$. Therefore, $d=B-e+f=B-0.7e$.

Also, $x_0 = m - d$

But $m = \sqrt{H^2 + d^2}$

$$\therefore\ x_0 = \sqrt{H^2 + d^2} - d$$

x_0=the distance between the focus F and the directrix.

The parabola intersects the x-axis at $\frac{x_0}{2}$.

Step 2

The equation of the parabola has to be formed in order to plot it, making use of the fact that any point on the parabola is at equal distance from its focus (F) and the directrix.

This is shown on the enlargement below.

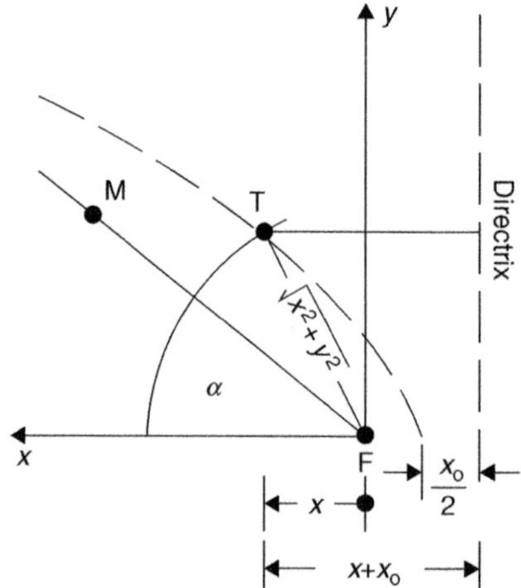

Figure 3.41

Equating the two distances and expressing the equation for the x-coordinate which is the formula for the parabola.

$$x + x_0 = \sqrt{x^2 + y^2}$$

$$(x + x_0)^2 = x^2 + y^2$$

$$x^2 + 2xx_0 + x_0^2 = x^2 + y^2$$

$$2xx_0 + x_0^2 = y^2$$

Therefore, the equation of the parabola: $$x = \frac{y^2 - x_0^2}{2x_0} \quad (3.47)$$

Knowing x_0, values of x can be plotted against y on the section of the dam.

Step 3

Next, the location of point N has to be found, that is the intersection of the parabola and slope FG.

Equation of the slope: $y = x \tan\alpha$

This is substituted into (3.47): $$x = \frac{(x\tan\alpha)^2 - x_0^2}{2x_0}$$

The result is a quadratic equation, whose solution is the x coordinate of point N.

$$2x_0 x = x^2 \tan^2\alpha - x_0^2 \qquad (\tan^2\alpha)x^2 - (2x_0)x - x_0^2 = 0$$

Therefore, $$x_N = \frac{2x_0 + \sqrt{4x_0^2 + 4x_0^2\tan^2\alpha}}{2\tan^2\alpha}$$

Cancelling, $$x_N = \frac{x_0 + x_0\sqrt{1+\tan^2\alpha}}{\tan^2\alpha}$$

$$x_N = x_0\left(\frac{1+\sqrt{1+\tan^2\alpha}}{\tan^2\alpha}\right) \quad (3.48)$$

Step 4

It is now possible to locate point M, the intersection to the phreatic line and slope FG, using Graph 3.5:

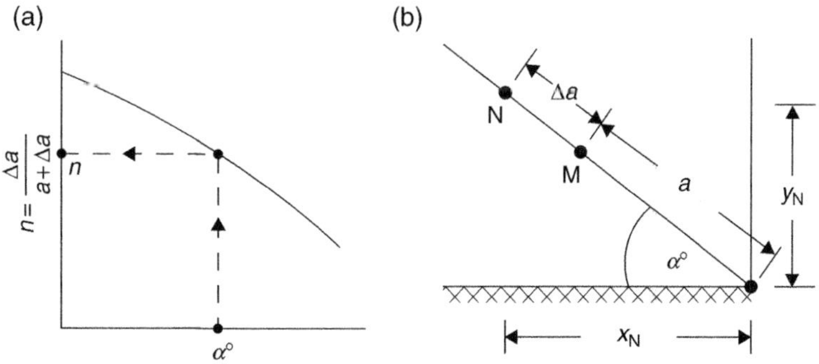

Figure 3.42

From Figure 3.42(b): $a + \Delta a = \sqrt{x_N^2 + y_N^2}$

But $y_N = x_N \tan\alpha$ (along the slope)

Substituting: $a + \Delta a = \sqrt{x_N^2 + x_N^2\tan^2\alpha}$

Hence, $\boxed{a+\Delta a=\sqrt{1+\tan^2\alpha}}$

From the graph: $\dfrac{\Delta a}{a+a}=n$

Expressing, $\Delta a=n(a+\Delta a)=nx_N\sqrt{1+\tan^2\alpha}$

Expressing distance FM: $a=(a+\Delta a)-\Delta a$

$$=x_N\sqrt{1+\tan^2\alpha}-nx_N\sqrt{1+\tan^2\alpha}$$

Therefore, point M is located by:

$$\boxed{a=(1-n)x_N\sqrt{1+\tan^2\alpha}\,\Big|\,\alpha>30^\circ} \qquad (3.49)$$

For $\alpha<30^\circ$

$$a=\frac{\alpha}{\cos\alpha}-\sqrt{\frac{d^2}{\cos^2\alpha}-\frac{H^2}{\sin^2\alpha}} \qquad (3.50)$$

Step 5
Sketch the phreatic line and continue to construct the flow net.

Example 3.11

The small earth dam, shown on Graph 3.6, retains 3 m deep water of a lagoon. It is underlain by impervious soil. The horizontal and vertical permeabilities within the structure are equal.

Draw: 1. The phreatic line
2. The flow net

Calculate the quantity of seepage at the toe per metre length of the dam.

Step 1
$f=0.3e=0.3\times3.4=1.02\,\text{m}$
$d=B-0.7e=11.5-0.7\times3.4=9.12\,\text{m}$

$x_0=\sqrt{H^2+d^2}-d=\sqrt{3^2+9.12^2}-9.12=0.48\,\text{m}$

Step 2
The equation of the parabola is:

$$x=\frac{y^2-0.48^2}{2\times0.48}=1.04y^2-0.24$$

Tabulate x for several values of y and plot the results.

Table 3.9

y	0	0.25	0.5	1	1.5	2	2.5	2.75	3
x	−0.24	−0.18	0.02	0.8	2.1	3.92	6.26	7.63	9.12

Step 3

From (3.48): $x_N = 0.48 \times \left(\frac{1+\sqrt{1+\tan^2 35}}{\tan^2 35} \right) = \frac{1.07}{0.49} = 2.18\,\text{m}$

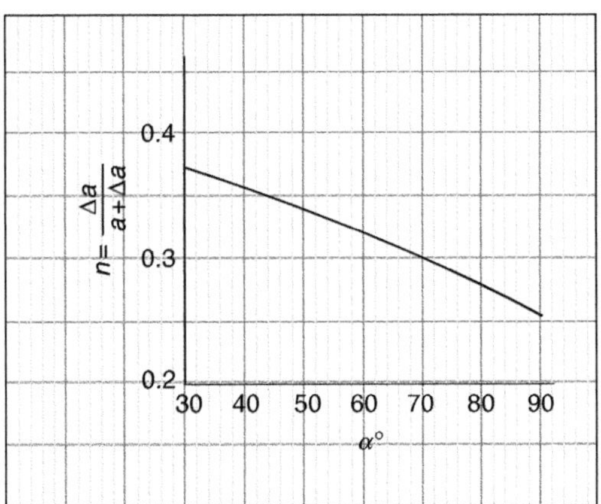

Graph 3.5

Step 4

From Graph 3.5 for $\alpha = 35°$ $n = 0.368$

And from (3.49): $a = (1 - 0.368) \times 2.18 \times \sqrt{1 + \tan^2 35}$

$= 1.378 \times 1.22 = 1.68\,\text{m}$

Also, $\frac{\Delta a}{a + \Delta a} = n\,\Delta a = an + n\Delta a$

$(1-n)\Delta a = an$

$\Delta a = \frac{an}{1-n} = \frac{1.68 \times 0.368}{1 - 0.368} = 0.98\,\text{m}$

Step 5

Measure lengths a and Δa along the slope to get points M & N. Draw the flow net.

Step 6

The flowrate at the toe of the dam is given by (3.26):

$Q = kH\frac{N_f}{N_e}$

From the net: $N_f = 5$, $N_e = 27$

$Q = 7.9 \times 10^{-4} \times 3 \times \frac{5}{27}$

$= 4.4 \times 10^{-4}\,\text{m}^3/\text{s}$

$= 4.4 \times 10^{-4} \times 60^2 \times 24\,\text{m}^3/\text{day}$

$= 38\,\text{m}^3/\text{day/metre length}$

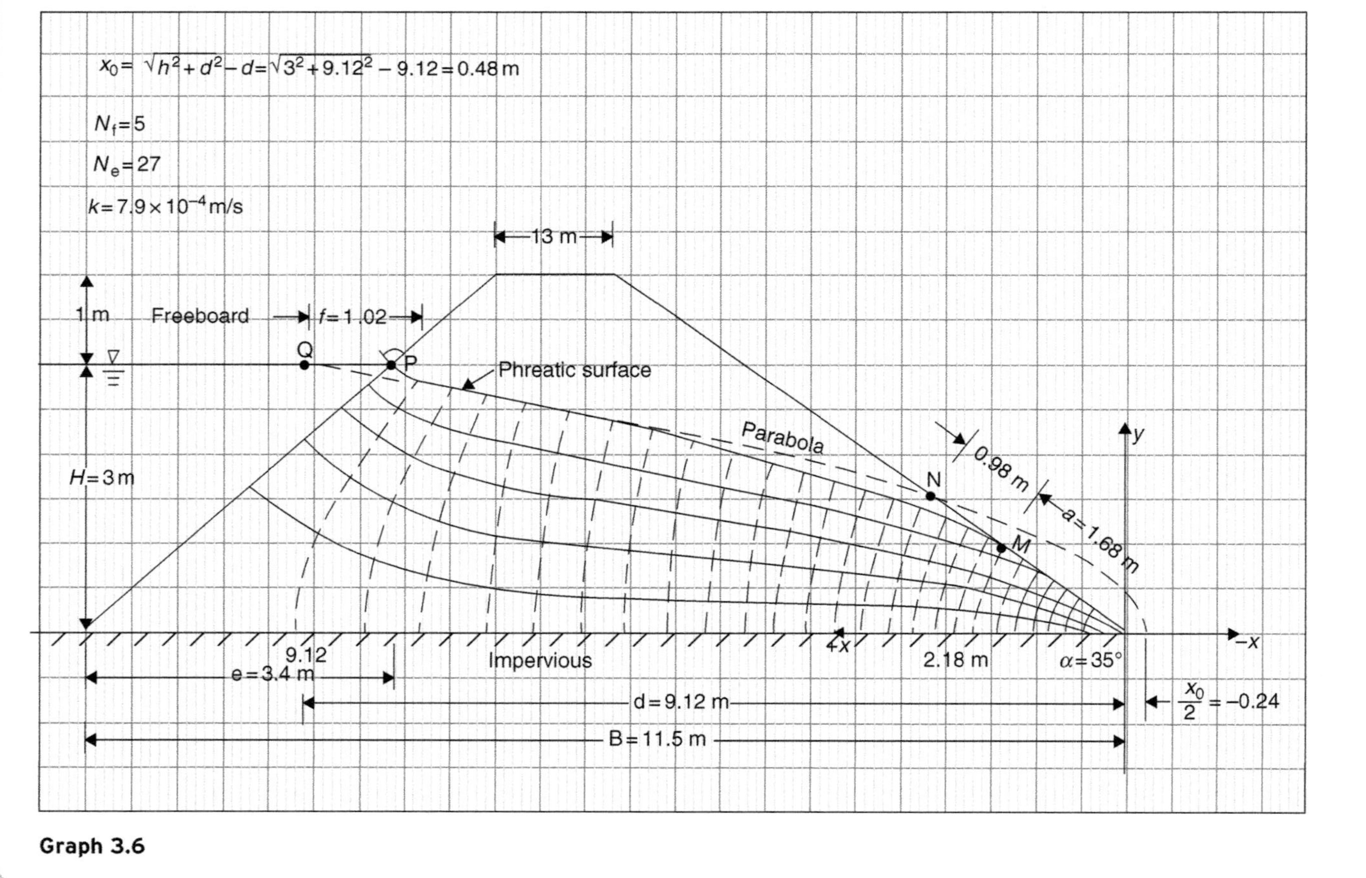

$x_0 = \sqrt{h^2 + d^2} - d = \sqrt{3^2 + 9.12^2} - 9.12 = 0.48\,m$
$N_f = 5$
$N_e = 27$
$k = 7.9 \times 10^{-4}\,m/s$
13 m
1 m
Freeboard
$f = 1.02$
Q
P
Phreatic surface
Parabola
0.98 m
N
$a = 1.68\,m$
M
y
$H = 3\,m$
9.12
$e = 3.4\,m$
Impervious
+x
2.18 m
$\alpha = 35°$
−x
$d = 9.12\,m$
$\frac{x_0}{2} = -0.24$
$B = 11.5\,m$

Graph 3.6

Problem 3.1

A permeability test was carried out on a specimen of sand and the results recorded as:

Cross-sectional area of the specimen	:	$A=150\,cm^2$
Length of the specimen	:	$L=30\,cm$
Porosity	:	$n=39\%$
Water collected in $t=22$ seconds	:	$q=83\,cm^3$
Pressure head	:	$h=19.2\,cm$
Specific gravity of sand	:	$G_s=2.66$

Determine
a) Coefficient of permeability of sand.
b) Discharge velocity.
c) Seepage velocity.
d) Critical hydraulic gradient.
e) Critical pressure head, at which the soil would fail.
f) Submerged unit weight of sand.

Problem 3.2

In order to extract water from the aquifer, shown in Figure 3.43, the water company intend to sink a 0.5 m diameter well as near to a protected building as possible. Because the structure is founded on shrinkable clay, it is imperative, that its water content should not be altered by the pumping operation. For this reason, the cone of depression is not allowed to encroach on a 200 m exclusion zone around the building. Calculate the yield from the well, in m³/s, if it is sunk 1200 m away from the protected structure. The water level in the central well must remain within the shrinkable clay layer. Assume that the shrinkable clay is saturated to the piezometric water level.

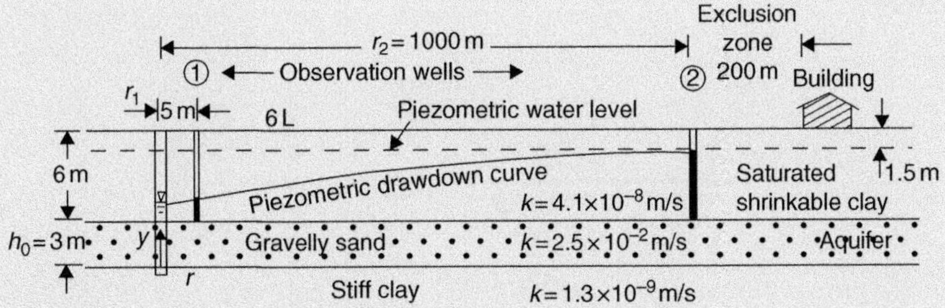

Figure 3.43

Ground water level is 1.5 m below the surface.
Depression of water in observation well No. 1 = 4 m.
Depression in observation well No. 2 = 0.

Problem 3.3

Referring to Example 3.6 of the main text, it is intended to reduce the uplift pressure on the dam, outlined in Figure 3.44, by placing an 11 m long sheet pile wall at the tail end. Determine the uplift pressures at corners D and E.

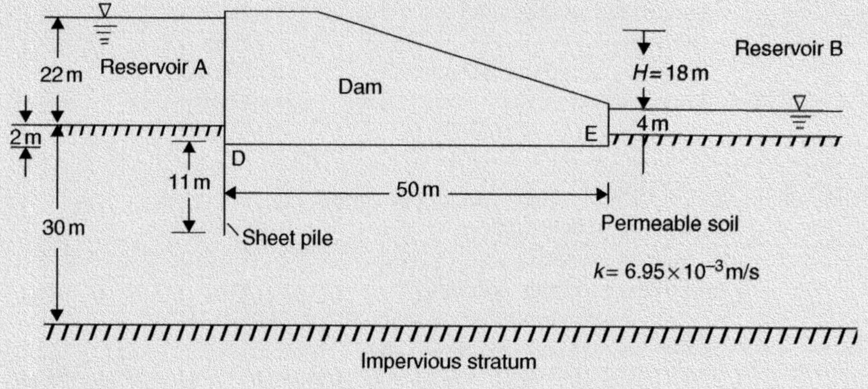

Figure 3.44

Problem 3.4

Steady seepage of water occurs through a sandy clay layer, underlain by coarse gravel under artesian pressure. The surface of sand slopes slightly, hence the emerging water flows downhill in a very thin layer. Piezometers placed into the gravel indicate a water level rise of 3 m above the ground surface.

The available information on the layer is shown in Figure 3.45.

Estimate the critical thickness (z) of the sandy clay layer, at which it would fail in shear.

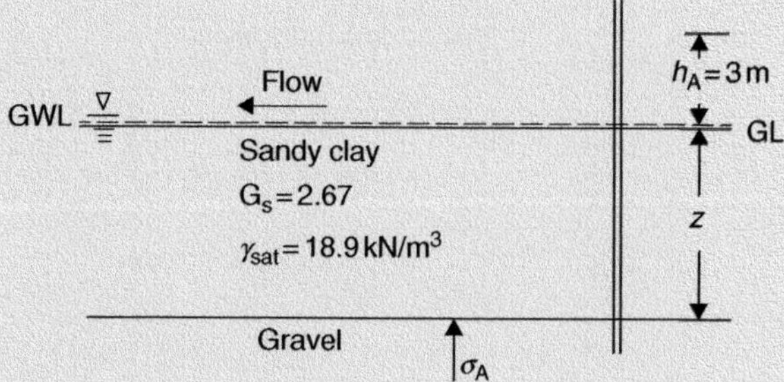

Figure 3.45

Problem 3.5

Starting from formula (1.43), derive the expressions for:

a) submerged density: $\gamma' = (G_s - 1)(1 - n)\gamma_w$

b) critical hydraulic gradient: $i_c = (G_s - 1)(1 - n)$

c) voids ratio: $e = \dfrac{\gamma_w}{\gamma_d - \gamma'} - 1$

Problem 3.6

The in-situ hydraulic gradient of sand ($G_s = 2.65$) below GWL was found to be 1.02. Determine its water content.

Problem 3.7

A 50 m wide concrete dam is shown on Graph 3.3 of Chapter 3. The diagram of uplift pressures is reproduced below.

Calculate the uplift force (S) per metre length of the dam.

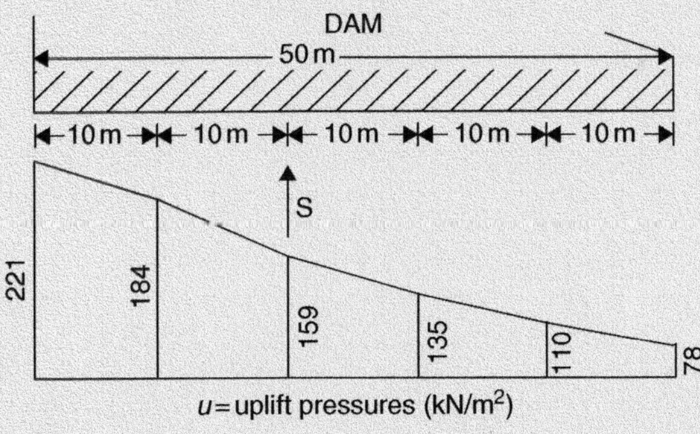

Figure 3.46

Problem 3.8

It is assumed that the pressure variation under the soil prizm, investigated for 'boiling' failure during seepage, adjacent to a sheet pile wall is parabolic, as shown in Figures 3.33 and 3.47. Prove formula (3.36), yielding the average pressure head h, as well as its distance from corner b. Draw the parabola as shown for ease of derivation.

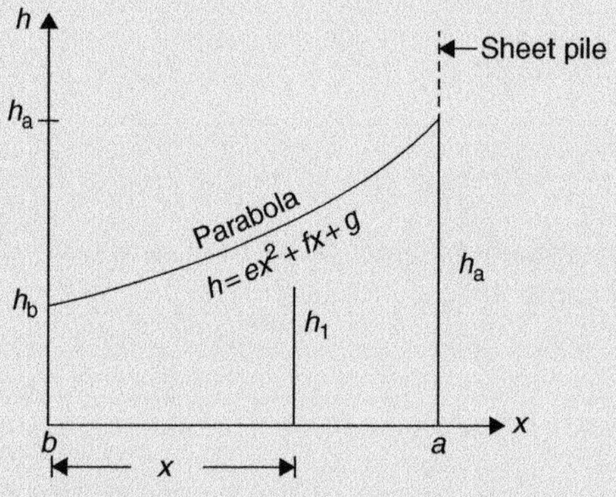

Figure 3.47

Chapter 4

Pressure at Depth Due to Surface Loading

When the ground surface is loaded, stresses are induced within the soil mass below. The theories evolved for the determination of these stresses assume that the soil is homogeneous and elastic. In addition, it is taken to be isotropic, that is, the stress at a point below ground level has the same value in all directions. There are two main types of loading:

1. Uniform overburden, such as compacted fill covering a large area around the point considered. The vertical stress induced equals to the weight of the deposited material, as shown in Figure 4.1:

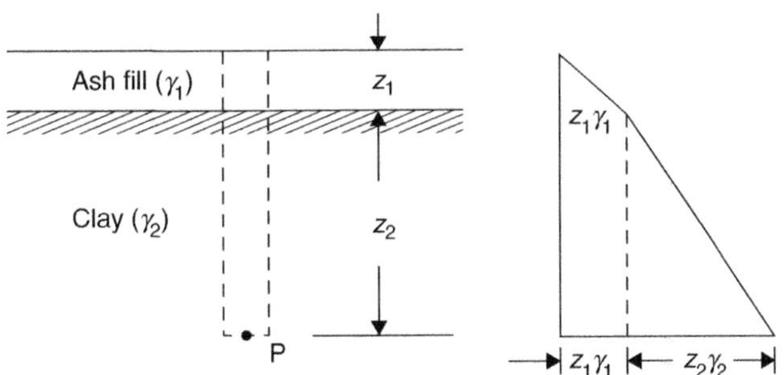

Figure 4.1

   The pressure distributed within a layer increases linearly with depth. The pressure of the ash fill surcharge remains constant ($z_1\gamma_1$) in the underlying layer. The total vertical pressure at point P is: $\sigma_v = z_1\gamma_1 + z_2\gamma_2$

2. Surface load of limited size e.g. a foundation. In this case, stresses are induced not only below, but also outside the base area as shown in Figure 4.2.

This second type of loading is of interest in this chapter. Practical problems, utilizing the formulae introduced, are:

1. Analysis of consolidation and settlement, when a change in vertical loading results in the compression of a soil layer.
2. Comparison of induced vertical stress with the bearing strength of a soil.

Introduction to Soil Mechanics, First Edition. Béla Bodó and Colin Jones.

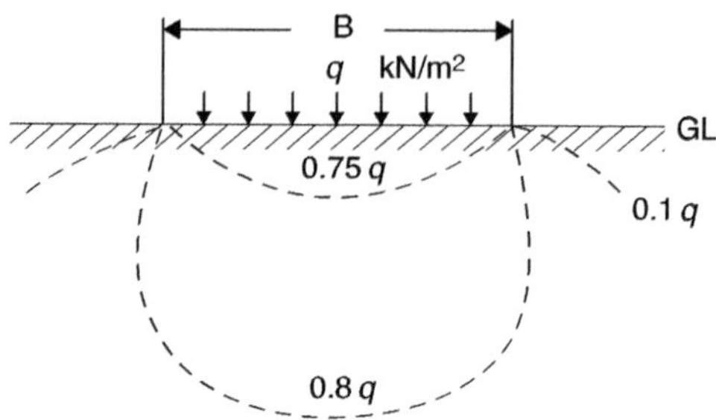

Figure 4.2

A number of formulae have been developed, depending on the configuration of the loading which were derived under assumptions not strictly true for soils. Their application should, therefore be taken as supplementary to engineering judgment.

It is not proposed to derive the formulae, in view of their complexity. Instead, they are quoted and their use illustrated by examples with the aid of nomograms, whenever possible. The following eight configurations are introduced as two-dimensional problems:

1. Concentrated point load
2. Concentrated line load
3. Uniform strip loading
4. Triangular strip loading
5. Superposition of strip loadings
6. Circular footing
7. Rectangular footing
8. Footing of irregular shape

4.1 Concentrated point load

Boussinesq (1885) solved this problem in three dimensions, which became the basis for all the other theories on this subject. Its two-dimensional aspects are of interest here, giving the vertical and horizontal pressures as well as the shear stress at a point (P) below a concentrated load (Q).

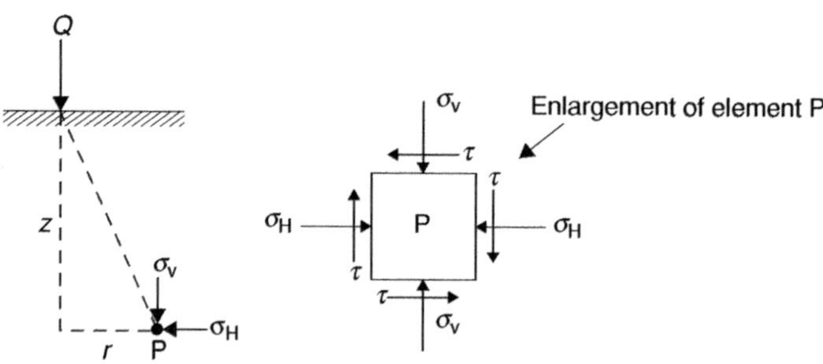

Figure 4.3

Influence factor:

$$I_0 = \left[\frac{3}{2\pi \left(1+\frac{r^2}{z^2}\right)^{5/2}} \right] \quad (4.1)$$

Vertical pressure at point P:

$$\sigma_v = I_0\left(\frac{Q}{z^2}\right)$$

The variation of I_0 with $\frac{r}{z}$ is given on Chart 4.1

Shear stress at P:

$$\tau = \sigma_v \frac{r}{z} \quad (4.2)$$

Horizontal pressure at P:

$$\sigma_H = \sigma_v\left(\frac{r}{z}\right)^2 \quad (4.3)$$

Example 4.1

A concentrated load of 1000 kN is placed on the surface. Estimate the vertical and horizontal pressures as well as the shear stress, at 6 m depth and 5 m away from the load, at point P.

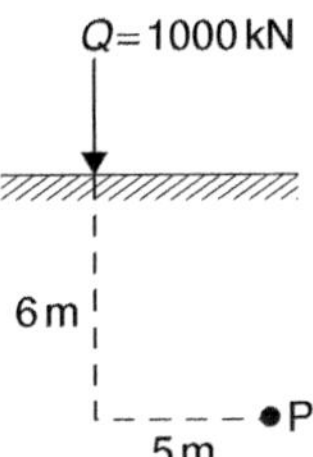

Figure 4.4

Influence factor: $I_0 = \dfrac{3}{2\pi\left(1+\frac{25}{36}\right)^{5/2}} = 0.128$

$$(4.1):\ \sigma_v = 0.128\times\left(\frac{1000}{36}\right) = 3.556\,\text{kN/m}^2$$

$$(4.2):\ \tau = 3.556\times\left(\frac{5}{6}\right) = 2.96\,\text{kN/m}^2$$

$$(4.3):\ \sigma_H = 3.556\times\left(\frac{5}{6}\right)^2 = 2.470\,\text{kN/m}^2$$

Note: Formula (4.3) is assumed to be valid if the soil is incompressible, that is, its volume does not change under the action of σ_H.

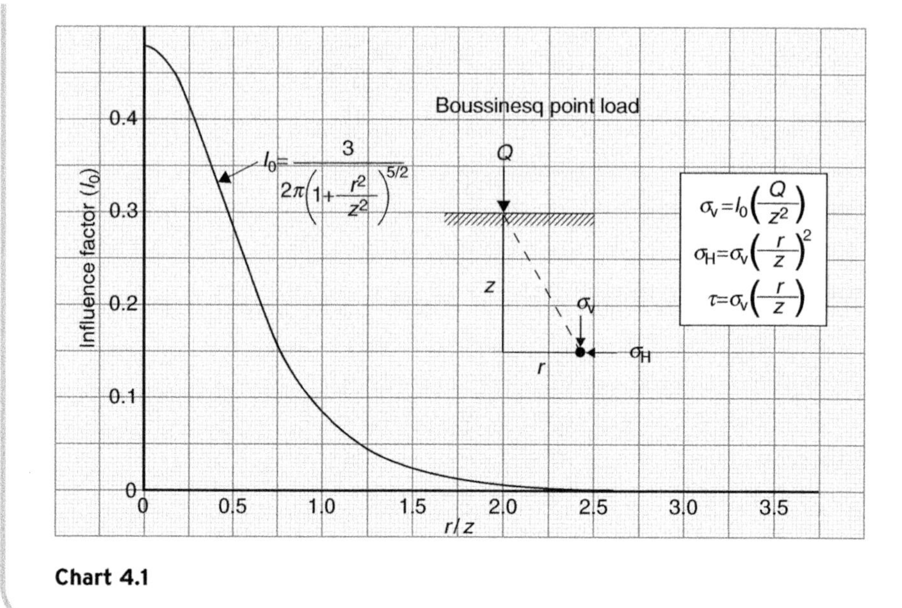

Chart 4.1

4.2 Concentrated line load

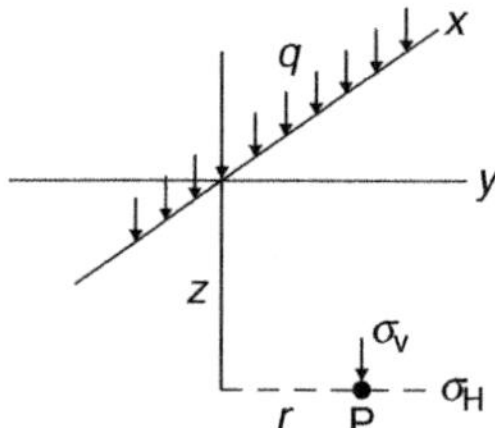

Figure 4.5

q = uniformly distributed load / metre

$$I_1 = \frac{2}{\pi\left(1+\frac{r^2}{z^2}\right)^2} \qquad \sigma_v = I_1\left(\frac{q}{z}\right) \tag{4.4}$$

Where I_1 is the influence factor for this case, it is given on Chart 4.2.

Shear stress at point P:

$$\tau = \sigma_v\left(\frac{r}{z}\right) \tag{4.5}$$

Horizontal pressure at P:

$$\sigma_H = \sigma_v\left(\frac{r}{z}\right)^2 \tag{4.6}$$

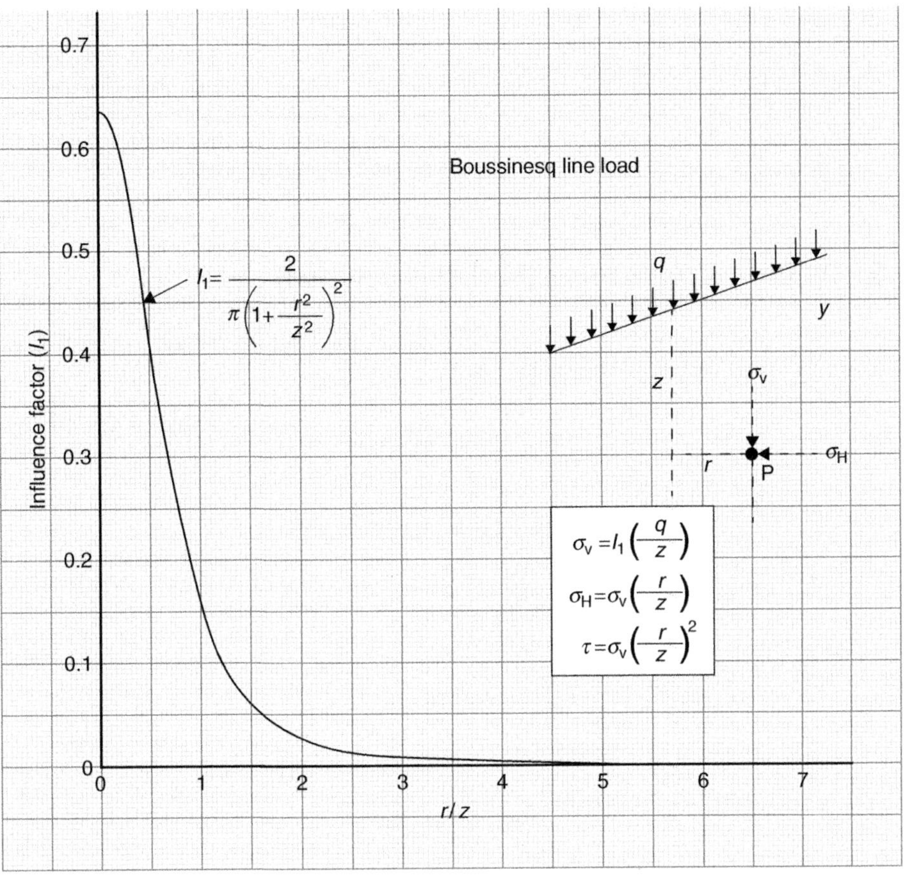

Chart 4.2

Example 4.2

Calculate σ_v, σ_H and τ at point P (shown in Figure 4.4) if instead of the point load, a line load of $q = 1000$ kN/m acts on the surface.

$$\left.\begin{array}{l} z = 6\text{m} \\ r = 5\text{m} \end{array}\right\} \quad I_1 = \frac{2}{\pi\left(1+\frac{25}{36}\right)^2} = 0.222$$

Vertical pressure at P: $\sigma_v = 0.222 \times \left(\frac{1000}{6}\right) = 37\,\text{kN/m}^2$

Horizontal pressure: $\sigma_H = 37 \times \left(\frac{5}{6}\right)^2 = 25.7\,\text{kN/m}^2$

Shear stress: $\tau = 37 \times \left(\frac{5}{6}\right) = 30.8\,\text{kN/m}^2$

Note: Point and line loading do not occur in reality, as pressure can only be imparted to the ground by footings having width as well as length. The concept, however, may be applied to foundations by dividing the base area into small sections, placing a point or line load at the centroid of each. The pressures at a point below the footing are then calculated by the principle of superposition, that is, the summation of the pressures induced at that point by the individual sections.

4.3 Uniform strip loading (Michell's solution)

The formulae were derived on the assumption, that the bearing pressure (q) is distributed evenly under an infinitely long footing.

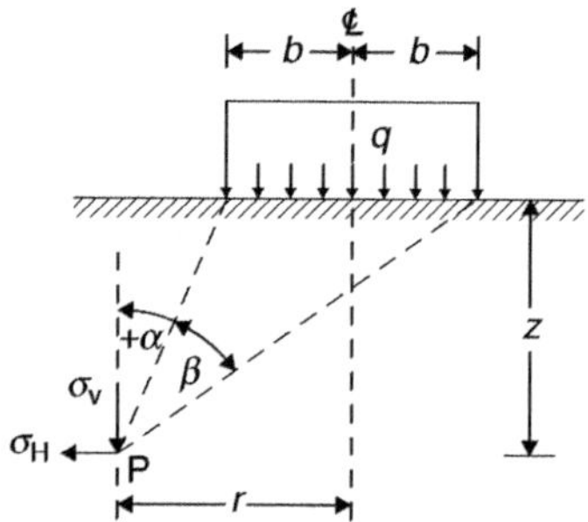

Figure 4.6

$$I_2 = \frac{\beta + \sin\beta\cos(2\alpha+\beta)}{\pi} \tag{4.7}$$

$$\sigma_v = \left[\frac{\beta + \sin\beta\cos(2\alpha+\beta)}{\pi}\right] q$$

$$\sigma_v = I_2 q \tag{4.8}$$

$$\sigma_H = \frac{2\beta}{\pi} q - \sigma_v \tag{4.9}$$

And the shear stress at P:

$$\tau = \sigma_v - \frac{\beta q}{\pi} \tag{4.10}$$

Where:

$$\alpha^\circ = \tan^{-1}\left(\frac{r-b}{z}\right) \tag{4.11}$$

and

$$\beta^\circ = \tan^{-1}\left(\frac{r+b}{z}\right) - \alpha^\circ \tag{4.12}$$

Under the centre line: $\alpha = -\frac{\beta}{2}$ and $r = 0$

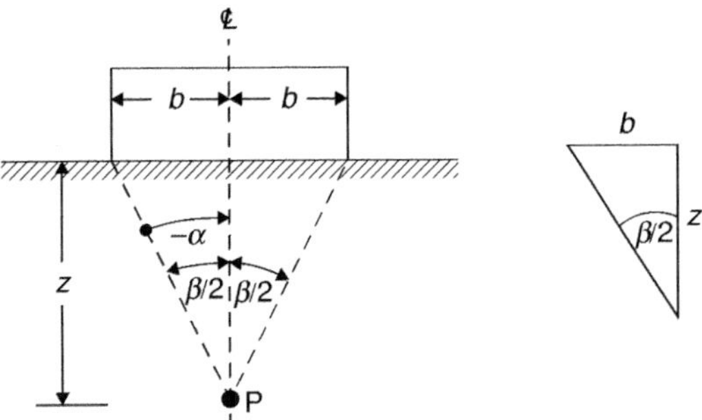

Figure 4.7

$$\tan\frac{\beta}{2} = \frac{b}{z}$$

$$\therefore \quad \boxed{\beta^\circ = 2\tan^{-1}\left(\frac{b}{z}\right)} \tag{4.13}$$

Therefore, under the centre line:

$$\boxed{I_3 = \frac{\beta + \sin\beta}{\pi}} \tag{4.14}$$

$$\boxed{\therefore \quad \sigma_V = I_3 q} \tag{4.15}$$

Example 4.3

Figure 4.8 shows a 1m wide strip footing transmitting a bearing pressure of 300 kN/m² to the soil. Calculate σ_V, σ_H and τ at a depth of 1m at 0.25 m intervals.

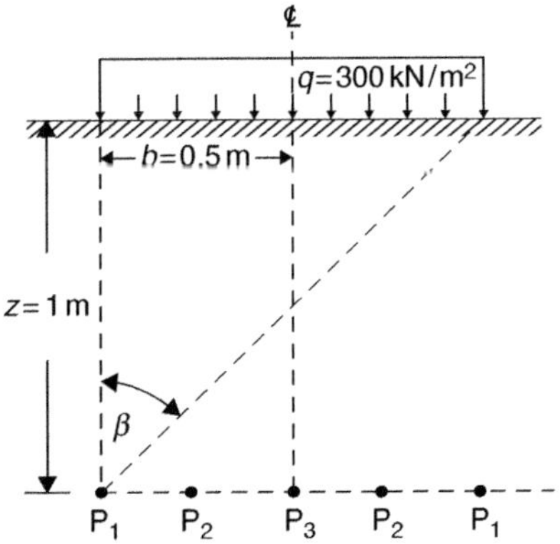

Figure 4.8

Below the edges (Point P_1)

$\alpha = 0 \quad r = b = 0.5\,\text{m}$

$$\beta = \tan^{-1}\left(\frac{2b}{z}\right) = \tan^{-1}(1) = 45^\circ$$

$$= \frac{\pi \times 45}{180}\text{Radian}$$

$$\therefore I_2 = \frac{1}{\pi}(\beta + \sin\beta\cos\beta)$$
$$= \frac{1}{\pi} \times \left(\frac{\pi \times 45}{180} + \sin 45 \times \cos 45\right)$$
$$= 0.409$$

From (4.8): $\sigma_v = 0.409 \times 300 = 122.7\ \text{kN/m}^2$

From (4.9): $\sigma_H = \frac{2}{\pi} \times \left(\frac{\pi \times 45}{180} \times 300\right) - 122.7 = 27.3\ \text{kN/m}^2$

From (4.10): $\tau = 122.7 - \left(\frac{\pi \times 45}{180}\right) \times \frac{300}{\pi} = 47.7\ \text{kN/m}^2$

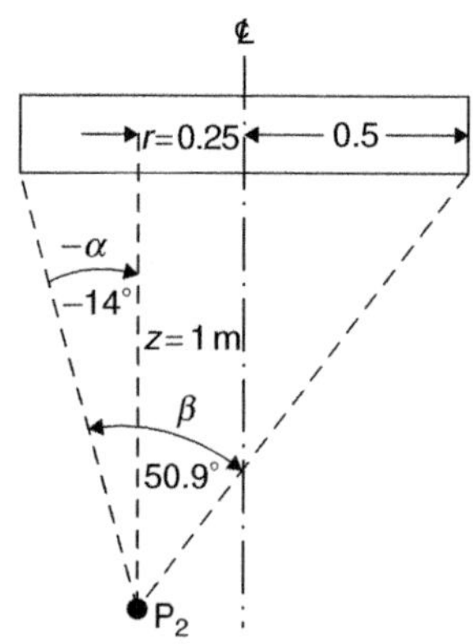

Figure 4.9

At point P_2

$$r = \frac{b}{z} = 0.25$$

$$\alpha = -\tan^{-1}\left(\frac{0.25}{1}\right) = -14°$$

$$\beta = \alpha + \tan^{-1}\left(\frac{0.25 + 0.5}{1}\right) = 50.9°$$

$$\therefore I_2 = \frac{1}{\pi} \times \left[\frac{\pi \times 50.9}{180} + \sin(50.9) \times \cos(-2.14 + 50.9)\right]$$
$$= \frac{1}{\pi} \times (0.888 + 0.776 \times 0.921) = 0.51$$

Vertical pressure at P_2: $\sigma_V = 0.51 \times 300 = 153\ \text{kN/m}^2$

Horizontal pressure at point P_2: $\sigma_H = \frac{2}{\pi}\left(\frac{\pi \times 50.9}{180}\right) \times 300 - 153 = 16.7\ \text{kN/m}^2$

Shear stress at P_2: $\tau = 153 - \frac{300}{\pi} \times \left(\frac{\pi \times 50.9}{180}\right) = 68.2\ \text{kN/m}^2$

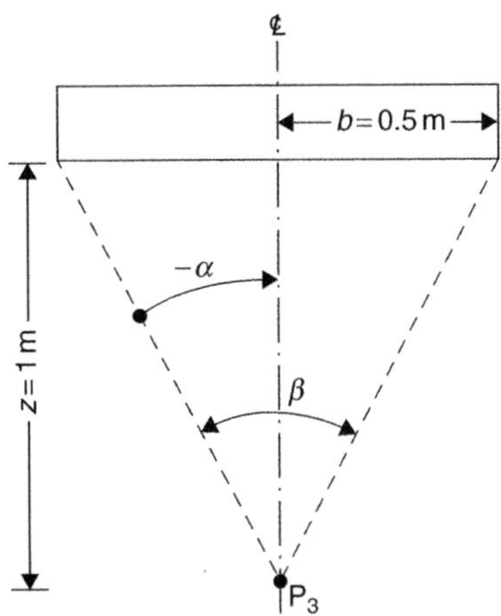

Figure 4.10

Under the centre (Point P_3)

$$\beta = 2 \times \tan^{-1}\left(\frac{0.5}{1}\right) = 53.12°$$

$$r = 0$$

$$I_3 = \frac{1}{\pi} \times \left(\frac{\pi \times 53.12}{180} + \sin 53.12\right) = 0.55$$

$$\sigma_V = 0.55 \times 300 = 165\ \text{kN/m}^2$$

Horizontal pressure: $\sigma_H = \frac{2}{\pi} \times \left(\frac{\pi \times 53.12}{180}\right) \times 300 - 165 = 12.1\ \text{kN/m}^2$

Shear stress: $\tau = 165 - \frac{300}{\pi} \times \left(\frac{\pi \times 53.12}{180}\right) = 76.47\ \text{kN/m}^2$

Figure 4.11 shows the variation of σ_V at 1m below the footing.

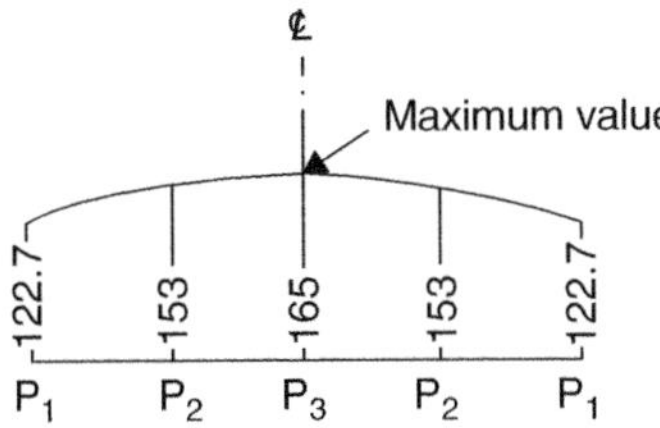

Figure 4.11

4.4 Bulb of pressure diagrams

See Charts 3.3 and 3.4. These are nomograms to estimate σ_V and τ under uniform strip loading. Although both can be found easily by calculation, nevertheless the charts are useful visual aid as to their spread.

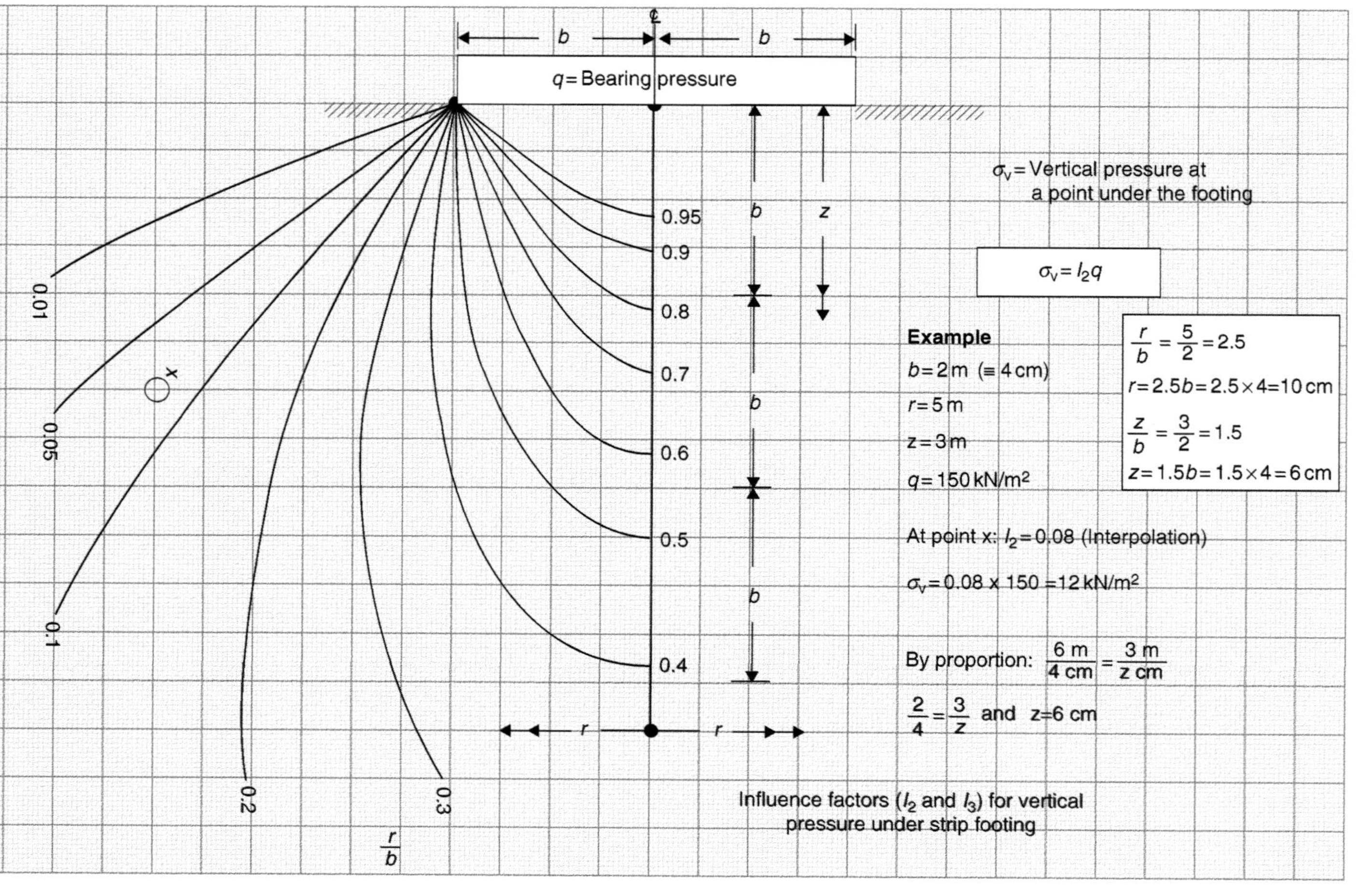

b
b
q = Bearing pressure
σv = Vertical pressure at a point under the footing
σv = I2q
0.95
0.9
0.8
0.7
0.6
0.5
0.4
b
b
b
z
r
r
Example
b = 2 m (≡ 4 cm)
r = 5 m
z = 3 m
q = 150 kN/m2
r/b = 5/2 = 2.5
r = 2.5b = 2.5 × 4 = 10 cm
z/b = 3/2 = 1.5
z = 1.5b = 1.5 × 4 = 6 cm
At point x: I2 = 0.08 (Interpolation)
σv = 0.08 x 150 = 12 kN/m2
By proportion: 6 m / 4 cm = 3 m / z cm
2/4 = 3/z and z = 6 cm
x
0.01
0.05
0.1
0.2
0.3
z/b
r/b
Influence factors (I2 and I3) for vertical pressure under strip footing

Chart 4.3

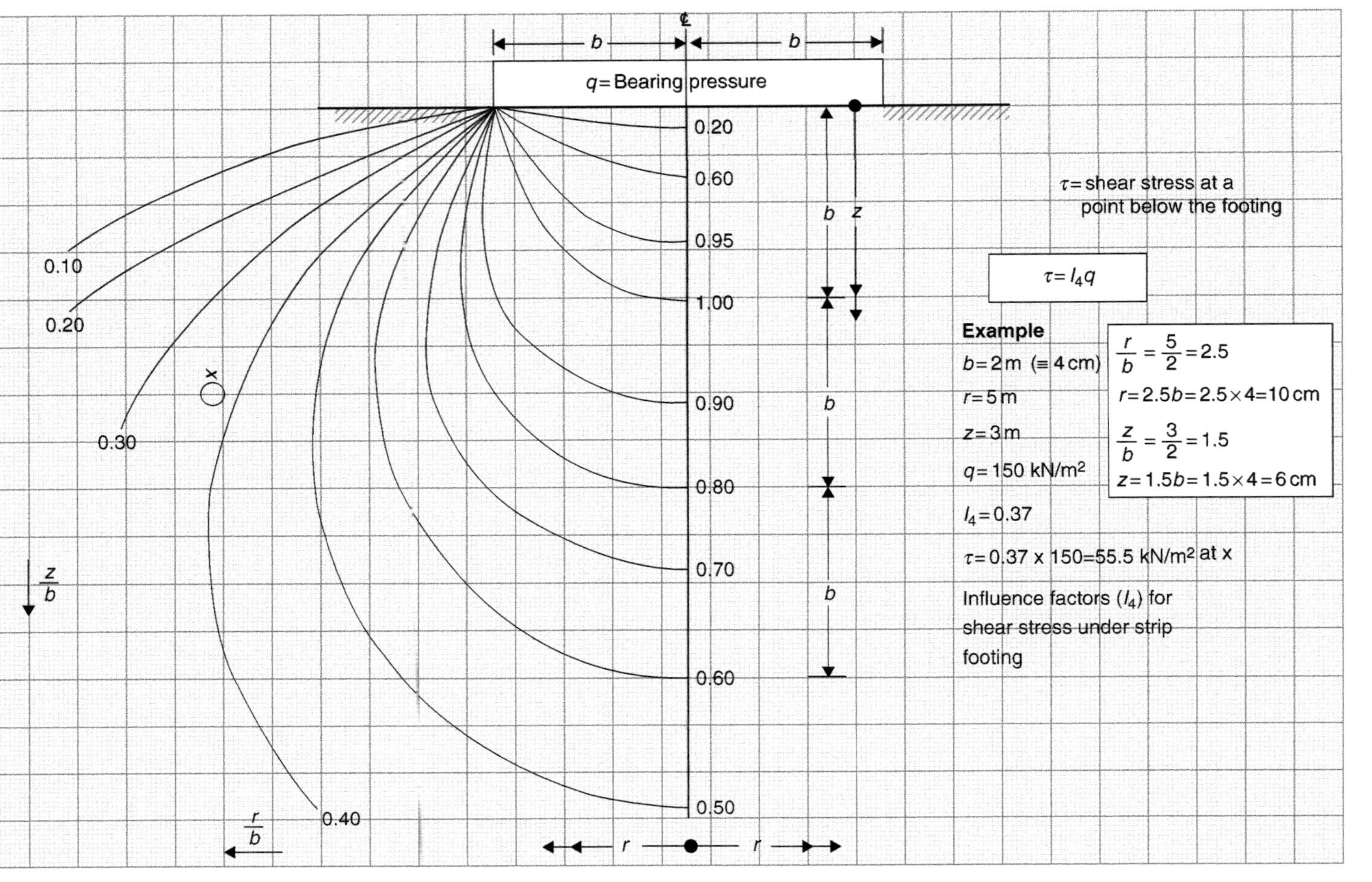

b
b
q= Bearing pressure
0.20
0.60
0.95
1.00
0.90
0.80
0.70
0.60
0.50
0.10
0.20
0.30
0.40
x
z/b
r/b
r
r
z
τ= shear stress at a point below the footing
τ= I4q
Example
b= 2 m (≡ 4 cm)
r= 5 m
z= 3 m
q= 150 kN/m2
I4= 0.37
τ= 0.37 x 150=55.5 kN/m2 at x
Influence factors (I4) for shear stress under strip footing
r/b = 5/2 = 2.5
r= 2.5b= 2.5×4=10 cm
z/b = 3/2 = 1.5
z= 1.5b= 1.5×4 = 6 cm

Chart 4.4

Example 4.4

This is to investigate the effect of a new footing placed adjacent to that in Example 4.3. The new base is 2 m wide, transmitting 250 kN/m² bearing pressure to the ground. Calculate the vertical pressure under footing A at 0.5 m intervals at 1 m depth.

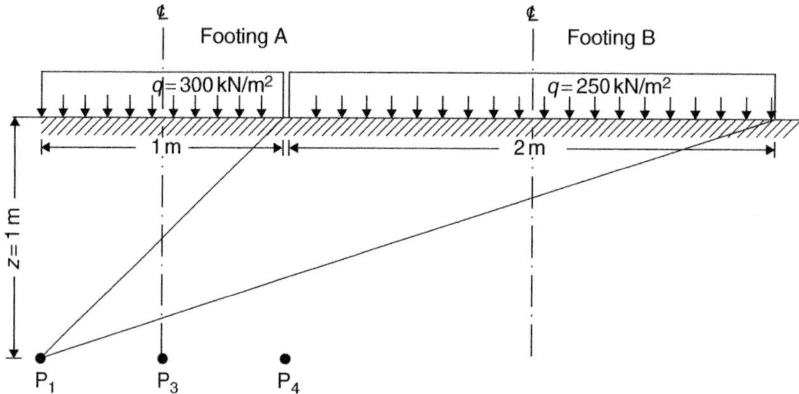

Figure 4.12

122.7	165	122.7	kN/m²	←	σ_v from Example 4.3
21.0	53.4	120		←	σ_v contributed by B
143.7	218.4	242.7		←	Total σ_v below footing A

Calculations for footing B

Pressure at P_1: $r = 2\,\text{m}$ $\therefore$ $\alpha = \tan^{-1}\left(\frac{2-1}{1}\right) = 45°$

$b = 1\,\text{m}$ $\beta = \tan^{-1}\left(\frac{2+1}{1}\right) - 45° = 26.6° = 0.464\,\text{radian}$

$$I_2 = \frac{1}{\pi}\times\left[0.464 + \sin 26.6 \times \cos(2\times 45 + 26.6)\right]$$

$$= \frac{1}{\pi}\times(0.264) = 0.0839$$

$$\therefore \quad \sigma_v = 122.7 + 0.0839\times 250 = 122.7 + 21 = 143.7\,\text{kN/m}^2$$

Pressure at P_3: $r = 1.5\,\text{m}$ $\therefore$ $\alpha = \tan^{-1}\left(\frac{1.5-1}{1}\right) = 26.6°$

$$\beta = \tan^{-1}\left(\frac{1.5+1}{1}\right) - 26.6° = 41.6° = 0.726\,\text{radian}$$

Influence factor: $$I_2 = \frac{1}{\pi} \times \left[0.726 + \sin 41.6 \times \cos(2 \times 26.6 + 41.6)\right]$$
$$= \frac{1}{\pi} \times (0.726 - 0.0556) = 0.213$$

$$\therefore \ \sigma_V = 165 + 0.213 \times 250 = 165 + 53.4 = 218.4 \text{ kN/m}^2$$

Pressure at P_4: $r = 1\text{m}$ $\therefore$ $$\alpha = \tan^{-1}\left(\frac{1-1}{1}\right) = 0°$$

$$\beta = \tan^{-1}\left(\frac{1+1}{1}\right) = 63.43° = 1.107 \text{ radian}$$

Influence factor: $$I_2 = \frac{1}{\pi} \times \left[1.107 + \sin 63.43 \times \cos 63.43\right] = 0.48$$

$$\therefore \sigma_V = 122.7 + 0.48 \times 250 = 122.7 + 120 = 242.7 \text{ kN/m}^2$$

Note: The results show that in the construction of new footings next or near to existing ones can affect adversely the latter. Footing A had obviously been constructed with consideration of the bearing strength as well as the consolidation characteristics of the soil. A large pressure increase due to footing B could, therefore, cause either bearing capacity failure or excessive settlement of the existing foundation. The sketch in Figure 4.13 depicts how differential settlement could affect a statically indeterminate structure, such as a rigid portal frame.

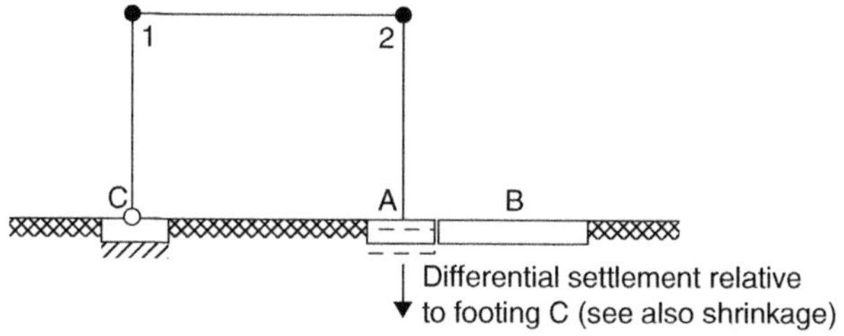

Figure 4.13

4.5 Vertical pressure under triangular strip load

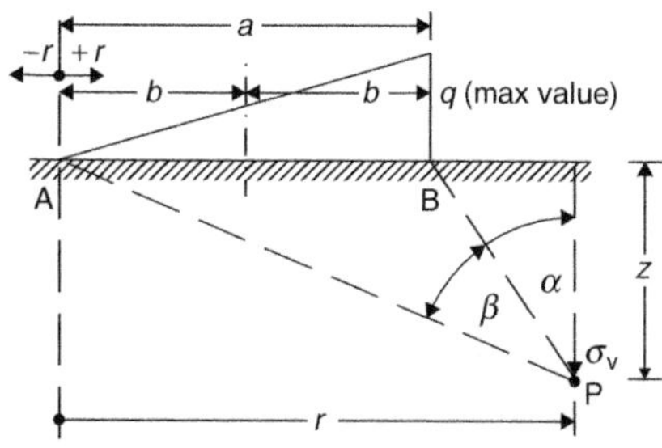

Figure 4.14

$$\alpha^\circ = \tan^{-1}\left(\frac{r-a}{z}\right) \tag{4.16}$$

$$\beta^\circ = \tan^{-1}\left(\frac{r}{z}\right) - \alpha^\circ \tag{4.17}$$

Vertical pressure:

$$\sigma_V = I_s q \tag{4.18}$$

$$I_s = \frac{1}{2\pi}\left(\frac{2r}{a}\beta - \sin 2\alpha\right) \tag{4.19}$$

Variation of α and β to be substituted into (4.19):

1. Outside edge A at point P:

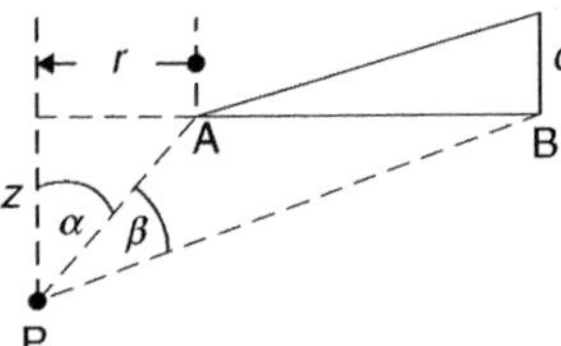

$$\alpha^\circ = \tan^{-1}\left(\frac{r-a}{z}\right)$$

$$\beta^\circ = \tan^{-1}\left(\frac{r}{z}\right) - \alpha^\circ$$

2. Under edge A ($r=0$):

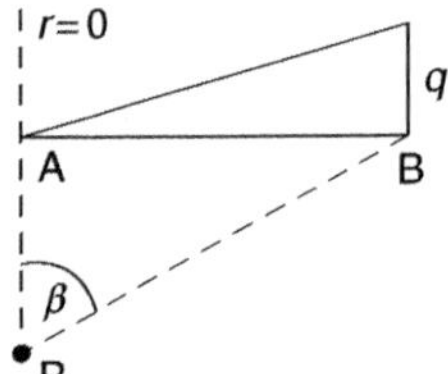

$$\alpha^\circ = \tan^{-1}\left(\frac{-a}{z}\right) \qquad \beta^\circ = -\alpha \tag{4.20}$$

$$\therefore \quad I_s = -\frac{\sin 2\alpha}{2\pi} \tag{4.21}$$

3. Under edge B (for maximum σ_V):

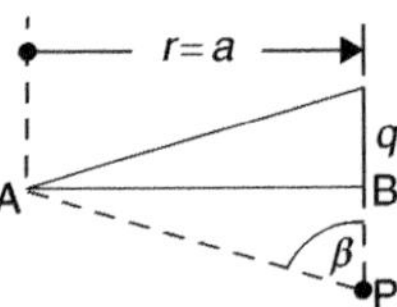

$$\alpha = 0 \qquad \beta^\circ = \tan^{-1}\frac{a}{z} \tag{4.22}$$

$$\therefore \quad I_s = -\frac{\beta}{\pi} \tag{4.23}$$

The triangular and uniform strip loadings may be combined into various shapes (Figure 4.15) and their contributions to the pressure at a point P are superimposed. The formulae to be used are indicated on the following diagrams:

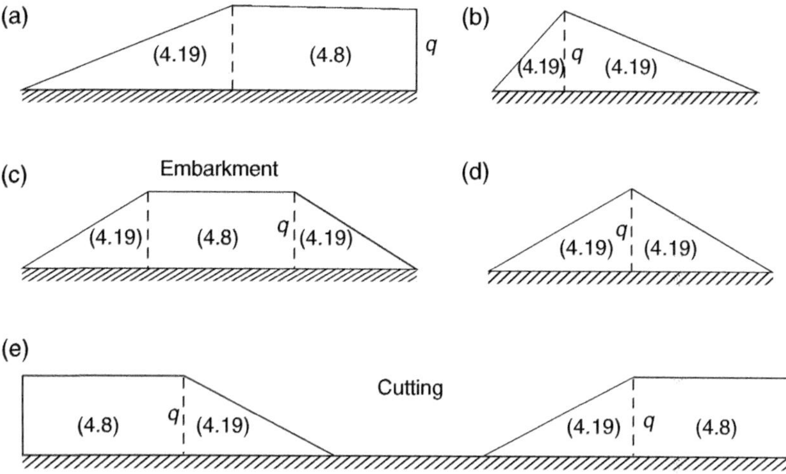

Figure 4.15

Example 4.5

A 4 m wide strip of ground is loaded by a triangle-shaped heap of material. The maximum applied pressure is $q = 80\,\text{kN/m}^2$. Calculate the vertical pressure, 1.5 metres below the surface, at points P_1, P_2 and P_3, located as shown in Figure 4.16.

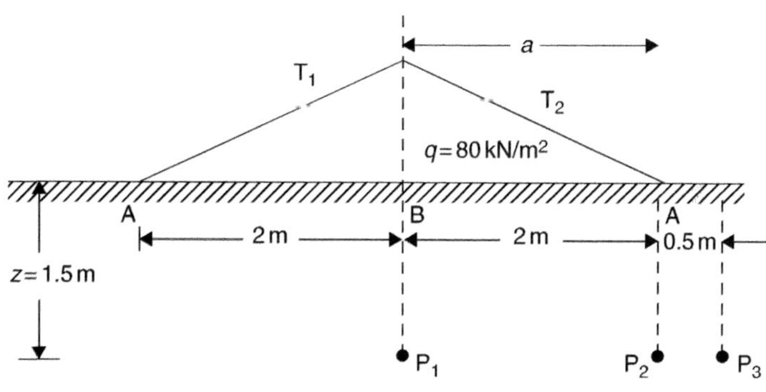

Figure 4.16

Pressures due to triangle T_1

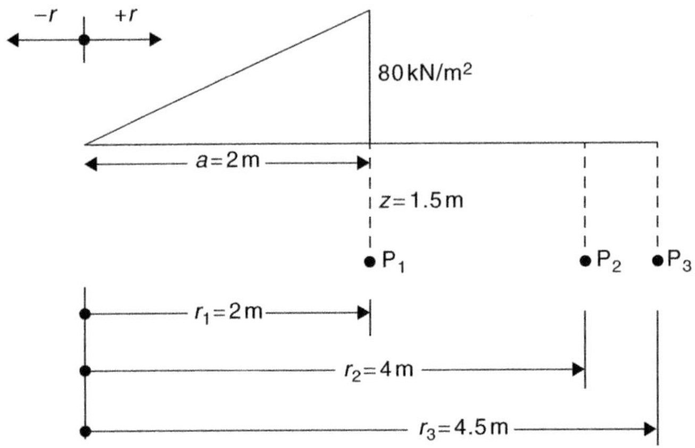

Figure 4.17

At point P_1

$\alpha = 0$

From (4.22): $\beta = \tan^{-1}\left(\frac{a}{z}\right) = \tan^{-1}\left(\frac{2}{1.5}\right)$

$= 53.13^\circ = 0.9273 \text{ radian}$

From (4.23): $I_s = \frac{\beta}{\pi} = \frac{0.9273}{\pi} = 0.295$

From (4.18): $\sigma_V = 0.295 \times 80 = 23.6 \text{ kN/m}^2$

At point P_2 ($r_2 = 4$ m)

From (4.16): $\alpha = \tan^{-1}\left(\frac{r_2 - a}{z}\right) = \tan^{-1}\left(\frac{4-2}{1.5}\right) = 53.13^\circ$

From (4.17): $\beta = \tan^{-1}\left(\frac{r_2}{z}\right) - \alpha = \tan^{-1}\left(\frac{4}{1.5}\right) - 53.13 = 16.31^\circ = 0.2847 \text{ radian}$

From (4.19): $I_s = \frac{1}{2\pi}\left(\frac{2 \times r_2}{a}\beta - \sin 2\alpha\right) = \frac{1}{2\pi}\left[\frac{4}{1} \times 0.2847 - \sin(106.23)\right] = 0.0285$

From (4.18): $\sigma_V = 0.0285 \times 80 = 2.3 \text{ kN/m}^2$

At point P_3 ($r_3 = 4.5$ m)

From (4.16): $\alpha = \tan^{-1}\left(\frac{4.5-2}{1.5}\right) = 59.04^\circ$

From (4.17): $\beta = \tan^{-1}\left(\frac{4.5}{1.5}\right) - 59.04 = 12.52^\circ = 0.2186 \text{ radian}$

From (4.19): $I_s = \frac{1}{2\pi} \times \left[\frac{4.5}{1} \times 0.2186 - \sin(118.08)\right] = 0.016$

From (4.18): $\sigma_V = 0.016 \times 80 = 1.3\,\text{kN/m}^2$

Pressures due to triangle T_2

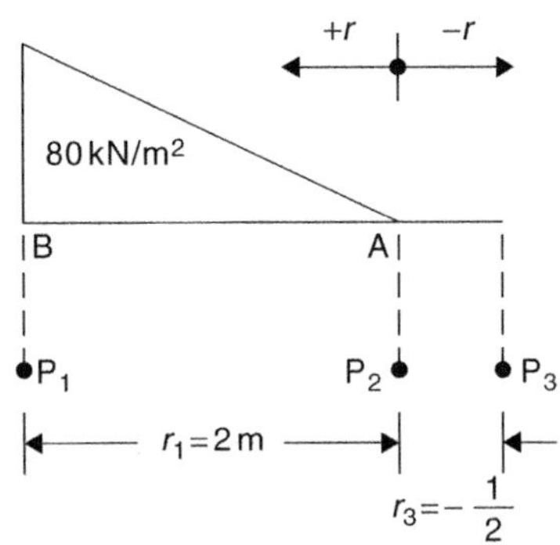

Figure 4.18

$r_1 = 2\,\text{m}$ $r_2 = 0$ $r_3 = -0.5\,\text{m}$	At point P1 $I_s = 0.295$ $\sigma_V = 23.6\,\text{kN/m}^2$ ∴ the same as for triangle T_1

At point P_2 (r_2=0)

From (4.20): $\alpha = \tan^{-1}\left(\frac{-a}{z}\right) = \tan^{-1}\left(\frac{-2}{1.5}\right) = -53.13^\circ$

From (4.21): $I_s = -\frac{\sin 2\alpha}{2\pi} = -\frac{\sin(-106.26)}{2\pi} = 0.153$

From (4.18): $\sigma_V = 0.153 \times 80 = 12.24\,\text{kN/m}^2$

At point P_3 (r_3=−0.5 m)

From (4.20): $\alpha = \tan^{-1}\left(\frac{r_3 - \alpha}{z}\right) = \tan^{-1}\left(\frac{-0.5-2}{1.5}\right) = -59.04^\circ$

From (4.21): $\beta = \tan^{-1}\left(\frac{r_3}{z}\right) - \alpha = \tan^{-1}\left(\frac{-0.5}{1.5}\right) + 59.04 = 40.6^\circ = 0.709\,\text{radian}$

From (4.19): $I_s = \frac{1}{2\pi} \times \left[\frac{-0.5}{1} \times 0.709 - \sin(-118.16)\right] \times = 0.084$

From (4.18): $\sigma_V = 0.084 \times 80 = 6.7\,\text{kN/m}^2$

Table 4.1 Summing

Point	I_5	σ_v kN/m²
P_1	2 x 0.295 = 0.59	2 x 23.6 = 47.20
P_2	0.0285 + 0.153 = 0.18	2.3 + 12.24 = 14.54
P_3	0.016 + 0.084 = 0.1	1.3 + 6.7 = 8.00

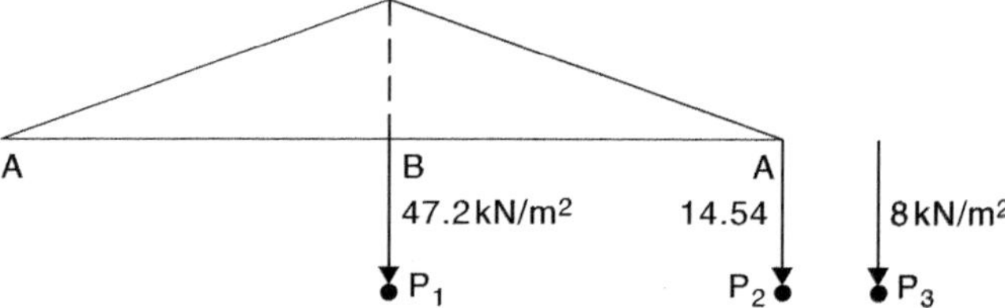

Figure 4.19

These results may be obtained from Chart 4.5.

4.6 Vertical pressure under circular area

Chart 4.6 contains the bulb of pressures, when the circular area (A) is loaded uniformly by either a distributed or a point load. The weight of the footing is assumed to be negligible.

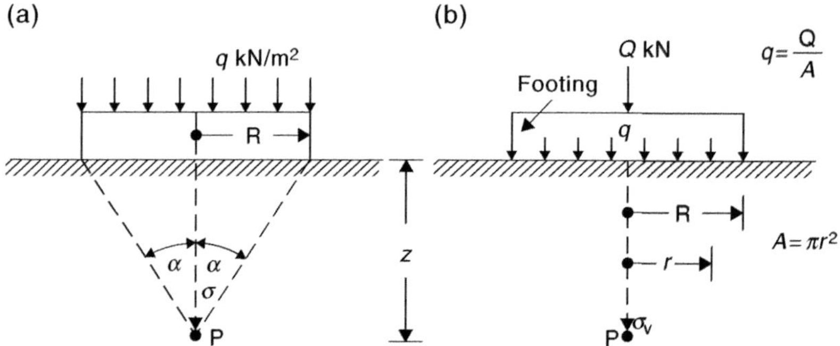

Figure 4.20

The influence coefficient for the vertical pressure under the centre of the area is given by:

Either

$$I_6 = 1 - \left[\frac{1}{1+\left(\frac{R}{z}\right)^2}\right]^{3/2} \tag{4.24}$$

or

$$I_6 = 1 - \cos^3\alpha \tag{4.25}$$

and

$$\sigma_v = I_6 q \tag{4.26}$$

The coefficient I_6 for various values of r, as given on Chart 4.6, has been found by numerical methods.

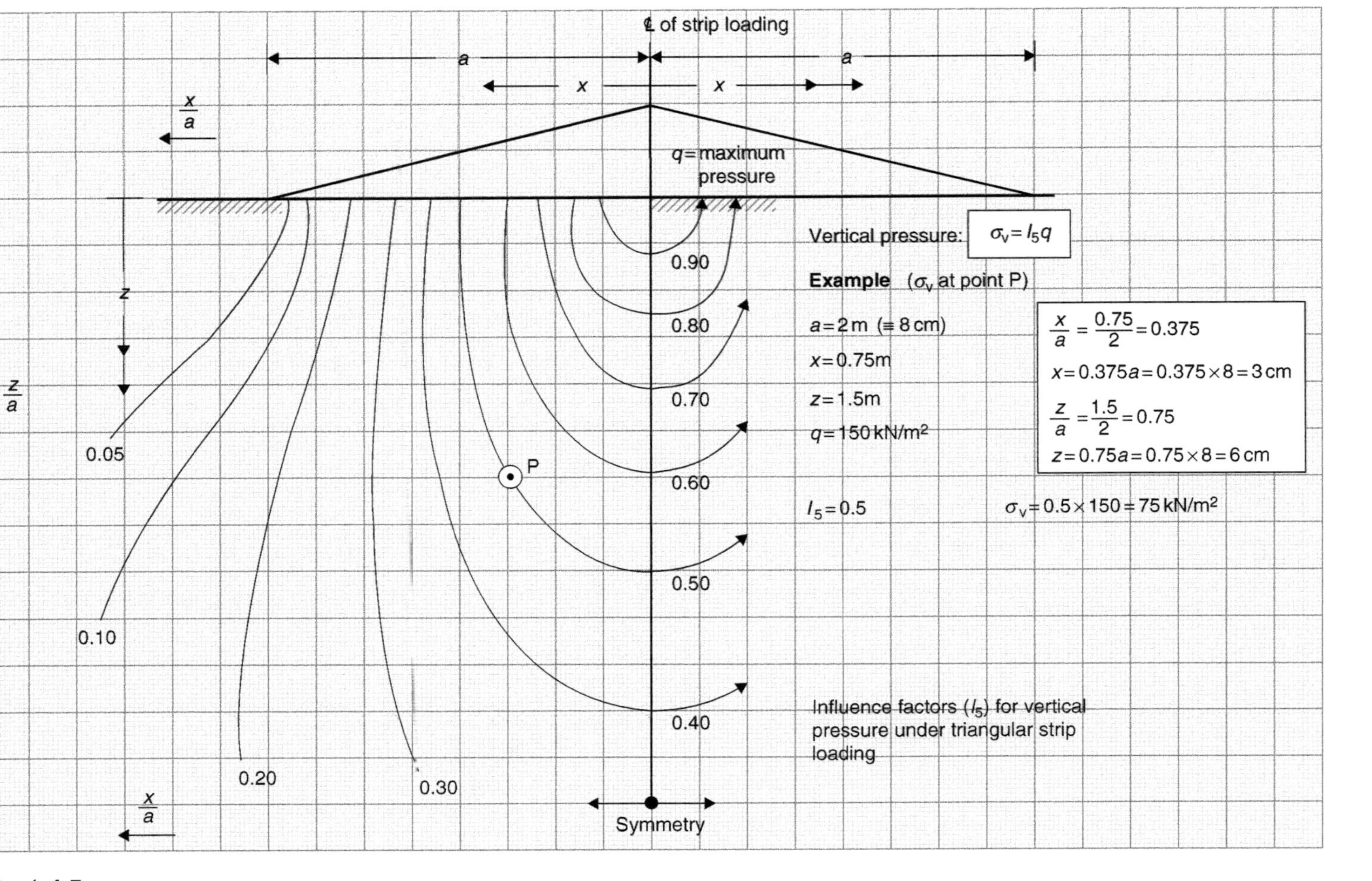

℄ of strip loading
a
a
x
x
x/a
q = maximum pressure
z
z/a
Vertical pressure: σv = I5q
Example (σv at point P)
a = 2 m (≡ 8 cm)
x = 0.75m
z = 1.5m
q = 150 kN/m2
I5 = 0.5
x/a = 0.75/2 = 0.375
x = 0.375a = 0.375 × 8 = 3 cm
z/a = 1.5/2 = 0.75
z = 0.75a = 0.75 × 8 = 6 cm
σv = 0.5 × 150 = 75 kN/m2
P
0.90
0.80
0.70
0.60
0.50
0.40
0.05
0.10
0.20
0.30
x/a
Symmetry
Influence factors (I5) for vertical pressure under triangular strip loading

Chart 4.5

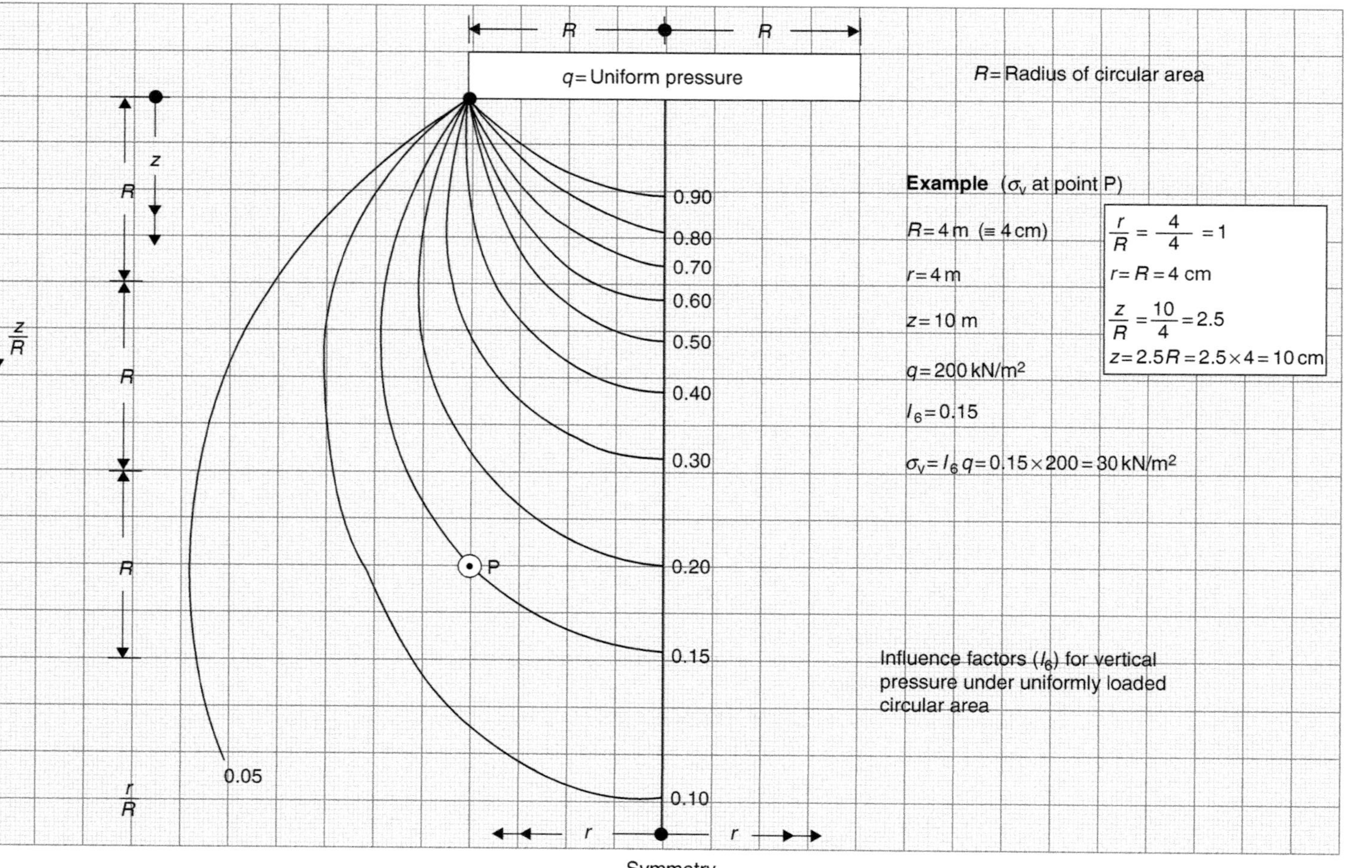

R
R
q = Uniform pressure
R = Radius of circular area
z
R
z/R
R
R
r/R
0.90
0.80
0.70
0.60
0.50
0.40
0.30
0.20
0.15
0.10
0.05
P
r
r
Symmetry
Example (σv at point P)
R = 4 m (≡ 4 cm)
r = 4 m
z = 10 m
q = 200 kN/m2
I6 = 0.15
σv = I6 q = 0.15 × 200 = 30 kN/m2
r/R = 4/4 = 1
r = R = 4 cm
z/R = 10/4 = 2.5
z = 2.5R = 2.5 × 4 = 10 cm
Influence factors (I6) for vertical
pressure under uniformly loaded
circular area

Chart 4.6

4.7 Rectangular footing

Steinbrenner proposed a method in 1934, for the determination of vertical pressure under the corners of the rectangular, uniformly loaded area. The method is based on Boussinesq's theory. Steinbrenner's influence factors (I_7) are given on Chart 4.7. The procedure can also be applied to any point on the rectangle, as shown in following example.

Example 4.6

A 2x6 m rectangular base is subjected to a uniformly distributed load of 400 kN/m², as shown in Figure 4.21. Calculate the vertical pressure (σ_V) 6 m below ground level, under points A, B, C and D.

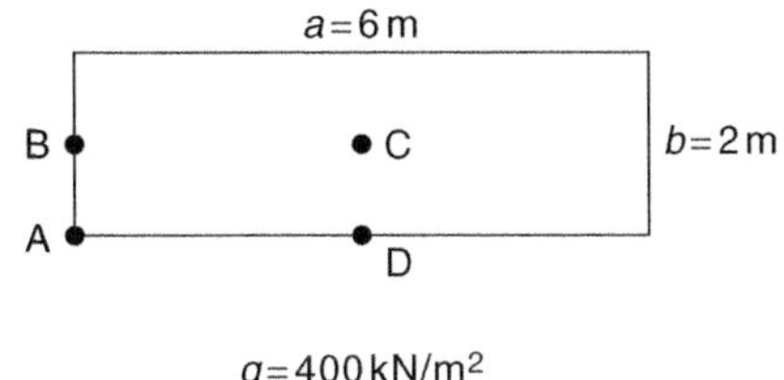

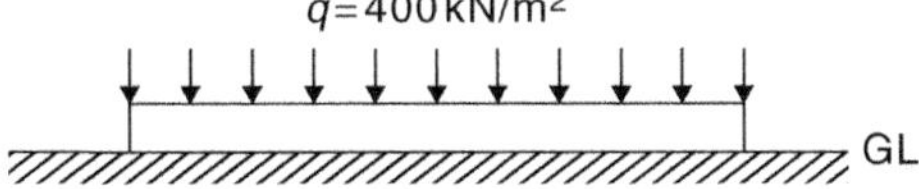

Figure 4.21

Under corner A: $\frac{a}{b}=\frac{6}{2}=3 \quad \frac{z}{b}=\frac{6}{2}=3$

From Chart 4.7: $I_7=0.086 \quad \therefore \quad \sigma_A=0.086\times 400=34.4\,\text{kN/m}^2$

Under point B: Before the method can be used, point B has to be made a corner point.

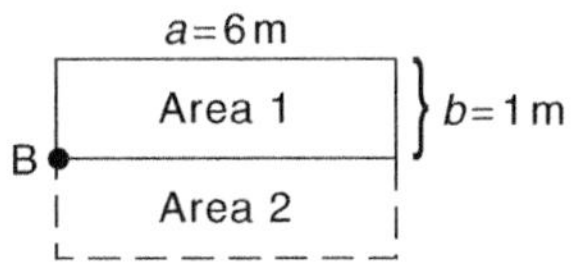

a=6 m

Area 1 } b=1 m

B

Area 2

Figure 4.22

$$\frac{a}{b}=\frac{6}{1}=6 \quad \frac{z}{b}=\frac{6}{1}=6$$

$$I_7=0.045$$

Point B is at the corner of two equal rectangles, each contributing the same pressure to B.

Hence, $\sigma_B = 2\,(I_7 q) = 2 \times 0.045 \times 400 = 36\,\text{kN/m}^2$

Under point D: After making it a corner point

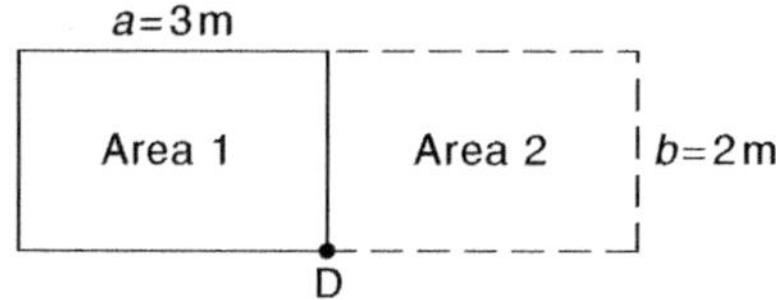

Figure 4.23

$$\frac{a}{b} = \frac{3}{2} = 1.5 \qquad \frac{z}{b} = \frac{6}{2} = 3$$

$$\therefore \quad I_7 = 0.061$$

Again, there are two areas, therefore:

$$\sigma_D = 2 \times (0.061 \times 400) = 48.8\,\text{kN/m}^2$$

Under the middle point C

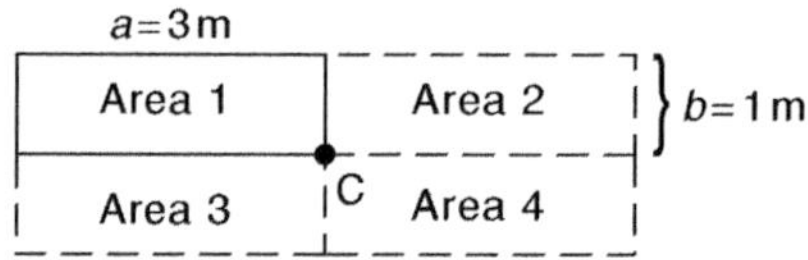

Figure 4.24

$$\frac{a}{b} = \frac{3}{1} = 3 \qquad \frac{z}{b} = \frac{6}{1} = 6$$

$$\therefore \quad I_7 = 0.032$$

There are four equal areas contributing, hence

$$\sigma_C = 4 \times (I_7 q) = 4 \times 0.032 \times 400 = 51.2\,\text{kN/m}^2$$

Fadum published a nomogram in 1948 for the influence factors, to be used with Steinbrenner's method, based on the formula:

$$I_8 = \frac{1}{4\pi}\left[\frac{2\,mn(F+1)\sqrt{F}}{(F+m^2n^2)F}\right] + \tan^{-1}\left(\frac{2mn\sqrt{F}}{F-m^2n^2}\right) \tag{4.27}$$

Where, $n = \frac{a}{z}$, $m = \frac{b}{z}$ and $F = m^2 + n^2 + 1$

The factors can only be evaluated by means of a programmable calculator. The figures are plotted on Chart 4.8 however.

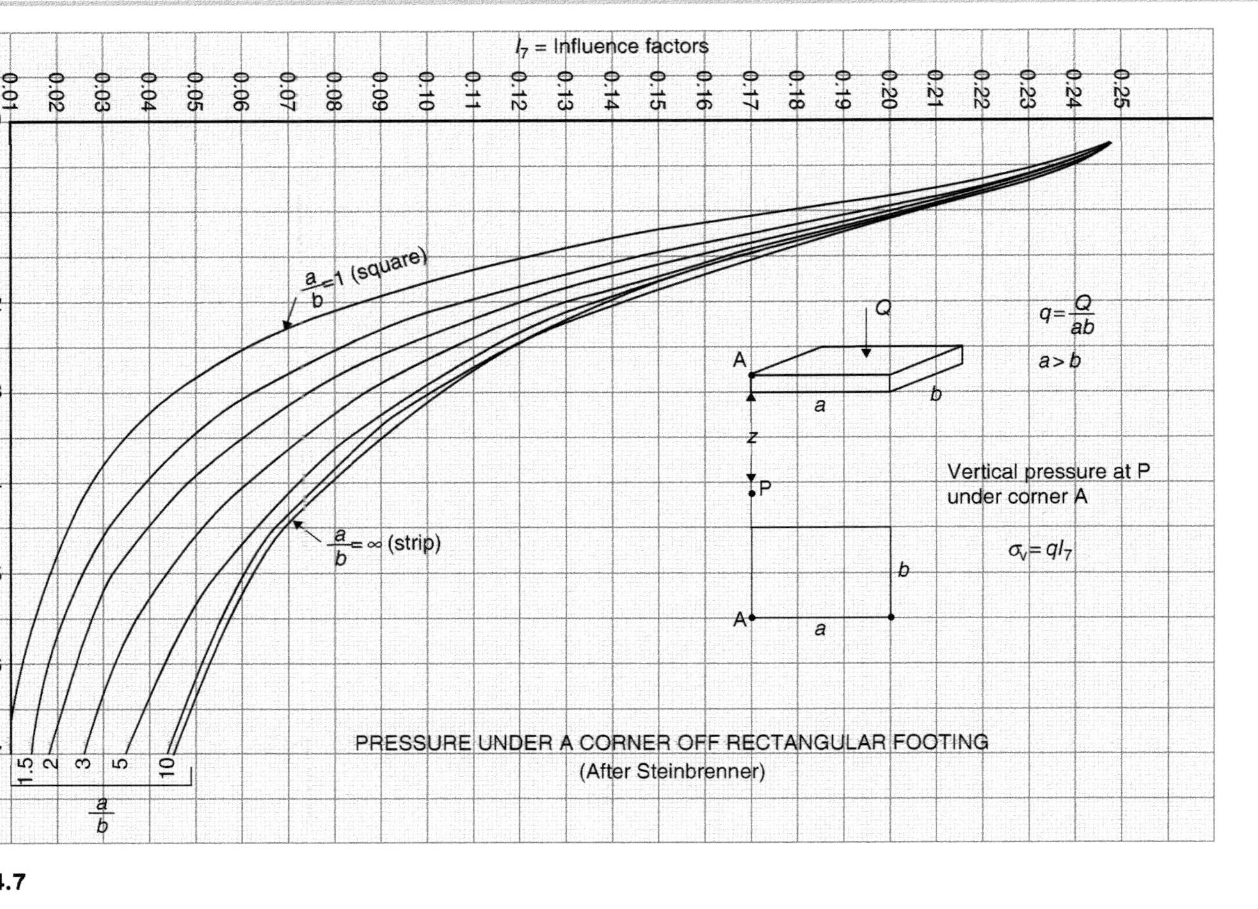

I_7 = Influence factors
0.01
0.02
0.03
0.04
0.05
0.06
0.07
0.08
0.09
0.10
0.11
0.12
0.13
0.14
0.15
0.16
0.17
0.18
0.19
0.20
0.21
0.22
0.23
0.24
0.25
0
1
2
3
4
5
6
7
$\frac{z}{b}$
$\frac{a}{b}$=1 (square)
$\frac{a}{b}=\infty$ (strip)
1.5
2
3
5
10
$\frac{a}{b}$
Q
A
a
b
z
P
$q=\frac{Q}{ab}$
$a > b$
Vertical pressure at P
under corner A
$\sigma_v = qI_7$
PRESSURE UNDER A CORNER OFF RECTANGULAR FOOTING
(After Steinbrenner)

Chart 4.7

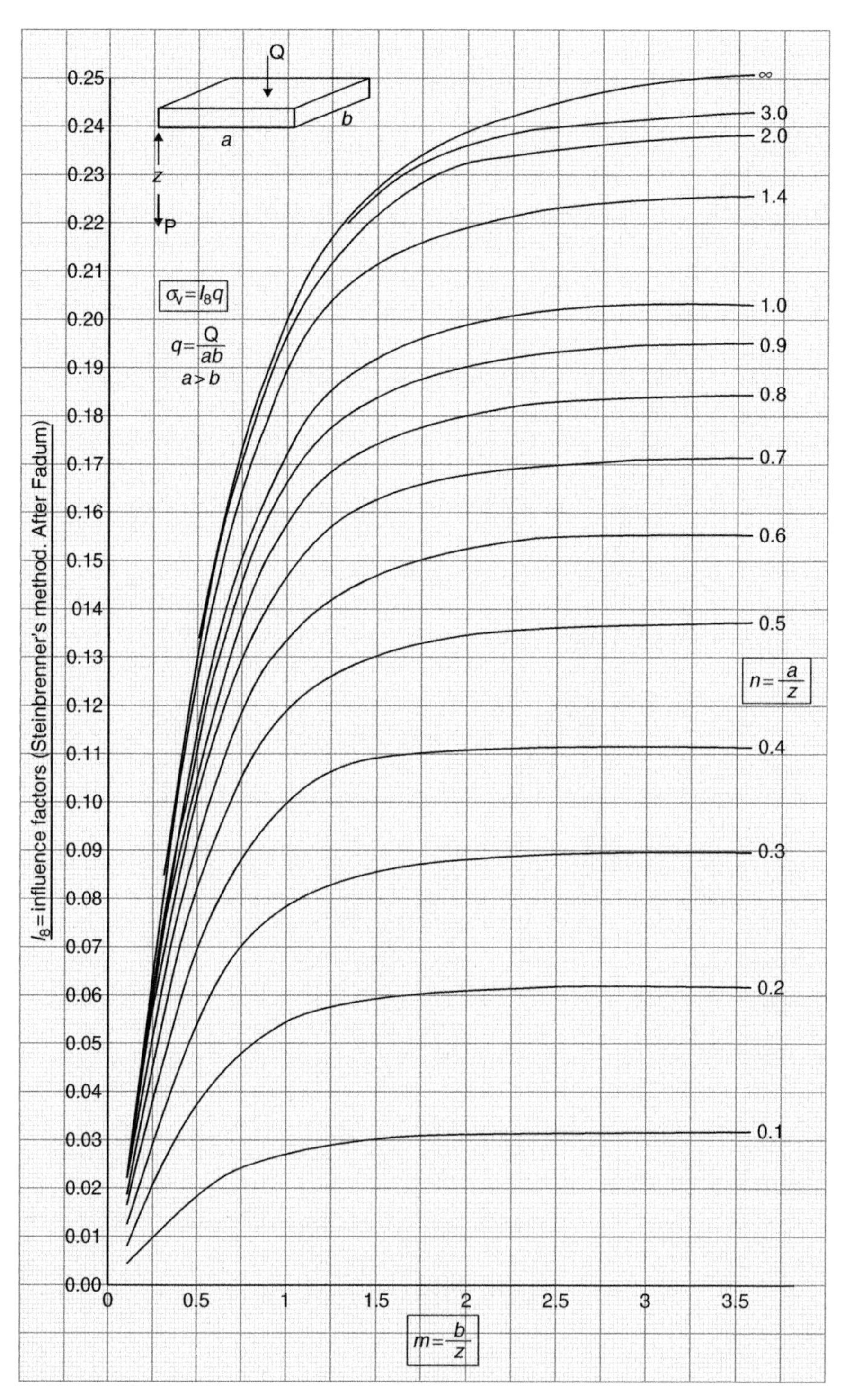
Q
a
b
z
P
$\sigma_v = I_8 q$
$q = \frac{Q}{ab}$
$a > b$
I_8 = influence factors (Steinbrenner's method. After Fadum)
0.25
0.24
0.23
0.22
0.21
0.20
0.19
0.18
0.17
0.16
0.15
014
0.13
0.12
0.11
0.10
0.09
0.08
0.07
0.06
0.05
0.04
0.03
0.02
0.01
0.00
0
0.5
1
1.5
2
2.5
3
3.5
$m = \frac{b}{z}$
$n = \frac{a}{z}$
∞
3.0
2.0
1.4
1.0
0.9
0.8
0.7
0.6
0.5
0.4
0.3
0.2
0.1

Chart 4.8

Example 4.7

Calculate the vertical pressure under point C in Example 4.6, using Forum's factors.

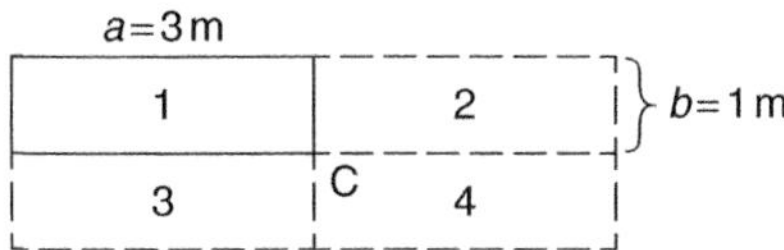

Figure 4.25

$$m=\frac{a}{z}=\frac{3}{6}=0.5 \qquad n=\frac{b}{z}=\frac{1}{6}=0.167$$

From Chart 4.8: $I_8 = 0.031$

$$\therefore\ \sigma_C = 4\times 0.031\times 400 = 49.6\ \text{kN/m}^2$$

4.8 Footings of irregular shape

Newmark introduced a graphical method for the evaluation of vertical pressure, anywhere under a flexible footing of arbitrary shape. It is based on formula (4.26), for vertical pressure induced underneath the centre of uniformly loaded circular area, that is:

$$\sigma_v = q\left[1-\frac{1}{\left(1+\frac{r^2}{z^2}\right)^{3/2}}\right]$$

Newmark constructed Chart 4.9 of concentric circles, after expressing the radius as:

$$r = z_0\sqrt{\left(1-\frac{\sigma_v}{q}\right)^{-2/3}-1}$$

Choosing an arbitrary value of $z_0 = 20$ m, the formula becomes:

$$\boxed{r = 20\sqrt{\left(1-\frac{\sigma_v}{q}\right)^{-2/3}-1}\ \text{m}} \qquad (4.28)$$

The concentric circles may be drawn to any desired scale, in this case to 1:470. Therefore the actual, drawn radii, demoted by ρ, are calculated from:

$$\boxed{\rho = \frac{r\times 10^3}{\eta_0} = \frac{r\times 10^3}{470} = \frac{r}{0.47}\ \text{mm}} \qquad (4.29)$$

where $\eta_0 = 470$ in the scale ratio $1{:}\eta_0$

Any number of concentric circles may be drawn by assigning values to $\frac{\sigma_v}{q}$ in formula (4.28), between zero and unity. Each circle represents a partial magnitude of pressure. The calculation for the radii of $C=10$ circles drawn on Chart 4.9 are tabulated.

Table 4.2

Circle number	0	1	2	3	4	5	6	7	8	9	10
Chosen $\frac{\sigma_v}{q}$	0	0.1	0.2	0.3	0.4	0.5	0.6	0.7	0.8	0.9	1
r from (4.28) (m)	0	5.4	8	10.4	12.7	15.3	18.4	22.2	27.7	38.2	∞
ρ from (4.29) (m)	0	11.5	17	22.0	27.1	32.6	39.0	47.2	59.0	81.2	∞

Note: The 10th circle cannot be drawn, as its radius is infinitely large.

Next, the circles are subdivided into a desired number of uniformly spaced sectors. In this case the number of sectors is $S=20$.

The influence value (I)

There are $C=10$ circles, or more precisely, 1 circle and 8 rings on the chart. Each ring contributes $\frac{\sigma_v}{10}$ to the pressure at the centre. Each sector contributes $\frac{\sigma_v}{20}$ to the pressure. There are 10 elements in each sector, contributing equally to the central pressure. There are 20 x 10 = 200 elements or fields, hence each contributes $\frac{\sigma_v}{200}=0.005\,\sigma_v$.

The influence value of each field is $I=0.005$. For a chart of E elements, it is given by

$$\boxed{I=\frac{1}{E}} \qquad \text{where } E=CS \tag{4.30}$$

Depth length $\overline{MN}$

This line is the scaled distance of the arbitrary depth ($z_o=20$ m), chosen in the design of the chart.

On this chart therefore, $\overline{MN}=\frac{z_0\times10^3}{\eta_0}=\frac{20\times10^3}{470}=42.5\,\text{mm}$

Application of the chart

Step 1: Equate the depth (z) at which the vertical pressure is required, to the depth length $\overline{MN}$. This determines the scale of the drawing.

Step 2: Draw the plan of the loaded area to this scale.

Step 3: Place the plan on the Newmark chart so that the point below which the pressure is to be determined is over the centre of the circles.

Step 4: Count the number (n) of elements covered by the loaded area, including portions of the partly-covered fields.

Step 5: Calculate the total pressure at z from:

$$\sigma_v=Inq=0.005nq \tag{4.31}$$

Note: Whilst the procedure is simple, there can be inaccuracies introduced in the estimation of the areas on the partially covered elements.

Example 4.8

Figure 4.26 shows the plan of an area loaded uniformly by $q = 350\,\text{kN/m}^2$. Estimate the pressure at 10 m below point A and check the result by Steinbrenner's method.

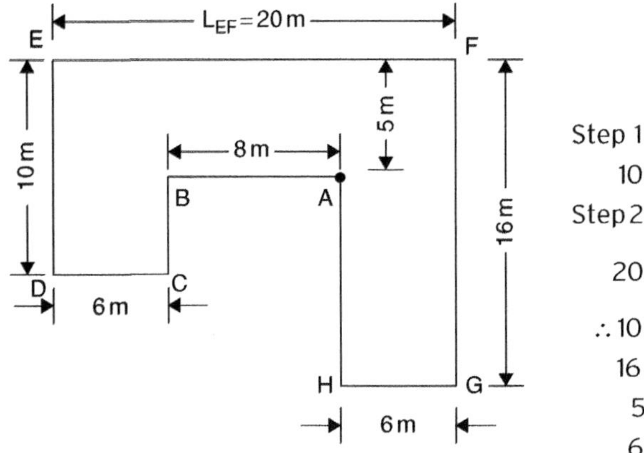

Figure 4.26

Step 1: $z = \overline{MN}$

$10\,\text{m} = 42.5\,\text{mm}$

Step 2:

$$20\,\text{m} = \frac{42.5}{10} \times 20 = 85\,\text{mm}$$

$$\therefore 10\,\text{m} = 42.5\,\text{mm}$$

$$16\,\text{m} = 4.25 \times 16 = 68\,\text{mm}$$

$$5\,\text{m} = 4.25 \times 5 = 21.25\,\text{mm}$$

$$6\,\text{m} = 4.25 \times 6 = 25.5\,\text{mm}$$

The loaded area can now be drawn on tracing paper.

Step 3: Position the plan on Chart 4.9 so that corner A is over the centre of the circles.

Step 4: The number of elements covered is approximately $n = 79.1$.

Step 5: The vertical pressure at $z = 10$ m is:

$$\sigma_v = 0.005 \times 79.1 \times 350 = 138.4\,\text{kN/m}^2$$

Checking: Using Fadum's influence factors (Chart 4.8):

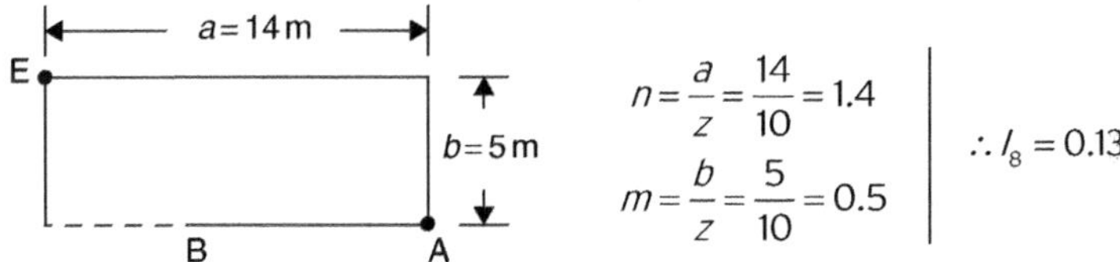

$$n = \frac{a}{z} = \frac{14}{10} = 1.4 \qquad m = \frac{b}{z} = \frac{5}{10} = 0.5 \qquad \therefore I_8 = 0.13$$

Figure 4.27a

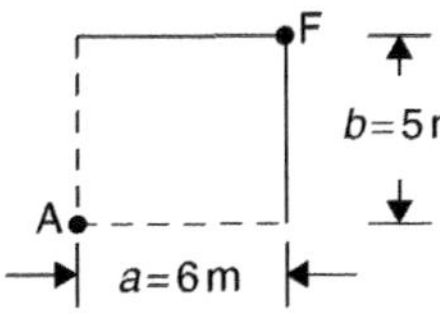

F
b = 5 m
A
a = 6 m

$$n = \frac{a}{z} = \frac{6}{10} = 0.6 \qquad m = \frac{b}{z} = \frac{5}{10} = 0.5 \qquad \therefore I_8 = 0.094$$

Figure 4.27b

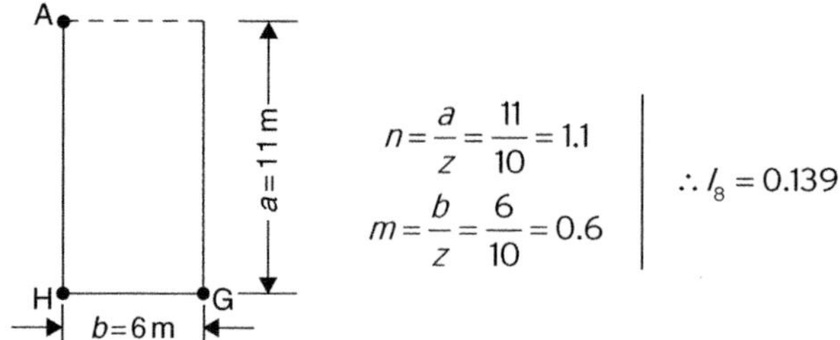

Figure 4.27c

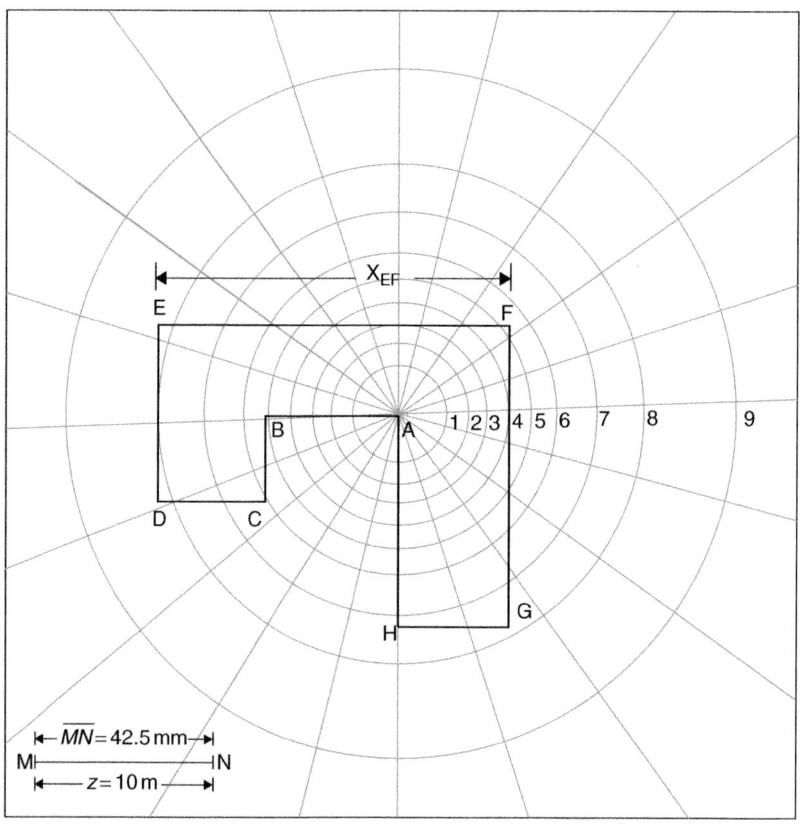

Influence value per field = 0.005

The chart was drawn to scale:- 1:470

The dimensions X of a structure are drawn to scale:-

$1{:}\eta$ or $1{:}\frac{z \times 10^3}{\overline{MN}}$ or $\frac{10 \times 10^3}{42.5}$ or 1:235

e.g. for $L_{EF} = 20$ m $\therefore X_{EF} = \frac{L_{EF} \times 10^3}{\eta} = \frac{20 \times 10^3}{235} = 85$ mm

or $X_{EF} = \frac{\overline{MN}}{z} L_{EF} = \frac{42.5}{10} \times 20 = 85$ mm

Chart 4.9

(a)

$a = 14\,\text{m}$ B A Area added $b = 5\,\text{m}$ D C

$$n = \frac{a}{z} = \frac{14}{10} = 1.4 \qquad m = \frac{b}{z} = \frac{5}{10} = 0.5 \qquad \therefore I_8 = 0.13$$

(b)

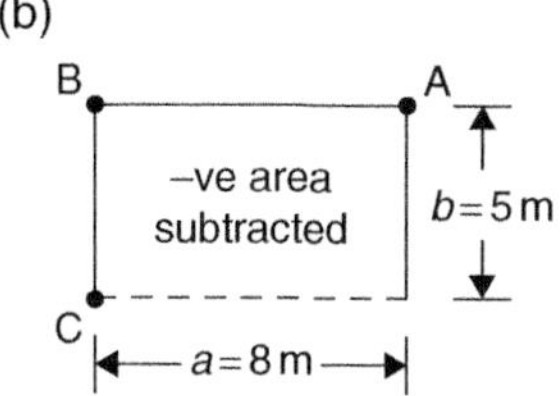

$$n = \frac{8}{10} = 0.8 \qquad m = \frac{5}{10} = 0.5 \qquad \therefore I_8 = -0.11$$

Figure 4.28

Total $I_8 = 2 \times 0.13 + 0.094 + 0.139 - 0.11 = 0.383$
The vertical pressure: $\sigma_A = 0.383 \times 350 = 134.1\,\text{kN/m}^2$
The two results are therefore comparable. See also Supplementary problem 4.3.

4.9 Pressure distribution under footings

It has been assumed in the foregoing that a footing transmits loading uniformly over the foundation area. In reality however, the distribution of contact pressure is influenced by several factors, associated with the characteristics of the following factors:

1. Footing
2. Soil
3. Loading

4.9.1 Influence of footing

It has been found from experiments and observation of actual structures that the contact pressure is influenced by:

a) The rigidity of the footing;
b) Its shape;
c) Its size;
d) Its depth below the surface;
e) The rigidity of the structure it supports.

a) Rigidity of footing
Footings can be either rigid or flexible. The contact pressure between these and the soil is different for each type. Further, the shape of the pressure distribution depends on whether the soil is cohesive or granular.

The significance of contact pressure distribution is in its effect on the immediate settlement just after the application of loading ($t = 0$), as well as on the design of the footing itself.

Rigid footing (Figure 4.29)
The settlement of a uniformly loaded rigid footing is uniform, whilst the distribution of contact pressure is not. The general reasoning is as follows:

a) Should the footing be flexible, than the settlement under its middle would be larger than at the edges.
b) This cannot occur as the footing is rigid.
c) Because of this, the settlement is uniform.
d) To achieve this, the pressure at the edges has to increase with a corresponding decrease at the middle.

Footing on cohesive soil:
In order that the pressure at the edges may increase, the soil has to have shear strength. This is necessary to prevent its outflow due to shear failure.

Footing on cohesionless soil:
As granular soils can 'flow' sideways at the edges, the contact pressure is zero there.

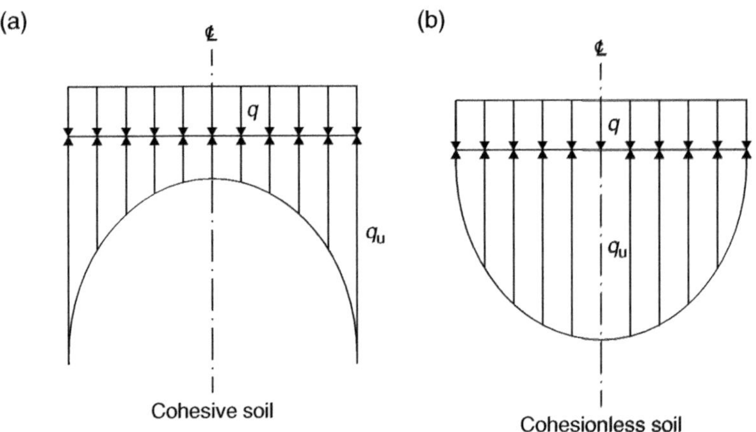

Figure 4.29

Flexible footing (Figure 4.30)
Under a rigid footing, the soil particles move downwards except possibly at the edges. Under a uniformly loaded flexible footing however, particles near the centre move downwards, whilst the rest move outwards. It follows, therefore, that there is maximum contact pressure under the centre and zero at the edges.

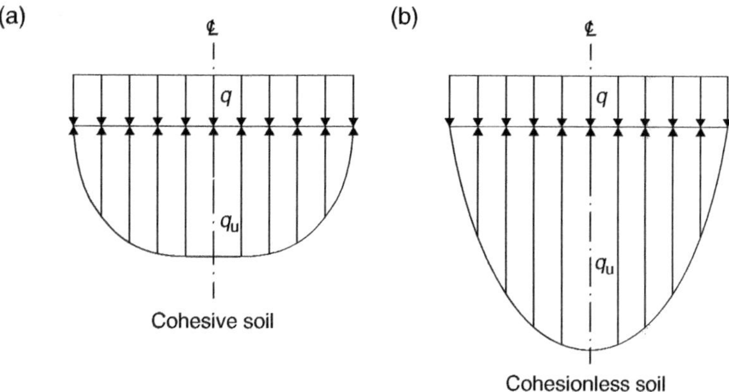

Figure 4.30

The contact pressure under very flexible footing is the mirror image of the external loading.

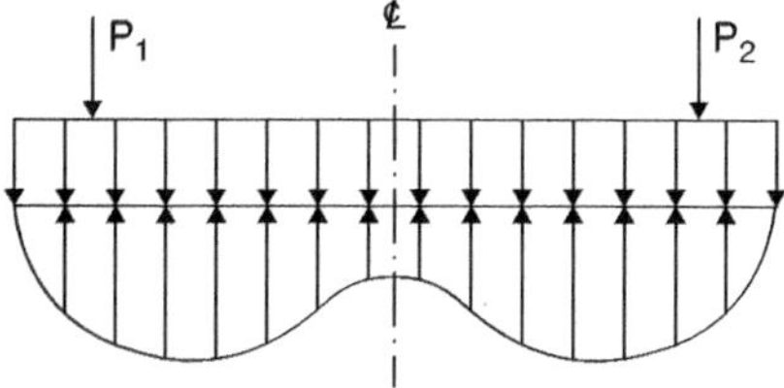

Figure 4.31

Note: In reality, there is no completely rigid or flexible footing. It is therefore, acceptable to assume uniform pressure distribution.

b) Shape of footing

The more closed is the plan area of the footing, the less uniform is the contact pressure. The pressure tends to increase, under the centre. For this reason, circular footings are more efficient in carrying load than strip ones.

c) Size of footing

The redistribution of contact pressure reduces with increasing width, as the disturbing effect at the edges are minimized.

d) Depth of footing

The large overburden above deep foundations prevents the outward movement of soil at the edges, hence the contact pressure becomes more uniform.

e) Rigidity of superstructure

Whilst flexible structure deflects with the consolidation of the soil, a rigid one does resist deformation. This increases the loading at the edges and releases it at the middle.

4.9.2 Influence of loading

Eccentric loading increases contact pressures at one edge of the rigid footing and decreases it under the other one (see section 8.13.1: Gravity Walls). In general, the positioning of loads affects the distribution of pressure mainly under flexible footings.

4.10 Linear dispersion of pressure

The methods described below are sometimes used to determine the pressure induces at a depth, by projecting the loaded areas onto the layer below. This has the effect of enlarging the base, thus decreasing the pressure. Three types of dispersion are introduced.

30° dispersion

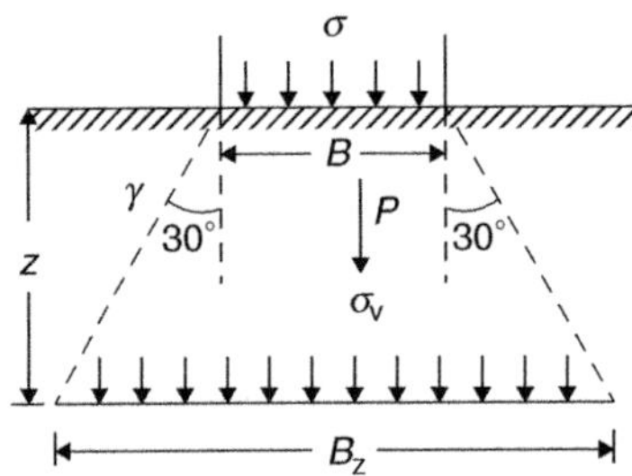

Figure 4.32

B = length of base of surface
B_z = length of enlarged base at z
P = force acting on the bases
σ = surface pressure
σ_z = induced pressure at z
σ_0 = overburden pressure = $z\gamma$
σ_v = total pressure at z

$$B_z = B + 2z\tan 30 = B + 1.15z$$
$$P = B\sigma = B_z\sigma_z$$
$$\therefore \sigma_z = \left(\frac{B}{B_z}\right)\sigma$$

Total pressure: $\sigma_v = \sigma_0 + \sigma_z$ or $\boxed{\sigma_v = z\gamma + \left(\dfrac{B}{B + 1.15z}\right)\sigma}$ (4.32)

45° dispersion

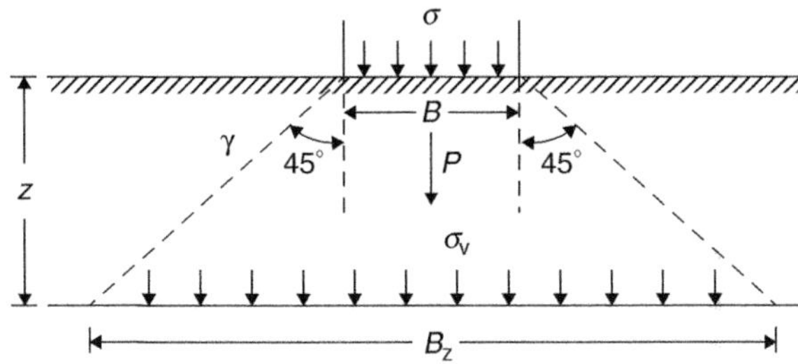

Figure 4.33

$$B_z = B + 2z\tan 45 = B + 2z$$

$$\sigma_z = \left(\frac{B}{B+2z}\right)\sigma$$

$$\boxed{\sigma_v = \sigma_0 + \left(\frac{B}{B+2z}\right)\sigma} \tag{4.33}$$

30°/45° dispersion

This is a combination of the above two transformations.

30° dispersion

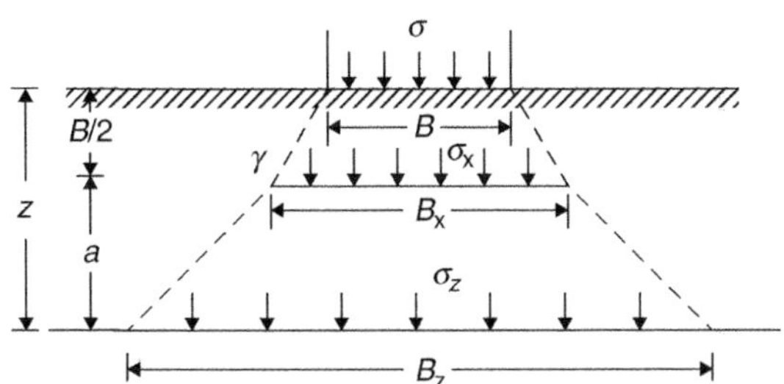

Figure 4.34

$$B_x = B + 1.15\left(\frac{B}{2}\right) = 1.575B$$

$$\sigma_x = \left(\frac{B}{B_x}\right)\sigma$$

$$= \frac{B\sigma}{1.575B} = 0.635\sigma$$

45° dispersion

$$B_z = B_x + 2a$$

But $a = z - \frac{B}{2}$

$$\therefore B_z = 1.575\text{B} + 2\times\left(z - \frac{B}{2}\right) = 0.575B + 2z$$

$$\sigma_z = \frac{B_x}{B_x + 2a}\sigma_x = \frac{B_x}{B_x + 2a}\left(\frac{B}{B_x}\right)\sigma = \frac{B}{B_x + 2a}\sigma$$

$$= \frac{B\sigma}{1.575B + 2a} = \frac{B\sigma}{1.575B + 2\times\left(z - \frac{B}{2}\right)}$$

Therefore, total pressure:
$$\boxed{\sigma_v = \sigma_0 + \left(\frac{B\sigma}{0.575B + 2z}\right)} \tag{4.34}$$

Example 4.9

A 2 m wide strip footing transmits 250 kN/m^2 to the soil of γ=18 kN/m. Calculate the total vertical pressure at 3.4 m depth, by the three methods.

Solution

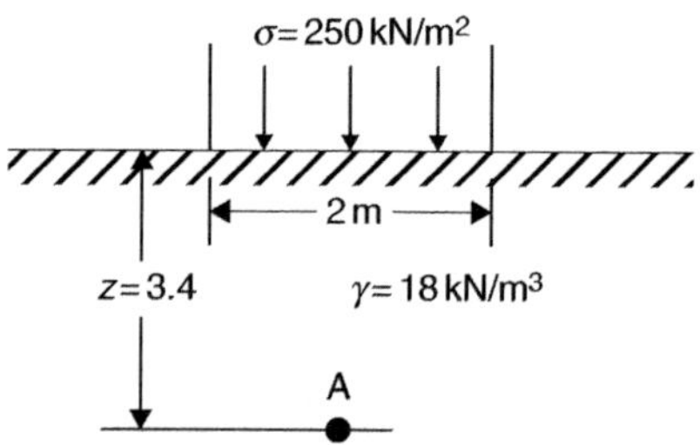

Figure 4.35

Overburden pressure: $\sigma_0 = 3.4 \times 18 = 61.2\ \text{kN/m}^2$

30°: $\sigma_v = 61.2 + \frac{2\times 250}{2 + 1.15\times 3.4} = 61.2 + 84.6 = 146\ \text{kN/m}^2$

45°: $\sigma_v = 61.2 + \frac{2\times 250}{2 + 2\times 3.4} = 61.2 + 56.8 = 118\ \text{kN/m}^2$

30°/45°: $\sigma_v = 61.2 + \frac{2\times 250}{0.575\times 2 + 2\times 3.4} = 61.2 + 62.8 = 124\ \text{kN/m}^2$

Note: These methods have no theoretical basis, hence the results should be considered indicative only. For footings founded below the surface: use net pressure (σ_n) instead of σ.

Problem 4.1

Figure 4.36 shows a section of ground and a shallow strip footing, seated at 1 m depth below ground level. Calculate the width of footing, which has to transmit a net pressure of 200 kN/m² (including self-weight) to the ground, without exceeding the bearing capacity of the clay.

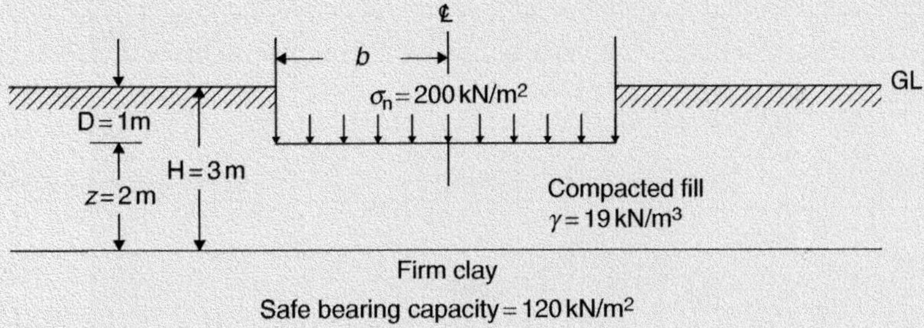

Figure 4.36

Problem 4.2

A long, reinforced concrete slab, 5 m wide, carrying 500 kN/m² uniformly distributed load, is placed 2 m away from an existing pile foundation, as shown in Figure 4.38. Estimate the average vertical and horizontal pressures acting on the surface of the pile embedded in the firm clay.

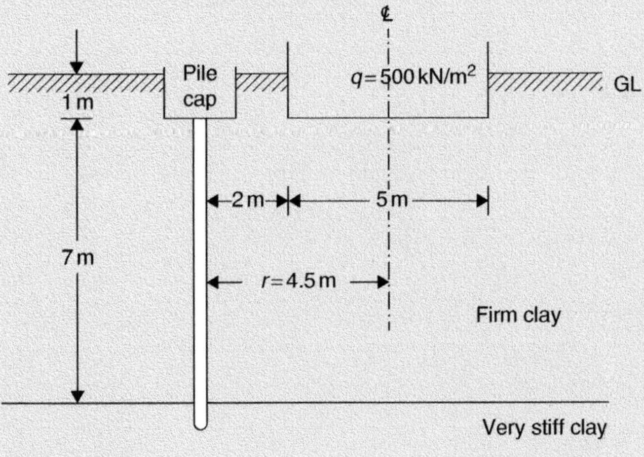

Figure 4.38

Problem 4.3

Construct a Newmark chart to the following specifications:

Number of circles = 8
Number of sectors = 12
Arbitrary depth = 15 m

Using this chart, estimate the vertical pressure 10 m below corner A of the plan shown.

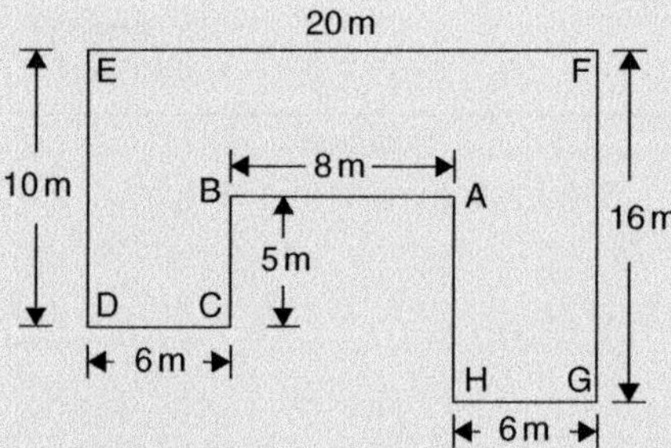

Figure 4.40

The plan area carries a uniformly distributed load of 350 kN/m². Compare the answer with that of Example 4.8 of the main text.

Problem 4.4

Applying formulae (4.7) to (4.10), prove that the horizontal and shear stresses a depth *z* below a uniform strip load can be expressed alternatively by:

a) $\sigma_H = \frac{1}{\pi}\left[\beta - \sin\beta\cos(2\alpha+\beta)\right]q$

b) $\tau = \frac{1}{\pi}\left[\sin\beta\cos(2\alpha+\beta)\right]q$

Chapter 5

Effective Pressure (σ')

An important problem in soil mechanics is the determination of pressures induced within fully saturated soil by either of two causes:

1. Change in the superimposed load
2. Variation of ground water conditions, e.g. when seepage occurs.

As a consequence of superimposed loading, the soil mass deforms, that is, consolidates by expelling some of the pore water, thus allowing the solid particles to pack together into a denser mass. There are three types of pressure acting on the soil at depth.

a) **Total pressure (σ)** of the superimposed load, such as foundation pressure, plus the weight of the overburden
b) **Pore water pressure (u)** in the voids, induced either by the weight of water or external load or both
c) **Effective pressure (σ')** between the soil grains. This is the actual cause of deformation, hence its name.

The relationship between the three pressures was introduced by Terzaghi as:

$$\boxed{\sigma' = \sigma - u} \tag{5.1}$$

The concept of effective pressure, also known as 'intergranular pressure' is best demonstrated by comparing the pressures within a small volume of soil, considered to be in one of the following three states:

1. Unloaded state
2. Loaded state
3. Flooded state.

5.1 Unloaded state

In Figure 5.1(a), the ground water level (GWL), hence the piezometric level, are assumed to coincide with the ground surface (GL) in the unloaded state. Therefore the water pressure (u) in the voids equals to the static water pressure, given by:

$$\boxed{u = z\gamma_w} \uparrow \ominus \tag{5.2}$$

Introduction to Soil Mechanics, First Edition. Béla Bodó and Colin Jones.

As the only load on the sample at depth z is the overburden of saturated density (γ_{sat}), the total pressure at sample level is given by:

$$\boxed{\sigma = z\gamma_{sat}} \downarrow \oplus \tag{5.3}$$

The effective pressure from (5.1):

$$\sigma' = \sigma - u = z\gamma_{sat} - z\gamma_w = z(\gamma_{sat} - \gamma_w)$$

But, submerged density is: $\gamma' = \gamma_{sat} - \gamma_w$

So, with GWL at GL:

$$\boxed{\sigma' = z\gamma'} \downarrow \oplus \tag{5.4}$$

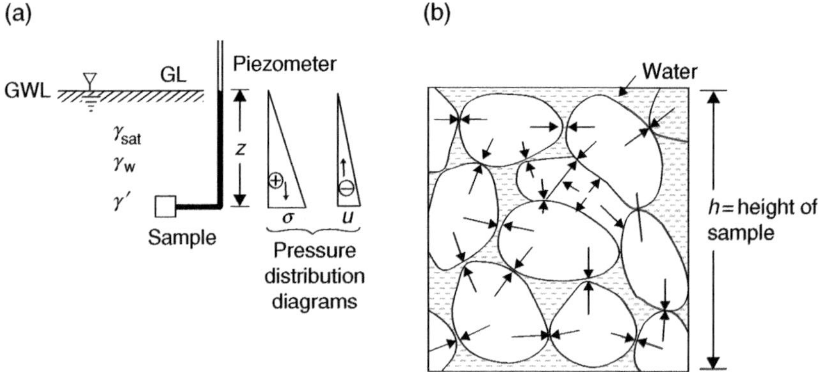

Figure 5.1

Notes:

1. The pressures can also be represented graphically by diagram as shown.
2. Both GWL and the piezometric level for the sample are under atmospheric pressure and coincide. This means that the water in the voids is in static equilibrium, hence there is no flow out of the sample.
3. Flow can only occur when the total pressure is increased by an external load, which in turn increases the pore pressure. This "excess" pore pressure induces seepage.
4. The GWL is normally below the surface, at some depth.

Example 5.1

Obtain the effective pressure at 3 m below the surface of saturated soil (γ_{sat}=19.69 kN/m³). The ground water level is at the surface, in this example.

Total pressure at depth z=3 m: $\sigma = z\gamma_{sat} = 3\times 19.69 = 59.07$ kN/m²

Pore pressure at 3 m: $u = z\gamma_w = 3\times 9.81 = 29.43$ kN/m²

Effective pressure therefore: $\sigma' = \sigma - u = 29.64$ kN/m²

Alternatively, $\sigma' = z\gamma' = z(\gamma_{sat} - \gamma_w) = 3\times(19.69 - 9.81) = 29.64$ kN/m²

Graphical representation of linear variation of pressure with depth:

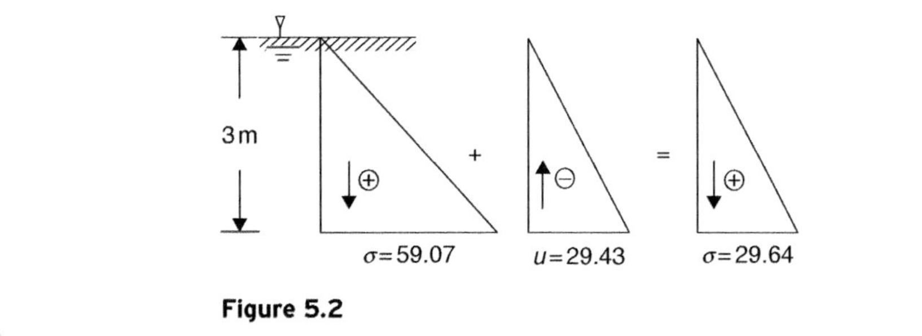

Figure 5.2

5.2 Loaded state

(See Figure 5.3). If the soil in its natural state is loaded by the construction of a rock embankment or a structure, for example, then initially the applied pressure (q) is carried by the pore water only and not the soil particles. The additional water pressure induced is called "Excess pore pressure", denoted by $\Delta u=q$. The change in u occurs immediately after the application of q at time $t=0$.

Figure 5.3(a) depicts the change in σ, u and σ' due to q, at the time of its application ($t=0$).

Total pressure: $\sigma = z\gamma_{sat} + q \quad \therefore \quad$ change in σ: $\Delta\sigma = q$

Pore pressure: $u = z\gamma_w + q \quad \therefore \quad$ change in u: $\Delta u = q$

The resulting change in the effective pressure is given by formula (5.1):

$$\Delta\sigma' = \Delta\sigma - \Delta u = q - q = 0$$

Therefore at time $t=0$ $\quad \begin{cases} \Delta u = \Delta\sigma = q \\ \Delta\sigma' = 0 \end{cases}$

The excess pore pressure induces flow of water from the voids, the rate of which depends on the permeability of soil. As a consequence of the outflow, the magnitude of Δu reduces progressively (Figure 5.3(b)), until it becomes zero. At that moment the flow stops. The considerable time taken to reach static equilibrium is usually indicated by $t=\infty$.

As Δu dissipates, load q is progressively transferred to the soil particles, that is the excess effective pressure becomes $\Delta\sigma'=q$ [Figure 5.3(c)].

Therefore at time $t=\infty$ $\quad \begin{cases} \Delta u = 0 \\ \Delta\sigma' = \Delta\sigma - \Delta u = q - 0 = q \end{cases}$

The excess effective pressure ($\Delta\sigma'$) reorientates the soil particles, pressing them into the voids left by the dissipated water, thus compressing the soil. The piezometric level coincides again with the initial water level. This process is called consolidation, caused by the excess effective intergranular pressure.

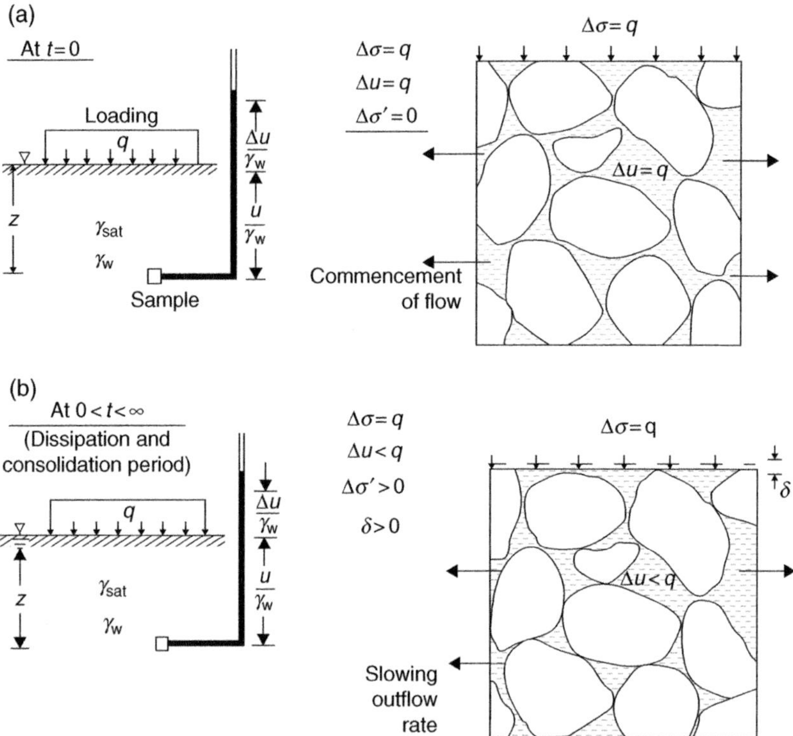

By now, most of the *excess* pore pressure has dissipated ($\Delta u < q$) with corresponding increase in $\Delta\sigma'$. The result is a deformation δ, which is a change in layer thickness due to consolidation.

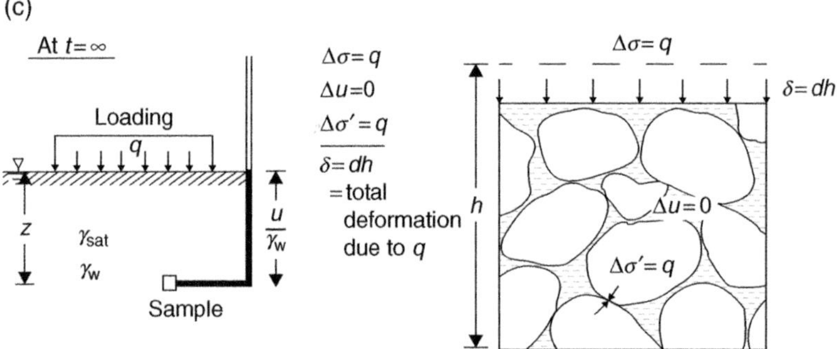

Figure 5.3

The pore water is now in static equilibrium, hence there is no outflow from the soil. The total deformation of the specimen is signified by *dh*.

Example 5.2

Calculate the pressures as well as any change in their magnitude, if the saturated soil in Example 5.1 is covered by hydraulic fill, weighing 20 kN/m². Obtain the values at point P, 3 metres below the original ground surface, at $t=0$ and $t=\infty$.

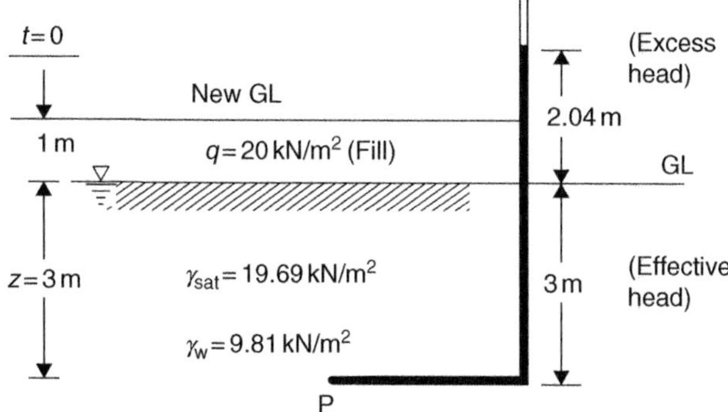

Figure 5.4

Increase in total pressure: $\Delta\sigma = q = 20\,\text{kN/m}^2$

$$\therefore \quad \sigma = z\gamma_{sat} + q = 3\times 19.69 + 20 = 79.07\,\text{kN/m}^2$$

Excess pore pressure: $\Delta u = \Delta\sigma = 20\,\text{kN/m}^2$

$$\therefore \quad u = z\gamma_w + \Delta u = 3\times 9.81 + 20 = 49.43\,\text{kN/m}^2$$

Equivalent piezometric pressure heads therefore are:

$$\text{Total head} = \frac{u}{\gamma_w} = \frac{49.43}{9.81} = 5.04\,\text{m}$$

$$\text{Excess head} = \frac{\Delta u}{\gamma_w} = \frac{20}{9.81} = 2.04\,\text{m}$$

$$\text{Effective head} = \frac{\sigma - u}{\gamma_w} = \frac{79.07 - 49.43}{9.81} = 3\,\text{m}$$

Excess effective pressure: $\Delta\sigma' = \Delta\sigma - \Delta u = 20 - 20 = 0$

Intergranular pressure at $t=0$: $\sigma' = \sigma - u = 79.07 - 49.43 = 29.64\,\text{kN/m}^2$

This is, of course, the same value as in Example 5.1, assuming that the fill is placed 'instantaneously'.

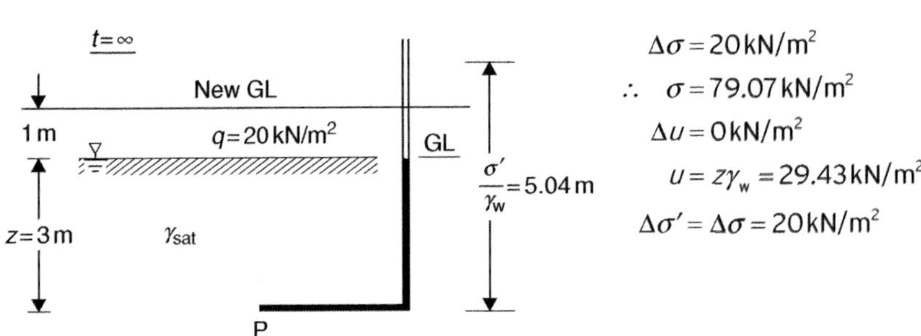

Figure 5.5

The final intergranular pressure: $\sigma' = 79.07 - 29.43 = 49.64\,\text{kN/m}^2$ which is equivalent to an effective head of 5.04 m.

5.3 Flooded state

If instead of the applied pressure q, the surface is flooded to such a depth that the hydrostatic pressure at ground level equals to q.

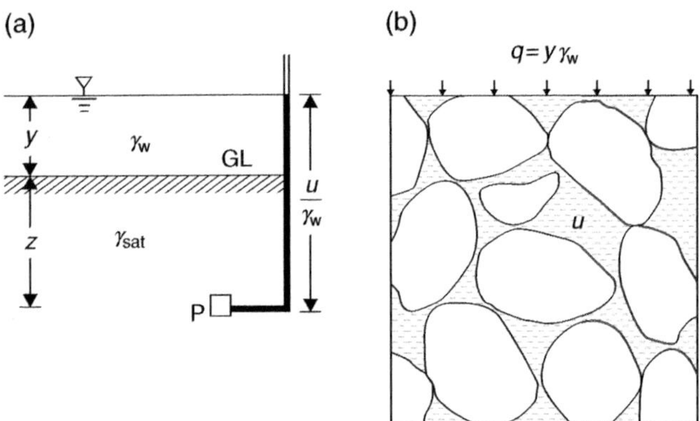

Figure 5.6

Total pressure at P: $\sigma = y\gamma_w + z\gamma_{sat}$

Pore pressure: $u = (y+z)\gamma_w = y\gamma_w + z\gamma_w$

Effective pressure: $\sigma' = \sigma - u = y\gamma_w + z\gamma_{sat} - y\gamma_w - z\gamma_w$

Cancelling $y\gamma_w$: $\sigma' = z(\gamma_{sat} - \gamma_w) = z\gamma'$

But, $\sigma' = z\gamma'$ is the intergranular pressure in the unloaded state (Figure 5.1). This implies, in this case, that water level above GL does not increase the intergranular pressure, hence has no consolidating effect.

The pore water pressure is increased, however, by the additional hydrostatic pressure $y\gamma_w$. Because u has no effect on consolidation, it is often called 'neutral pressure'.

Example 5.3

Calculate the pressures at 3 m below ground surface level, if the saturated soil in Example 5.1 is flooded to a depth of 2.04 m, as shown in Figure 5.7.

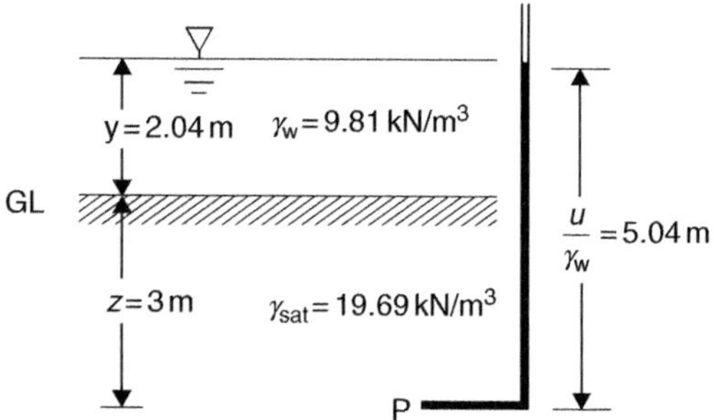

Figure 5.7

At P: $\sigma = 2.04 \times 9.81 + 3 \times 19.69$

$= 79.08\ \text{kN/m}^2$

$u = 5.04 \times 9.81 = 49.44\ \text{kN/m}^2$

$\therefore \quad \sigma' = 79.08 - 49.44 = 29.64\ \text{kN/m}^2$

Comparing this result with the value of the effective pressure in Example 5.1, it is obvious that the increase in water level has not changed its magnitude. There is no excess effective pressure.

Graphical representation

In graphical solutions to this type of problems, one must remember, that:

1. Pressure of overburden and water increases linearly between the top and bottom of the layer considered.
2. Pressure in a stratum due to the weight of the layer above it is constant throughout.

The above problem can therefore be drawn as shown in Figure 5.8 and the diagrams summed.

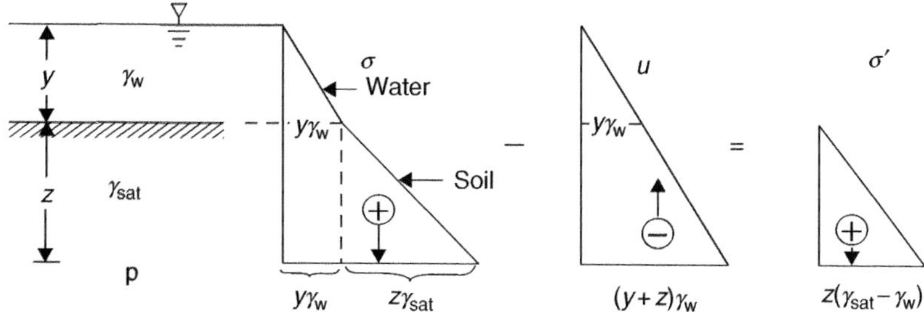

Figure 5.8

5.4 Types of problem

Having clarified the meaning and significance of the effective pressure, the idea can now be extended to five practical problems in connection with:

1. Stratification of the soil (Problem 1)
2. Excavation (Problem 2)
3. Artesian pressure (Problem 3)
4. Seepage pressure/piping (Problem 4)
5. Pumping of ground water (Problem 5)

Problem 1

Figure 5.9 shows a cross-section through the strata underlying the site, which is loaded by deposited soil, imposing a uniform pressure of $q=50\,\text{kN/m}^2$ at surface level. The ground water level is at a depth of 4 m. The bulk densities of the three layers are also indicated.

Assuming, that the surface load is applied 'instantaneously' (at time $t=0$), calculate at the top of each stratum:

1. The pressures σ, $\Delta\sigma$, u, Δu, σ' and $\Delta\sigma'$ at time $t=0$.
2. The pressures σ, u and σ' after the excess pore pressure had dissipated from the two clay layers (i.e. when $\Delta u=0$), after a considerable period of time, normally indicated as $t=\infty$.

Notes: The assumption of instantaneous loading means that during the period of construction, there is imperceptible dissipation of excess pore pressure from the clay strata.

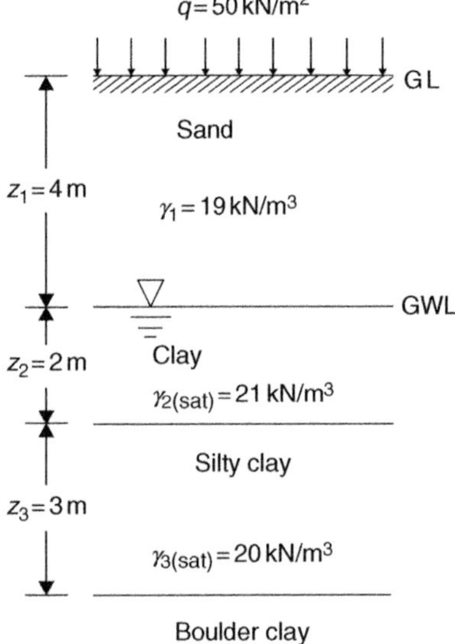

Figure 5.9

At $t=0$

$z=0\,\text{m}$

Total pressure:	$\sigma=q$	$=50\,\text{kN/m}^2$
Excess total pressure:	$\Delta\sigma=q$	$=50\,\text{kN/m}^2$
Pore pressure:	u	$=0$
Excess pore pressure:	Δu	$=0$
Effective pressure:	$\sigma'=\sigma-u$	$=50\,\text{kN/m}^2$
Excess effective pressure:	$\Delta\sigma'=\Delta\sigma-\Delta u$	$=50\,\text{kN/m}^2$

$z=4\,\text{m}$

$$\sigma=q+z_1\gamma_1=50+4.19 \quad =126\,\text{kN/m}^2$$
$$\Delta\sigma=q \quad =50\,\text{kN/m}^2$$
$$\Delta u=q \quad =50\,\text{kN/m}^2$$
$$u=\Delta u \quad =50\,\text{kN/m}^2$$
$$\sigma'=\sigma-u=126-50 \quad =76\,\text{kN/m}^2$$
$$\Delta\sigma'=\Delta\sigma-\Delta u=50-50 \quad =0\,\text{kN/m}^2$$

$z=6\,\text{m}$

$$\sigma=q+z_1\gamma_1+z_2\gamma_{2(\text{sat})}=126+2\times21 \quad =168\,\text{kN/m}^2$$
$$\Delta\sigma=q \quad =50\,\text{kN/m}^2$$
$$\Delta u=q \quad =50\,\text{kN/m}^2$$
$$u=z_2\gamma_w+\Delta u=2\times9.81+50 \quad =69.4\,\text{kN/m}^2$$
$$\sigma'=\sigma-u=168-69.6 \quad =98.4\,\text{kN/m}^2$$
$$\Delta\sigma'=\Delta\sigma-\Delta u=50-50 \quad =\text{kN/m}$$

$z=9\,\text{m}$

$$\sigma=q+z_1\gamma_1+z_2\gamma_{2(\text{sat})}+z_3\gamma_{3(\text{sat})}=168+3\times20 \quad =228\,\text{kN/m}^2$$
$$\Delta\sigma=q \quad =50\,\text{kN/m}^2$$
$$\Delta u=q \quad =50\,\text{kN/m}^2$$
$$u=(z_2+z_3)\gamma_w+\Delta u=5\times9.81+50 \quad =99\,\text{kN/m}^2$$
$$\sigma'=\sigma-(u+\Delta u)=228-99 \quad =129\,\text{kN/m}^2$$
$$\Delta\sigma'=50-50 \quad =0\,\text{kN/m}^2$$

At $t=\infty$

By this time, the excess pore pressure had been dissipated and the excess total pressure $q=50\,\text{kN/m}^2$ is carried entirely by the consolidated soil skeleton. The values of the pore pressure are now the same as existed prior to the application of q. The final effective pressures (when $\Delta u=0$), to be verified by the reader, are given in Table 5.1.

Table 5.1

Depth below GL (m)	**0**	**4**	**6**	**9**
Total pressure: σ kN/m²	50	126	168	228
Pore pressure: u kN/m²	0	0	19.6	49
Effective pressure: σ' kN/m²	50	126	148.4	179

Graphical representation ($t=\infty$)

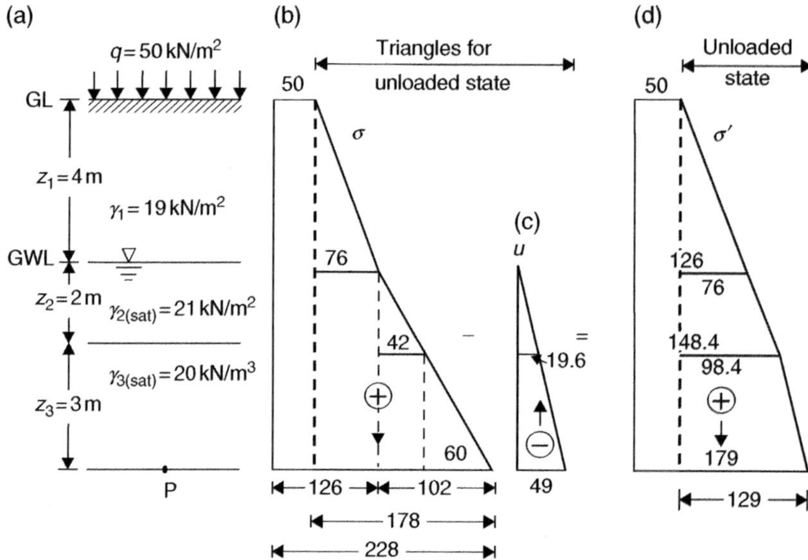

Figure 5.10

Notes:

i. Parts of the diagrams are highlighted by thicker lines. These are the stress diagrams for the soil in its unloaded state, that is when $q=50\,\text{kN/m}^2$ is not applied.

ii. Downward pressure is +ve
Upward pressure is −ve (buoyancy)

Problem 2

If instead of placing the uniform load onto the ground surface, some of the overburden is removed – e.g. for the foundation of a large building – then there are two cases to be considered:

1. The base of excavation is above GWL.
2. The base is below GWL.

Base above GWL

In this case, the pressure of the excavated material may be considered as negative surface load ($-q$), hence its value is simply subtracted from the pressures in the unloaded state. The relevant pressure diagrams for a 1.5 m deep excavation on the site of Problem 1 are drawn in Figure 5.11. The decrease in total and effective stress is therefore $-q=-1.5\times 19=-28.5\,\text{kN/m}^2$.

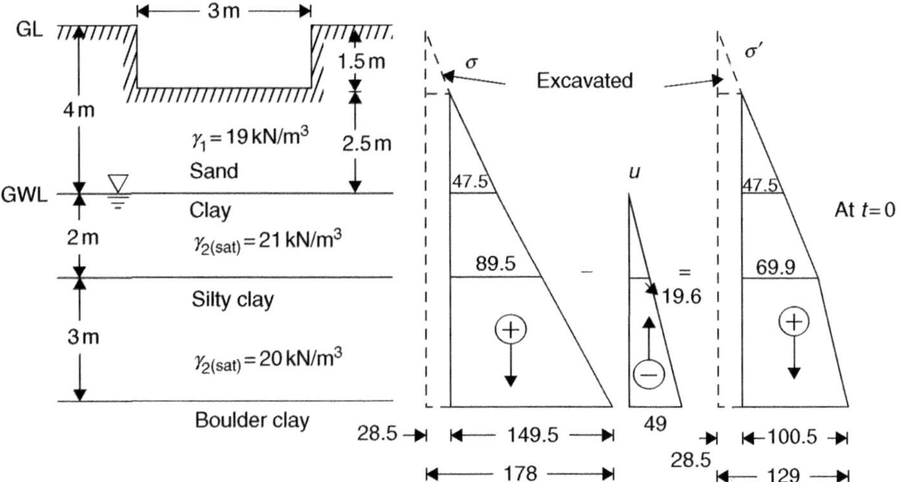

Figure 5.11

If now a 3m-wide strip footing is placed into the excavation, transmitting 128.5kN/m² to the soil, then the actual 'net' bearing pressure applied at the depth of 1.5m is σ_n=128.5−28.5=100kN/m². The maximum vertical pressure induced by σ_n is given by the Boussinesq-Michell formula (4.14) at any point below the centre of the foundation. The results are added to the total pressure triangle of Figure 5.11. Therefore, the value of σ_v at the top of each layer is calculated from:

$$\text{(4.14)} \quad \sigma_v = (\beta + \sin\beta)\frac{\sigma_n}{\pi}$$

$$\text{(4.13)} \quad \beta = 2\tan^{-1}\left(\frac{b}{z}\right)$$

Chapter 4

At $z=2.5\,\text{m}$ $\quad \beta = 2\times\tan^{-1}\left(\dfrac{1.5}{2.5}\right) = 61.93° = 1.081$ radian

$$\therefore \quad \sigma_v = (1.081 + \sin 61.93)\times\frac{100}{\pi} = 62.5\,\text{kN/m}^2$$

At $z=4.5\,\text{m}$ $\quad \beta = 2\times\tan^{-1}\left(\dfrac{1.5}{4.5}\right) = 36.87° = 0.644$ radian

$$\therefore \quad \sigma_v = (0.644 + \sin 36.87)\times\frac{100}{\pi} = 39.6\,\text{kN/m}^2$$

At $z=7.5\,\text{m}$ $\quad \beta = 2\times\tan^{-1}\left(\dfrac{1.5}{7.5}\right) = 22.62° = 0.395$ radian

$$\therefore \quad \sigma_v = (0.395 + \sin 22.62)\times\frac{100}{\pi} = 24.8\,\text{kN/m}^2$$

The final values of σ′ in Figure 5.12 can now be used to calculate the consolidation of the clay layers. It is again assumed that the rate of excavation as well as the construction

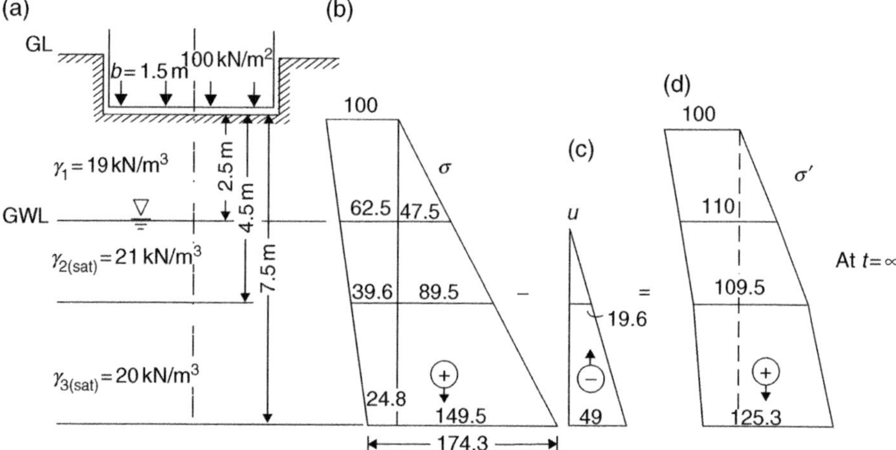

Figure 5.12

of the base are 'instantaneous', hence no pore water has dissipated from the clay layers during the construction period.

Base below GWL

The extent of problems encountered during excavation of open cuts, below the ground water level, depends on whether the soil is sand, gravel or clay. In sand and gravel, the GWL would be re-established quickly because of high permeability. Also any change in pore pressure will be dissipated almost 'instantaneously' at $t=0$, hence no long-term consolidation has to be considered. The effective pressure variation in granular soils will be discussed later.

The excavation of open cuts in clays is completely different matter. In this case, the removal of soil is rapid enough to prevent the dissipation of excess pore pressure from the underlying clay strata, because of their low permeability, hence assume the unloading to be 'instantaneous'.

Suppose the depth of excavation for a large structure is 6 m, instead of 1.5 m, then the variation of pressures at $z=9$ m, or at any other depth, may still be evaluated at $t=0$.

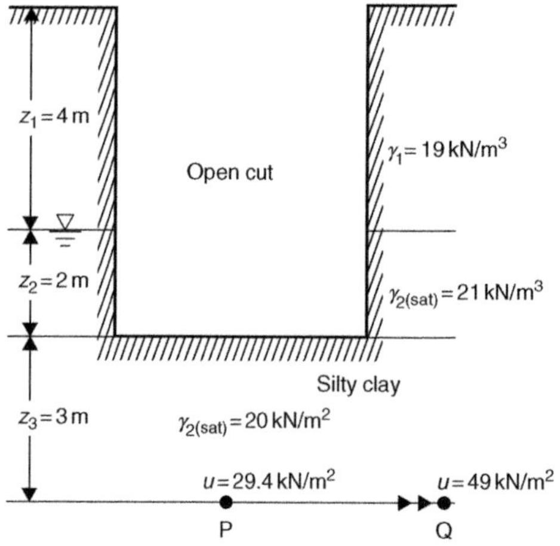

Figure 5.13

Pressure at points P and Q.
Soil is in natural, undisturbed state.
From Figure 5.10:

$$\sigma_1 = 178\,\text{kN/m}^2$$
$$u_1 = -49\,\text{kN/m}^2$$
$$\sigma'_1 = 129\,\text{kN/m}^2$$

Pressures at the end of excavation period are different at points P and Q, the latter being, some way away from the open cut whilst the pore pressure at Q remains 49 kN/m^2, its value is diminished at P, because of the reduction in the water level. As the excavation is assumed to be rapid, Δu_2 has no time to dissipate and equalize the pore pressure at P and Q.

At P:

	$\Delta\sigma_2 = z_1\gamma_1 + z_2\gamma_{2(sat)}$	$= 4\times19 + 2\times21$	$= 118$ N/m^2	
Either	$\sigma_2 = z_3\gamma_{3(sat)}$	$= 3\times20$	$= 60$ kN/m^2	
or	$\sigma_2 = \sigma_1 - \Delta\sigma_2$	$= 178 - 118$	$= 60$ kN/m^2	
Also,	$\Delta\sigma_2 = \sigma_1 - \sigma_2$	$= 178 - 60$	$= 118$ kN/m^2	
	$\Delta u_2 = z_2\gamma_w$	$= 2\times9.81$	$= 19.6$ kN/m^2	
Either	$u_2 = z_3\,\gamma_w$	$= 3\times9.81$	$= 29.4$ kN/m^2	
Or	$u_2 = u_1 - \Delta u_2$	$= 49\times19.6$	$= 29.4$ kN/m^2	
Finally,	$\sigma'_2 = \sigma_2 - u_2$	$= 60 - 29.4$	$= 30.6$ kN/m^2	< 129 kN/m^2

Decrease:

Either	$\Delta\sigma'_2 = \sigma'_1 - \sigma'_2$	$= 129 - 30.6$	$= 98.4$ kN/m^2
Or	$\Delta\sigma'_2 = \Delta\sigma_2 - \Delta u_2$	$= 118 - 19.6$	$= 98.4$ kN/m^2

Note: The decrease in σ' indicates that the soil becomes weaker because of the diminished interparticle pressure. If however, the structure is now placed 'instantaneously' into the cut, then its weight would increase the total, hence the effective pressure. For instance, calculate the pressure at P, taking the net weight of the structure as $\sigma_n = 200$ kN/m^2 using formulae (4.13) and (4.14).

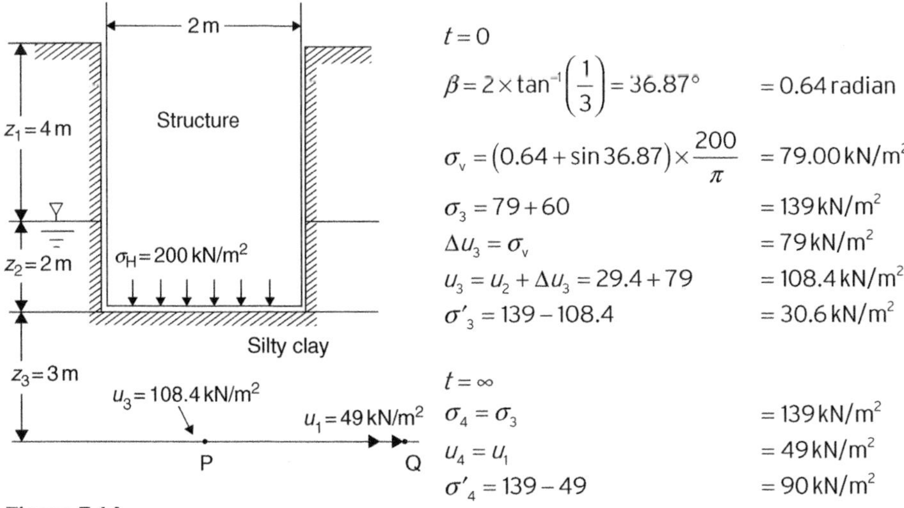

Figure 5.14

Notes:

a) The purpose of these calculations is to assess how much the effective pressure has changed at various depths, due to the excavation and the construction of the structure. Table 5.2 summarizes the deviations, at depth 6 and 9 metres, from the value of σ_1' prior to the start of excavation.

Table 5.2

σ'	Depth (m)		6	9
kN/m²	Before excavation	: σ'_1	98.4	129.0
	After excavation	: σ'_2	0	30.6
	At $t=0$	: σ'_3	0	30.6
	At $t=\infty$	: σ'_4	180.4	90.0

b) The increased average effective pressure within the silty clay layer indicates its slow consolidation and the consequent settlement of the structure depending on the characteristics of the soil. This subject will be dealt with in a later chapter on consolidation and settlement. For settlement analysis, the average positive value of the effective pressure within a layer has to be estimated. Negative value would suggest possible swelling of the clay at that depth.

Problem 3

Artesian pressure σ_A is encountered in permeable layers, such as gravel, underlying non-permeable stratum e.g. clay. The water in the gravel layer is under pressure, caused by a hydrostatic head as shown, or by the overburden.

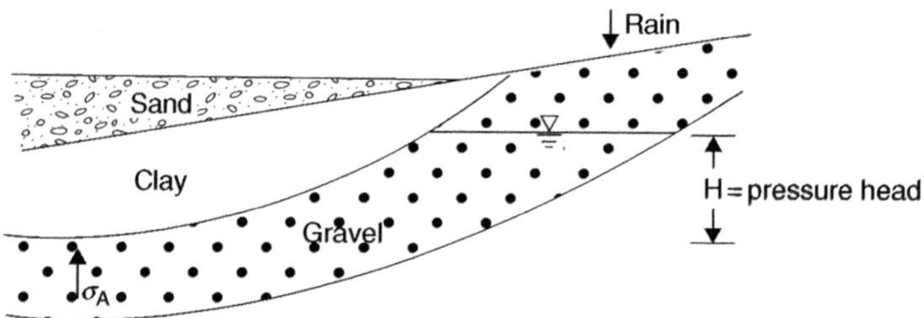

Figure 5.15

If a borehole is sunk through the clay layer, the artesian pressure released would indicate significant problems during construction or excavation of the soil.

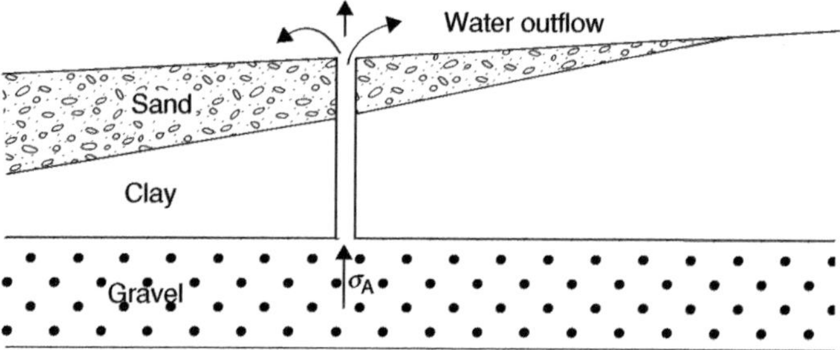

Figure 5.16

The influence of artesian pressure on the effective one is shown below.

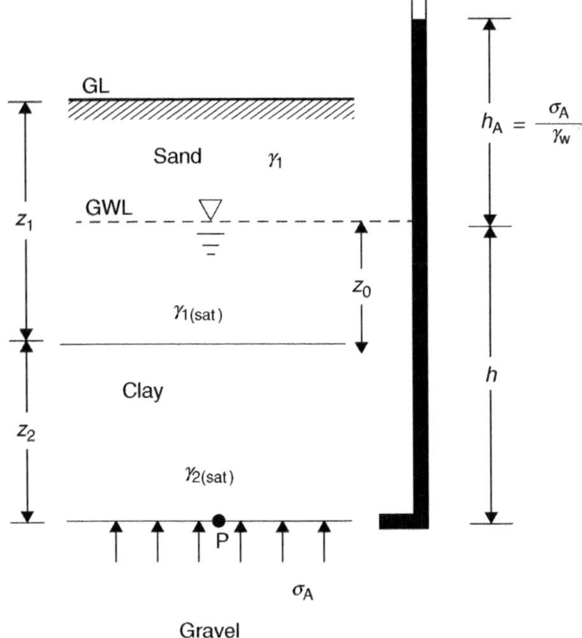

Figure 5.17

(Artesian pressure head)
Pressures at the bottom of clay layer (at P).

$$\sigma = (z_1 - z_0)\gamma_1 + z_0\gamma_{1(sat)} + z_2\gamma_{2(sat)}$$

$$u = h\gamma_w + \sigma_A = (z_0 - z_2)\gamma_w + \sigma_A$$

$$\sigma' = \sigma - u$$

$$\therefore \quad \sigma' = z_1\gamma_1 - z_0\gamma_1 + z_0\gamma_{1(sat)} + z_2\gamma_{2(sat)} - z_0\gamma_w - z_2\gamma_w - h_A\gamma_w$$

$$= (z_1 - z_0)\gamma_1 + z_0\left[\gamma_{1(sat)} - \gamma_w\right] + z_2\left[\gamma_{2(sat)} - \gamma_w\right] - h_A\gamma_w$$

$$\therefore \quad \sigma' = (z_1 - z_0)\gamma_1 + z_0\gamma_1' + z_2\gamma_2' - h_A\gamma_w$$

The effect of artesian pressure is to:

1. increase the pore pressure
2. decrease the effective pressure

Notes:

a) The artesian pressure is zero at the top of the clay layer, where it can dissipate quickly within the sand.
b) The distribution of σ_A is linear within the clay layer.
c) If the total pressure (σ) of the layers of soil, at the bottom of the clay is smaller than the total upward pressure, then the ground could fail in uplift. In this case $\sigma \leq u + \sigma_A$ when the clay could fail in shear (heaving). (See Chapter 3).

Problem 4

The effective pressure was estimated in the previous three problems under static groundwater conditions. If for some reason water movement occurs through the soil, the resulting seepage pressure will alter the value of the effective pressure. Two types of seepage have to be considered:

1. Downward seepage
2. Upward seepage.

Downward seepage

The downward pressure reduces the pore pressure (u), hence increases σ' at a depth z. Consider the sheet pile wall of a cofferdam, driven into the sandy-gravel layer of a river. Because of seepage, the water level rises inside the dam as shown.

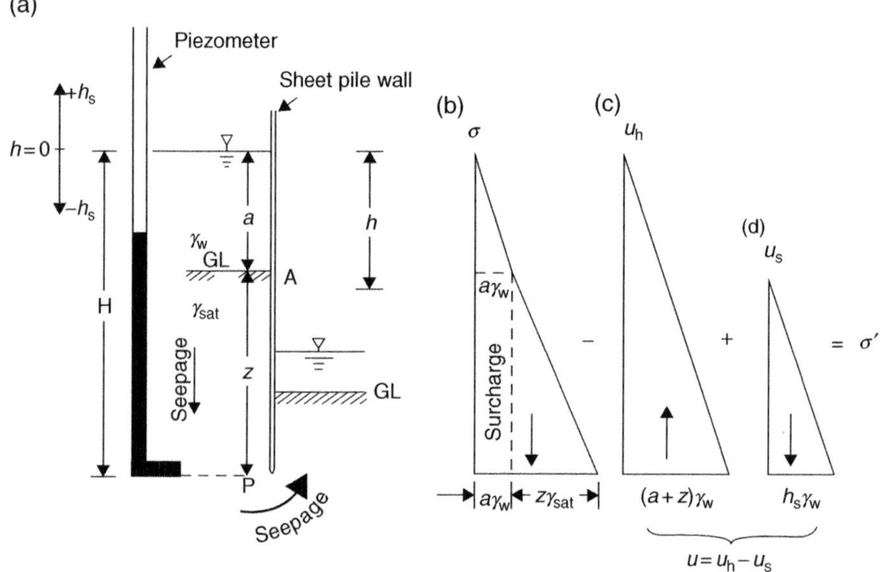

Figure 5.18

where $-h_s$ = seepage pressure head
u_s = seepage pore pressure
u = pore pressure due to γ_w and u_s
u_h = pore pressure at depth h due to γ_w only.

From Figure 5.18, the pressures at depth H are:

$$\left.\begin{aligned} u_h &= (a+z)\gamma_w \\ u_s &= -h_s\gamma_w \end{aligned}\right\} \therefore u = a\gamma_w + z\gamma_w - h_s\gamma_w$$

$$\sigma = a\gamma_w + z\gamma_{sat}$$

Hence, $$\sigma' = \sigma - u = a\gamma_w + z\gamma_{sat} - a\gamma_w - z\gamma_w + h_s\gamma_w$$

Cancelling $a\gamma_w$: $$\sigma' = z(\gamma_{sat} - \gamma_w) + h_s\gamma_w$$

$$\therefore \boxed{\sigma' = z\gamma' + h_s\gamma_w}$$

It is customary to express h_s in terms of the hydraulic gradient, given by:

$$i = \frac{\text{Head loss}}{\text{Length of flow path}} \quad \text{or} \quad \boxed{i = \frac{h_s}{z}} \qquad (3.2)\ \&\ (5.5)$$

Hence the seepage pressure is $\boxed{u_s = iz\gamma_w \,\text{kN/m}^2}$ (5.6)

Therefore, effective pressure: $\boxed{\sigma' = z(\gamma' + i\gamma_w)\,\text{kN/m}^2}$ (5.7)

Suppose the quantities in Figure 5.18 are given as

$$a = 2\,\text{m} \qquad \gamma_{sat} = 19\,\text{kN/m}^3$$
$$z = 4\,\text{m} \qquad \gamma_w = 10\,\text{kN/m}^3$$

Piezometric level below water surface: $h_s = -0.3\,\text{m}$
Calculate the pressures at point P ($h = 6\,\text{m}$):

Pore pressure:	$u_h = (a+z)\gamma_w = 6 \times 10$	$= 60\,\text{kN/m}^2 \uparrow$
See page pressure:	$u_s = -h_s\gamma_w = -0.3 \times 10$	$= -3\,\text{kN/m}^2 \downarrow$
Final pore pressure:	$u = u_h + u_s = 60 - 3$	$= 57\,\text{kN/m}^2 \uparrow$
Total pressure:	$\sigma = a\gamma_w + z\gamma_{sat} = 2 \times 10 + 4 \times 19$	$= 96\,\text{kN/m}^2 \downarrow$
Effective pressure:	$\sigma' = \sigma - u = 96 - 57$	$= 39\,\text{kN/m}^2 \downarrow$
Hydraulic gradient:	$i = \frac{h_s}{z} = \frac{0.3}{4}$	$= 0.075$

Alternatively,

From (5.6):	$u_s = iz\gamma_w = 0.075 \times 4 \times 10$	$= 3\,\text{kN/m}^2$
But	$\gamma' = \gamma_{sat} - \gamma_w = 19 - 10$	$= 9\,\text{kN/m}^2$
From (5.7):	$\sigma' = z(\gamma' + i\gamma_w) = (9 + 0.0075 \times 10) \times 4 = 39\,\text{kN/m}^2$	

The calculations show that downward seepage increases the intergranular pressure. At the surface of the soil $u_s = 0$ as it is completely dissipated there.

Upward seepage
(See also section 3.7). In this case, the upward pressure increases u and decreases σ'. Because of the decreased intergranular pressure, the soil could become weak to such an extent, that it would fail in piping. The type of failure when granular material, mainly sand, 'boils up' is also called 'quick condition'. For this reason, the upward seepage is of more engineering interest than the downward seepage. Figure 5.19 shows the pressures inside the coffer dam. In this case the seepage pressure head is positive ($+h_s$).

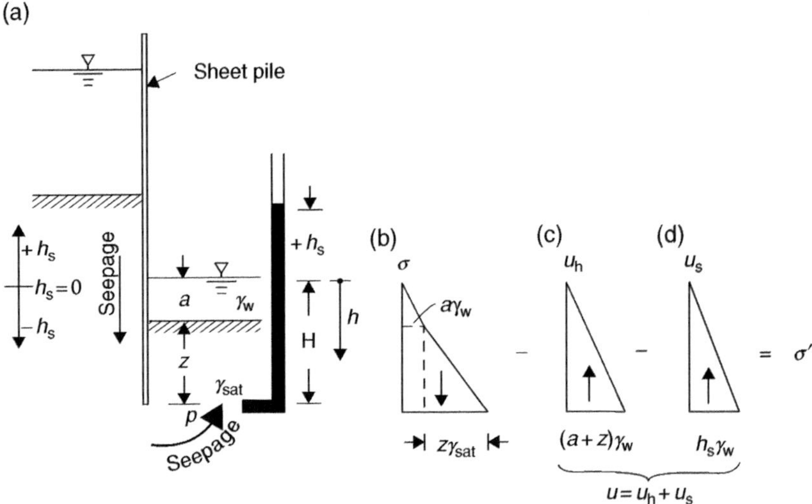

Figure 5.19

$$\left.\begin{aligned} u_h &= (a+z)\gamma_w \\ u_s &= h_s\gamma_w \end{aligned}\right\} \qquad \left.\begin{aligned} u &= (a+z)\gamma_w + h_s\gamma_w \\ \sigma &= a\gamma_w + z\gamma_{sat} \end{aligned}\right\} \quad \sigma' = \sigma - u$$

$$\therefore \quad \sigma' = a\gamma_w + z\gamma_{sat} - a\gamma_w - z\gamma_w - h_s\gamma_w$$

Simplifying $$\boxed{\sigma' = z\gamma' - h_s\gamma_w \,\text{kN/m}^2} \tag{5.8}$$

or $$\boxed{\sigma' = z(\gamma' - i\gamma_w) \,\text{kN/m}^2}$$

Quick condition occurs when the effective pressure becomes zero, i.e.

$$\sigma' = z\gamma' - h_s\gamma_w = 0$$

From which, $$\boxed{z\gamma' = h_s\gamma_w}$$

Note: $+h_s$ is equivalent to the artesian pressure head h_A, except that, whilst h_s could cause boiling failure in sand, h_A induces shear failure (heaving) in clays.

The formula shows that if the seepage pressure represented by h_s is large enough, it could equal or exceed the overburden pressure $z\gamma'$ thus forcing the soil upwards. This phenomenon is also called 'piping' or 'boiling'.

Piping depends largely on the length of the path (z) through which water flows upwards. Failure occurs when z has a certain critical value (z_c), expressed from:

$$z_c\gamma' = h_s\gamma_w \qquad \therefore \boxed{z_c = \left(\frac{\gamma_w}{\gamma'}\right) h_s \,\text{m}} \tag{5.9}$$

This indicates that the sheet pile must be driven deeper into the sand than length z_c. The critical hydraulic gradient (i_c) corresponding to z_c is given by:

$$\boxed{i_c = \frac{h_s}{z_c} = \frac{\gamma'}{\gamma_w}} \qquad (3.2)\ \&\ (5.10)$$

But, for saturated soil: $$\gamma' = \left(\frac{G-1}{1+e}\right)\gamma_w$$

So, $$\boxed{i_c = \frac{G_s - 1}{1+e}} \qquad (3.42)$$

The critical hydraulic gradient, therefore, is a function of the soil's structure.

Suppose the quantities in Figure 5.19 are given as:

$a = 1\text{m}$ $\quad \gamma_{sat} = 19\,\text{kN/m}^3$ $\quad h_s = 0.25\text{m}$

$z = 1.5\,\text{m}$ $\quad \gamma_w = 10\,\text{kN/m}^3$

Also, $G_s = 2.65$
$e = 0.86$

Calculate the pressures at point P: $h = 2.5\,\text{m}$:

Pore pressure: $u_h = (a+z)\gamma_w \quad = 2.5 \times 10 \quad = 25\,\text{kN/m}^2$

See page pressure: $u_s = h_s\gamma_w \quad = 0.25 \times 10 \quad = 2.5\,\text{kN/m}^2$

Final pore pressure: $u = u_h + u_s \quad = 25 + 2.5 \quad = 27.5\,\text{kN/m}^2$

Total pressure: $\sigma = a\gamma_w + z\gamma_{sat} \quad = 10 + 1.5 \times 19 \quad = 38.5\,\text{kN/m}^2$

Effective pressure: $\sigma' = \sigma - u \quad = 38.5 - 27.5 \quad = 11\,\text{kN/m}^2$

Hydraulic gradient: $i = \frac{h_s}{z} = \frac{0.25}{1.5} = 0.167$

From (5.8): $\sigma' = (\gamma' - i\gamma_w)z = (9 - 0.167 \times 10) \times 1.5 = 11\,\text{kN/m}^2$ (as before)

Critical path length: $z_c - \left(\frac{\gamma_w}{\gamma'}\right)h_s - \left(\frac{10}{9}\right) \times 0.25 = 0.28\,\text{m}$

Therefore, the actual path length of $z = 1.5\,\text{m}$ is satisfactory against piping. See Chapter 8 for the estimation of sheet pile stability, however;

Critical hydraulic gradient: $i_c = \frac{h_s}{z_c} = \frac{0.25}{0.28} = 0.893$

Alternatively: $$i_c = \frac{G_s - 1}{1+e} = \frac{2.65-1}{1+0.86} = 0.887 \approx 0.893$$

Check effective pressure at critical stage:

$$\sigma_c' = (\gamma' - i_c\gamma_w)z_c = (9 - 0.893 \times 10) \times 0.28 = 0.02 \approx 0$$

The deviation of σ_c' from zero is due to cumulative arithmetic errors.

Note: In this section, only the pressures caused by seepage have been discussed. Consult Chapter 3 on Permeability, for the evaluation of forces due to movement of water in soils.

The development of seepage pressures was demonstrated by Peck (1953), as shown diagrammatically in Figure 5.19. In this, a cylinder containing saturated soil is connected to a water-tank. Initially, the water in both containers is at the same level [Figure 5.20(a)], hence there is no seepage ($h_s = 0$).

If the water-tank is raised [Figure 5.20(b)], then upward flow is induced through the soil in order to equalize the water levels. The seepage pressure head ($+h_s$) is above the water level in the cylinder, increasing the pore-water pressure within the soil.

If, however, the water-tank is lowered (Figure 5.20(c)), then downward flow is induced in order to equalize the water levels. The seepage pressure-head ($-h_s$) is below the water level in the cylinder, decreasing the pore-pressure within the soil. The relevant pressures at point P are summarized:

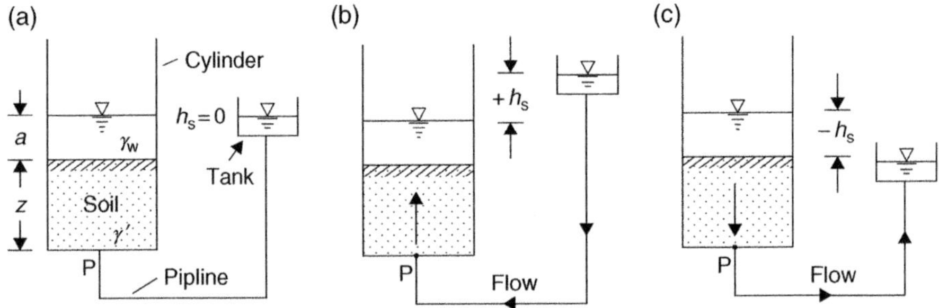

Figure 5.20

$$
\begin{array}{lll}
\sigma = a\gamma_w + z\gamma_{sat} & \sigma = a\gamma_w + z\gamma_{sat} & \sigma = a\gamma_w + z\gamma_{sat} \\
u = (a + z)\gamma_w & u = (a + z)\gamma_w & u = (a + z)\gamma_w \\
u_s = 0 & u_s = h_s\gamma_w & u_s = -h_s\gamma_w \\
\sigma' = z\gamma' & \sigma' = (\gamma' - i\gamma_w)z & \sigma' = (\gamma' - i\gamma_w)z
\end{array}
$$

5.5 Effect of seepage on shallow footings

Footings constructed on granular soil may experience upward seepage force should the soil be flooded, or the water table otherwise raised rapidly below them. In sand, this could cause washouts, that is, the removal of foundation material. To obviate this type of base failure it is advisable to use piles to support the structure [see Chapter 9. Figure 5.33(d)].

The effect of upward seepage, of hydraulic gradient i, is to decrease the submerged density of the soil.

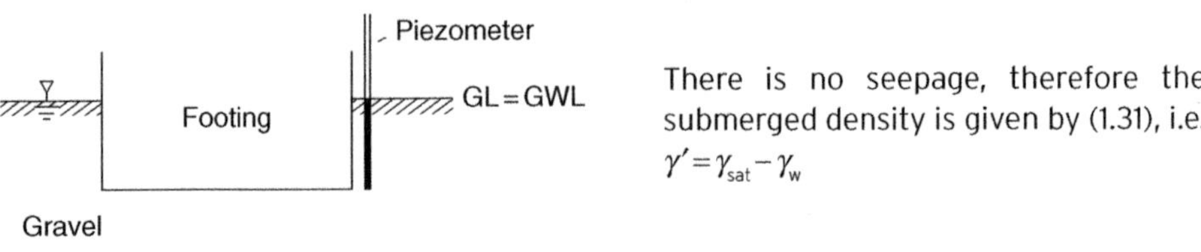

There is no seepage, therefore the submerged density is given by (1.31), i.e. $\gamma' = \gamma_{sat} - \gamma_w$

Figure 5.21

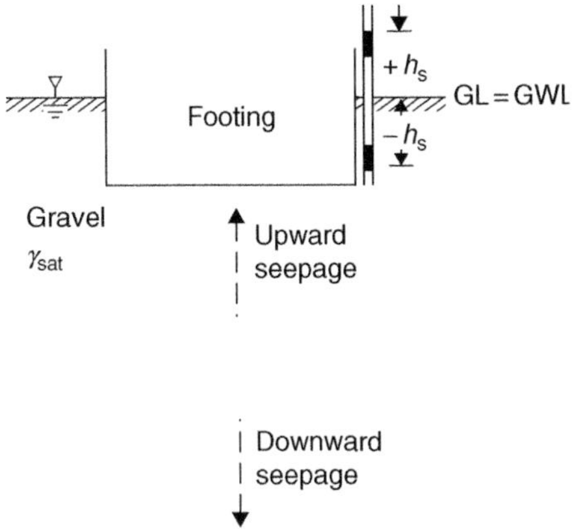

i = hydraulic gradient

Submerged density for upward seepage:

From (5.8) $$\gamma'' = (\gamma_{sat} - \gamma_w) - i\gamma_w \qquad (5.11)$$

Or $$\gamma'' = \gamma' - i\gamma_w$$

Submerged density for downward seepage:

From (5.7) $$\gamma'' = (\gamma_{sat} - \gamma_w) + i\gamma_w \qquad (5.12)$$

Or $$\gamma'' = \gamma' + i\gamma_w$$

Figure 5.22

Problem 5

Pumping is carried out for various reasons. The variation of pore pressure during the following two situations are introduced in this section:

1. Ground water lowering
2. Reduction of artesian pressure

5.6 Ground water lowering (at atmospheric pressure)

Figure 5.23 shows a section of the ground. The ground water level has to be lowered to the top of the clay layer in order to allow excavation for a foundation.

When the water is suddenly removed from above the clay, the pore pressure is not altered immediately in its voids. Eventually, sometime after pumping, equilibrium is attained due to positive capillary tension, resulting in increased effective pressure.

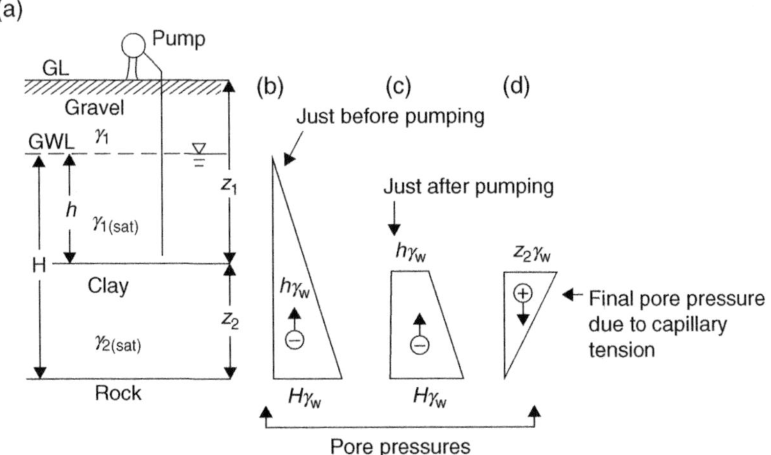

Figure 5.23

5.7 Reduction of artesian pressure

The result of decreased pressures (σ_A) is an increased effective one.

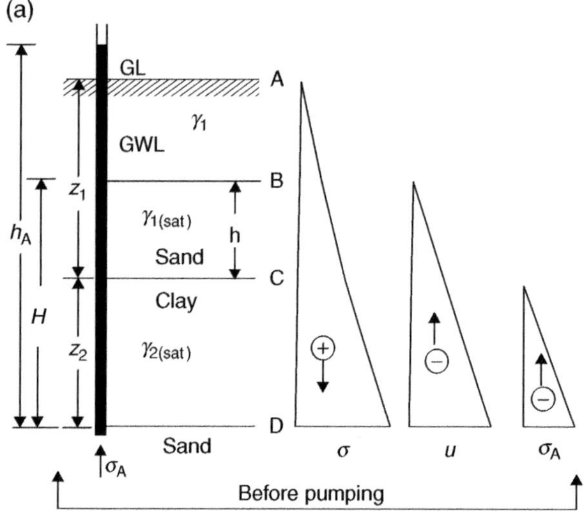

$$\sigma = \gamma_1(z_1 - h) + \gamma_{1(sat)}h + \gamma_{2(sat)}z_2$$

$$u = H\gamma_w$$

$$\sigma_A = h_A\gamma_w$$

(b)

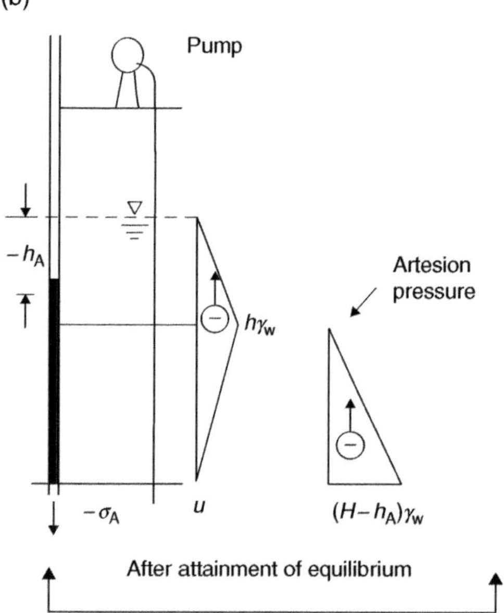

Pore pressure is zero at *B* and *D* because of dissipation.

Figure 5.24

A section of ground is shown in Figure 5.25. The water level in the piezometer, installed in the gravel, is 3 m above, whilst the ground-water table is 1.5 m below the ground surface.

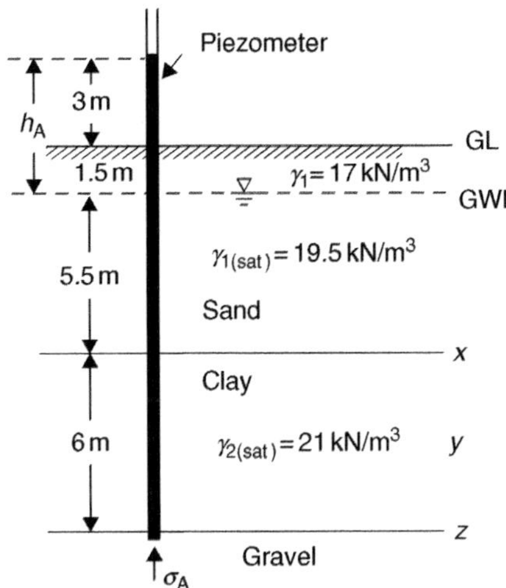

Figure 5.25

Calculate the effective pressures at points *x*, *y* and *z*:

1. before pumping
2. after the GWL is lowered by 5 m
3. after the piezometric level is lowered by 9.5 m.

Before pumping $\quad h_A = 4.5\text{m} \quad \therefore \quad \sigma_A = 4.5 \times 9.81 = 44.1\text{kN/m}^2$

At *x*:

$$\sigma_x = 1.5 \times 17 + 5.5 \times 19.5 \qquad = 132.8\,\text{kN/m}^2$$
$$u_x = 5.5 \times 9.81 + \sigma_A = 54 + 44.1 \quad = 98.1\,\text{kN/m}^2$$
$$\sigma'_x = \sigma_x - u_x = 132.8 - 98.1 \qquad = 34.7\,\text{kN/m}^2$$

At *y*:

$$\sigma_y = 132.8 + 3 \times 21 \quad = 195.8\,\text{kN/m}^2$$
$$u_y = 8.5 \text{ x } 9.81 + 44.1 \quad = 127.5\,\text{kN/m}^2$$
$$\sigma'_y = 195.8 - 127.5 \quad = 68.3\,\text{kN/m}^2$$

At *z*:

$$\sigma_z = 195.8 + 3 \times 21 \quad = 258.8\,\text{kN/m}^2$$
$$u_z = 11.5 \times 9.81 + 44.1 \quad = 157.0\,\text{kN/m}^2$$
$$\sigma'_z = 258.8 - 157 \quad = 101.8\,\text{kN/m}^2$$

Lowering the GWL only, by 5 m

As the water level in the piezometer is unaltered the artesian pressure head difference becomes: $h_A = 4.5 + 5 = 9.5\,\text{m} \quad \therefore \quad \sigma_A = 9.5 \text{ x } 9.81 = 93.2\text{ kN/m}^2$

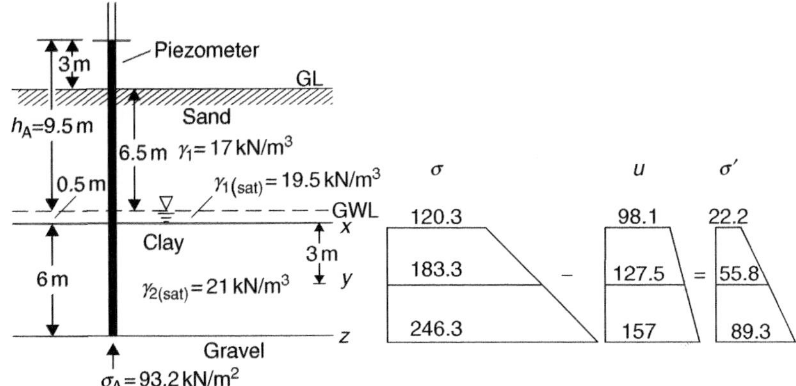

Figure 5.26

At x: $\sigma_x = 6.5\times17 + 0.5\times19.5 = 120.3\,\text{kN/m}^2$

$u_x = 0.5\times9.81 + 93.2 = 98.1\,\text{kN/m}^2$ (unaltered)

$\sigma'_x = 120.3 - 98.1 = 22.2\,\text{kN/m}^2$

At y: $\sigma_y = 120.3 + 3\times21 = 183.3\,\text{kN/m}^2$

$u_y = 3.5\times9.81 + 93.2 = 127.5\,\text{kN/m}^2$ (unaltered)

$\sigma'_y = 183.3 - 127.5 = 55.8\,\text{kN/m}^2$

At z: $\sigma_z = 183.3 + 3\times21 = 246.3\,\text{kN/m}^2$

$u_z = 6.5\times9.81 + 93.2 = 157.0\,\text{kN/m}^2$ (unaltered)

$\sigma'_z = 246.3 - 157 = 89.3\,\text{kN/m}^2$

The resulting pressure diagrams are shown on Figure 5.26. Note that the resultant pore pressure has not changed, due to the increased value of h_A.

Lowering the piezometric level by 9.5 m
As the two levels coincide $h_A = 0 \quad \therefore \quad \sigma_A = 0$

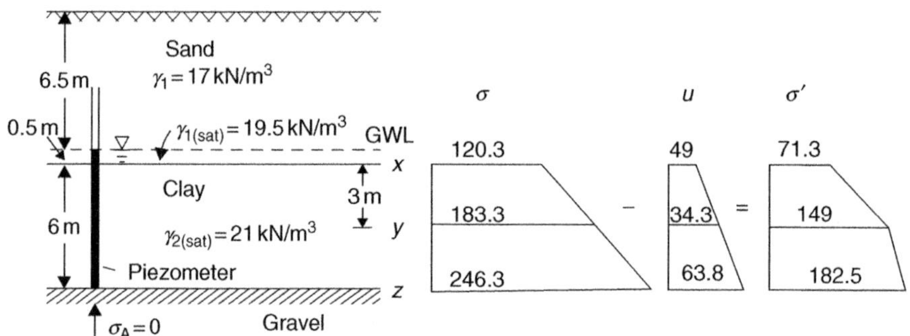

Figure 5.27

At x: $\sigma_x = 120.3\,\text{kN/m}^2$

$u_x = 0.5\times9.81 = 49\,\text{kN/m}^2$

$\sigma'_x = 120.3 - 49 = 71.3\,\text{kN/m}^2$

At y: $\sigma_y = 183.3\,\text{kN/m}^2$

$u_y = 3.5\times9.81 = 34.3\,\text{kN/m}^2$

$\sigma'_y = 183.3 - 34.3 = 149\,\text{kN/m}^2$

At z: $\sigma_z = 246.3\,\text{kN/m}^2$

$u_z = 6.5\times9.81 = 63.8\,\text{kN/m}^2$

$\sigma'_z = 246.3 - 63.8 = 182.5\,\text{kN/m}^2$

Note:

The artesian piezometric level could be theoretically below the gravel. In that case the artesian pressure has a negative value.

5.8 Capillary movement of water

Water surface exposed to the atmosphere is under tension. For example, this so called 'capillary tension' formed the meniscus at the stem of the hydrometer in Chapter 2. The phenomenon is more pronounced at the water surface in a very small diameter tube. Figure 5.28 demonstrates that if one end of a small-bore tube is inserted into water, then the fluid rises to some height (h_c).

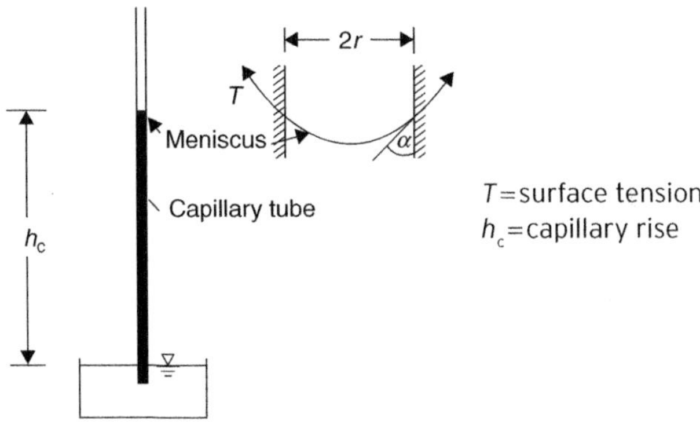

Figure 5.28

The average value is given by: $$h_c = \frac{0.15}{r}\cos\alpha \qquad (5.13)$$

Water also rises above the ground water table because of surface tension, although the voids do not form straight capillaries. The speed of rise depends on the soil types:

a) In clay the capillary rise is slow due to the very small pore size as well as to the presence of water bonded to the clay particles.
b) In sand and silty sand, the rise depends on the:
   - pore size
   - particle shape and distribution density
   - original water content
   - viscosity of water

The upper boundary of the zone affected by surface tension is called the 'capillary fringe'. The zone may be divided into two regions:

1. Closed capillary fringe, where the soil may be considered full saturated (S_r=1)
2. Open capillary fringe, where the soil is only partially saturated (S_r<1).

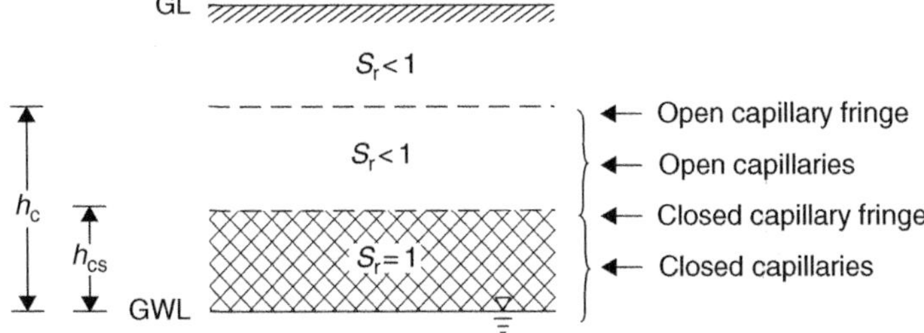

Figure 5.29

The determination of the capillary heads is somewhat unreliable. The Figures contained in Table 5.3, obtained by Lane and Washburn (1946), provide comparative Figures for guidance.

Table 5.3

Soil	Particle size D_{10} (mm)	Voids ratio e	Capillary head (cm) h_c	h_{cs}
Coarse gravel	0.82	0.27	5.4	6
Sandy gravel	0.20	0.45	28.4	20
Fine gravel	0.30	0.29	19.5	20
Silty gravel	0.06	0.45	106.0	68
Coarse sand	0.11	0.27	82.0	60
Medium sand	0.02	0.48–0.66	239.6	120
Fine sand	0.03	0.36	165.5	112
Silt	0.006	0.95–0.93	359.2	180

where h_c=height of open capillary fringe above the water table.
h_{cs}=height of closed, saturated fringe.

The effect of capillary water is to increase the unit weight of soil. The increase depends on its porosity as well as whether:

1. the soil is originally dry or partially saturated;
2. the capillary region is closed or open.

Dry soil

1. γ_d=dry unit weight
2. γ_{sat}=saturated unit weight
3. $\Delta\gamma$=change in unit weight

After saturation

$$\boxed{\Delta\gamma = \gamma_{sat} - \gamma_d} \tag{5.14}$$

But, from (1.42):

$$\gamma_{sat} = \left(\frac{G_s + e}{1+e}\right)\gamma_w = \left(\frac{G_s}{1+e} + \frac{e}{1+e}\right)\gamma_w$$

$$= \left(\frac{G_s}{1+e}\right)\gamma_w + \left(\frac{e}{1+e}\right)\gamma_w$$

From (1.41): $\gamma_d = \left(\frac{G_s}{1+e}\right)\gamma_w$

From (1.12): $n = \left(\frac{e}{1+e}\right)$

$$\boxed{\gamma_{sat} = \gamma_d + n\gamma_w} \tag{1.64}$$

Hence,

$$\boxed{\Delta\gamma = \gamma_{sat} - \gamma_d = n\gamma_w} \tag{5.15}$$

Partially saturated soil
Unit weight (γ)
Degree of saturation (s_r)

After partial saturation: $\Delta\gamma = \gamma_{sat} - \gamma$

From (1.38): $\gamma_{sat} = \left(\frac{G_s + S_r e}{1+e}\right)\gamma_w$

$$\Delta\gamma = \left(\frac{G_s + e}{1+e}\right)\gamma_w - \left(\frac{G_s + S_r e}{1+e}\right)\gamma_w$$

Therefore, $= \frac{\gamma_w}{1+e}(G_s + e - G_s - S_r e) = \left(\frac{e}{1+e}\right)(1 - S_r)\,\gamma_w$

But $$n = \frac{e}{1+e} \quad \therefore \quad \Delta\gamma = \gamma_{sat} - \gamma = (1 - S_r)n\gamma_w \qquad (5.16)$$

Example 5.4

A 5 m thick fine sand layer was deposited over an area underlain by coarse gravel. The ground-water table was at the surface of the gravel. Subsequent observations indicated that the sand became saturated by capillary action to a height of 1.2 m above the gravel surface. Calculate the densities of sand in the closed capillary region if:

a) it is dry initially ($S_r = 0$)
b) it is partially saturated initially at $S_r = 0.28$

The known characteristics of the compacted sand are:

$e = 0.52$

$\gamma_d = 17\ \mathrm{kN/m^2}$

$G_s = 2.64$

a) Dry sand

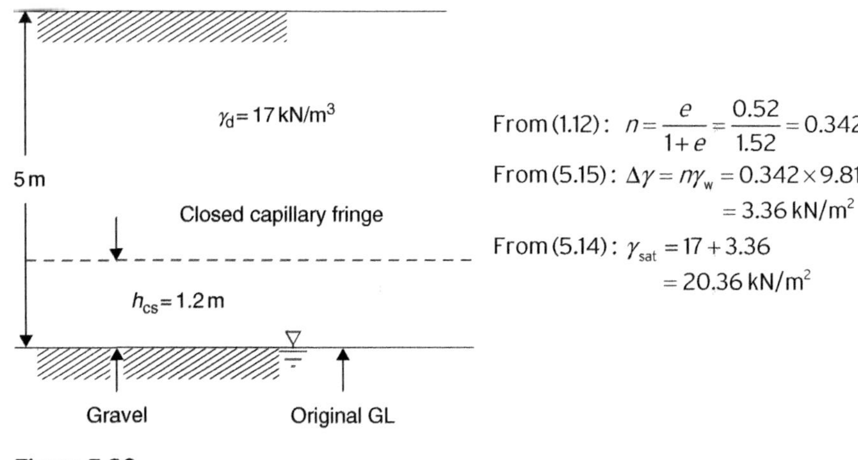

From (1.12): $n = \frac{e}{1+e} = \frac{0.52}{1.52} = 0.342$

From (5.15): $\Delta\gamma = n\gamma_w = 0.342 \times 9.81 = 3.36\ \mathrm{kN/m^2}$

From (5.14): $\gamma_{sat} = 17 + 3.36 = 20.36\ \mathrm{kN/m^2}$

Figure 5.30

(b) Partially saturated sand (S_r=0.28)

From (5.16): $\Delta\gamma=(1-S_r)n\gamma_w=(1-0.28)\times0.342\times9.81=2.42\,\text{kN/m}^2$

Also, $\Delta\gamma=\gamma_{sat}-\gamma=2.42$

Hence, $\gamma=\gamma_{sat}-2.42=20.36-2.42=17.94\,\text{kN/m}^2$

Summary of densities:

For $S_r=0$	Dry	$\gamma_d=17.00\,\text{kN/m}^2$
	Increase	$\Delta\gamma=3.36\,\text{kN/m}^2$
	Saturated ($S_r=1$)	$\gamma_{sat}=20.36\,\text{kN/m}^2$
For $S_r=0.28$	Increase	$\Delta\gamma=2.42\,\text{kN/m}^2$
	Partially saturated	$\gamma=17.94\,\text{kN/m}^2$

Effective pressure and capillary action

The effect the closed capillary region on the pore pressure, hence on the effective stress is twofold:

1. An increase in the density of the soil within the region, as discussed above. This increases the total stress (σ)
2. The pore pressure is negative (Below atmospheric) throughout the region. The value at the closed capillary fringe is given by:

$$\boxed{u_{cs}=-\gamma_w h_{cs}} \tag{5.17}$$

This pressure varies linearly with depth, becoming zero at the ground water table, where the pressure is atmospheric. The capillary action has no effect on the pore water pressure below the GWL.

Example 5.5

The gravel layer in Example 5.4 is 3 m thick and has a saturated density of 22 kN/m³. Draw the effective pressure diagram, taking 1.2 m capillary region into account.

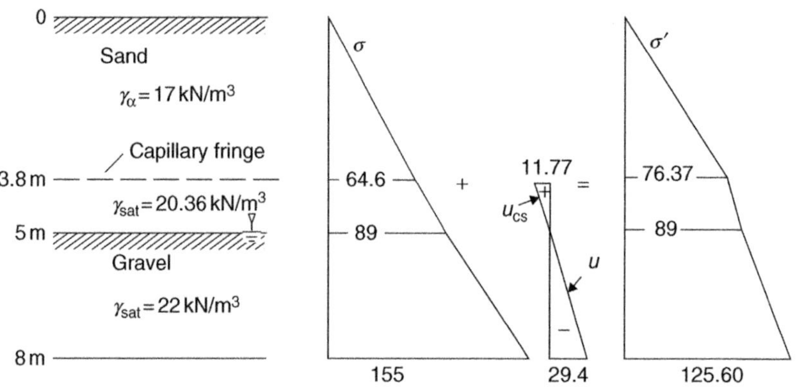

Figure 5.31

At 3.8 m $\sigma = 17 \times 3.8 = 64.6\ \text{kN/m}^2$

$u_{cs} = -9.81 \times 1.2 = -11.77\ \text{kN/m}^2$

$\sigma' = \sigma - u_{cs} = 64.6 - (-11.77) = 76.37\ \text{kN/m}^2$

At 5 m $\sigma = 17 \times 3.8 + 1.2 \times 20.36 = 89.03\ \text{kN/m}^2$

$u_{cs} = u = 0$

$\sigma' = 89.03\ \text{kN/m}^2$

At 8 m $\sigma = 89.03 + 22 \times 3 = 155.03\ \text{kN/m}^2$

$u = 9.81 \text{x} 3 = 29.4\ \text{kN/m}^2$

$\sigma' = \sigma - u = 125.6\ \text{kN/m}^2$

Note: If capillary action is ignored, than the effective pressure is underestimated below and overestimated within the region as shown in Figure 5.32.

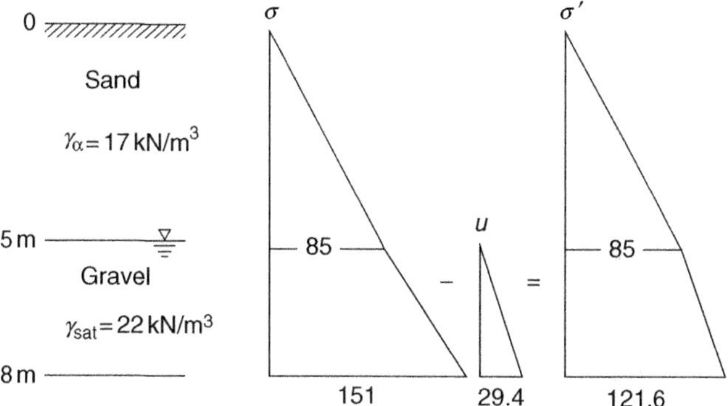

Figure 5.32

At 5 m $\sigma = 17 \times 5 = 85\ \text{kN/m}^2$

$u = 0$

$\sigma' = \sigma = 85\ \text{kN/m}^2$

At 8 m $\sigma = 85 + 22 \times 3 = 151.0\ \text{kN/m}^2$

$u = 9.81 \times 3 = -29.4\ \text{kN/m}^2$

$\sigma' = \sigma - u = 121.6\ \text{kN/m}^2$

Thus in the capillary region, the water is in tension, whilst the soil skeleton is under increased effective compression. In cohesive soil, therefore, the additional effective pressure contributes to its consolidation.

In clay, the capillary action is slower and the capillary head is much larger than for coarser soils listed in Table 5.3. The maximum value of h_{cs} for clay is assumed to be 10 m however. When a saturated sample of fine-grained soil is first taken from the ground, it does not fall apart as some or all of its shear strength (called apparent cohesion) is due to existing capillary tension.

Determination of capillary effect on site

In order to obtain reasonable estimates of h_c and h_{cs}, the soil above the groundwater table should be sampled and tested for water content and degree of saturation, at frequent intervals. The depth of each sample is plotted against its water content and the resulting curve should give an indication of the position of each fringe.

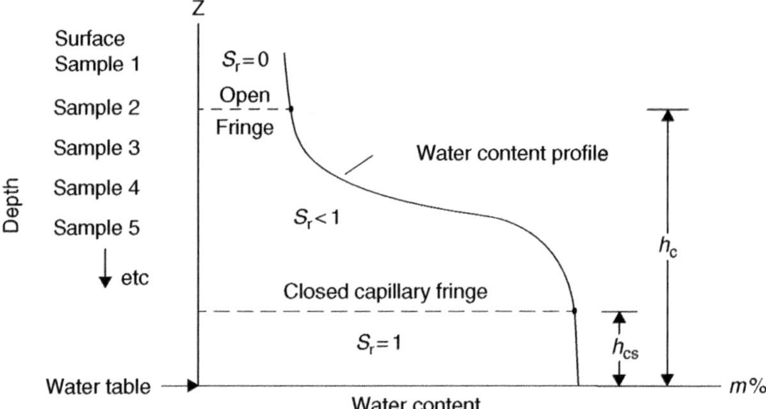

Figure 5.33

5.8.1 Equilibrium moisture content (m_E)

When saturated fine-grained soil is subjected to constant pressure, at constant drainage condition, the resulting excess pore pressure causes;

1. The outflow of some of the pore water. This flow continues, until all of the excess pressure is dissipated and the remaining water in the voids is in hydrostatic equilibrium. This water, at any depth, is called the 'equilibrium moisture content' associated with the particular applied pressure and drainage condition.
2. A decrease in the voids ratio, hence in the volume of the soil (see also consolidation).

Conversely, removal of loading would allow the ingress of water and an increase in the volume (swelling) of soil, until new equilibrium water content profile is attained eventually.

The variation of moisture content in the subgrade of roads or industrial pavements is of some importance, because of possible damage due to:

1. Swelling and shrinkage
2. Freezing and thawing.

Determination of m_E

As fine-grained soil may be considered saturated above the water table, the equilibrium moisture content may be estimated at any depth from the effective pressure–voids ratio curve of the oedometer test (see Consolidation – Chapter 7).

Step 1: Carry out the consolidation test on the soil and draw the σ'–ecurve.

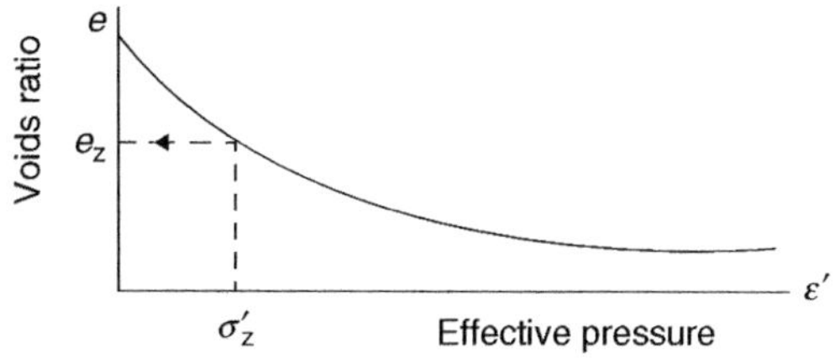

Figure 5.34

Step 2: Determine the effective pressure (σ'_z) at the depth (z) considered and obtained e_z as shown.

Step 3: Calculate the moisture content (m_z) for soil from formula (1.36), taking $S_r=1$.

$$\therefore \quad \boxed{m_z = \frac{e_z}{G_s}} = (m_E \text{ at depth } z)$$

Example 5.6

It is proposed that a concrete road should be constructed on homogeneous clay. The groundwater table is 2 m below subgrade level. The total weight of pavement and sub-base is 10.8 kN/m².

An oedometer consolidation test carried out on a clay specimen, taken from 1 m depth, yielded the following results, plotted on Graph 5.1(a):

Table 5.4a

σ' (kN/m²)	0	10	20	30	40	50	60	65
e	0.720	0.676	0.657	0.631	0.616	0.598	0.588	0.576

Calculate: the average value of saturated density, using $G_s=2.75$, from formula (1.42):

$$\gamma_{sat} = \left(\frac{G_s + e}{1+e}\right)\gamma_w$$

Table 5.4b

γ_{sat} (kN/m³)	19.79	20.05	20.17	20.34	20.43	20.55	20.62	20.7

Average density: $\gamma_{sat} = \dfrac{162.65}{8} = 20.33\,\text{kN/m}^3$

This value is used in the calculation of effective pressures. The clay above the water table is assumed to be saturated, because of capillary action.

Determine: The equilibrium moisture content profile for two cases:

1. Prior to the construction of the road
2. A long period after the end of road construction, when all excess pore pressure had dissipated.

The results of calculations are summarized in Table 5.4b for both cases. The effective pressure at depth z, for case 1, is given by:

$$\sigma' = \gamma_{sat} z = 20.33\,\text{kN/m}^2$$

For case 2, the surcharge weight ($q = 10.8\,\text{kN/m}^2$) of the pavement has to be added. The effective pressure at depth z is now given by:

$$\sigma' = 20.33z + 10.8\,\text{kN/m}^2$$

Table 5.5

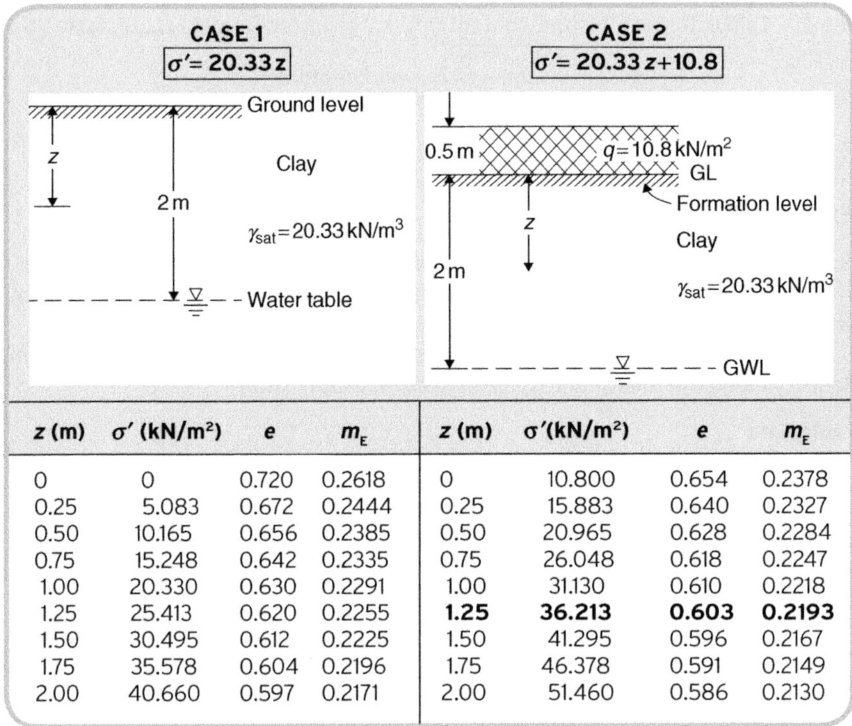

z (m)	σ' (kN/m²)	e	m_E	z (m)	σ'(kN/m²)	e	m_E
0	0	0.720	0.2618	0	10.800	0.654	0.2378
0.25	5.083	0.672	0.2444	0.25	15.883	0.640	0.2327
0.50	10.165	0.656	0.2385	0.50	20.965	0.628	0.2284
0.75	15.248	0.642	0.2335	0.75	26.048	0.618	0.2247
1.00	20.330	0.630	0.2291	1.00	31.130	0.610	0.2218
1.25	25.413	0.620	0.2255	**1.25**	**36.213**	**0.603**	**0.2193**
1.50	30.495	0.612	0.2225	1.50	41.295	0.596	0.2167
1.75	35.578	0.604	0.2196	1.75	46.378	0.591	0.2149
2.00	40.660	0.597	0.2171	2.00	51.460	0.586	0.2130

Calculations for $z = 1.25$ m Case 2:

$$\sigma' = 20.33 \times 1.25 + 10.8 = 36.213\,\text{kN/m}^2$$

From Graph 5.1(a): $e = 0.603$

From (1.36): $$m_E = \frac{e}{G_s} = \frac{0.603}{2.75} = 0.2193$$

The equilibrium moisture content profiles are drawn on Graph 5.1(b). These verify that increased load decreases the water content, until the excess pore pressure dissipates and the moisture content profile reaches equilibrium.

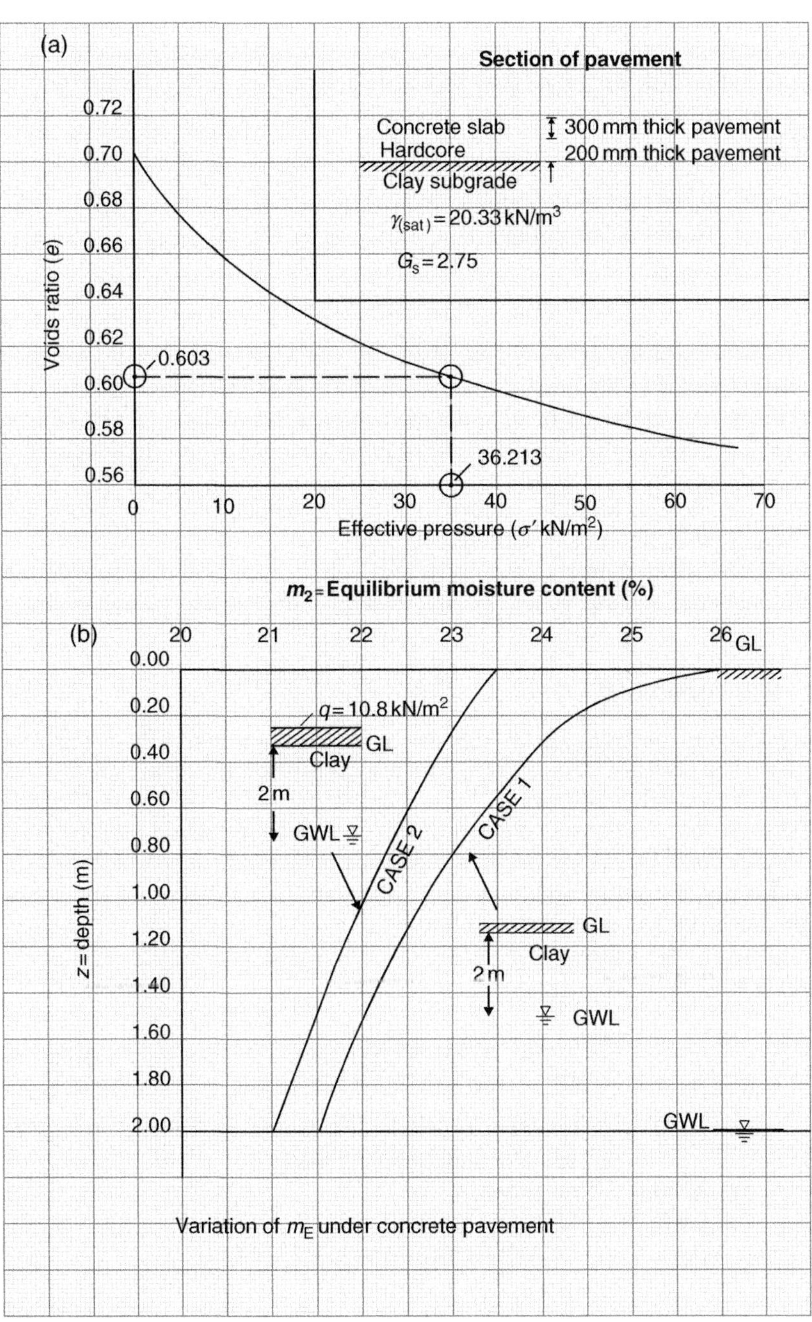
(a)
Section of pavement
Concrete slab
Hardcore
300 mm thick pavement
200 mm thick pavement
Clay subgrade
$\gamma_{(sat)}$ = 20.33 kN/m³
G_s = 2.75
Voids ratio (e)
0.72
0.70
0.68
0.66
0.64
0.62
0.60
0.58
0.56
0.603
36.213
0
10
20
30
40
50
60
70
Effective pressure (σ′ kN/m²)
m_2 = Equilibrium moisture content (%)
(b)
20
21
22
23
24
25
26
GL
z = depth (m)
0.00
0.20
0.40
0.60
0.80
1.00
1.20
1.40
1.60
1.80
2.00
q = 10.8 kN/m²
GL
Clay
2 m
GWL
CASE 2
CASE 1
GL
Clay
2 m
GWL
GWL
Variation of m_E under concrete pavement

Graph 5.1

Compaction and m_E

It was pointed out in Chapter 1, that the deposited soil is normally compacted near to its optimum moisture content (m_0). However, if a pavement is to be placed on the compacted surface, then the thickness of the road should be determined by the eventual equilibrium moisture content profile, as well as by the traffic to be carried and the CBR results. In general if:

1. $m_0 > m_E$, the soil will loose water;
2. $m_0 < m_E$, the soil will gain water.

Notes:

1. Concrete surfacing prevents evaporation of soil water as well as the entry of rain water. This allows the gradual attainment of the equilibrium profile.
2. If the ground surface is open to the elements, as it is at the edges of a pavement, then m_E varies with climatic and seasonal changes to an approximate depth of 1 m.
3. The variation of moisture content can cause damage to a structure built on cohesive soils, because of subsequent:
   (a) freezing and thawing;
   (a) shrinkage and swelling, due to desiccation and absorption respectively.

5.8.2 Soil suction (S_s)

Capillary action is caused by surface tension acting on the water at the menisci formed between the soil particles. The negative pore pressure, below the open capillary fringe of height h_c, is an indication of suction.

The strength of suction is a function of the degree of saturation, that is, dryer soil sucks up water faster than wet soil. It is analogous to blotting paper, which draws up ink faster when dry and slower when partially saturated. This implies that suction is high when the soil is dry and zero when saturated.

Pore pressure was expressed in terms of total stress and suction by Croney and Coleman in 1953 as:

$$\boxed{u = \alpha\sigma - S_s} \tag{5.18}$$

Where, $\alpha = 0$ for incompressible soil, where no volume change occurs upon the application of load
$\alpha = 0.5$ for silty clay
$\alpha = 0.15$ for sandy clay
$\alpha = 1$ for compressible, saturated soil e.g. clay

The compressibility factor (α) may be obtained in terms of the plasticity index, by the empirical, therefore approximate formula:

$$\boxed{\alpha = \frac{\text{PI}}{40}} \quad \text{for} \quad \text{PI} \leq 40\% \tag{5.19}$$

Effective stress

It can be expressed at any depth in terms of soil suction.

From (5.1): $\sigma' = \sigma - u$ $\quad\therefore\; \sigma' = \sigma - (\alpha\sigma - S_s)$

From (6.18): $u = \alpha\sigma - S_s$ $\quad = \sigma - \alpha\sigma - S_s$

Therefore for $0 \le \alpha \le 1$ $\quad \boxed{\sigma' = (1-\alpha)\sigma + S_s}$ (5.20)

Two particular cases can be derived from this general formula.

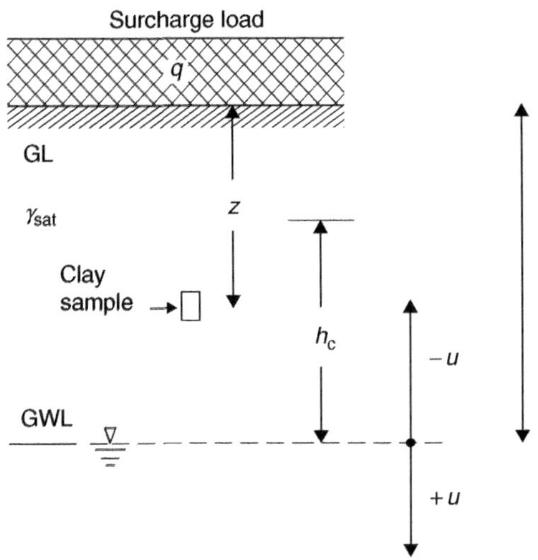

Capillary region. The pore pressure is −ve above and+ve below the water table.
Total pressure: $\sigma = z\gamma_{sat} + q$

Figure 5.35

Case 1: $\boxed{\alpha = 0}$ $\quad \sigma' = (1-0)\sigma + S_s$

$\quad = \sigma + S_s$

But $\quad \sigma' = \sigma - u \quad \therefore \quad \sigma - u = \sigma - S_s$ $\quad \boxed{\therefore\; S_s = -u \approx \gamma_w h_c}$

This shows that if the soil sample is removed from the ground, there are no external pressures exerted on it and the pore water pressure in the specimen is balanced by the capillary action.

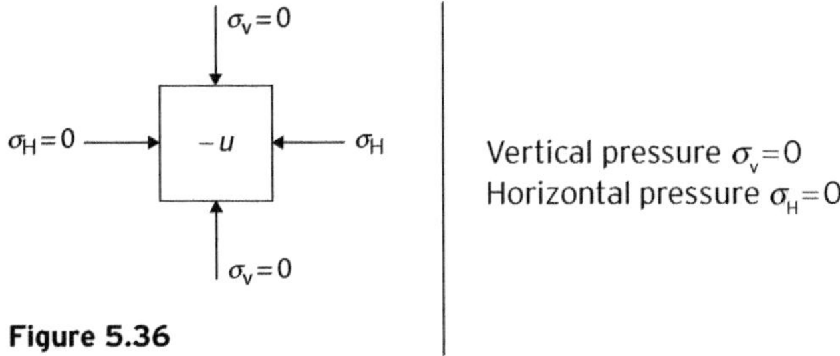

Vertical pressure $\sigma_v = 0$
Horizontal pressure $\sigma_H = 0$

Figure 5.36

Note that the surface tension at the surface of the sample gives material its 'apparent cohesion', which would disappear if the soil is immersed in water.

Also, if the total stress is not increased by a surcharge load ($q=0$), hence there is no excess pore pressure, then the suction equals to the pore pressure.

$$\boxed{S_s=-u\approx\gamma_w h_c}$$

This also applies to incompressible materials such as dense, compact sand and rock, where the application of surcharge load does not increase the pore pressure. The load is balanced by the intergranular pressure, hence no dissipation or consolidation can occur.

Case 2: $\boxed{\alpha=1}$ $\sigma'=(1-1)\sigma+S_s$ $\therefore\boxed{\sigma'=S_s}$

So $\sigma'=\sigma-u=S_s$ $\therefore\boxed{u=\sigma-S_s}$

This means that at the end of the consolidation process, after the dissipation of the excess pore pressure, the soil suction equals to the effective pressure.

Soil suction index (pF)

The magnitude of suction is indicated by this index. It is defined as the logarithm (base 10) of suction expressed in terms of the height of water in centimeters.

That is, $\boxed{pF=\log_{10}(S_s)}$ $0\le pF\le 6$ (5.21)

Therefore, if pF of a soil is known from tests, then the suction is given by:

$$\boxed{S_s=10^{(pF)}\,\Big|\,\text{cm}} \tag{5.22}$$

Expressing S_s in metres: $S_s=\dfrac{10^{(pF)}}{10^2}$

Therefore, $\boxed{S_s=10^{(pF-2)}\,\Big|\,\text{m}}$ (5.23)

In pressure units: $\boxed{S_s=\gamma_w\times10^{(pF-2)}\,\Big|\,\text{kN/m}^2}$ (5.24)

Variation of suction pressure over the range of pF is listed below.

Table 5.6

Index	Suction		
	$S_s=10^{(pF)}$	$S_s=10^{(pF-2)}$	$S_s=9.81\times10^{(pF-2)}$
pF	**cm**	**m**	**kN/m²**
0	1	0.01	9.81×10^{-2}
1	10	0.10	9.81×10^{-1}
2	10^2	1.00	9.81
3	10^3	10	9.81×10
4	10^4	10^2	9.81×10^2
5	10^5	10^3	9.81×10^3
6	10^6	10^4	9.81×10^4

The value of pF for a particular soil can be determined over the range $0 \le \text{pF} \le 6$, by the 'High pressure membrane method', as described in laboratory manuals. The results are plotted on the $m-$pF graph.

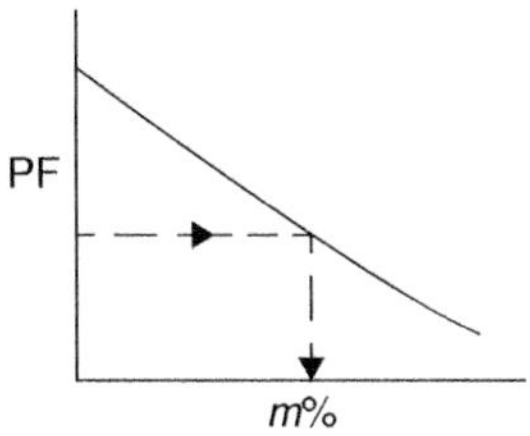

Figure 5.37

Suction and m_E

It has been pointed out that at the completion of consolidation the pore water is in hydrostatic equilibrium and that

$$\alpha = 1$$
$$S_s = \sigma'$$

This provides an alternative method for the determination of the equilibrium moisture content. The procedure to calculate m_E at a depth z is as follows:

Step 1: Calculate the total stress (σ) at z.

Step 2: Determine the pore pressure (u) at z, remembering that u is negative above and positive below the ground water table.

Step 3: Calculate the effective pressure at z by:

$$\sigma' = \sigma - (\pm u) = S_s (\text{kN/m}^2)$$

Step 4: Index pF is evaluated after expressing it from formula (5.24):

$$S_s = \gamma_w \times 10^{(\text{pF}-2)} \quad \text{or} \quad \frac{S_s}{\gamma_w} = 10^{(\text{pF}-2)}$$

$$\therefore \quad \log\frac{S_s}{\gamma_w} = \text{pF} - 2 \quad \text{and} \quad \boxed{\text{pF} = 2 + \log\left(\frac{S_s}{\gamma_w}\right)} \qquad (5.25)$$

Note that the unit if S_s is kN/m².

Step 5: Using the given m-pF diagram (see Graph 5.2a), the moisture content m%, for the evaluated pF index, is read off. This is the equilibrium moisture content at depth z.

Example 5.7

Referring to Example 5.6, determine the equilibrium moisture content variation to 3 m depth below formation level, using the m–pF curve of Graph 5.2a.

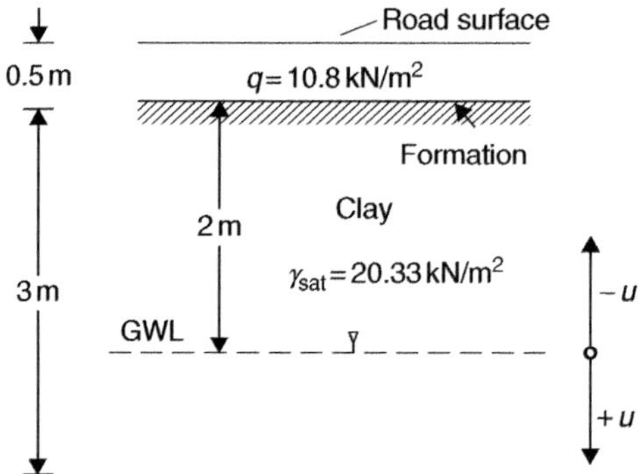

Figure 5.38

For $z = 1.25\,\text{m}$

$$\sigma = 10.8 + 20.33 \times 1.25 = 36.21\,\text{kN/m}^2$$
$$u = (2 - 1.25) \times 9.81 = 7.36\,\text{kN/m}^2$$
$$\sigma' = \sigma - u$$
$$= 36.21 - (-7.36) = 43.57\,\text{kN/m}^2$$
$$\therefore\ S_s = \sigma' = 43.57\,\text{kN/m}^2$$

From (5.25): $\text{pF} = 2 + \log\left(\dfrac{43.57}{9.81}\right) = 2 + 0.6475 = 2.65$

Marking pF = 2.65 on Graph 5.2(a), the equilibrium moisture content is read off as $m_E = 21.9\%$ at depth $z = 1.25$ m. Similar calculations are to be made for each depth and the results tabulated.

Table 5.7

	z	$\sigma = 10.8 + 20.33z$	$u = 9.81\,(2 - z)$	$\sigma' = \sigma - u$	$\text{pF} = 2 + \log\left(\dfrac{S_s}{9.81}\right)$	m_E
	m		kN/m²			%
	0	10.80	−19.62	30.42	2.49	23.8
	0.25	15.88	−17.17	33.05	2.53	23.3
	0.50	20.97	−14.72	35.68	2.56	22.8
	0.75	26.05	−12.26	38.31	2.59	22.5
	1.00	31.13	−9.81	40.94	2.62	22.2
	1.25	**36.21**	**−7.36**	**43.57**	**2.65**	**21.9**
−u	1.50	41.30	−4.91	46.20	2.67	21.7
↑	1.75	46.38	−2.45	48.83	2.70	21.5
	2.00	51.46	0	51.46	2.72	21.3
↓	2.50	61.63	4.91	56.72	2.76	21.1
+u	3.00	71.79	9.81	61.98	2.80	21.0

The equilibrium moisture content profile is plotted on Graph 5.2(b).

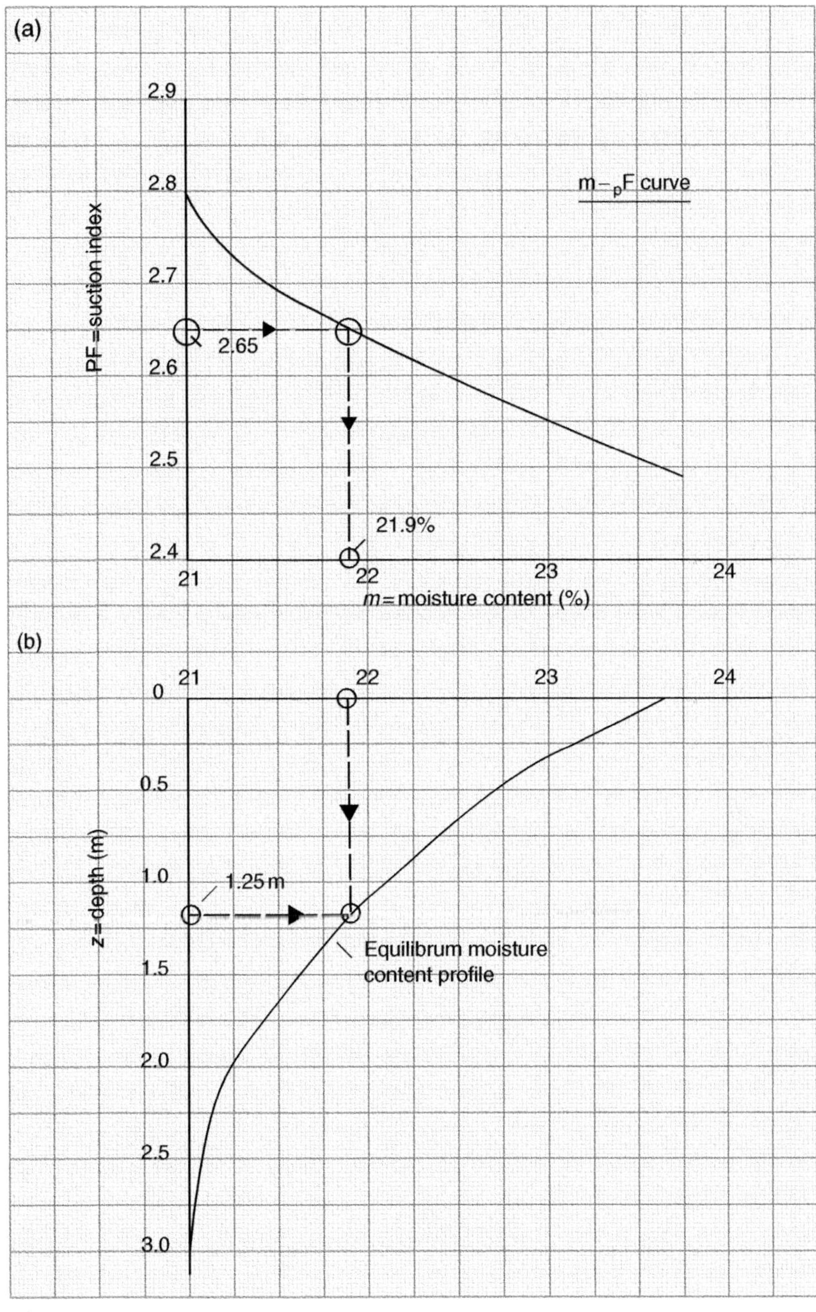

Graph 5.2

Problem 5.1

Excavation is carried out for the purpose of road construction. The formation is designed to be 6 m below existing ground level. There had been no site investigation to ascertain the soil profile, shown in Figure 5.39. The contractor is unaware that the groundwater table is at the top of the stiff clay and there is an artesian pressure of 45 kN/m² in the gravel layer.

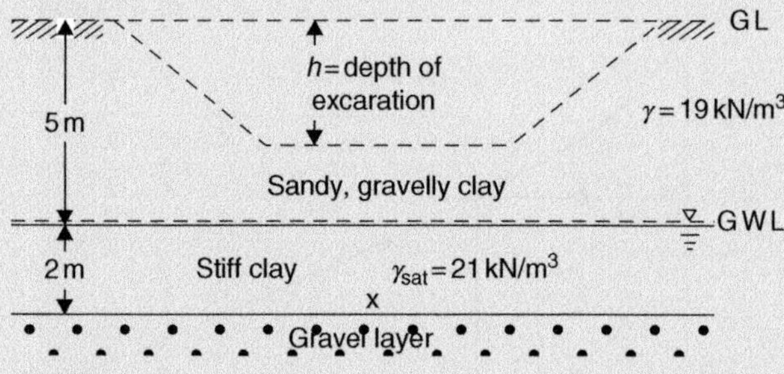

Figure 5.39

Artesian pressure: $\sigma_n = 45\,\text{kN/m}^2$

Estimate that depth of excavation at which the stiff clay fails under the artesian pressure, flooding the trench, thus causing disruption to the earthworks.

Problem 5.2

A stream is to be diverted through a new culvert, before a 20 m-thick compacted fill is placed over it in a land reclamation scheme. The maximum depth of water flow in the conduit is not expected to exceed 1.5 metres. The final ground profile is to be as shown:

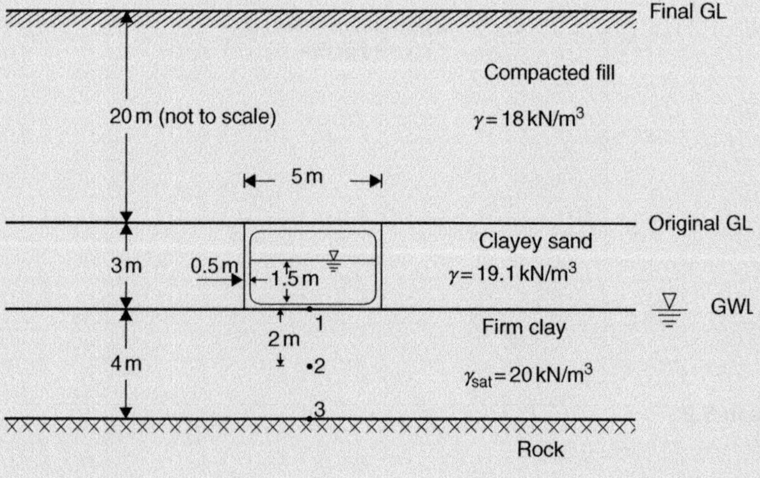

Figure 5.40

The external dimensions of the reinforced concrete culvert section is 3 m × 5 m, having 500 m thick walls. The mass density of reinforced concrete may be assumed as 2446 kg/m^3.

Calculate the total pore water and effective pressures at points 1, 2 and 3, before the commencement and after the completion of the scheme.

Problem 5.3

Figure 5.43 shows the cross-section of the ground and the available information on the two soil layers overlying the stiff clay. Calculate the total and effective pressures at the top and bottom of the silty clay layer.

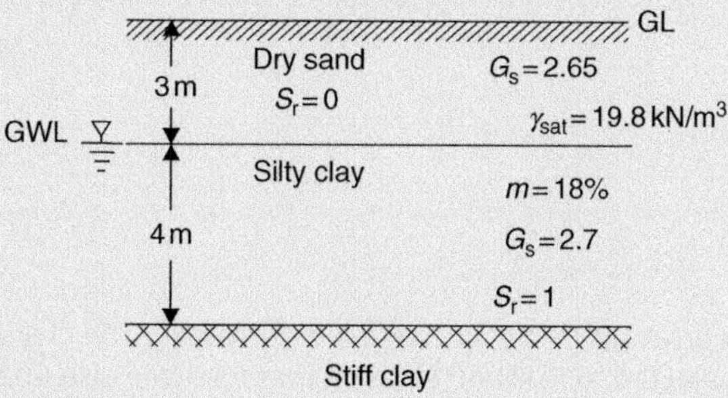

Figure 5.43

Problem 5.4

A jetty is to be constructed at a lake. The available information on the ground conditions is shown in Figure 5.45. Only the dry unit weight and the specific gravity of each layer are known.

Determine for each layer:

1. porosity
2. voids ratio
3. saturated unit weight
4. submerged unit weight.

Calculate the effective pressure at the boundaries of each layer in terms of:

a) saturated unit weight
b) submerged unit weight.

Draw the total pore and effective pressure diagrams.

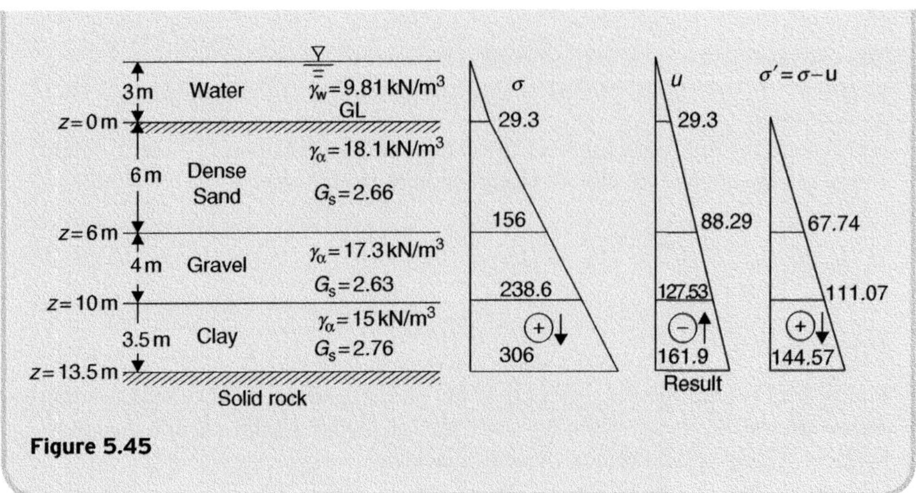

Figure 5.45

Problem 5.5

A 3 m thick clay layer forms the bed of a 4 m deep lake. The clay is underlain by coarse gravel subjected to 18 kN/m² artesian pressure (see also Chapter 3).

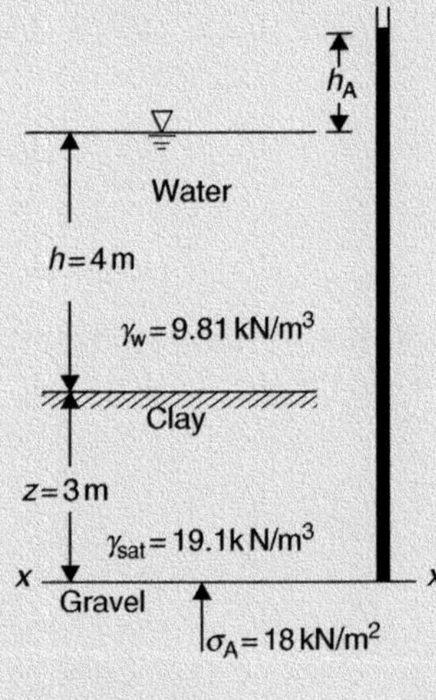

Figure 5.46

1. Derive an expression for the effective pressure (σ') at the top of the gravel in terms of:
   a) z, γ' and σ_A
   b) z and the modified submerged density γ''.
2. Derive a formula for the critical thickness (z_c) of the clay layer, at which it fails under σ_A=18 kN/m².
3. Calculate z_c, γ'', σ' and h_A.

Problem 5.6

Artesian pressure σ_A exists in the gravel layer below several layers of soil, shown in Figure 5.49. Show that the effective pressure at the top of the gravel (point x) can be expressed by:

1. Either $\sigma'_x = z_1\gamma_1 + z_2\gamma'_2 + z_3\gamma'_3 - \sigma_A$
2. Or $\sigma'_x = z_1\gamma_1 + z_2\gamma''_2 + z_3\gamma''_3 + \sigma_A$

where from (5.11): $\gamma'' = \gamma'_1 - i\gamma_w$

but,

$$i = \frac{h_A}{z}, \quad h_A = \frac{\sigma_A}{\gamma_w} \quad \therefore \; i = \frac{\sigma_A}{z}$$

For a layer therefore,

$$\boxed{\gamma'' = \gamma' - \frac{\sigma_A}{z}}$$

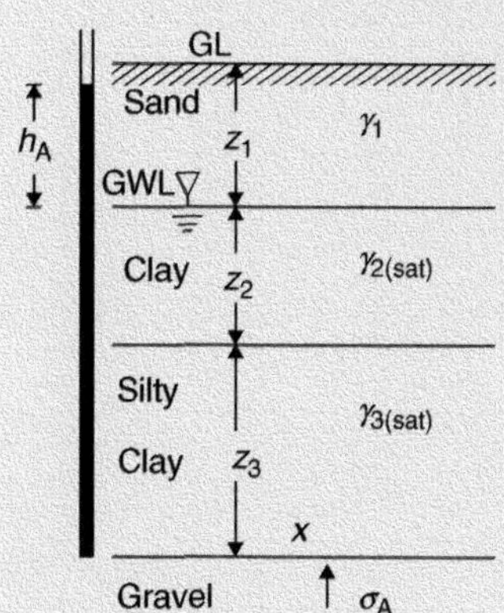

Figure 5.47

The submerged and modified submerged densities are:

$$\gamma'_2 = \gamma_{2(sat)} - \gamma_w \quad \text{and} \quad \gamma''_2 = \gamma'_2 - \frac{\sigma_A}{z_2}$$

$$\gamma'_3 = \gamma_{3(sat)} - \gamma_w \quad \text{and} \quad \gamma''_3 = \gamma'_3 - \frac{\sigma_A}{z_3}$$

Problem 5.7

Figure 5.48 shows a section of the ground, indicating artesian pressure of 11.8 kN/m^2 within the gravel layer as well as the unit weights of the layers above it. Estimate the effective pressure at points *x*, *y* and *z*, in terms of:

a) Natural bulk densities
b) Submerged densities
c) Modified submerged densities

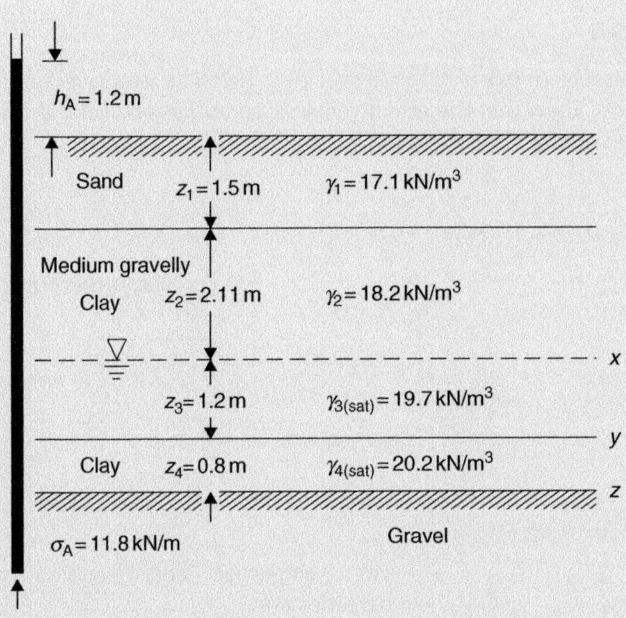

Figure 5.48

Submerged densities:

$$\gamma_3' = 9.89\,\text{kN/m}^3 \quad \gamma_3'' = 9.89 - \frac{11.8}{1.2} = 0.06\,\text{kN/m}^3$$

$$\gamma_4' = 10.39\,\text{kN/m}^3 \quad \gamma_4' = 10.39 - \frac{11.8}{0.8} = -4.36\,\text{kN/m}^3$$

Problem 5.8

A 944 cm^3 dry soil sample is gradually flooded. It has 45% porosity and 15.1 kN/m^3 dry density. Estimate the mass of added water and unit weight when the soil is:

a) Partially saturated to $S_r = 0.69$
b) Fully saturated

Chapter 6
Shear Strength of Soils

Definition

The shear strength of a soil is its maximum resistance to shearing stresses.

When soil is subjected to vertical loading, it fails in shear, that is it deforms plastically by sliding over a slip surface as shown in Figure 6.1(a).

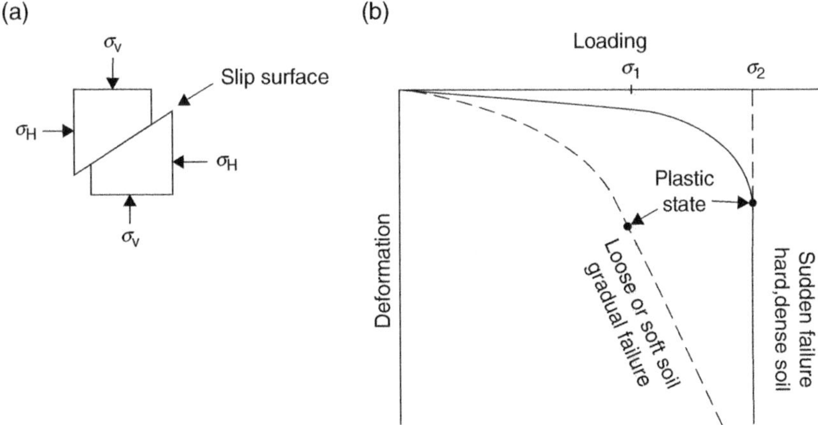

Figure 6.1

Plastic deformation means, that it cannot be reversed by the removal of loading.

The soil enters plastic state, when the shearing stress acting on the slip surface, reaches its shear strength. Figure 6.1(b) depicts the two ways, in which the soil can fail:

1. Suddenly, in hard cohesive soils. In this case, local failure at one point precipitates general rupture of the entire slip surface, followed by large deformation.
2. Gradually, in soft clayey and loose, granular soils. In this case the deformation progresses slowly over the slip surface.

Normally the mode of failure would fall between these two limiting cases.

Introduction to Soil Mechanics, First Edition. Béla Bodó and Colin Jones.

6.1 Coulomb-Mohr Theory

There are several theories concerning the failure of loaded soil. Of these, the combination of Coulomb's and Mohr's are sufficiently accurate for engineering purposes. More so, since they were modified to take into account of the effect of pore pressure on the shear strength.

Coulomb (1776) suggested, that soil fails because the strength provided by inter-particle friction and cohesion is exceeded by the applied shearing stress on the slip surface. The stresses acting on the surface, inclined at an angle α to the horizontal, are shown on Figure 6.2(a). Coulomb related the quantities present by the linear equation:

$$\boxed{\tau = c + \sigma_n \tan\phi} \tag{6.1}$$

This is represented graphically in Figure 6.2(b).

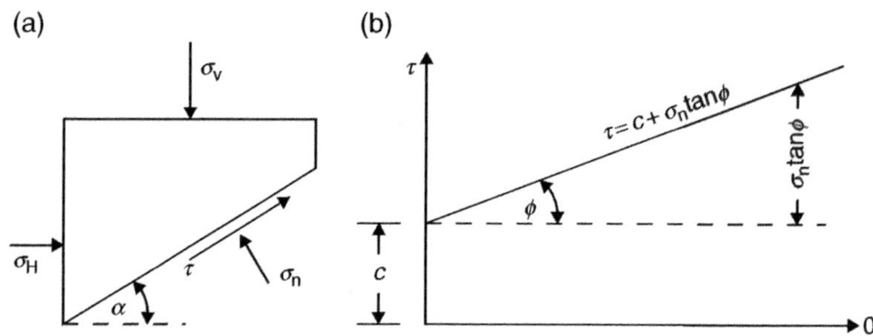

Figure 6.2

Where: τ = shear stress on any plane
σ_n = normal pressure on the plane
σ_v = vertical (major) principal stress
σ_H = horizontal (minor) principal stress
ϕ = angle of friction
c = cohesion
α = inclination of the plane considered

Mohr (1910) developed the graphical representation of the possible stresses at a point on a plane, within a material. The derivation of his graphical construction can be found in books on statics or strength at materials. Basically, if a soil specimen is subjected to an all-round horizontal pressure (σ_H) - as it would be in-situ - and a vertical pressure (σ_v), then the shear and normal stresses on any plane within the soil can be represented by points on the circumference of a circle, as shown.

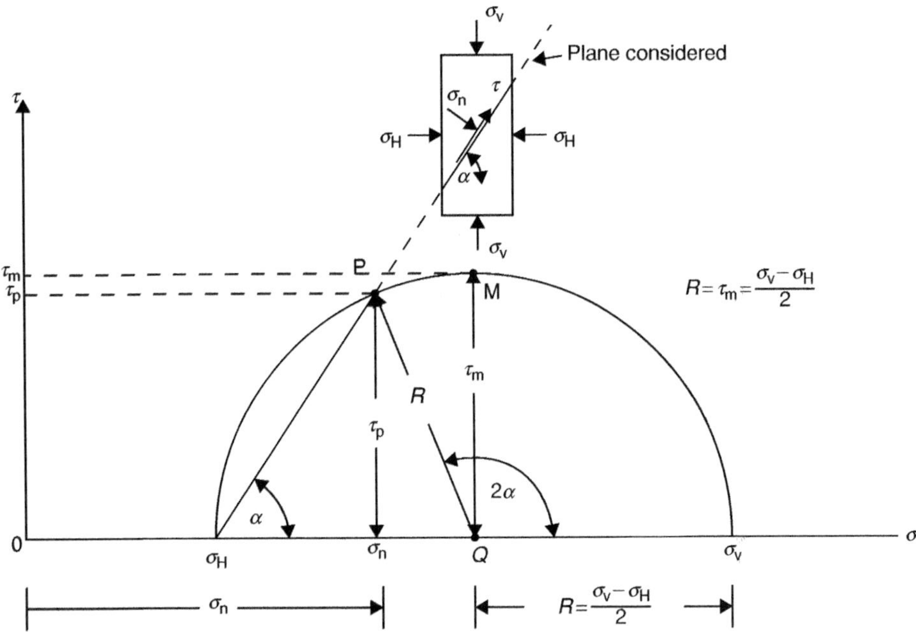

Figure 6.3

Where: R = radius of the circle

τ_p = shear stress on plane inclined at angle α

σ_n = normal pressure on the same plane

τ_m = maximum possible value of shear stress

Note that σ_v and σ_H are also called "principal stresses".

These quantities can be calculated from the following expressions:

$$\tau_p = R \sin 2\alpha = \frac{\sigma_v \quad \sigma_H}{2} \sin 2\alpha \tag{6.2}$$

$$\sigma_n = \frac{\sigma_v + \sigma_H}{2} + \frac{\sigma_v - \sigma_H}{2} \cos 2\alpha \tag{6.3}$$

6.1.1 Stresses on the plane of failure

Mohr's method can now be applied to find the straight line, which satisfies Coulomb's equation, yielding numerical values for the cohesion and the angle of friction.

According to the theory, if a number of specimens from the same material are tested to failure, under different principal stresses, then point P of each circle has to lie on a common tangent (Figure 6.4). The equation of this line, often called 'failure envelope' is that given by Coulomb.

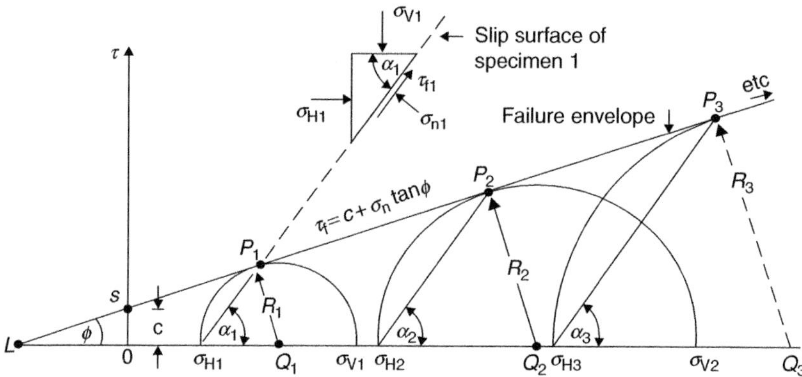

Figure 6.4

Where τ_f = shear strength or maximum resistance to shear } at failure
σ_n = normal pressure on the same surface

The shear strength parameters c and ϕ as well as the failure shear stress τ_f can be derived from Figure 6.5.

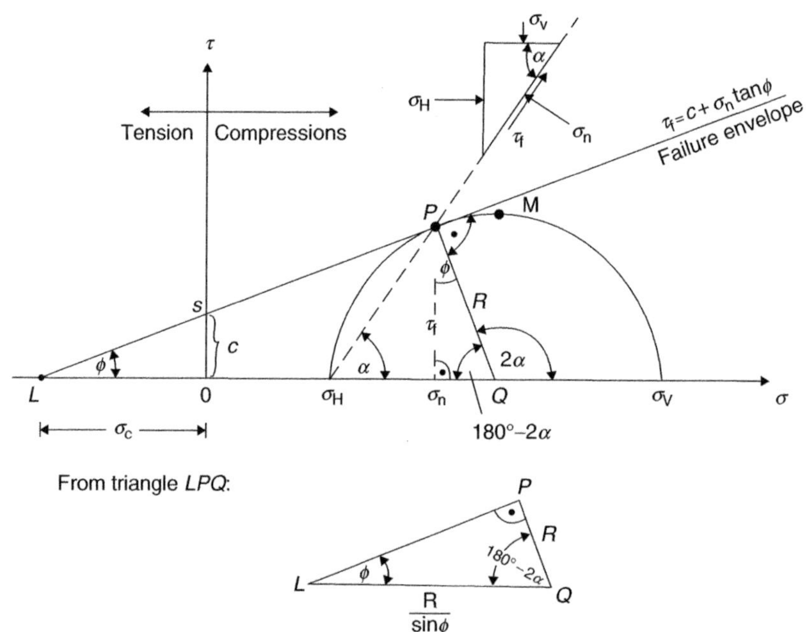

Figure 6.5

$$180° = \phi + 90° + 180° - 2\alpha \quad \therefore \quad \boxed{\phi = 2\alpha - 90°} \tag{6.4}$$

$$\tan\phi = \frac{c}{\sigma_i} \quad \therefore \quad \boxed{c = \sigma_i \tan\phi} \tag{6.5}$$

Where σ_i may be considered as an internal tensile stress.

$$\cos\phi = \frac{\tau_f}{R} \text{ but } R = \frac{\sigma_v - \sigma_H}{2} \quad \therefore \quad \boxed{\tau_f = \frac{\sigma_v - \sigma_H}{2}\cos\phi} \tag{6.6}$$

Also, from (6.2):

$$\boxed{\tau_f = \frac{\sigma_v - \sigma_H}{2}\sin 2\alpha}$$

From (6.1):

$$\boxed{\tau_f = c + \sigma_n \tan\phi}$$

Note: Figure 6.6 shows that the soil does not fail on the plane of the maximum shear stress (τ_{max}) through point M, but on a steeper one:

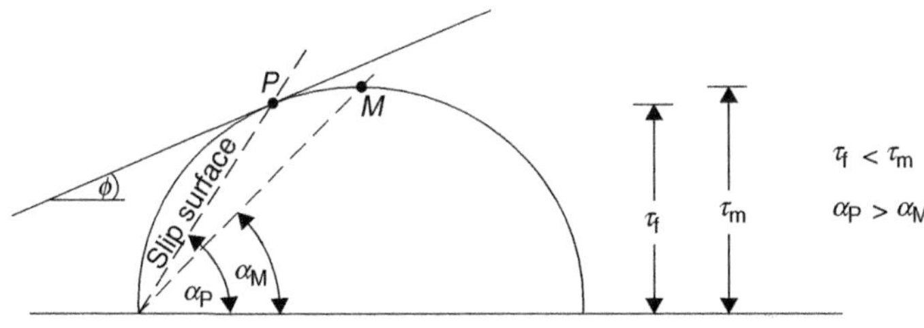

Figure 6.6

6.1.2 Friction and cohesion

Coulomb's equation defines the shear stress in terms of friction angle ϕ and the cohesion intercept c. These are empirical constants only, depending on the natural state of the soil as well as on the method of testing. They indicate, however, that the shear strength of soils largely depends on friction between the particles and on their cohesion.

Friction

In coarse-grained soils, the shear strength depends largely on:

1. Surface roughness of the grains.
2. Interlocking of grains and state of compaction.
3. Magnitude of contact pressures.
4. Adhesion in finer-grained wet soils due to thin water layer between the contact points.

Cohesion

In fine-grained soil the shear strength is assumed to be dependent largely on:

1. Water content of the soil
2. Shape, size and packing of the particles
3. Adhesion due to the thin film of water between the contact surfaces.

6.1.3 Apparent cohesion

This is caused by surface tension, acting at the ends of the thin film of water between the contact surfaces of fine-grained, moist soils. The apparent cohesion disappears if the soil is flooded or completely dried. This is why sand particles stick together, whilst partially saturated.

Notes: Soils are often referred to as:
a) ϕ – soil when $c = 0$
b) c – soil when $\phi = 0$
c) $c - \phi$ – soil when neither ϕ nor c is zero.

6.2 Stress path (Lambe, 1964)

This is another graphical method to represent the failure envelope. In addition, the variation of stress within a specimen during test can be visualized by drawing the path of change (see also Chapter 8).

The *p*–*q* diagram

Figure 6.7 shows this diagram as well as Mohr's.

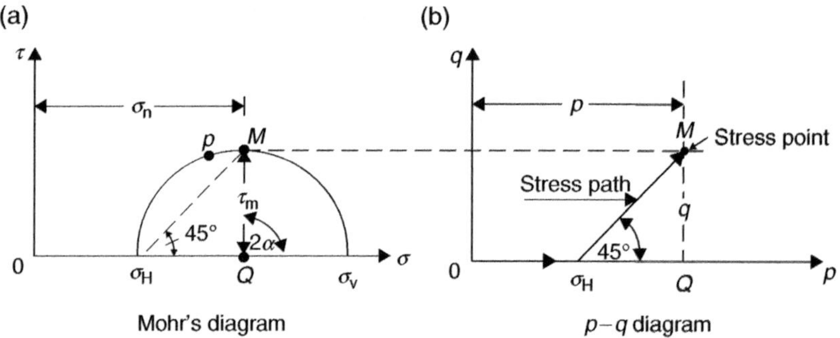

Figure 6.7

In this method of presentation, it is not the tangent point of the Mohr-circle, but its peak point (M) is drawn, having p and q as coordinates.

From (6.3): $$\sigma_n = \frac{\sigma_V + \sigma_H}{2} + \frac{\sigma_V - \sigma_H}{2}\cos 2\alpha$$

But $2\alpha = 90°$ $$\therefore \sigma_n = \frac{\sigma_V + \sigma_H}{2} = p \quad \therefore \boxed{p = \frac{\sigma_V + \sigma_H}{2}} \tag{6.7}$$

From (6.2): $$\tau_m = \frac{\sigma_V - \sigma_H}{2}\sin 2\alpha$$

$$q = \tau_m$$

and $$\sin 90 = 1$$

$$\therefore \boxed{q = \frac{\sigma_V - \sigma_H}{2}} \tag{6.8}$$

Notes: a) The stress path extends as the vertical pressure is increased.
b) Both $\sigma-\tau$ and $p-q$ coordinates may be drawn on the same diagram as shown.

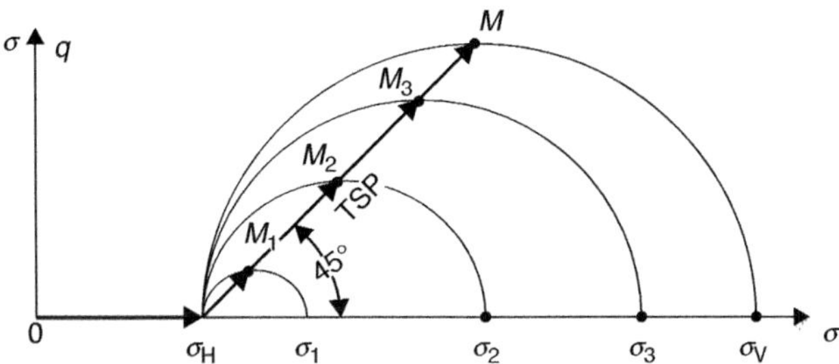

Figure 6.8

Where σ_1, σ_2, and σ_3 are the increasing values of vertical pressure, defining the intermediate Mohr-circles before failure is reached at σ_v.

and TSP = total stress path. The soil is assumed to be either partially saturated or loaded rapidly.

Note: The influence of water content and saturation on Mohr's circle and stress paths will be considered shortly.

6.2.1 Stress path failure envelope

Figure 6.9 shows the stress path in three triaxial tests carried out on a material and the resulting failure envelope.

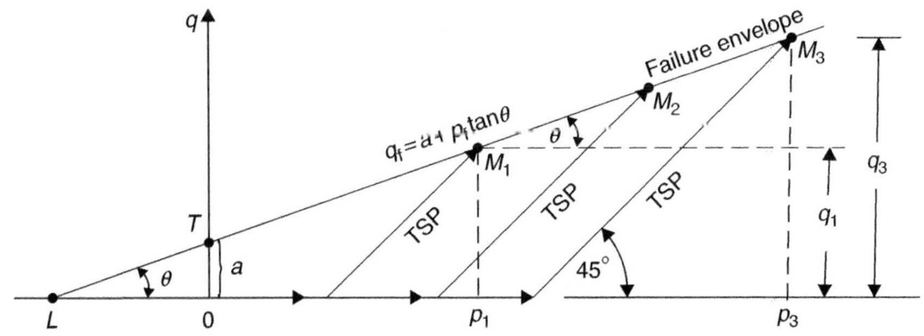

Figure 6.9

The equation of the failure envelope is defined as:

$$\boxed{q_f = a + p_f \tan\theta} \tag{6.9}$$

The total stress paths are parallel for the same material. Also, they are inclined at 45°, when tested triaxially (see also Figure 6.13).

The shear strength parameters, c and ϕ, can now be derived in terms of a and θ.

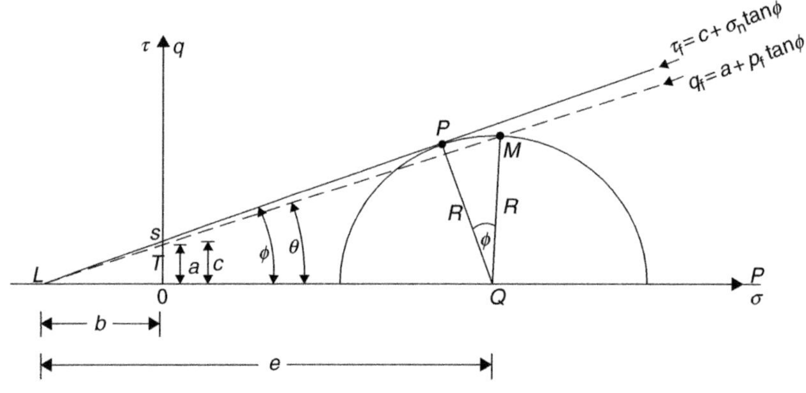

Figure 6.10

From triangle PLQ: $\sin\phi = \dfrac{R}{e}$

From triangle LMQ: $\tan\theta = \dfrac{R}{e}$

$\sin\phi = \tan\theta$

$$\therefore \boxed{\phi = \sin^{-1}(\tan\theta)} \tag{6.10}$$

From triangle LOS: $\tan\phi = \dfrac{c}{b}$

From triangle LOT: $\tan\phi = \dfrac{a}{b}$

$b = \dfrac{c}{\tan\phi} = \dfrac{a}{\tan\theta}$

$\tan\phi = \dfrac{\sin\phi}{\cos\phi} = \dfrac{\tan\theta}{\cos\phi}$

$$\text{Substituting } c = \frac{a}{\tan\theta} \times \frac{\tan\theta}{\cos\phi} \quad \therefore \boxed{c = \frac{a}{\cos\phi}} \tag{6.11}$$

Note: Angle θ can be calculated directly from two test results as shown in Figures 6.9 and 6.11.

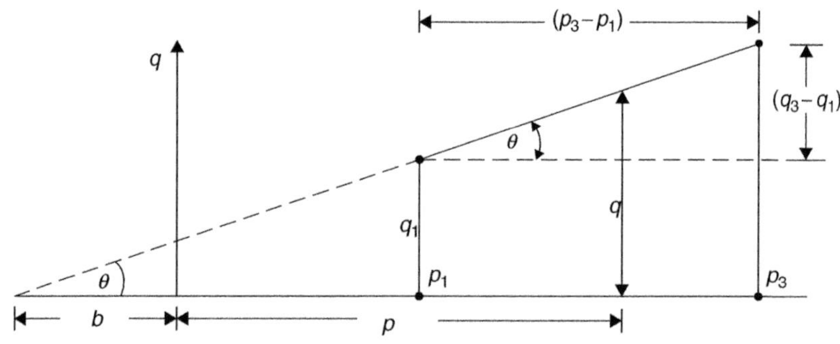

Figure 6.11

$$\tan\theta = \frac{q_3 - q_1}{p_3 - p_1} \qquad \boxed{\theta = \tan^{-1}\frac{q_3 - q_1}{p_3 - p_1}} \tag{6.12}$$

$$b + p = \frac{q}{\tan\theta} = \frac{q}{\sin\phi} \qquad \therefore\ b = \frac{q}{\sin\phi} - p$$

$$c = b\tan\phi = \left(\frac{q}{\sin\phi} - p\right)\tan\phi \quad \therefore \quad \boxed{c = \left(\frac{q}{\sin\phi} - p\right)\tan\phi} \tag{6.13}$$

Summary

The shear strength parameters c and ϕ may be evaluated from test results in three ways:

1. Graphically, by Mohr's circles
2. Semi-graphically, by stress paths
3. Analytically, by means of stress path coordinates.

All three methods are illustrated in Example 6.1.

Example 6.1

Three specimens of the same soil were subjected to the pressures given below:

Table 6.1

Specimen	1	2	3
σ_H (kN/m²)	100	200	300
σ_v (kN/m²)	400	600	800

Obtain c and ϕ as well as the stresses on the three slip surfaces from the relevant diagram.

Method 1

The two shear strength parameters, measured from the Mohr-circle diagram, drawn on Graph 6.1 are:

$$c = 70\,\text{kN/m}^2$$
$$\phi = 19.5° \qquad \therefore\ \tan\phi = 0.354$$
$$\alpha = 55°$$

Table 6.2

τ_f (kN/m²)	140	190	235
σ_n (kN/m²)	200	335	465
$R = \tau_m$ (kN/m²)	150	200	250

Equation of the failure envelope: $\tau_f = 70 + 0.354\,\sigma_n$

Method 2

The stress path envelope is drawn on Graph 6.2, by calculating the relevant values of p and q. The TSP arrows may be omitted.

Table 6.3

p (kN/m^2)	250	400	550
q (kN/m^2)	150	200	250

From Graph 6.2: $a = 66\,\text{kN/m}^2$

$\theta = 18°$ $\quad\therefore\quad$ $\tan\theta = 0.325$

The equation of the envelope is: $q_f = 66 + 0.325p_f$

From (6.10): $\phi = \sin^{-1}(\tan\theta) = \sin^{-1}(0.325) = 18.96\ (\approx 19.5)$

From (6.11): $$c = \frac{a}{\cos\phi} = \frac{66}{\cos 18.66} = 69.7\,\text{kN/m}^2\ (\approx 70)$$

Method 3

Substituting two of the stress path coordinates, already calculated in Method 2, into (6.12) for angle θ:

$q_1 = 150\,\text{kN/m}^2$

$q_3 = 250\,\text{kN/m}^2$

$p_1 = 250\,\text{kN/m}^2$

$p_3 = 550\,\text{kN/m}^2$

$q_3 - q_1 = 100$

$p_3 - p_1 = 300$

$$\tan\theta = \frac{100}{300} = 0.333$$

$$\theta = \tan^{-1}\frac{100}{300} = 18.4°$$

For angle ϕ, apply (6.10): $$\phi = \sin^{-1}(\tan\theta) = \sin^{-1}(0.333) = 19.4°$$

For cohesion c, apply (6.13): $$c = \left(\frac{q}{\sin\phi} - p\right)\tan\phi$$

$\sin\phi = \sin 19.4 = 0.332$

$\tan\phi = \tan 19.4 = 0.352$

Taking $p = 550\,\text{kN/m}^2$

and $q = 250\,\text{kN/m}^2$

$$c = \left(\frac{250}{0.332} - 550\right) \times 0.352 = 71.5\,\text{kN/m}^2$$

Therefore, c and ϕ can be determined accurately by calculation only.

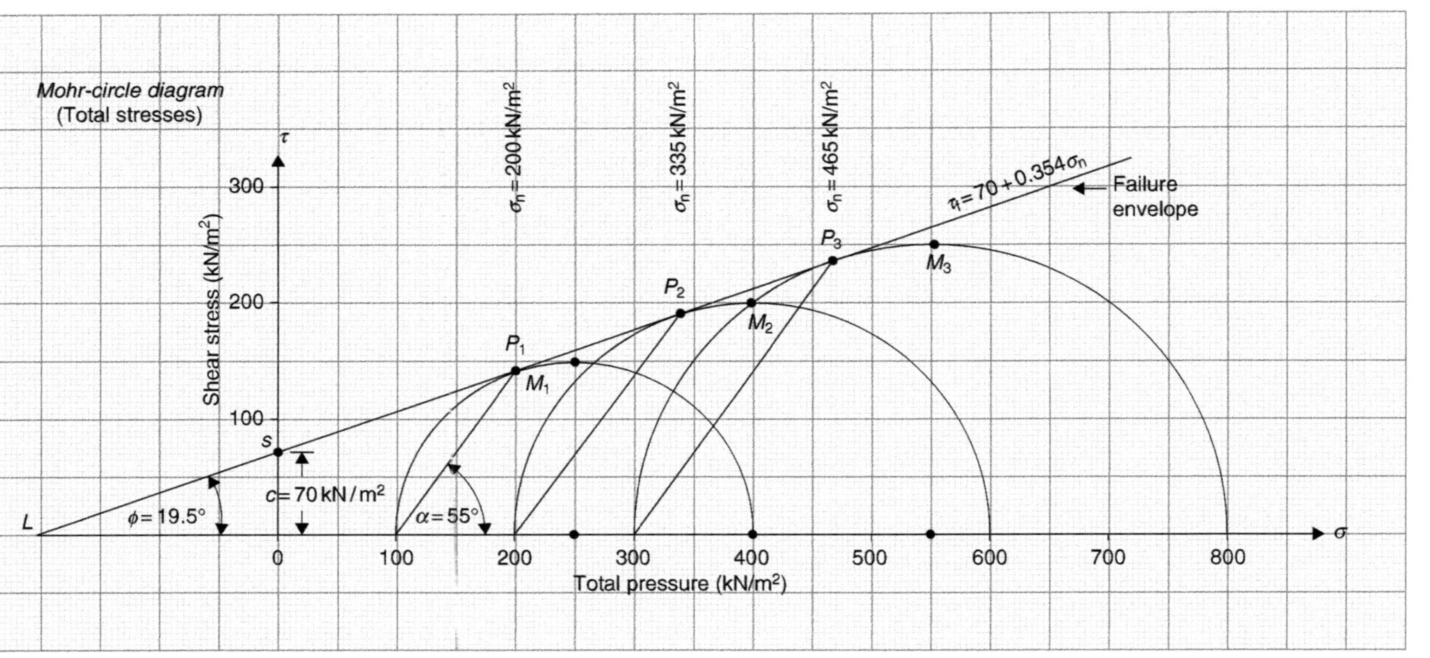

Mohr-circle diagram
(Total stresses)
τ
300
200
100
Shear stress (kN/m²)
σn = 200 kN/m²
σn = 335 kN/m²
σn = 465 kN/m²
τf = 70 + 0.354σn
Failure envelope
P1
P2
P3
M1
M2
M3
s
L
c = 70 kN/m²
φ = 19.5°
α = 55°
0
100
200
300
400
500
600
700
800
σ
Total pressure (kN/m²)

Graph 6.1

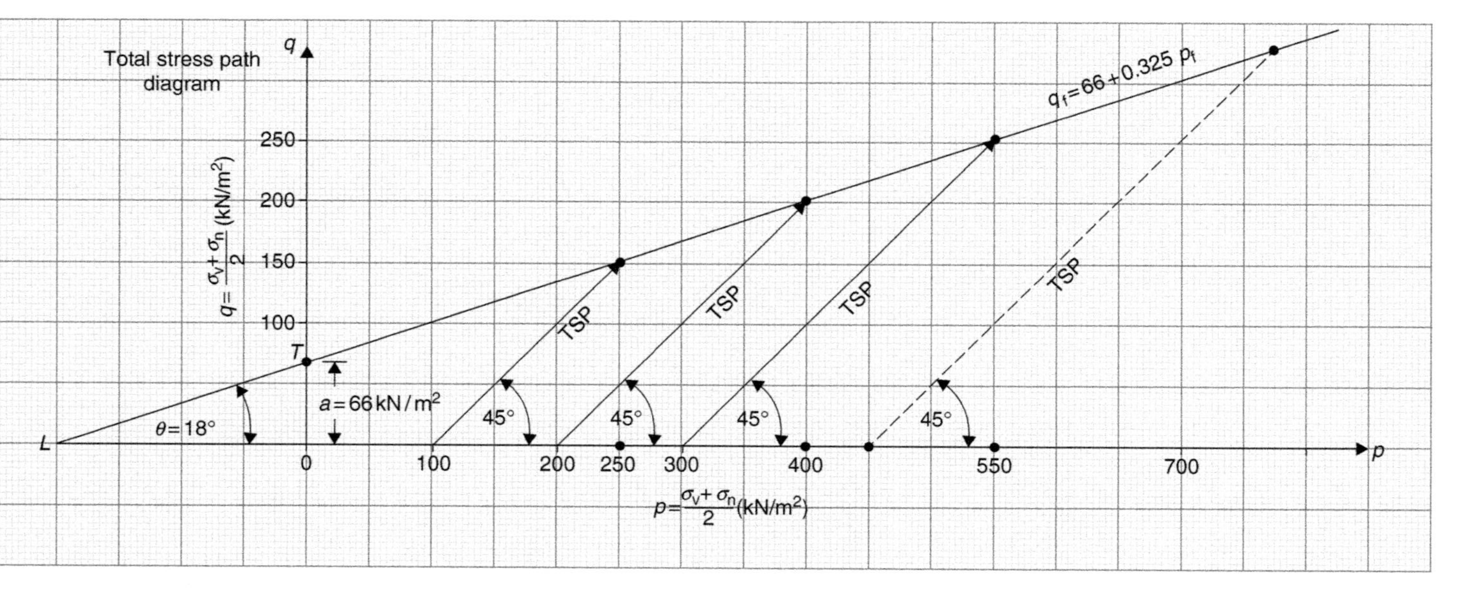
Total stress path diagram
q
250
200
150
100
T
L
0
100
200
250
300
400
550
700
p
$q = \frac{\sigma_v + \sigma_n}{2}$ (kN/m²)
$p = \frac{\sigma_v + \sigma_n}{2}$ (kN/m²)
θ = 18°
a = 66 kN/m²
45°
45°
45°
45°
TSP
TSP
TSP
TSP
$q_f = 66 + 0.325\ p_f$

Graph 6.2

6.2.2 Variation of stress path

An advantage of the stress path method is that stress increases can be visualized. Because of this, the direction of the path may be altered at will by varying the external pressures σ_V and σ_H. Eight possibilities are shown below, in terms of total stress path (TSP), by means of formulae:

$$(6.7): \quad p = \frac{\sigma_V + \sigma_H}{2}$$

$$(6.8): \quad q = \frac{\sigma_V - \sigma_H}{2}$$

Case 1 $\sigma_V = \sigma_H = 40\,\text{kN/m}^2$

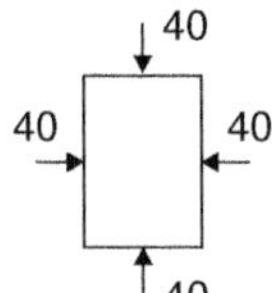

$$p = \frac{40+40}{2} = 40\,\text{kN/m}^2$$

$$q = \frac{40-40}{2} = 0\,\text{kN/m}^2$$

(a)

q

Stress point

p

40

Case 2 $\sigma_V > \sigma_H$

$\sigma_V = 40\,\text{kN/m}^2$

$\sigma_H = 20\,\text{kN/m}^2$

$$p = \frac{40+20}{2} = 30\,\text{kN/m}^2$$

$$q = \frac{40-20}{2} = 10\,\text{kN/m}^2$$

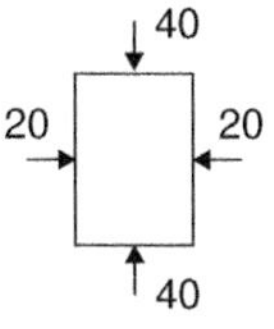

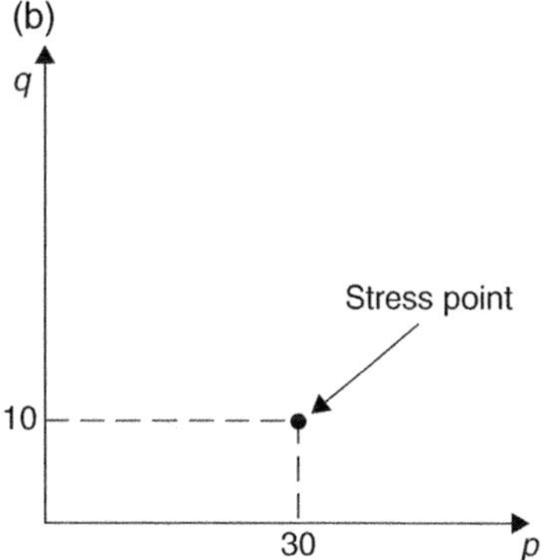

Figure 6.12

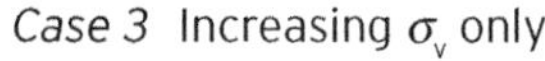

Case 3 Increasing σ_V only

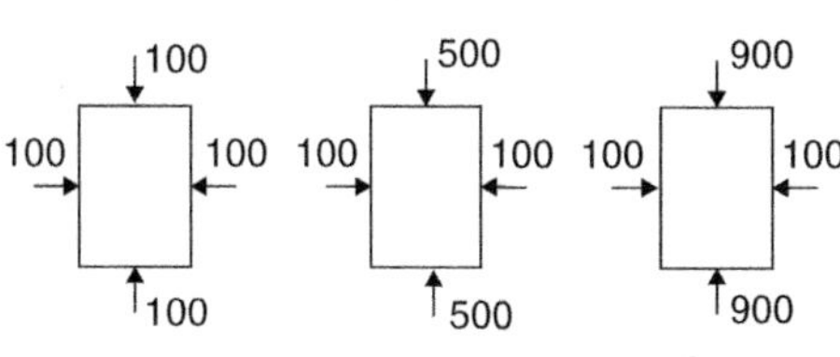

p	100	300	500
q	0	200	400

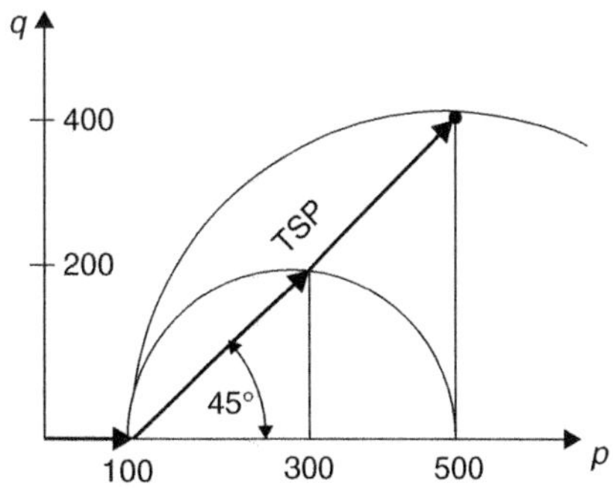

Figure 6.13

Note: This case is similar to that illustrated in Figure 6.8, which is the usual procedure in the triaxial test.

Case 4 Decreasing σ_H only whilst σ_v is kept constant.

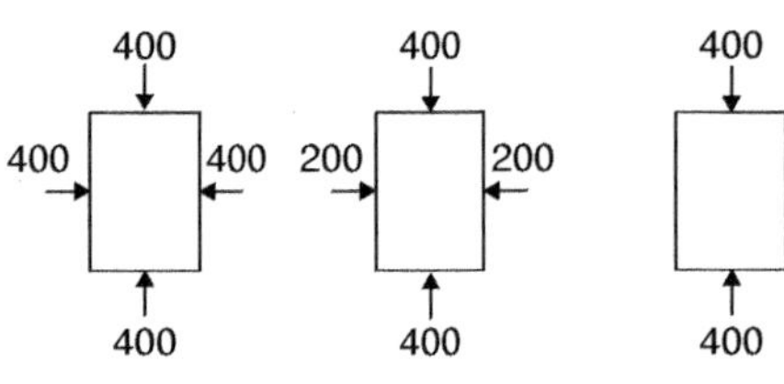

p	400	300	200
q	0	100	200

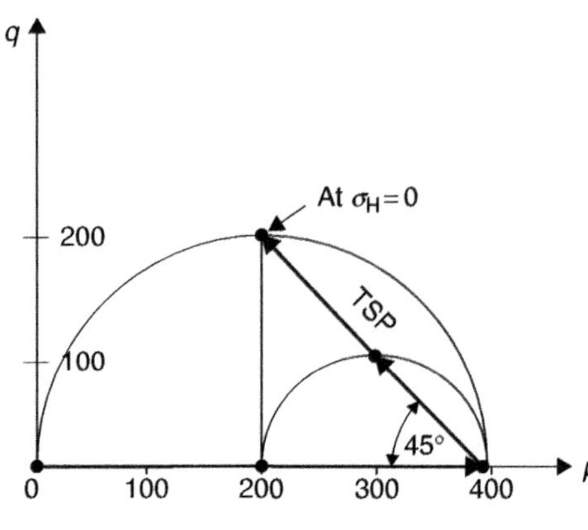

Figure 6.14

These calculations are facilitated by formulae (6.14) and (6.15), derived below.

From (6.7): $\sigma_v = 2p - \sigma_H$

From (6.8): $-\sigma_v = -2q \pm \sigma_H$

Subtracting: $0 = 2p - 2q - 2\sigma_H$

$$\therefore \sigma_H = p - q \tag{6.14}$$

Substitute (6.14): $\sigma_v = 2p - (p - q)$

$$\therefore \ \sigma_v = p + q \tag{6.15}$$

Case 5 Increasing σ_v and decreasing σ_H by equal amounts.

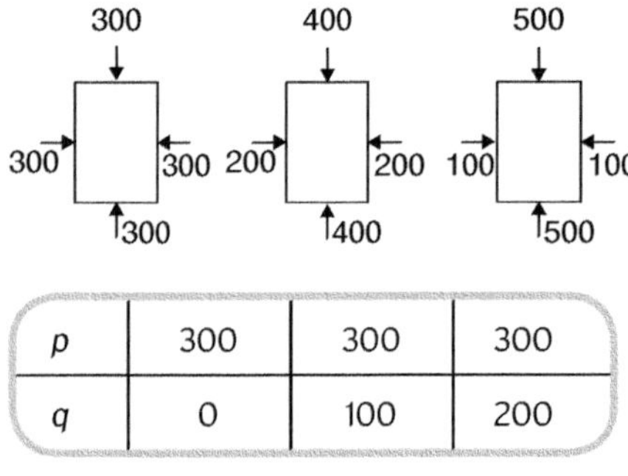

p	300	300	300
q	0	100	200

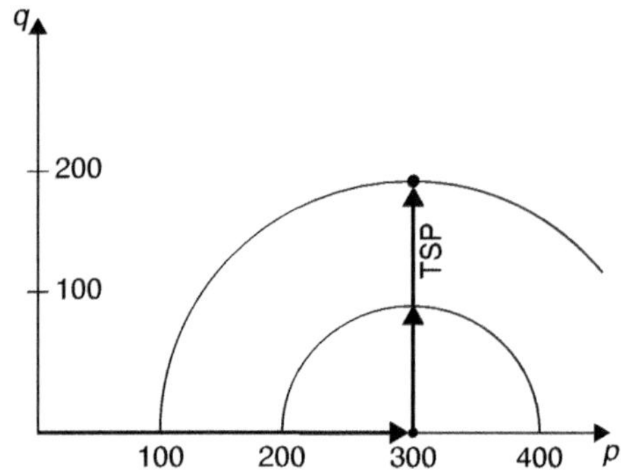

Figure 6.15

Case 6 Increasing both σ_v and σ_H by equal amounts.

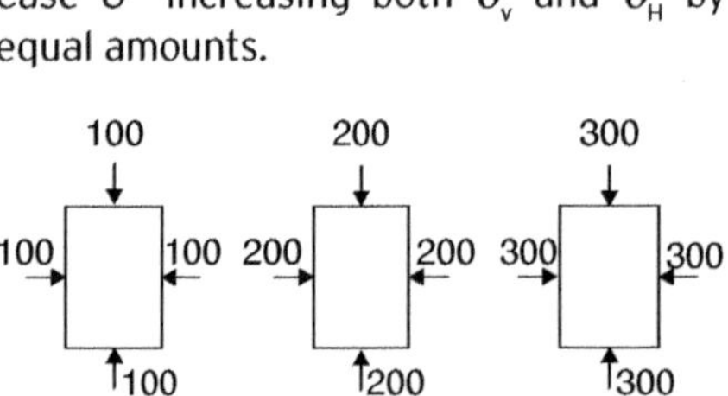

p	100	200	300
q	0	0	0

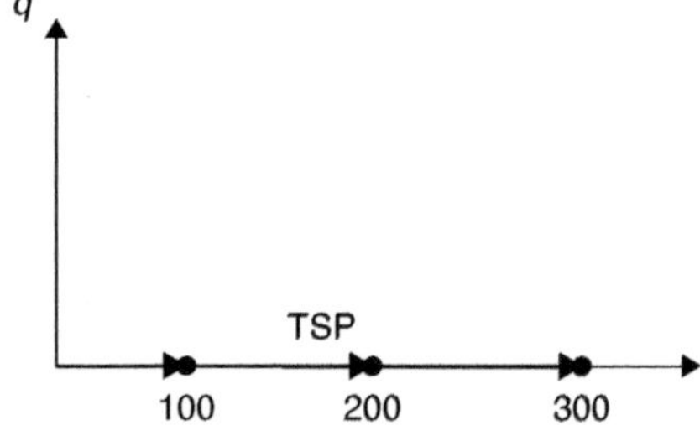

Figure 6.16

Case 7 Decreasing both σ_V and σ_H by equal amounts.

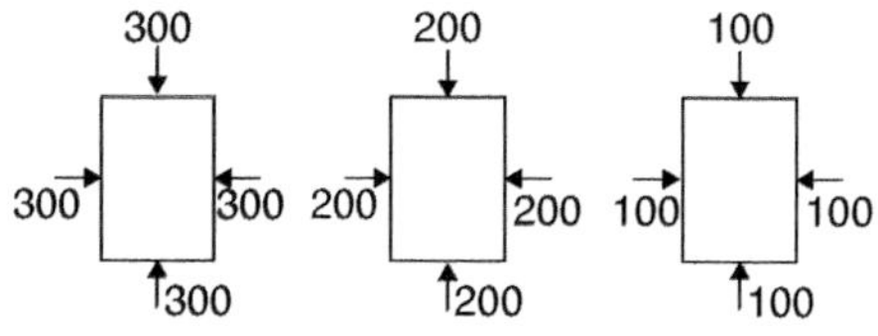

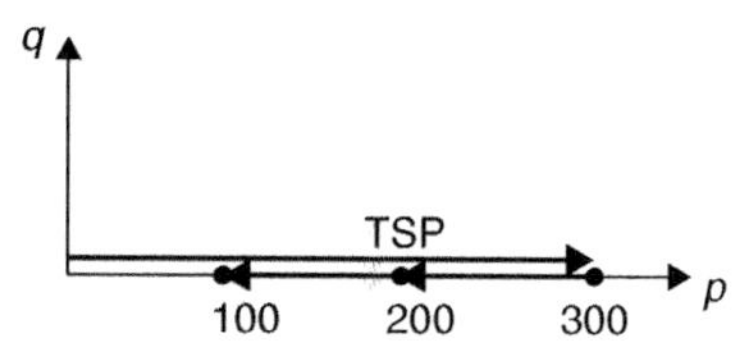

Figure 6.17

p	300	200	100
q	0	0	0

Case 8 Increasing both σ_V and σ_H. Also, $\sigma_H = 0.5\ \sigma_V$

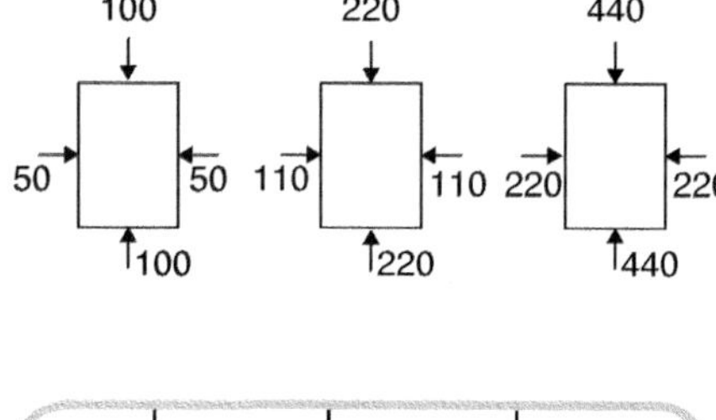

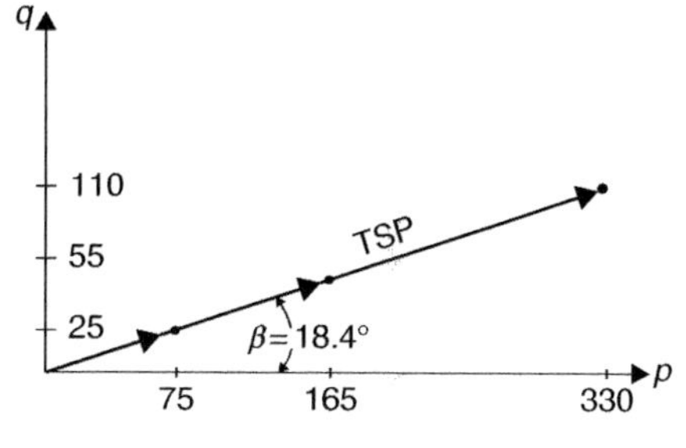

p	75	165	330
q	25	55	110

Figure 6.18

Note: The inclination of TSP is not 45° in this case.

From (6.7): $p = \dfrac{\sigma_V + 0.5\sigma_V}{2} = 0.75\sigma_V$

From (6.8): $q = \dfrac{\sigma_V + 0.5\sigma_V}{2} = 0.25\sigma_V$

$\tan\beta = \dfrac{q}{p} = \dfrac{0.25}{0.75} = 0.333$

$\beta = 18.4°$

Also, $\dfrac{q}{p} = 0.333 \qquad \therefore \quad p = 3q$

Theoretically therefore, any stress path can be chosen, provided there is a test equipment; capable of changing σ_V and σ_H during a test. A hydraulically loaded triaxial cell (Bishop-Wesley, 1975) has been designed for this purpose at Imperial College, London.

6.3 Effect of saturation

The Mohr's circle representation discussed so far is applicable to homogeneous, isotropic material in general. Soil, however, contains water. If the voids are fully saturated then external loading induces the already much discussed pore water pressure (u). This means, that the soil is subjected to an effective pressure (σ'), rather than a total pressure (σ), as expressed by:

$$\boxed{\sigma' = \sigma - u} \tag{5.1}$$

As the shear strength of a soil now depends on σ', Mohr's circles have to be modified to reflect this fact. The equation of the failure envelope, therefore, has the form:

$$\boxed{\tau_f = c' + \sigma_n' \tan\phi'} \tag{6.16}$$

$$\text{or} \quad \boxed{\tau_f = c' + (\sigma - u)\tan\phi'}$$

Where, c' = cohesion intercept in terms of effective stress σ'
ϕ' = angle of shearing resistance in terms of σ'.

6.3.1 Effective Mohr's circle

The graphical effect of u is to shift the total pressure circle towards the origin. Figure 6.19 shows the effective pressure circle relative to the total one.

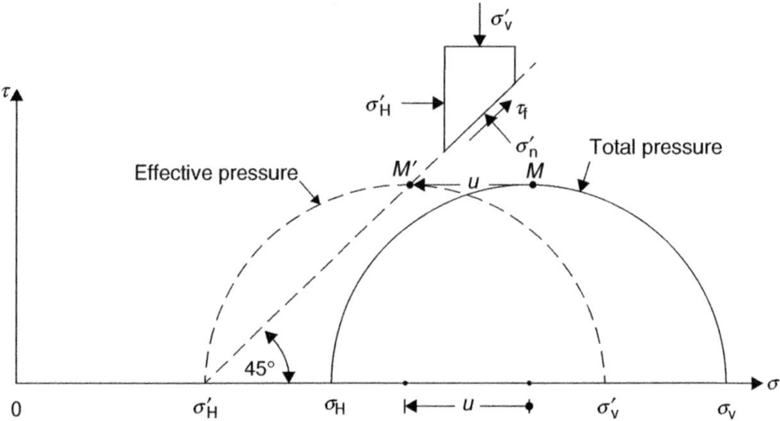

Figure 6.19

It is assumed, in this case, that the pore pressure cannot dissipate, that is water cannot flow out of the soil specimen, during the application of the loading.

6.3.2 Effective stress path (ESP)

It has to be appreciated that the pore pressure increases gradually from zero to its maximum value (u), upon the application of the total pressure. The ESP is obtained by subtracting the increments of pore pressure (Δu) at each point along the total stress path (TSP). The ESP is a path, curved towards the origin.

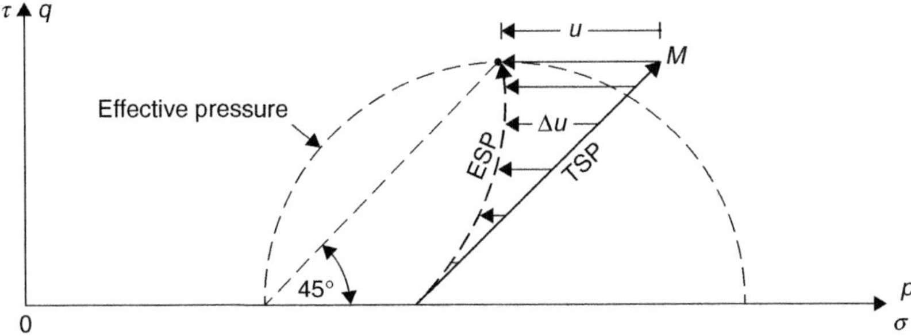

Figure 6.20

Note: Formulae (6.2) to (6.15) are still valid, but with effective values. The total pressure in increased gradually during a test.

Example 6.2

Suppose the final pore pressure for the three specimens, in Example 6.1, are also given as tabulated. Calculate c' and ϕ'.

Table 6.4

Specimen	1	2	3
σ_H (kN/m^2)	100	200	300
σ_v (kN/m^2)	400	600	800
u (kN/m^2)	20	90	150

The effective pressures σ'_H and σ'_v as well as p' and q' are:

Table 6.5

Specimen	1	2	3	
$\sigma'_H = \sigma_H - u$	80	110	150	
$\sigma'_v = \sigma_v - u$	380	510	650	
p'	230	310	400	From (6.7)
q'	150	200	250	From (6.8)

Method 1

Draw Graph 6.3 and measure the shear strength parameters as:

$$c' = 20\,\text{kN/m}^2$$
$$\phi' = 36°$$
$$\therefore \quad \tau_f = 20 + 0.727\sigma'_n$$

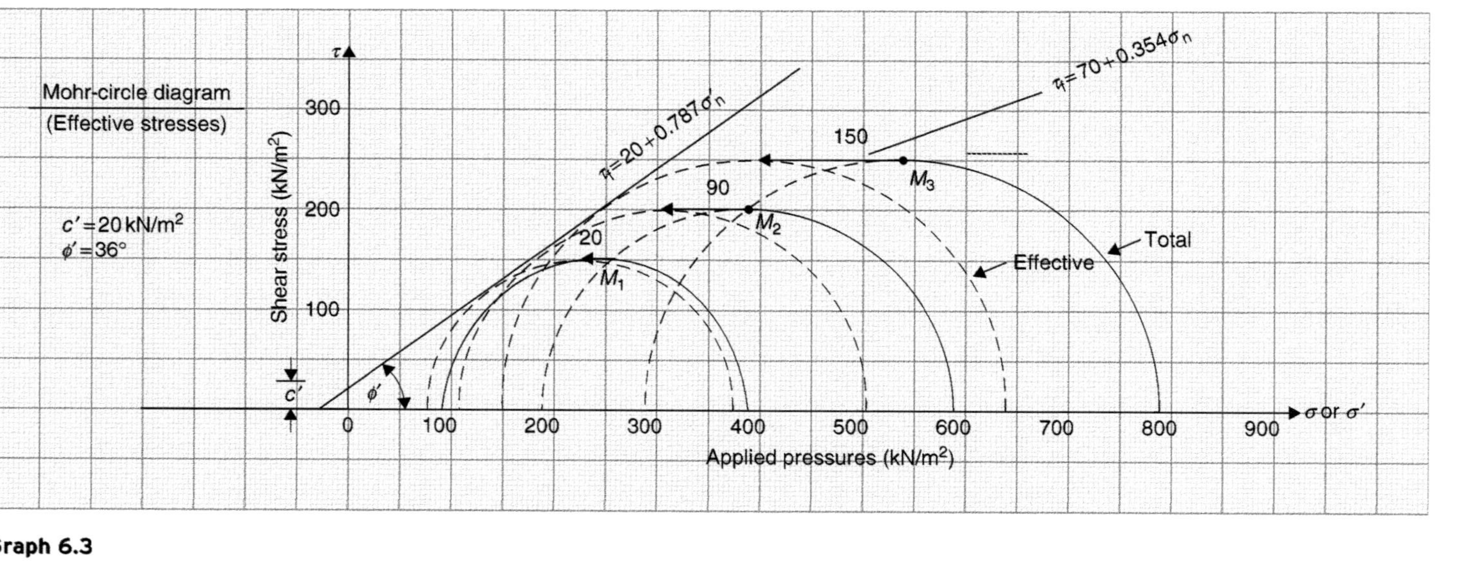

Mohr-circle diagram
(Effective stresses)
c′ = 20 kN/m²
φ′ = 36°
τ
Shear stress (kN/m²)
300
200
100
c′
φ′
0
100
200
300
400
500
600
700
800
900
σ or σ′
Applied pressures (kN/m²)
τf = 20 + 0.787σ′n
τf = 70 + 0.354σn
150
90
20
M1
M2
M3
Effective
Total

Graph 6.3

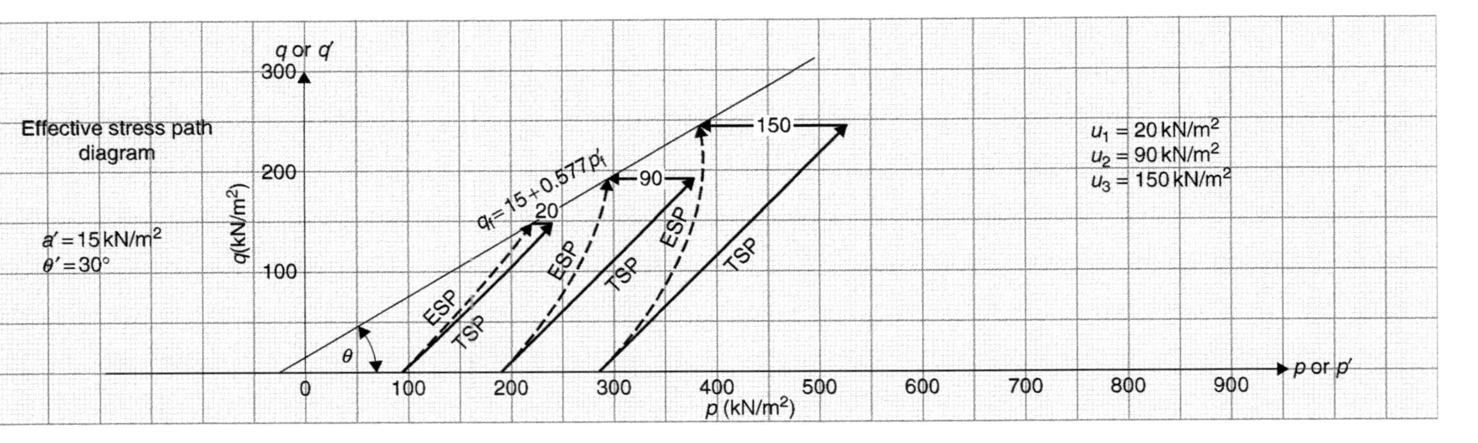
Effective stress path diagram
a′ = 15 kN/m²
θ′ = 30°
q or q′
300
200
100
q(kN/m²)
0
100
200
300
400
500
600
700
800
900
p or p′
p (kN/m²)
θ
q_f = 15 + 0.577 p′_f
20
90
150
ESP
TSP
ESP
TSP
ESP
TSP
u₁ = 20 kN/m²
u₂ = 90 kN/m²
u₃ = 150 kN/m²

Graph 6.4

Method 2

Draw Graph 6.4 and measure the effective stress parameters as:

$$a' = 15\,\text{kN/m}^2 \qquad \theta = 30^\circ$$

$$\therefore \quad q'_f = 15 + 0.577 p'_f$$

From (6.10): $\phi' = \sin^{-1}(\tan\theta') = \sin^{-1}(0.577) = 35.3^\circ$

From (6.11): $c' = \dfrac{a'}{\cos\phi'} = \dfrac{15}{\cos 35.3} = 18.4\,\text{kN/m}^2$

Method 3

$$q'_3 - q'_1 = 250 - 150 = 100 \qquad p'_3 - p'_1 = 400 - 230 = 170 \qquad \therefore \quad \tan\theta' = \frac{100}{170} = 0.588 \quad \text{from (6.12)}$$

From (6.10): $\phi' = \sin^{-1}(0.588) = 36.03^\circ$

From (6.13):

$$c' = \left(\frac{q'_3}{\sin 36.03} - p'_3\right) \times \tan 36.03 = \left(\frac{250}{0.588} - 400\right) \times 0.727 = 18.3\,\text{kN/m}^2$$

6.4 Measurement of shear strength

It was explained in the last section how to determine, theoretically, the shear strength parameters of a particular type of soil, subjected to known horizontal and vertical pressures. In this section various tests are introduced, whose main purpose is to obtain the magnitude of applied pressures at failure. There are four types of tests normally used for this purpose:

1. Standard triaxial
2. Unconfined compression
3. Shear box
4. Vane

The standard triaxial tests are used most extensively; hence, these are to be dealt with in some detail.

6.4.1 Triaxial tests

The purpose of these tests is to establish shear strength by simulating as accurately as possible, the ground condition encountered in a particular engineering problem. In practice, one of the following four triaxial tests is chosen to achieve comparability:

a) Unconsolidated undrained: (UU)
b) Quick undrained: (QU)
c) Consolidated undrained: (CU)
d) Consolidated drained: (CD)

The triaxial apparatus

Figure 6.21(a) shows the outline of Bishop's (1957) triaxial cell and loading arrangement.

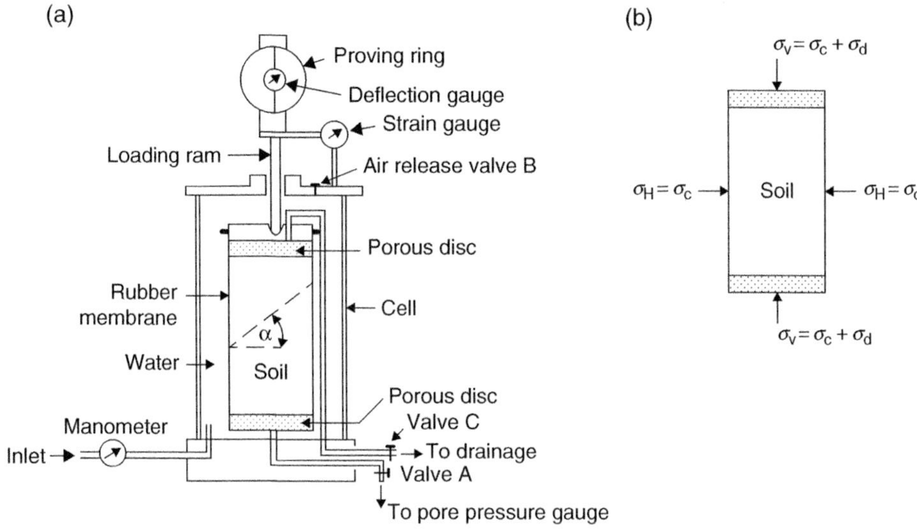

Figure 6.21

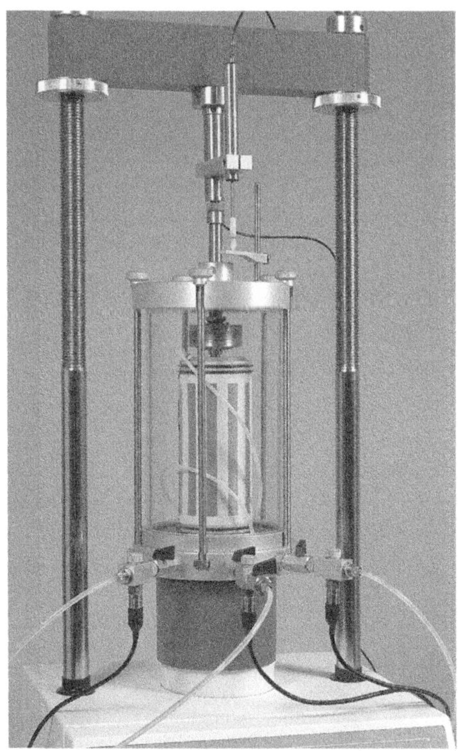

Tritest 50 complete with triaxial cell and transducers. Reproduced by permission of ELE International.

The specimen is tested by first applying an all-round cell pressure σ_c and then increasing the so called *deviator stress* σ_d by means of the loading ram, until the sample fails.

Notes:

1. The horizontal pressure equals to the cell pressure during the test.
2. The vertical pressure changes with the deviator stress.

   From Figure 6.21(b), the principal stresses are given by: $\sigma_H = \sigma_c$, $\sigma_v = \sigma_c + \sigma_d$

3. Alternative designation, in the literature for the horizontal and vertical principal stresses is σ_3 and σ_1 respectively.

6.4.2 Variation of pore pressure

The gradual increase in the total pressures σ_c and σ_v, during a triaxial test, induces corresponding increases (Δu) in the pore pressure within the soil. The correspondence between these quantities was derived by Skempton (1954), in terms of two coefficients *A* and *B*. The effect of σ_c and σ_d on the variation of *u* is considered separately.

Application of σ_c
Figure 6.22 depicts the application of hydrostatic pressure only.

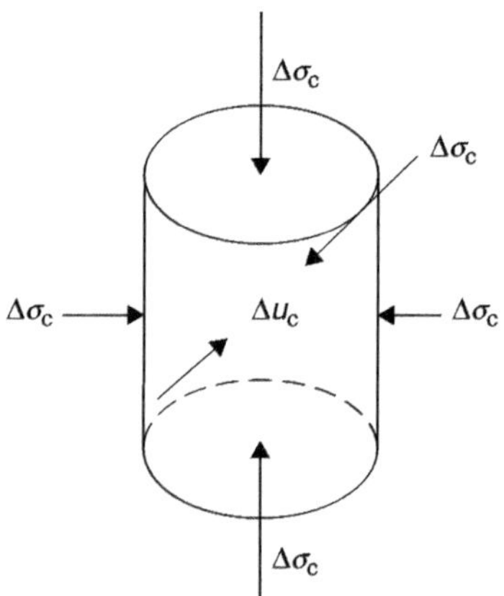

$\Delta\sigma_c$ = Small increment in the cell-water pressure σ_c. It induces a small increase (Δu_c) in the pore water pressure. The magnitude of Δu_c depends also on the degree of saturation. The increase is given by:

$$\boxed{\Delta u_c = B\Delta\sigma_c} \qquad (6.17)$$

Figure 6.22

where *B* is called the *pore pressure parameter*. It is found by increasing the cell pressure to any desired value $\Delta\sigma_c$ and measuring the corresponding pore pressure Δu_c. Its magnitude is an indication of the degree of saturation:

$B = 1$ - Saturated soil
$0 < B < 1$ - Partially saturated
$B = 0$ - Dry

Therefore, this first stage of a triaxial test may be used to determine the state of saturation.

Application of σ_d
If now, the vertical pressure is increased without allowing water outflow from the specimen, then it induces further pressure increase. Figure 6.23 shows the addition of the deviator stress to the cell pressure.

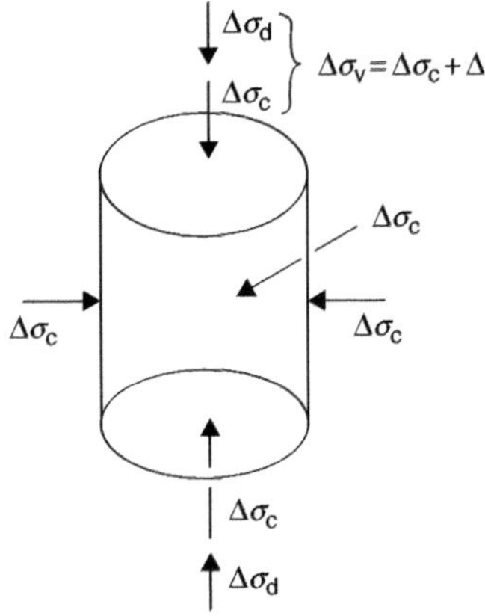

Figure 6.23

$\Delta\sigma_d$ = Small increment in the deviator stress, applied by the loading ram in the triaxial test. The induced pore pressure is given by:

$$\boxed{\Delta u_d = \bar{A}\Delta\sigma_d} \quad (6.18)$$

Where $\bar{A}$ is another pore pressure parameter. It is calculated from the pore pressure measured for the range of $\Delta\sigma_d$.

6.4.3 Total excess pore pressure

Combining the above two increments, the overall change in the pore pressure, during a triaxial test, can be determined from:

$$\Delta u = \Delta u_c + \Delta u_d$$
$$\Delta u = B\Delta\sigma_c + \bar{A}\,\Delta\sigma_d$$
$$\Delta u = B\left(\Delta\sigma_c + \frac{\bar{A}}{B}\Delta\sigma_d\right)$$

Let

$$\boxed{A = \frac{\bar{A}}{B}} \quad (6.19)$$

Then finally

$$\boxed{\Delta u = B(\Delta\sigma_c + A\,\Delta\sigma_d)} \quad (6.20)$$

Where A is also a *pore pressure parameter*. The value of parameter A depends largely on the stress history of the soil. Some typical values are:

$-0.5 < A < 0$ for heavily over-consolidated clay
$0 < A < 0.5$ for lightly over-consolidated clay
$0.5 < A < 1.5$ for normally consolidated clay

See definitions in Sections 7.3.1 and 7.3.2.

Example 6.3

A clay sample was tested in an undrained triaxial test. The applied cell pressure was 100 kN/m^2 and the corresponding induced pore pressure observed as 55 kN/m^2. The deviator stress was applied in two 200 kN/m^2 increments and the pore pressures noted. The data are summarized below. Calculate the pore pressure parameters.

Table 6.6

	1	2
$\Delta\sigma_d$	200	400
Δu_d	95	120
Δu	150	175

$\Delta\sigma_c = 100\,\text{kN/m}^2$
$\Delta u_c = 55\,\text{kN/m}^2$
$\Delta u = \Delta u_c + \Delta u_d$

From (6.17): $B = \dfrac{\Delta u_c}{\Delta\sigma_c} = \dfrac{55}{100} = 0.55 < 1$ | $\therefore$ Partially saturated clay

From (6.18): $\bar{A} = \dfrac{\Delta u_d}{\Delta\sigma_d}$

From (6.19): $A = \dfrac{\bar{A}}{B}$

	1	2
$\bar{A}$	0.475	0.3
A	0.863	0.545

From (6.20): check: $\Delta u = B(\Delta\sigma_c + A\Delta\sigma_d)$

Increment 1. $\Delta u = 0.55 \times (100 + 0.863 \times 200) = 150\,\text{kN/m}^2$

Increment 2. $\Delta u = 0.55 \times (100 + 0.545 \times 400) = 175\,\text{kN/m}^2$

Note: The pore pressure parameters are used to determine the variation of u in the field.

6.4.4 Unconsolidated-undrained tests

During this test, the soil specimen is sheared without allowing the outflow of pore water. This is to simulate ground conditions in actual engineering problems, where there is no time for pore pressure dissipation, during the construction of structures in cohesive soils of low permeability. There are two standard tests of this type:

1. Undrained test on partially saturated soil, denoted as UU-test.
2. Undrained test on saturated cohesive soils, usually referred to as 'Quick' or QU-test, applicable to both normally and overconsolidated clays.

UU-test
In the UU-test both the total as well as the pore pressures are measured. The presentation of the results are in terms of effective stresses, hence the Mohr's circle are drawn as in Example 6.2, resulting in:

Notes:

1. The failure envelope is normally curved when expressed in terms of total stresses, and approximates a straight line, when drawn in terms of effective ones. For this reason the use of the effective stress parameters are preferred.
2. c'_u and ϕ'_u are effective undrained cohesion and angle of friction respectively.

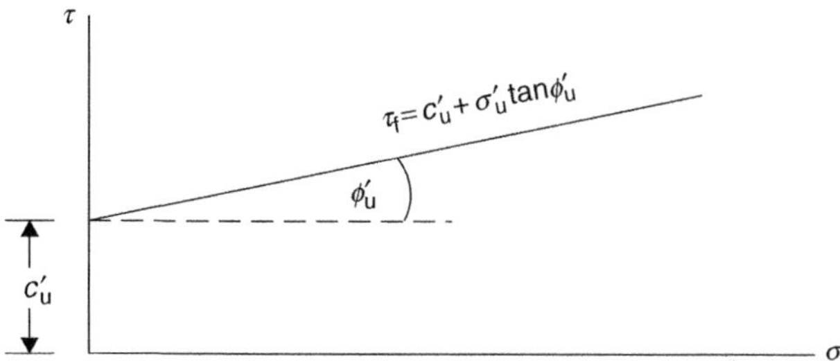

Figure 6.24

Example 6.4

Table 6.7 contains the results of a UU-test carried out on three 38/76 mm, partially saturated clay samples. The proving ring constant was $n=4.5$ N/division.

Determine the equation for the failure envelope in terms of effective stresses.

Table 6.7

σ_c (kN/m²)	**100**			**175**			**250**			
u_f (kN/m²)	**40**			**45**			**85**			**At failure**
$\sigma'_c = \sigma_c - u_f$	**60**			**130**			**165**			**At failure**
	x (mm)	r_x (div)	ε%	x (mm)	r_x (div)	ε%	x (mm)	r_x (div)	ε%	
	0.7	4	0.9	1.0	11.0	1.3	0.6	9.0	0.8	
	2.0	11	2.6	1.9	20.5	2.5	1.2	16.0	1.6	
	3.4	16.5	4.5	3.0	29.0	3.9	2.0	26.1	2.6	
	5.3	23.5	7.0	4.3	37.4	5.7	2.9	34.4	3.8	
	7.8	28.0	10.3	6.0	46.6	7.9	4.1	44.4	5.4	
	9.6	28.9	12.6	7.6	52.8	10.0	5.3	52.6	7.0	
	11.2	26.9	14.7	9.4	57.2	12.4	7.0	61.9	9.2	
	12.0	26.0	15.8	10.4	58.4	13.7	8.4	67.8	11.0	
				12.2	56.6	16.0	9.5	71.9	12.5	
				12.6	54.1	16.6	11.4	74.1	15.0	
							12.2	73.9	16.1	
							13.8	70.7	18.2	
Mode of Failure	Failure at $x=9.6$			Failure at $x=10.4$			Failure at $x=11.4$			
h_0=76mm	$r_x=28.9$ $\varepsilon=12.6\%$			$r_x=58.4$ $\varepsilon=13.7\%$			$r_x=74.1$ $\varepsilon=15\%$			

where x = strain dial gauge reading (mm), indicating the shortening of the sample.
r_x = reading of the proving ring dial gauge, indicating the force (P_x), applied by the loading ram to the sample. The deviator stress is determined from x and r_x.

$$\varepsilon = \text{strain} = \frac{\text{shortening of sample}}{\text{initial length}} \quad \boxed{\varepsilon = 100\frac{x}{h_o} \;\Big|\; \%} = \frac{100\,x}{76}\%$$

Area correction in undrained test

Applied force: $P_x = nr_x = 4.5\, r_x$ Newton

Initial cross-sectional area of the specimen:

$$A_o = \frac{38^2\pi}{4} = 1134\text{ mm}^2$$

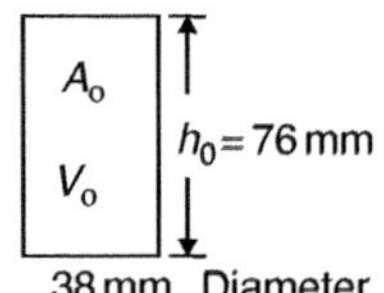

Initial volume: $V_o = h_o A_o = 76 \times 1134 = 86184\text{ mm}^3$

Initial height: $h_o = 76\text{ mm}$

As the sample is compressed, its volume remains practically the same, but its height, and hence the cross-sectional area, changes with x.

Changed height at x: $h_x = h_o - x = h_o\left(1 - \frac{x}{h_o}\right) = h_o(1-\varepsilon)$

Volume at x: $V_o = A_x h_x$

Where A_x = cross-sectional area at x.

Equating $\quad V_o = A_o h_o = A_x h_x \quad \boxed{A_x = \frac{A_o h_o}{h_x} = \frac{V_o}{h_x} = \frac{A_o}{1-\varepsilon}}$ (6.21)

Deviator stress at x: $\quad \boxed{\sigma_x = \frac{P_x}{A_x} = \frac{nr_x}{A_x}}$ (6.22)

Where r_x = proving ring dial reading at x:

$$\therefore \quad \sigma_x = \frac{nr_x}{\frac{V_o}{h_x}} = \frac{nr_x h_x}{V_o} \quad \text{and} \quad \sigma_x = \frac{nr_x(h_o - x)}{V_o}\ (\text{N/mm}^2)$$

or $\quad \boxed{\sigma_x = \frac{1000nr_x(h_o - x)}{V_o} \;\Big|\; (\text{kN/m}^2)}$ (6.23)

This is the deviator stress, corrected for changing area. The quantities h_o, x, V_o are in millimeters and n in newtons. In this example,

$$\sigma_x = \frac{1000 \times 4.5 \times r_x(76 - x)}{86184} = 0.0522 \times r_x(76 - x)\,\text{kN/m}^2$$

Deviator stresses at failure:

Test 1: $x = 9.6\,\text{mm}$, $r_x = 28.9$ | $\sigma_1 = 0.0522 \times 28.9 \times (76 - 9.6) = 100\,\text{kN/m}^2$

Test 2: $x = 10.4\,\text{mm}$, $r_x = 58.4$ | $\sigma_2 = 0.0522 \times 58.4 \times (76 - 10.4) = 200\,\text{kN/m}^2$

Test 3: $x = 11.4\,\text{mm}$, $r_x = 74$ | $\sigma_3 = 0.0522 \times 74.1 \times (76 - 11.4) = 250\,\text{kN/m}^2$

The results are drawn on Graph 6.5 and the shear stress parameter measured as:

$c'_u = 10\,\text{kN/m}^2$, $\phi'_u = 25°$ | $\therefore \quad \tau_f = 10 + \sigma'_n \times \tan 25 = 10 + 0.466 \times \sigma'_n$

u_f = pore pressure at failure.

Alternative determination of the deviator stresses may be made by plotting the x and r_x dial reading on Chart 6.1. The advantage of this is not only that σ_d, at failure, can be read off directly from the chart, but that the point of failure becomes obvious.

If a sample does not exhibit clear shear failure but slumps, then it is assumed to occur at ε=20%, which is at x=15.2 mm on this chart.

Construction of Chart 6.1

It is designed for n=4.5 N/div and sample size of 38 × 76 mm, using formula (6.23), therefore any of the diagonal lines represented by r_x can be drawn by: $\sigma_x = 0.0522 \times r_x\,(76 - x)$.

Line r_x=100 is drawn between x=0, x=16 for example:

$$\text{If } x = 0: \quad \sigma_x = 0.0522 \times 100 \times (76 - 0) = 396.7\,\text{kN/m}^2$$
$$\text{If } x = 16: \quad \sigma_x = 0.0522 \times 100 \times (76 - 16) = 313.2\,\text{kN/m}^2$$

These figures indicate the two ends of line $r_x = 100$ as shown:

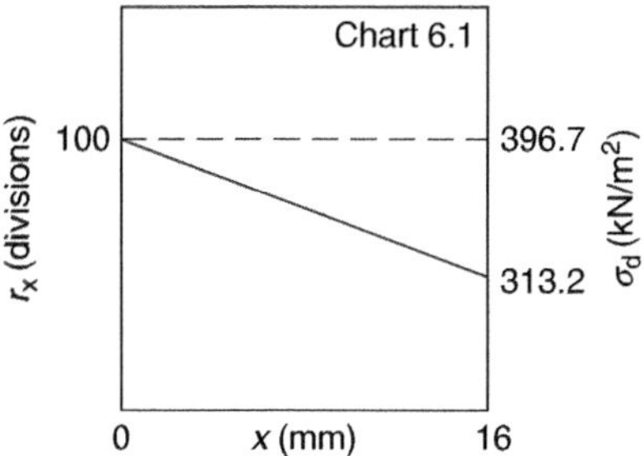

All the other lines can be drawn in this manner with fewer lines for clarity.

Figure 6.25

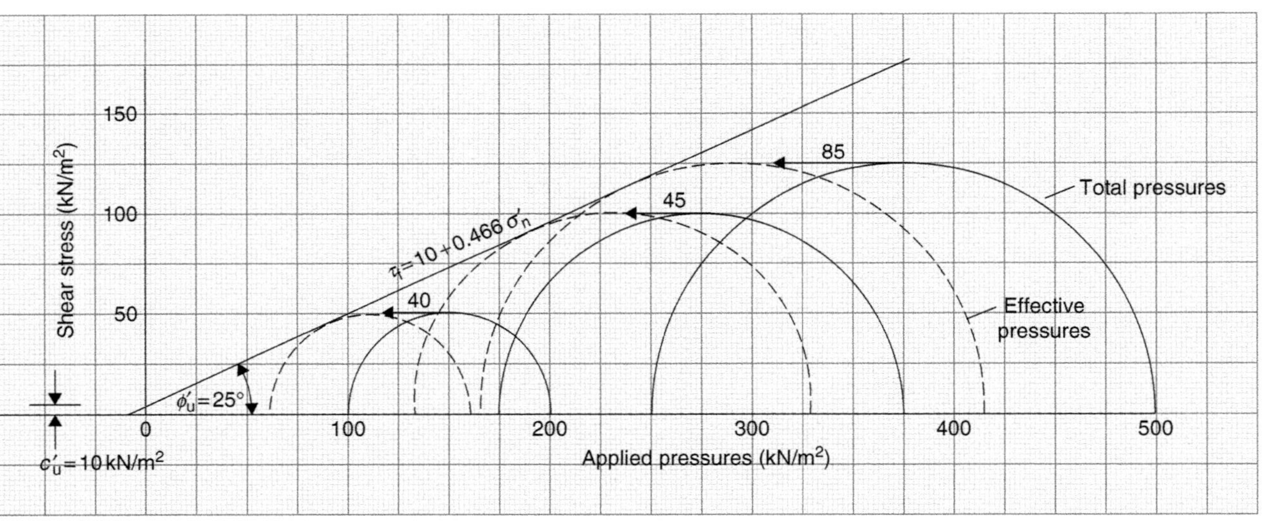

150
100
50
Shear stress (kN/m²)
0
100
200
300
400
500
Applied pressures (kN/m²)
τ = 10 + 0.466 σ′n
85
45
40
Total pressures
Effective pressures
φ′u = 25°
c′u = 10 kN/m²

Graph 6.5

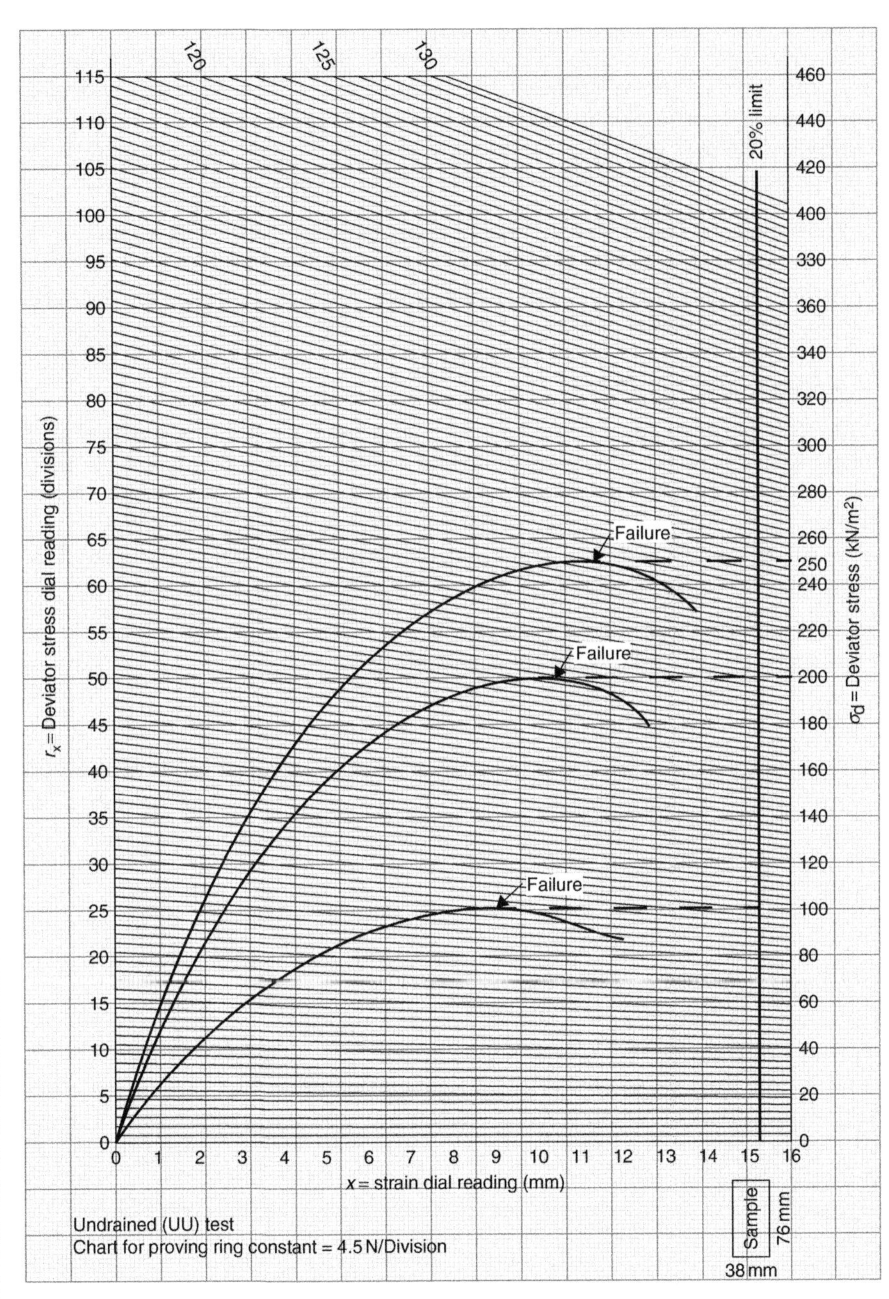

120
125
130
20% limit
Failure
Failure
Failure
r_x = Deviator stress dial reading (divisions)
σ_d = Deviator stress (kN/m²)
x = strain dial reading (mm)
Undrained (UU) test
Chart for proving ring constant = 4.5 N/Division
Sample
76 mm
38 mm

Chart 6.1

Some applications of the UU-test

It can be used to simulate the soil conditions in practical problems, where there is imperceptible pore water pressure dissipation during the imposition of external loading. In other words, the loads are placed over such a short period that there is only negligible outflow of water from soils of low permeability, hence there is not time for consolidation and strength increase. The soil is therefore weakest at this juncture, but gets stronger with the passage of time. Some of the practical problems are:

1. Excavation and end of construction of foundations in partially saturated soil.
2. Stability of compacted fill of low permeability.
3. During the construction of earth dam, where large pore pressures can develop, owing to the speed of construction.

6.4.5 Quick-undrained test

The QU-test is so called because the specimen is sheared, usually over a period of 15 to 20 minutes. Pore pressure is not measured and no drainage is allowed from the sample. The test is carried out usually on saturated, cohesive soils and the results are presented in terms of total stresses and shear strength parameters c_u and ϕ_u. The failure envelope depends not only on the degree of saturation, but on the stress history of the soil. Typical Coulomb envelopes are as follows.

1. *Saturated, normally consolidated clay (NCC)*
   Where the voids ratio (e) has not changed after deposition.

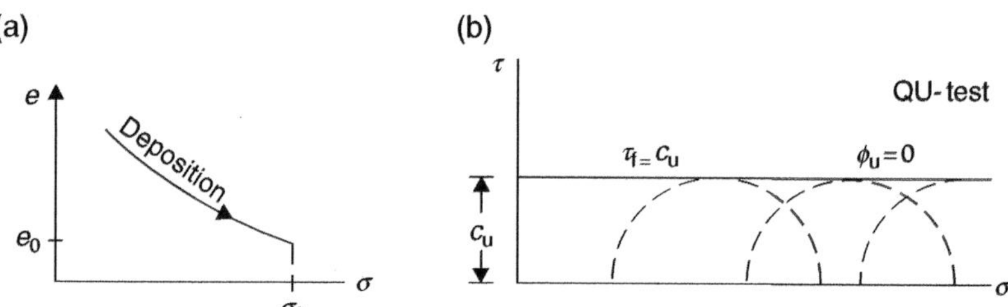

Figure 6.26

Where σ_0 = existing overburden pressure.

In this case, each Mohr-circle is of the same diameter, because any increase in the cell pressure induces pore pressure of equal magnitude. As a consequence, the effective pressure remains constant, as shown below. This applies equally to saturated, overconsolidated clay (OCC).

In Test 1, the cell pressure σ_c induces a pore pressure (u_1) in the sample.

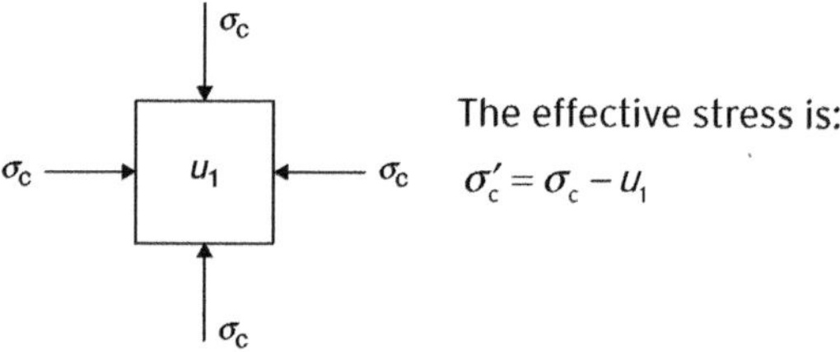

The effective stress is:

$$\sigma_c' = \sigma_c - u_1$$

Figure 6.27

In Test 2, the cell pressure is increased to $(\sigma_c + \Delta\sigma_c)$ and $\Delta\sigma_c$ induces additional pore pressure (Δu) in the specimen.

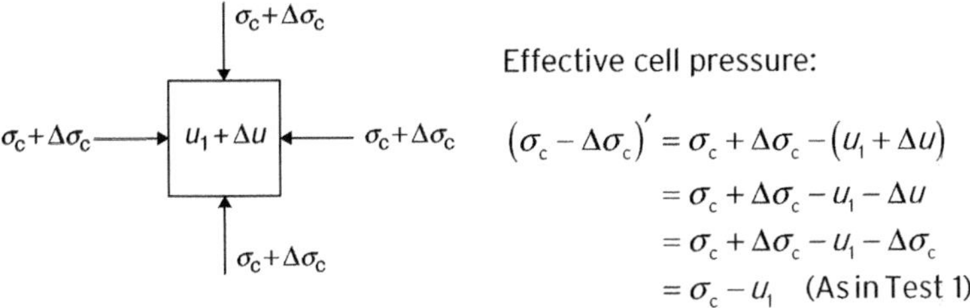

Figure 6.28

Therefore, the effective cell pressure is the same for both samples of the same material, hence they fail at the same deviator stress and the failure envelope becomes horizontal, for both NCC and OCC.

2. *Partially saturated NCC*
The failure envelope is curved, until the stresses become large enough to compress the air voids and cause saturation.

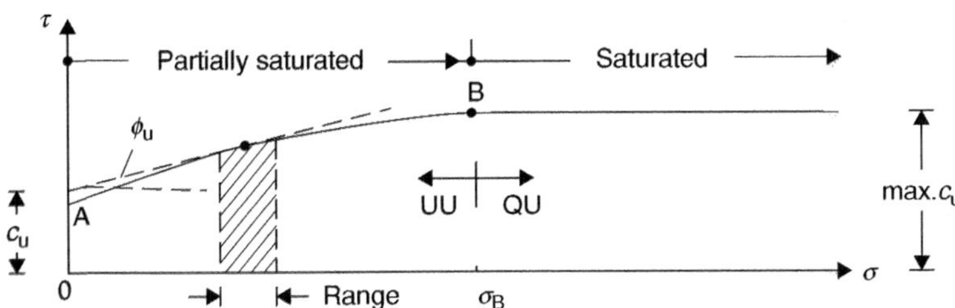

Figure 6.29

Either UU-test is carried out up to pressure σ_B, to get c'_u and σ'_u or apply QU-test for a limited range on the curved part, to get c_u and ϕ_u. QU-test is applicable beyond point B.

3. *Partially saturated overconsolidated clay (OCC)*
The voids ratio has changed after deposition, because of subsequent removal of overburden.

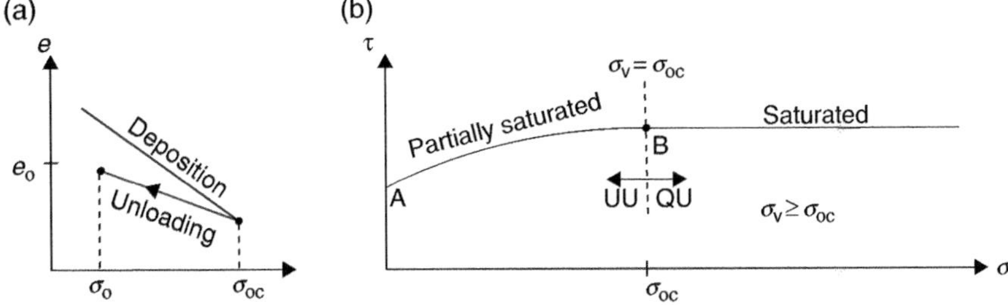

Figure 6.30

Where σ_{oc} = original pressure, prior to the removal of overburden (overconsolidation pressure).
σ_o = existing overburden pressure.
e_o = existing voids ratio.

If the test pressure is smaller than σ_{oc}, then negative pore pressure develops, due to the swelling of the clay. When the vertical load equals or exceeds σ_{oc}, the failure envelope becomes horizontal. Beyond point B, the procedure is the same as for NCC.

Some applications of the QU test

Similarly to the UU-test, the QU-test applied to the determination of short-term stability of structures, constructed in soils of low permeability. The test is not applicable, when either the period of construction is exceedingly long or the drainage path from the soil is too short as these allow faster outflow of water than assumed and simulated.

Retaining walls: Total stress analysis applies when:

a) The soil is saturated and little or no drainage occurs during construction.
b) Temporary excavation is supported by a structure and insignificant pore water outflow is expected during its short life-time.

Excavation for foundation: The clay is weakest at the end of excavation, when the removal of the overburden could result in ground heave, due to the decrease in the effective pressure. After the construction of the footing, the applied load, which is usually larger than the removed overburden reduces the uplift and then strengthens the soil by consolidating it. Short-term stability of natural ground under embankment, yet to be constructed, may be analysed in terms of total stresses, proving the period of construction is too short for drainage to occur.

6.4.6 Consolidated-undrained (CU) test

The difference between this and the UU-test is that in the CU-test, the water is allowed to flow out of the consolidating specimen as the pore pressure due to σ_c dissipates completely. After this, the outflow valve C (Figure 6.21) is closed and the deviator stress is applied slowly in order to equalize the pore water pressure, which is then measured.

As in the UU-test, the shear strength parameters may be expressed in terms of both total and effective stresses.

c_{cu}	In terms of total stress
ϕ_{cu}	for saturated soil
c'	in terms of effective stresses
ϕ'	for saturated and partially saturated soil

Some applications of CU-test

a) Foundations for structures, where the weight of the structure consolidates the supporting soil and then, there is a sudden increase in the loading. This occurs repeatedly in water tanks and grain silos.
b) Foundation of earth dams and embankments, where some consolidation could occur over extended construction period.

c) To determine the effective stress parameters c' and ϕ' of saturated soil. This is not possible in the QU-test (see Figure 6.28). In the CU-test however, the pore pressure induced by σ_c is zero at the beginning of the undrained compression stage as in Figure 6.31.

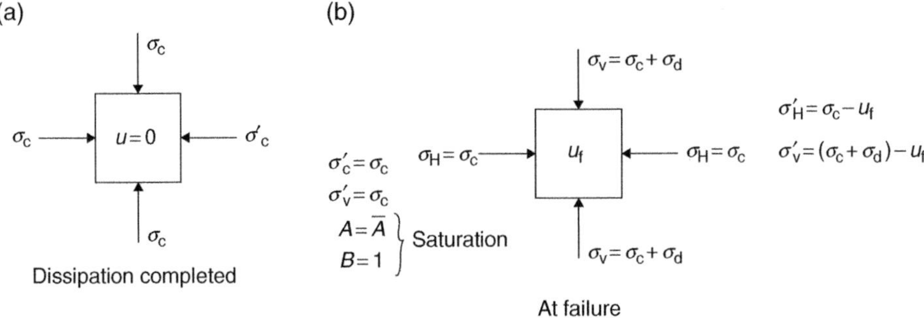

Figure 6.31

Figure 6.31(b) shows, that if several specimens of the same soil are sheared at different cell pressures, then the failure envelope can be drawn in terms of effective stresses as well as total ones. Typical envelopes are:

1. Sand and NCC

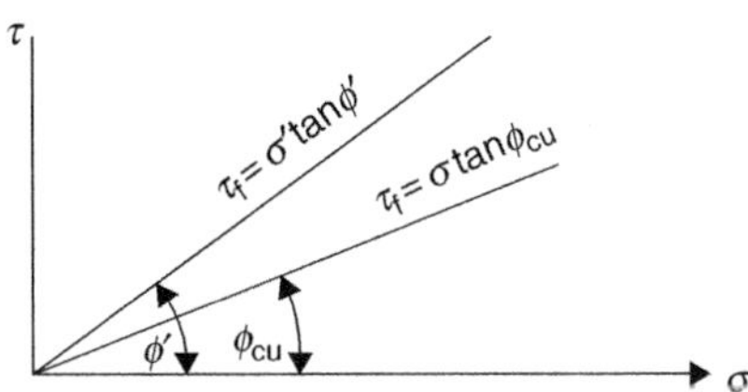

Figure 6.32

2. Overconsolidated clay (OCC)

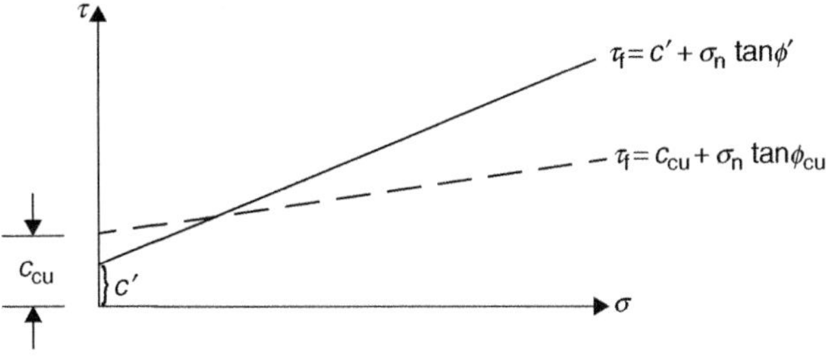

Figure 6.33

6.4.7 Consolidated-drained (CD) test

In this test, the sample is consolidated, as in the CU-test under cell pressure and allowed to drain until $u=0$. Deviator stress is then applied at a slow rate so, that any pore pressure induced can dissipate. As the pore pressure is zero at failure, the total applied pressures are effective in precipitating failure i.e. $\sigma'_v = \sigma_c + \sigma_d$

$$\sigma'_H = \sigma_H = \sigma_c$$

Typical failure envelopes

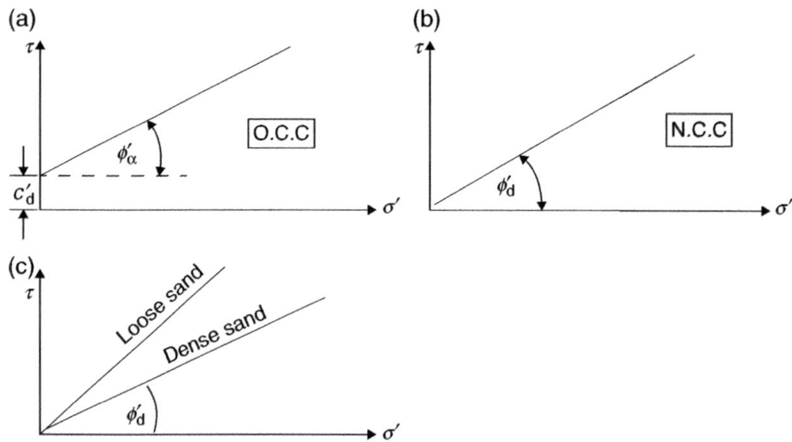

Figure 6.34

Area correction in drained tests

During an undrained test, the cross-sectional area changes, whilst the volume remains constant. In drained test, however, the volume also changes as a consequence of water loss. In order to determine the deviator stress at failure, formula (6.23) has to be modified to take this into account. The cross-sectional area A_x is derived in terms of volume.

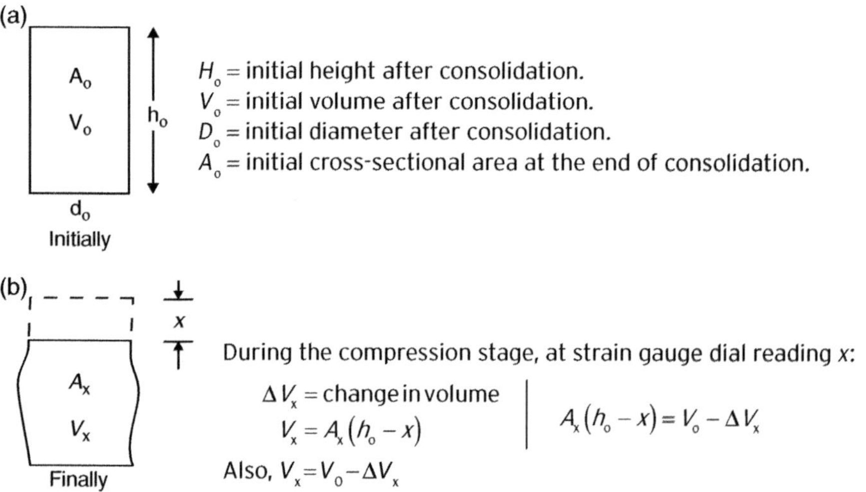

H_o = initial height after consolidation.
V_o = initial volume after consolidation.
D_o = initial diameter after consolidation.
A_o = initial cross-sectional area at the end of consolidation.

During the compression stage, at strain gauge dial reading x:

ΔV_x = change in volume

$V_x = A_x(h_o - x)$

Also, $V_x = V_0 - \Delta V_x$

$$A_x(h_o - x) = V_o - \Delta V_x$$

Figure 6.35

$$A_x = \frac{V_o - \Delta V_x}{h_o - x} = \frac{V_o\left(1 - \frac{\Delta V_x}{V_o}\right)}{h_o\left(1 - \frac{x}{h_o}\right)} \qquad \text{But,} \quad A_0 = \frac{V_0}{h_0}$$

$$\therefore A_x = A_o\left(\frac{1 - \frac{\Delta V_x}{V_o}}{1 - \frac{x}{h_o}}\right)$$

Deviator stress at x:

From (6.22): $\sigma_x = \dfrac{nr_x}{A_x}$ $\quad\therefore\quad$ $$\boxed{\sigma_x' = \frac{nr_x}{A_o}\left(\frac{1 - \frac{x}{h_o}}{1 - \frac{\Delta V_x}{V_o}}\right)} \tag{6.24}$$

Note: For saturated soil, both V_o and ΔV_x are determined from the volume of water dissipated and collected in a burette. For partially saturated soils, a twin-burette arrangement is used in laboratories.

Some applications of the (CD) test

In general, the test is used in problems, where water can drain under loading e.g:

a) Foundations and piles in sand or gravel.
b) Retaining walls in sand or gravel.
c) Sudden variation of water level in sand or gravel slopes of rivers or reservoirs e.g. sudden drawdown.
d) Earth retaining structures and fills.
e) Foundation of earth or other structures, where some consolidation could occur due to slow progress of construction.
f) Stability during construction in fissured clay.

Note: The test may be carried out on soil samples of all types such as: disturbed, remoulded, compacted or re-deposited. In granular soil, either drained or undrained test may be used to the same effect.

6.4.8 Unconfined compression strength of clays

The consistency of clays may be related to the results of unconfined compression tests ($\tau_f = c_u$).

Table 6.8

Consistency	$\sigma_u = 2c_u$ (kN/m²)
Very soft	< 25
Soft	25–50
Medium	50–100
Stiff	100–200
Very stiff	200–400
Hard	> 400

The unconfined compression test

If the laboratory triaxial test is carried out without surrounding the soil specimen with rubber membrane and cell pressure, then it is said to be tested under unconfined conditions. The usual application of this undrained test is in the field, using the apparatus shown in Figure 6.36 on cohesive soil only.

The compressive strength is obtained by applying an axial load to an undisturbed specimen and measuring the resulting deformation.

During the test, the volume (V) of the sample remains the same, but its cross-sectional area changes.

Unconfined compression apparatus

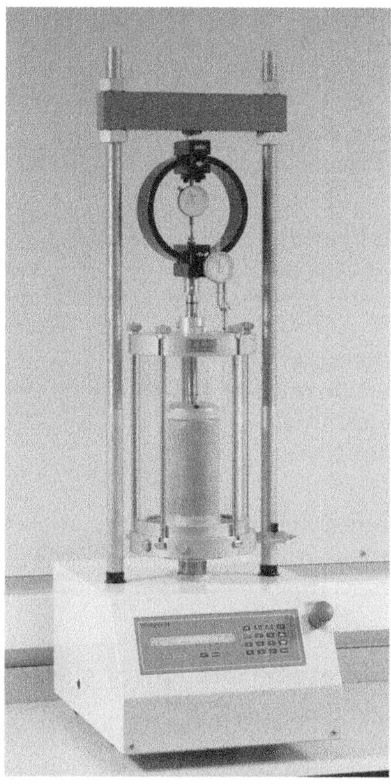

Figure 6.36 EL25-3700 series MultiPlex50 for quick undrained triaxial test. Reproduced by Permission of ELE International.

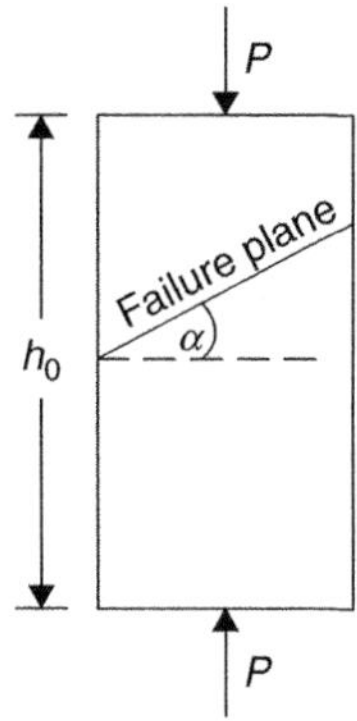

Figure 6.37

k = proving ring constant
x = strain dial gauge reading
$P = kx$

Initially: A_o = cross-sectional area
H_o = height
V = volume
$\therefore V = A_o h_o$

At failure: A_x = cross-sectional area
h_x = height

$$V = A_x(h_o - h_x) \quad \therefore \quad A_x = \frac{V}{h_o - h_x}$$

$$\sigma_u = \text{compressive stress} = \frac{P}{A_x}$$

Substituting, $\sigma_u = \dfrac{kx(h_o - h_x)}{V}$

Or $\boxed{\sigma_u = \dfrac{kx(h_o - h_x)}{A_o h_o}}$

Unconfined compression strength (6.25)

Mohr-circle representation:

$$\boxed{c_u = \frac{\sigma_u}{2}}$$

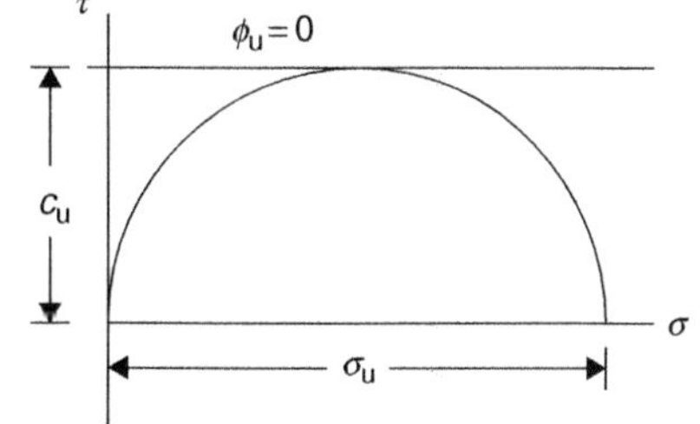

Figure 6.38

If the angle (α) of the failure plane can be measured then, theoretically, ϕ_u and c_u can be determined.

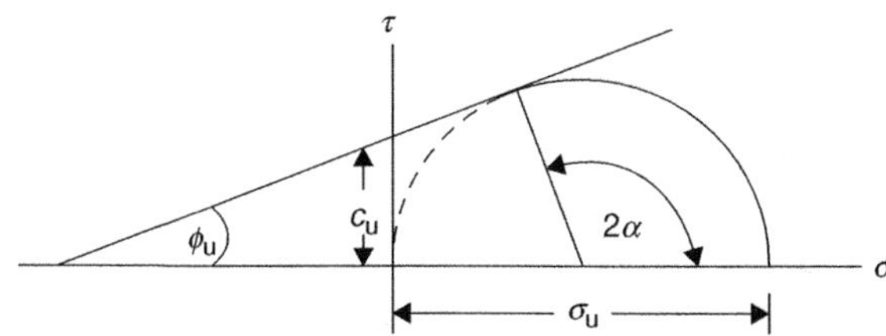

Figure 6.39

Where c_u = undrained shear strength

6.4.9 Standard shear box test

The shear box test is used for the determination of shear strength of soils, such as sand, by direct shear. The test can be drained or undrained and may be used for clay and remoulded soil, although the triaxial apparatus is more versatile in the testing of these materials. Figure 6.40 shows the schematic cross-section of the apparatus:

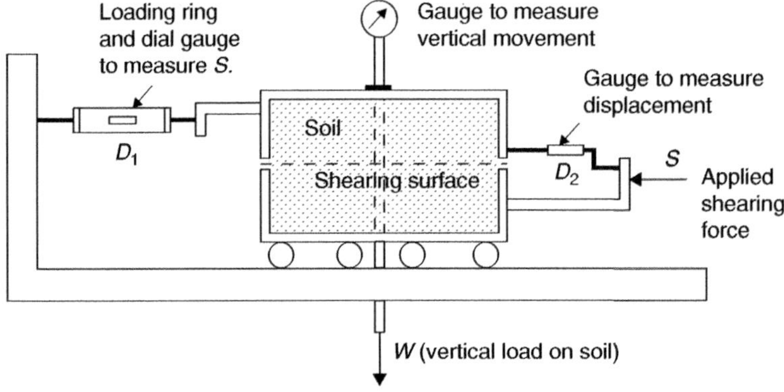

Figure 6.40

Outline of the undrained test-procedure:

Step 1: Place the soil into the box, levelling off the surface, which should be about 5 mm below the top of the box.

Step 2: Place the upper toothed grid, face downwards on the soil as well as the loading pad on top of the assembly.

Step 3: Apply the loading (W).

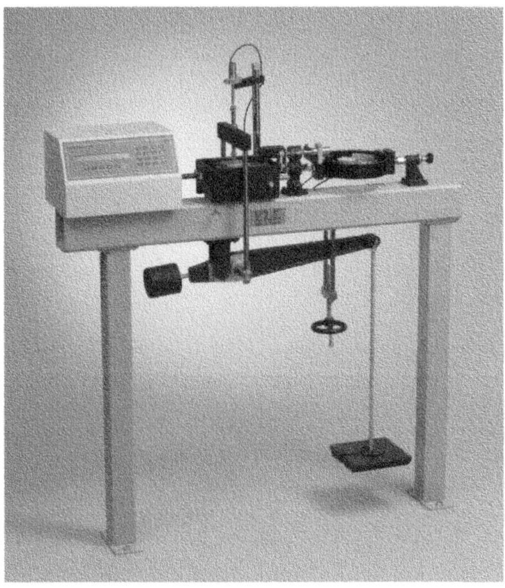

Figure 6.41 EL26-2114 series Digital Direct/Residual Shear Apparatus. Reproduced by permission of ELE International.

Step 4: Switch on the motor and read the dial gauges at regular intervals, say every 15 seconds, until failure occurs.

Step 5: Calculate the applied shear force (s) from the proving ring constant and the load gauge reading.

Step 6: Repeat steps 1-5 at least four times.

Step 7: Calculate the compressive (σ) and shear stresses (τ) and plot τ against ϕ to obtain the coulomb envelope, hence c and ϕ.

Example 6.5

The result of the shear box test on clayey sand is tabulated below.

Proving ring constant: $n = 0.0102$ kN/div

Area of sample: $A = 3600\,\text{mm}^2 = 3.6 \times 20^{-3}\,\text{m}^2$

Table 6.9

$$S = nD_1 \text{ and } \tau = \frac{S}{A}$$

		Sample A			Sample B			Sample C		
		$\sigma_n = 42\,\text{kN/m}^2$			$\sigma_n = 78\,\text{kN/m}^2$			$\sigma_n = 130\,\text{kN/m}^2$		
Time (Sec)	**Dial (D_2) readings (mm)**	**Dial (D_1) (div)**	**Shear force S (kN)**	**Shear stress τ (kN/m²)**	**D_1 (div)**	**S (kN)**	**τ (kN/m²)**	**D_1 (div)**	**S (kN)**	**τ (kN/m²)**
0	0	0	0	0	0	0	0	0	0	0
15	0.3	31.8	0.02	9	63.5	0.065	18.0	91.8	0.094	26.0
30	0.6	54.7	0.056	15.5	102.4	0.104	29.0	141.2	0.144	40.0
45	0.9	74.1	0.076	21.0	129.9	0.132	36.8	183.5	0.187	52.0
60	1.2	90.0	0.092	25.5	148.2	0.151	42.0	215.3	0.220	61.0
75	1.5	104.0	0.106	29.5	170.8	0.174	48.4	240.0	0.245	68.0
90	1.8	114.7	0.117	32.5	186.4	0.19	52.8	259.0	0.264	73.5
105	2.1	123.5	0.126	35.0	198.7	0.203	56.3	277.4	0.283	78.6
120	2.4	130.4	0.133	37.0	209.6	0.214	59.4	292.2	0.298	82.8
135	2.7	137.6	0.140	39.0	217.4	0.222	61.6	302.4	0.308	85.7
150	3.0	139.4	0.142	39.5	218.8	0.223	62.0	315.9	0.322	89.5
165	3.3	135.9	0.139	38.5	217.0	0.221	61.5	324.7	0.331	92.0
180	3.6	123.5	0.126	35.0	197.6	0.202	56.0	332.4	0.339	94.2
195	3.9							338.8	0.346	96.0
210	4.2							343.8	0.351	97.4
225	4.5							348.0	0.355	98.6
240	4.8							351.0	0.358	99.4
255	5.1							347.0	0.354	98.3
270	5.4							338.8	0.346	96.0

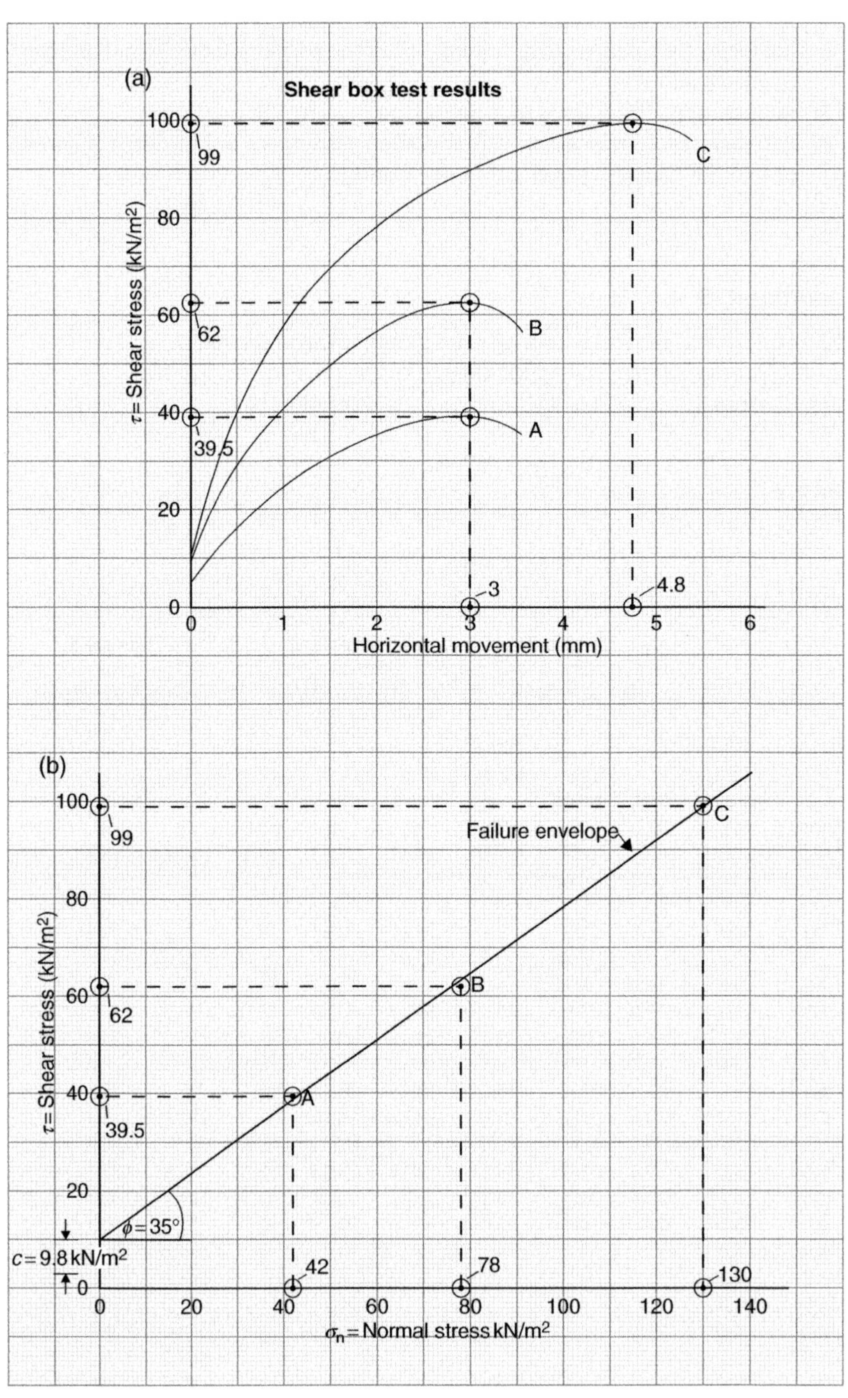
(a)
Shear box test results
τ = Shear stress (kN/m²)
100
99
80
60
62
40
39.5
20
0
C
B
A
3
4.8
0
1
2
3
4
5
6
Horizontal movement (mm)
(b)
100
99
Failure envelope
C
80
τ = Shear stress (kN/m²)
60
62
B
40
39.5
A
20
φ = 35°
c = 9.8 kN/m²
0
42
78
130
0
20
40
60
80
100
120
140
σn = Normal stress kN/m²

Graph 6.6

By plotting the tabulated figures on Graph 6.6(a) the maximum values of the shear stress are determined.

Table 6.10

Sample	τ_{max} (kN/m^2)
A	39.5
B	62.0
C	99.0

By plotting these figures in turn, against their respective normal stress on Graph 6.6(b), the failure envelope can be drawn and the shear strength parameters determined as:

$$C = 9.8\,\text{kN/m}^2$$
$$\phi = 35°$$

6.4.10 The Vane shear test

It happens sometimes, that the clay to be tested is so plastic, that it cannot be extruded from the ground, without causing extensive disturbance to its structure. Because of its softness, the soil cannot be tested in the triaxial apparatus. This applies especially to sensitive clays. In these circumstances, the in-situ Vane test is used to obtain fairly reliable values for the undrained shear strength (c_u) of the clay. In general, the test is used for intact soft to firm clays, having shear strength less than 100 kN/m^2.

Sensitivity of clays (S_t)

The shear strengths of clays are adversely affected by disturbance to a varying degree. The sensitivity of a particular soil to the destruction of its structure is given by:

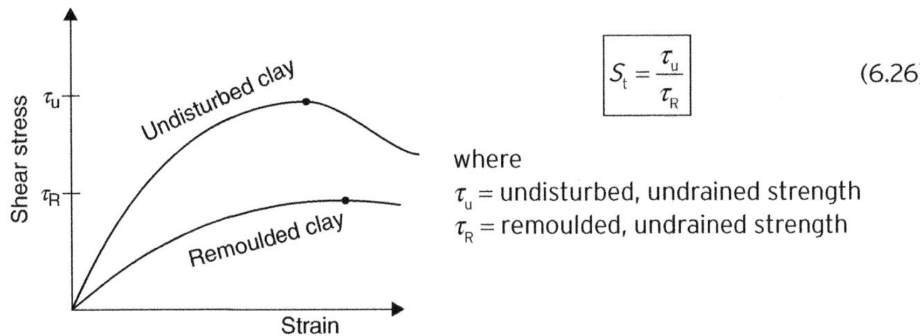

$$S_t = \frac{\tau_u}{\tau_R} \quad (6.26)$$

where
τ_u = undisturbed, undrained strength
τ_R = remoulded, undrained strength

Figure 6.42

In order to determine the sensitivity of a type of clay therefore, the shear strength of an undisturbed sample is first obtained from an undrained test. After this, the specimen is remoulded at the same moisture content and its unit weight as well as shear strength found. The clay is then classified, using Table 6.11.

Table 6.11

Sensitivity (S_t)	Classification
1	Insensitive
1–2	Low sensitivity
2–4	Medium sensitivity
4–8	Sensitive
8–16	Extra sensitive
>16	Quick

Outline of the field test

The test is most frequently carried out in a borehole. The apparatus consist of a four-bladed vane attached to a torque-measuring apparatus, by means of extension rods, as shown.

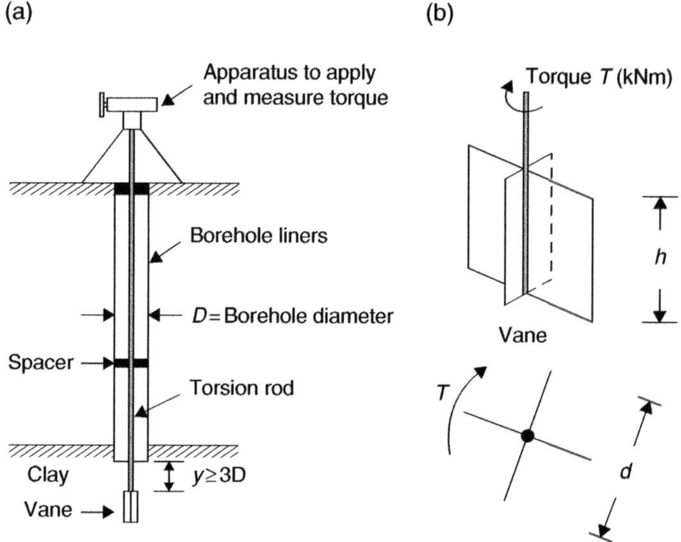

Figure 6.43

Normally, there are two vane sizes to choose from, depending on the strength of the soil.

Table 6.12

Consistency	Shear strength c_u (kN/m²)	Vane size (mm) d	h
Soft to firm	<50	75	150
Firm to stiff	50–100	50	100

Step 1: Push the vane into the clay to a depth of $y \geq 3D$, below the bottom of the borehole.

Step 2: Apply a steadily increasing torque to the rod, until failure occurs, that is when sudden loss of resistance is noted.

Step 3: Record the applied torque (T) at failure.

Step 4: Calculate the undrained shear strength from:

$$c_u = \frac{T}{\pi d^2 \left(\frac{h}{2} + \frac{d}{6}\right)} \tag{6.27}$$

6.4.11 Residual shear strength (Skempton 1964)

Clay in its undisturbed state is made up of flat, flaky particles, orientated in a random manner. This contributes to its shear strength.

Figure 6.44

If the clay is tested and strained in a drained shear box test, the particles re-orientate somewhat under the effective pressure such that the shearing resistance increases.

Figure 6.45

At failure strain, the shear strength reaches its maximum value. Beyond this point, the orientation of the particles gradually becomes parallel, accompanied by a decrease in strength. At large strain, the particles are found to be parallel along the failure plane.

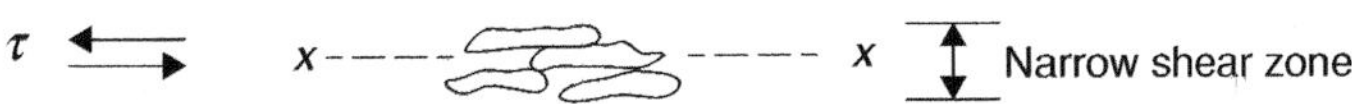

Figure 6.46

The shear strength of the clay along the plane of failure remains constant, if the sample is subject to even larger strains. At this stage, the soil is said to have reached its *residual strength*.

It must be emphasized, that the re-orientation of particles occurs in a thin layer along the shear zone only. The variation of shear stress with displacement under constant effective pressure is shown below.

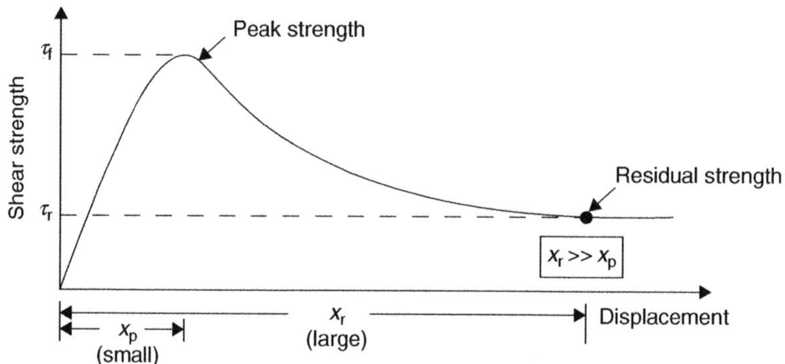

Figure 6.47

The Coulomb envelopes for peak and residual strength, obtained from several tests on the same material, show marked difference between their shear strength parameters.

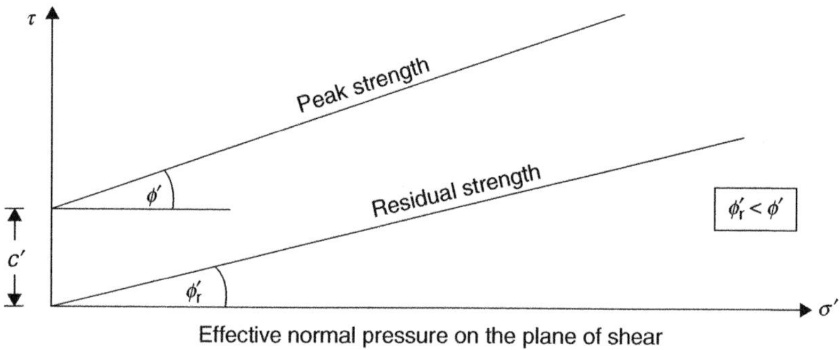

Figure 6.48

For practical purposes, the residual cohesion intercept is taken to be zero, hence the equation for:

1. Peak strength: $\tau_f = c' + \sigma' \tan\phi'$
2. Residual strength: $\tau_r = \sigma' \tan\phi'_r$

Notes:

a) As the specimen is in fact remoulded within the shear zone, the residual strength of any remoulded sample can be determined directly by a suitable shear test.
b) The residual strength does not depend on the consolidation history of a clay. Subjected to the same effective pressure, the clay exhibits the same residual strength in both normally and overconsolidated state.
c) The shear test must be carried out slowly to allow time for the dissipation of excess pore pressure.
d) Displacement (x_r) could be quite large. This can be achieved in the standard shear box by repeating the test, on the same specimen, as many times as necessary to reach a steady value of shear strength. After the split box reaches the end of its travel, it is returned to the original position and the sample is sheared again.

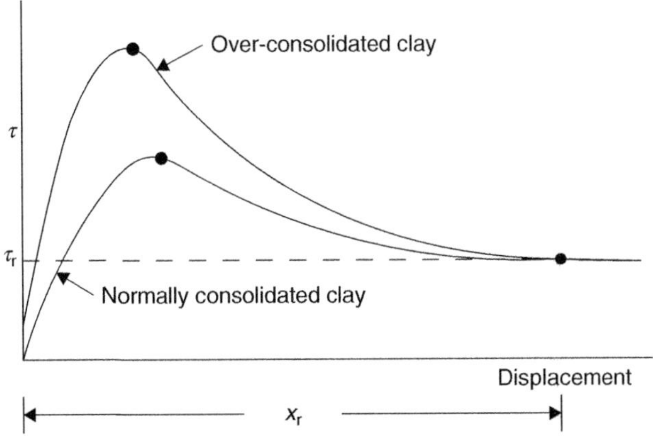

Figure 6.49

Alternatively, a reversible shear box can be used. In this apparatus, the box can travel back and forth, thus allowing the displacement to be as long as required to reach residual strength, without repeatedly starting from the beginning.

Alternatively, a ring shear apparatus is used for the above purpose, but it is not yet standard laboratory equipment.

6.5 Thixotropy of clay

Clay loses strength when remoulded. The magnitude of loss depends on its sensitivity [see formula (6.26)]. If, however, its water content remains unchanged, it regains some or most of its strength with time. The *phenomenon* is called *thixotropy*. Bentonite, mainly composed of the clay mineral montmorillonite, is such a thixotropic material, which expands with increasing water content. A suspension of it will eventually become a gel, which can be re-converted into suspension by mixing. It is often used to support the sides of temporary trenches for diaphragm walls. Subsequent placing of concrete displaces the reconstituted slurry.

6.6 Undrained cohesion and overburden pressure

The value of c_u varies with depth. Skempton (1957) introduced the following approximate relationship for normally consolidated clays.

$$\left(\frac{c_u}{\sigma'_0}\right)_{NCC} = 0.11 + 0.0037 \times (PI\%) \tag{6.28}$$

Where PI = Plasticity Index
σ'_0 = effective overburden pressure

For overconsolidated clays, Ladd and Foott (1974) purposed the empirical formula:

$$\left(\frac{c_u}{\sigma'_0}\right)_{OCC} = (OCR)^{0.8}\left(\frac{c_u}{\sigma'_0}\right)_{NCC} \tag{6.29}$$

where, OCR = Overconsolidation ratio by formula (7.9).

Example 6.6

The clay in Figure 6.50 has the following Atterberg Limits: LL = 66.4% and PL = 29% From (2.12), the Plasticity Index is: PI = LL − PL

$$= 66.4 - 29 = 37.4\%$$

The clay is overconsolidated, having OCR = 1.8. Estimate the undrained cohesion 14 m below ground level at point *P*.

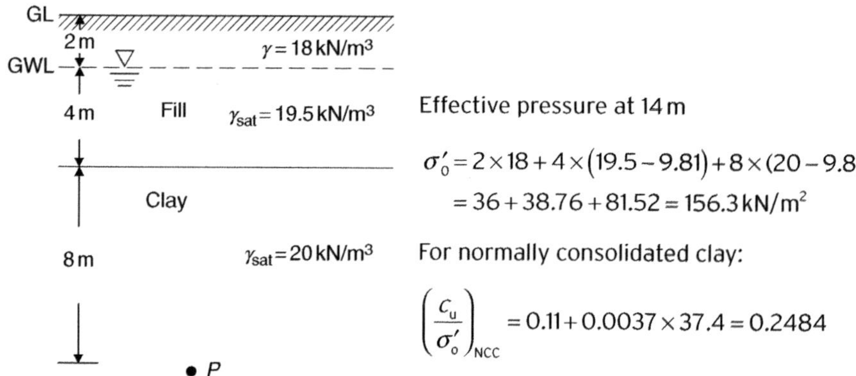

Effective pressure at 14 m

$$\sigma'_o = 2\times 18 + 4\times(19.5 - 9.81) + 8\times(20 - 9.81)$$
$$= 36 + 38.76 + 81.52 = 156.3\,\text{kN/m}^2$$

For normally consolidated clay:

$$\left(\frac{c_u}{\sigma'_o}\right)_{NCC} = 0.11 + 0.0037\times 37.4 = 0.2484$$

Figure 6.50

For the overconsolidated clay therefore, from (6.29):

$$\left(\frac{c_u}{\sigma'_o}\right)_{OCC} = (1.8)^{0.8}\times 0.2484 = 0.3975$$

$$c_u = 0.3975\times 156.3 = 62.1\,\text{kN/m}^2$$

For direct application to overconsolidated clays, the two formulae are combined:

$$\boxed{c_u = (\text{OCR})^{0.8}\left[0.11 + 0.0037\,(\text{PI}\,\%)\right]\sigma'_o} \qquad (6.30)$$

And $c_u = 1.8^{0.8}\times(0.11 + 0.0037\times 37.4)\times 156.3 = 62.1\,\text{kN/m}^2$

Problem 6.1

A clay specimen was tested in a triaxial apparatus at zero cell pressure, until brittle failure occurred at 150 kN/mm^2 deviator stress. The following details were recorded:

Initial height of the specimen = 76 mm
Initial diameter = 38 mm
Loading gauge reading at failure = 41 divisions
Strain gauge reading at failure = 5.2 mm
Angle of the failure plane = 57°

Estimate the:
a) proving ring constant
b) Cross-section of the specimen at failure
c) Theoretical values of undrained cohesion and angle of friction.

Problem 6.2

The results of triaxial tests, carried out on three clay samples, are tabulated below. Each test was performed by varying the pressures during the process.

Test 1: Decreasing the cell pressure (σ_c), whilst increasing the deviator stress (σ_d) so, that the total vertical pressure (σ_v) remains constant.

Table 6.13

$\sigma_H = \sigma_c$ (kN/m^2)	500	400	300	200
σ_d (kN/m^2)	0	100	200	300

Test 2: Decreasing the cell pressure and increasing the total vertical pressure by the same amount.

Table 6.14

$\sigma_H = \sigma_c$ (kN/m^2)	500	450	400	350	300
σ_v (kN/m^2)	500	550	600	650	700

Test 3: The cell pressure is kept constant, whilst increasing the deviator stress, as in the standard triaxial test.

Table 6.15

$\sigma_H = \sigma_c$ (kN/m^2)	500	500	500	500
σ_d (kN/m^2)	0	200	400	600

a) Draw the stress path diagram and obtain the equation of its failure envelope.
b) Draw the Mohr circle diagram and obtain the equation of its failure envelope.

Problem 6.3

A 4 m thick layer of sandy gravel alluvial deposit is underlain by saturated, normally consolidated clay, as shown in Figure 6.52. A clay sample was extracted from a depth of 9 m below ground surface, where laboratory tests indicate the following soil characteristics:

Coefficient of earth pressure at rest = 0.7
Pore pressure parameters: A = 0.65
B = 1

Estimate the pore pressure in the sample just after its removal from the ground.

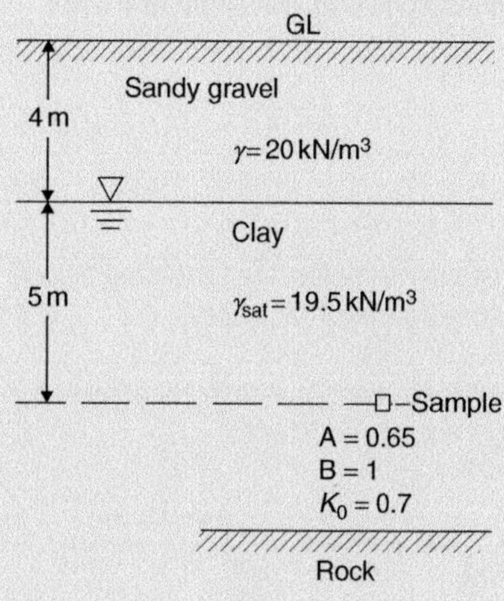

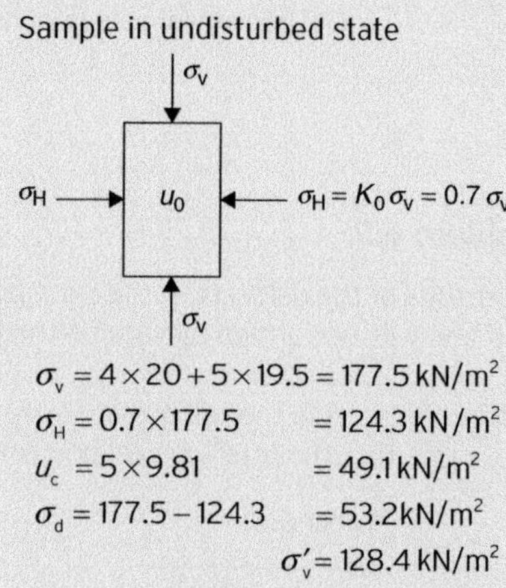

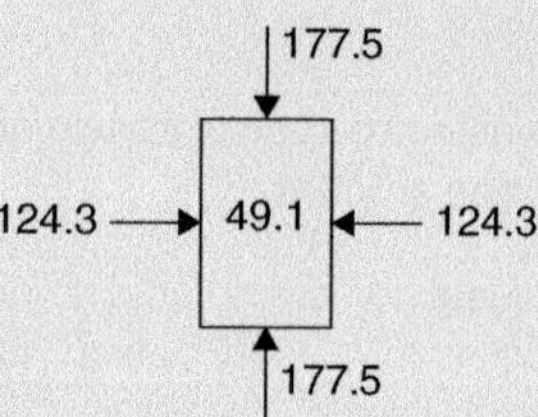

Figure 6.52

Problem 6.4

The residual pore pressure in each of the three samples taken from 9 m below ground surface, as in problem 8.3, is $-110\,kN/m^2$. Each sample is tested triaxially under different conditions, that is in a:

i. Quick test
ii. Consolidated, undrained (CU) test.
iii. Consolidated, drained (CD) test.

Estimate the effective stresses in the samples after the application of:

a) cell pressure $= 100\,kN/m^2$
b) deviator stress $= 120\,kN/m^2$

Assume $B = 1$ and $A = 0.65$

Chapter 7
Consolidation and Settlement

When fine-grained, cohesive, saturated soil is in its natural state, the water in the voids (pores) is under hydrostatic pressure. If a surcharge load (q) is placed upon the soil layer, it induces a pressure ($\Delta u = q$) in the pore water in excess of the hydrostatic one. It is explained in Chapter 5 that as the excess pore pressure dissipates, the soil particles reorientate, thus decreasing the volume of voids. The long-term effect of external load, is the compression (consolidation) of the layer. As a consequence of consolidation, there is a decrease in total volume of a layer and the load is subjected to downward movement (settlement). Its strength increases however.

There are two problems to solve in this chapter:

1. The magnitude of consolidation of soil and consequent settlement of a structure.
2. The length of time taken to reach maximum settlement of a structure.

7.1 Consolidation

Any change in total volume (V) means corresponding change in voids ratio (e) as well as in the moisture content. It is, therefore, convenient to predict the magnitude of consolidation in terms of voids ratio. This is done by means of the oedometer test, in terms of effective pressure and voids ratio. A brief outline of the test procedure is as follows:

Step 1: Determine the initial moisture content (m_0), of the soil, as well as its specific gravity G_s.
Step 2: Cut a sample, using a cutter of the required size. Trim off the top and bottom surfaces, until they are flush with the edges of the cutter (Figure 7.1).
Step 3: Weigh the cutter and sample in order to determine the density.
Step 4: Saturate the porous discs and place them into the ring with the cutter plus sample sandwiched between the two.
Step 5: Place the pressure pad onto the specimen and load it with the lowest pressure required (Figure 7.2).
Step 6: Flood the cell with water up to the top of the sample.

Introduction to Soil Mechanics, First Edition. Béla Bodó and Colin Jones.

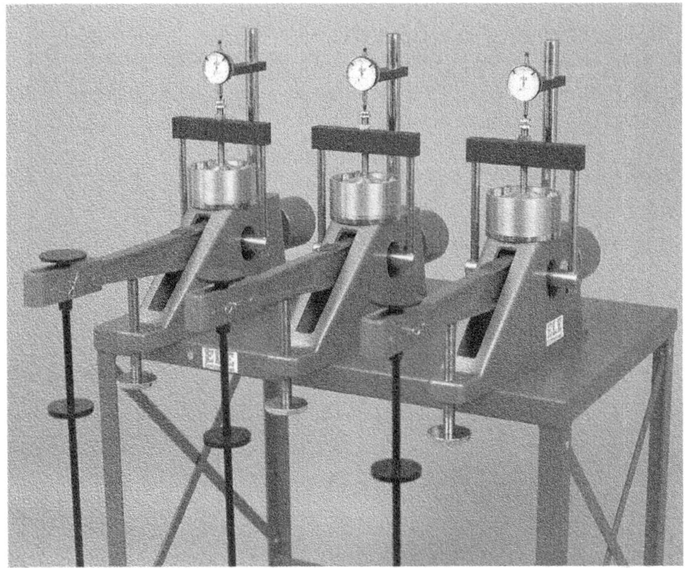

Figure 7.1 EL25-0402 Consolidation Frames. Reproduced by permission of ELE International.

Step 7: Read the micrometer dial (D_x) frequently at first, gradually increasing the time interval until there is no perceptible change from the previous reading.

Step 8: Add the next load-increment and repeat steps 7 and 8 as required.

Step 9: After the last reading was taken for the heaviest load, remove all weights from the hanger. The resulting swelling is observed on the dial gauge.

Step 10: Release the water from the cell and the sample from the cutter. Determine its moisture content at the time of removal and calculate the voids ratio after swelling from: $e_s = m_s G_s$.

Figure 7.2 Consolidation Cell Components. Reproduced by permission of ELE International.

7.2 The pressure-voids ratio curve

The results of an oedometer test are normally presented by this curve. Two procedures are introduced in this section for the determination of the voids ratio corresponding to each load increment. These are:

1. Analytical evaluation
2. Graphical construction on Chart 7.1

7.2.1 Analytical solution

Figures 7.3, 7.4 and 7.5 illustrate diagrammatically the variation of volume and voids ratio, due to increments in the effective pressure during the oedometer test.

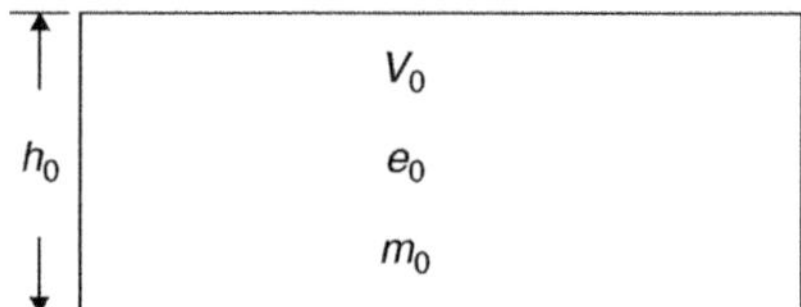

Figure 7.3

m_0 = Initial moisture content
h_0 = Initial height
e_0 = Initial voids ratio
V_0 = Initial volume

The sample is loaded

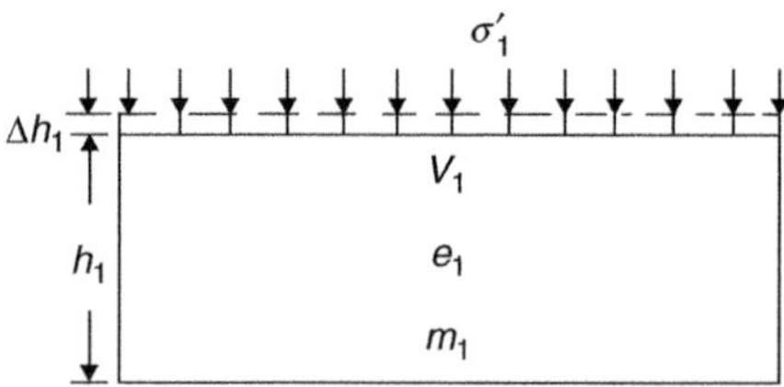

Figure 7.4

σ'_1 = initial applied effective pressure
Δh_1 = decrease in height due to σ'_1
h_1 = height of sample at the end of stage 1

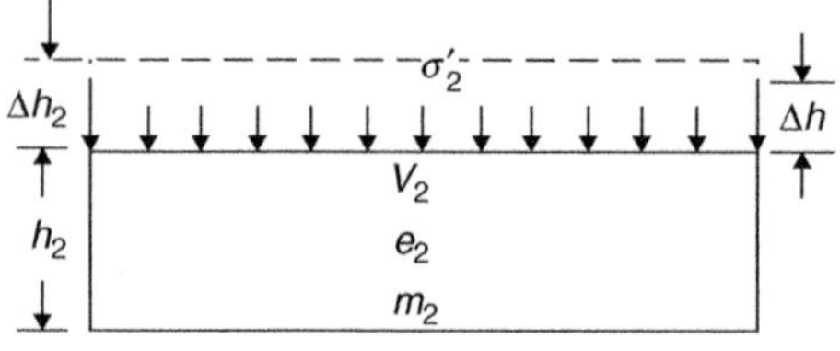

Figure 7.5

$$\sigma'_2 = \sigma'_1 + \Delta\sigma'_2$$

$\Delta\sigma'_2$ = pressure increment
Δh = decrease in height due to $\Delta\sigma'_2$
Δh_2 = decrease in height due to σ'_2
h_2 = height of sample at the end of stage 2

The formula for the voids ratio is derived from the volumetric changes observed between stages 1 and 2, caused by the load increment $\Delta\sigma'_2$.

Volume of specimen at stage 1: $V_1 = Ah_1$
where A is its cross-sectional area in plan.

Volume of specimen at stage 2: $V_2 = Ah_2$

From Chapter 1, formula (1.10): $V = (1 + e)\ V_s$

Therefore: $V_1 = (1+e_1)V_s$ from which $\boxed{V_s = \dfrac{V_1}{1+e_1}}$

$V_2 = (1+e_2)V_s$

Change in volume: $\Delta V = V_1 - V_2 = (1+e_1)V_s - (1+e_2)V_s$

$= (e_1 - e_2)V_s$

Eliminating the volume of solids,

$$\Delta V = (e_1 - e_2)\frac{V_1}{1+e_1} \qquad \therefore \quad \boxed{\Delta V = \left(\frac{e_1 - e_2}{1+e_1}\right)V_1}$$

Change in height: $\Delta h = h_1 - h_2$
Change in voids ratio: $\Delta e = e_1 - e_2$
Change in volume: $\Delta V = A\Delta h$

Eliminating area from $V_1 = Ah_1$

But, $\Delta V = A\Delta h$

$$A = \frac{V_1}{h_1} = \frac{\Delta V}{\Delta h}$$

$$\therefore \boxed{\Delta V = \frac{V_1}{h_1}\Delta h}$$

Equating $$\Delta V = \frac{V_1}{h_1}\Delta h = \frac{e_1 - e_2}{1+e_1}V_1$$

Eliminating V_1 $$\frac{\Delta h}{h_1} = \frac{e_1 - e_2}{1+e_1} \qquad \boxed{\frac{\Delta h}{h_1} = \frac{e}{1+e_1}} \qquad (7.1)$$

Substituting for Δh: $$\boxed{\frac{h_1 - h_2}{h_1} = \left(\frac{e_1 - e_2}{1+e_1}\right)} \qquad (7.2)$$

In general terms, the voids ratio at any stage can be calculated from the:

a) Height and voids ratio at the previous stage
b) Height at the stage considered.

7.2.2 Equation of the σ'–e curve

The shape of the curve in both compression and swelling stage is drawn in Figure 7.6. Its equation is derived from (7.2).

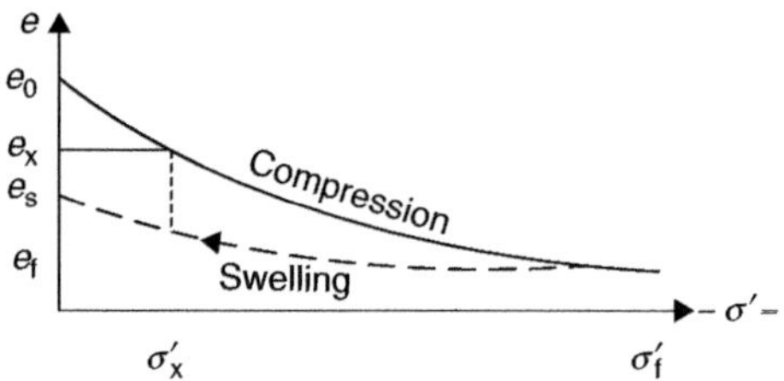

Figure 7.6

Where e_0 = initial voids ratio, before loading
e_x = voids ratio at an intermediate stage
e_f = final voids ratio just before unloading
e_s = voids ratio after swelling
Also, h_0 = initial height of specimen
h_x = height at an intermediate stage
h_f = final, minimum height
h_s = height after swelling is completed, usually 24 hours later
σ'_x = effective pressure at an intermediate stage
σ'_f = final effective pressure
m_0 = initial moisture content
m_s = moisture content after swelling
D_x = dial reading at any stage

The voids ratio for saturated soil is given by $e = mG_s$. The specific gravity is either determined or assumed to be in the region of $G_s = 2.7$.

In step 1 of the test, m_0 is measured, hence the initial voids ratio can be calculated as:

$$e_0 = m_0 G_s (S_r = 1)$$

In formula (7.2), choose:

$$e_1 = e_0 \text{ (known)}$$
$$e_2 = e_x$$
$$h_1 = h_0 \text{ (known)}$$
$$h_2 = h_x \text{ (known)}$$

Substituting,

$$\frac{h_0 - h_x}{h_0} = \frac{e_0 - e_x}{1 + e_0}$$

Expressing e_x:

$$h_0 + h_0 e_0 - h_x(1 + e_0) = h_0 e_0 - h_0 e_x$$
$$h_0 - h_x(1 + e_0) = -h_0 e_x$$

From which,

$$\boxed{e_x = \left(\frac{1 + e_0}{h_0}\right) h_x - 1} \qquad (7.3)$$

This is the desired equation of the curve, which can be simplified further by expressing h_x in terms of oedometer dial readings D_x.

$$h_x = h_0 - D_x$$

Substituting,
$$e_x = \left(\frac{1+e_0}{h_0}\right)(h_0 - D_x) - 1$$
$$= 1 + e_0 - \left(\frac{1+e_0}{h_0}\right)D_x - 1$$

From which,
$$\boxed{e_x = e_0 - \left(\frac{1+e_0}{h_0}\right)D_x} \quad (7.4)$$

The application of (7.4) means that there is no need to calculate heights, as the void ratios can be found directly from the oedometer dial readings as the test progresses, provided e_0 is known at the start. If, however, e_s and h_s are determined after swelling, then (7.3) and (7.4) are transformed into:

$$\boxed{e_x = \left(\frac{1+e_s}{h_s}\right)h_x - 1} \quad (7.5)$$

$$\boxed{e_x = \left(\frac{1+e_s}{h_s}\right)(h_0 - D_x) - 1} \quad (7.6)$$

Example 7.1

The results of an oedometer test carried out on a sample of clay are given below. The initial moisture content was 27.8%. The initial height and diameter of the specimen were 19 mm and 76 mm respectively. Assume $G_s = 2.7$.

Calculate the void ratios by both formulae (7.3) and (7.4). Draw the pressure-voids ratio curve.

Table 7.1

x	σ'_x kN/m^2	D_x (mm)	h_x (mm)	$e_x = \left(\frac{1.751}{19}\right)h_x - 1$	$e_x = 0.751 - \left(\frac{1.751}{19}\right)D_x$
0	0	0.00	19.00	0.751	0.751
1	25	0.22	18.78	0.731	0.731
2	50	0.40	18.60	0.714	0.714
3	100	0.72	18.28	0.685	0.685
4	200	1.12	17.88	0.648	0.648
5	400	1.57	17.43	0.606	0.606
6	600	1.82	17.18	0.583	0.583
7	750	1.90	17.10	0.576	0.576
8	0	0.49	18.51	0.705	0.706

Where $e_0 = 0.278 \times 2.7 = 0.751$

$$\text{Hence}\,(7.3):\; e_x = \frac{h_x}{19} \times (1 + 0.751) - 1 = \left(\frac{1.751}{19}\right) \times h_x - 1$$

$$\text{And}\,(7.4):\quad e_x = 0.751 - \left(\frac{1.751}{19}\right) D_x$$

Note: Whilst h_x has to be calculated step-by-step by formula (7.3), the application of (7.4) yields the answer directly from D_x.

e.g. for stage $x = 5$, $D_x = 1.57$

$$\therefore e_5 = 0.751 - \left(\frac{1.751}{19}\right) \times 1.57 = 0.606$$

The result in Table 7.1 is plotted on Graph 7.1.

7.2.3 Alternative conventional procedure

The voids ratio is normally calculated in the literature by means of differences as given by the generalized form of (7.1) that is: $\dfrac{\Delta h}{h_0} = \dfrac{e}{1 + e_0}$

The procedure is detailed in Table 7.2.

Table 7.2

	x	σ'_x kN / m^2	D_x (mm)	h_x (mm)	Δh_x (mm)	$\Delta e_x = \left(\frac{1+e_0}{h_0}\right)\Delta h_x$	e_x
$x = 0$	0	0	0.00	19.00	–	–	0.751
	1	25	0.22	18.78	–0.22	–0.020	0.731
	2	50	0.40	18.60	–0.18	–0.017	0.714
	3	100	0.72	18.28	–0.32	–0.029	0.685
	4	200	1.12	17.88	–0.40	–0.037	0.648
	5	400	1.57	17.43	–0.45	–0.041	0.607
	6	600	1.82	17.18	–0.25	–0.023	0.584
x_f	7	750	1.90	17.10	–0.08	–0.007	0.577
x_s	8	0	0.49	18.51	+1.41	+0.130	0.707

Note: It must be remembered that e_0 is the voids ratio of saturated soil. Therefore, unless the specimen is fully saturated ($S_r = 1$) initially, then the moisture content and voids ratio after step 10, that is after saturation and swelling, should be applied to calculate e_x.

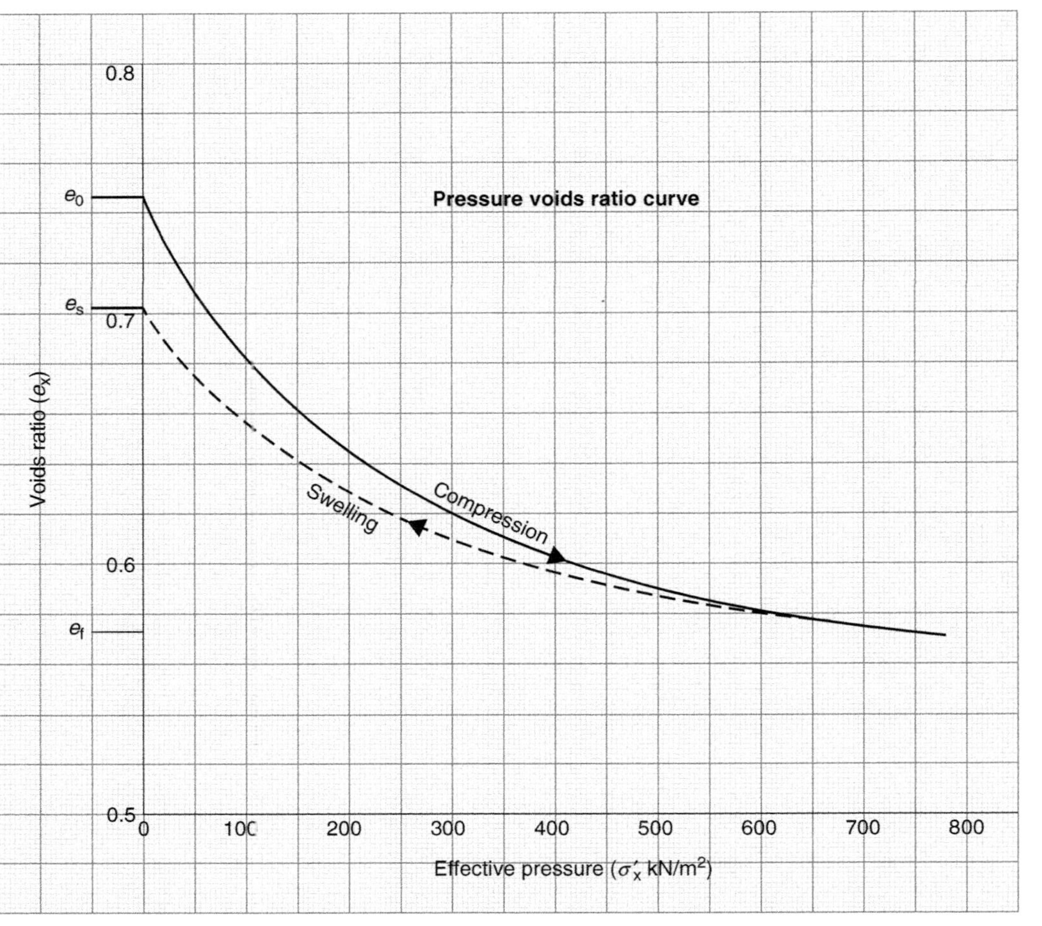

Pressure voids ratio curve
Voids ratio (e_x)
0.8
e_0
e_s
0.7
0.6
e_f
0.5
0
100
200
300
400
500
600
700
800
Compression
Swelling
Effective pressure (σ'_x kN/m^2)

Graph 7.1

7.2.4 Graphical solution

The range of voids ratio may be obtained without calculations, by means of Chart 7.1.

Step 1: If e_s and D_s are known, then plot point C.
Step 2: Draw the voids ratio line so that every point on it is at the same proportional distance from the two adjacent construction lines as point C. Note, that these lines are not parallel. In this example, C is at equal distance from both lines. For this reason, draw the lines through C so that points A and B are also at equal distance from each line. Reading D_f defines point B.

Points A and B represent the initial and final void ratios respectively. Any other value, corresponding to a dial reading can now be read-off directly, e.g.

$$\text{For } D_4 = 1.12\,\text{mm} \quad e_4 = 0.648$$

If the initial voids ratio (e_0) is determined in the beginning, then the voids ratio line may be drawn prior to the commencement of the test, without knowing the actual positions of points B and C. Therefore,

Step 1: Plot point A at $D_x = 0$ on Chart 7.2
Step 2: Draw the voids ratio line.
Step 3: Choose a suitable scale for σ'_x along the top of the nomogram. In this example: $20\,\text{mm} = 100\,\text{kN/m}^2$.
Step 4: Plot the test results (σ'_x and D_x) after each reading during the test.
Step 5: Plot points B and C at the end of the test.

This nomogram can now be used to find the voids ratio, corresponding to any effective pressure, directly, as shown on Chart 7.2.

$$\text{For } \sigma'_1 = 200\,\text{kN/m}^2 \qquad e_1 = 0.650$$

$$\text{For } \sigma'_2 = 400\,\text{kN/m}^2 \qquad e_2 = 0.605$$

$$\therefore \text{ Change } \Delta\sigma' = 200\,\text{kN/m}^2 \qquad \Delta e = 0.045$$

It will be obvious from the Direct Method in Example 7.2, that Chart 7.2 could be applied to the estimation of foundation settlement, well before the completion of the test.

Construction of Chart 7.1

This nomogram is constructed for a 19/76 mm cutter ring. Should these dimensions be different, a new chart can easily be drawn by means of formula (7.4).

For example, in this case: $h_0 = 19\,\text{mm}$, hence (7.4) becomes: $e_x = e_0 - \left(\dfrac{1+e_0}{19}\right)D_x$

Choose a suitable value for D_x, preferably near the top of the chart, say $D_x = 2\,\text{mm}$ and substitute: $e_x = e_0 - \dfrac{2}{19} - \dfrac{2}{19} \times e_0 = \dfrac{17}{19} \times e_0 - \dfrac{2}{19}$

From which, $\boxed{e_x = 0.895 \times e_0 - 0.1053}$

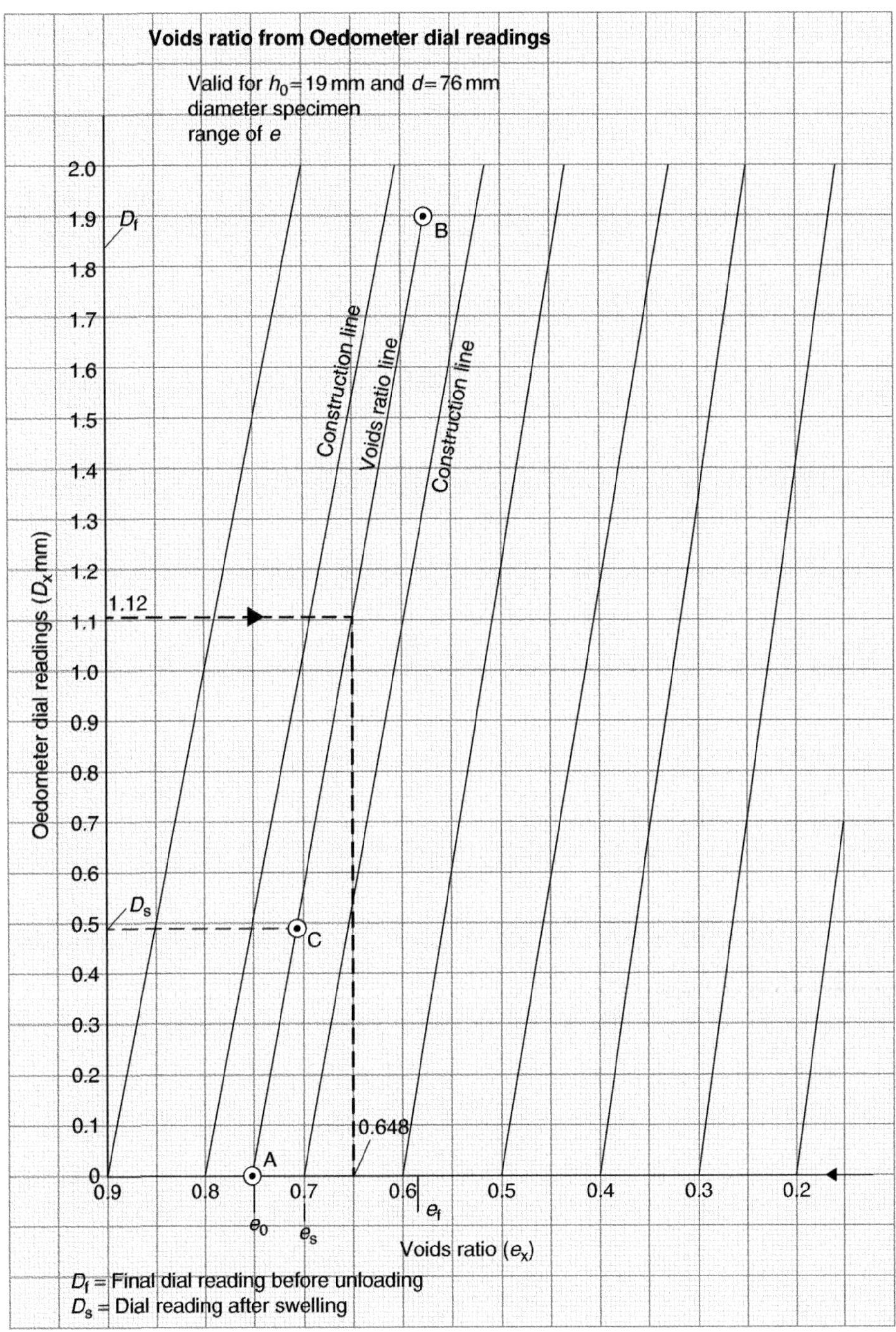

Voids ratio from Oedometer dial readings
Valid for h_0 = 19 mm and d = 76 mm diameter specimen
range of e
Oedometer dial readings (D_x mm)
2.0
1.9
1.8
1.7
1.6
1.5
1.4
1.3
1.2
1.1
1.0
0.9
0.8
0.7
0.6
0.5
0.4
0.3
0.2
0.1
0
D_f
D_s
1.12
B
C
A
0.648
Construction line
Voids ratio line
Construction line
0.9
0.8
0.7
0.6
0.5
0.4
0.3
0.2
e_0
e_s
e_f
Voids ratio (e_x)
D_f = Final dial reading before unloading
D_s = Dial reading after swelling

Chart 7.1

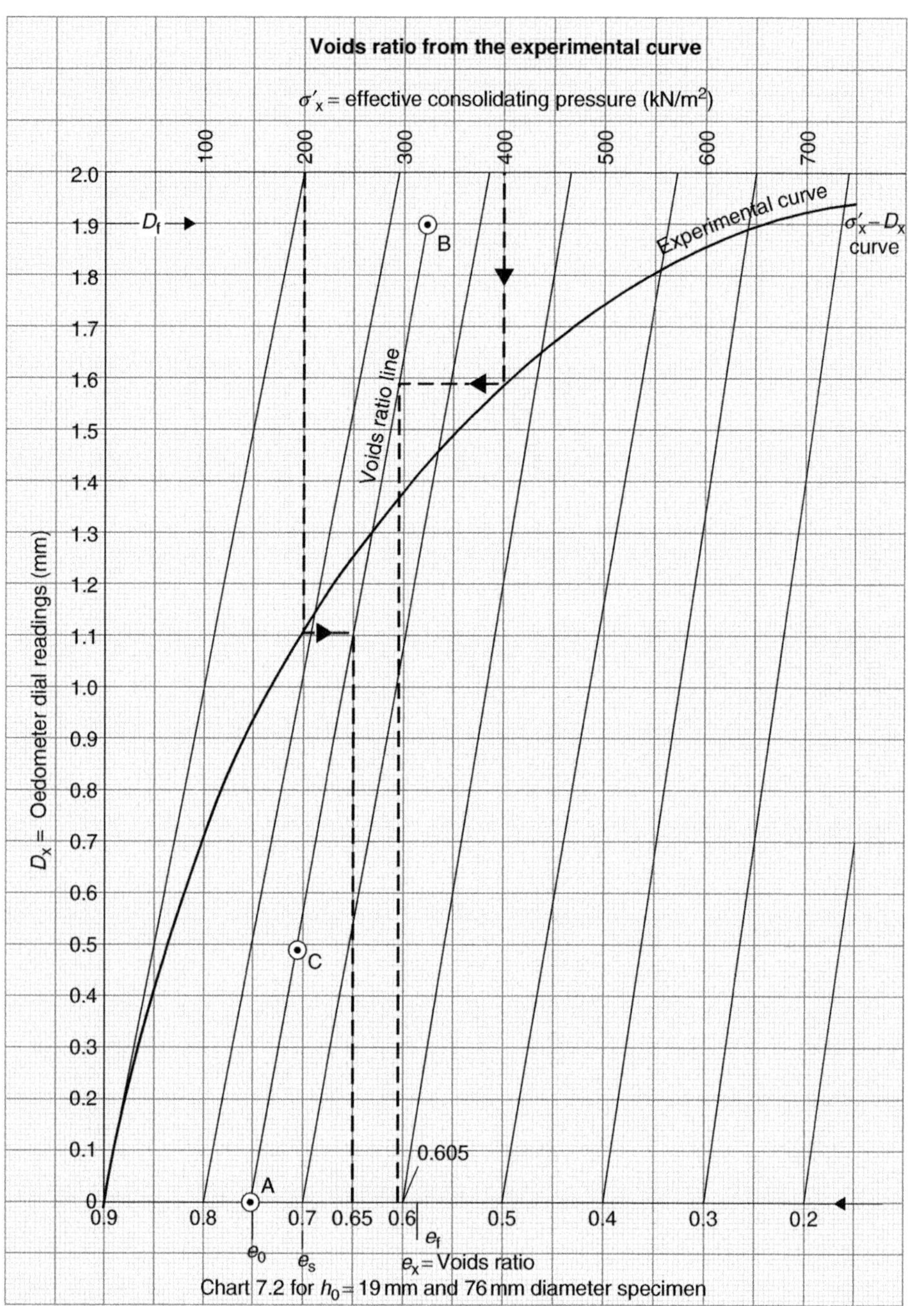
Voids ratio from the experimental curve
σ'x = effective consolidating pressure (kN/m²)
100
200
300
400
500
600
700
2.0
1.9
1.8
1.7
1.6
1.5
1.4
1.3
1.2
1.1
1.0
0.9
0.8
0.7
0.6
0.5
0.4
0.3
0.2
0.1
0
Df
B
Experimental curve
σ'x – Dx curve
Voids ratio line
Dx = Oedometer dial readings (mm)
C
A
0.605
0.9
0.8
0.7
0.65
0.6
0.5
0.4
0.3
0.2
ef
e0
es
ex = Voids ratio
Chart 7.2 for h0 = 19 mm and 76 mm diameter specimen

Chart 7.2

This equation gives the values of e_x along the horizontal line at D_x=2 mm for each value of e_o, situated on the horizontal line at D_x=0. The construction is shown in Figure 7.7.

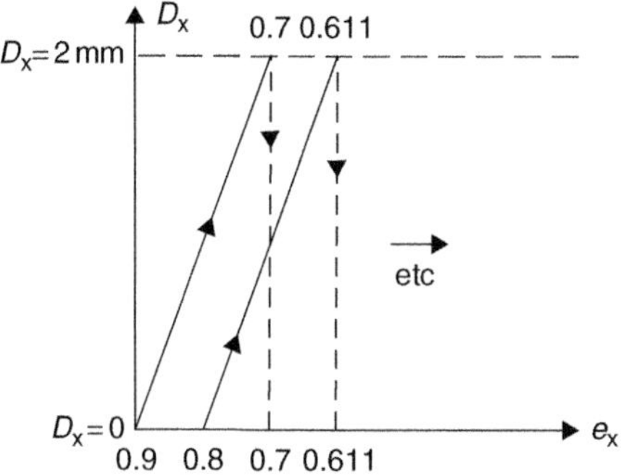

Figure 7.7

Table 7.3

e_o	e_x
0.9	0.700
0.8	0.611
0.7	0.521
0.6	0.432
0.5	0.342
↓etc.	↓etc.

The lines, of course, may be extended for larger values of D_x.

7.3 Forms of the σ'–e curve

The shape of the curve depends largely on the geological history of the soil. From this point of view, there are two main types of clay:

1. Normally consolidated
2. Overconsolidated.

The former is more compressible than the latter. Cohesive soils, such as clay, are formed by sedimentation, that is, the particles are gradually deposited and compressed by the weight of increasingly thick overburden. The σ'–e curve for this material – approximately straight line when drawn on a semi-logarithmic scale – is called the *Virgin Consolidation Curve*.

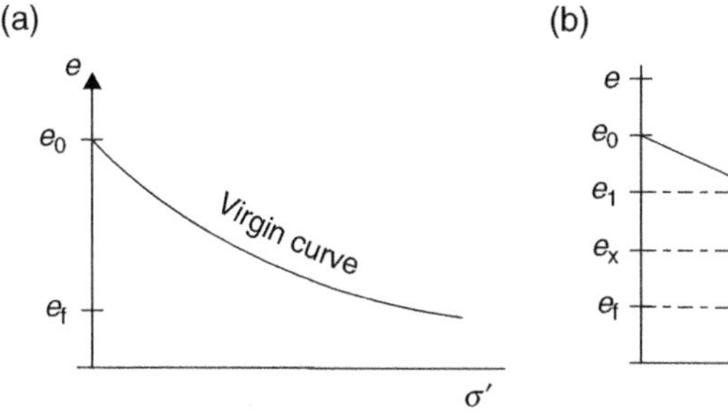

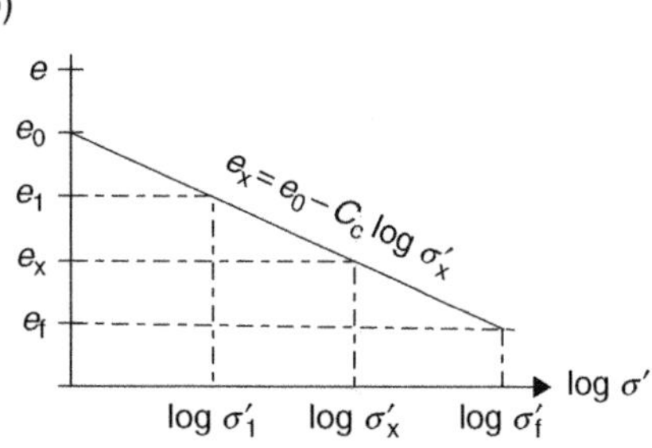

Figure 7.8

7.3.1 Normally consolidated clay

This clay is formed as described above. However, as the sample is taken from a depth, it swells because of the removed overburden pressure. Upon recompression, in the oedometer, the log σ'–e curve deviates from the straight line until the original overburden pressure is reached. It follows the virgin curve beyond this point.

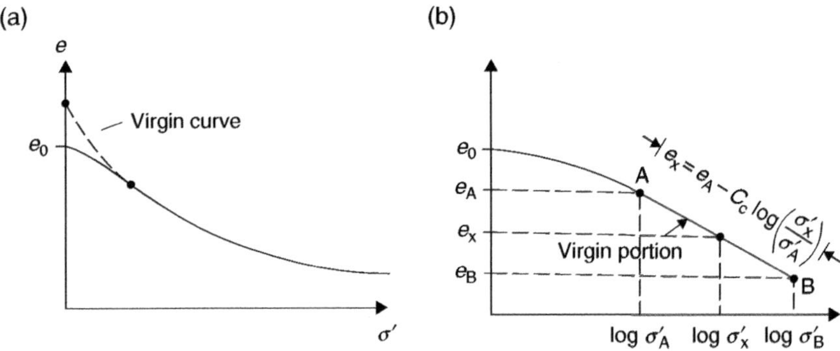

Figure 7.9

Equation of the straight line from Figure 7.8b:
The slope is called *Compression index* C_c.

$$C_c = -\frac{e_x - e_A}{\log \sigma'_x - \log \sigma'_1} = -\frac{e_x - e_A}{\log\left(\frac{\sigma'_x}{\sigma'_A}\right)}$$

Expressing e_x :

$$\boxed{e_x = e_A - C_c \log\frac{\sigma'_x}{\sigma'_A}} \quad (7.7)$$

In Figure 7.8(b): $e_A = e_0$

$\sigma'_A = 0$

$$\boxed{\therefore\ e_x = e_0 - C_c \log \sigma'_x} \quad (7.8)$$

7.3.2 Overconsolidated clays

In this case, some of the overburden had been removed by erosion or otherwise. The most obvious example is the retreating ice after the Ice Age, leaving the well consolidated clay under reduced geostatic pressure. Consequently, the clay is denser than expected from its overburden and less compressible than normally consolidated clay. Removal from the ground does not produce appreciable swelling of a sample.

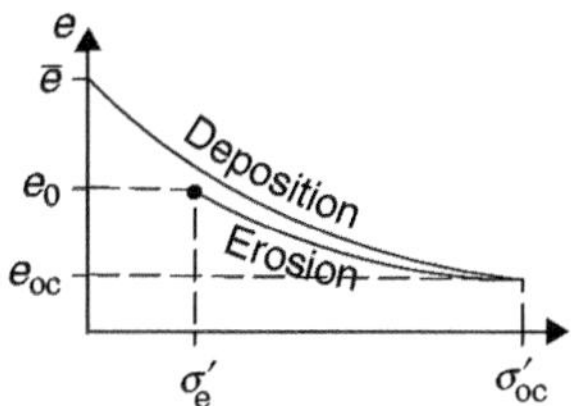

$\bar{e}$ = voids ratio before deposition
e_0 = voids ratio after erosion
OCR = overconsolidation ratio
e_{oc} = overconsolidated voids ratio

Figure 7.10

$$\text{OCR} = \frac{\text{max. effective pressure}}{\text{existing effective pressure}} = \boxed{\frac{\sigma'_{oc}}{\sigma'_0}} \tag{7.9}$$

For normally consolided clays: OCR = 1
For overconsolidated clays: OCR > 1

Casagrande proposed a graphical construction for the estimation of the overconsolidation pressure (σ'_{oc}). There are four steps to follow:

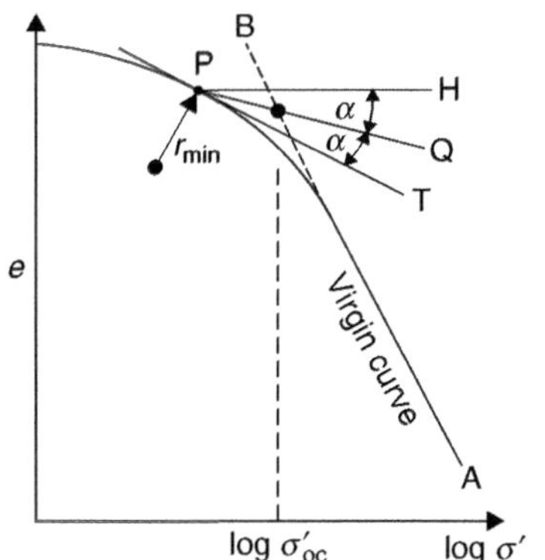

Step 1: Extend the virgin curve (A–B)
Step 2: Locate point P, the locus of of minimum curvature (r_{min})
Step 3: Draw horizontal and tangent lines P–H and P–T respectively
Step 4: Draw line P–Q as shown.

σ'_{oc} is given by the intersection of lines A–B and P–Q.

Figure 7.11

7.4 Coefficient of compressibility (a_v)

It is the slope of the σ'–e curve at a point, that is for *small* increments of pressure:

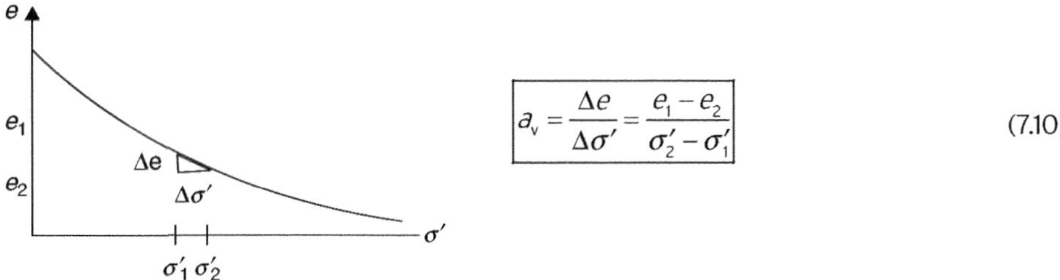

$$\boxed{a_v = \frac{\Delta e}{\Delta \sigma'} = \frac{e_1 - e_2}{\sigma'_2 - \sigma'_1}} \tag{7.10}$$

Figure 7.12

7.5 Coefficient of volume change (m_v)

It is the change in unit volume, caused by unit change in the effective stress. The coefficient is applied to the estimation of consolidation and settlement.

The coefficient m_v may be formulated in two ways:

1. Voids ratio method
2. Direct method, using the oedometer reading D_x

7.5.1 Voids ratio method

$$\boxed{m_v = \frac{a_v}{1+e_1}} \tag{7.11}$$

But from (7.10) $\quad a_v = \dfrac{e_1 - e_2}{\sigma_2' - \sigma_1'} \quad \therefore \quad \boxed{m_v = \dfrac{e_1 - e_2}{(1+e_1)(\sigma_2' - \sigma_1')}}$ (7.12)

7.5.2 Direct method

It is possible to derive a formula for m_v which does not include voids ratio. This means that it not necessary to use void ratios at all for the estimation of settlement.

From (7.10): $a_v = \dfrac{e_1 - e_2}{\sigma_2' - \sigma_1'}$

From (7.4): $e_1 = e_0 - \left(\dfrac{1+e_0}{h_0}\right) D_1$

And: $e_2 = e_0 - \left(\dfrac{1+e_0}{h_0}\right) D_2$

$$\therefore \; e_1 - e_2 = \left(\frac{1+e_0}{h_0}\right)(D_2 - D_1)$$

Also, $1+e_1 = 1+e_0 - (1+e_0)\dfrac{D_1}{h_0}$ $\quad \therefore \; 1+e_1 = \left(\dfrac{1+e_0}{h_0}\right)(h_0 - D_1)$

Therefore, $\boxed{a_v = \left(\dfrac{1+e_0}{h_0}\right)\left(\dfrac{D_2 - D_1}{\sigma_2' - \sigma_1'}\right) = \dfrac{1+e_0}{h_0}\left(\dfrac{\Delta D}{\Delta \sigma'}\right)}$ (7.13)

From (7.11): $m_v = \dfrac{\dfrac{1+e_0}{h_0}\left(\dfrac{D_2 - D_1}{\sigma_2' - \sigma_1'}\right)}{\dfrac{1+e_0}{h_0}(h_0 - D_1)}$

Therefore, $\boxed{m_v = \dfrac{D_2 - D_1}{(h_0 - D_1)(\sigma_2' - \sigma_1')}} = \boxed{\dfrac{\Delta D}{(h_0 - D_1)\Delta \sigma'}}$ (7.14)

where ΔD = the difference between two dial readings.

Note: It is not required therefore to know the values of e_0 or e_s in order to calculate m_v. Chart 7.3 is constructed from (7.14), thus m_v can be read off directly from the experimental curve.

Example 7.2

Using the Oedometer test results of Example 7.1, calculate a_v and m_v for the pressure range between $\sigma'_1 = 90\,kN/m^2$ and $\sigma'_2 = 200\,kN/m^2$, by the two methods:

1. Voids method ratio

From Graph 7.1:

$$\begin{array}{llll} e_1 = 0.69 & \text{for} & \sigma'_1 = 90\,kN/m^2 \\ e_2 = 0.648 & \text{for} & \sigma'_2 = 200\,kN/m^2 \\ \hline \Delta e = 0.042 & & \Delta\sigma' = 110\,kN/m^2 \end{array}$$

From (7.10): $$a_v = \frac{\Delta e}{\Delta\sigma'} = \frac{0.042}{110} = 0.000382\,m^2/kN = 0.382\,m^2/MN$$

From (7.11): $$m_v = \frac{a_v}{1+e_1} = \frac{0.382}{1.69} = 0.226\,m^2/MN$$

2. Direct method

$e_0 = 0.751$ and $h_0 = 19\,mm$

From Chart 7.2:

$$\begin{array}{llll} D_1 = 0.66 & \text{for} & \sigma'_1 = 90\,kN/m^2 \\ D_2 = 1.12\,mm & \text{for} & \sigma'_2 = 200\,kN/m^2 \\ \hline \Delta D = 0.46\,mm & & \Delta\sigma' = 110\,kN/m^2 \end{array}$$

From (7.13): $$a_v = \frac{1+e_0}{h_0}\left(\frac{\Delta D}{\Delta\sigma'}\right) = \frac{1.751}{19}\left(\frac{0.46}{110}\right) = 0.000385\,m^2/MN$$

$$= 0.385\,m^2/MN$$

From (7.14): $$m_v = \frac{\Delta D}{(h_0 - D_1)\Delta\sigma'} = \frac{0.46}{(19-0.66)\times 110} = 0.000228\,m^2/MN$$

$$= 0.228\,m^2/MN$$

The nomogram on Chart 7.3 can be used for the evaluation of m_v directly from the oedometer readings. The procedure is indicated for the solution if this example in Figure 7.13.

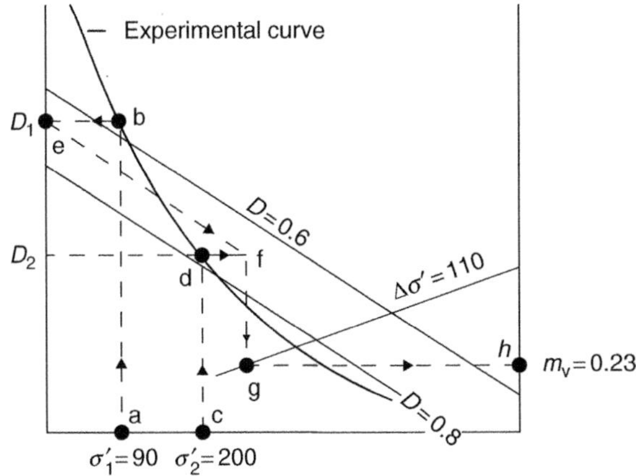

Sequence of construction

1. a – b – e
2. c – d
3. e – f, d – f
4. f – g
5. g – h = m_v

Figure 7.13

Note that line e–f is drawn between the two not-quite parallel *D*-lines.

Construction of Chart 7.3

The nomogram is constructed for a $h_0 = 19$ mm ring from formula (7.14):

$$m_v = \frac{1}{\Delta\sigma'}\left(\frac{D_2 - D_1}{h_0 - D_1}\right) \quad \text{or} \quad m_v = \frac{x}{\Delta\sigma'} \quad \text{where} \quad x = \frac{D_2 - D_1}{h_0 - D_1}$$

The *D*-lines are plotted by assuming a convenient value of D_2 and substituting it into the expression for the auxiliary variable *x*.

$$\text{Taking,:} \quad D_2 = 2.2 \quad \text{getting:} \quad x = \frac{2.2 - D_1}{19 - D_1}$$

$$\text{Taking,} \quad D_1 = 1.8 \quad \text{then} \quad x = \frac{2.2 - 1.8}{19 - 1.8} = 0.0233$$

$$\text{Or} \quad D_1 = 1 \quad \text{then} \quad x = \frac{2.2 - 1}{19 - 1} = 0.0667$$

These results are drawn in Figure 7.14 as diagonal *D*-lines.

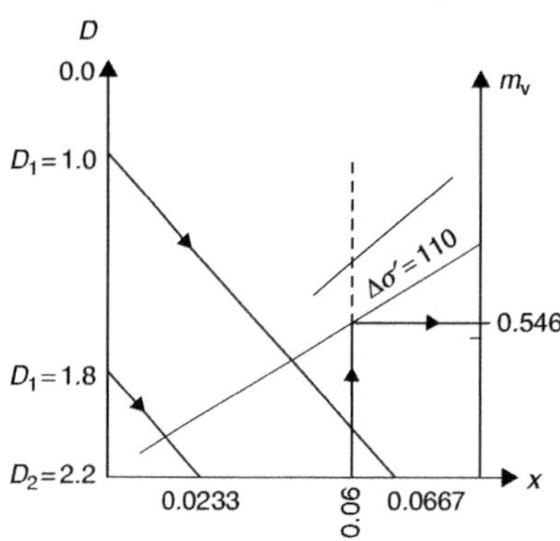

Figure 7.14

All the other lines can be drawn this way, provided the graph paper is large enough in the *x*-direction. If not, then choose, say $D_2 = 1.2$, so that the $D_x = 0$ line falls within the nomogram.

The $\Delta\sigma'$-lines are constructed from $x = m_v\Delta\sigma'$ by assuming a suitable value for *x*, e.g. $x = 0.06$ and expressing the coefficient of volume compressibility as

$$m_v = \frac{0.06}{\Delta\sigma'}$$

$$\text{Choosing } \Delta\sigma' = 110 \text{ getting: } m_v = \frac{0.06}{110} = 0.000546\,\text{m}^2/\text{kN}$$

$$= 0.546\,\text{m}^2/\text{MN}$$

The line, connecting this point and the origin, corresponds to $\Delta\sigma' = 110\,\text{kN/m}^2$. When all the other $\Delta\sigma'$-lines are drawn, then the auxiliary variable should be ignored. Its use was only to aid the construction. Instead, the scale for σ'_x may be chosen along the *x*-axis. In this chart 20 mm represents 100 kN/m^2.

7.6 Estimation of settlement

Once the oedometer results are available, the probable settlement of a structure placed on the soil tested can be assessed. The vertical movements of its footings are evaluated by assuming one-dimensional consolidation that is any horizontal volume change is not considered. Three methods of settlement evaluation are to be introduced:

1. Voids ratio method
2. The method using m_v
3. Direct method.

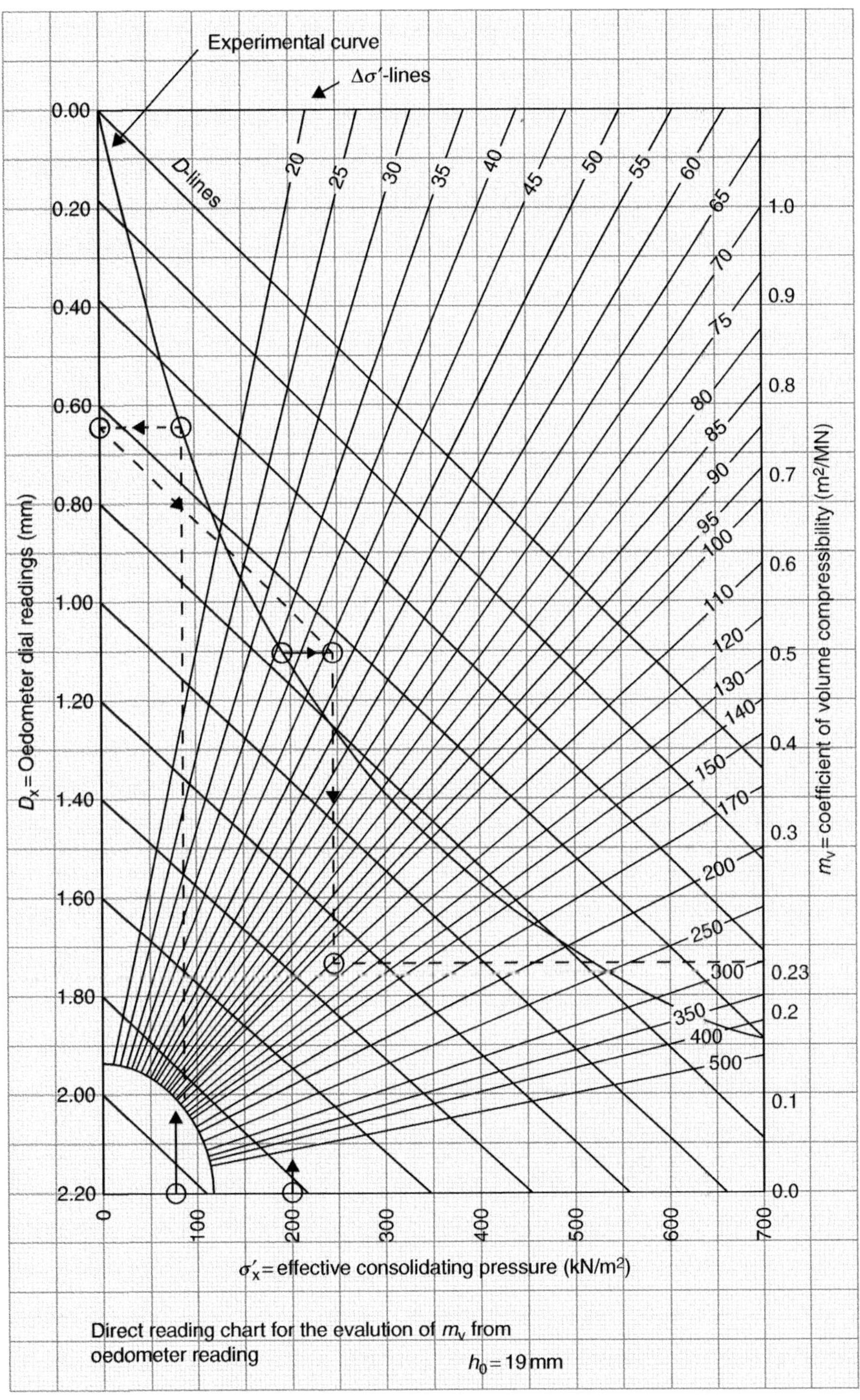
Experimental curve
Δσ′-lines
D-lines
D_x = Oedometer dial readings (mm)
σ'_x = effective consolidating pressure (kN/m²)
m_v = coefficient of volume compressibility (m²/MN)
Direct reading chart for the evalution of m_v from oedometer reading
h_0 = 19 mm

Chart 7.3

Figure 7.15 shows the outline of the problem in general. The quantities indicated are common to all three methods.

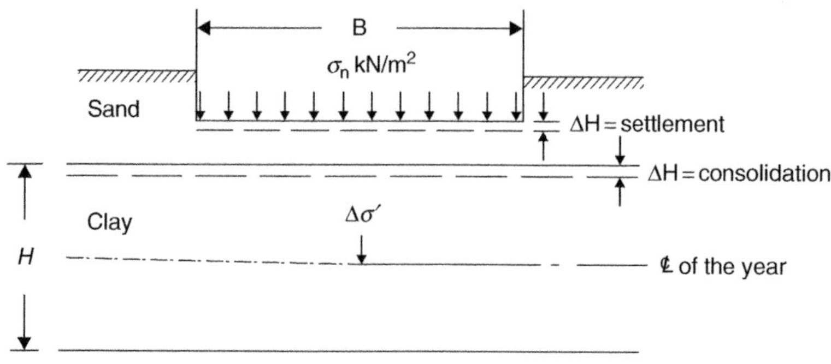

Figure 7.15

where σ_n = Net bearing pressure (kN/m^2)
H = Thickness of the layer (m)
ΔH = Long-term settlement or consolidation (mm)
B = Width of footing
$\Delta\sigma'$ = Effective consolidating pressure at the "point" considered.

General procedure
In order to estimate the magnitude of consolidation of a clay layer, calculate:

1. The effective pressure ($\Delta\sigma'_1$) of the undisturbed soil, at the centre-line of the layer.
2. The effective pressure at the same depth after excavation to foundation level.
3. The pressure induced by the net foundation loading at the middle of the layer, using the appropriate Boussinesq-type formula or bulb of pressure diagram.
4. The sum of the above two pressures to get $\Delta\sigma'_2$.
5. The excess effective pressure: $\Delta\sigma' = \Delta\sigma'_2 - \Delta\sigma'_1$. It is then substituted into the method applied.

Note: If the clay layer is very thick, it may be subdivided into thinner layers. The above procedure is then applied to the centre of each layer and the calculated amounts of consolidation summed, to get the total change in thickness (ΔH).

See Section 7.6 for the definition of net foundation pressure.

7.6.1 Voids ratio method

For the oedometer specimen: $$\frac{\Delta h}{h_1} = \frac{\Delta e}{1+e_1} \qquad (7.1)$$

For the clay layer: $$\frac{\Delta H}{H} = \frac{e_2 - e_1}{1+e_1}$$

Hence the change in thickness off the layer is given by:

$$\boxed{\Delta H = \frac{e_2 - e_1}{1+e_1} H} \qquad (7.15)$$

where H = initial thickness of the clay layer
e_1 = initial voids ratio of the clay layer in its unloaded state.
e_2 = voids ratio at the end of the consolidation period ($t = \infty$), after the dissipation of the excess pore pressure (Δu).

Example 7.3

Figure 7.16 shows a 4 m wide strip footing, transmitting a uniformly distributed net load of 200 kN/m². The ground water level is at 2.5 m below ground surface. Calculate the settlement of the base, using the test results of Example 7.1.

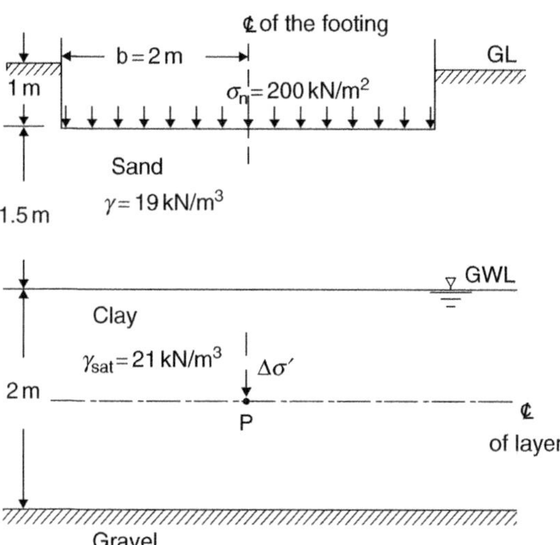

Figure 7.16

Step 1: Effective pressure in natural, undisturbed state at point P: $\sigma_1' = 2.5 \times 19 + 1 \times (21 - 9.81) = 58.69\,\text{kN/m}^2$

Step 2: Effective pressure at P after excavation. $\sigma' = \sigma_1' - 1 \times 19 = 58.69 - 19 = 39.69\,\text{kN/m}^2$

Step 3: Net pressure at P due to $\sigma_n' = 200\,\text{kN/m}^2$ is obtained by means of the bulb of pressure diagram (Chapter 4) for strip foating at its centre line, for maximum value. The influence factor at 2.5 m below footing level is read-off as 0.75 (Chart 4.3).

$$\text{Hence, } \sigma_v = 0.75\,\sigma_n = 0.75 \times 200 = 150\,\text{kN/m}^2$$

Step 4: Effective pressure immediately after construction ($t = 0$) is $\sigma_2' = \sigma' + \sigma_v = 39.69 + 150 = 189.69\,\text{kN/m}^2$

Step 5: Excess effective pressure at P is:

$$\Delta\sigma' = \sigma_2' - \sigma_1' = 189.69 - 58.69 = 131\,\text{kN/m}^2$$

Step 6: The voids ratios corresponding to the clay in Example 7.1 is found on Graph 7.1:

$$\begin{array}{ll} \text{For } \sigma'_1 = 58.69\,\text{kN/m}^2 & e_1 = 0.705 \\ \text{For } \underline{\sigma'_2 = 189.69\,\text{kN/m}^2} & \underline{e_2 = 0.652} \\ \Delta\sigma' = 131.00\,\text{kN/m}^2 & \Delta e = 0.053 \end{array}$$

Step 7: The magnitude of settlement is given by formula (7.15) for $H = 2\,\text{m}$

$$\Delta H = \left(\frac{\Delta e}{1+e_1}\right)\text{H} = \left(\frac{0.053}{1.705}\right) \times 2000 = 62\,\text{mm}$$

7.6.2 Method using m_v

The settlement formula is derived in terms of m_v and $\Delta\sigma'$ from:

$$(7.15)\ \Delta H = \left(\frac{\Delta e}{1+e_1}\right)H \qquad \therefore \left(\frac{\Delta e}{1+e_1}\right) = \frac{\Delta H}{H}$$

$$(7.10)\ \Delta e = a_v \Delta\sigma' \qquad \therefore \frac{a_v}{1+e_1}\Delta\sigma' = \frac{\Delta H}{H} \qquad \therefore \boxed{\Delta H = (m_v \Delta\sigma')H} \qquad (7.16)$$

$$(7.11)\ m_v = \frac{a_v}{1+e_1} \qquad \therefore m_v \Delta\sigma' = \frac{\Delta H}{H}$$

Example 7.4

Calculate ΔH for the footing in Example 7.3,

Where $\Delta\sigma' = 131\,\text{kN/m}^2$ (Step 6)

$$\begin{array}{l} e_1 = 0.705 \\ \underline{e_2 = 0.652} \\ \therefore\ \Delta e = 0.053 \end{array}$$

From (7.10): $a_v = \dfrac{\Delta e}{\Delta\sigma'} = \dfrac{0.053}{131}$

From (7.11): $m_v = \dfrac{a_v}{1+e_1} = \dfrac{0.053}{131 \times 1.705} = 0.000237\ \text{m}^2/\text{kN}$

Or from (7.14): $m_v = \dfrac{D_2 - D_1}{(h_0 - D_1)\Delta\sigma'} = \dfrac{\Delta D}{(h_0 - D_1)\Delta\sigma'}$

$$\begin{array}{ll} \text{Where, for } \sigma'_1 = 58.69\,\text{kN/m}^2 & D_1 = 0.48\,\text{mm} \\ \underline{\sigma'_2 = 189.69\,\text{kN/m}^2} & \underline{D_2 = 1.08\,\text{mm}} \\ \Delta\sigma' = 131\,\text{kN/m}^2 & \Delta D = 0.6\,\text{mm} \end{array}$$

$$\therefore m_v = \frac{0.6}{(19-0.48)\times 131} = 0.000247\ \text{m}^2/\text{kN}$$

The third option is to read off the value of m_v, directly from Chart 7.3.

The magnitude of settlement is given by (7.16):

$$\Delta H = (m_v\sigma')H = 0.000237\times 131\times 2 = 0.062\ \text{m}$$
$$= 62\ \text{mm}$$

This is the same figure as in Example 7.3.

7.6.3 Direct method

It has already been stated that it is not necessary to calculate the void ratios for the estimation of consolidation settlement.

Note, that $\Delta H = 62$ mm has been calculated from m_v expressed in terms of dial reading (D_x) and applied pressure in the oedometer test. The formula can be simplified further to eliminate $\Delta\sigma'$.

From (7.16): $\Delta H = m_v \Delta\sigma' H$

From (7.14): $m_v = \dfrac{D_2 - D_1}{(h_0 - D_1)\Delta\sigma'}$

$$\therefore \quad \Delta H = \frac{D_2 - D_1}{(h_0 - D_1)\Delta\sigma'}\Delta\sigma' H$$

Eliminating $\Delta\sigma'$

$$\therefore \quad \boxed{\Delta H = \frac{D_2 - D_1}{(h_0 - D_1)}H} \qquad (7.17)$$

Therefore, settlement calculations do not require the knowledge of voids ratio.

Chart 7.4 has been constructed by means of (7.17), from which ΔH can be obtained directly, even during the oedometer test.

Example 7.5

Calculate ΔH for the footing in Example 7.3 from (7.17). Check the results by means of the nomogram on Chart 7.4.

From the pressure-dial reading (experimental) curve:

$$D_1 = 0.48\ \text{mm} \quad \text{for} \quad \sigma'_1 = 58.69\ \text{kN/m}^2$$
$$D_2 = 1.08\ \text{mm} \quad \text{for} \quad \sigma'_2 = 189.69\ \text{kN/m}^2$$
$$D_2 - D_1 = 0.6\ \text{mm}$$

$$\therefore \Delta H = \frac{D_2 - D_1}{(h_0 - D_1)}H = \left(\frac{0.6}{19-0.48}\right)\times 2000 = 65\ \text{mm}$$

The procedure for reading Chart 7.4 is similar to that of Chart 7.3. The result is: $\Delta H = 66$ mm. Any deviation from the calculated results is due to graphical inaccuracy.

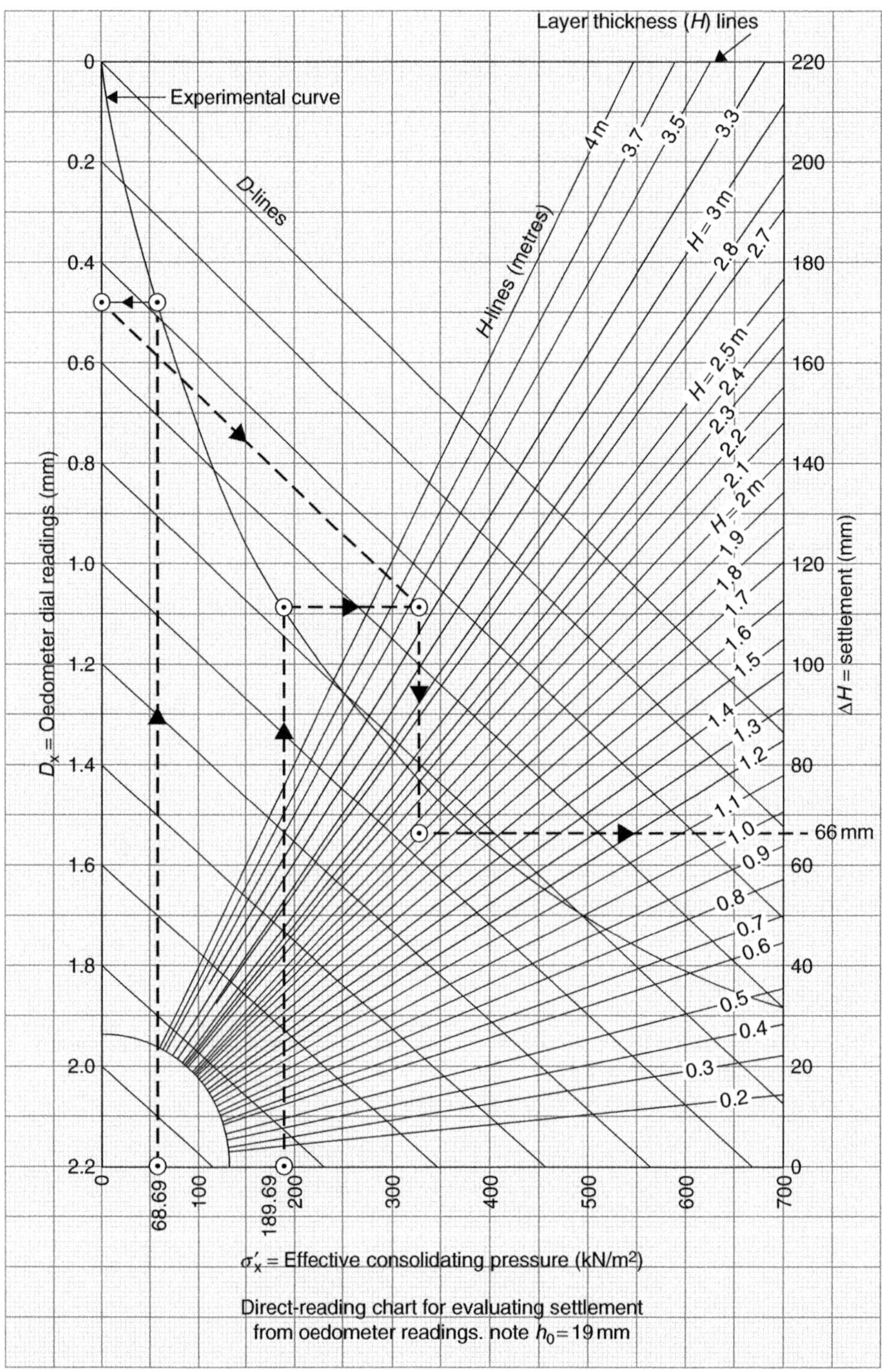

Layer thickness (H) lines
Experimental curve
D-lines
H-lines (metres)
4 m
3.7
3.5
3.3
H = 3 m
2.8
2.7
H = 2.5 m
2.4
2.3
2.2
2.1
H = 2 m
1.9
1.8
1.7
1.6
1.5
1.4
1.3
1.2
1.1
1.0
0.9
0.8
0.7
0.6
0.5
0.4
0.3
0.2
66 mm
0
0.2
0.4
0.6
0.8
1.0
1.2
1.4
1.6
1.8
2.0
2.2
D_x = Oedometer dial readings (mm)
220
200
180
160
140
120
100
80
60
40
20
0
ΔH = settlement (mm)
0
68.69
100
189.69
200
300
400
500
600
700
σ'_x = Effective consolidating pressure (kN/m²)
Direct-reading chart for evaluating settlement
from oedometer readings. note h_0 = 19 mm

Chart 7.4

Construction of Chart 7.4

The nomogram is constructed in similar manner to Chart 7.3 for a $h_0 = 19\,\text{mm}$ ring, but from formula (7.17).

$$\Delta H = \frac{D_2 - D_1}{(h_0 - D_1)} H \quad \text{or} \quad \Delta H = xH \quad \text{where } x = \frac{D_2 - D_1}{h_0 - D_1}$$

The *D*-lines are again plotted as on Chart 7.3, relative to the auxiliary *x*-axis. The *H*-lines are then constructed from $\Delta H = xH$, by assuming a suitable value for *x* as before, say $x = 0.06$ and expressing the change in height as: $\Delta H = 0.06H$.

By choosing suitable values for *H*, the nomogram can be completed. For example:

$$\text{If } H = 2\,\text{m} = 2000\,\text{mm}, \quad \text{then} \quad \Delta H = 0.06 \times 2000 = 120\,\text{mm}$$

The construction is shown below:

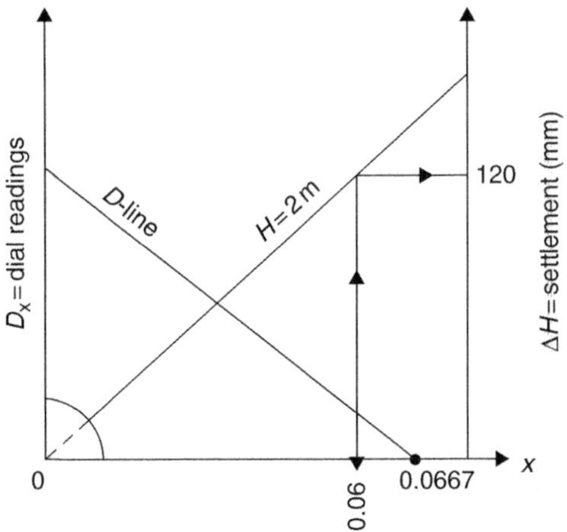

Figure 7.17

Again, upon the completion of the nomogram, the auxiliary variable is ignored and the scale of σ_x' chosen along the *x*-axis as before.

7.7 Rate of consolidation

It has been explained in some detail that consolidation occurs during the dissipation of excess pore pressure, induced by external loading. The process is very slow in fine-grained soil and could last for several years, or decades, after the completion of the structure. It is possible, therefore, that unacceptably large settlement will occur years later. Also, a certain amount of settlement has to be taken into account during the construction period, should it be prolonged. An outstanding example of this is the leaning tower of Pisa. Its construction started in the late 12th century and, because of settlement problems, it was completed in the 14th century. The movement has continued for 700 years, but recent successful remedial measures prevented eventual collapse.

7.7.1 Variation of excess pore pressure with time

Terzaghi's theory of one-dimensional consolidation is simulated in the oedometer test, where the soil can drain in two directions, similarly to an 'open' clay layer confined between two permeable ones as shown:

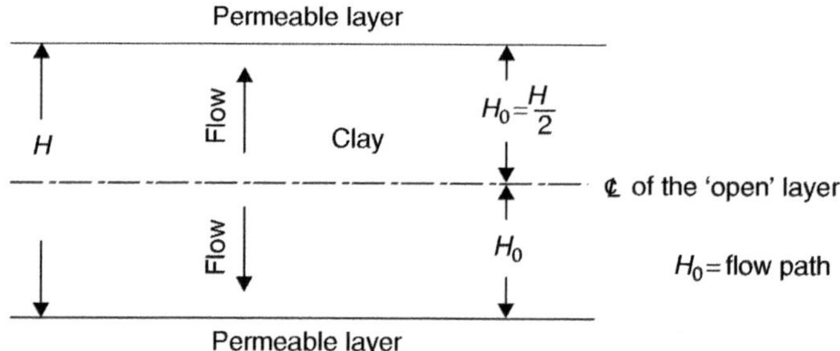

Figure 7.18

The speed of water, hence pore pressure, dissipation depends largely on:

1. H_0 = length of the flow path
2. k = permeability of the soil
3. m_v = compressibility of the soil
4. The type and magnitude of loading
5. The shape of the initial excess pore pressure (u_0) distribution through the layer

Although, there is two-way drainage in the oedometer, nevertheless the theory applies equally to soil, which can drain one way only, through a half-open layer as shown:

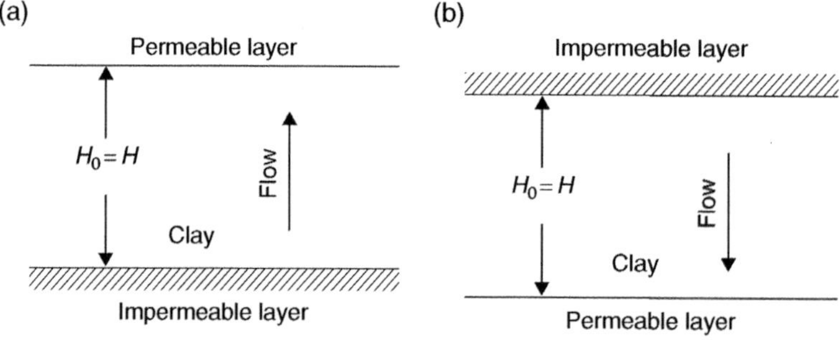

Figure 7.19

In these cases, the flowpath (H_0) equals to the thickness (H) of the clay layer.

7.7.2 Typical pore pressure distributions

The initial excess pore pressure ($\Delta u = u_0$) distribution with depth in a layer depends on the type of loading placed on it 'instantaneously' at the time $t=0$. The shape of the distribution influences the time taken in the dissipation of the excess pressure. The most typical cases of pressure distribution are:

Case 1: Rectangular, that is uniform distribution with depth. It occurs under very wide surcharge (q), slabs or when the layer is so thin, that the pressure difference between top and bottom is negligible.

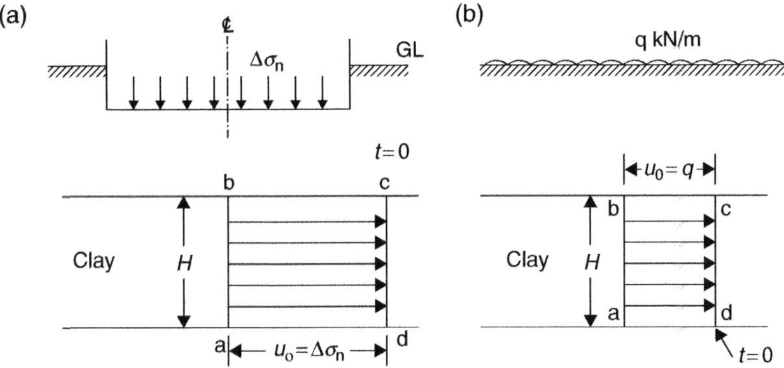

Figure 7.20

Case 2: Triangular with apex at the top of the layer. It occurs within earthworks of compacted cohesive soil constructed on impervious layer, or in hydraulic fills placed on impervious base. Pressure is due to self-weight.

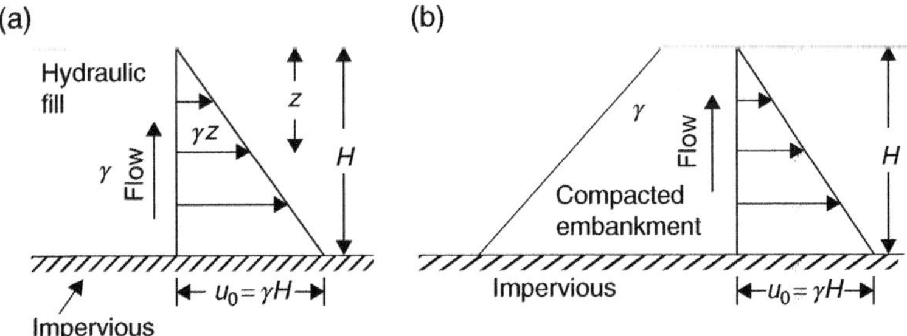

Figure 7.21

Case 3: Triangular with apex at the bottom of the layer. This occurs, when the layer is so thick that the pressure induced by a surface load - estimated by a Boussinesq-based formula - at the bottom is negligible. It is assumed, that the loading is transmitted to the ground by impermeable concrete foundation.

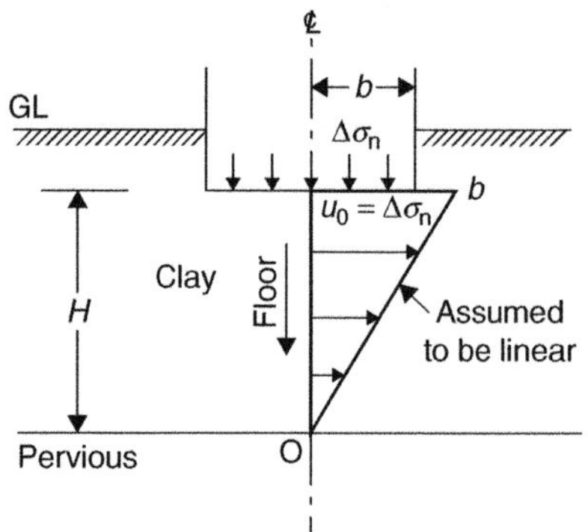

Note:
The variation of pressure under the centreline of a strip footing, for example, varies nearly linearly with depth.

Figure 7.22

Case 4: Trapezoidal, when a clay layer is contained by two permeable ones at a depth below a footing. The (Boussinesq) pressure difference at the top and bottom of the layer is large enough to form a trapezoidal pressure diagram.

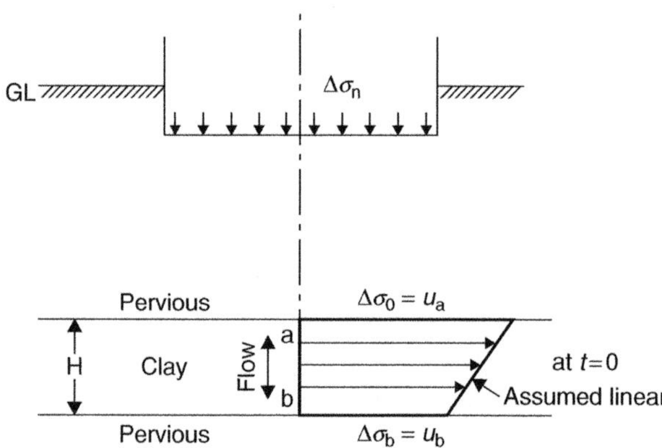

Figure 7.23

7.7.3 Estimation of time

It is not proposed to detail the mathematical justification of the method to estimate time. Instead, the procedure will be demonstrated for the rectangular (Case 1) distribution, which can be extended to the other cases. In order to do this, some basic quantities to be used have to be defined. Their application will be made clear shortly.

Coefficient of consolidation:

$$\boxed{c_v = \frac{k}{\gamma_w m_v}} \tag{7.18}$$

Time factor:

$$\boxed{T_v = \frac{c_v t}{H_0^2}} \tag{7.19}$$

where t = time

Degree of consolidation at any depth z is defined as:

$$U_z = \frac{\text{Excess pore pressure dissipated after time } t}{\text{Initial excess pore pressure}}$$

$$U_z = \frac{u_0 - u_t}{u_0} \quad \text{or} \quad \boxed{U_z = 1 - \frac{u_t}{u_0}} \tag{7.20}$$

where, u_0 = initial excess pore pressure at $t = 0$

u_t = pore pressure at time t

U_z can also be expressed alternatively as the percentage consolidation achieved at a depth z at time t.

Therefore,

$$\boxed{U_z = 100 \times \left(1 - \frac{u_t}{u_0}\right) \%} \tag{7.21}$$

Normally, the estimation of the average percentage consolidation (U) of an entire layer is required. Chart 7.5 should be used for this purpose.

7.7.4 Coefficient of consolidation (c_v)

It is apparent from formula (7.18) and from the presence of the coefficient of compressibility m_v, that c_v can be obtained from the oedometer test. As each value of m_v is applied to a range of pressure, so is c_v. The graphical determination of c_v was introduced by Taylor, as described below:

Step 1: Place the load increment, for which the value of c_v is required, on the specimen in the oedometer and record both the dial readings and time at intervals, until negligible consolidation is observed.

Step 2: Plot the square root of time against the dial readings (D_t) and draw the experimental curve as shown in Figure 7.24.

Step 3: Extend the straight portion of the experimental curve, until it cuts the vertical axis at point D_0.

Step 4: Draw a straight line from point D_0 such that the abscissa at each dial reading is 1.15 times that of the experimental curve, along its straight portion. This line cuts the curve at point x, which represents 90% consolidation in the pressure range.

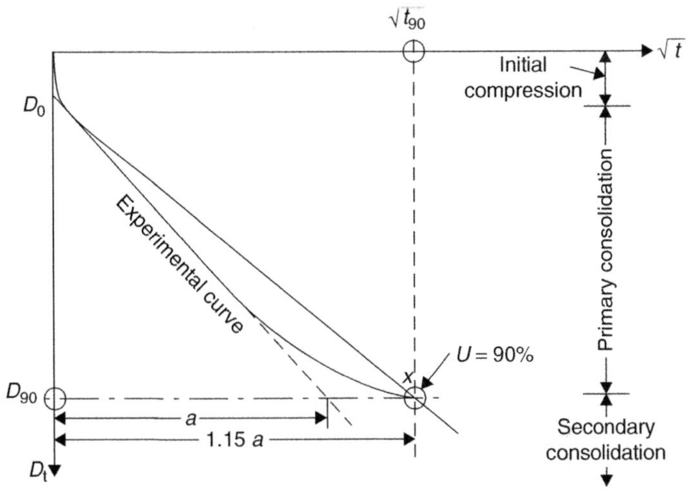

Figure 7.24

Step 5: Estimate c_v from (7.18): $$c_v = \frac{0.212\,h^2}{t_{90}} \quad (7.22)$$

where h = average height of the specimen in the oedometer for the load increment. Also taking $T_v = 0.848$ for U_{90} on Chart 7.5 (Curve B for open layer).

Example 7.6

A settlement of 62 mm was estimated for the clay in Example 7.3, induced by the load increment:

$\sigma'_1 = 58.69\ \text{kN/m}^2$
$\sigma'_2 = 189.69\ \text{kN/m}^2$
$\Delta\sigma' = 131.00\ \text{kN/m}^2$

The coefficient of consolidation for this loading is found by averaging the c_v, determined for load increments 100 and 200 kN/m².

Table 7.4

(Extracts from Table 7.2)

x	$\Delta\sigma'_x$ (kN/m²)	D_x (mm)	h_x (mm)	e_x
1	50	0.40	18.60	0.714
2	100	0.72	18.28	0.685
3	200	1.12	17.88	0.648

Test results for load increment $\Delta\sigma'_2 = 100\ \text{kN/m}^2$

Table 7.5

t (min)	0	0.2	0.5	1	5	10	30	60	120	240	360	480	1440
$\sqrt{t}$	0	0.45	0.71	1	2.2	3.2	5.5	7.7	11	15.5	19	21.9	37.9
D_x (mm)	0.4	0.43	0.44	0.455	0.48	0.515	0.545	0.585	0.635	0.673	0.69	0.7	0.72

$$\text{Average height}: h = \frac{1}{2} \times (18.28 + 18.60) = 18.44\ \text{mm}$$

$$\text{From Graph 7.2A}: \sqrt{t_{90}} = 13.6 \quad \therefore \quad t_{90} = 185\ \text{minutes}$$

$$\text{From (7.22)} \quad c_v = \frac{0.212 \times 18.44^2}{185} = 0.39\ \text{mm}^2/\text{min}$$

(a)

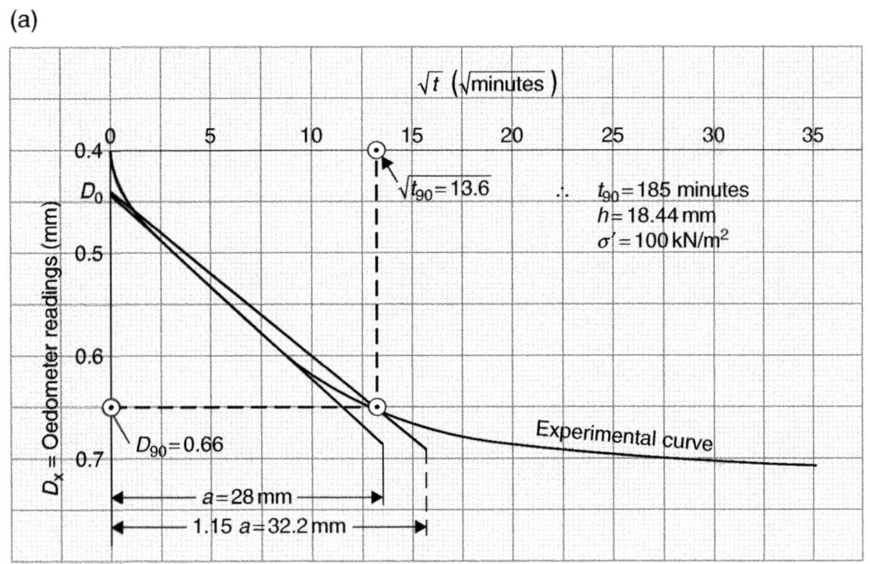

√t (√minutes)
0
5
10
15
20
25
30
35
0.4
D0
0.5
0.6
0.7
Dx = Oedometer readings (mm)
√t90 = 13.6
∴ t90 = 185 minutes
h = 18.44 mm
σ' = 100 kN/m²
D90 = 0.66
Experimental curve
a = 28 mm
1.15 a = 32.2 mm

(b)

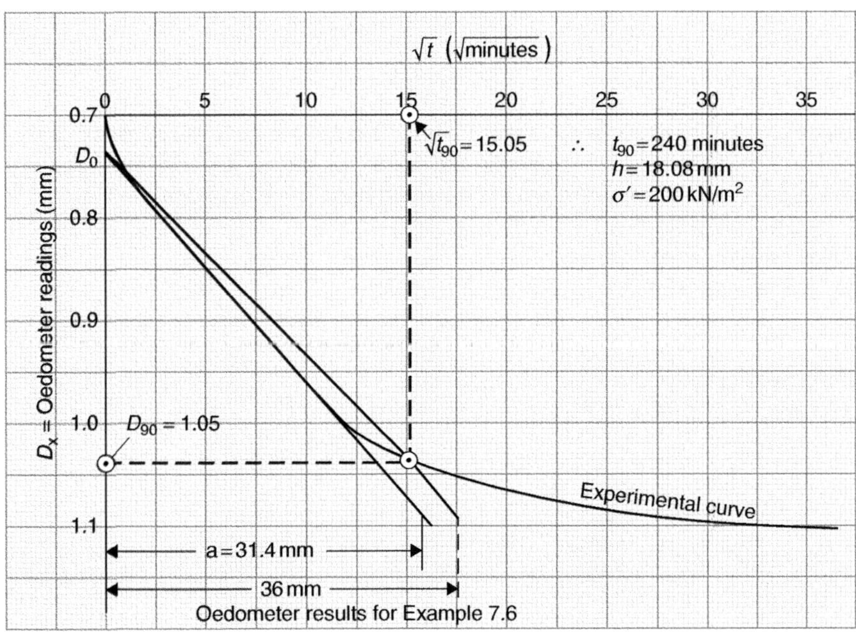

√t (√minutes)
0
5
10
15
20
25
30
35
0.7
D0
0.8
0.9
1.0
1.1
Dx = Oedometer readings (mm)
√t90 = 15.05
∴ t90 = 240 minutes
h = 18.08 mm
σ' = 200 kN/m²
D90 = 1.05
Experimental curve
a = 31.4 mm
36 mm
Oedometer results for Example 7.6

Graph 7.2

Test results for load increment $\Delta\sigma'_3 = 200\,kN/m^2$

Table 7.6

t (min)	0	0.2	0.5	1	5	10	30	60	120	240	360	480	1440
$\sqrt{t}$	0	0.45	0.71	1	2.2	3.2	5.5	7.7	11	15.5	19	21.9	37.9
D_x (mm)	0.72	0.73	0.75	0.76	0.795	0.84	0.887	0.94	1.0	1.05	1.072	1.087	1.12

$$\text{Average height}: h = \frac{1}{2} \times (17.88 + 18.28) = 18.08\,\text{mm}$$

$$\text{From Graph 7.2B}: \sqrt{t_{90}} = 15.5 \quad \therefore \quad t_{90} = 240\,\text{minutes}$$

$$\text{and} \quad c_v = \frac{0.212 \times 18.08^2}{240} = 0.289\,\text{mm}^2/\text{min}$$

$$\text{The average values is: } c_v = \frac{1}{2} \times (0.39 \times 0.289) = 0.34\,\text{mm}^2/\text{min, for these increments.}$$

Example 7.7

Using $c_v = 0.277\,\text{mm}^2/\text{min}$, estimate the time taken to reach 50%, 70% and 90% of the total theoretical consolidation of a 2 m thick clay layer, subjected to an initial uniform, excess pore pressure distribution of 131 kN/m². Assume:

1. Two-way drainage
2. One-way drainage

$$\textit{Two-way drainage}: H_0 = \frac{H}{2}$$

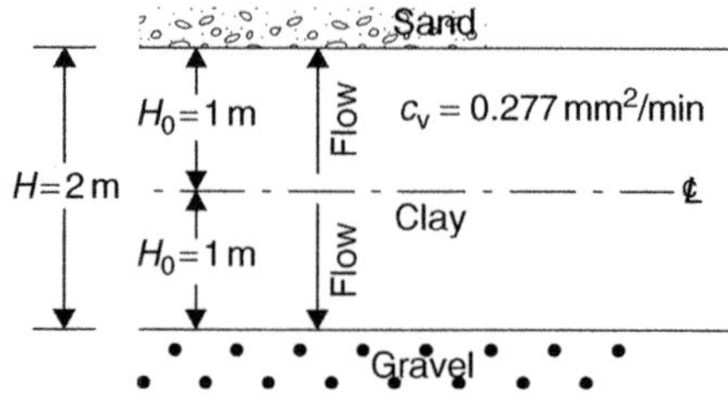

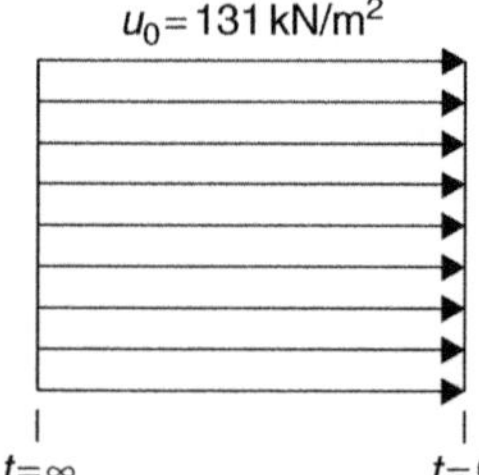

Figure 7.25

Curve B in Chart 7.5 is applicable to uniform pressure distribution for both one-way and two-way drainage.

From (7.19): $t = \frac{H_0^2}{C_v} T_v = \frac{1000^2}{0.277} T_v = \boxed{3.61 \times 10^6 T_v}$ minutes

For $U = 50\%$, $T_v = 0.195$ $\therefore$ $t_{50} = 3.61 \times 0.195 \times 10^6 = 0.704 \times 10^6$ minutes $= 489$ days

For $U = 70\%$, $T_v = 0.403$ $\therefore$ $t_{70} = 3.61 \times 0.403 \times 10^6 = 1455 \times 10^6$ minutes $= 1010$ days

For $U = 90\%$, $T_v = 0.848$ $\therefore$ $t_{90} = 3.61 \times 0.848 \times 10^6 = 3.06 \times 10^6$ minutes $= 2126$ days $= 5$ years 10 months

One-way drainage $H_0 = H$

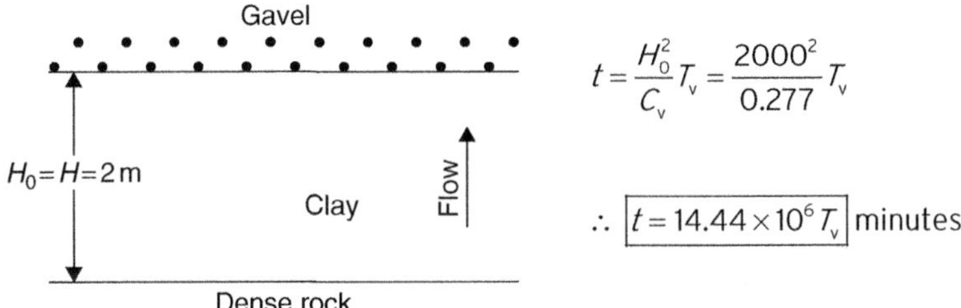

$$t = \frac{H_0^2}{C_v} T_v = \frac{2000^2}{0.277} T_v$$

$\therefore$ $\boxed{t = 14.44 \times 10^6 T_v}$ minutes

Figure 7.26

For $U = 50\%$, $T_v = 0.195$ $\therefore$ $t_{50} = 14.44 \times 0.195 \times 10^6 = 2.82 \times 10^6$ minutes $= 1958$ days

For $U = 70\%$, $T_v = 0.403$ $\therefore$ $t_{70} = 14.44 \times 0.403 \times 10^6 = 5.82 \times 10^6$ minutes $= 4041$ days

For $U = 90\%$, $T_v = 0.848$ $\therefore$ $t_{90} = 14.44 \times 0848.10 \times 10^6 = 12.24 \times 10^6$ minutes $= 8504$ days $= 23\frac{1}{2}$ years

Note: Because of the assumptions made in the theoretical derivation of the formulae, as well as due to the uncertainties of the in-situ drainage conditions, the results are not exact, and should be considered as indicative information only.

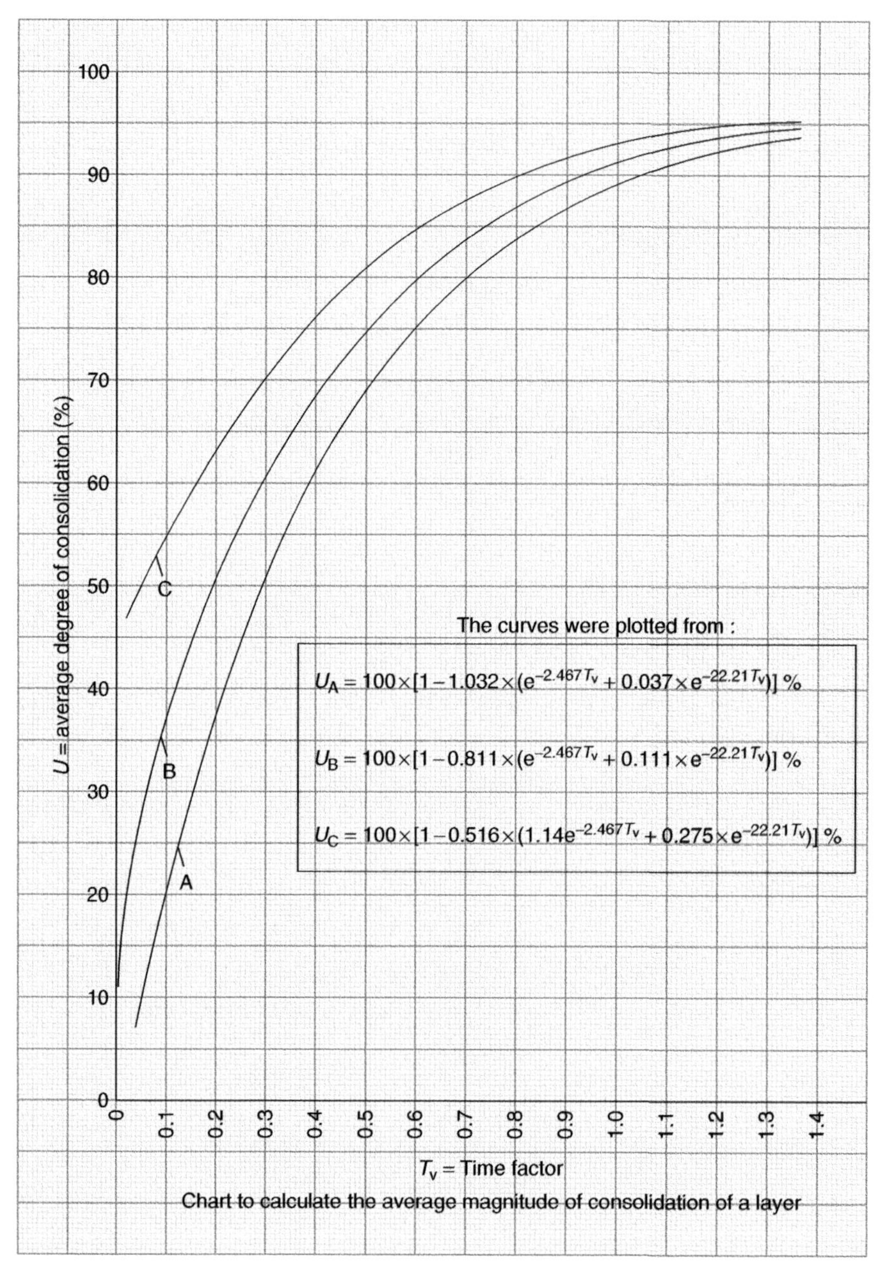
100
90
80
70
60
50
40
30
20
10
0
U = average degree of consolidation (%)
C
B
A
The curves were plotted from :
$U_A = 100\times[1-1.032\times(e^{-2.467T_v}+0.037\times e^{-22.21T_v})]$ %
$U_B = 100\times[1-0.811\times(e^{-2.467T_v}+0.111\times e^{-22.21T_v})]$ %
$U_C = 100\times[1-0.516\times(1.14e^{-2.467T_v}+0.275\times e^{-22.21T_v})]$ %
0
0.1
0.2
0.3
0.4
0.5
0.6
0.7
0.8
0.9
1.0
1.1
1.2
1.3
1.4
T_v = Time factor
Chart to calculate the average magnitude of consolidation of a layer

Chart 7.5

7.8 Pore pressure isochrones

It is theoretically possible to visualize the progress of pore pressure dissipation with time, by means of isochrones. These are sinusoidal curves, indicating the variation of pressure with depth, at any time t_x between $t = 0$ and $t = \infty$.

Remember: $t = 0$ assumes instantaneous application of loading.
$t = \infty$ indicates that all excess pore pressure had dissipated.

The actual shape of an isochrones depends on:

a) the initial distribution of excess pore pressure
b) the drainage conditions (one or two-way).

The various relationships during the consolidation process are shown below for uniform initial excess pore pressure distribution and two-way drainage.

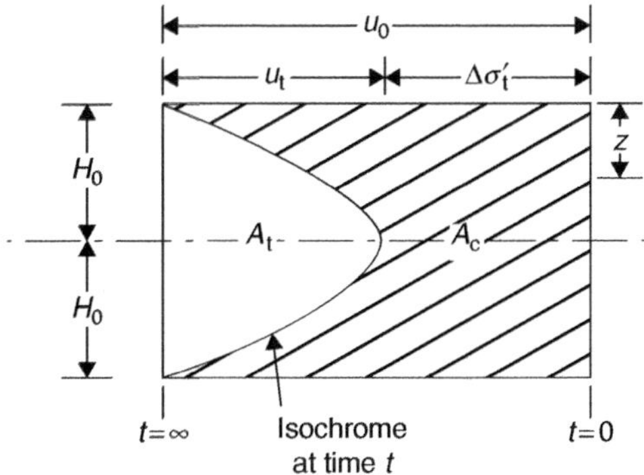

The tth isochrone is constructed by means of Chart 7.6. From this figure:

$$u_t = u_0 - \Delta\sigma'_t$$

Figure 7.27

Where A_t = area under an isochrone. It equals to the amount of consolidation yet to occur.
A_c = area indicating the amount of consolidation completed so far.
A = area of pressure diagram.

$$\boxed{A = A_t + A_c} \tag{7.23}$$

$\Delta\sigma'_t$=increasing effective pressure at time t, as the excess pore pressure dissipates.
u_t = pore pressure at time t.

Application of Chart 7.6

Only one half of the curves are drawn as the rest of the isochrones are symmetric. The procedure is best introduced by an example.

Example 7.8

Figure 7.25 shows an open, that is two-way-draining, clay layer under 131 kN/m² uniform excess pore pressure. The coefficient of consolidation of the clay is 0.277 mm²/min. Draw the isochrones for the excess pore pressure existing at 376, 752 and 1500 days after the loading of the clay.

From (7.19): $T_v = \frac{C_v}{H_0^2} = \frac{0.277t}{(1000)^2} t = 0.277 \times 10^{-6} t$

For $t = 376 \times 24 \times 60 = 541440$ minutes
For $T_v = 0.277 \times 541440 \times 10^{-6} = 0.14997$ (say $T_v = 0.15$)
For $t = 752 \times 1440 = 1082880$ minutes
For $T_v = 0.277 \times 1082880 \times 10^{-6} = 0.29995$ (say $T_v = 0.3$)
For $t = 1500 \times 1440 = 2160000$ minutes
For $T_v = 0.277 \times 2.16 = 0.598$ (say $T_v = 0.6$)

The flow path is $H_0 = 1$ m, therefore the depth factor: $\delta = \frac{z}{H_0} = \frac{z}{1} = z$ m

Also the pore pressure u_t can be expressed from (7.20) in terms of u_0 and u_z.

$$U_z = 1 - \frac{u_t}{u_0} \quad \therefore \quad \boxed{u_t = u_0(1 - U_z)} \tag{7.24}$$

Step 1: Read-off the value of U_z for various depth factors along the curve of T_v, from Chart 7.6.
Step 2: Calculate $u_t = 131 \times (1 - U_z)$ for each depth factor.
Step 3: Draw u_t for various depths. In this example z happens to be the same as δ, because $H_0 = 1$.

The results are tabulated and the three isochrones drawn on Graph 7.3.

The plotted isochrones can be used for three purposes:

1. To visualize the progress of excess pore pressure dissipation with time.
2. To calculate the average percentage consolidation to time t.
3. To calculate and plot curve B on Chart 7.5.

7.8.1 Average percentage consolidation

The area (A_t) under an isochrone can be calculated by means of Simpson's Rule. As the total area (A) is known, the amount of consolidation to time t is given by (7.23).

$$\boxed{A_c = A - A_t}$$

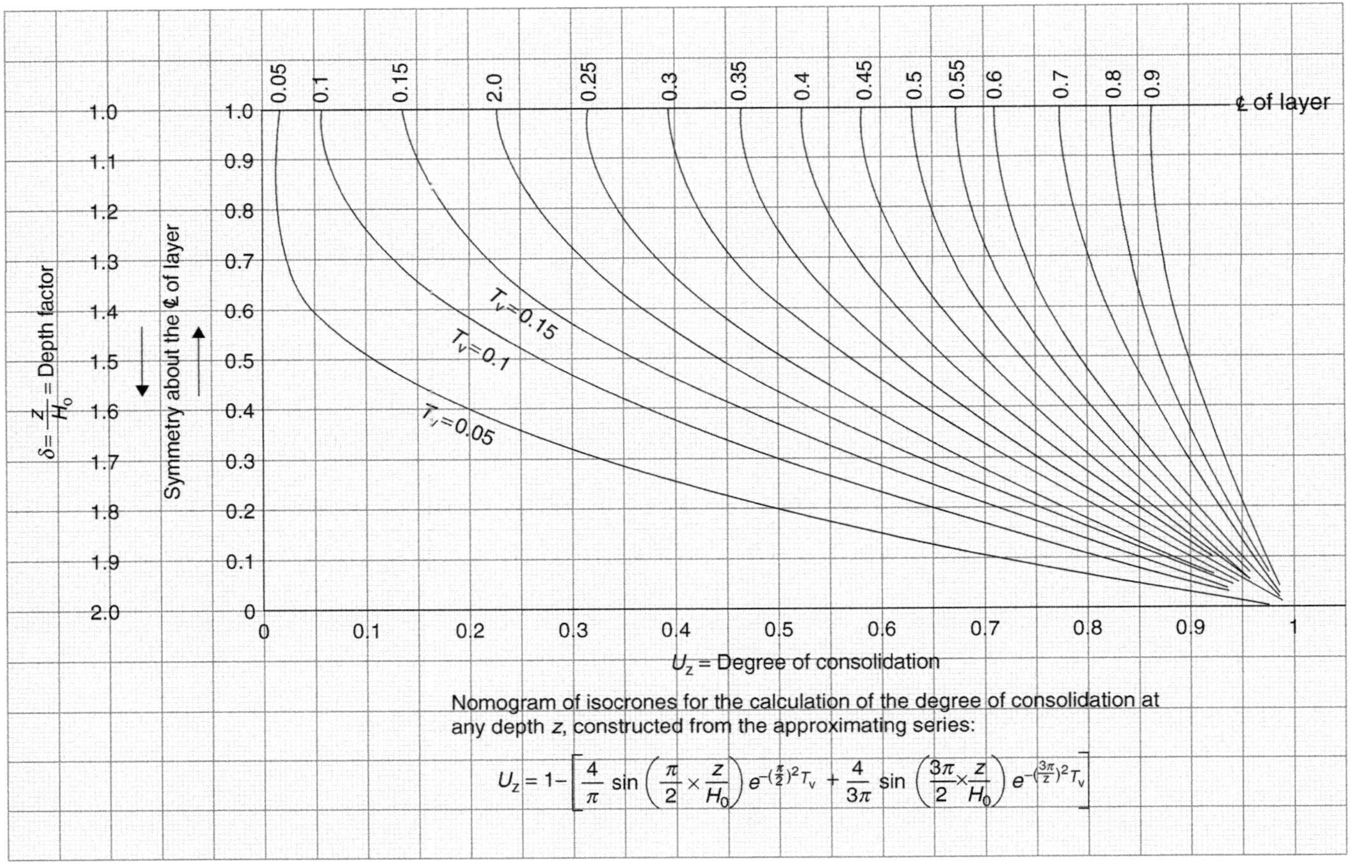

$\delta = \frac{z}{H_0}$ = Depth factor

Symmetry about the ℄ of layer

℄ of layer

$T_v = 0.15$

$T_v = 0.1$

$T_v = 0.05$

U_z = Degree of consolidation

Nomogram of isocrones for the calculation of the degree of consolidation at any depth z, constructed from the approximating series:

$$U_z = 1 - \left[\frac{4}{\pi} \sin\left(\frac{\pi}{2} \times \frac{z}{H_0}\right) e^{-(\frac{\pi}{2})^2 T_v} + \frac{4}{3\pi} \sin\left(\frac{3\pi}{2} \times \frac{z}{H_0}\right) e^{-(\frac{3\pi}{2})^2 T_v} \right]$$

Chart 7.6

Table 7.7

	$\delta = z$ (m)	0.1	0.2	0.3	0.4	0.5	0.6	0.7	0.8	0.9	1
$T_v = 0.15$	U_z	0.855	0.715	0.585	0.467	0.367	0.285	0.217	0.173	0.144	0.135
	$u_t = 131\,(1 - U_z)$	19	37.3	54.4	69.8	82.9	93.7	102.6	108.3	112.1	113.3
$T_v = 0.3$	U_z	0.906	0.816	0.725	0.643	0.573	0.508	0.457	0.423	0.4	0.394
	u_t (kN/m^2)	12.3	24.1	36	46.8	55.9	64.5	71.1	75.6	78.6	79.4
$T_v = 0.6$	U_z	0.955	0.912	0.872	0.83	0.795	0.767	0.743	0.726	0.715	0.71
	u_t (kN/m^2)	5.6	11.5	16.8	22.3	26.9	30.5	33.7	35.9	37.3	38

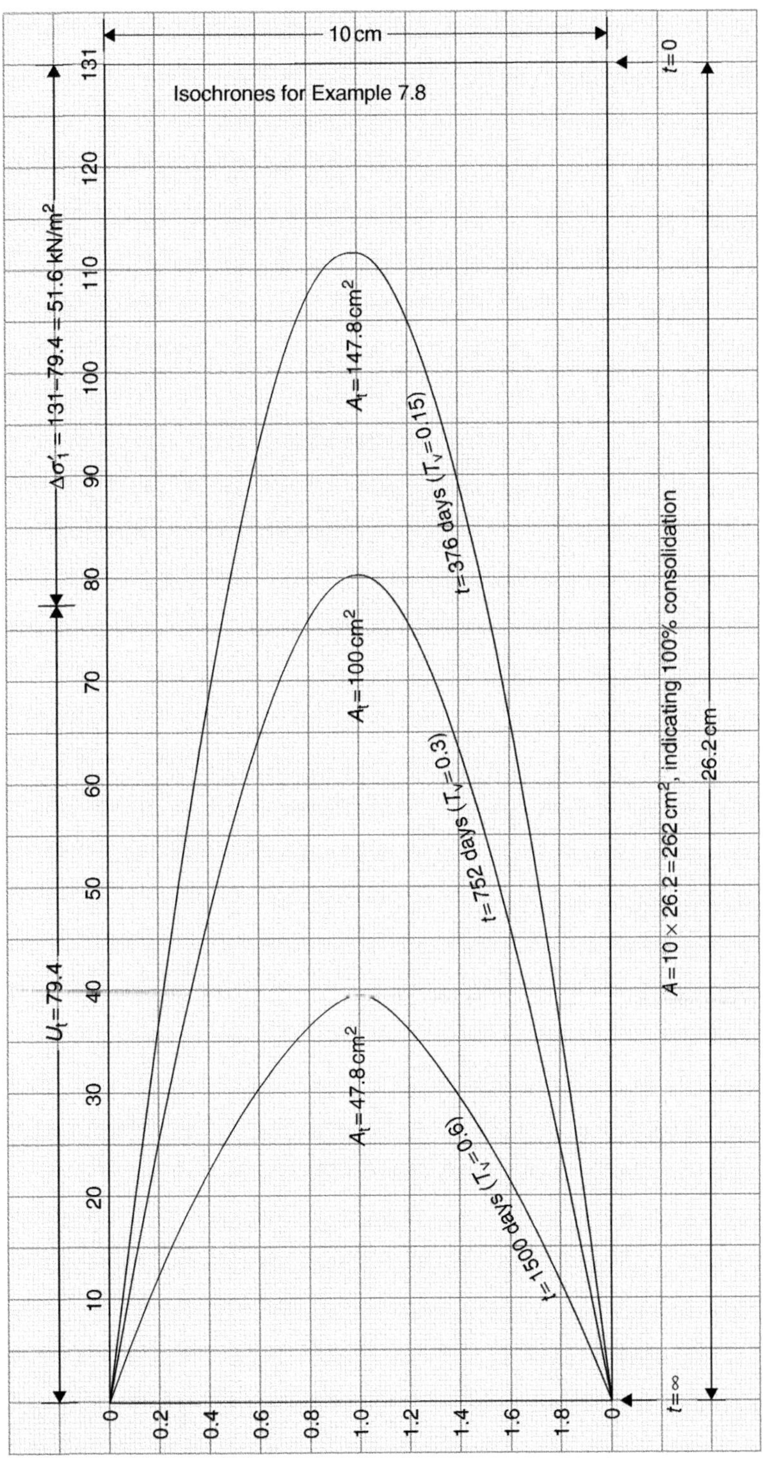
10 cm
Isochrones for Example 7.8
$\Delta\sigma'_1$ = 131−79.4 = 51.6 kN/m²
U_t = 79.4
0
10
20
30
40
50
60
70
80
90
100
110
120
131
0
0.2
0.4
0.6
0.8
1.0
1.2
1.4
1.6
1.8
0
A_t = 147.8 cm²
t = 376 days (T_v = 0.15)
A_t = 100 cm²
t = 752 days (T_v = 0.3)
A_t = 47.8 cm²
t = 1500 days (T_v = 0.6)
A = 10 × 26.2 = 262 cm², indicating 100% consolidation
26.2 cm
t = 0
t = ∞

Graph 7.3

If A represents 100% consolidation at time $t=\infty$, then A_c is equivalent to U% average consolidation, that is:

$$\boxed{U = 100\frac{A_c}{A}\,\%} \tag{7.25}$$

Example 7.9

The areas A_t for the three isochrones are given in Graph 7.3. Determine U% for each and compare the results with the coordinates of curve B on Chart 7.5. The calculations are tabulated below:

Table 7.8

From Graph 7.3			
T_v	A_t (cm²)	$A_c = A - A_t = 262 - A_t$ (cm²)	$U = 100\frac{A_c}{262}\%$
0.15	147.8	114.2	43.6
0.3	100	162	61.8
0.6	47.8	214.2	81.8

Check for curve B, Chart 7.5:

For $T_v = 0.15$ $U = 44\%$
For $T_v = 0.3$ $U = 61.4\%$ Correct
For $T_v = 0.6$ $U = 81.2\%$

Application of Chart 7.5

Figures 7.20–7.23 show various types of initial pore pressure distributions. Of these, only uniform distribution in an open layer and the application of curve B were discussed in the foregoing examples. In this section the most typical cases in relation to curves A, B and C will be summarized for open as well as half-open drainage.

Open layer

In this case, curve B is to be used for linear pore pressure distributions of whatever shape, taking $H_0 = \frac{H}{2}$, that is half of the layer thickness.

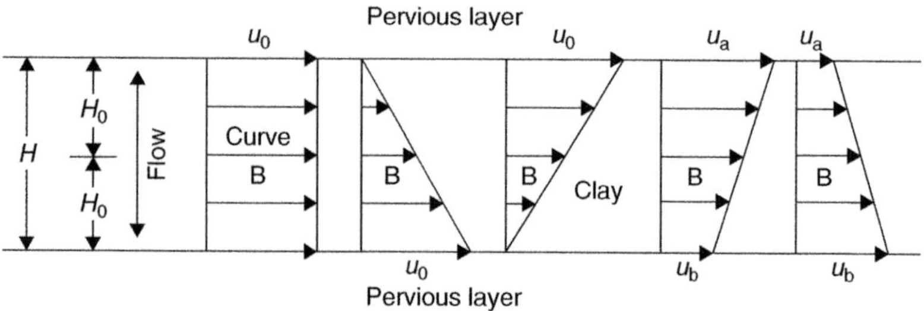

Figure 7.28

Half-open layer (Either at its top or at its bottom)
In this case, the flow path equals to the layer thickness, that is $H_0 = H$.

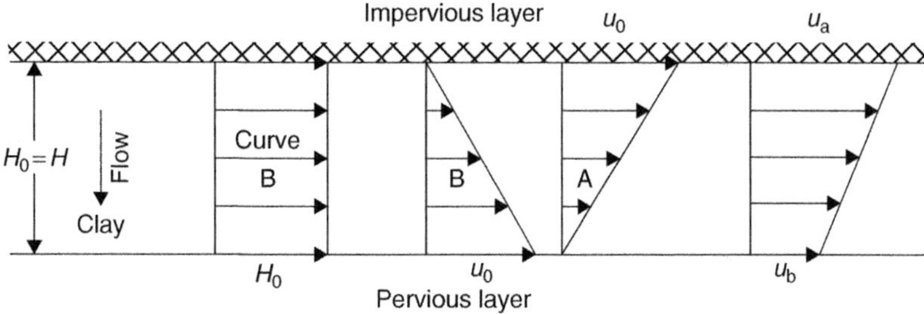

Figure 7.29

Note that for trapezoidal distribution, the $U\%$ for each value of T_v has to be calculated from:

$$\boxed{U_T = U_B - \frac{r-1}{r+1}(U_B - U_A)\ \%} \tag{7.26}$$

and

$$\boxed{r = \frac{u_a}{u_b}} > 1 \tag{7.27}$$

where
$U_T = U\%$ for the trapezoidal distribution
$U_B = U\%$ obtained from curve B
$U_A = U\%$ obtained from curve A
u_a = initial pore pressure at the top
u_b = initial pore pressure at the bottom

Example 7.10

Calculate the percentage consolidation for the given trapezoidal distribution at $T_v = 0.5$.

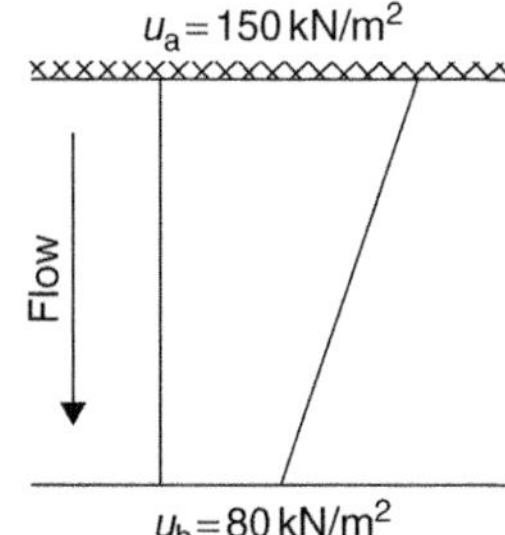

Figure 7.30

$$r = \frac{150}{80} = 1.875$$

From Chart 7.5: For $T_v = 0.5$
$U_B = 76.4\%$
$U_A = 70\%$

$$\therefore\ U_T = 76.4 - \frac{0.875}{2.875} \times (76.4 - 70) = 74.45\%$$

In this way, a curve can be constructed for each particular ratio r.

Should u_a be smaller than u_b, then the following formula would yield U_T %:

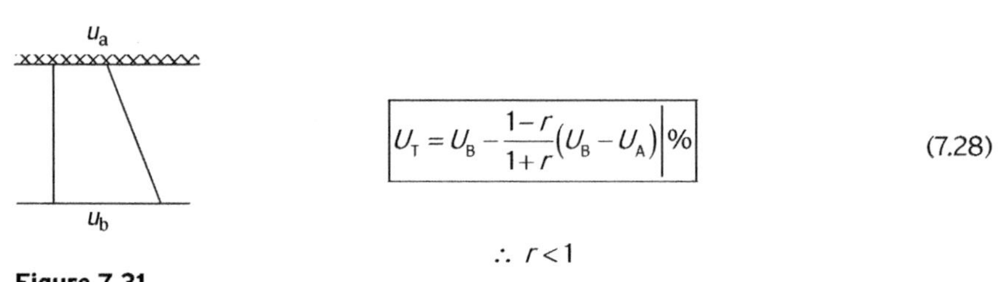

$$U_T = U_B - \frac{1-r}{1+r}(U_B - U_A)\Big|\% \qquad (7.28)$$

$$\therefore \ r < 1$$

Figure 7.31

Variation of isochrones - typical cases

The way in which the pore pressure dissipates depends on the shape of the excess pore pressure diagram and, whether the layer is open or half-closed. The variation of isochrones for the distribution shapes shown in Figure 7.29 are depicted below.

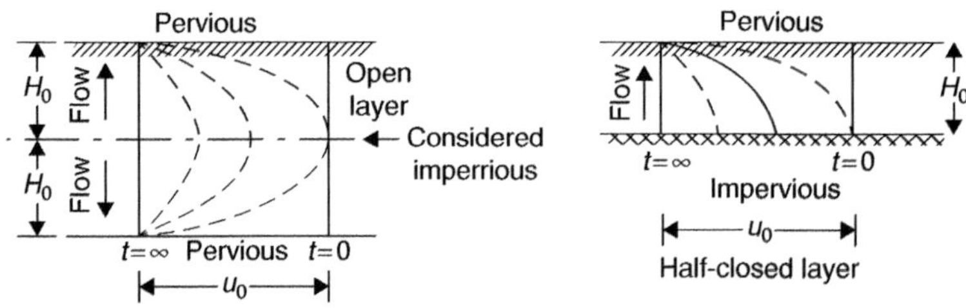

Figure 7.32

For rectangular initial distribution curve B (Chart 7.5) may be used for both open and half-closed layer.

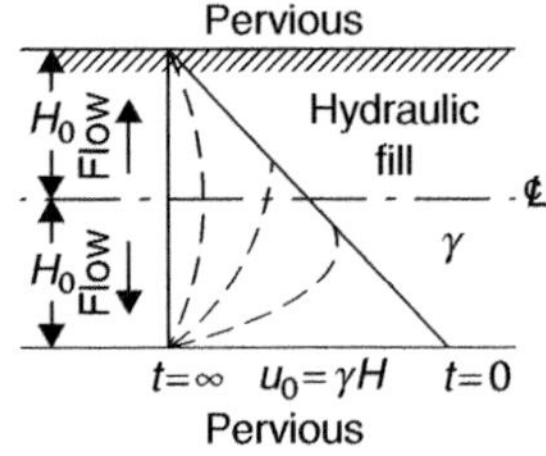

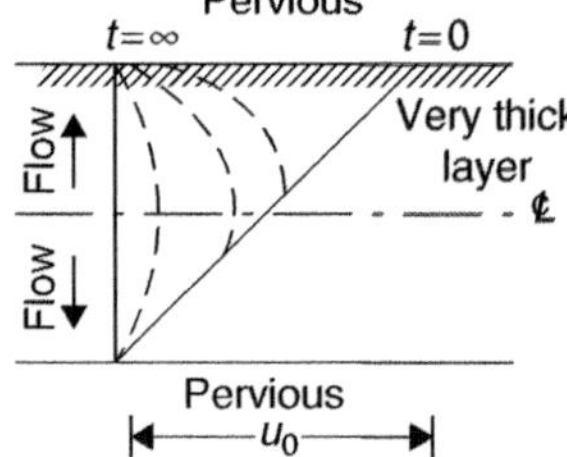

Figure 7.33

Curve B can be used, when the open layer is hydraulic fill or very thick so, that the pressure induced at the base may be assumed zero.

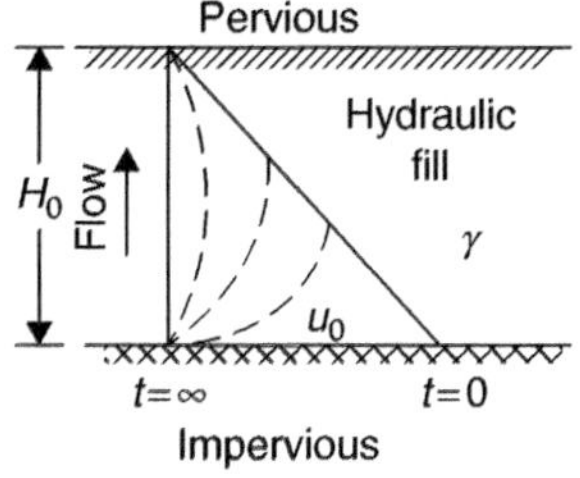

Figure 7.34

In half-closed layer curve A may be used.

$$u_0 = \gamma H$$

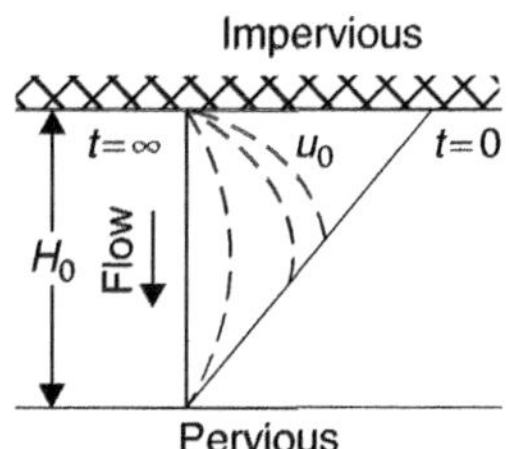

Figure 7.35

Half-closed, thick layer subjected to foundation pressure. Use curve A.

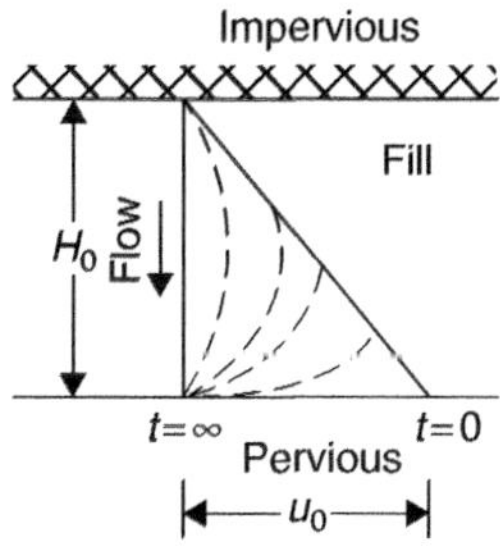

Figure 7.36

Half-closed layer. Excess pressure is due to self-weight. Use curve C.

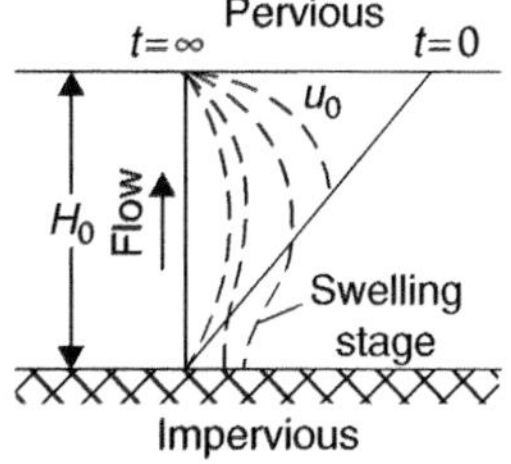

Figure 7.37

Very thick half-closed layer. Initially, water flows in both directions and the lower part of the layer swells. Eventually, only upward flow occurs. Use curve C.

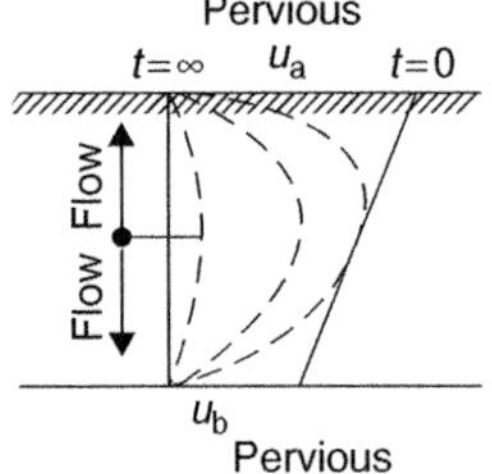

Thick open layer. Excess pressure is due to a combination of self weight and superficial loading. Use curve B.

Figure 7.38

7.9 Coefficient of permeability (*k*)

Once the coefficients of consolidation (c_v) and volume compressibility have been determined, then k can be calculated from (7.18).

Example 7.11

$$\text{For a given clay} \quad m_v = 0.000546\ \text{m}^2/\text{kN}$$
$$c_v = 0.34\ \text{mm}^2/\text{min}$$
$$= 5.7\times 10^{-9}\ \text{m}^2/\text{s}$$

Calculate the coefficient of permeability in metre/second units.

$$\text{From (7.18)}: k = c_v m_v \gamma_w$$
$$= 5.7\times 10^{-9}\times 5.46\times 10^{-4}\times 9.81$$
$$= 305\times 10^{-13}$$
$$= 0.305\times 10^{-10}\ \text{m/s}$$

7.10 Time from similarity

Suppose a clay layer of thickness H is proposed to be loaded by effective pressure σ'.

If a sample from this clay is subjected to the same pressure in the oedometer, then its c_v and h can be calculated. This value of c_v applies to the clay of thickness H equally. Now for the degree of consolidation, T_v is the same for both layers, therefore formula (7.19) may be written as:

$$\frac{T_v}{c_v} = \frac{t}{h^2} = \frac{t}{h^2} \qquad \boxed{t_x = \frac{tH^2}{h^2}} \tag{7.29}$$

where t = time taken in the oedometer to reach U% consolidation.
t_x = time required for U% consolidation of the clay layer of thickness H.

7.11 Total settlement

Figure 7.24 indicates that the total settlement has three components:

1. Initial compression
2. Primary consolidation
3. Secondary consolidation

7.11.1 Initial compression

Initial compression or immediate settlement is a rapid elastic deformation of saturated cohesive soil, under suddenly applied load. The magnitude of settlement may be estimated from the general formula for rigid footing placed on the surface.

$$dH_i = \frac{qB}{E}(1-\mu)I_p \tag{7.30}$$

where B = width of footing
q = bearing pressure
E = modulus of elasticity
μ = Poisson's Ratio = 0.5
I_p = influence factor
L = length of footing

Skempton's influence factors

Table 7.9

Rectangle L/B	1	1.5	2	3	4	5	10	100	Circle
I_p	0.82	1.06	1.20	1.42	1.58	1.70	2.10	3.47	0.79

The modulus of elasticity is found from the triaxial compression test for each soil.
Average ranges: Soft clays: $1400 < E < 3500\ \text{kN/m}^2$
Hard clays: $5500 < E < 14000\ \text{kN/m}^2$

7.11.2 Primary consolidation

This is assumed to occur when practically all of the excess pore water had been dissipated. In theory, 100% primary consolidation is completed at this stage. Figure 7.24 shows how 90% consolidation is estimated. Once U = 90% is known, U = 100% is determined by proportion. Thus, by Taylor's 'square root of time' method in Graph 7.2.

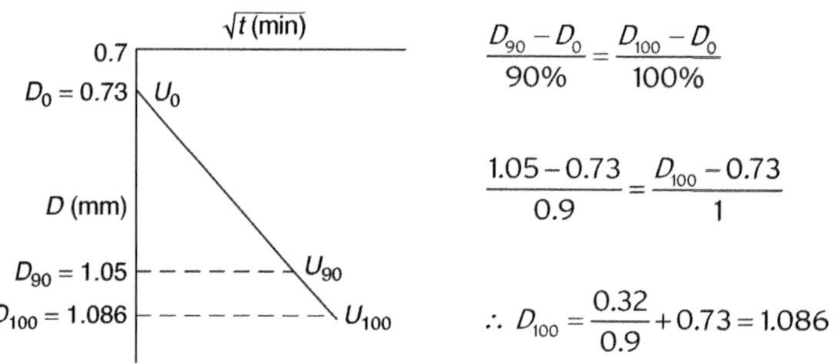

$$\frac{D_{90} - D_0}{90\%} = \frac{D_{100} - D_0}{100\%}$$

$$\frac{1.05 - 0.73}{0.9} = \frac{D_{100} - 0.73}{1}$$

$$\therefore\ D_{100} = \frac{0.32}{0.9} + 0.73 = 1.086\ \text{mm}$$

Figure 7.39

Alternatively, the percentage consolidation ($U\%$), hence the coefficient of consolidation can be determined graphically by Casagrande's log of time method as shown in Figure 7.40.

Step 1: Plot the dial readings against the logarithm of time (t).

Step 2: Locate D_0 (for U_0=0%), by selecting two points (A and B) on the first, parabolic part of the curve at times t_1 and $t_2=4t_1$.

Step 3: Measure the vertical distance between points A and B.

Step 4: Draw length AB vertically from A to locate D_0 (for U_0).

Step 5: Extend the two straight portions of the experimental curve until they intersect at point X. This locates D_{100} for U_{100}=100%.

Step 6: Scale D_0 to D_{100}

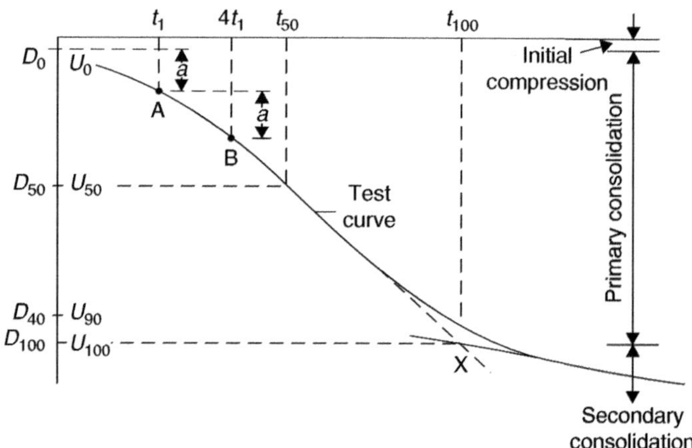

Figure 7.40

7.11.3 Secondary consolidation

This is indicated by the continuation of the oedometer test curve beyond the primary stage, that is, after the near-complete dissipation of excess pore pressure. The reason for it is assumed to be the viscosity of water flowing very slowly through the, by now, denser soil. Secondary compression is smaller in overconsolidated clays than in normally consolidated ones. The extent of the consolidation may be estimated from:

$$\boxed{\Delta H = c_\alpha \left(\frac{\text{Log}\, t_2 - \log t_1}{H} \right)} \tag{7.31}$$

where c_α = coefficient of secondary consolidation given by

$$\boxed{c_\alpha = \frac{\Delta e}{\log t_2 - \log t_1}} = \frac{\Delta e}{\log\left(\frac{t_2}{t_1}\right)} \tag{7.32}$$

where Δe is as shown below.

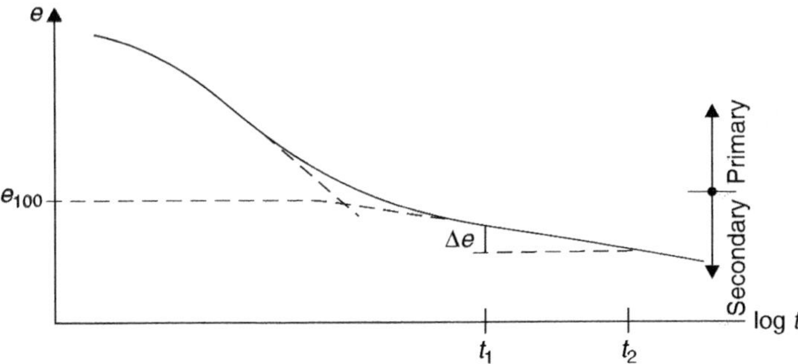

Figure 7.41

Organic soils and plastic clays have low values of c_α, hence their secondary consolidation is high. Conversely, overconsolidated clays have high values of c_α. Soils were classified by G. Mesri, according to their secondary compressibility:

Table 7.10

c_α	Secondary compressibility	
<0.002	Very low	Over consolidated clays
0.004	Low	
0.008	Medium	Normally consolidated clays
0.016	High	
0.032	Very high	
0.064	Extremely high	Organic soils

Problem 7.1

A consolidation test has been carried out on a standard 19 mm thick clay sample. The oedometer's deflection gauge indicated 1.66 mm, just before the removal of the last load, that is, no swelling was allowed. The voids ratio was found to be 0.55 at this stage.

Determine:

a) The initial voids ratio of the saturated specimen
b) The height of the specimen for the voids ratio of 0.62

Problem 7.2

Superficial deposits were removed some years ago from a site, to be used for housing development. The section of the ground in Figure 7.42 shows the known properties of two soil layers overlying solid rock. The overconsolidation ratio of the stiff clay layer was found to be 5.

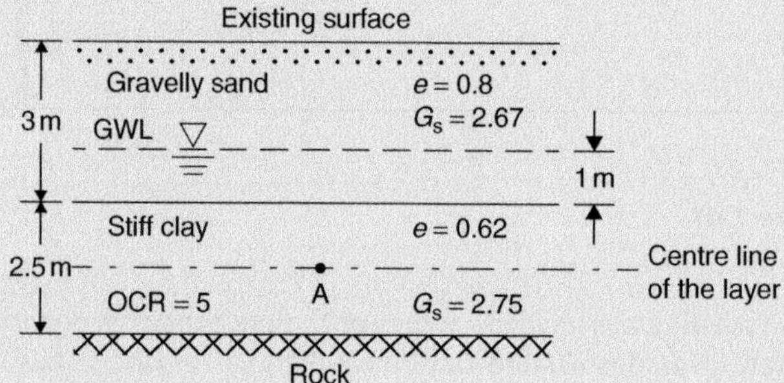

Figure 7.42

Estimate:

a) The pressure exerted originally by the removed overburden on the top of the gravelly sand layer.
b) The thickness of the superficial deposit, assuming its mass density as 1988 kg/m³.

Problem 7.3

An oedometer test has been carried out on a clay specimen, taken from a 2.5 m thick layer. The results recorded were:

Table 7.12

									Swelling stage ↓
σ'_x = pressure (kN/m^2)	0	50	100	200	300	400	500	600	0
D_x = dial reading (mm)	0	0.24	0.46	0.80	1.11	1.31	1.44	1.50	0.66

The voids ratio, 24 hours after the removal of the last, 600 kN/m^2 load, was found to be 54.2%.

The original effective pressure at the centre of the 2.5 m thick clay layer is 100 kN/m^2.

The final effective pressure, after the construction of the structure, will be 200 kN/m^2, at the same depth.

Estimate:

a) The voids ratio for each load increment
b) The coefficient of volume compressibility for $\Delta\sigma' = 200 - 100 = 100$ kN/m^2
c) The settlement of the structure due to $\Delta\sigma' = 100$ kN/m^2, in millimetres

Problem 7.4

Site and laboratory investigations indicate that a 2 m thick sandy gravel surface layer of 19.8 kN/m^3 density is underlain by 4.4 m thick medium clay of 20.4 kN/m^3 unit weight. The clay itself is underlain by very stiff clay. The ground water table is 2 m below the surface. It is proposed that:

a) Either the entire area is covered by compacted fill, surcharging the ground by 360 kN/m^2
b) Or constructing a 6 m wide, long, rigid, concrete slab on the surface, imparting 360 kN/m^2 foundation pressure to the ground.

Estimate the effective and hence the excess effective pressures, induced by the two types of loading of the centre ($x-x$) of the layer at time:

1. $t = 0$, that is at the start of consolidation
2. $t = \infty$, that is at the completion of consolidation.

Compare the two results.

Problem 7.5

The oedometer consolidation test results, carried out on a normally consolidated clay, are given in Table 7.14. Plot log σ'_x against e_x on Graph 7.4 and determine the Compression Index, hence express the voids ratio in terms of formula (7.8). Indicate the extent of validity of this expression.

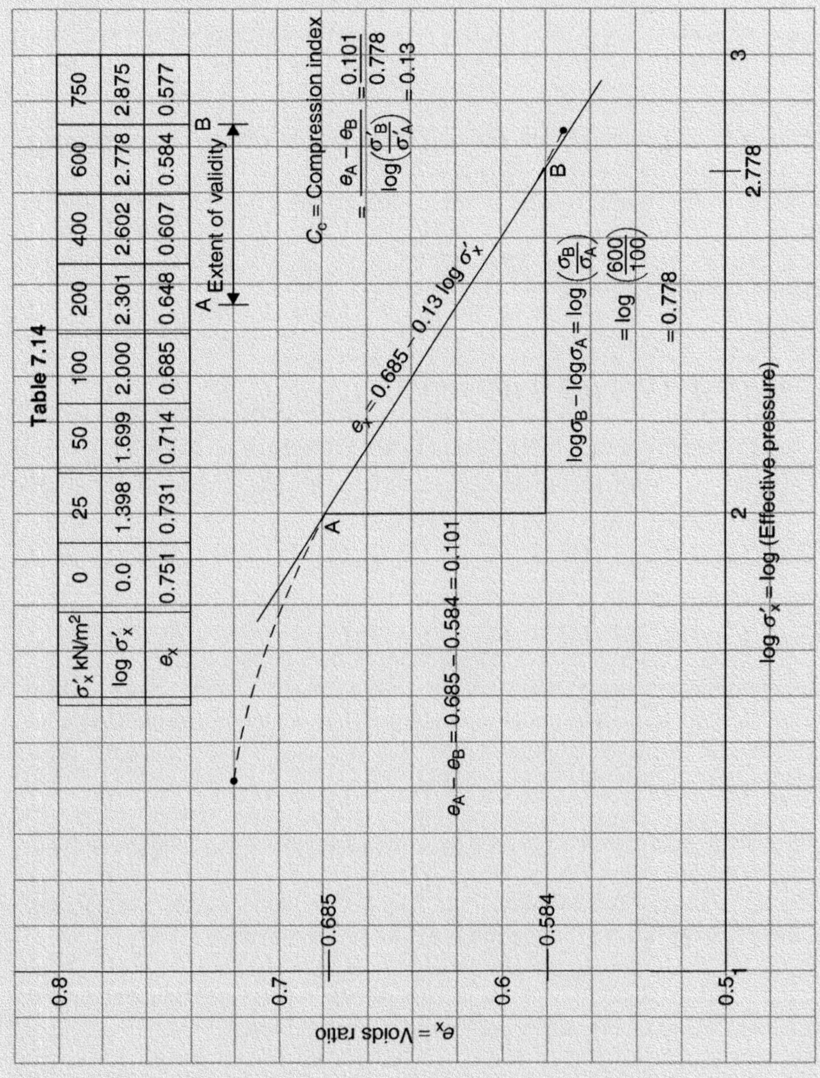

Table 7.14

σ'_x kN/m^2	0	25	50	100	200	400	600	750
log σ'_x	0.0	1.398	1.699	2.000	2.301	2.602	2.778	2.875
e_x	0.751	0.731	0.714	0.685	0.648	0.607	0.584	0.577

Graph 7.4

Problem 7.6

A 4 m wide strip footing is to support part of a settlement-sensitive structure. It is proposed to be based 0.5 m below the surface in a 3.3 m thick, dense, coarse sand layer. The sand is underlain by normally consolidated clay, 4.6 m thick, below which compact gravel is found. The ground water table is 3.3 m below the surface. The footing transmits a net pressure of 250 kN/m^2, including self weight to the soil. Soil characteristics:

Sand:	Unit weight	= 18.2 kN/m^3
Clay:	Saturated density	= 20.1 kN/m^3
	Equation of void ratio:	$e_x = 0.707 - 0.083 \log(\sigma_x)$
	Valid for:	$50 \le \sigma'_x \le 500$ kN/m^2
	Coefficient of consolidation	= 4.14 mm^2/min
	Coefficient of volume change	= 0.1812×10^{-3} m^2/kN

Calculate the voids ratio in the unexcavated state and just after the construction of footing at:

Point A: Top of the clay layer
Point C: Centre line of the clay layer
Point B: Bottom of the clay layer

Estimate the primary consolidation of the clay layer.

Problem 7.7

Calculate the time taken for the structure in Problem 7.6 to settle 78 mm.

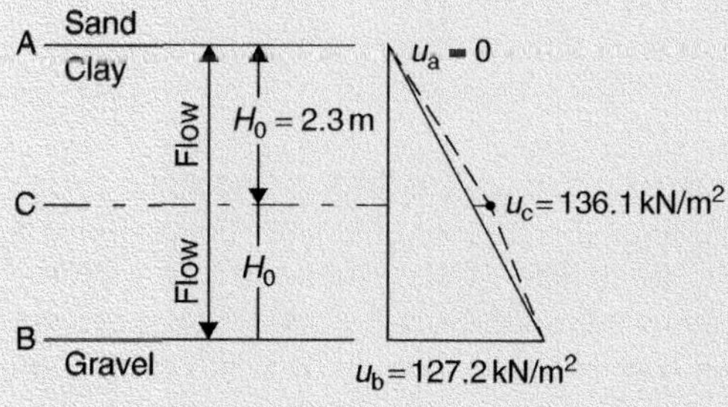

The pore pressure distribution is nearly triangular, thus curve B (Chart 7.5) may be applied.

Figure 7.50

Problem 7.8

The structure in Problems 7.6 and 7.7 was indicated to be sensitive to settlement. However, it was estimated that the 4.6m thick clay layer would consolidate about 78mm in 2.1 years. In order to prevent, or at least minimize settlement, the ground is proposed to be pre-loaded by metal kentledge of 43kN/m^3 unit weight, for one year.

Calculate the height of kentledge layer required to consolidate the clay by 78mm, using the time-consolidation curve on Graph 7.5 for the pressure range 100–200kN/m^2 at $m_v = 0.1812 \times 10^{-3}$ m^2/kN

Problem 7.9

The magnitude of consolidation of the clay layer in Problem 7.6 was estimated to be 87mm, by using the maximum pressure below the centre of the footing (See Figure 7.49). Show that the result would not be much more different, should the settlement be determined by means of the average induced pressure.

Problem 7.10

Starting from formula (7.11), show that $m_v = \dfrac{\Delta h}{h_1 \Delta \sigma'}$

Problem 7.11

A long 4m wide concrete slab is constructed in an 11m thick, homogeneous clay layer, one metre below the ground surface to minimize seasonal effects. There is no evidence of ground water table. Oedometer test results indicate that the coefficient of volume change may be taken as 2.47×10^{-4} m^2/kN for the pressure range 50–300kN/m^2. It is specified that the slab should not settle more than 250mm.

Determine: the required net and total bearing pressures to produce consolidation of this magnitude.

Check, whether the bearing capacity of the clay is exceeded by the estimated bearing pressure. Adopt 2.5 as thefactor of safety.

Chapter 8
Lateral Earth Pressure

Vertical pressure (σ_v) at a point below ground surface is normally induced in four ways:

a) By an imposed structural load, as given by one of the Boussinesq-based formulae. It decreases with depth in a non-linear manner.
b) By a surcharge of infinite extent on the ground surface. This pressure does not vary with depth, but remains constant.
c) By the overburden. This geostatic pressure increases linearly with depth.
d) By the ground water. This hydrostatic pressure also increases linearly with depth.

Each of these induces horizontal pressure (σ_H) at the point considered. The magnitude of σ_H is governed by a constant of proportionality (K) normally called the 'coefficient of lateral earth pressure. This relationship is expressed in general as:

$$\sigma_H \propto \sigma_v \qquad \text{that is} \qquad \boxed{\sigma_H = K\sigma_v} \tag{8.1}$$

Figure 8.1 shows these four types of pressure.

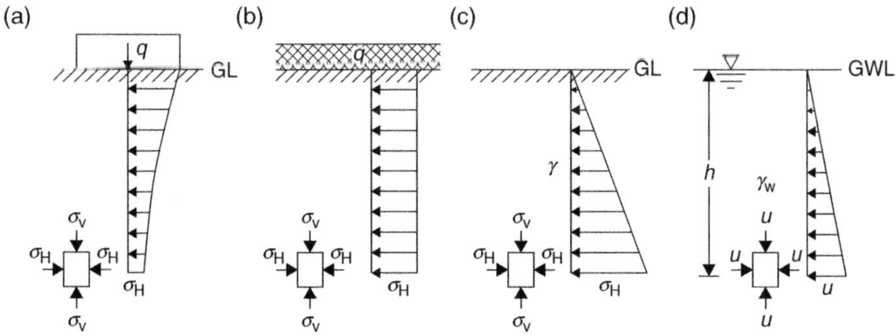

Figure 8.1

Introduction to Soil Mechanics, First Edition. Béla Bodó and Colin Jones.

	$\sigma_V = lq$	$\sigma_V = q$	$\sigma_V = h\gamma$	$\sigma_V = u = h\gamma_W = \sigma_H$
Active:	$\sigma_H = K_a\sigma_V$	$\sigma_H = K_a q$	$\sigma_H = K_0\sigma_V$	$K = 1$
Passive:	$\sigma_H = K_p\sigma_V$	$\sigma_H = K_p q$		

Where, K_0 = coefficient of earth pressure at rest, as the soil is in its natural undisturbed state when only vertical strain can occur during deposition.

K_a = coefficient of active earth pressure, when the soil can expand horizontally due to load σ'_v and yielding lateral support.

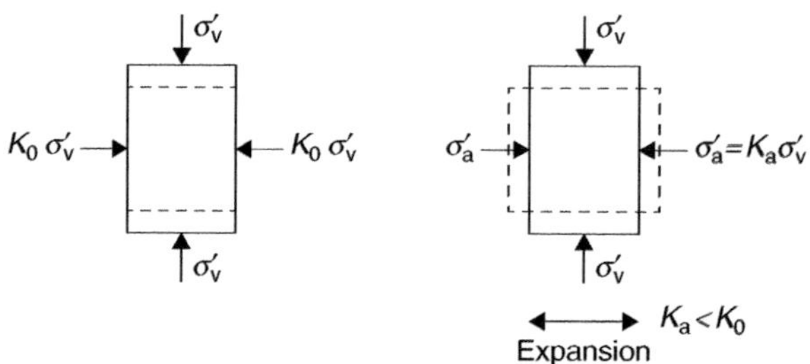

Figure 8.2

Note, that the pressures are expressed as effective ones, when ground water is present. In these terms, the 'active' pressure is given by:

$$\boxed{\sigma'_a = K_a\sigma'_v} \tag{8.2}$$

8.1 Resistance to active expansion

If the expanding soil is supported by some sort of structure, which in turn is supported by soil on its other side, resistance to this expansion develops as the soil in front of the structure is compressed. The available resisting horizontal pressure depends also on the vertical soil pressure on the other side.

$$\boxed{\sigma'_p = K_p\sigma'_v} \tag{8.3}$$

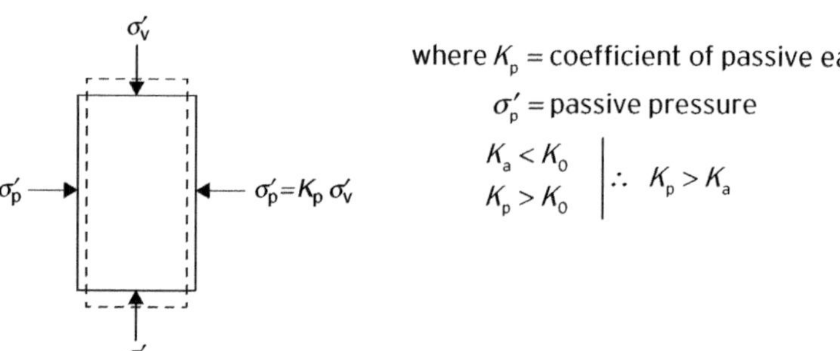

Figure 8.3

where K_p = coefficient of passive earth pressure.

σ'_p = passive pressure

$$\left.\begin{aligned} K_a &< K_0 \\ K_p &> K_0 \end{aligned}\right| \therefore\ K_p > K_a$$

Figure 8.4 shows a typical occurrence of active and passive pressures.

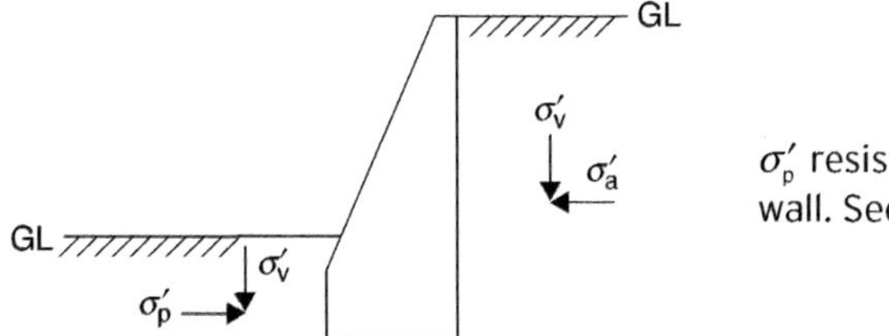

σ'_p resists the action of σ'_a on the retaining wall. See also Figure 8.7.

Figure 8.4

Note: Values of K_0, K_a, K_p and their application were evolved in the following theories:

1. Rankine's for cohesionless soil (ϕ-soil)
2. Rankine-Bell theory for c-ϕ soil
3. Coulomb's wedge theory
4. Culmann's construction

When there is no evidence of groundwater table, then $\sigma'_a = \sigma_a = K_a \sigma_v$ and $\sigma'_p = \sigma_p = K_p \sigma_v$.

8.2 The value of K_0

It is determined in triaxial tests, during which the lateral strain is kept zero by the synchronized increase of cell pressure and deviator stress. The coefficient is applied in the design of structures constructed so, that the soil cannot deform, hence may be considered at rest. This type of structures are:

a) Braced excavations
b) Basement walls
c) Culverts and underpaths
d) Cantilever retaining walls
e) Abutments of rigid portal frame bridges.

The magnitude of K_0 depends largely on the:

1. Density of the soil
2. Stress history of the soil.

There are several empirical formulae for the determination of K_0, but typical ranges are listed in Table 8.1.

Table 8.1

Soil	K_0
Compact sand	0.4–0.6
Loose sand	0.45–0.5
Normally consolidated clay	0.5–0.75
Overconsolidated clay	1–4

Empirical formulae

a) Jáky (1944): For normally consolidated sand and clay:

$$\boxed{K_0 = 1 - \sin\phi} \tag{8.4}$$

b) Alpan (1967): For normally consolidated clay:

$$\boxed{K_0 = 0.19 + 0.233\log(\text{PI})} \tag{8.5}$$

Where, PI = Plasticity index

c) Mayne-Kulhawy (1982) for overconsolidated soil:

$$\boxed{K_0 = (1 - \sin\phi)(\text{OCR})^{\sin\phi'}} \tag{8.6}$$

Where, OCR = overconsolidation ratio

8.3 Stress path representation (Lambe 1967)

During the sedimentation process, normal consolidation occurs and there is no perceptible horizontal compression, hence $\sigma'_H < \sigma'_v$ and $K_0 < 1$. If the value of K_0 is known, then the stress path for points at increasing depth below ground level can be plotted. The soil is said to be in its K_0-state and the line connecting the plotted points is called the K_0-line. The slope angle θ_0 can be determined directly from K_0, using the formulae derived below (see also chapter 6):

From (6.14): $\sigma'_H = p - q = \left(1 - \frac{q}{p}\right)p$

From (6.15): $\sigma'_v = p + q = \left(1 + \frac{q}{p}\right)p$

$\therefore$ $\sigma'_H = (1 - \tan\theta_0)p$; $\sigma'_v = (1 + \tan\theta_0)p$

From Graph 8.1: $\tan\theta_0 = \dfrac{q}{p}$

From (8.1): $K_0 = \dfrac{\sigma'_H}{\sigma'_v}$ $\quad\therefore$

$$\boxed{K_0 = \frac{1 - \tan\theta_0}{1 + \tan\theta_0}} \tag{8.7}$$

From (8.7): $K_0 + K_0\tan\theta_0 = 1 - \tan\theta_0$

$(1 + K_0)\tan\theta_0 = 1 - K_0$

$$\tan\theta_0 = \frac{1 - K_0}{1 + K_0} \qquad \therefore \boxed{\theta_0 = \tan^{-1}\left(\frac{1 - K_0}{1 + K_0}\right)} \tag{8.8}$$

The significance of the K_0 line

The following sections will show that the K_0 line provides a visual dividing line between active expansion and passive compression. The positions of the K_a and K_p lines, relative to K_0 are shown in Figure 8.5.

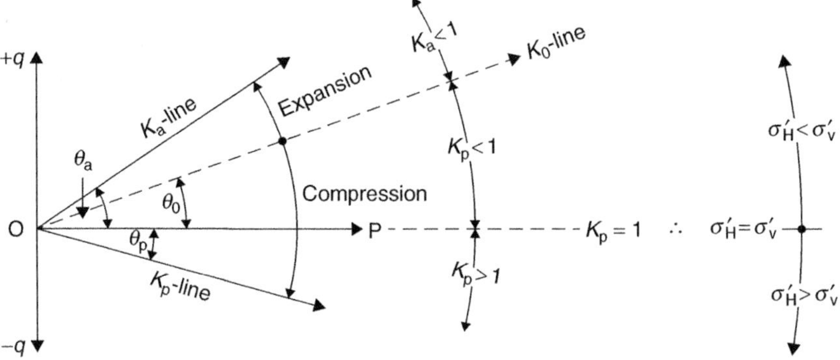

Figure 8.5

The departure of K_a and K_p from the K_0 line depends on the relative magnitudes of σ'_H and σ'_V as indicated. See also Example 8.2.

Example 8.1

Figure 8.6 shows the profile of uniform, normally consolidated clay with ground water level at a depth of 2 m. The coefficient of lateral earth pressure at rest is $K_0=0.6$.

Calculate σ'_V, σ'_H as well as the stress path co-ordinates p' and q' at 1, 2, 3 and 4 m below ground level. Draw the K_0 line on Graph 8.1.

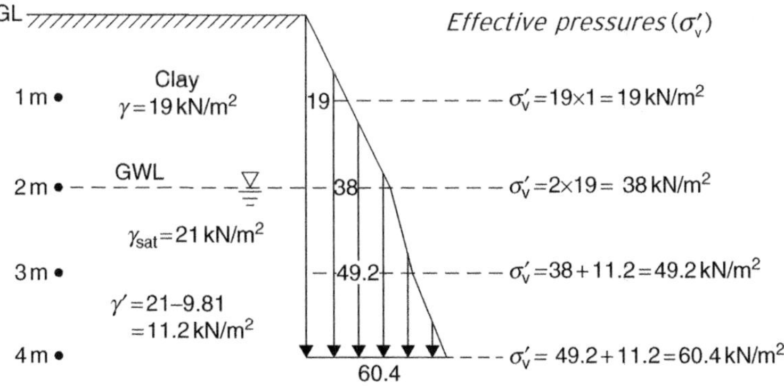

Figure 8.6

Table 8.2 contains the rest of the calculations, using the stress path formulae introduced in Chapter 6.

Table 8.2

Point	σ'_v (kN/m²)	$\sigma'_H=0.6\,\sigma'_v$ (kN/m²)	$q=\frac{1}{2}(\sigma'_v-\sigma'_H)$ (kN/m²)	$p=\sigma'_H+q=\frac{1}{2}(\sigma'_v-\sigma'_H)$ (kN/m²)
1	19	11.4	3.8	15.2
2	38	22.8	7.6	30.4
3	49.2	29.5	9.8	39.4
4	60.4	36.2	12.1	48.3

From Graph 8.1, the slope angle of K_0 line is $\theta_0 = 14°$. Graph 8.2 shows the Mohr-circle representation of the pressures in the soil at rest, the envelope lies practically along the K_0 line, at angle $\phi'_0=14.4°$.

8.4 Rankine's theory of cohesionless soil

The theory was proposed by Rankine (1857) for homogeneous ϕ-soil having *horizontal* surface. Its development can be illustrated by Mohr's diagram. The basis of the proposal was, that the soil changes from a state of elastic equilibrium into a plastic one, when the entire mass is on the point of imminent shear failure. Any further change in the applied pressure would cause continuous deformation (plastic flow). This state is reached on the Mohr-diagram, when the circle of principal stresses (σ_V & σ_H) reaches the Coulomb failure envelope. As the shear stress is zero on the plane of principal stresses, they are plotted along the horizontal (σ) axis at $\tau=0$.

Figure 8.7 depicts the movement of soil mass after plastic failure. The surface of the retaining wall is assumed to be smooth.

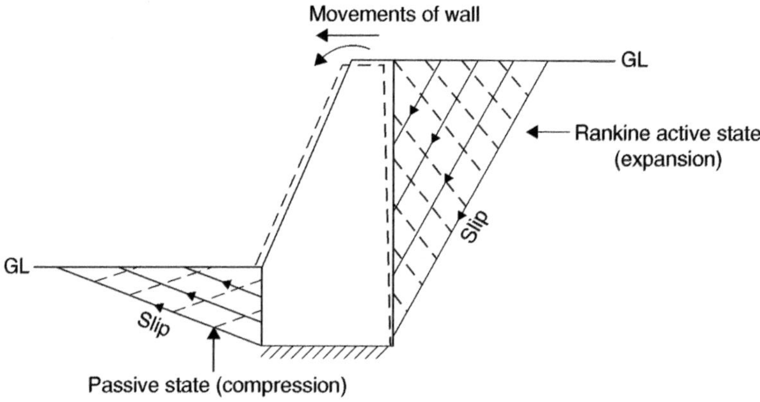

Figure 8.7

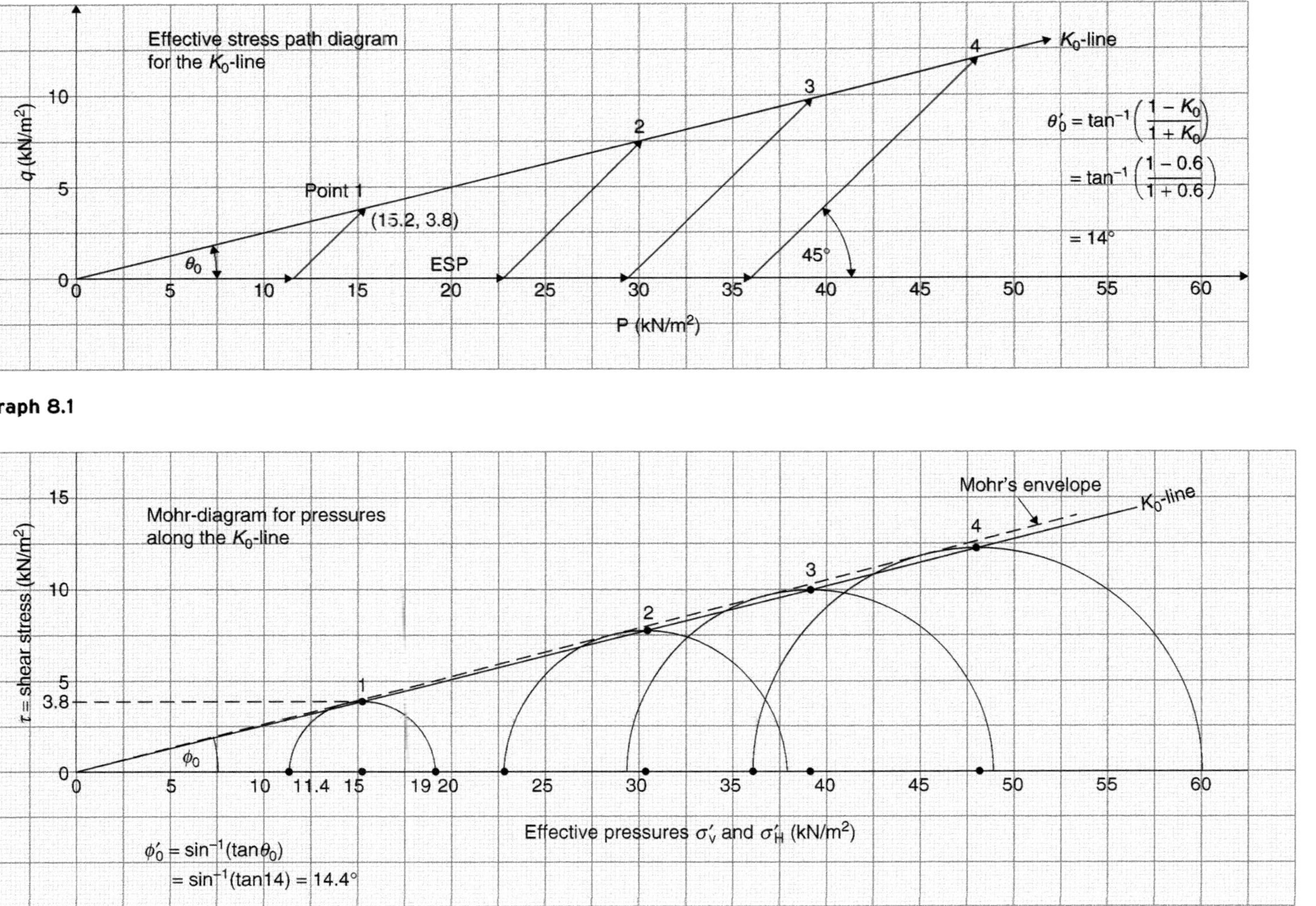

Graph 8.1

Graph 8.2

Active rankine state

The development of this state is shown in Figure 8.8.

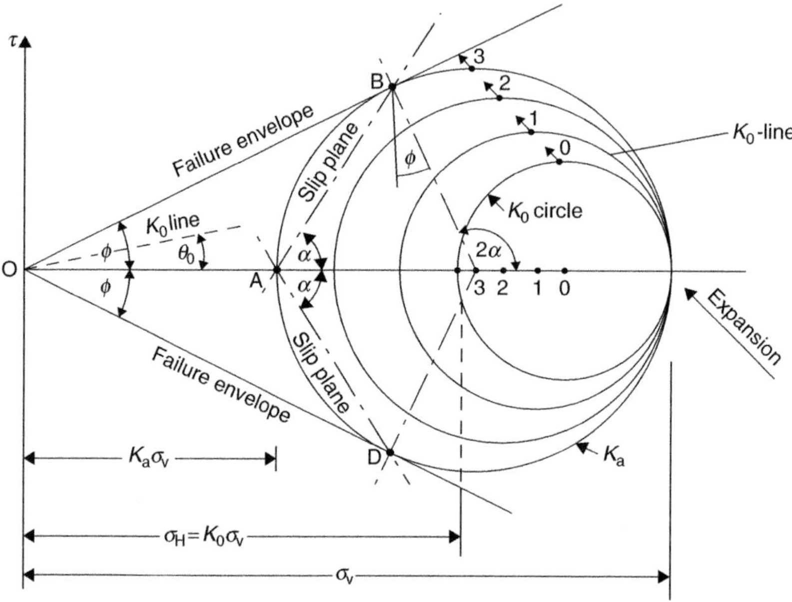

Figure 8.8

a) The K_0 circle is drawn from σ_V and $\sigma_H = K_0\sigma_V$ to signify the pressures in natural state. The circle is far from touching the failure envelope, hence no shear deformation can occur.
b) As the wall in Figure 8.7 moves forward slowly, the soil expands towards it. σ_H decreases whilst σ_V remains the same. This can only mean that the value of K gets smaller as the diameters of the circles increase (indicated by circles 1 and 2).
c) When circle 3 touches the failure surface, then K reaches its smallest value and the full shear strength of the soil is mobilized so, that:

$$\boxed{\sigma_H = K_a\sigma_V} \tag{8.9}$$

d) The rupture or slip surfaces AB and AD are inclined at $\left(45° + \frac{\phi}{2}\right)$ to σ_V as shown below.

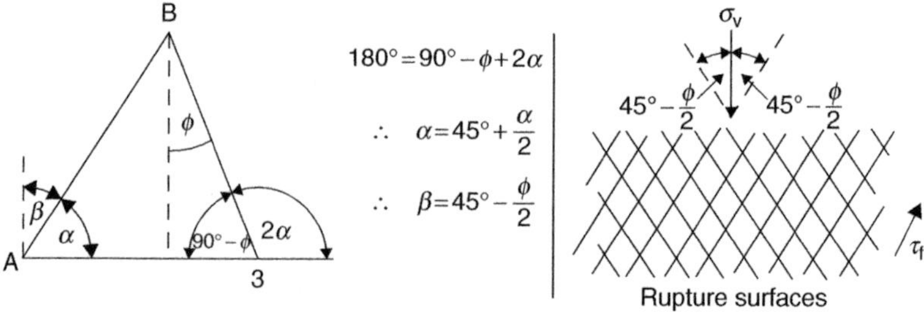

Figure 8.9

e) The shear stress τ_f acts upwards, against the direction of slipping.
f) Derivation of the active pressure formula:

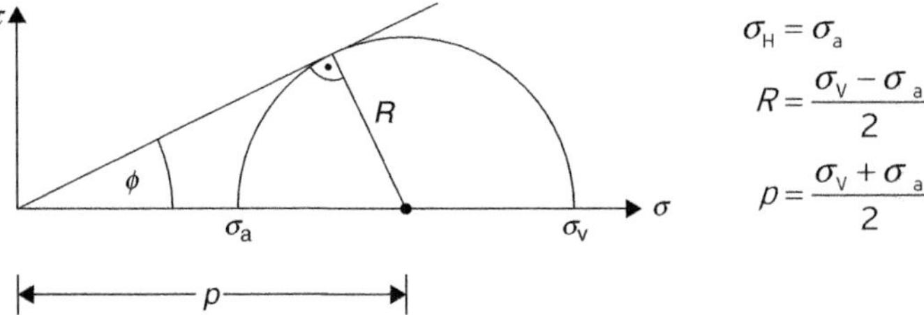

$$\sigma_H = \sigma_a$$
$$R = \frac{\sigma_V - \sigma_a}{2}$$
$$p = \frac{\sigma_V + \sigma_a}{2}$$

Figure 8.10

$$\sin\phi = \frac{R}{p} = \frac{\frac{1}{2}(\sigma_V - \sigma_a)}{\frac{1}{2}(\sigma_V + \sigma_a)} = \frac{\sigma_V - \sigma_a}{\sigma_V + \sigma_a}$$

$$\sigma_V \sin\phi + \sigma_a \sin\phi = \sigma_V - \sigma_a$$

Expressing the active pressure: $\boxed{\sigma_a = \left(\frac{1-\sin\phi}{1+\sin\phi}\right)\sigma_V}$ (8.10)

Hence the coefficient: $\boxed{K_a = \frac{1-\sin\phi}{1+\sin\phi}}$ (8.11)

Alternatively, $\boxed{K_a = \tan^2\left(45° - \frac{\phi}{2}\right)}$ (8.12)

Proof: $\frac{1-\sin\phi}{1+\sin\phi} = \tan^2\left(45° - \frac{\phi}{2}\right)$ can be derived by basic trigonometry, using a circle of unit radius:

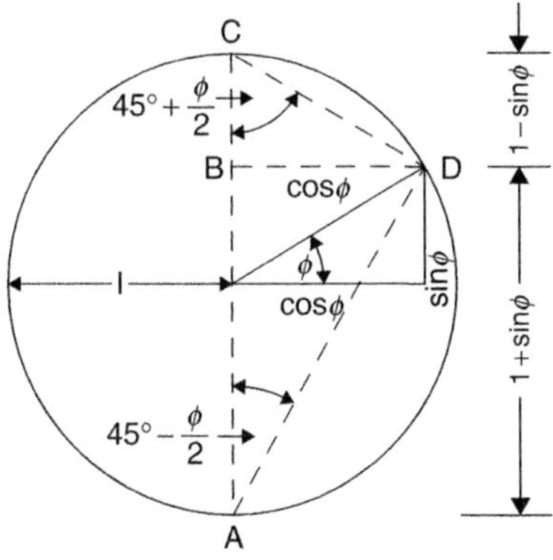

Figure 8.11

From triangle BAD:

$$\tan\left(45° - \frac{\phi}{2}\right) = \frac{\cos\phi}{1+\sin\phi}$$

From triangle BCD:

$$\tan\left(45° + \frac{\phi}{2}\right) = \frac{\cos\phi}{1+\sin\phi}$$

But $\tan\left(45° - \frac{\phi}{2}\right) = \frac{1}{\tan\left(45° + \frac{\phi}{2}\right)}$

$$\frac{1-\sin\phi}{1+\sin\phi} = \frac{\cos\phi}{\tan\left(45^\circ+\frac{\phi}{2}\right)} \times \frac{\tan\left(45^\circ-\frac{\phi}{2}\right)}{\cos\phi} = \frac{\tan\left(45^\circ-\frac{\phi}{2}\right)}{\tan\left(45^\circ+\frac{\phi}{2}\right)}$$

$$= \frac{\tan\left(45^\circ-\frac{\phi}{2}\right)}{\frac{1}{\tan\left(45^\circ-\frac{\phi}{2}\right)}} = \tan^2\left(45^\circ-\frac{\phi}{2}\right)$$

Sometimes these expressions are signified by N_ϕ

eg. $$N_\phi = \frac{1+\sin\phi}{1-\sin\phi} = \tan^2\left(45^\circ+\frac{\phi}{2}\right) \qquad \therefore \quad K_a = \frac{1}{N_\phi} \tag{8.13}$$

and $$\frac{1}{N_\phi} = \frac{1-\sin\phi}{1+\sin\phi} = \tan^2\left(45^\circ-\frac{\phi}{2}\right) \qquad \sigma_a = \frac{\sigma_V}{N_\phi} \tag{8.14}$$

Passive Rankine state
(See Figure 8.12.)

a) Starting again from K_0 state, circles of progressively smaller radii are drawn, this time, to simulate compression. This is, because σ_H increases.
b) It is indicated in Figure 8.12, that $\sigma_H < \sigma_V$ at first, although it is increasing as the radii are decreasing and σ_H approaches the value of σ_V.
c) At some stage $\sigma_H = \sigma_V$. The radius becomes zero, $\phi = 0$ hence $K = 1$.
d) After this $\sigma_H > \sigma_V$ and increasing,

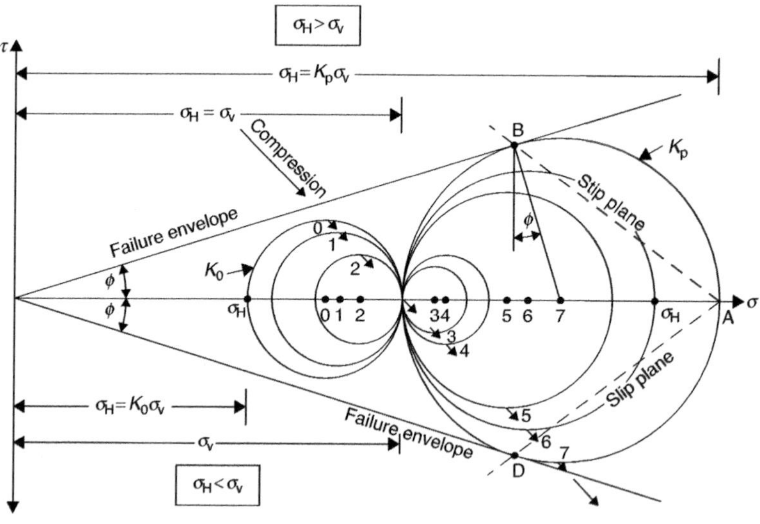

Figure 8.12

e) Circle 7 touches the failure envelope, giving the maximum value of σ_H:

$$\boxed{\sigma_p = K_p \sigma_V} \tag{8.15}$$

f) The rupture surfaces AB and AD are inclined $\left(45° - \dfrac{\phi}{2}\right)$ to the line of action of $\sigma_p = K_p \sigma_V$ as shown.

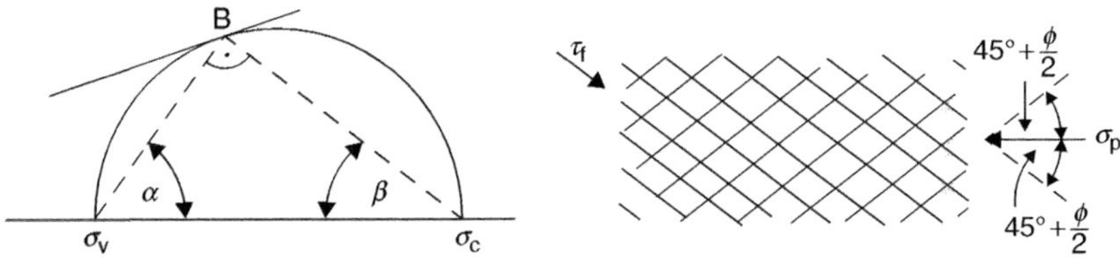

Figure 8.13

$$\alpha = 45° - \frac{\phi}{2}$$

$$B = 90° - \alpha = 45° + \frac{\phi}{2}$$

g) The shear stress acts downwards

h) Derivation of the passive pressure formula:

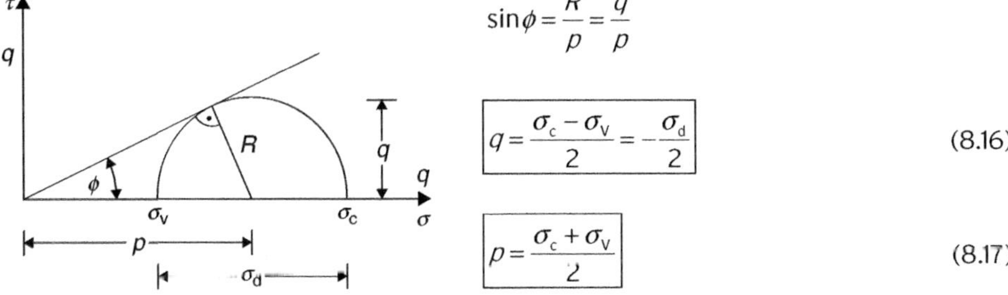

$$\sin\phi = \frac{R}{p} = \frac{q}{p}$$

$$\boxed{q = \frac{\sigma_c - \sigma_V}{2} = \frac{\sigma_d}{2}} \tag{8.16}$$

$$\boxed{p = \frac{\sigma_c + \sigma_V}{2}} \tag{8.17}$$

Figure 8.14

$$\sin\phi = \frac{\frac{1}{2}(\sigma_p - \sigma_v)}{\frac{1}{2}(\sigma_p + \sigma_v)} = \frac{\sigma_p - \sigma_v}{\sigma_p + \sigma_v}$$

$$\sigma_p \sin\phi + \sigma_v \sin\phi = \sigma_p - \sigma_v$$

$$\sigma_v(1 + \sin\phi) = \sigma_p(1 - \sin\phi)$$

$$\boxed{\sigma_p = \left(\frac{1 + \sin\phi}{1 - \sin\phi}\right)\sigma_v \qquad \sigma_p = \tan^2\left(45° + \frac{\phi}{2}\right)\sigma_v} \tag{8.18}$$

$$\boxed{K_p = \frac{1 + \sin\phi}{1 - \sin\phi} = N_\phi} \tag{8.19}$$

But $\quad K_a = \dfrac{1}{N_\phi} \qquad \therefore \; \boxed{K_p = \dfrac{1}{K_a}}$ (8.20)

8.4.1 Stress path representation (Lambe)

The information found on Figures 8.8 and 8.12 is summarized in Figure 8.15. See Chapter 6 for explanation.

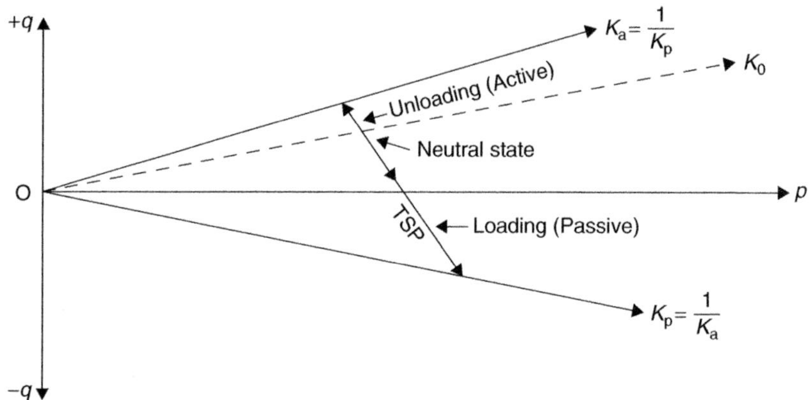

Figure 8.15

At failure	$: K = K_f$	
At the unloading stage	$: K_f = K_a$	$K > K_a$
At the loading stage	$: K_f = K_p$	$K < K_p$

Example 8.2

Triaxial tests carried out on three compact sand specimens yielded $\phi = 25°$, from which the value of K_0 and the inclination of the corresponding failure K_f-line were calculated theoretically as:

$$K_0 = 1 - \sin\phi = 1 - \sin 25 = 0.58 \quad (8.4)$$

and

$$\theta_f = \tan^{-1}(\sin\phi) = \tan^{-1}(\sin 25) = 23° \quad (6.10)$$

Two more specimens were tested in order to draw the stress path for the active and passive Rankine states, starting from the K_0 state. The test was carried out in two stages, keeping the vertical pressure σ_v constant at 660 kN/m². The results are tabulated below.

Step 1: Choose an arbitrary value for the cell pressure in the K_0 state and calculate the relevant values of σ_d, p, q, K_0 and θ_0 as shown in Table 8.3. The results are plotted on Graph 8.3 as point "A".

Table 8.3 (K_0 state)

$K_0 \frac{\sigma_c}{\sigma_v}$	0.576	–
σ_v	660	kN/m^2
$\sigma_c = K_0\sigma_v$	380	kN/m^2
$\sigma_d = \sigma_v - \sigma_c$	280	kN/m^2
$q = \frac{\sigma_d}{2}$	140	kN/m^2
$p = q + \sigma_c$	520	kN/m^2
$\theta° = \tan^{-1}\left(\frac{q}{p}\right)$	15	Degree

Step 2: The procedure during the first test was to decrease σ_c by (say) 55 kN/m^2 and increase σ_d by the same amount so, that σ_v remained 660 kN/m^2. The workings in this unloaded (active) stage were again tabulated, starting from the K_0 state.

Table 8.4 (Active state)

	Point A		Point B
σ_v kN/m^2	660	660	660
σ_c N/m^2	380	325	270
σ_d kN/m^2	280	335	390
q kN/m^2	140	167.5	195
p kN/m^2	520	492.5	465
K	0.576	0.49	0.41
$\theta°$	15	18.8	22.8

Step 3: During the second test (Passive stage) σ_c was increased arbitrarily by 140 kN/m^2 and σ_d changed, bearing in mind (from Figure 8.12), that

a) If $\sigma_H = \sigma_c < \sigma_v$, then for each circle:

$+q$, σ_c, σ_v

$$\sigma_v = \sigma_c + \sigma_d \qquad K = \frac{\sigma_c}{\sigma_v} \qquad q = \frac{\sigma_v - \sigma_c}{2} = \frac{\sigma_d}{2} \qquad p = q + \sigma_c$$

Therefore 140 kN/m^2 is subtracted from σ_d until it becomes zero at $\sigma_v = \sigma_H = \sigma_c$.

b) When $\sigma_c > \sigma_v$, then for each circle:

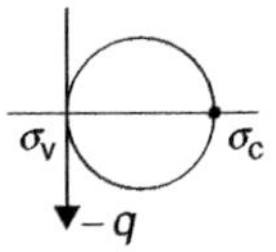

$$\sigma_v = \sigma_c - \sigma_d \quad K = \frac{\sigma_c}{\sigma_v} \quad q = -\left(\frac{\sigma_v - \sigma_c}{2}\right) = -\frac{\sigma_d}{2} \quad p = \sigma_v - q$$

Therefore, as soon as $\sigma_c \geq \sigma_v$, the deviator stress increases instead of decreases. In this example: σ_d is increased by 140 kN/m^2.

The calculations commenced from K_0 state and the results tabulated.

Table 8.5 (Passive state)

	$\sigma_v=\sigma_c+\sigma_d$					$\sigma_v=\sigma_c-\sigma_d$	
σ_v kN/m^2	660	660	660	660	660	660	660
σ_c kN/m^2	380	520	660	800	940	1080	1220
σ_d kN/m^2	280	140	0	–140	–280	–420	–560
q kN/m^2	140	70	0	–70	–140	–210	–280
p kN/m^2	520	590	660	730	800	870	940
K	0.576	0.79	1	1.21	1.42	1.64	1.85
$\theta°$	15	6.8	0	–5.5	–9.9	–13.6	–16.6
	Point A		Point C				Point D

Step 4: The content of Tables 8.4 and 8.5 are plotted on Graph 8.3 along line BD, with arrows to indicate expansion or compression.

Notes:

a) It is not necessary to calculate many points in order to draw line BE. Because point C is given by $p=\sigma_c=\sigma_v$ at $K=1$ and A can be calculated, as in Table 8.3, the two points define the line.

b) Once the TSP has been drawn, p and q can be found for any value of σ_c chosen, as the stress path from it to the TSP is inclined to the horizontal at 45°. Also, the centre of the passive failure circle can be read-off at point F. In this case from Graph 8.3:

$$\left.\begin{aligned} p &= 1135\,\text{kN/m}^2 \\ q &= -475\,\text{kN/m}^2 \\ K_p &= 2.44 \end{aligned}\right\} \text{Point E}$$

From $K = \frac{\sigma_c}{\sigma_v}$ $\quad 2.44 = \frac{\sigma_c}{660}$ $\quad \therefore \sigma_c = 1610$ kN/m^2 (cell pressure)

From $\sigma_d = \sigma_c - \sigma_v$ $\quad \sigma_d = 1610 - 660 = 950$ kN/m^2 (Deviator stress)

Alternatively $\sigma_d = 2q = 2 \times 475 = 950$ kN/m^2

c) Much smaller strain is required to fail in active expansion than in passive compression.

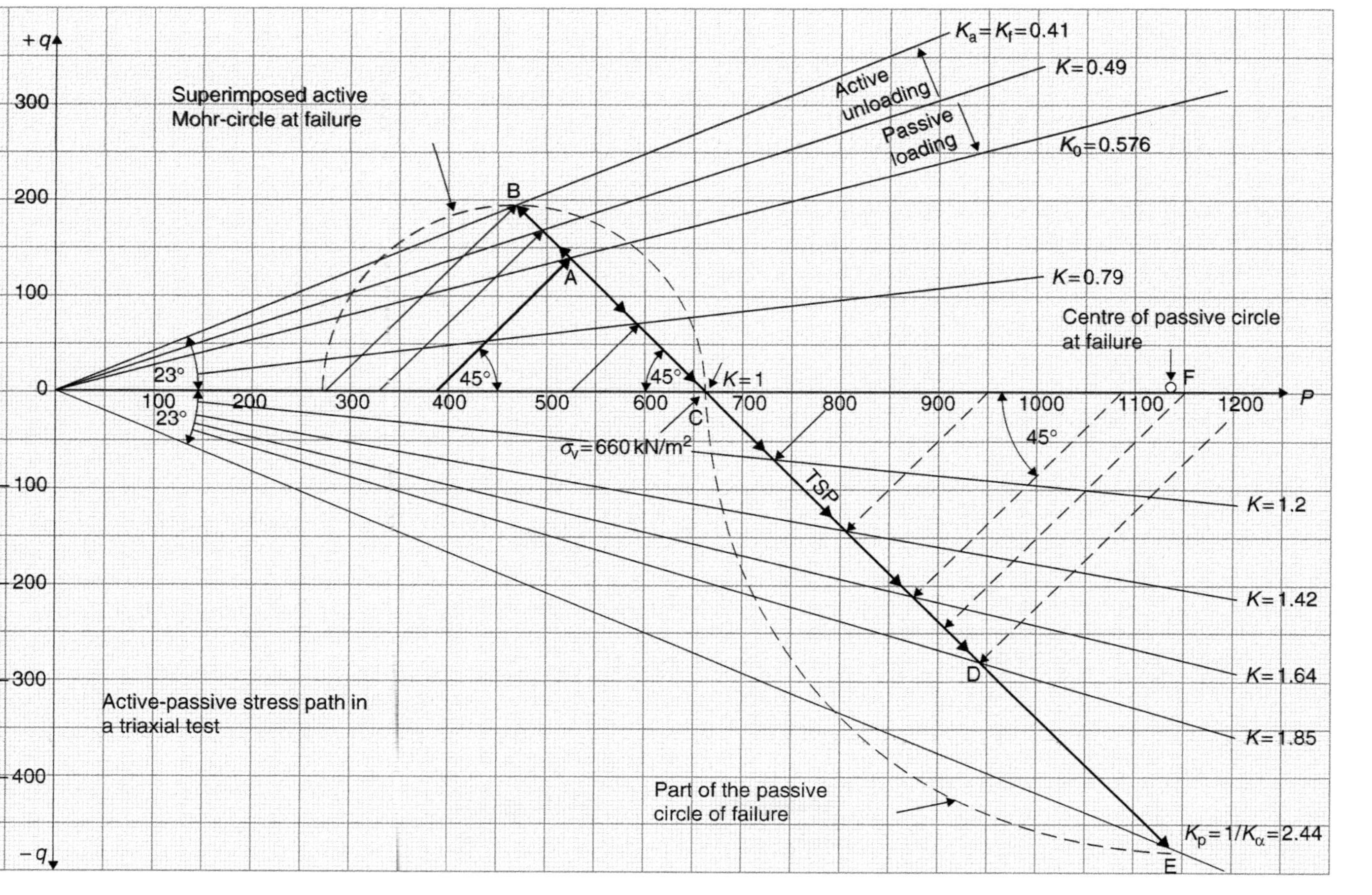
+q
−q
P
300
200
100
0
−100
−200
−300
−400
100
200
300
400
500
600
700
800
900
1000
1100
1200
Superimposed active
Mohr-circle at failure
$K_a = K_f = 0.41$
$K = 0.49$
$K_0 = 0.576$
$K = 0.79$
$K = 1$
$K = 1.2$
$K = 1.42$
$K = 1.64$
$K = 1.85$
$K_p = 1/K_\alpha = 2.44$
Active
unloading
Passive
loading
Centre of passive circle
at failure
F
A
B
C
D
E
23°
23°
45°
45°
45°
$\sigma_v = 660\ kN/m^2$
TSP
Active-passive stress path in
a triaxial test
Part of the passive
circle of failure

Graph 8.3

8.5 Rankine–Bell theory for c–ϕ soil

Bell extended Rankine's theory to c–ϕ soils, applicable to retaining structures having smooth, vertical surfaces. The active and passive pressures are derived by means of Mohr's diagram.

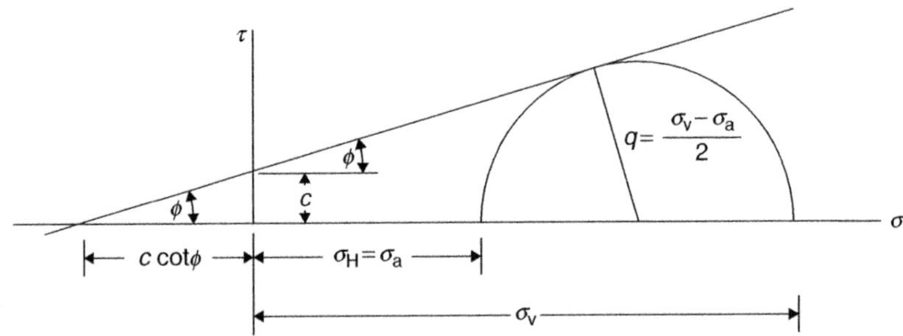

Figure 8.16

$$\sin\phi = \frac{\frac{1}{2}(\sigma_v - \sigma_a)}{c \times \cot\phi + \sigma_a + \frac{1}{2}(\sigma_v - \sigma_a)} = \frac{\sigma_v - \sigma_a}{2c \times \cot\phi + \sigma_a + \sigma_v}$$

$$2c \times \sin\phi \cot\phi + \sigma_a \sin\phi + \sigma_v \sin\phi = \sigma_v - \sigma_a$$

$$2c \times \sin\phi \times \frac{\cos\phi}{\sin\phi} + \sigma_a(1+\sin\phi) = \sigma_v(1-\sin\phi)$$

Expressing:
$$\sigma_a = \left(\frac{1-\sin\phi}{1+\sin\phi}\right)\sigma_v - \frac{2c - \cos\phi}{1+\sin\phi}$$

From $\sin^2\phi + \cos^2\phi = 1$

$$\cos\phi = \sqrt{1-\sin^2\phi} = \sqrt{(1-\sin\phi)(1+\sin\phi)}$$

$$\frac{\cos\phi}{1+\sin\phi} = \frac{\sqrt{(1-\sin\phi)(1+\sin\phi)}}{1+\sin\phi} = \sqrt{\frac{1-\sin\phi}{1+\sin\phi}}$$

But from (8.11):
$$K_a = \frac{1-\sin\phi}{1+\sin\phi} \quad \therefore \frac{\cos\phi}{1+\sin\phi} = \sqrt{K_a}$$

$$\therefore \sigma_a = \left(\frac{1-\sin\phi}{1+\sin\phi}\right)\sigma_v - 2c\sqrt{\frac{1-\sin\phi}{1+\sin\phi}}$$

Therefore the active pressure is: $\boxed{\sigma_a = K_a\sigma_v - 2c\sqrt{K_a}}$ (8.21)

Similarly, the formula for the passive pressure is: $\boxed{\sigma_p = K_p\sigma_v + 2c\sqrt{K_p}}$ (8.22)

8.5.1 Tension cracks

The active pressure formula is made up by a positive term ($K_a\sigma_v$), which increases with depth and a negative constant number $\left(2c\sqrt{K_a}\right)$. At a certain depth, usually denoted by z_0, the two terms become equal in magnitude, resulting in zero active pressure. The soil is in tension from the surface down to depth z_0. The formula for z_0 is derived with reference to Figure 8.17.

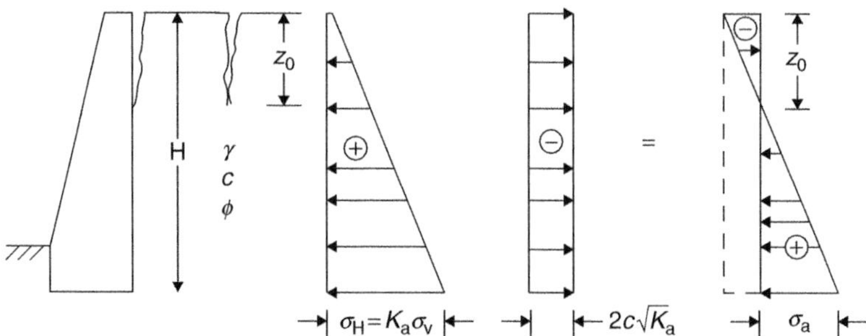

Figure 8.17

When $\sigma_a = 0$ then $K_a\sigma_v - 2c\sqrt{K_a} = 0$

But $\sigma_v = z_0\gamma$

$$K_a z_0 \gamma = 2c\sqrt{K_a}$$

$$z_0 = \frac{2c}{\gamma}\frac{\sqrt{K_a}}{K_a}$$

$$= \frac{2c}{\gamma}\sqrt{\frac{K_a}{K_a^2}}$$

$$\boxed{z_0 = \frac{2c}{\gamma\sqrt{K_a}}} \qquad (8.23)$$

For pure clay: $\phi = 0$ and $K_a = 1$

$$\therefore \boxed{z_0 = \frac{2c}{\gamma}} \qquad (8.24)$$

Note: If ϕ is small or zero, in a fully saturated soil, the shear strength parameters are obtained by undrained test, hence c_u and ϕ_u are used in the formulae. If, however, ϕ is large and the soil is very permeable, then drained test results ϕ'_d and c'_d.

The formula for z_0 can also be derived from Mohr's circle taking $\sigma_a = 0$.

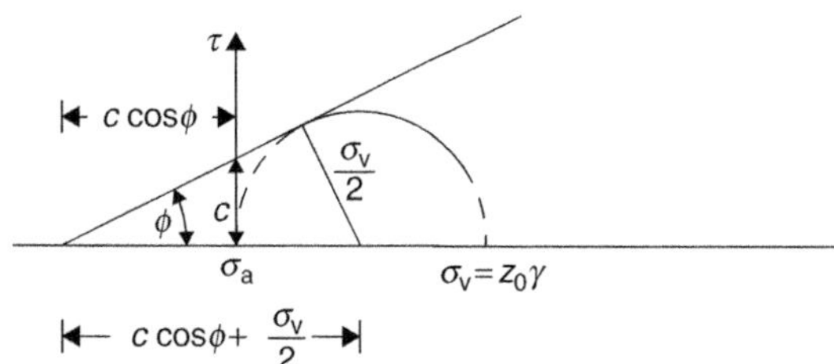

Figure 8.18

The derivation starts from:

$$\sin\phi = \frac{\frac{\sigma_v}{2}}{c\cot\phi + \frac{\sigma_v}{2}}$$

8.5.2 Effect of surcharge (*q* kN/m) on z_0

This decreases the value of z_0. In any case, the actual value could be a third lower then the theoretical one.

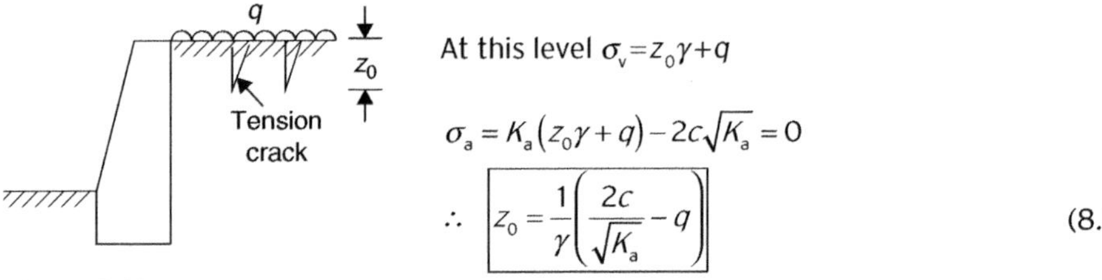

At this level $\sigma_v = z_0\gamma + q$

$$\sigma_a = K_a(z_0\gamma + q) - 2c\sqrt{K_a} = 0$$

$$\therefore \quad \boxed{z_0 = \frac{1}{\gamma}\left(\frac{2c}{\sqrt{K_a}} - q\right)} \tag{8.25}$$

Figure 8.19

8.5.3 Water in the cracks only

When water drains into the cracks of impermeable soil, it does not alter the pore pressure at depth in the short term. However, it increases the hydrostatic pressure on the wall.

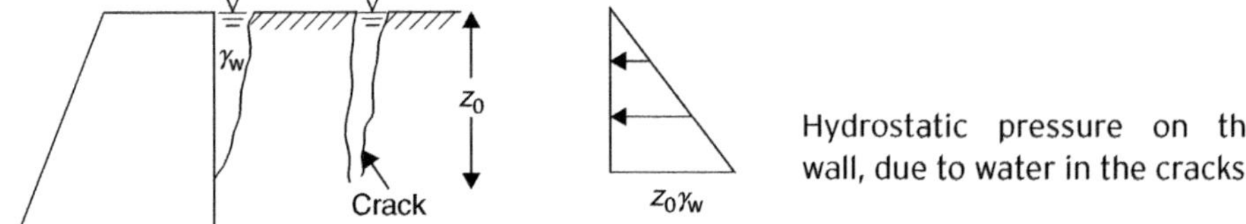

Hydrostatic pressure on the wall, due to water in the cracks.

Figure 8.20

8.6 Rankine–Bell theory for *c*-soil

Formulae (8.21) to (8.25) are simplified by using the undrained cohesion c_u, taking $\phi_u = 0$.

From (8.11): $K_a = 1$ hence $K_p = \dfrac{1}{K_a} = 1$

From (8.21): $$\boxed{\sigma_a = \sigma_v - 2c_u} \tag{8.26}$$

From (8.22): $$\boxed{\sigma_p = \sigma_v + 2c_u} \tag{8.27}$$

From (8.23) and (8.24): $$\boxed{z_0 = \frac{2c_u}{\gamma}} \tag{8.28}$$

From (8.25): $$\boxed{z_0 = \frac{1}{\gamma}(2c_u - q)} \tag{8.29}$$

8.7 Pressure–force and its line of action

The usual shape of a pressure diagram is either triangular or rectangular. The pressure-force *P* can easily found from these, bearing in mind that:

1. $\left| \begin{matrix} P = A\bar{\sigma}_H \\ A = H \end{matrix} \right|$ where P = force over 1 m length of wall
A = surface area over 1 m length
H = height of the wall over which the diagram is drawn
$\bar{\sigma}_H$ = average active or passive pressure of the diagram
2. Force P acts through the centroid (C.G) of a pressure-diagram at distance y from its base.

8.7.1 Triangular diagram for uniform soil

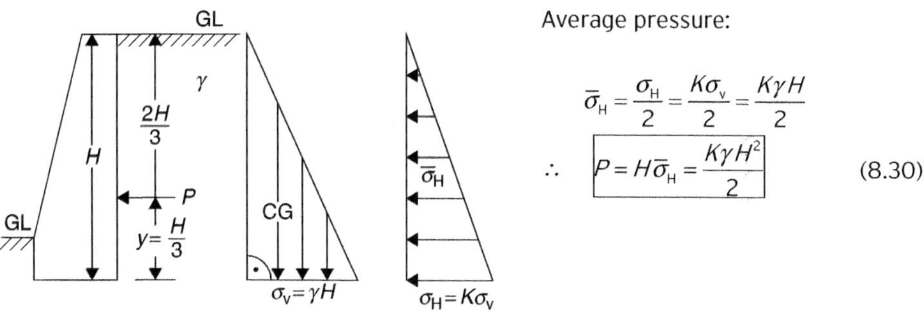

Average pressure:

$$\bar{\sigma}_H = \frac{\sigma_H}{2} = \frac{K\sigma_v}{2} = \frac{K\gamma H}{2}$$

$$\therefore \quad \boxed{P = H\bar{\sigma}_H = \frac{K\gamma H^2}{2}} \tag{8.30}$$

Figure 8.21

The centroid of a triangle is located at one-third of its sides, measured from the right angle.

$$\therefore \quad \boxed{y = \frac{H}{3}} \tag{8.31}$$

For the active case:

$$P_a = \frac{K_a \gamma H^2}{2}$$

For the passive case:

$$P_p = \frac{K_p \gamma H^2}{2} = \frac{\gamma^2}{2K_a}$$

Force P can be determined either from the diagrams or directly by the formula as shown in Example 8.3.

8.7.2 Triangular diagram for water

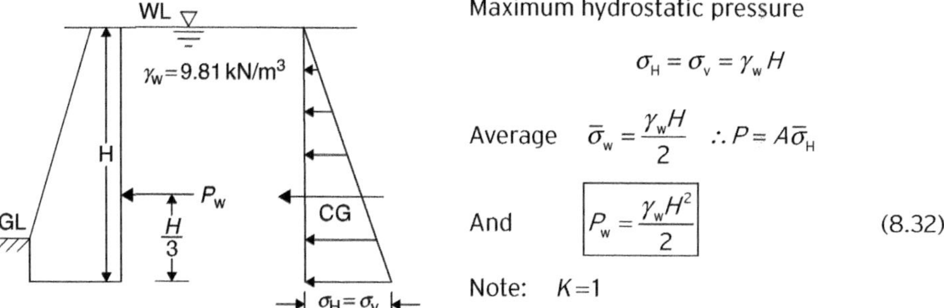

Maximum hydrostatic pressure

$$\sigma_H = \sigma_v = \gamma_w H$$

Average $\quad \bar{\sigma}_w = \frac{\gamma_w H}{2} \quad \therefore P = A\bar{\sigma}_H$

And $\quad \boxed{P_w = \frac{\gamma_w H^2}{2}} \tag{8.32}$

Note: $K = 1$

Figure 8.22

8.7.3 Rectangular diagram for surcharge only

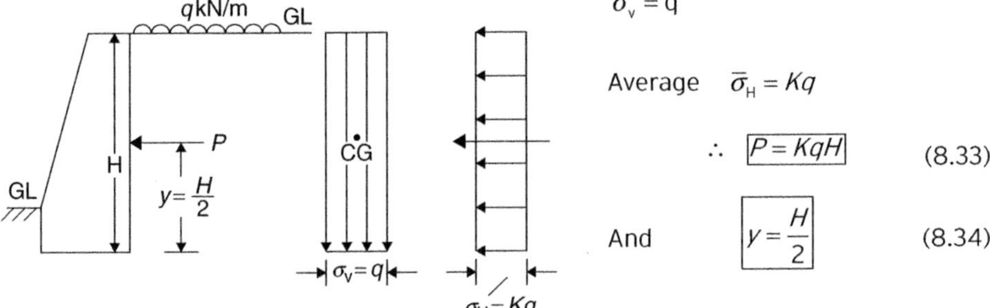

$\sigma_v = q$

Average $\bar{\sigma}_H = Kq$

$\therefore \boxed{P = KqH}$ (8.33)

And $\boxed{y = \frac{H}{2}}$ (8.34)

Figure 8.23

For the active case: $P_a = K_a qH$

For the passive case: $P_p = K_p qH = \frac{qH}{K_a}$

The diagrams are combined in most problems in various ways and the line of action of resultant force is found by taking moments about the base, as shown in the next example.

Example 8.3

For the smooth retaining wall, shown in Figure 8.24, find the active force and its line of action in two cases:

1. There is sufficient drainage layer behind the wall to ensure, that water level remains below the base.
2. The drainage layer and weepholes are blocked and the ground water level is at 4 m below the surface.

Case 1

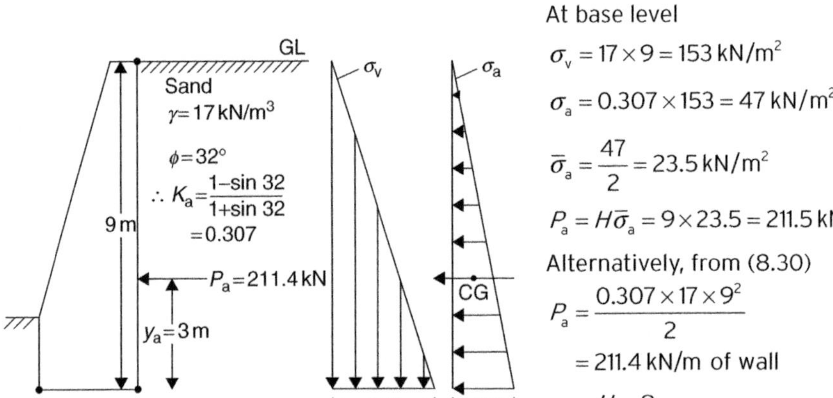

At base level

$\sigma_v = 17 \times 9 = 153\ \text{kN/m}^2$

$\sigma_a = 0.307 \times 153 = 47\ \text{kN/m}^2$

$\bar{\sigma}_a = \frac{47}{2} = 23.5\ \text{kN/m}^2$

$P_a = H\bar{\sigma}_a = 9 \times 23.5 = 211.5\ \text{kN}$

Alternatively, from (8.30)

$P_a = \frac{0.307 \times 17 \times 9^2}{2}$

$= 211.4\ \text{kN/m of wall}$

$y_a = \frac{H}{3} = \frac{9}{3} = 3\ \text{m}$

Figure 8.24

Case 2

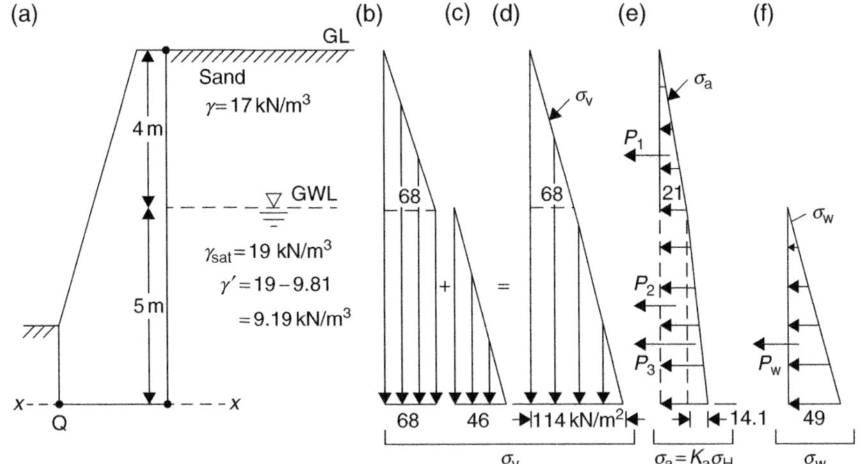

Figure 8.25

Diagram (b): At 4 m, $\sigma_v = 17 \times 4 = 68\,\text{kN/m}^2$

At 9 m, $\sigma_v = 68\,\text{kN/m}^2$ as the dry sand acts as a surcharge at 4 m.

Diagram (c): At 9 m, $\sigma_v = 5 \times 9.19 = 46\,\text{kN/m}^2$

Diagram (d): At 9 m, $\sigma_v = 68 + 46 = 114\,\text{kN/m}^2$

Diagram (e): At 4 m, $\sigma_a = 0.307 \times 68 = 21\,\text{kN/m}^2$ (rounded up)

$$\bar{\sigma}_a = \frac{21}{2} = 10.5\,\text{kN/m}^2 \quad \therefore P_1 = 4 \times 10.5 = 42\,\text{kN}$$

$$y_1 = 5 + \frac{4}{3} = 6.33\,\text{m} \quad \text{from base}$$

At 9 m, Rectangular and triangular diagrams

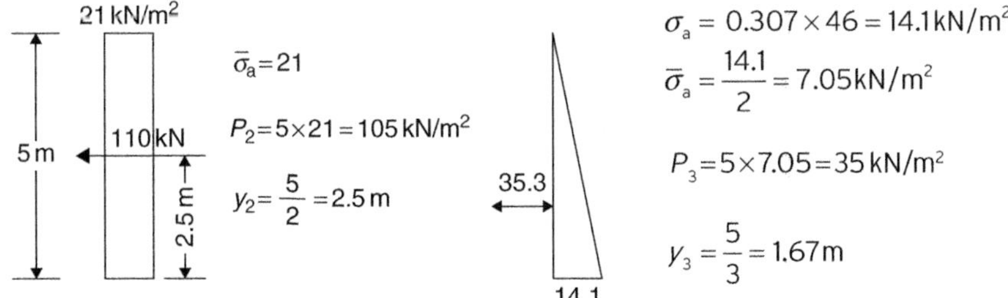

$$\sigma_a = 0.307 \times 46 = 14.1\,\text{kN/m}^2$$

$$\bar{\sigma}_a = \frac{14.1}{2} = 7.05\,\text{kN/m}^2$$

$$P_3 = 5 \times 7.05 = 35\,\text{kN/m}^2$$

$$y_3 = \frac{5}{3} = 1.67\,\text{m}$$

Figure 8.26

Diagram (f): At 9 m, Hydrostatic pressure: $\sigma_w = 5 \times 9.81 = 49\,\text{kN/m}^2$

Hydrostatic Force: $P_w = \frac{49}{2} \times 5 = 122.5\,\text{kN}$

$$y_w = \frac{5}{3} = 1.67\,\text{m}$$

The resultant of these four forces is their sum:

$$P_a = \overleftarrow{42} + \overleftarrow{105} + \overleftarrow{35.3} + \overleftarrow{122.5} = \overleftarrow{305}\ \text{kN}$$

Sum of moments of the four forces about toe Q:

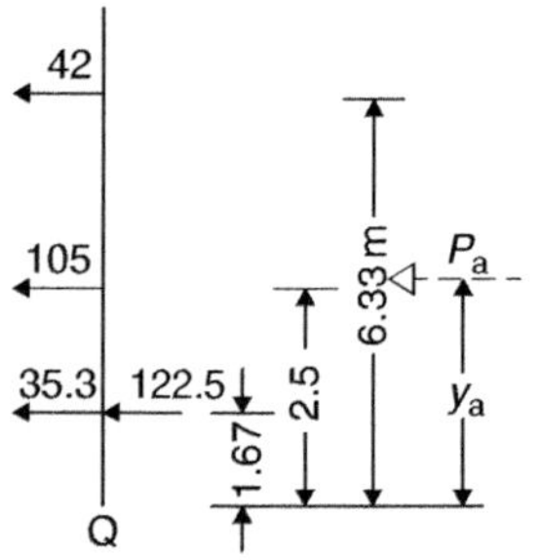

$$\overset{+}{\curvearrowleft}\ M_Q = \Sigma M_Q = 42\times 6.33 + 105\times 2.5 + (35.3+122.5)\times 1.67$$

$$= 265.9 + 262.5 + 263.5 = 792\ \text{kNm}$$

This equals to the moment of the resultant about the toe Q:

$$M_Q = P_a y_a = 305 y_a$$

Figure 8.27

Equating to get the line of action of the resultant:

$$305 y_a = 792 \qquad \therefore \qquad y_a = \frac{792}{305} = 2.6\ \text{m}$$

Comparing cases:
Case 1: $P_a = 211.4\ \text{kN}$ $y_a = 3\ \text{m}$ $M_Q = 211.4\times 3 = 634\ \text{kNm}$
Case 2: $P_a = 305\ \text{kN}$ $y_a = 2.6\ \text{m}$ $M_Q = 793\ \text{kNm}$

These moments try to overturn the wall about its tow (Q).

As the overturning moment is larger in Case 2, it is prudent to design the wall on the assumption, that the drainage is blocked.

Example 8.4

If a surcharge of $q = 100\ \text{kN/m}^2$ is placed on the surface behind the wall in Example 8.3, Case 1, then determine the value of the active force and the line of action.

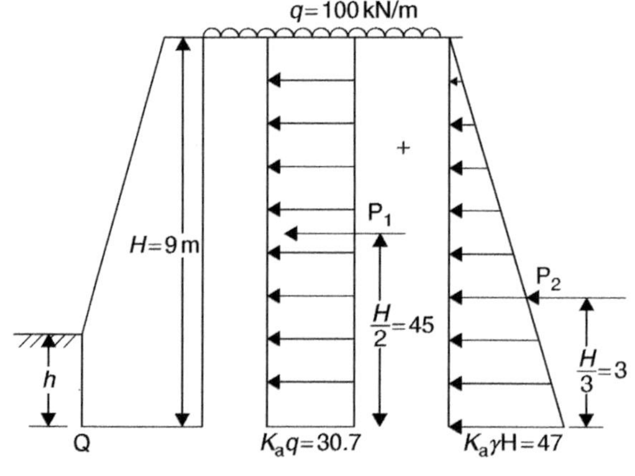

Figure 8.28

From Example 8.3:

$\gamma = 17\ \text{kN/m}^3$

$\phi = 32°$

$K_a = 0.307$

Active pressures at base level:

1. Due to surcharge:
$\sigma_1 = K_a q = 0.307\times 100$
$= 30.7\ \text{kN/m}^2$
2. Due to overburden:
$\sigma_2 = K_a \gamma H = 0.307\times 17\times 9$
$= 47\ \text{kN/m}^2$

Active force due to surcharge: $P_1 = \sigma_1 H = 30.7 \times 9 = 276.3\,\text{kN}$

Acting at: $y_1 = \frac{H}{2} = 4.5\,\text{m}$

Active force due to overburden: $P_2 = \frac{\sigma_2}{2} H = \frac{47}{2} \times 9 = 211.5\,\text{kN}$

Acting at: $y_2 = \frac{H}{3} = 3\,\text{m}$

Total active force: $\overleftarrow{P_a} = \overleftarrow{P_1} + \overleftarrow{P_2} = \overleftarrow{276.3} + \overleftarrow{211.5} = \overleftarrow{487.8}\,\text{kN}$

Taking moments about Q to determine its line of action.

$$\overset{\curvearrowleft}{\Sigma M_Q} = \overset{\curvearrowleft}{P_1 y_1} + \overset{\curvearrowleft}{P_2 y_2} = \overset{\curvearrowleft}{P_a y_a} \qquad y_a = \frac{1}{P_a}(P_1 y_1 + P_2 y_2)$$

$$y_a = \frac{1}{487.8}(276.3 \times 4.5 + 211.5 \times 3) = \frac{1877.9}{487.8} = 3.85\,\text{m}$$

Alternatively:

The values of P_a and y_a can be calculated directly from the general formulae derived from:

$$\sigma_1 = K_a q \qquad \bar{\sigma}_1 = K_a q \quad P_1 = K_a q H \quad y_1 = \frac{H}{2}$$

$$\sigma_2 = K_a \gamma H \qquad \bar{\sigma}_2 = \frac{K_a \gamma H}{2} \quad P_2 = \frac{K_a \gamma H^2}{2} \quad y_2 = \frac{H}{3}$$

$$P_a = P_2 + P_1 = \frac{K_a \gamma H^2}{2} + K_a q H$$

From which,
$$\boxed{P_a = \frac{K_a H}{2}(\gamma H + 2q)} \tag{8.35}$$

$$= \frac{0.307 \times 9}{2} \times (17 \times 9 + 2 \times 100) = 487.7$$
$$\approx 488\,\text{kN}$$

Also,
$$y_a = \frac{1}{P_a}(P_1 y_1 + P_2 y_2)$$
$$= \frac{1}{P_a}\left(K_a q H \times \frac{H}{2} + \frac{K_a \gamma H^2}{2} \times \frac{H}{3}\right)$$
$$= \frac{K_a H^2}{2P_a}\left(q + \frac{\gamma H}{3}\right) = \frac{K_a H^2}{6P_a}(3q + \gamma H)$$
$$= \frac{K_a H^2 (3q + \gamma H)}{6 \times \frac{K_a H}{2}(2q + \gamma H)} \quad \therefore \quad \boxed{y_a = \frac{H}{3}\left(\frac{3q + \gamma H}{2q + \gamma H}\right)} \tag{8.36}$$

$$\therefore y_a = \frac{9}{3} \times \left(\frac{3 \times 100 + 17 \times 9}{2 \times 100 + 17 \times 9}\right) = 3 \times \left(\frac{453}{353}\right) = 3.85\text{m}$$

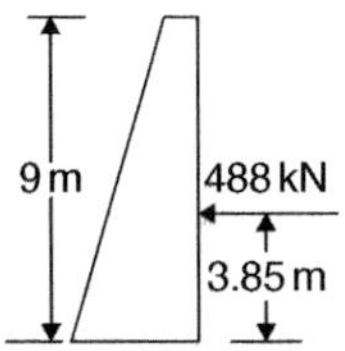

Figure 8.29

Similarly for passive force: $$P_p = \frac{K_p h}{2}(\gamma h + 2q) \quad (8.37)$$

$$y_p = \frac{h}{3}\left(\frac{3q + \gamma h}{2q + \gamma h}\right) \quad (8.38)$$

8.8 Wall supporting sloping surface

The extension of Rankine's theory to sloping surfaces applies only to cohesionless soils. The orientation of the principal stresses are as shown in Figure 8.30.

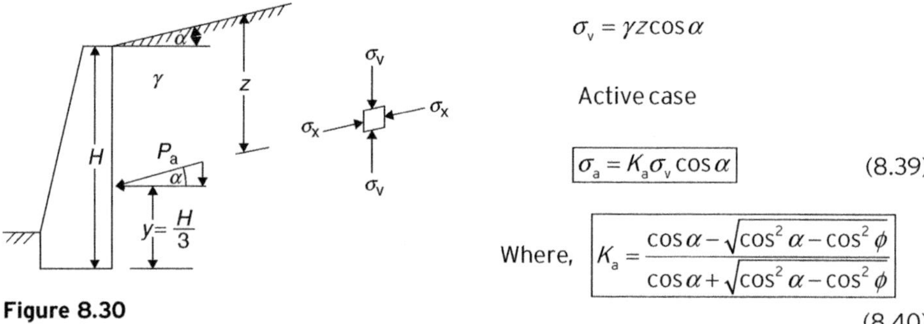

$$\sigma_v = \gamma z \cos\alpha$$

Active case

$$\sigma_a = K_a \sigma_v \cos\alpha \quad (8.39)$$

Where, $$K_a = \frac{\cos\alpha - \sqrt{\cos^2\alpha - \cos^2\phi}}{\cos\alpha + \sqrt{\cos^2\alpha - \cos^2\phi}} \quad (8.40)$$

Figure 8.30

Notes:

1. If $\alpha = 0$, the surface is horizontal and $K_a = \frac{1 - \sqrt{1 - \cos^2\phi}}{1 + \sqrt{1 - \cos^2\phi}} = \frac{1 - \sin\phi}{1 + \sin\phi}$ which is (8.11)
2. The surface at the front of the wall is normally horizontal, hence passive resistance is calculated by (8.20). If there is a slope however, then apply:

$$K_p = \frac{1}{K_a} = \frac{\cos\alpha + \sqrt{\cos^2\alpha - \cos^2\phi}}{\cos\alpha - \sqrt{\cos^2\alpha - \cos^2\phi}} \quad \text{and} \quad \sigma_p = K_p \sigma_v \cos\alpha$$

3. If then $\alpha = \phi$ then $K_a = K_p = 1$ thus the slope cannot be steeper than the soils angle of friction.
4. Active force: $P_a = H\bar{\sigma}_a$

$$\bar{\sigma}_a = \frac{K_a \sigma_v \cos\alpha}{2}$$

$$\sigma_v = \gamma H$$

$$P_a = \frac{K_a \gamma H^2 \cos\alpha}{2} \quad (8.41)$$

$$y = \frac{H}{3}$$

8.9 General formulae for c–ϕ soil

The derivation of the expressions for active and passive forces P_a and P_p is in terms of the Rankine-Bell Theory.

8.9.1 Active case

The active force (P_a) is determined by ignoring the length of possible tension crack given by:

$$z_0 = \frac{2c}{\gamma\sqrt{K_a}} \tag{8.23}$$

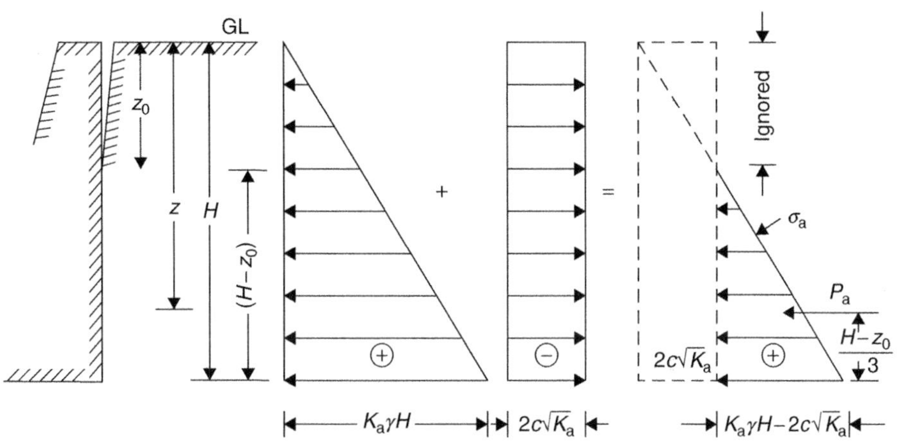

Figure 8.31

Active pressure: $$\sigma_a = K_a\gamma H - 2c\sqrt{K_a}$$

Average value: $$\bar{\sigma}_a = \frac{K_a\gamma H}{2} c\sqrt{K_a}$$

Active force:

$$\begin{aligned} P_a = \bar{\sigma}_a(H-z_0) &= \bar{\sigma}_a\left(H - \frac{2c}{\gamma\sqrt{K_a}}\right) \\ &= \left(\frac{K_a\gamma H}{2} - c\sqrt{K_a}\right)\left(H - \frac{2c}{\gamma\sqrt{K_a}}\right) \\ &= \frac{K_a\gamma H^2}{2} - cH\sqrt{K_a} - \frac{K_a cH}{\sqrt{K_a}} + \frac{2c^2}{\gamma} \\ &= \frac{K_a\gamma H^2}{2} - cH\sqrt{K_a} - cH\sqrt{K_a} + \frac{2c^2}{\gamma} \end{aligned}$$

Therefore,

$$\boxed{P_a = \frac{K_a\gamma H^2}{2} - 2cH\sqrt{K_a} + \frac{2c^2}{\gamma}} \tag{8.42}$$

The active force is located at $y_a = \dfrac{H-z_0}{3}$

Notes:

a) Water pressure in the cracks should also be taken into account, as it exerts an overturning effect on a retaining wall.

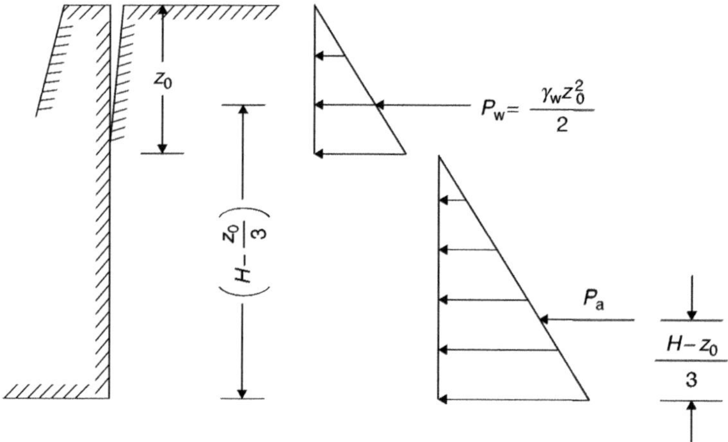

Figure 8.32

b) Surcharge q alters the depth of crack given by: $z_0 = \frac{1}{\gamma}\left(\frac{2c}{\sqrt{K_a}} - q\right)$ (8.25)

The active force also includes q.

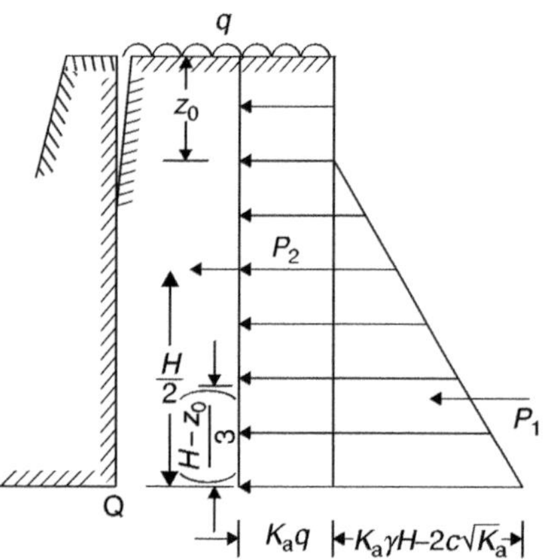

Figure 8.33

$$P_1 = \frac{K_a \gamma H^2}{2} - 2cH\sqrt{K_a} + \frac{2c^2}{\gamma}$$

$$P_2 = K_a q H$$

$$\therefore\ P_a = \frac{K_a \gamma H^2}{2} + K_a q H - 2cH\sqrt{K_a} + \frac{2c^2}{\gamma}$$

$$\boxed{P_a = K_a H\left(\frac{\gamma H}{2} + q\right) - 2cH\sqrt{K_a} + \frac{2c^2}{\gamma}} \qquad (8.43)$$

Its point of application is obtained from:

$$P_1\left(\frac{H - z_0}{3}\right) + P_2\frac{H}{2} = P_a y_a$$

or $$\boxed{y_a = \frac{1}{P_a}\left[P_1\left(\frac{H - z_0}{3}\right) + \frac{P_2 H}{2}\right]} \qquad (8.44)$$

8.9.2 Passive case (with surcharge)

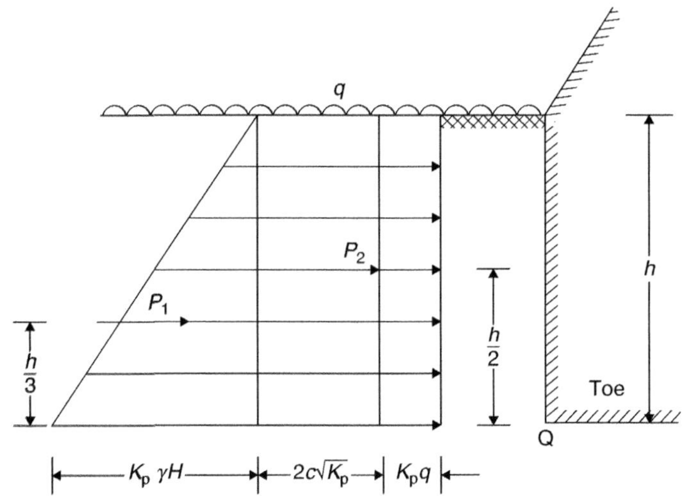

Passive pressure:
$\sigma_p = K_p\gamma h + 2c\sqrt{K_p}$

Surcharge pressure:
$\sigma_s = K_p q$

Figure 8.34

Average passive pressure: $\bar{\sigma}_p = \dfrac{K_p\gamma h}{2} + 2c\sqrt{K_p} + K_p q$

Passive force: $P_1 = \dfrac{K_p\gamma h^2}{2}$ acting at $y_1 = \dfrac{h}{3}$

Passive force: $P_2 = 2ch\sqrt{K_p} + K_p qh$

Total passive force: $P_p = P_1 + P_2 = \dfrac{K_p\gamma h^2}{2} + 2ch\sqrt{K_p} + K_p qh$

$$\therefore \boxed{P_p = K_p h\left(\frac{\gamma h}{2} + q\right) + 2ch\sqrt{K_p}} \qquad (8.45)$$

For line of action take moments about the toe:

$$\overset{+}{\curvearrowright}\sum M_Q = P_1 y_1 + P_2 y_2 = P_p y_p$$

$$P_p y_p = \frac{K_p\gamma h^2}{2}\times\frac{h}{3} + \left(2ch\sqrt{K_p} + K_p qh\right)\frac{h}{2}$$

$$= \frac{K_p\gamma h^3}{6} + ch^2\sqrt{K_p} + \frac{K_p qh^2}{2}$$

$$= h^2\left[K_p\left(\frac{\gamma h}{6} + \frac{q}{2}\right) + c\sqrt{K_p}\right]$$

$$\therefore \boxed{y_p = \frac{h^2}{P_p}\left[K_p\left(\frac{\gamma h}{6} + \frac{q}{2}\right) + c\sqrt{K_p}\right]} \qquad (8.46)$$

Example 8.5

The retaining wall, shown in Figure 8.35, having smooth surface is built on clayey soil. The surface carries a surcharge of $q = 35\,\text{kN/m}^2$. Calculate:

1. P_a, assuming that no tension cracks develop.
2. P_a after the development of cracks.
3. Passive force P_p.

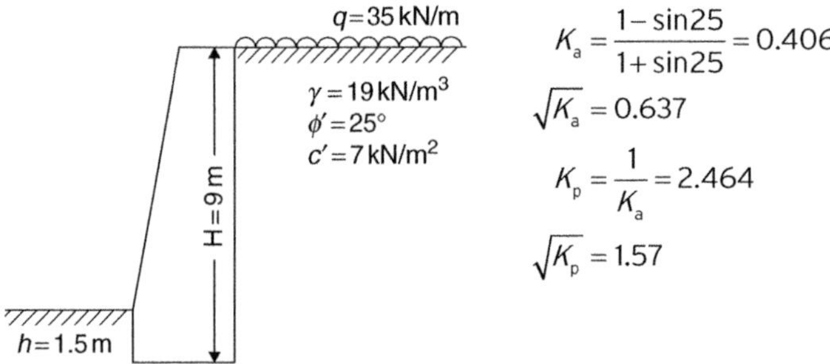

Figure 8.35

1. Assuming no tension cracks

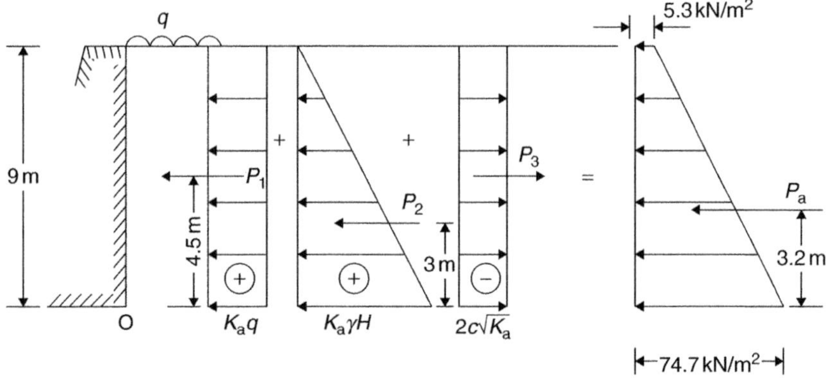

Figure 8.36

Pressures at base level

$\bar{\sigma}_1 = K_a q = 14.2\,\text{kN/m}^2$

$\bar{\sigma}_2 = K_a \gamma H = 69.4\,\text{kN/m}^2$

$\bar{\sigma}_3 = -2c\sqrt{K_a} = -8.9\,\text{kN/m}^2$

Average pressure due to q: $\bar{\sigma}_1 = K_a q = 0.406 \times 35 = 14.2\,\text{kN/m}^2$
Average force due to q: $P_1 = 14.2 \times 9 = 128\,\text{kN}$
Its line of action: $y_1 = 4.5\,\text{m}$

Average pressure due to γ: $\bar{\sigma}_2 = \frac{K_a \gamma H}{2} = \frac{0.406 \times 19 \times 19}{2} = 34.7\,\text{kN/m}^2$

Active force due to γ: $P_2 = 34.7 \times 9 = 312\,\text{kN}$

Its line of action: $y_2 = 3\,\text{m}$

Average pressure due to c: $\bar{\sigma}_3 = 2c\sqrt{K_a} = 2 \times 7 \times 0.637 = 8.9\,\text{kN/m}^2$

Active force due to c: $P_3 = 8.9 \times 9 = 80$

Its line of action: $y_3 = 4.5\,\text{m}$

Total active force: $\overleftarrow{P_a} = \overleftarrow{P_1} + \overleftarrow{P_2} - \overrightarrow{P_3} = 128 + 312 - 80 = 360\,\text{kN}$

Alternatively, using the resultant pressure diagram.

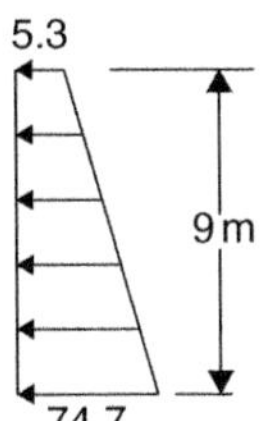

Pressure of top $= 5.3\,\text{kN/m}^2$

Pressure of base $= 74.7\,\text{kN/m}^2$

Average pressure $= \frac{5.3 + 74.7}{2} = 40\,\text{kN/m}^2$

Active force $P_a = 40 \times 9 = 360\,\text{kN}$

Figure 8.37

For its line of action: $\Sigma M_O = 4.5P_1 + 3P_2 - 4.5P_3 = P_a y_a$

Hence, $y_a = \frac{1}{360}(4.5 \times 128 + 3 \times 312 - 4.5 \times 80) = \frac{1152}{360} = 3.2\text{m}$

Alternatively, determine P_a from formula derived from Figure 8.36.

Average pressures: $\bar{\sigma}_1 = K_a q \quad \therefore \quad P_1 = K_a qH$

$$\bar{\sigma}_2 = \frac{K_a \gamma H}{2} \quad \therefore \quad P_2 = \frac{K_a \gamma H^2}{2}$$

$$\bar{\sigma}_3 = -2c\sqrt{K_a} \quad \therefore \quad P_3 = -2cH\sqrt{K_a}$$

Total active force:

$$\overleftarrow{P_a} = \overleftarrow{P_1} + \overleftarrow{P_2} + \overleftarrow{P_3} = K_a qH + \frac{K_a \gamma H^2}{2} - 2cH\sqrt{K_a}$$

$$\therefore \quad \boxed{P_a = K_a H\left(\frac{\gamma H}{2} + q\right) - 2cH\sqrt{K_a}} \qquad (8.47)$$

Hence, $P_a = 0.406 \times 9 \times \left(\frac{19.9}{2} + 35\right) - 2 \times 7 \times 9 \times 0.637 = 360\,\text{kN}$ (as before)

2. Assuming tension cracks

From (8.25): $z_0 = \frac{1}{\gamma}\left(\frac{2c}{\sqrt{K_a}} - q\right) = \frac{1}{19} \times \left(\frac{2 \times 15}{0.637} - 35\right) = 0.64\text{m}$

Water pressure in the crack: $\sigma_w = \gamma z_0 = 19 \times 0.64 = 12\,\text{kN/m}^2$

Hydrostatic force in the crack: $P_w = \frac{12}{2} \times 0.64 = 3.8\,\text{kN}$

From (8.43): $P_a = K_a H\left(\frac{\gamma H}{2} + q\right) - 2cH\sqrt{K_a} + \frac{2c^2}{\gamma}$

$$= 0.406 \times 9 \times \left(\frac{19.9}{2} + 35\right) - 2 \times 7 \times 9 \times 0.637 + \frac{2 \times 7^2}{19}$$

$$= 440 - 80 + 5 = 365\,\text{kN}$$

From (8.44): $y_a = \frac{1}{P_a}\left[P_1\left(\frac{H - z_0}{3}\right) + \frac{P_2 H}{3}\right]$

$$P_1 = \frac{K_a \gamma H^2}{2} - 2cH\sqrt{K_a} + \frac{2c^2}{\gamma}$$

$$= \frac{0.406 \times 19 \times 9^2}{2} - 80 + 5 = 237.4\text{kN}$$

$$P_2 = K_a qH = 0.406 \times 35 \times 9 = 127.9\text{kN}$$

$$\therefore P_a = P_1 + P_2 = 365.3\,\text{kN} \approx 365\,\text{kN}$$

and $y_a = \frac{1}{365}\left[\frac{237.4 \times (9 - 0.64)}{3} + \frac{127.9 \times 9}{2}\right]$

$$\therefore y_a = \frac{1237}{365} = 3.39\,\text{m}$$

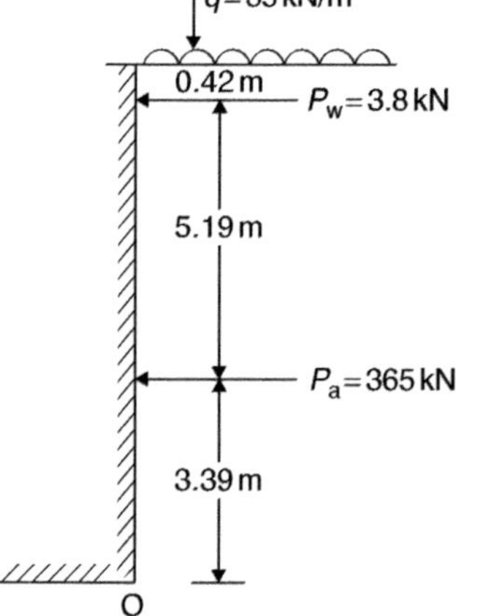

Figure 8.38

Notes:

a) The term: $\frac{2c^2}{\gamma}$ increases P_a due to the neglected tensile stresses over length z_0.

b) z_0 decreases as q increases.

3. Passive resistance

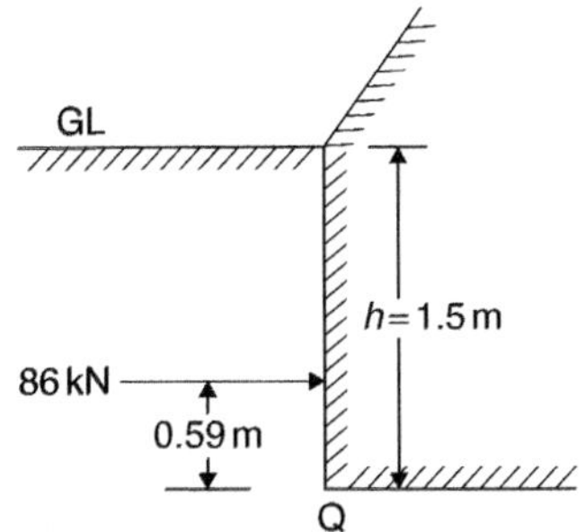

Figure 8.39

Disregarding surcharge q in formula (8.45):

$$P_p = \frac{K_p \gamma h^2}{2} + 2ch\sqrt{K_p}$$

$$= \frac{2.464 \times 19 \times 1.5^2}{2} + 2 \times 7 \times 1.5 \times 1.57$$

$$= 52.7 + 33 = 86\,\text{kN}$$

Its line of action from (8.46):

$$Y_p = \frac{h^2}{P_p}\left(\frac{K_p \gamma h}{6} + c\sqrt{K_p}\right)$$

$$= \frac{1.5^2}{86}\left(\frac{2.464 \times 19 \times 15}{6} + 7 \times 1.57\right)$$

$$= \frac{2.25 \times 22.7}{86} = 0.59\,\text{m}$$

8.10 Formulae for pure clay ($\phi = 0$)

As the short-term stability is the most critical for saturated soils, apply undrained cohesion c_u.

$$\phi = 0 \quad \therefore \quad K_a = K_p = 1$$

Wall without surcharge

From (8.28): Crack: $z_0 = \dfrac{2c_u}{\gamma}$

From (8.42): Active: $P_a = \dfrac{\gamma H^2}{2} - 2c_u H + \dfrac{2c_u^2}{\gamma}$

From (8.45): Passive: $P_p = \dfrac{\gamma h^2}{2} + 2c_u h$

From (8.46): $Y_p = \dfrac{h^2}{P_p}\left(\dfrac{\gamma h}{6} + c_u\right)$

Wall with surcharge

From (8.29): Crack: $z_0 = \dfrac{1}{\gamma}(2c_u - q)$

From (8.43): Active: $P_a = H + \left(\dfrac{\gamma H}{2} + q - 2c_u\right) + \dfrac{2c_u^2}{\gamma}$

From (8.45): Passive: $P_p = h\left(\dfrac{\gamma h}{2} + q + 2c_u\right)$

8.11 Height of unsupported clay

Vertical cuts may be excavated in stiff clays, without supporting them by walls, sheet piles or bracing, in the short term. The height (H_0) that may be left exposed is obtained from formula (8.47) ignoring surcharge q.

$$P_a = \frac{K_a \gamma H^2}{2} - 2cH\sqrt{K_a} = 0$$

For the soil to stand, P_a has to be zero, hence,

$$\frac{K_a \gamma H_0^2}{2} - 2cH_0\sqrt{K_a}$$

$$K_a \gamma H_0 = 4c\sqrt{K_a}$$

Expressing,

$$H_0 = \frac{4c\sqrt{K_a}}{\gamma K_a}$$

But $\frac{\sqrt{K_a}}{K_a} = \sqrt{\frac{K_a}{K_a^2}} = \frac{1}{\sqrt{K_a}}$

$$\therefore \boxed{H_0 = \frac{4c}{\gamma\sqrt{K_a}}} \quad (8.48)$$

$$\boxed{H_0 = 2z_0} \quad (8.49)$$

This critical height reduces with time due to change in the pore pressure, hence in the effective stress. The clay becomes weaker as its cohesion decreases due to softening.

For pure clay $\phi = 0 \quad \therefore K_a = 1$

$$\therefore \boxed{H_0 = \frac{4c}{\gamma}} \quad (8.50)$$

8.12 Wedge theories

The analytical theory, proposed by Coulomb in 1776, was applicable to walls:

a) Supporting cohesionless or cohesive soil.
b) Constructed with either vertical or inclined surfaces.
c) Supporting soil, having either horizontal or uniformly sloping ground surface.
d) Having either smooth or rough surfaces.

It is assumed in the theory, that the soil fails along a plane surface and the wedge thus formed is pressing against the wall.

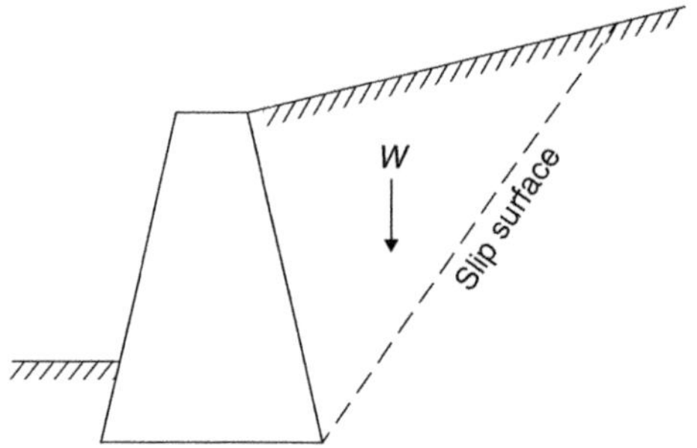

W = weight of the wedge.

Figure 8.40

The problem was to find that slip surface, or wedge, which exerted the maximum force on the wall. Each wedge had to be analysed separately, until the critical one was found. As the calculations were somewhat tedious, Culmann devised a simpler, graphical method in 1875, which was applicable also to irregular ground surfaces as well as all types of surcharge loading. The Culmann-procedure is described in supplementary Problem 8.9.

Because of its simplicity and versatility, the graphical method will be discussed fairly extensively in this section mainly for the active case.

The passive case can be treated in the same way (see Supplementary Problem 8.7).

8.12.1 Procedure for cohesionless soil

It often occurs that cohesionless soil is used as backfill to a wall after its construction. It is assumed, that there is friction between the soil and the wall surface. The angle of wall friction (δ) is either determined by laboratory experiment or its value arbitrarily assumed to be within the range $0.5\ \phi < \delta < \frac{2}{3}\phi$.

Figure 8.41 shows the forces acting on the wall and on a trial wedge. Also, the force polygon is drawn for the determination of active force (P_a), acting on the wall, due to this particular wedge. It is not known, at this stage, whether P_a is maximum value, corresponding to the actual slip surface, or not.

(a)

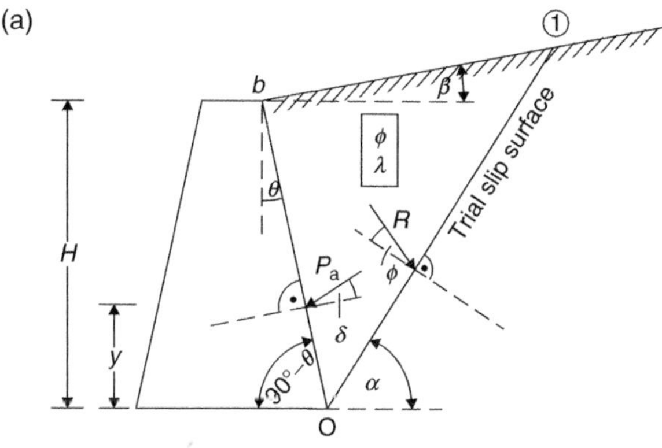

Forces acting on the wall and the slip surface due to the wedge.

(b)

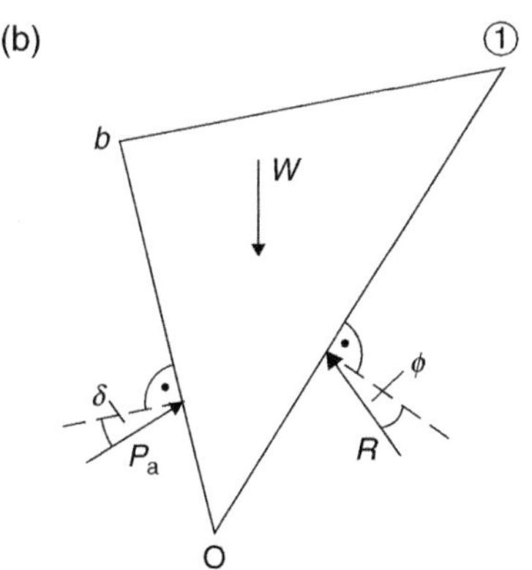

Equal and opposite forces acting on the wedge. The position of R is arbitrary on the diagram.

Figure 8.41

where W = weight of the wedge. This is the only force to be determined prior to the drawing of the polygon. Its direction is of course vertical

R = Reaction of soil on the wedge. It value is measured from the polygon.

ϕ = angle of friction for the soil. Force R is drawn at this angle to the normal on the slip surface.

δ = the arbitrarily chosen wall friction angle. Active force P_a is drawn at this angle to the normal on the walls surface.

The polygon of forces acting on the wedge can now be drawn to scale (See example 8.6).

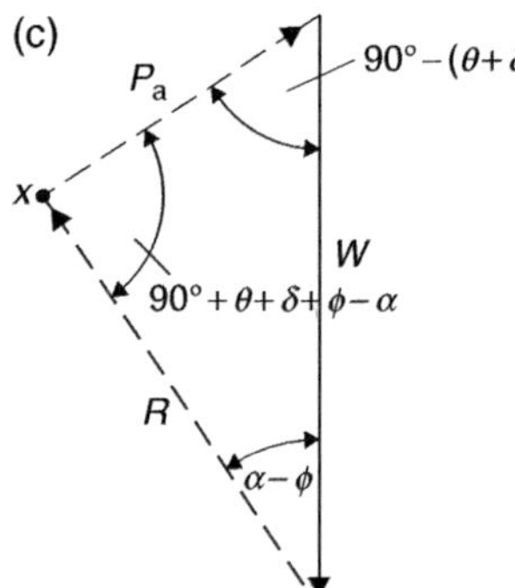

A polygon like this is drawn for several slip surfaces and point x on each is connected by a curve, called Culmann line. The maximum value of P_a can be located from the curve. Its reverse is then acting on the wall.

Figure 8.41 (continued)

Example 8.6

The retaining wall shown in Figure 8.42 supports cohesionless soil as well as a light foundation, transmitting uniform pressure of 50 kN/m². The soil characteristics are:

Unit weight	$\gamma = 19\,\text{kN/m}^3$
Angle of friction	$\phi = 32°$

The wall friction is assumed to be $\delta = 0.55\phi$
$= 17.6°$

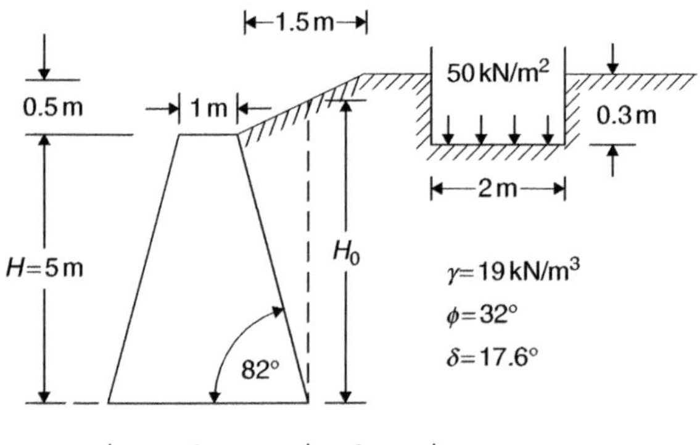

Determine the active force on the wall and the inclination of the corresponding failure surface.

Figure 8.42

Construction of Graph 8.4

Step 1: The configuration of Figure 8.42 is drawn to a scale of 1 cm = 0.5 m.

Step 2: Five trial slip surfaces are drawn at arbitrary angles. Calculate the weight of each wedge (Table 8.6).

Step 3: Draw the R-forces at $\phi=32°$ to the normal to each plane, anywhere on each surface. Only their directions are important.

Step 4: Draw the direction of P_a at $\delta=17.6°$ to the normal to the surface of the wall at height $y=\frac{H_0}{3}$ from its base. However, see the discussion as to the point of application (Section 8.12.3).

Step 5: Draw the polygon for each wedge on diagram (b), thus locating points x_1 to x_5.

Step 6: Draw the Culmann line through these points.

Step 7: Find the maximum value of P_a. In this case, it coincides with the force induced by the total weight of wedge 3.

Step 8: The angle of inclination of the slip surface is calculated as shown on diagram (b).

Table 8.6 Weight of 1 m wide wedges and, from Graph 8.4, the active forces acting on each.

Wedge	1	2	3	4	5
Volume: V (m^3)	4.5	8.23	10.61	12.63	16.6
Weight: W (kN)	19×4.5 = 85.5	19×8.23 = 156.4	19×10.61+50 = 251.5	19×12.63+100 = 340.0	19×16.6+100 = 415.4
P_a (kN)	67.5	92.5	110.0 ↗	105.0	85.0

Resulting active force acting on the wall is: $P_a=110$ kN ↙

The horizontal component F_a tries to overturn the wall. The vertical component W_a increases the bearing pressure of the base and counters the overturning moment.

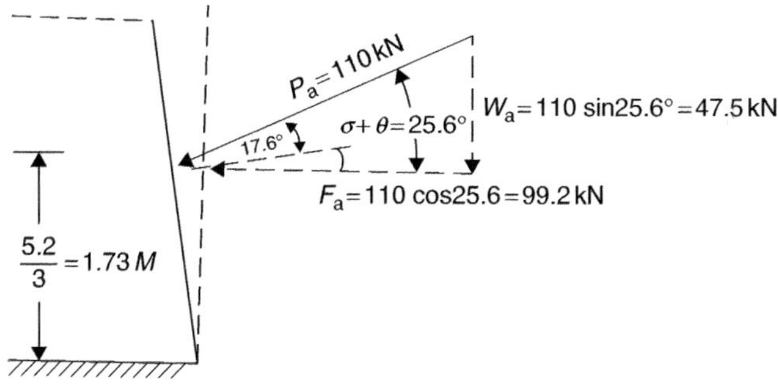

Figure 8.43

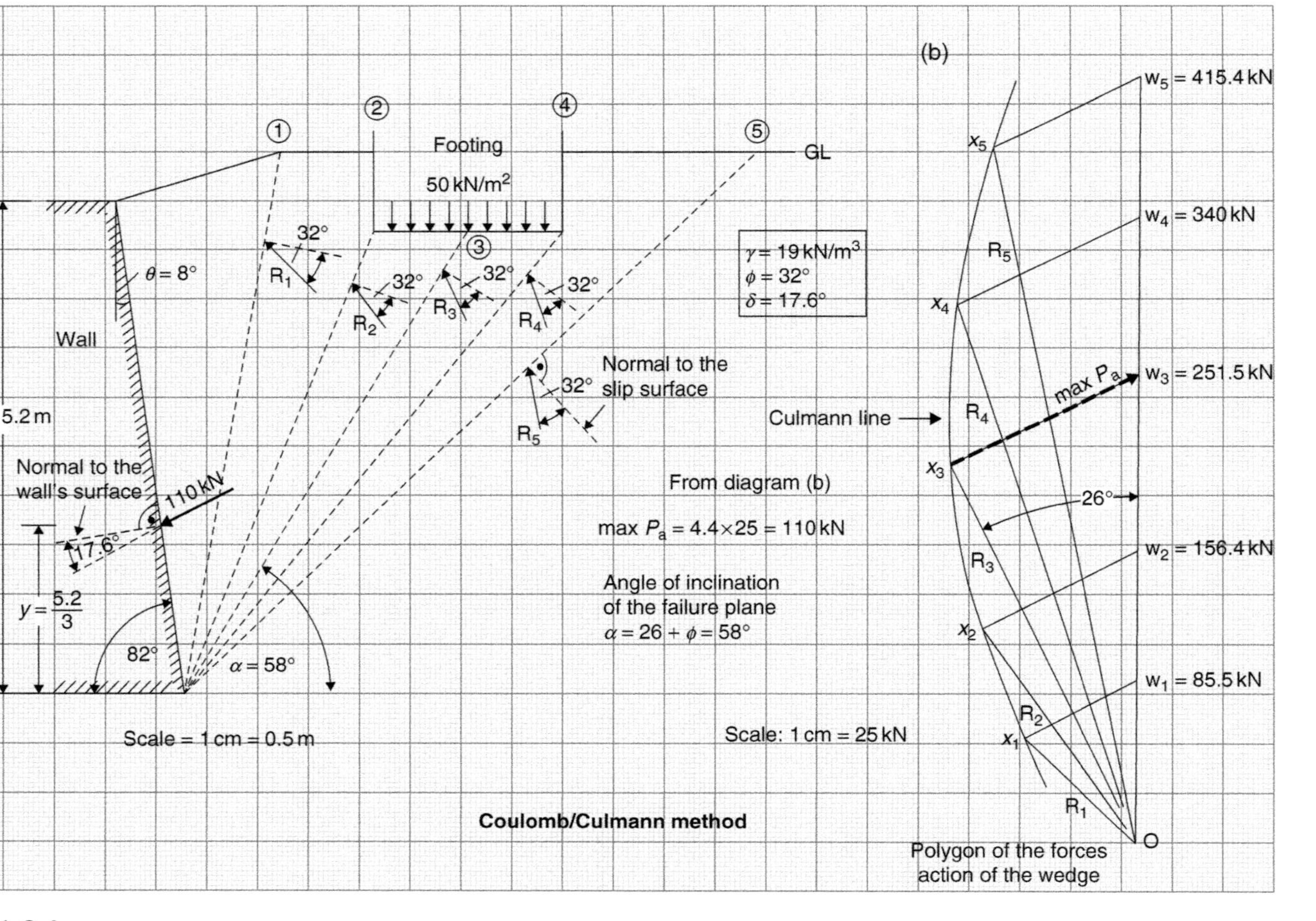

(a)
(b)
①
②
③
④
⑤
Footing
50 kN/m²
GL
32°
R1
R2
R3
R4
R5
θ = 8°
Wall
γ = 19 kN/m³
ϕ = 32°
δ = 17.6°
Normal to the
slip surface
H0 = 5.2 m
Normal to the
wall's surface
110 kN
17.6°
y = 5.2/3
82°
α = 58°
Scale = 1 cm = 0.5 m
From diagram (b)
max Pa = 4.4×25 = 110 kN
Angle of inclination
of the failure plane
α = 26 + ϕ = 58°
Scale: 1 cm = 25 kN
Coulomb/Culmann method
w5 = 415.4 kN
w4 = 340 kN
w3 = 251.5 kN
w2 = 156.4 kN
w1 = 85.5 kN
x5
x4
x3
x2
x1
max Pa
Culmann line
26°
O
Polygon of the forces
action of the wedge

Graph 8.4

8.12.2 Procedure for cohesive soil

If the backfill is cohesive (c–ϕ) soil, then in addition to the wall friction, the adhesion between the wall and the soil must be taken into account. Also, it is on the safe side to assume the presence of tension cracks, calculated from formula (8.23):

$$z_0 = \frac{2c}{\gamma\sqrt{K_a}}$$

And from (8.11):

$$K_a = \frac{1-\sin\phi}{1+\sin\phi}$$

Figure 8.44 depicts the forces acting on the wall and on a trial wedge, in this case.

(a)

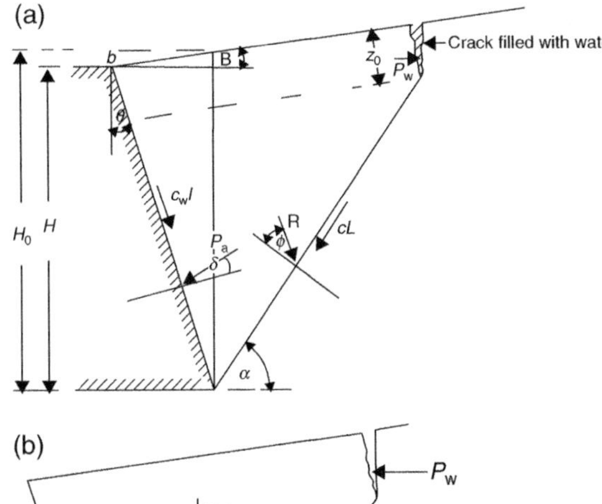

Forces acting on the wall and on the slip surface.
P_w = force of water in the tension crack

$$P_w = \frac{\gamma_w z_0^2}{2} \qquad (8.51)$$

(b)

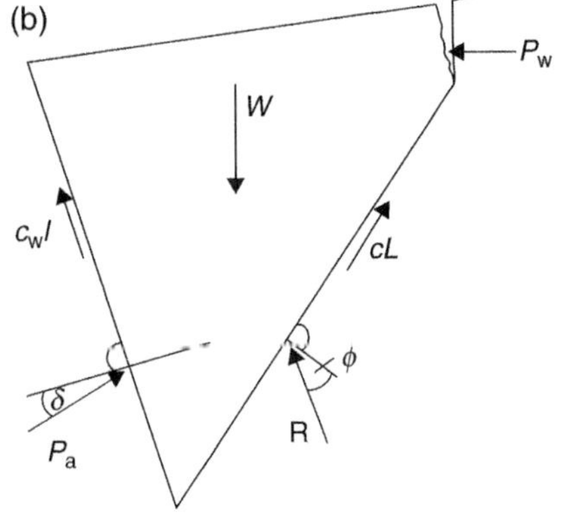

Equal and opposite forces acting on the wedge. These forces are drawn on the polygon

Figure 8.44

where: c = cohesion of the soil

c_w = adhesion between soil and wall, chosen arbitrarily. Recommended values are:

$$c_w = c \quad \text{if} \quad c \le 50\,\text{kN/m}^2$$
$$c_w = 50\,\text{kN/m}^2 \quad \text{if} \quad c > 50\,\text{kN/m}^2$$

L = Length of the slip surface

l = Length of the back face (o-b)

$c_w l$ = Adhesive force, resisting shear

cL = Cohesive force, resisting shear.

The polygon of forces, acting on the wedge can now be drawn. Its shape depends on the included forces.

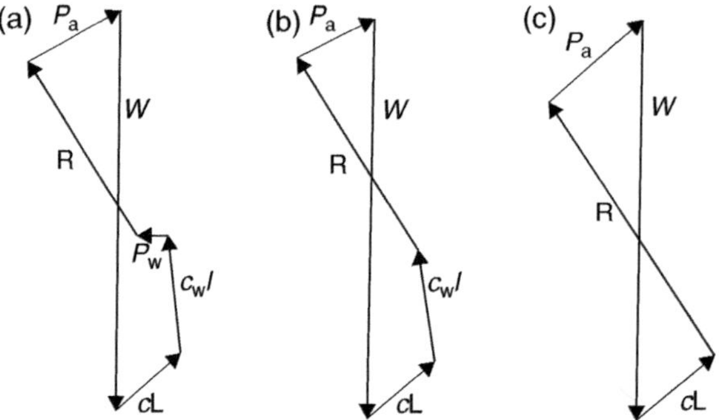

Figure 8.45

Figure 8.45(a): If all forces are included.
Figure 8.45(b): P_w is not included. The cracks are assumed to be dry.
Figure 8.45(c): P_w and $c_w l$ are not included. No adhesion between the wall and soil is assumed.

Example 8.7

The retaining wall of Example 6.6 supports cohesive soil, having strength parameters as shown.

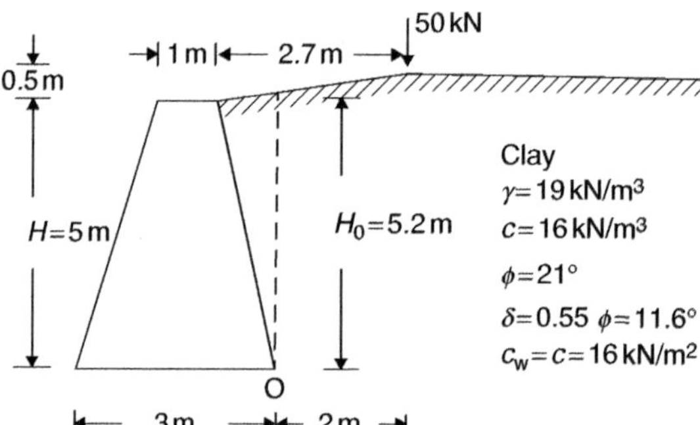

Figure 8.46

A point load is placed on the surface at 2.7m from the crest of the wall. Determine the magnitude of the active force acting on the wall, assuming the tension cracks full of water.

Coefficient of active pressure: $K_a = \dfrac{1-\sin 21}{1+\sin 21} = 0.47$

Depth of tension cracks: $z_0 = \dfrac{2\times16}{19\times\sqrt{0.47}} = 2.5\text{m}$

Hydrostatic force in cracks: $P_w = \dfrac{9.81\times2.5^2}{2} = 31\text{kN}$

Length l=5.2 m ∴ Adhesive force=$c_w l$=16×5.2=83 kN

Table 8.7 Weights of 1m wide wedges and, from Graph 8.5, the active force acting on each.

Wedge number	1	2	3	4
Length L (m)	3.5	4.25	5	5.8
cL=16 L (kN)	56	68	80	93
Volume V (m³)	9.42	14.22	18.22	22.17
Weight W (kN)	179+50=229	270+50=320	346+50=396	421+50=471
P_a (kN)	75 ↗	67	50	40

The resultant maximum active force acting on the wall is P_a=75 kN ↙

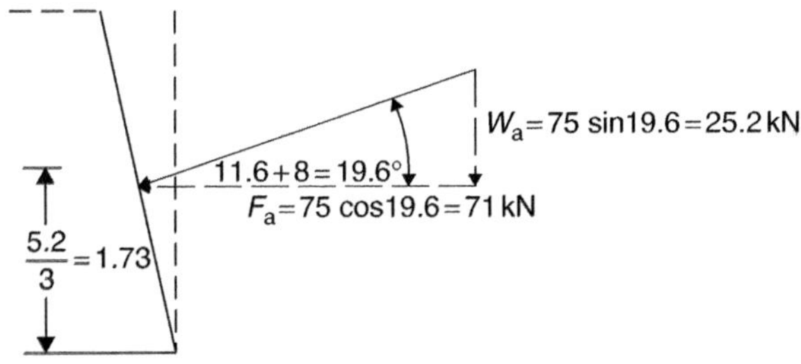

Figure 8.47

Notes:

1. The procedure applies equally to walls having vertical, smooth or rough surfaces.
2. Code of Practice CP2 (1951) recommended, that the depth of tension cracks should be calculated from:

$$z_0 = \frac{2c}{\gamma}\sqrt{\frac{1+\frac{c_w}{c}}{K_a}} \qquad (8.52)$$

As $c_w = c$ for range $0 \le c \le 50\,\text{kN/m}^2$, this formula becomes:

$$z_0 = \frac{2.82c}{\gamma\sqrt{K_a}} \qquad (8.53)$$

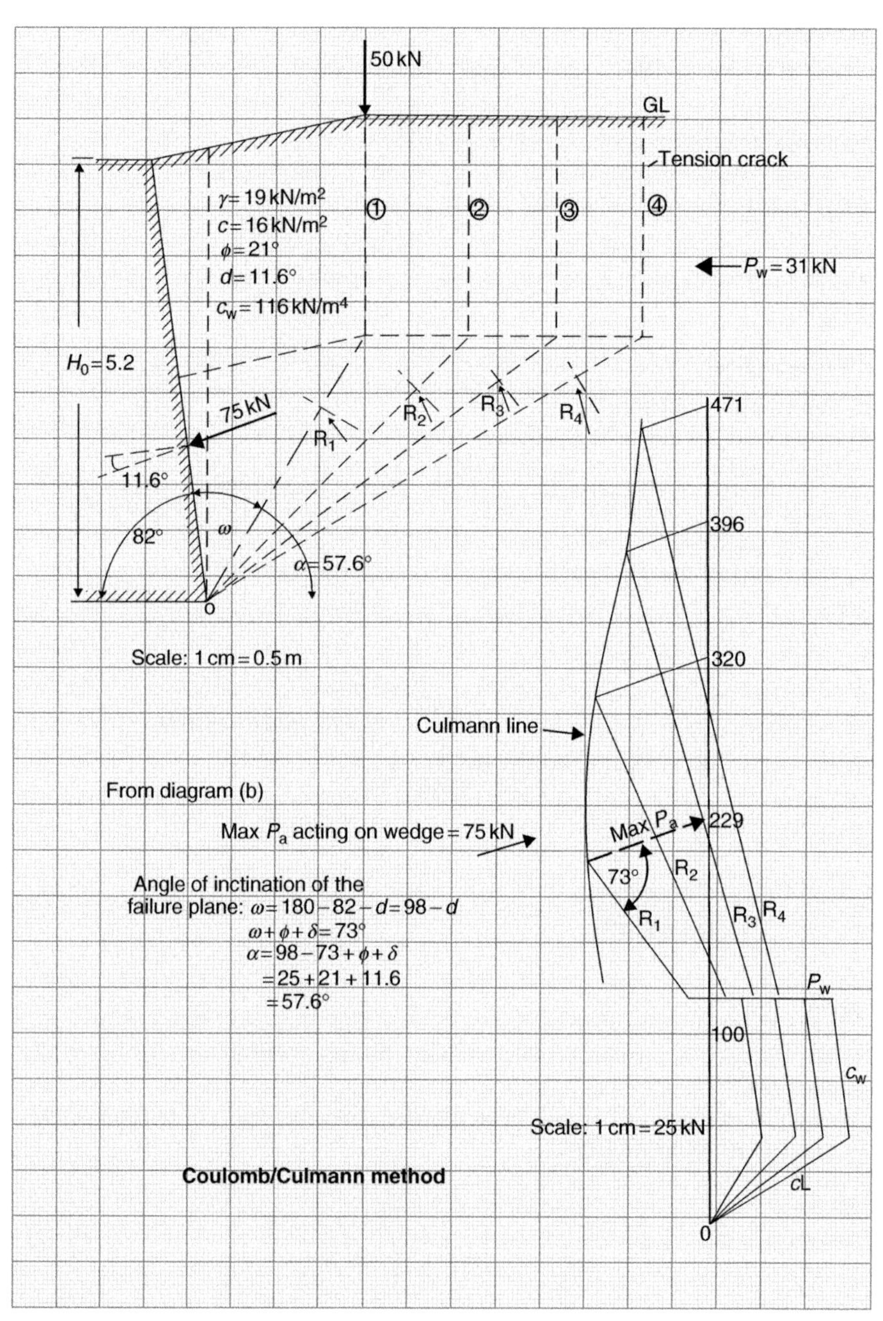
50 kN
GL
Tension crack
γ = 19 kN/m²
c = 16 kN/m²
φ = 21°
d = 11.6°
c_w = 116 kN/m⁴
①
②
③
④
P_w = 31 kN
H_0 = 5.2
75 kN
R_1
R_2
R_3
R_4
11.6°
82°
ω
α = 57.6°
o
471
396
320
229
100
0
Scale: 1 cm = 0.5 m
Culmann line
From diagram (b)
Max P_a acting on wedge = 75 kN
Max P_a
73°
Angle of inctination of the failure plane: ω = 180 − 82 − d = 98 − d
ω + φ + δ = 73°
α = 98 − 73 + φ + δ
= 25 + 21 + 11.6
= 57.6°
P_w
c_w
cL
Scale: 1 cm = 25 kN
Coulomb/Culmann method

Graph 8.5

Using the expression, the depth of cracks in Example 8.7 is found to be:

$$z_0 = \frac{2.82 \times 16}{19\sqrt{0.47}} = 3.46\,\text{m}$$

This means:

i. Shorter slip surfaces, hence smaller value of cohesive force cL.
ii. Smaller weight of each 'wedge'.
iii. Hydrostatic force in the crack.

The effect of (i) and (iii) is an increased overturning moment on the wall.

8.12.3 Point of application of P_a(x)

Whenever several forces act on a wall, their resultant many be obtained by taking moments as shown in Example 8.3.However, for the irregular ground surface behind the wall, the point of application (x) may be found by the following procedure:

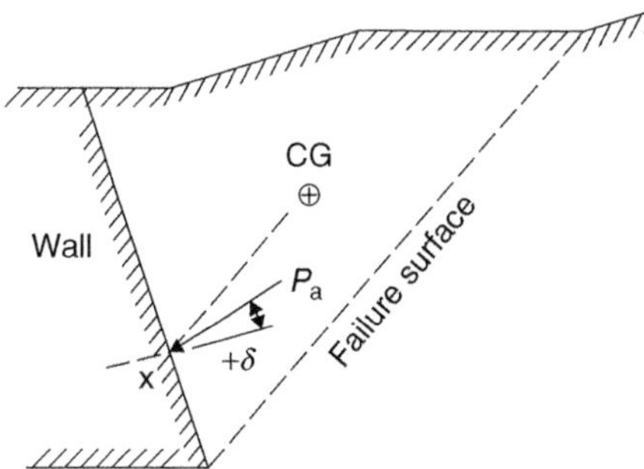

Figure 8.48

Step 1: Find the centroid (CG) of the critical wedge.

Step 2: Draw a line from CG parallel with the failure surface to intersect this wall at point x.

In most cases, however, the line of action is at $y = \frac{H}{3}$ as for hydrostatic pressure e.g.

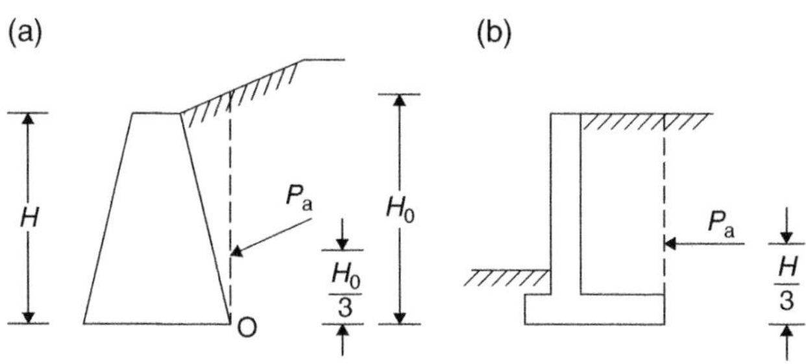

Figure 8.49

See also Section 8.7 on 'Pressure force and its line of action' in this chapter.

8.12.4 Effect of static water table

If there is no drainage, or the filter layer is blocked behind the wall, than the force polygon for a wedge has to take the hydrostatic force into account.

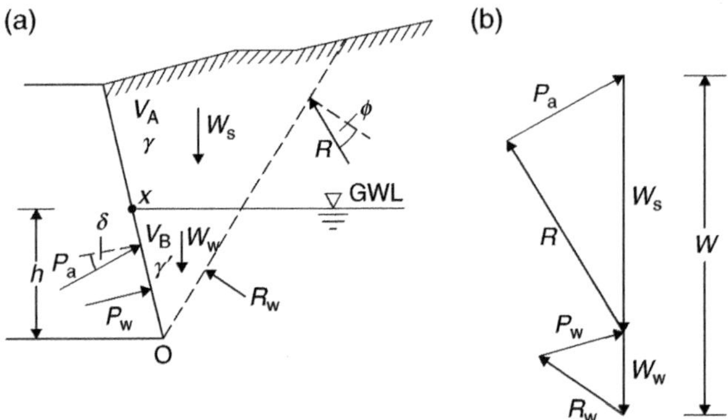

Figure 8.50

where

V_A and V_B are the volume of wedge and water respectively

Weight of water: $W_w = V_B \gamma_w$

Weight of soil: $W_s = V_A \gamma + V_B \gamma'$

Hydrostatic force: $P_w = \dfrac{\gamma_w h^2}{2}$

8.13 Stability of retaining walls

The purpose of a retaining wall is to support solid materials, earth or otherwise. There are four main types:

1. Gravity
2. Cantilever
3. Counterfort
4. Buttress.

8.13.1 Gravity walls

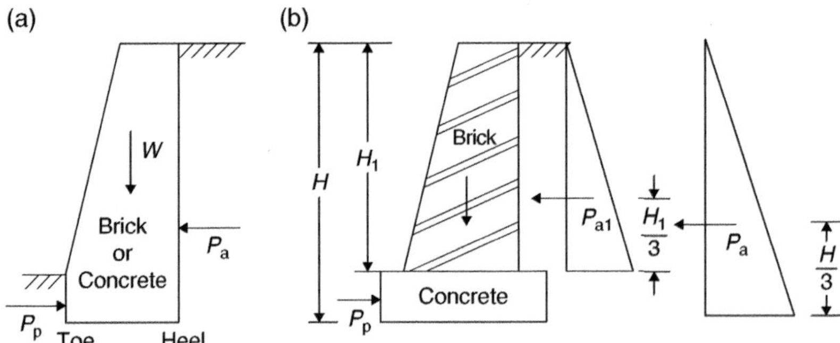

Figure 8.51

Notes:

1. Tension in brick wall must be avoided as it could open up the joints at the heel.
2. Unless the brick wall is firmly keyed into the concrete base, separate check should be made as to its stability.

8.13.2 Cantilever walls

These are constructed from reinforced concrete.

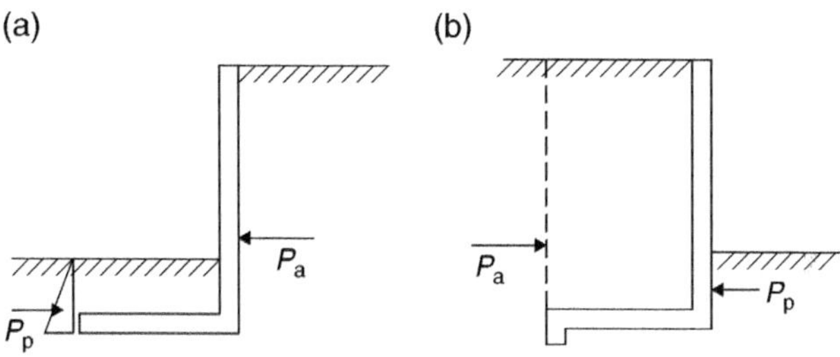

Figure 8.52

8.13.3 Buttress and counterfort walls

These are also constructed from reinforced conrete.

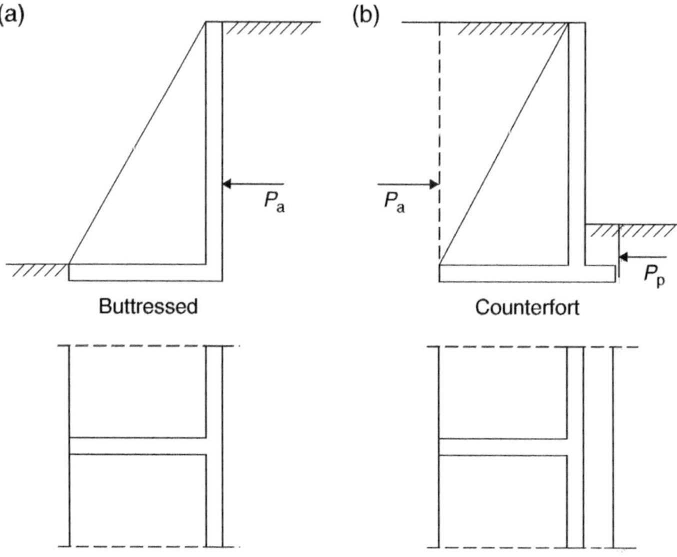

Figure 8.53

8.13.4 Stability check

After the determination of all forces acting on the wall, there are four checks to be made against its failure by:

1. Overturning
2. Sliding
3. Overstressing the foundation soil
4. Tension in brickwork.

8.13.4.1 Overturning

The moment of the active force P_a, about the toe, tries to rotate the wall anticlockwise, in Figure 8.54 and overturn it, hence this moment is called 'Overturning moment (M_0)'.

The moments of the wall's weight (W) and the passive force (P_p) try to rotate the wall clockwise about the toe, thus resisting the action of M_0. For this reason, it is called 'Resisting moment (M_R)'.

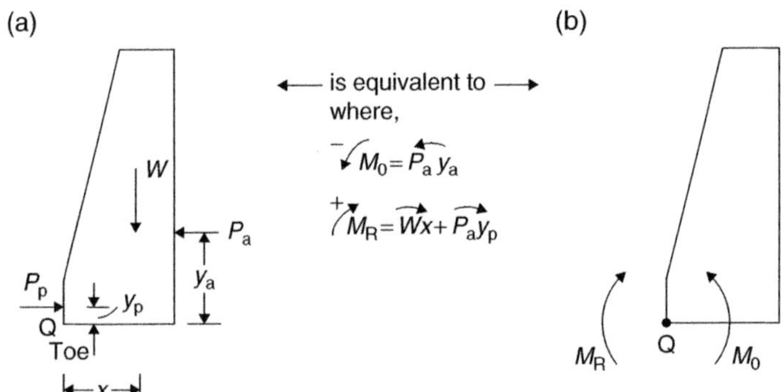

Figure 8.54

For equilibrium: $M_R = M_0$.

It is necessary to achieve not only equilibrium, but additional safety, hence the resisting moment is required to be at least twice as large as the overturning one i.e.

$$\boxed{M_R \geq 2M_0}$$

This means, that the factor of safety give by:

$$\boxed{F_S = \frac{M_R}{M_0} \geq 2} \tag{8.54}$$

8.13.4.2 Sliding

The horizontal component of the active force tries to push the wall forward. This is restricted by the passive force, if any, as well as the friction force (F) between the base of the wall and the soil. Factor of safety against sliding is also required to be at least 2.

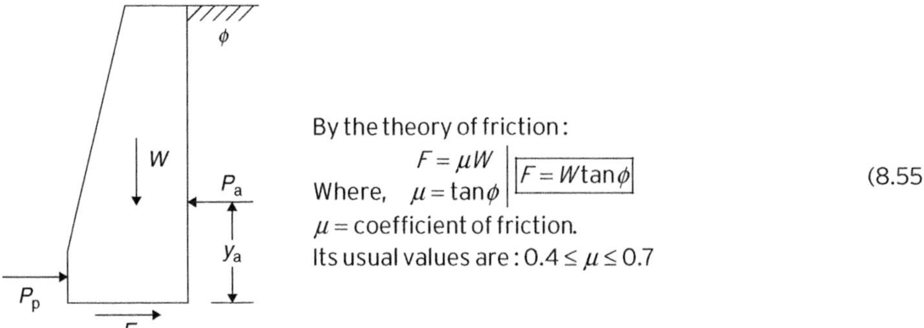

By the theory of friction:

$$F = \mu W \quad \text{Where,} \quad \mu = \tan\phi \qquad \boxed{F = W\tan\phi} \tag{8.55}$$

μ = coefficient of friction.

Its usual values are: $0.4 \leq \mu \leq 0.7$

Figure 8.55

The factor of safety is given by:
$$F_S = \frac{F + P_p}{P_a}$$

Therefore,
$$\boxed{F_S = \frac{W\tan\phi + P_p}{P_a} \geq 2} \tag{8.56}$$

8.13.4.3 Overstressing of soil

This problem occurs, when the compressive stress under the base of the wall exceeds the bearing strength of the soil. Depending on the position of the resultant of all forces acting on the wall, the pressure at the toe could be much larger than at the heel. It is possible to attain zero or negative stress (i.e. tension) at the heel.

However, the *Middle Third Rule* states that: 'If the resultant of forces (R) acts within the middle third of the base area, then the stress everywhere over it is compressive'.

Note: Even if the factors of safety for sliding and overturning are satisfactory, the wall could fail, if the soil is too weak below the base (see bearing capacity, Chapter 9).

Figure 8.56 depicts the occurrence of compression and tension.

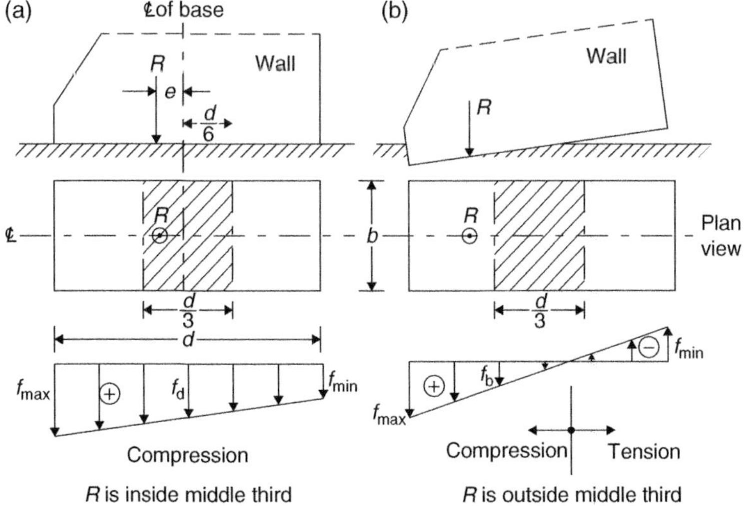

Figure 8.56

f_{max} = maximum compressive stress. It should be smaller than the bearing capacity of the soil.

e = eccentricity of R, measured from the centre line of the base, which is the neutral axis of its area.

Stress f is made up of two components:

1. Direct pressure due to R.

$$f_d = \frac{R}{A}$$

where $A = bd$ (basearea) $\therefore f_d = \dfrac{R}{bd}$

2. Bending stress (the reader is recommended to consult works on the theory of bending in Strength of Materials).

$$f_b = \pm\frac{M}{Z} \quad \begin{pmatrix} \text{compression}: +\text{ve} \\ \text{tension} \quad : -\text{ve} \end{pmatrix}$$

Where, $M = Re$

And, $Z = \dfrac{bd^2}{6}$ (Section Modulus of the base area)

The formula for the combined pressures can now be derived from:

$$f_{max} = f_d \pm f_b$$

$$= \frac{R}{bd} \pm \frac{Re}{\frac{bd^2}{6}}$$

$$\therefore \boxed{f_{\substack{max \\ min}} = \frac{R}{bd}\left(1 \pm \frac{6e}{d}\right)} \tag{8.57}$$

8.13.4.4 Tension in Brickwork

This only occurs when $e > \frac{d}{6}$, as indicated in Figure 8.56(b). Brickwork has little tensile strength, hence the wall should be re-designed in order to make $e < \frac{d}{6}$.

Example 8.8

The active thrust on a 9m high wall was determined for cases 1 and 2 in Example 8.3. Design the wall to support the active force in case 1, with adequate factors of safety.

Available information:

Soil: sand | $\phi = 32°$

$\gamma = 17\,\text{kN/m}^3$

Bearing capacity $= 200\,\text{kN/m}^2$

$P_a = 211.4\,\text{kN} \quad y_a = 3\,\text{m}$

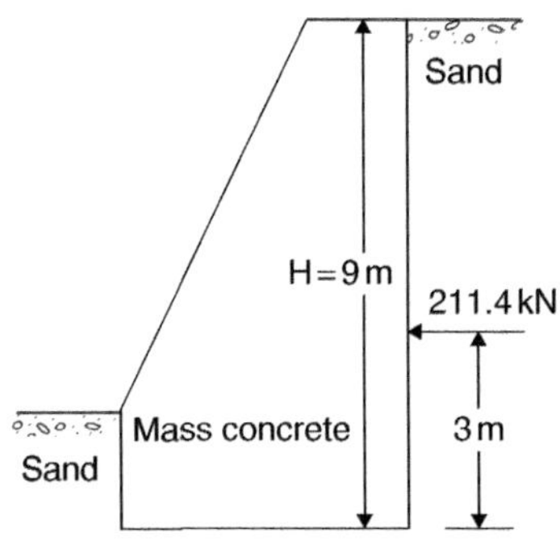

Specified shape for the wall.

Unit weight of concrete $\gamma_c = 24\,\text{kN/m}^3$

Figure 8.57

Step 1: Choose trial dimensions for the wall, say:

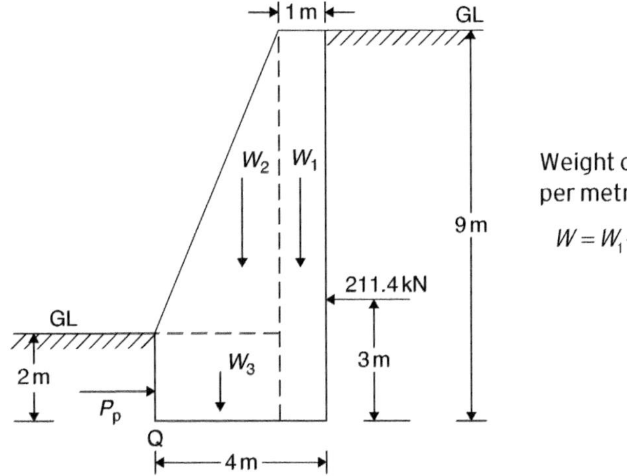

Weight of the wall per metre length:

$$W = W_1 + W_2 + W_3$$

Figure 8.58

Step 2: Determine all forces acting on the wall, by dividing it into standard shapes as shown: $W_1 = 1 \times 9 \times 24 = 216\,\text{kN}$

$$W_2 = \frac{3.7}{2} \times 24 = 252\,\text{kN}$$

$$W_3 = 2 \times 3 \times 24 = 144\,\text{kN}$$

$$\therefore \quad W = 612\,\text{kN}$$

For passive force:

$$K_p = \frac{1+\sin 32}{1-\sin 32} = \frac{1.53}{0.47} = 3.25$$

$$\sigma_v = 2 \times 17 = 34\,\text{kN/m}^2$$

$$\sigma_p = 3.25 \times 34 = 110.5\,\text{kN/m}^2$$

Average pressure: $\bar{\sigma}_p = \dfrac{110.5}{2} = 55.3\,\text{kN/m}^2$

$P_p = 2 \times 55.3 = 110.5\,\text{kN}$

Alternatively from (8.30): $P_p = \dfrac{K_p \gamma H^2}{2} = \dfrac{3.25 \times 17 \times 2^2}{2} = 110.5\,\text{kN}$

Its line of action: $y_p = \dfrac{2}{3} = 0.67\,\text{m}$

Step 3: Check the safety factor against overturning, by taking moments about the toe.

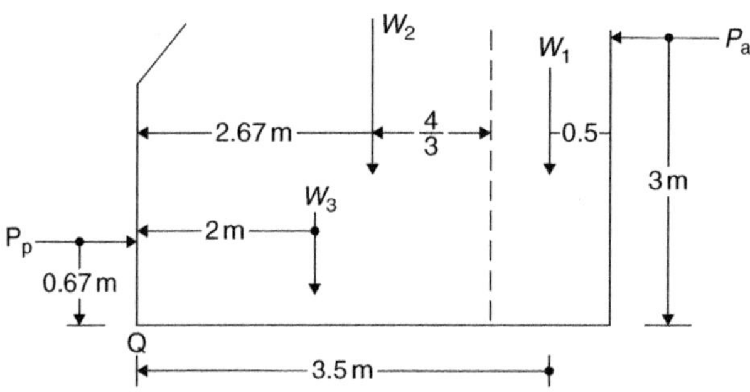

Figure 8.59

Moments about Q:

Overturning moment: $M_O = 3 \times P_a = 3 \times 211.4 = 634\,\text{kNm}$

Resisting moment: $M_R = 3.5W_1 + 2.67W_2 + 2W_3 + 0.67\,P_p$

$= 3.5 \times 216 + 2.67 \times 252 + 2 \times 144 + 0.67 \times 110.5$

$= 756 + 673 + 288 + 74 = 1791\,\text{kNm}$

Therefore, $F_s = \dfrac{M_R}{M_o} = \dfrac{1791}{634} = 2.82$ is satisfactory

Step 4: Forces to be taken into account for F_s against sliding are shown below:

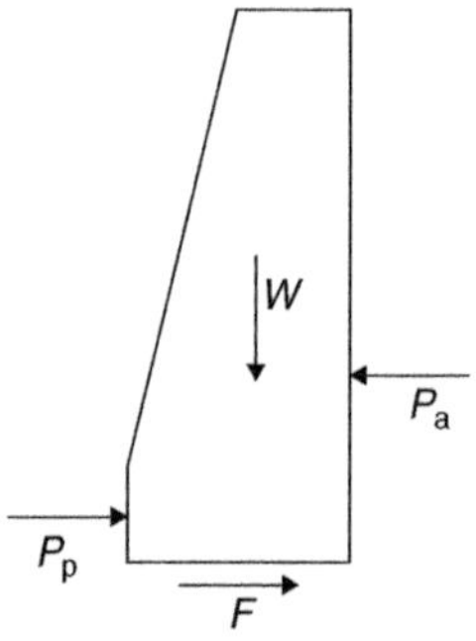

$\mu = \tan\phi = \tan 32 = 0.625$

$F = W \tan\phi = 612 \times 0.625 = 383\,\text{kN}$

$\therefore\ F_s = \dfrac{F + P_p}{P_a} = \dfrac{383 + 110.5}{211.4} = 2.33$

It is just satisfactory, assuming that P_p is not diminished later by erosion or excavation.

Figure 8.60

Step 5: Determine the resultant and its line of action relative to the toe by summing the moments of all forces acting on the wall about Q.

The resultant of vertical forces is the total weight i.e. $R = W = 612$ kN.

Taking moments about Q, using the lever arms shown.

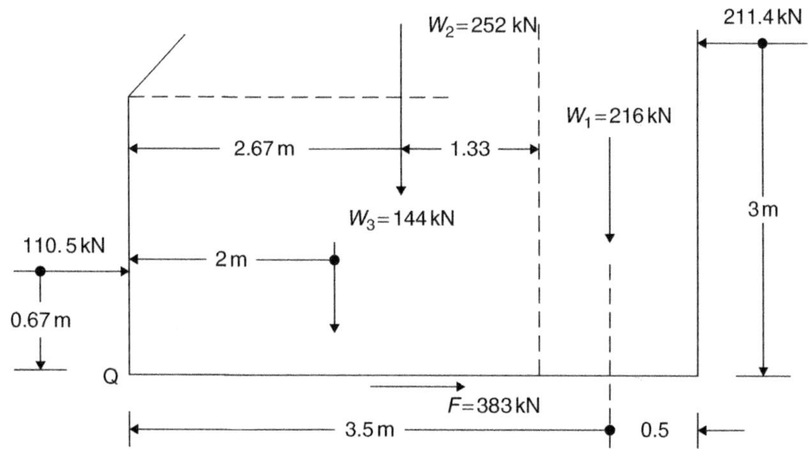

Figure 8.61

$$\overset{+}{\curvearrowright}\Sigma M_Q = 3.5W_1 + 2.67W_2 + 2 \times W_3 + 110.5 \times 0.67 - 211.4 \times 3$$

$$= 3.5 \times 216 + 2.67 \times 252 + 2 \times 144 + 74 - 634 = 1791 - 634 = 1157 \text{ kNm}$$

Note that $\boxed{\Sigma M_Q = M_R - M_O}$

But the moment of the resultant has to equal ΣM_Q for equilibrium. It is assumed to act at some distance (x_R) from the toe.

Moment of R about $Q = Rx_R = 612x_R$

For equilibrium therefore, $\Sigma M_Q = 1157 = 612x_R$

From which, $x_R = \dfrac{1157}{612} = 1.89$ m

Now, that its point of application on the base is known, R can replace the entire force system. Force F may be ignored as it has no moment about Q.

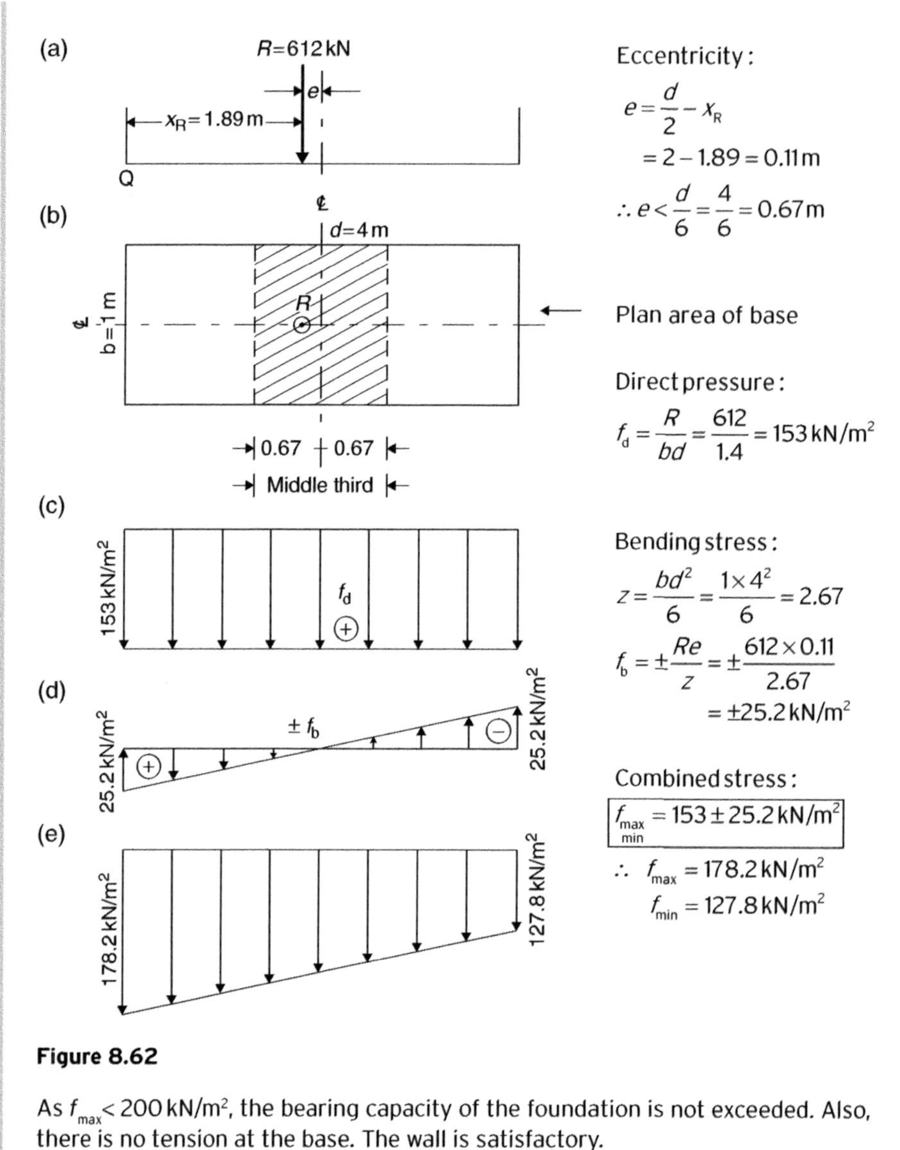

Figure 8.62

As $f_{max} < 200\,kN/m^2$, the bearing capacity of the foundation is not exceeded. Also, there is no tension at the base. The wall is satisfactory.

8.14 Sheet piles

These are 'weightless' retaining walls, usually made of interlocking steel sections to make them watertight. They have smoother surfaces than gravity or concrete retaining structures, hence wall friction is negligible. Their flexibility is ignored in the stability calculations for simplicity. There are two main types of sheet pile walls:

1. Cantilever (Maximum $H = 4\,m$)
2. Anchored

8.14.1 Cantilever sheet pile walls

These may be used as either temporary or permanent supports to granular soil and perhaps short-term ones to clays and silts. The supported height is comparatively small, due to the flexibility of the wall. The piles are driven into the soil to a depth necessary for stability, which depends completely on the passive resistance of the soil.

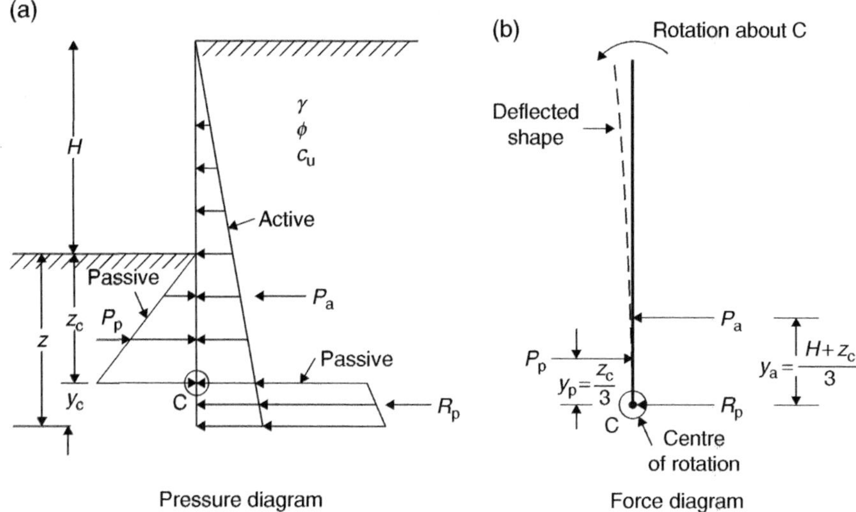

Figure 8.63

The pile is assumed to rotate about point C, a little distance above its lower end, under the action of the active force P_a. The rotation is resisted by two passive forces P_p and R_p. As the determination of R_p is complicated, the stability calculations are simplified by:

a) assuming that R_p acts at the centre of rotation (C) as shown in Figure 8.63(b)
b) lengthening the theoretical penetration depth (z_c) by 20% so, that

$$\left.\begin{aligned} y_c &= 0.2z_c \\ z &= z_c + y_c \end{aligned}\right| \quad \therefore \boxed{z = 1.2z_c} \tag{8.58}$$

Also, the passive resistance is decreased by a factor of safety F_s as described later.

Determination of z for φ-soils

$$\text{Active force:} \quad P_a = \frac{K_a\gamma(H+z_c)^2}{2} \quad \Bigg| \quad \text{acting at } y_a = \frac{H+z_c}{3}$$

$$\left.\begin{aligned} &\text{Passive force:} \quad P_p = \frac{K_a\gamma z_c^2}{2} \\ &\text{Factored} \quad P_p = \frac{K_a\gamma z_c^2}{2F_s} \end{aligned}\right| \quad \text{acting at } y_p = \frac{z_c}{3}$$

Note: If moment of forces are taken about the centre of rotation, then R_p can be eliminated, thus it need not be taken into consideration any more.

For equilibrium, the sum of moments about c=0

$$\overset{+}{\curvearrowright}\Sigma M_C = P_p y_p - P_a y_a = 0$$

Or $\overset{+}{\curvearrowright}\Sigma M_c = \frac{K_p \gamma z_c^2}{2} \times \frac{z_c}{3} - \frac{K_a \gamma (H+z_c)^2}{2} \times \frac{H+z_c}{3} = 0$

$$= \frac{K_p \gamma z_c^3}{6} - \frac{K_p \gamma (H+z_c)^3}{6} = 0$$

Cancelling γ and the divisor 6.

$$= K_p z_c^3 - K_a (H+z_c)^3 = 0$$

$$K_a (H+z_c)^3 = K_p z_c^3$$

$$\left(\frac{H+z_c}{z_c}\right)^3 = \frac{K_p}{K_a}$$

$$\frac{H}{z_c} + 1 = \sqrt[3]{\frac{K_p}{K_a}}$$

$$\frac{H}{z_c} = \sqrt[3]{\frac{K_p}{K_a}} - 1$$

From which,

$$z_c = \frac{H}{\sqrt[3]{\frac{K_p}{K_a}} - 1} \tag{8.59}$$

Total depth:

$$z = \frac{1.2H}{\sqrt[3]{\frac{K_p}{K_a}} - 1} \tag{8.60}$$

8.14.2 Factor of safety

Before the above formulae can be applied, the value of K_p has to be calculated so, that it includes a factor of safety. Two methods are introduced below.

Method 1

Both K_p and K_a are modified by the factor $F_\phi = F_s$ where F_ϕ is the factor of safety with respect to friction. The passive Rankine coefficient is calculated first from the friction mobilized (ϕ'_m), instead of the friction available (ϕ'). The active coefficient is then determined from K_p.

For cohesionless soil, the factor of safety may be defined as:

$$F_\phi = F_s = \frac{\tan\phi'}{\tan\phi'_m} \tag{8.61}$$

From which, $$\tan\phi'_{m} = \frac{\tan\phi'}{F_{s}}$$

Passive coefficient
$$\boxed{K_{p} = \frac{1+\tan\phi'_{m}}{1-\tan\phi'_{m}}} \quad (8.62)$$

Active Coefficient
$$K_{a} = \frac{1}{K_{p}} \quad (8.20)$$

Substituting K_{a} into formula (8.60) for z:

$$\boxed{z = \frac{1.2H}{\sqrt[3]{K_{p}^{2}}-1}} \quad (8.63)$$

Method 2

Only K_{p} is divided by the factor of safety and K_{a} remains unaltered. The value of F_{s} is usually 2 or more.

Formula (8.60) is expressed as:

$$\boxed{z = \frac{1.2H}{\sqrt[3]{\frac{K_{p}}{F_{s}K_{a}}}-1}} \quad (8.64)$$

Example 8.9

A 3.5 m deep excavation in cohesionless soil is to be supported by cantilever sheet piles. The soil characteristics are:

$$\gamma = 19\,\text{kN/m}^{3}$$
$$\phi' = 34^{\circ}$$
$$c' = 0$$

Estimate: a) The total length of pile to be driven into the ground by the two methods taking $F_{s}=2$

b) The factor of safety if $z=4\,\text{m}$

a) Method 1: $\tan\phi'_{m} = \frac{\tan 34}{2} = 0.3373 \quad \therefore \quad \phi'_{m} = 18.63^{\circ}$

$$K_{p} = \frac{1+\sin 18.63}{1-\sin 18.63} = \frac{1.3195}{0.6805} = 1.94$$

$$K_{a} = \frac{1}{K_{p}} = \frac{1}{1.94} = 0.52$$

From (8.63): $$z = \frac{1.2\times 3.5}{\sqrt[3]{1.94^{2}}-1}$$

$$= \frac{4.2}{0.554} = 7.6\,\text{m}$$

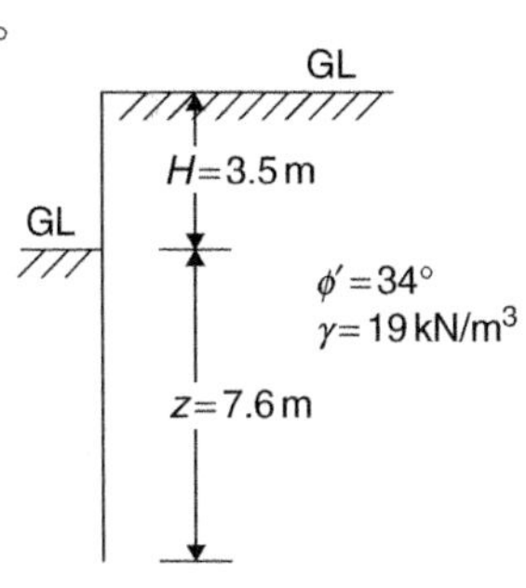

Figure 8.64

Method 2: $K_a = \dfrac{1-\sin 34}{1+\sin 34}$

$= \dfrac{0.441}{1.559} = 0.283$

$K_p = \dfrac{1}{K_a} = \dfrac{1}{0.283} = 3.54$

From (8.64): $z = \dfrac{1.2 \times 3.5}{\sqrt[3]{\dfrac{3.54}{2 \times 0.233}} - 1} = \dfrac{4.2}{0.841} = 5\,\text{m}$

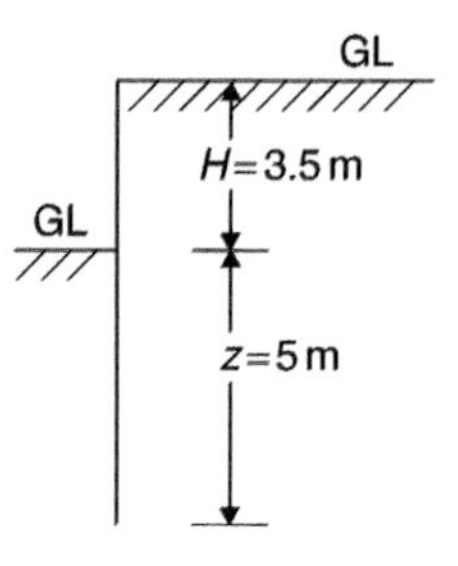

Figure 8.65

Note: There is a large discrepancy between the two results:

Method 1: $z=7.6\,\text{m}$
Method 2: $z=5\,\text{m}$

Check: In order to decide which value to accept, comparison is made between the theoretical passive resistance required and available one, for both methods. Referring to Figure 8.63(b).

$$\overset{+}{\curvearrowleft}\ \Sigma M_C = P_p y_p - P_a y_a = 0$$

$$\therefore \quad \boxed{P_p = \left(\frac{y_a}{y_p}\right) P_a}$$

Method 1: $K_a = 0.52 \qquad y_a = \dfrac{7.6+3.5}{3} = 3.7\,\text{m}$

$K_p = 1.94 \qquad y_p = \dfrac{7.6}{3} = 2.53\,\text{m}$

Active: $P_a = \dfrac{K_a \gamma (H+z)^2}{2} = \dfrac{0.52 \times 19 \times (3.5+7.6)^2}{2} = 609\,\text{kN}$

Required: $P_p = \left(\dfrac{y_a}{y_p}\right) P_a = \dfrac{3.7}{2.53} \times 609 = 891\,\text{kN}$

Available: $P_p = \dfrac{K_a \gamma z^2}{2} = \dfrac{1.94 \times 19 \times 7.6^2}{2} = 1065\,\text{kN}$

Ratio: $r_1 = \dfrac{\text{Available } P_p}{\text{Required } P_p} = \dfrac{1065}{891} = 1.2$

Method 2: $K_a = 0.283 \qquad y_a = \dfrac{3.5+5}{3} = 2.83\,\text{m}$

$K_p = 3.54 \qquad y_p = \dfrac{5}{3} = 1.67\,\text{m}$

Active: $P_a = \dfrac{K_a \gamma (H+z)^2}{2} = \dfrac{0.283 \times 19 \times (3.5+5)^2}{2} = 194\,\text{kN}$

Required: $P_p = \left(\frac{y_a}{y_p}\right)P_a = \frac{2.83}{1637}\times 194 = 329\,\text{kN}$

Available: $$P_p = \frac{\frac{K_p}{F_s}\gamma z^2}{2} = \frac{K_p\gamma z^2}{2F_s} = \frac{3.54\times 19\times 5^2}{2\times 2} = 420\,\text{kN}$$

Ratio: $$r_2 = \frac{\text{Available } P_p}{\text{Required } P_p} = \frac{420}{329} = 1.28$$

Conclusions:
The ratios r_1 and r_2 are practically equal, but Method 2 yielded a more economical driving length ($z=5$ m) than Method 1. For this reason, Method 2 will be applied in the rest of this chapter.

b) If $z = 4$ m then F_s is expressed from (8.64):

$$\sqrt[3]{\frac{K_p}{F_s K_a}} - 1 = \frac{1.2H}{z} \qquad \frac{K_p}{F_s K_a} = \left(\frac{1.2H}{z}+1\right)^3$$

$$F_s = \frac{K_p}{K_a\left(\frac{1.2H}{z}+1\right)^3}$$

$$= \frac{3.54}{0.283\times\left(\frac{1.2\times 3.5}{z}+1\right)^3} = 1.45 < 2$$

Check:

Active: $$P_a = \frac{0.283\times 19\times(3.5+4)^2}{2} = 151\,\text{kN}$$

$$y_a = \frac{3.5+4}{3} = \frac{7.5}{3} = 2.5\,\text{m} \quad y_p = \frac{4}{3} = 1.33\,\text{m}$$

Required: $$P_p = \frac{7.5}{4}\times 151 = 283\,\text{kN}$$

Available: $$P_p = \frac{3.54\times 19\times 4^2}{2\times 2} = 269\,\text{kN}$$

Ratio: $$r = \frac{269}{283} = 0.95 < 1$$

Therefore longer depth of penetration is required.

8.14.3 Bending of sheet piles

Sheet pile wall could fail in bending, because of its flexibility. It is necessary, therefore, to determine the position and magnitude of the maximum bending moment (M_{max}) so, that a suitability stiff pile can be chosen from the manufacturer's catalogue.

Position of M_{max}

It occurs at a depth h at which the active and passive forces are equal in magnitude, that is, where the shear force in the cantilever is zero. Applying Method 2, consider point X:

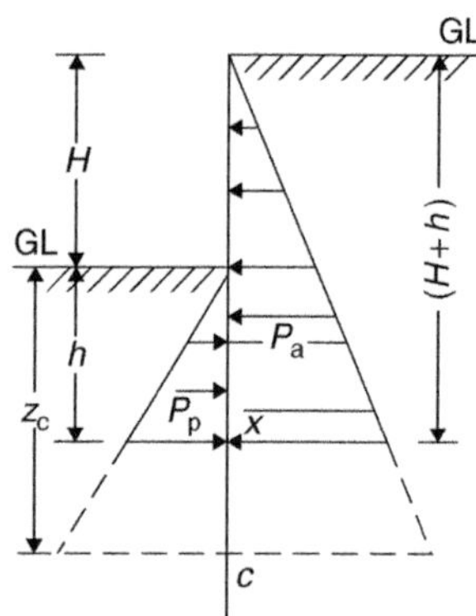

Figure 8.66

Active force: $P_a = \dfrac{K_a\gamma(H+h)^2}{2}$

Passive force: $P_p = \dfrac{K_a\gamma h^2}{2F_s}$

Equating forces: $P_a = P_p$

or $$\frac{K_a\gamma(H+h)^2}{2} = \frac{K_p\gamma h^2}{2F_s}$$

$$K_a(H+h)^2 = \frac{K_p h^2}{F_s}$$

$$\left(\frac{H+h}{h}\right)^2 = \frac{K_p}{K_a F_s}$$

Rearranging, $$\left(\frac{H}{h}+1\right) = \sqrt{\frac{K_p}{K_a F_s}}$$

From which, $\dfrac{H}{h} = \dfrac{K_p}{K_a F_s} - 1$, hence the position of M_{max} is given by: $h = \dfrac{H}{\dfrac{K_p}{K_a F_s} - 1}$ (8.65)

For M_{max}, take moments about point X.

P_a is acting at: $y_a = \dfrac{H+h}{3}$

P_p is acting at: $y_p = \dfrac{h}{3}$

$$M_{max} = \Sigma M_x = P_p y_p - P_a y_a$$

$$\therefore M_{max} = \frac{K_p\gamma h^3}{6F_s} - \frac{K_a\gamma(H+h)^3}{6}$$

$$M_{max} = \frac{\gamma}{6}\left[\frac{K_p h^3}{F_s} - K_a(H+h)^3\right] \quad (8.66)$$

For Example 8.9, Method 2: $K_p = 3.537$ and $F_s = 2$
$K_a = 0.283$

$$\therefore h = \frac{H}{\dfrac{K_p}{\sqrt{F_s}} - 1} = \frac{3.5}{\dfrac{3.537}{\sqrt{2}} - 1} = \frac{3.5}{1.5} = 2.33\text{m}$$

$$\text{And } M_{max} = \frac{19}{6} \times \left[\frac{3.537 \times 2.33^3}{2} - 0.283 \times (3.5 + 2.33)^3 \right]$$

$$= \frac{19}{6} \times (22.37 - 56.08) = -106.6 \text{ kNm}$$

$$= 107 \text{ kNm} \circlearrowleft$$

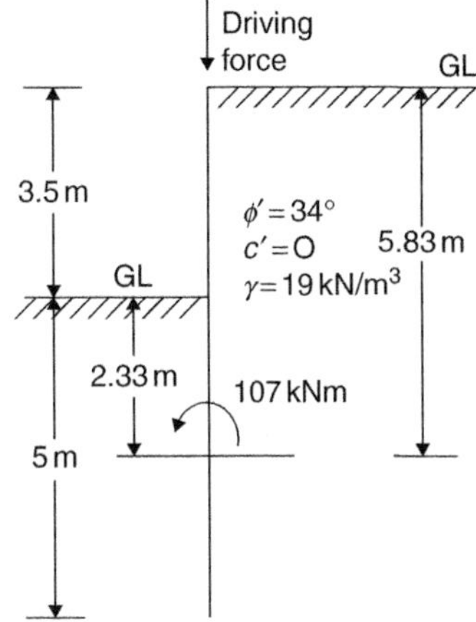

Figure 8.67

The cross-section has to be chosen from the section table, supplied by manufacturers, to carry 107 kNm as well as to withstand the driving force without buckling.

8.14.4 Sheet pile in cohesive soils

It has been shown that the height of free-standing cuts in cohesive soil is given by:

Either for c-ϕ soil as:

$$H_0 = \frac{4c}{\gamma \sqrt{K_a}} \tag{8.48}$$

or for c-soil as:

$$H_0 = \frac{4c}{\gamma} \tag{8.50}$$

As a small vertical cut can stand unsupported at height H_0 for a reasonable length of time, it is generally considered uneconomical to retain it by cantilever sheet piles in the short term.

8.15 Anchored sheet pile walls

These are suitable for permanent supports to both cohesionless and cohesive soils.

Advantages:

1. Reduction in the depth of penetration
2. Increase in the height to be supported
3. Allowing the use of lighter sections, due to the reduction of bending moments and deflections

Unknowns:

a) Depth of penetration (z)
b) Tension (T) in the tie rods
c) Maximum bending moments (M_{max})

Methods of analysis:

i. Free-earth support method
ii. Fixed-earth support method

8.15.1 Free-earth support method

It is assumed, in this analysis, that:

1. The wall is hinged at its base, hence it can rotate about this point.
2. the passive resistance (R_p) against the rotation of the base is negligible.

Sheet pile in cohesionless soil

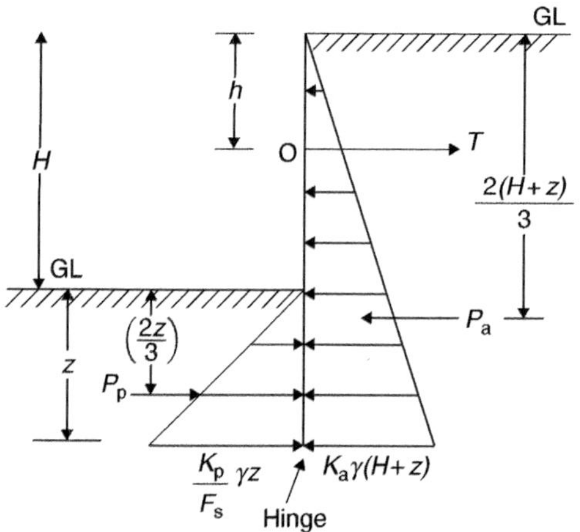

Active: $P_a = \dfrac{K_a\gamma(H+z)^2}{2}$

Passive: $P_p = \dfrac{K_p\gamma z^2}{2F_s}$

Tension force

For equilibrium, the sum of horizontal forces = 0

$$\overset{+}{\rightarrow}\ \Sigma F_x = \vec{T} + \vec{P_p} - \overleftarrow{P_a} = 0$$

$$\therefore \quad T = P_a - P_p$$

Figure 8.68

$$T = \frac{K_a\gamma(H+z)^2}{2} - \frac{K_a\gamma z^2}{2F_s}$$

∴ Tension force:

$$\boxed{T = \frac{\gamma}{2}\left[K_a(H+z)^2 - \frac{K_p z^2}{F_s}\right]} \tag{8.67}$$

For z, take moments about the tie rod.

$$\overset{+}{\curvearrowleft}\ \Sigma M_0 = P_a\left[\frac{2}{3}(H+z) - h\right] - \frac{P_p}{F_s}\left(\frac{2z}{3} + H - h\right) = 0$$

Equating,

$$\boxed{P_a\left[\frac{2}{3}(H+z) - h\right] = \frac{P_p}{F_s}\left(\frac{2z}{3} + H - h\right)} \tag{8.68}$$

As P_a and P_p are also expressed in terms of z, there is no advantage in deriving a complicated cubic formula in order to evaluate it. Rather, an easier application of 8.68 is now illustrated by an example.

Example 8.10

With reference to the vertical cut in Example 8.9, determine the depth of penetration (z) required if the pile is anchored at 1.2 m below ground level, as well as the tension (T) in the tie rod per metre run of wall. Adopt $F_s = 2$.

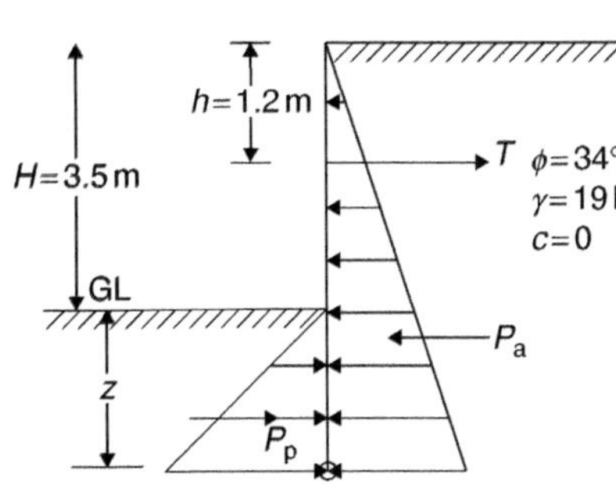

Figure 8.69

$$K_a = \frac{1-\sin 34}{1+\sin 34} = 0.283$$

$$K_p = \frac{1}{K_a} = 3.537$$

Active force:

$$P_a = \frac{K_a\gamma(H+z)^2}{2} = \frac{0.283\times 19\times(3.5+z)^2}{2}$$
$$= 2.69\times(3.5+z)^2$$

Passive force: $$P_p = \frac{K_p\gamma z^2}{2} = \frac{3.537\times 19\times z^2}{2} = 33.6z^2$$

Applying formula (8.68):

$$2.69\times(3.5+z)^2\left[\frac{2}{3}(3.5+z)-1.2\right] = \frac{33.6z^2}{2}\left(\frac{2z}{3}+3.5-1.2\right)$$

$$\frac{2}{3}(3.5+z)^3 - 1.2(3.5+z)^2 = 6.25z^2\left(\frac{2z}{3}+2.3\right)$$

Expanding the cubic and quadratic terms:

$$\frac{2}{3}\times\left(42.88+36.75z+10.5z^2+z^3\right) - 1.2\times(12.25+7z+z^2) = 6.25z^2\left(\frac{2z}{3}+2.3\right)$$

$$28.6+24.5z+7z^2+0.67z^3-14.7-8.4z-1.2z^2 = 4.17z^3+14.4z^2$$

Collecting similar terms,

$$(4.17-0.67)z^3+(14.4+1.2-7)z^2+(8.4-24.5)z+(14.7-28.9)=0$$

$$3.5z^3+8.6z^2-16.1z-14.2=0$$

$$\boxed{z^3+2.46z^2-4.6z-4.1=0}$$

This cubic equation is solved here graphically.

Rearranging, $z^3+2.46z^2=4.6z+4.1$

Let $L=z^3+2.46z^2$ | $L=R$

And $R=4.6z+4.1$ | Tabulating L and R for several vales of z.

Table 8.8

z m	1	1.5	1.6	1.7	1.8	2
$L=z^3+2.46z^2$	3.46	8.91	10.40	12.02	13.80	17.84
$R=4.6z+4.1$	8.70	11.00	11.46	11.92	12.38	13.30

It appears from the table that L and R are equal somewhere between $z=1.6$ and $z=1.7$. To be more precise, the tabulated values are plotted on Graph 8.6 and the intersection of the two curves yields $z=1.69$ m. The tension in the anchoring rod can now be calculated.

$$T=\frac{\gamma}{2}\left[K_a(H+z)^2-\frac{K_p z^2}{F_s}\right]=\frac{19}{2}\times\left[0.283\times(3.5+1.69)^2-\frac{3.537\times1.69^2}{2}\right]$$

$$=9.5\times(7.62-5.05)=24.4\text{kN/m length of wall.}$$

This value is usually increased by at least 15% to take account of anchorage stresses.

Therefore, $T=1.15\times24.4=28$ kN/m

Comparison: In Example 8.9: $z=5$ m

In Example 8.10: $z=1.7$ m

Sheet piles in cohesive soils

Figure 8.70 shows the pressure diagram, assuming full tension crack development. The coefficient of passive pressure is divided by the factor of safety as before in Method 2.

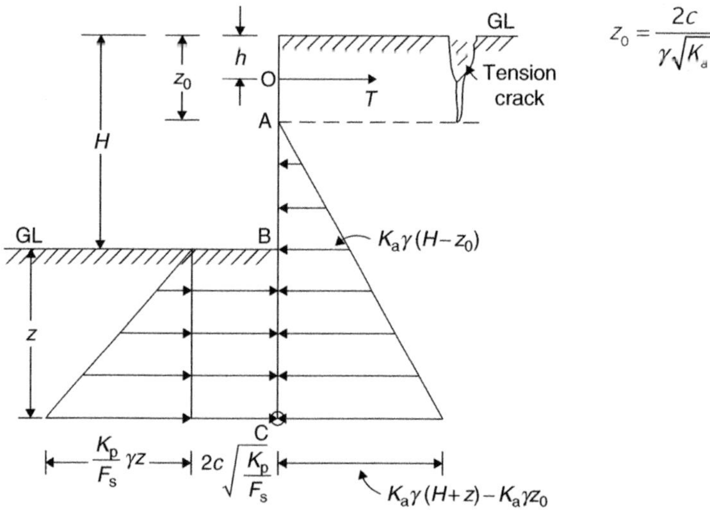

Figure 8.70

The pressure diagram is simplified by indicating net pressures, as in Figure 8.71.

Just above point B: $\sigma_a = K_a\gamma(H - z_0)$

Just below point B: $$\sigma_B = 2c\sqrt{\frac{K_p}{F_s}} K_a\gamma(H - z_0) \qquad (8.69)$$

At hinge c: $$\sigma_c = \frac{K_p}{F_s}\gamma z + 2c\sqrt{\frac{K_p}{F_s}} - K_a\gamma(H + z) + K_a\gamma z_0$$

Rearranging, $$= 2c\sqrt{\frac{K_p}{F_s}} - K_a\gamma(H - z_0) + \gamma z\left(\frac{K_p}{F_s} - K_a\right)$$

Therefore, $$\sigma_C = \sigma_B + \gamma\left(\frac{K_p}{F_s} - K_a\right)z$$

The net pressure diagram is drawn in Figure 8.71.

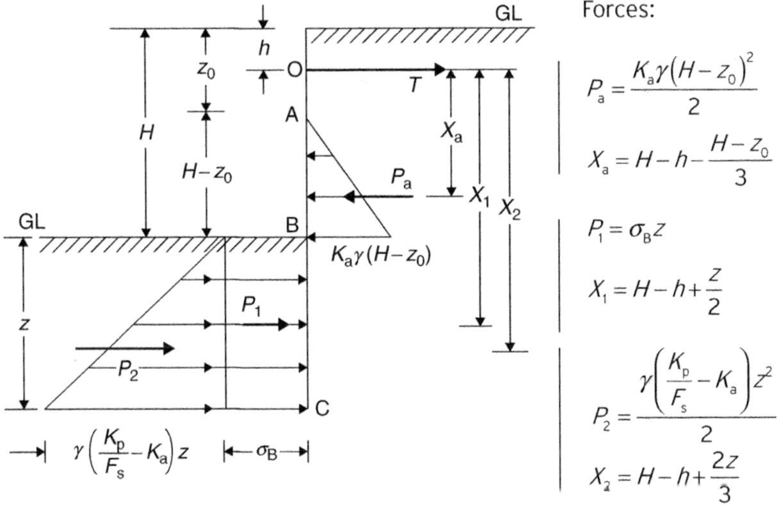

Figure 8.71

In order to derive a cubic equation for z, take moments about the tie.

$$\overset{+}{\curvearrowleft}\ \Sigma M_O = P_a x_a + P_1 x_1 - P_2 x_2 = 0$$

$$P_a\left(H - h - \frac{H - z_0}{3}\right) - \sigma_B z\left(H - h + \frac{z}{2}\right) - \frac{\gamma\left(\frac{K_p}{F_s} - K_a\right)z^2}{2}\left(H - h + \frac{2z}{3}\right)$$

$$\frac{P_a}{3}(2H - 3h + z_0) - \sigma_B(H - h)z - \frac{\sigma_B}{2}z^2 - \frac{\gamma(H - h)\left(\frac{K_p}{F_s} - K_a\right)}{2}z^2 - \frac{\gamma\left(\frac{K_p}{F_s} - K_a\right)}{3}z^3 = 0$$

Therefore, the cubic equation is expressed as:

$$\left[\frac{\gamma\left(\frac{K_p}{F_s}-K_a\right)}{3}\right]z^3+\left[\frac{\sigma_B}{2}+\frac{\gamma(H-h)\left(\frac{K_p}{F_s}-K_a\right)}{2}\right]z^2+\sigma_B(H-h)z-\frac{P_a}{3}(2H-3h+z_0)=0 \tag{8.70}$$

Inspite of its complicated look, the cubic can be solved easily, as shown in Example 8.11. For the equilibrium of horizontal forces, their sum must be zero.

$$\overset{+}{\rightarrow}\ \Sigma F_x = \overrightarrow{T}+\overrightarrow{P_1}+\overrightarrow{P_2}-\overleftarrow{P_a}=0$$

Expressing the force in the tie:

$$\boxed{T=P_a-P_1-P_2}$$

Or substituting the expressions of these forces + 15%:

$$T=\frac{1.15}{2}\left[K_a\gamma(H-z_0)^2-2\sigma_B z-\gamma\left(\frac{K_p}{F_s}-K_a\right)z^2\right] \tag{8.71}$$

Example 8.11

A sheet pile wall is supporting a 3.5 m cut in clayey, soil, as shown in Figure 8.72. Calculate the depth of penetration required and the tension in the tie rod. Factor of safety = 2.

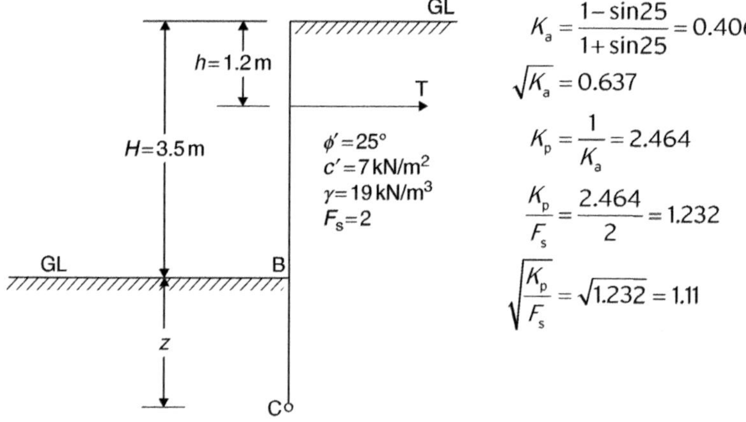

Figure 8.72

$$K_a=\frac{1-\sin 25}{1+\sin 25}=0.406$$

$$\sqrt{K_a}=0.637$$

$$K_p=\frac{1}{K_a}=2.464$$

$$\frac{K_p}{F_s}=\frac{2.464}{2}=1.232$$

$$\sqrt{\frac{K_p}{F_s}}=\sqrt{1.232}=1.11$$

$$Z_0 = \frac{2c'}{\gamma\sqrt{K_a}} = \frac{2.7}{19 \times 0.637} = 1.16\,\text{m} \qquad \frac{K_p}{F_s} - K_a = 1.232 - 0.406 = 0.826$$

$$H - z_0 = 3.5 - 1.16 = 2.34\,\text{m} \qquad H - h = 3.5 - 1.2 = 2.3\,\text{m}$$

Active force: $$P_a = \frac{K_a\gamma(H - z_0)^2}{2} = \frac{0.406 \times 19 \times 2.34^2}{2} = 21.1\,\text{kN}$$

Pressure: $$\sigma_B = 2c\sqrt{\frac{K_p}{F_s}} - K_a\gamma(H - z_0) = 2.7 \times 1.11 - 0.406 \times 19 \times 2.34 = -2.51\,\text{kN/m}^2$$

Coefficients: $$\frac{\gamma}{3}\left(\frac{K_p}{F_s} - K_a\right) = \frac{19}{3} \times 0.826 = 5.23$$

$$\frac{1}{2}\left[\sigma_B + \gamma(H - h)\left(\frac{K_p}{F_s} - K_a\right)\right] = \frac{1}{2} \times (-2.51 + 19 \times 2.3 \times 0.826) = 16.8$$

$$\sigma_B(H - h) = -2.51 \times 2.3 = -5.77$$

$$\frac{P_a}{3}(2H - 3h + z_0) = \frac{21.1}{3} \times (2 \times 3.5 - 3 \times 1.2 + 1.16) = 32$$

Substituting into (8.70)

$$5.23z^3 + 16.8z^2 - 5.77z - 32 = 0$$

or

$$\boxed{z^3 + 3.21z^2 - 1.1z - 6.12 = 0}$$

Graphical solution: $$z^3 + 3.21z^2 = 1.1z + 6.12$$

Let $L = z^3 + 3.21z^2$, $R = 1.1z + 6.12$ $\qquad L = R$

Tabulating the calculations:

As L becomes larger than R at $z=1.3$, the intersection point of the two curves is some-where between $z=1.2$ and $z=1.3$. It is seen on Graph 8.7, that $z=1.29$ m.

Table 8.9

z (m)	1	1.1	1.2	1.3	1.4	1.5
$L = z^3 + 3.21z^2$	4.21	5.22	6.35	7.62	9.03	10.6
$R = 1.1z + 6.12$	7.22	7.33	7.44	7.55	7.66	7.77

Tension in the tie:

$$T = \frac{1.15}{1}\left[K_a\gamma(H - z_0)^2 - 2\sigma_B z - \gamma\left(\frac{K_p}{F_s} - K_a\right)z^2\right]$$

$$= \frac{1.15}{2} \times (0.406 \times 19 \times 2.34^2 + 2 \times 2.51 \times 1.29 - 19 \times 0.826 \times 1.29^2)$$

$$= \frac{1.15}{2} \times (42.24 + 6.48 - 26.12) = 13\,\text{kN/m of wall}$$

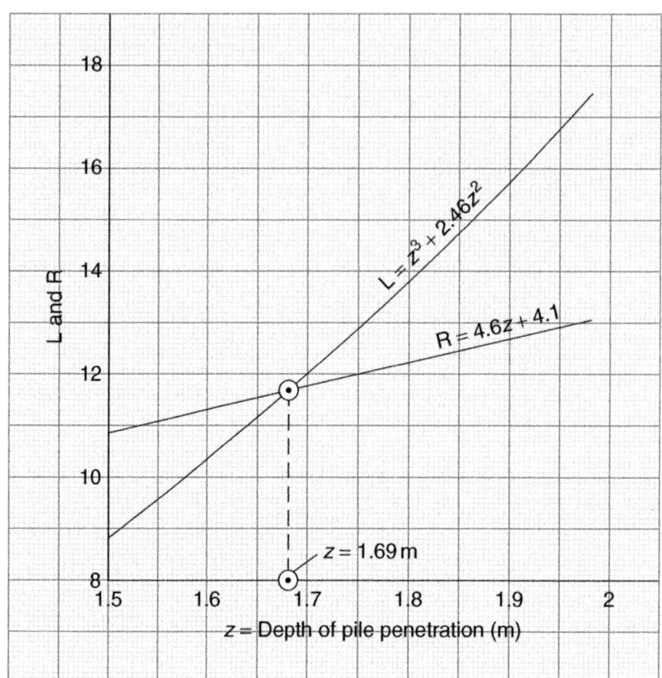

Graph 8.6

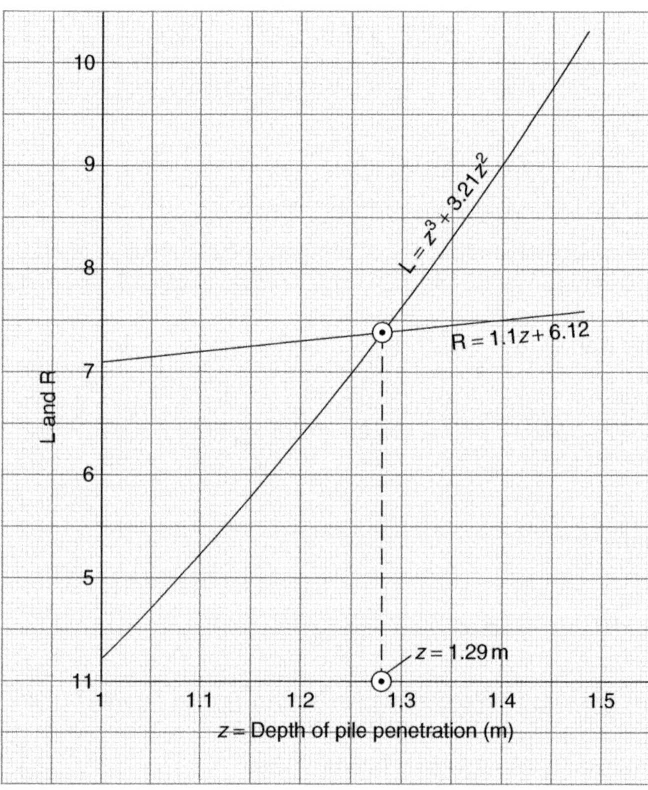

Graph 8.7

Therefore, the tie is subjected to small tension in this example. Alternatively, z and T can be determined from the net pressure diagram (Figure 8.73).

Pressures: $\gamma\left(\frac{K_p}{F_s}-K_a\right)z = 19\times 0.826z = \overrightarrow{15.69z}\,\text{kN/m}^2$

$$\sigma_B = -2.51 = \overleftarrow{2.51}\,\text{kN/m}^2$$

$$K_a\gamma(H-z_0) = 0.406\times 19\times 2.34 = \overleftarrow{18}\,\text{kN/m}^2$$

The pressure diagram may now be drawn.

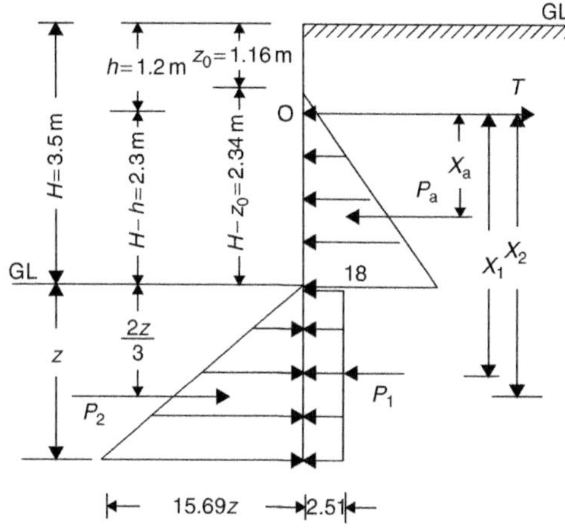

Note that σ_B is negative, hence it is drawn in the opposite sense to P, as shown.

Figure 8.73

Forces and their distances from the tie.

$P_a = 21.1\,\text{kN}$ $\quad x_a = H-h-\frac{H-z_0}{3}$

$$= 2.3-\frac{2.34}{3} = 1.52\,\text{m}$$

$P_1 = 2.51z$ $\quad x_1 = H-h+\frac{z}{2} = 2.3+\frac{z}{2}$

$P_2 = \frac{15.69z^2}{2}$ $\quad x_2 = H-h+\frac{2z}{3} = 2.3+\frac{2z}{3}$

Sum the moment of forces about T to get z.

$$\overset{+}{\curvearrowright}\sum M_O = \overset{\curvearrowright}{P_a x_a}+\overset{\curvearrowright}{P_1 x_1}-\overset{\curvearrowleft}{P_2 x_2} = 0$$

$$= 21.1\times 1.52+2.51z\left(2.3+\frac{z}{2}\right)-\frac{15.69z^2}{2}\times\left(2.3+\frac{2z}{3}\right) = 0$$

$$32.1+5.77z+1.25z^2-18z^2-5.23z^3 = 0$$

Rearranging,

$$5.23z^3 + 16.8z^2 - 5.77z - 32.1 = 0$$

This is the same cubic equation as before, yielding $z = 1.29$ m.

Tension T is determined by equating the sum of horizontal forces to zero.

$$\overset{+}{\rightarrow}\ \Sigma F_x = \vec{T} - \overset{\leftarrow}{P_a} - \overset{\leftarrow}{P_1} + \overset{\rightarrow}{P_2} = 0$$

or $\quad T = P_a + P_1 - P_2$

But $\quad P_1 = 2.51z = 2.51 \times 1.29 = 3.24$ kN

$$P_2 = \frac{15.69z^2}{2} = 7.85 \times 1.29^2 = 13.1\,\text{kN}$$

$\therefore \quad T = 21.1 + 3.24 - 13.1 = 11.2$ kN

Increasing T by 15%: $T = 1.15 \times 11.2 = 13$ kN/m length of wall.

Note: The theoretical penetration depth is not increased by 20% in this method as $R_p \approx 0$, unless erosion of the lower ground is expected.

8.15.2 Fixed-earth support method

In this analysis the lower end of the pile is assumed to be fixed and cannot rotate. The fixity is provided by the passive resistance of the soil. Because the achievement of perfect fixity is doubtful in most soils, the method is applied only to dense sands and gravels. The solution is again derived by means of the net pressure diagram. Assumed pressure diagram:

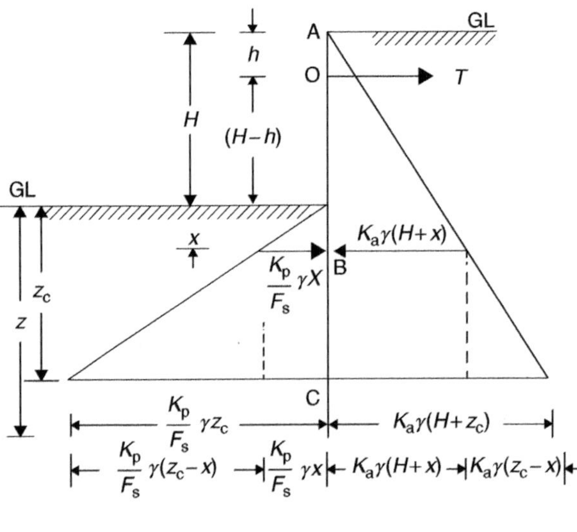

Net pressures

(K_p is reduced by F_s)

$$\text{At B}: \sigma_B = K_a\gamma(H+x) - \frac{K_p}{F_s}\gamma x$$

$$\text{At C}: \sigma_C = \frac{K_p}{F_s}\gamma(z_c - x)$$

$$+\frac{K_p}{F_s}\gamma x$$

$$-K_a\gamma(H+x)$$

$$-K_a\gamma(z_c - x)$$

$$\sigma_C = \left(\frac{K_p}{F_s} - K_a\right)\gamma(z_c - x)$$

$$-K_a\gamma(z_c - x) + K_p\gamma x$$

$$\therefore \sigma_C = \left(\frac{K_p}{F_s} - K_a\right)\gamma(z_c - x) - \sigma_B$$

Figure 8.74

Net pressure diagram

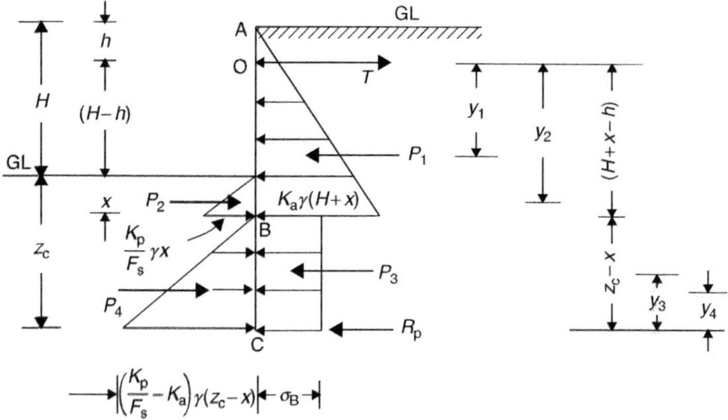

Figure 8.75

In order to solve this statically indeterminate problem, the sheet pile is divided into two equivalent 'beams' AB and BC. It is assumed, that the point of contraflexure is x m below the lower ground level. The length x depends on the friction angle (ϕ'), according to experiments by Blum/Terzaghi, as long as there is no additional pressures on the wall due to surcharge and high water table. Approximate values are:

Table 8.10

ϕ'	15°	20°	25°	30°	35°	40°
x	$0.37H$	$0.25H$	$0.15H$	$0.08H$	$0.035H$	$-0.007H$

Forces acting

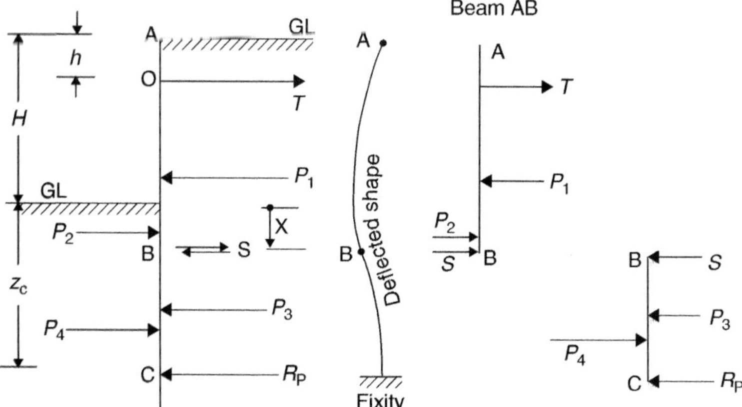

Figure 8.76

where S = shear force in the pile at point B.

Considering beam AB to determine S and T.

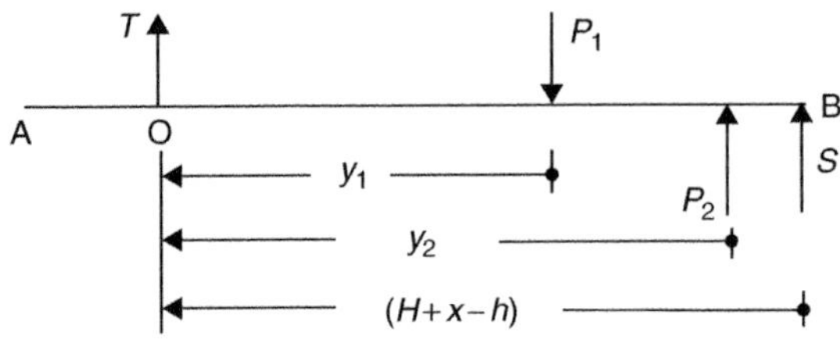

Figure 8.77

$$P_1 = \frac{K_a \gamma (H+x)^2}{2} \qquad Y_1 = H + x - h - \frac{H+x}{3} = \frac{2}{3}(H+x) - h$$

$$P_2 = \frac{1}{2}\left(\frac{K_p}{F_s}\gamma x^2\right) \qquad Y_2 = H - h + \frac{2x}{3}$$

Moments about T:

$$\overset{+}{\curvearrowright} \Sigma M_O = P_1 y_1 - P_2 y_2 - S(H+x-h) = 0$$

$$\therefore \quad \boxed{S = \frac{P_1 Y_1 - P_2 Y_2}{H+x-h}} \qquad (8.72)$$

Sum of forces equated to zero for equilibrium:

$$+\uparrow \Sigma F = T - P_1 + P_2 + S = 0$$

From which $\boxed{T = P_1 - P_2 - S}$ or $\boxed{T = 1.15\,(P_1 - P_2 - S)}$ (8.73)

Considering beam BC to determine z_c

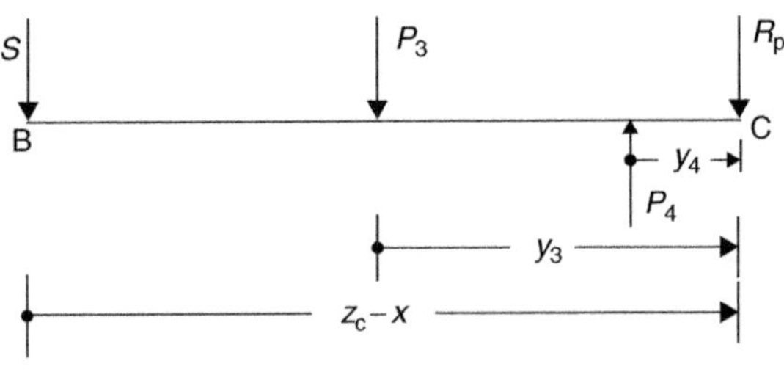

Figure 8.78

$$P_3 = \sigma_B (z_c - x) \qquad y_3 = \frac{z_c - x}{2}$$

$$P_4 = \frac{\gamma}{2}\left(\frac{K_p}{F_s} - K_a\right)(z_c - x)^2 \qquad y_4 = \frac{z_c - x}{3}$$

Taking moments about R_p to determine z_c.

$$\overset{+}{\curvearrowright}\Sigma M_c = P_4 y_4 - P_3 y_3 - S(z_c - x) = 0$$

$$\frac{\gamma}{2}\left(\frac{K_p}{F_s} - K_a\right)(z_c - x)^2 \times \frac{z_c - x}{3} - \sigma_B (z_c - x) \times \frac{(z_c - x)}{2} - S(z_c - x) = 0$$

$$\frac{\gamma}{3}\left(\frac{K_p}{F_s} - K_a\right)(z_c - x)^2 - \sigma_B (z_c - x) - 2S = 0 \qquad \text{[a quadratic]}$$

From which,

$$z_c - x = \frac{\sigma_B \pm \sqrt{\sigma_B^2 + \dfrac{8\gamma S}{3}\left(\dfrac{K_p}{F_s} - K_a\right)}}{\dfrac{2\gamma}{3}\left(\dfrac{K_p}{F_s} - K_a\right)}$$

Or

$$\boxed{z_c = \frac{\sigma_B + \sqrt{\sigma_B^2 + \dfrac{8\gamma S}{3}\left(\dfrac{K_p}{F_s} - K_a\right)}}{\dfrac{2\gamma}{3}\left(\dfrac{K_p}{F_s} - K_a\right)} + x} \qquad (8.74)$$

Increasing z_c by 20%: $\boxed{z = 1.2 z_c}$

Maximum bending occurs at distance y from T.

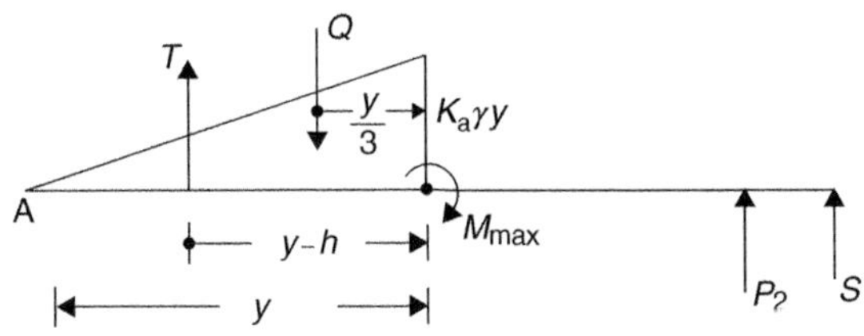

Figure 8.79

Force on length y:

$$Q = \frac{K_a \gamma y^2}{2}$$

For equilibrium of forces at y: $$T = Q = \frac{K_a \gamma y^2}{2}$$

Expressing: $$\boxed{y = \sqrt{\frac{2T}{K_a \gamma}}}$$

Therefore, $$M_{max} = T(y - h) - \frac{Qy}{3} = T(y - h) - \frac{Ty}{3} = T\left(y - h - \frac{y}{3}\right)$$

Hence, $$\boxed{M_{max} = T\left(\frac{2}{3}y - h\right)} \qquad (8.75)$$

Example 8.12

Design the sheet pile wall in Example 8.10 by the fixed-earth method and determine the maximum bending moment.

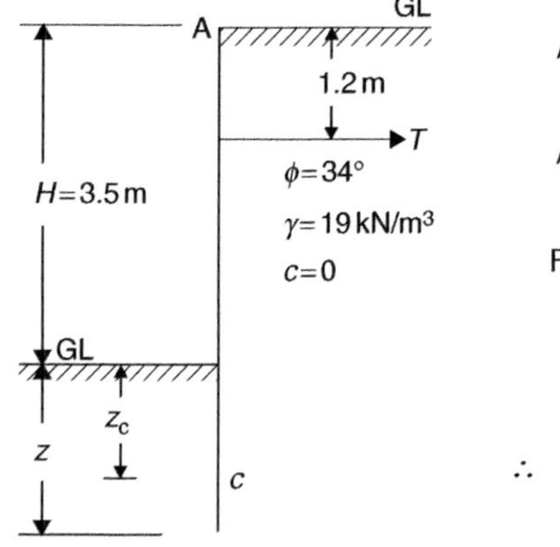

Figure 8.80

$$K_a = 0.283 \qquad \frac{K_p}{F_s} = \frac{3.537}{2} = 1.77$$

$$K_p = 3.537 \qquad \frac{K_p}{F_s} - K_a = 1.49$$

From Table 8.10 by interpolation for $\phi' = 34°$

30°	34°	35°
0.08	0.044	0.035

$$\therefore \quad x = 0.044H = 0.044 \times 3.5 = 0.15\,\text{m}$$

Pressures:

$$K_a\gamma(H+x) = 0.283 \times 19 \times (3.5+0.15) = 19.63$$

$$\frac{K_p}{F_s}\gamma x = 1.77 \times 19 \times 0.15 = 5$$

$$\therefore \sigma_B = 19.63 - 5 = 14.63\,\text{kN/m}^2$$

$$\left(\frac{K_p}{F_s} - K_a\right)\gamma(z_c - x) = 1.49 \times 19(z_c - x) = 28(z_c - 0.15)$$

The net factored pressure distribution is used to solve the problem.

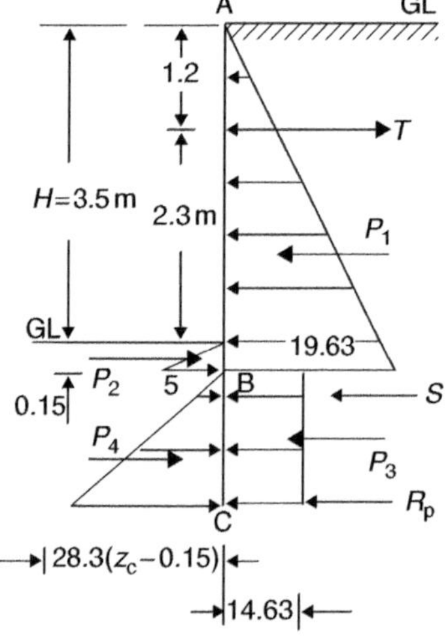

Figure 8.81

Beam AB

$$P_1 = \frac{19.63 \times 3.65}{2} = 35.8\,\text{kN}$$

$$y_1 = \frac{2}{3}(H+x) - h = \frac{2}{3} \times 3.65 - 1.2 = 1.23\,\text{m}$$

$$P_2 = \frac{5 \times 0.15}{2} = 0.38\,\text{kN}$$

$$y_2 = H - h + \frac{2x}{3} = 2.3 + \frac{2 \times 0.15}{3} = 2.4\,\text{m}$$

Shear force:

$$S = \frac{P_1 y_1 - P_2 y_2}{H + x - h} \tag{8.72}$$

$$= \frac{35.8 \times 1.23 - 0.38 \times 2.4}{3.65 - 1.2} = 17.6\,\text{kN}$$

From (8.73) $T=1.15\,(P_1-P_2-S)=1.15\times(35.8-0.38-17.6)=20.5\,\text{kN/m}$

Beam BC

$$P_3 = 14.63(z_c - 0.15) \qquad y_3 = \frac{z_c - 0.15}{2}$$

$$P_4 = \frac{28.3}{2}(z_c - 0.15)^2 = 14.15(z_c - 0.15)^2 \qquad y_4 = \frac{z_c - 0.15}{3}$$

$$\overset{+}{\curvearrowright}\ \Sigma M_c = P_4 y_4 - P_3 y_3 - S(z_c - 0.15) = 0$$

$$\frac{28.3}{6}(z_c - 0.15)^3 - \frac{14.63}{2}(z_c - 0.15)^2 - 17.6(z_c - 0.15) = 0$$

$$4.72(z_c - 0.15)^2 - 7.32(z_c - 0.15) - 17.6 = 0$$

$$(z_c - 0.15)^2 - 1.55(z_c - 0.15) - 3.73 = 0$$

Let $t = z_c - 0.15$ and solve the quadratic

$t^2 - 1.55t - 3.73 = 0$

$$t = \frac{1.55 \pm \sqrt{2.4 + 4\times 3.73}}{2} = \frac{1.55 \pm 4}{2}$$

$$= \frac{1.55 + 4}{2} = 2.78$$

Therefore, $z_c = t + 0.15 = 2.9\,\text{m}$

Alternatively by (8.74): $$Z_c = \frac{14.63 + \sqrt{214 + \dfrac{8\times 19\times 17.6\times 1.49}{3}}}{2\times 19\times 1.49/3} + x$$

$$= \frac{14.63 + 39.28}{18.87} + 0.15 = 3\,\text{m} \approx 2.9\,\text{m}$$

And $z = z_c + 0.2z_c = 3 + 0.2\times 3 = 3.6\,\text{m}$ (Taking $z_c = 3\,\text{m}$)

Max. BM: $$y = \sqrt{\frac{2T}{K_a\gamma}} = \sqrt{\frac{2\times 20.5}{0.283\times 19}} = 2.76\text{m}$$

$$\therefore \quad M_{max} = T\left(\frac{2}{3}y - h\right) = 20.5\times\left(\frac{2}{3}\times 2.76 - 1.2\right) = 13\,\text{kNm/m}$$

Table 8.11

Comparison:	**Example 8.10**	**Example 8.12**
T (kN)	28	20.5
z (m)	1.7	3.6

It appears from these results that a longer pile is required to fix the end, thus increasing the resistance against rotation. Corresponding tension (T) is smaller, which means smaller maximum bending moment in the pile, hence smaller required cross section.

8.15.3 Anchorage

There are two points to consider in the design of support to the tie rods.

1. The distance of the anchor from the sheet piles, that is the length of tie rod.
2. The prevention of bearing failure of soil supporting the anchor.

8.15.4 Length of tie rod (L)

It is important to place the anchor well away from the sheet pile wall that is outside the zone of possible failure. Figure 8.82 shows the safe zone suggested by Lohmeyer (1934).

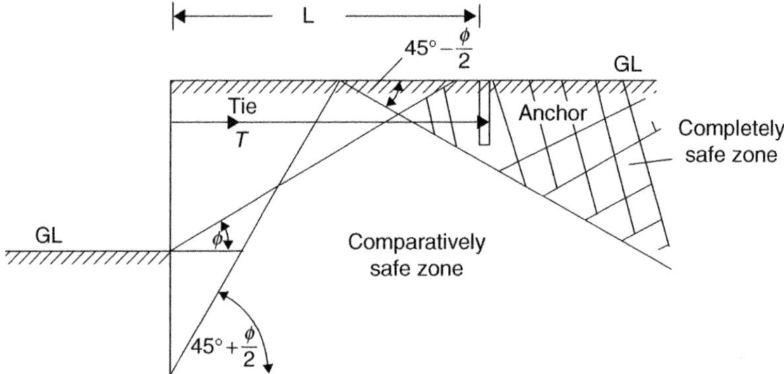

Figure 8.82

8.15.5 Stability of anchors

Anchors may be constructed in various ways:

1. Sheet pile anchor wall (Figure 8.83a)
2. Raking pile anchor (Figure 8.83b)
3. Ground anchor (Figure 8.83c)

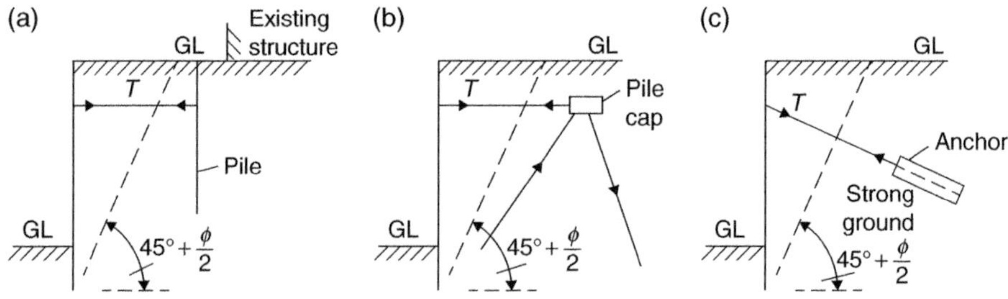

Figure 8.83

4. Anchor wall lying parallel to the sheet pile wall, extending from the ground surface. Its height (b) is determined from the equilibrium of forces acting on it.

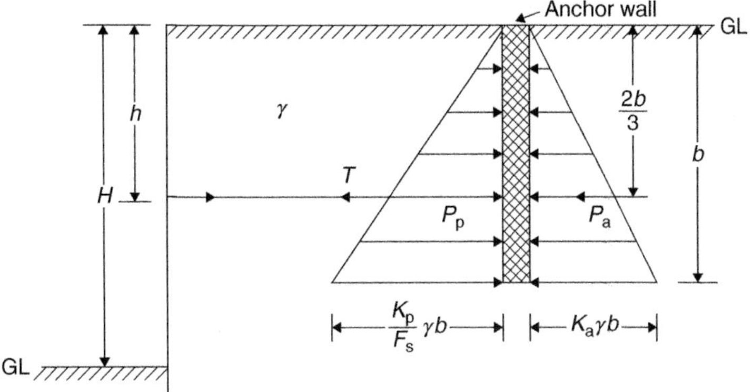

Figure 8.84

$$P_a = \frac{K_a \gamma b^2}{2} \quad P_p = \frac{K_p \gamma b^2}{2F_s}$$

Horizontal equilibrium of forces:

$$T + P_a = P_p$$

$$T = \frac{K_p \gamma b^2}{2F_s} - \frac{K_a \gamma b^2}{2}$$

$$T = \frac{\gamma}{2}\left(\frac{K_p}{F_s} - K_a\right) b^2 \qquad \text{from which:} \quad \boxed{b = \sqrt{\frac{2T}{\gamma\left(\frac{K_p}{F_s} - K_a\right)}}} \tag{8.76}$$

The depth of tie rod is:

$$\boxed{h = \frac{2b}{3}} \tag{8.77}$$

5. Sunken, parallel anchor wall. Again, its size is determined by the equilibrium of forces acting on it.

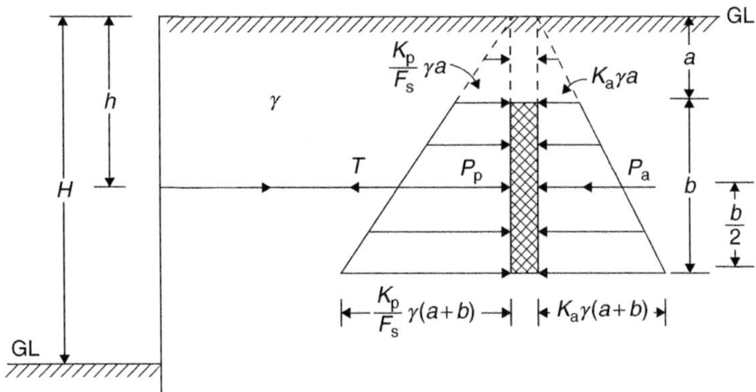

Figure 8.85

$$P_a = \frac{1}{2}\left[K_a\gamma a + K_a\gamma(a+b)\right]b$$
$$= \frac{K_a\gamma}{2}(2a+b)b$$
$$P_p = \frac{1}{2}\left[\frac{K_p}{F_s}\gamma a + \frac{K_p}{F_s}\gamma(a+b)\right]b$$
$$= \frac{K_p\gamma}{2F_s}(2a+b)b$$

Equilibrium of forces: $\overleftarrow{T} + \overleftarrow{P_a} = \overrightarrow{P_p}$

or $T = P_p - P_a$

Substituting: $$T = \frac{K_p\gamma}{2F_s}(2a+b)b - \frac{K_a\gamma}{2}(2a+b)b$$
$$= \frac{\gamma}{2}\left(\frac{K_p}{F_s} - K_a\right)\left(2ab + b^2\right)$$

From Figure 8.86: $$h = a + \frac{b}{2} \qquad \therefore \boxed{a = h - \frac{b}{2}} \qquad (8.78)$$

Substituting, $$T = \frac{\gamma}{2}\left(\frac{K_p}{F_s} - K_a\right)\left[2\left(h - \frac{b}{2}\right)b + b^2\right]$$
$$= \frac{\gamma}{2}\left(\frac{K_p}{F_s} - K_a\right)\left(2hb - b^2 + b^2\right)$$
$$\frac{2T}{\gamma\left(\frac{K_p}{F_s} - K_a\right)} = 2hb$$

$$\therefore \boxed{b = \frac{T}{\gamma\left(\frac{K_p}{F_s} - K_a\right)h}} \qquad (8.79)$$

All variables on the right-hand side are known as h has to be pre-determined in order to calculate T.

Example 8.13

Determine the height of the sunken anchor wall and the length of tie rod for the sheet pile wall in Example 8.12.

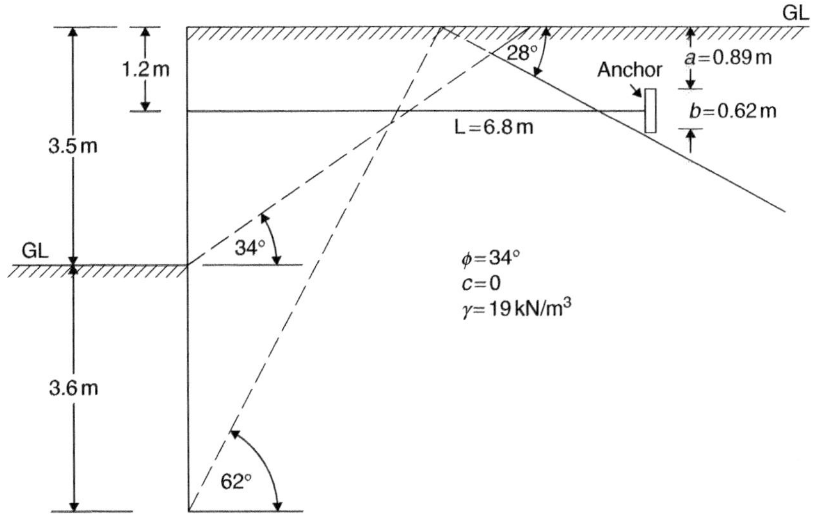

Figure 8.86

From Example 8.12:

$$K_a = 0.283 \qquad \frac{K_p}{F_s} = \frac{3.537}{2} = 1.77 \qquad \frac{K_p}{F_s} - K_a = 1.49$$

$$T = 20.5\,\text{kN/m of wall length} \qquad h = 1.2\,\text{m}$$

From (8.79): $$b = \frac{20.5}{19 \times 1.44 \times 1.2} = 0.62\,\text{m}$$

From (8.78): $$a = h - \frac{b}{2} = 1.2 - \frac{0.62}{2} = 0.89\,\text{m}$$

After drawing the anchor on the scaled section, within the safe zone, the length of the tie rod is measured as: $L = 6.8$ m (say).

8.16 Effect of ground water

The effect of water level difference on sheet pile walls, driven into cohesionless material, was discussed in Chapter 3, with regard to seepage only. However, the driving length and the tension in the anchor rod are also affected by the presence of ground water. The pressure distribution below the water level is in terms of the submerged density of the soil. The water pressure diagram depends on the relative positions of the upper and lower water levels. There are three cases to consider:

1. If water level on one side of the piles, due to flooding

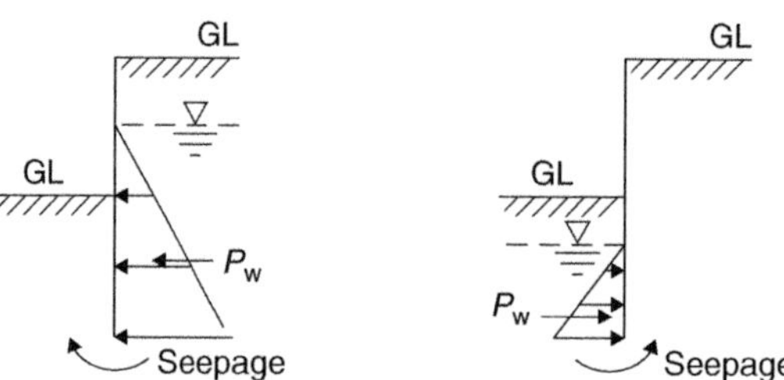

Figure 8.87(a) **Figure 8.87(b)**

2. Water levels at the same height.

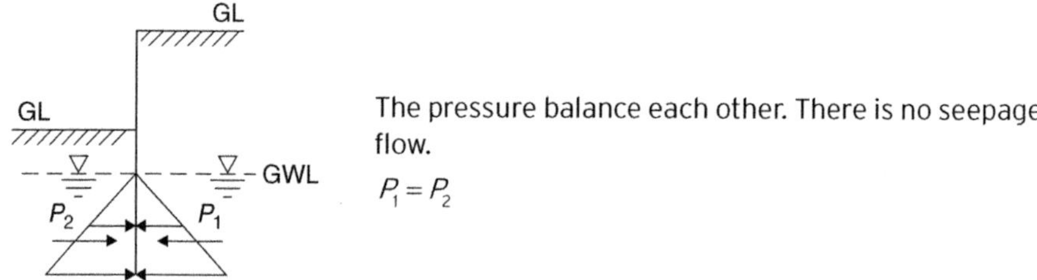

The pressure balance each other. There is no seepage flow.

$P_1 = P_2$

Figure 8.87(c)

3. Water is at different levels, ignoring seepage initially.

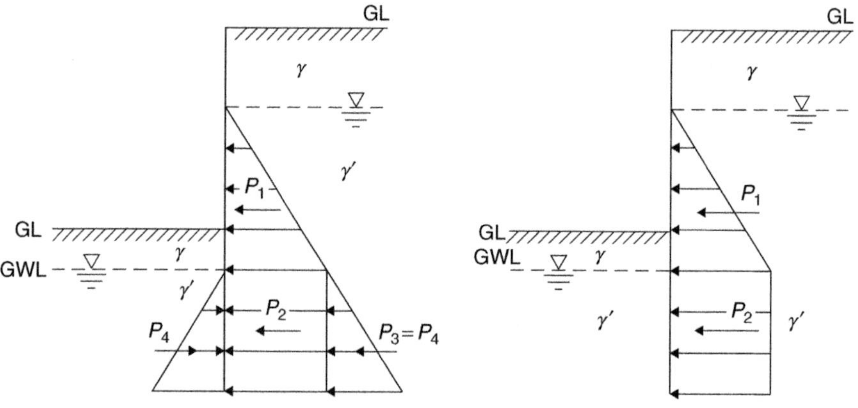

Figure 8.87(d) **Figure 8.87(e)**

The net pressure diagram (e) is to be used in a design and the result is modified for seepage, as necessary.

Example 8.14

A 3.5 m deep excavation is cohesionless soil is to be supported by sheet pile walls so, that works may be carried out at the lower level. Any ground water inside the excavation is pumped out from a sump. The water level at the other side of the sheet piles coincides with the proposed position of the tie rod at 1.2 m depth. Determine:

1. The depth of penetration (z) of the pile.
2. The force in the tie rod.

The problem is to be solved by the free-earth support method.

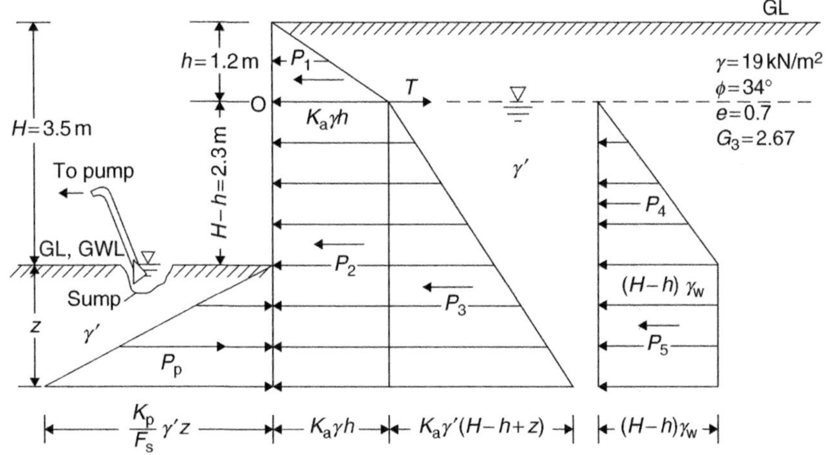

Figure 8.88

Pressure and force diagrams

$K_a = 0.283$

$K_p = 3.537 \quad \left| \dfrac{K_p}{F_s} = 1.769 \right.$

$F_s = 2$

$$\text{From (1.43)}: \gamma' = \frac{(G_s - 1)\gamma_w}{1+e} = \frac{(2.67-1)\times 9.81}{1+0.7} = 9.64\,\text{kN/m}^3$$

Pressures:

$$\frac{K_p}{F_s}\gamma' z = 1.769 \times 9.64z = 17z$$

$$K_a\gamma h = 0.283 \times 19 \times 1.2 = 6.45\,\text{kN/m}^2$$

$$K_a\gamma'(H-h+z) = 0.283 \times 9.64 \times (3.5-1.2+z) = 2.73 \times (2.3+z)$$

$$(H-h)\gamma_w = 2.3 \times 9.81 = 22.56\,\text{kN}$$

Forces and their distances from the tie rod:

$$P_1 = \frac{K_a \gamma h}{2} \times h = \frac{6.45}{2} \times 1.2 = 3.87\,\text{kN}$$

$$y_1 = \frac{h}{3} = \frac{1.2}{3} = 0.4\,\text{m}$$

$$P_2 = K_a \gamma h (H - h + z) = 6.45 \times (2.3 + z)$$

$$y_2 = \frac{1}{2}(H - h + z) = 0.5 \times (2.3 + z)$$

$$P_3 = \frac{K_a \gamma' (H - h + z)}{2}(H - h + z) = \frac{2.73}{2} \times (2.3 + z)^2 = 1.37 \times (2.3 + z)^2$$

$$y_3 = \frac{2}{3}(H - h + z) = 0.67 \times (2.3 + z)$$

Water:

$$P_4 = \frac{(H - h)\gamma_w}{2}(H - h) = \frac{22.56}{2} \times 2.3 = 25.9\,\text{kN}$$

$$y_4 = \frac{2}{3}(H - h) = \frac{2 \times 2.3}{3} = 1.53\,\text{m}$$

$$P_5 = (H - h)\gamma_w z = 22.56z$$

$$y_5 = H - h + \frac{z}{2} = 2.3 + \frac{z}{2}$$

Passive:

$$P_p = \frac{K_p \gamma' z}{2F_s} \times z = \frac{17z}{2} \times z = 8.5z^2$$

$$y_p = H - h + \frac{2}{3}z = 2.3 + 0.67z$$

To evaluate tension T, equate the moments of forces about the tie to zero.

$$\overset{+}{\curvearrowleft}\Sigma M_O = P_p y_p + P_1 y_1 - P_2 y_2 - P_3 y_3 - P_4 y_4 - P_5 y_5 = 0$$

$$8.5z^2 \times (2.3 + 0.67z) + 3.87 \times 0.4 - 6.45 \times 0.5 \times (2.3 + z)^2$$
$$- 1.37 \times 0.67\,(2.3 + z)^3 - 25.9 \times 1.53 - 22.56z \times (2.3 + 0.5z) = 0$$

$$19.55z^2 + 5.7z^3 + 1.5 - 3.22 \times (2.3 + z)^2 - 0.92 \times (2.3 + z)^3 - 39.6 - 51.9z - 11.3z^2 = 0$$

Quadratic terms: $(2.3+z)^2 = 5.29 + 4.6z + z^2$
Cubic terms: $(2.3+z)^3 = 12.2 + 15.9z + 6.9z^2 + z^3$

Collecting similar terms and expanding the exponential ones:

$$5.7z^3 + 8.25z^2 - 51.9z - 38.1 - \left(17 + 14.8z + 3.22z^2\right) - \left(11.2 + 14.6z + 6.3z^2 + 0.92z^3\right) = 0$$

Collecting similar terms:

$$(5.7-0.92)z^3+(8.25-3.22-6.3)z^2-(51.9+14.8+14.6)z-(38.1+17+11.2)=0$$

$$\boxed{4.78z^3-1.27z^2-81.3z-66.3=0}$$

The solution of this cubic equation by graphics or otherwise is $z=4.61\,\text{m}$

Tension T:

$$P_1=3.87\,\text{kN}$$
$$P_2=6.45\times(2.3+4.61)=6.45\times6.91=44.6\,\text{kN}$$
$$P_3=1.37\times(6.91)^2=65.4\,\text{kN}$$
$$P_4=25.9\,\text{kN}$$
$$P_5=22.56\times4.61=104\,\text{kN}$$
$$P_p=8.5\times(4.61)^2=180.6$$

For horizontal equilibrium $\Sigma F_x=0$

$$\overset{+}{\rightarrow}\quad \Sigma F_x=\overrightarrow{P_p}+\overrightarrow{T}-\overleftarrow{P_1}-\overleftarrow{P_2}-\overleftarrow{P_3}-\overleftarrow{P_4}-\overleftarrow{P_5}=0$$

$$\therefore\quad T=P_1+P_2+P_3+P_4+P_5-P_p$$
$$=3.87+44.6+65.4+25.9+104-180.6$$
$$=243.77-180.6=63\,\text{kN/m of wall length}$$

Increasing this by 15% $T=1.15\times63=72.5\,\text{kN}$

If the ties are placed at 3 meters apart, then each tie has to be designed to carry $3\times72.5\approx218\,\text{kN}$.

Note: $z=4.61\,\text{m}$ should be checked for adequacy against 'boiling' failure. See Example 8.15.

Example 8.15

The driving depth in example 8.14 was calculated to be $z=4.61\,\text{m}$, assuming no seepage. This would only occur when the pile penetrates an underlying impervious layer. Assuming that the impervious layer is at 15m depth, determine:

a) The factor of safety against boiling failure, using the partial flow net on Graph 8.8.
b) The driving length required, due to the modification of submerged densities for seepage, at $F_s=2$.

Solution

a) The factor of safety against piping is found by the methods of Chapter 3.

From Graph 8.8: $N_e=19$ $\qquad \Delta h=\dfrac{H-h}{N_e}=\dfrac{2.3}{19}=0.121$

$N_a=11$

$N_b=15.7$

$H_T=z+H=4.61+2.3=6.91\,\text{m}$

$H_2=z=4.61\,\text{m}$

Average pressure head (h_1) acting on the base (a-b) of prism:

$$h_a = H_T - \Delta h N_a = 6.91 - 0.121 \times 11 = 5.58\,\text{m}$$
$$h_b = H_T - \Delta h N_B = 6.91 - 0.121 \times 15.7 = 5\,\text{m}$$

From (3.36): $h_1 = \dfrac{h_a + 2h_b}{3} = \dfrac{5.58 + 2.5}{3} = 5.2\,\text{m}$

From (3.37): $h_s = h_1 - h_2 = 5.2 - 4.61 = 0.59\,\text{m}$

Average hydraulic gradient: $i_{av} = \dfrac{h_s}{z} = \dfrac{0.59}{4.61} = 0.128$

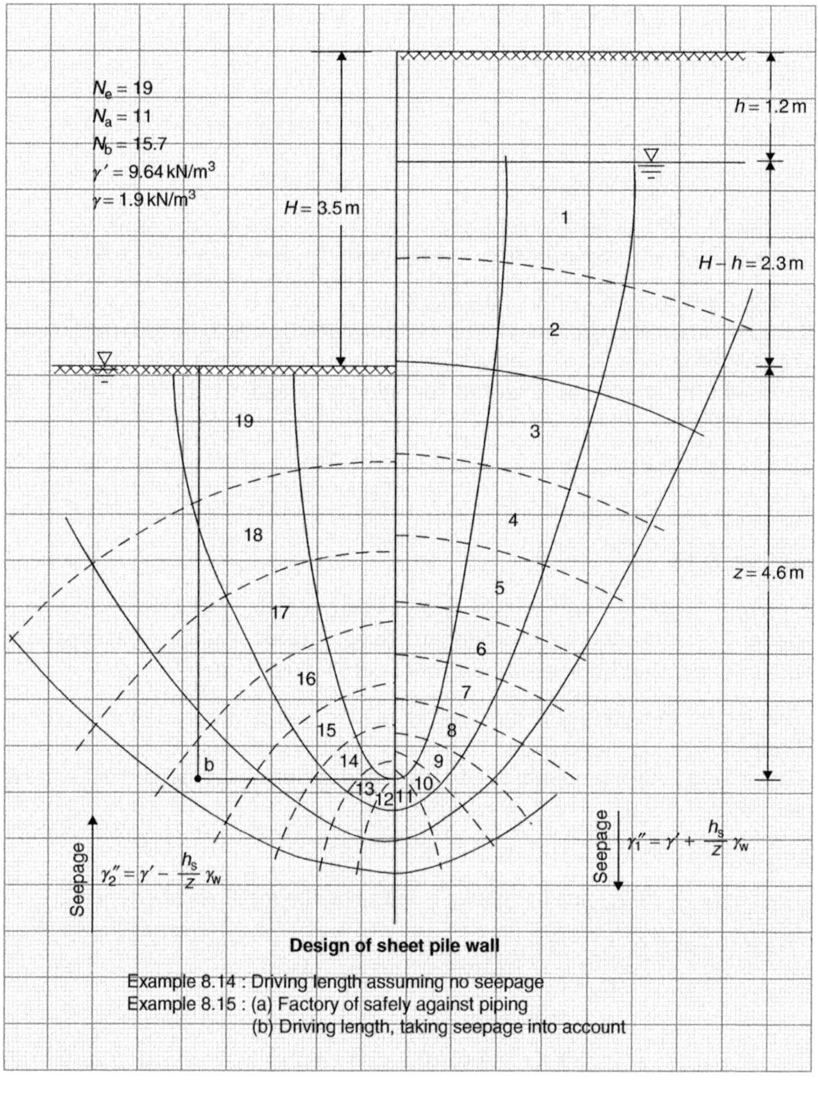

Graph 8.8

Critical hydraulic gradient: $$i_c = \frac{\gamma'}{\gamma_w} = \frac{9.64}{9.81} = 0.983$$

Factor of safety against piping failure from (3.42).

$$F_s = \frac{i_c}{i_{av}} = \frac{0.983}{0.128} = 7.7$$

b) In order to determine the driving depth and retain 2 as the factor of safety, the submerged density has to be modified for upward and downward seepage (see also chapter 5)

Downward seepage: $$\gamma'' = \gamma' + \frac{h_s}{z}\gamma_w = 9.64 + \frac{0.59 \times 9.81}{4.61}$$
$$= 9.64 + 1.26 = 10.9\,\text{kN/m}^2$$

Upward seepage: $$\gamma_2'' = \gamma' - \frac{h_s}{z}\gamma_w = 9.64 - 1.26 = 8.4\,\text{kN/m}^2$$

This means that on the upstream side of the pile, the submerged density becomes heavier, thus exerting increased active pressure on the structure. On the other side, however, the submerged weight is decreased, resulting in smaller passive pressure. As a consequence, the driving depth has to be increased, in order to keep 2 as the factor of safety. The procedure is exactly the same as in Example 8.14, using the pressure diagram in Figure 8.88.

Pressures

$$\frac{K_p}{F_s}\gamma_2''\, z = \frac{3.537}{2} \times 8.4z = 14.86z$$

$K_a\gamma h = 6.45\text{kN/m}^2$ (Unchanged)

$K_a\gamma_1''\,(2.3 + z) = 0.283 \times 10.9 \times (2.3 + z) = 3.1 \times (2.3 + z)$

$(H - h)\gamma_w = 22.56\text{kN/m}^2$ (unchanged)

Forces

$$P_1 = \frac{K_a\gamma h^2}{2} = 3.87\text{kN} \quad y_1 = 0.4\text{m (unchanged)}$$

$$P_1 y_1 = 3.87 \times 0.4 = 1.5\,\text{kNm}$$

$$P_2 = K_a\gamma h(2.3 + z) = 6.45 \times (2.3 + z) \qquad y_2 = 0.5 \times (2.3 + z)$$

$$P_2 y_2 = 6.45 \times 0.5 \times (2.3 + z)^2 = 3.23 \times \left(5.29 + 4.6z + z^2\right)$$
$$= 17.1 + 14.9z + 3.23z^2$$

$$P_3 = \frac{K_a\gamma_2''}{2} \times (2.3 + z)^2 = \frac{3.1}{2} \times (2.3 + z)^2 = 1.55 \times (2.3 + z)^2 \qquad y_3 = 0.67 \times (2.3 + z)$$

$$P_3y_3 = 1.55 \times 0.67 \times (2.3 + z)^3 = 1.04 \times (12.2 + 15.9z + 6.9z^2 + z^3)$$
$$= 12.7 + 16.5z + 7.2z^2 + 1.04z^3$$

$P_4 = 25.9\text{kN}$ (unchanged) $\quad y_4 = 1.53\text{m}$

$$P_4y_4 = 25.9 \times 1.53 = 39.6\text{kNm}$$

$P_5 = 22.56z \quad y_5 = 2.3 + \frac{z}{2}$ (Unchanged)

$$P_5y_5 = 22.56z \times \left(2.3 + \frac{z}{2}\right) = 52z + 11.3z^2$$

$$P_p = \frac{K_p\gamma_2'' z}{F_s} \times \frac{z}{2} = \frac{14.86z^2}{2} = 7.43z^2 \quad y_p = 2.3 + 0.67z$$

$$P_py_p = 7.43z^2 \times (2.3 + 0.67z) = 17.1z^2 + 5z^2$$

$$+\curvearrowleft \Sigma M_O = 17.1z^2 + 5z^3 + 1.5 - 17.1 - 14.9z - 3.23z^2 - 12.7 - 16.5z$$
$$-7.2z^2 - 1.04z^3 - 39.6 - 52z - 11.3z^2 = 0$$

Summing like terms to get the cubic equation

$$3.96z^3 - 4.63z^2 - 83.3z - 67.9 = 0$$

From which $z = 5.5\,\text{m}$

Therefore, the effect of seepage is to increase the length of sheet piles by 0.9 m, at F_s=2.

8.17 Stability of deep trenches

Narrow, deep trenches - excavated to accommodate pipelines, diaphragm walls etc. - have to be supported temporarily to prevent collapse prior or during construction. Two methods of support are introduced below:

1. Horizontal bracing
2. Bentonite slurry support

8.17.1 Horizontal bracing

This method of support is applied to vertical cuts, deeper than 1.2 m. The material used is either timber or steel.

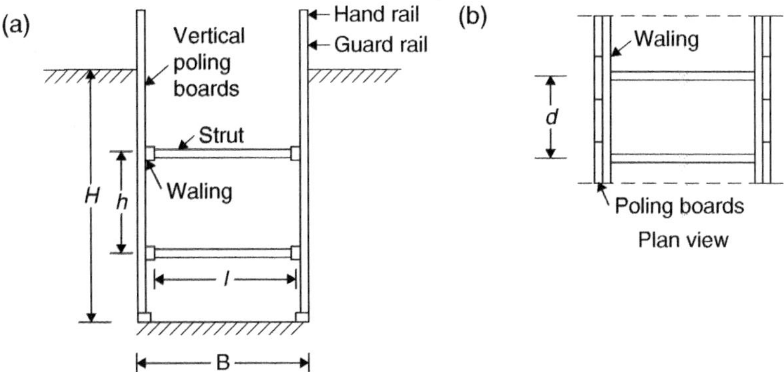

Figure 8.89

Timber support

For cohesive soils, one or several poling board may be omitted (open timbering), whilst closed timbering is appropriate in loose and wet soils. At least 1m high handrail should be constructed on both sides for safety.

The vertical and horizontal distances between the struts (or braces) depend on their strength to carry the estimated pressures acting on the poling boards and the waling. Also, the cross-section of the strut depends on its material, length (l) and the axial force transmitted by the waling.

The supporting arrangement depicted in Figure 8.89 may be altered in various ways by:

a) Using adjustable steel struts as long as $l<1.67$ m
b) Using steel column or suitable beam section as strut
c) Using sheet piles instead of poling board driven into the ground for deeper cuts.
d) Using standard steel I-beam or column sections driven into the ground and placing timber sheeting horizontally behind the flanges. The sheeting should be continuous for cuts of $H>7$ m.

8.17.1.1 Pressure distribution against the sheeting

In contrast to retaining wall problems, there is no theoretical solution for the soil-pressure distribution over the depth of cut. The reason for this is twofold:

1. Whilst a retaining wall is homogeneous, stiff structure, the sheeting is flexible. The flexibility depends on the positioning of the struts during excavation. Moreover, any non-uniformity in the supported soil can greatly influence the magnitude of force in each strut, but has no effect on a rigid wall.
2. The development and magnitude of force in the struts depend largely on the sequence of their installation, hence on the method of excavation.

Notes: Should one strut fail, then the adjacent members would carry its load. This cannot occur, when the support is a solid wall.

Because of uncertainties involved, the pressure distribution, in this case, can only be approximated on the bases of actual measurement of strut loads during and after construction. Terzaghi and Peck suggested the distributions, for dense and, loose sand as shown in Figure 8.90. These are drawn, empirically to envelope the experimental measurements of pressures developed in actual structures.

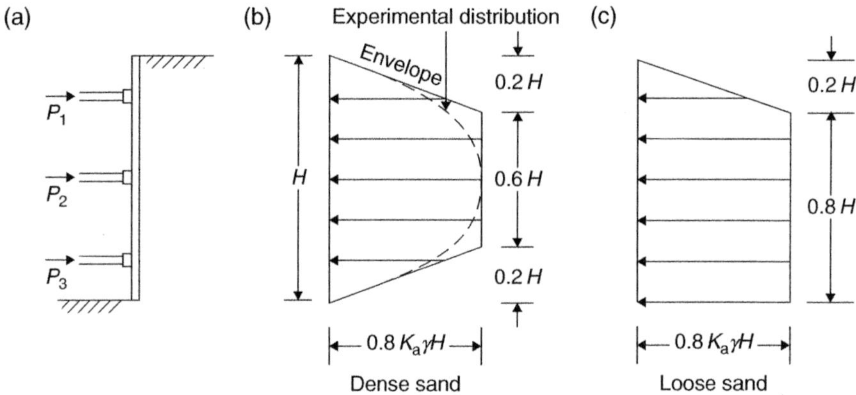

Figure 8.90

where

$$K_a = \frac{1-\sin\phi}{1+\sin\phi} \tag{8.11}$$

The suggested maximum pressure, acting on the sheeting, is 0.8 times the active value determined from Rankine's theory.

Note: The diagrams show idealized, empirical pressure distributions. In reality, the actual, measured shapes can be somewhat different, depending on the method of strut installation. However, as the envelopes indicate larger than the in-situ test results, these may be used to estimate the forces acting on the member of the structure.

The pressure diagrams suggested by Terzaghi and Peck for cohesive soils are shown in Figure 8.91.

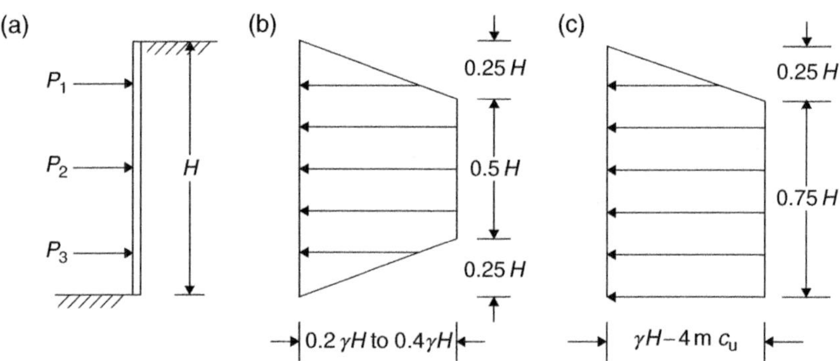

Figure 8.91

where c_u = unconfined compressive strength.

The choice of envelope depends on the stability number of the clay:

If $\frac{\gamma H}{c_u} \leq 4$ then envelope (b) is to be used

$\frac{\gamma H}{c_u} > 4$ and $(\gamma H - 4\text{m}\, c_u) > 0.4\,\gamma H$, then use (c)

$\frac{\gamma H}{c_u} > 4$ and $(\gamma H - 4\text{m}\, c_u) < 0.4\,\gamma H$, then use (b)

$\frac{\gamma H}{c_u} > 4$ then $0.4 \leq m < 1$ for normally consolidated clays. Otherwise $m = 1$

Evaluation of forces in the struts

The pressure envelopes indicate the approximate loading on the sheeting, hence on the struts or braces. Any additional surface loads must be taken into account. However, it is best to keep all superimposed loads, such as construction materials at least 2 m away from the trench.

In order to determine the compression in each strut, the poling is divided into "simply supported beams" by assuming hinges at the strut supports, as shown in the following example.

Example 8.16

A trench 2.5 m wide and 5.5 m deep is to be excavated in compact dry sand, in order to place a 500 mm dia. pipeline with centre at 5 m below ground level. The characteristics of the sand are:

$$\gamma = 19.4\,\text{kN/m}^3$$
$$\phi = 32°$$

Determine the forces in three horizontal timber struts placed at 1.1 m, 2.75 m and 4.4 m from the surface at 1.8 m intervals.

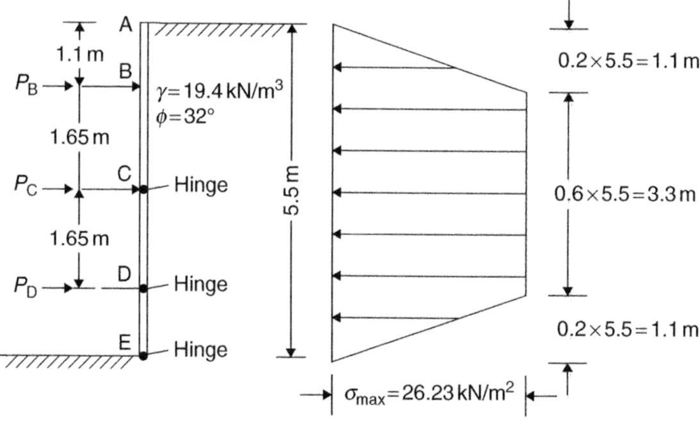

Figure 8.92

The maximum pressure of compact, dense sand is given by

$$\sigma_{max} = 0.8K_a\gamma H$$
$$= 0.8\times\left(\frac{1-\sin 32}{1+\sin 32}\right)\times 19.4\times 5.5$$
$$= 0.8\times\left(\frac{0.47008}{1.5299}\right)\times 106.7 = 26.23\,\text{kN/m}^2$$

This completes the pressure envelope for this problem and the simply supported beams can now be analysed.

Beam ABC is a simple supported cantilever.

26.23 kN/metre width of valing
A
B
C
1.1 m
1.65 m
PB
PCB

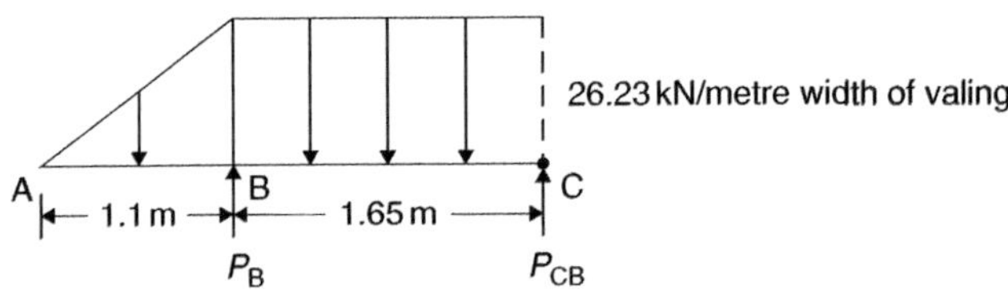

Figure 8.93

Force acting on cantilever AB is given by the area of the triangle, multiplied by the spacing ($d = 1.8$ m)

$$F_1 = 1.8\times\left(\frac{1.1\times 26.23}{2}\right) = 26\,\text{kN acting at } \frac{1.1}{3} = 0.37\,\text{m from B}$$

Similarly, the force acting on span BC is given by:

$$F_2 = 1.8\times(1.65\times 26.23) = 77.9\,\text{kN acting at centre span}$$

The equivalent force diagram can now be drawn and forces in the struts determined.

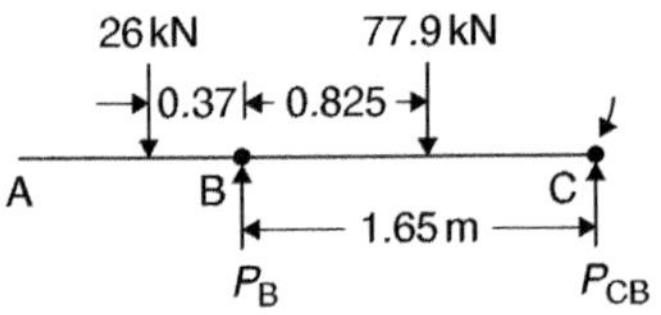

26 kN
77.9 kN
0.37
0.825
A
B
C
1.65 m
PB
PCB

Sum of moments about B=0

$$\sum M_B = 77.9\times 0.825 - 26\times 0.37 - 1.65P_{CB} = 0$$
$$54.65 - 1.65P_{CB} = 0$$

Figure 8.94

From which the force in strut C, contributed by the load on length AC is:

$$P_{CB} = \frac{54.65}{1.65} = 33.1\,\text{kN}$$

Summing the vertical forces to get P_B

$$+\uparrow \Sigma V = \overset{\uparrow}{P_B} + \overset{\uparrow}{P_{CB}} - \overset{\downarrow}{26} - \overset{\downarrow}{77.9} = 0$$

$$P_B = 103.9 - P_{CB} = 103.9 - 33.1 = 70.8\,\text{kN}$$

Beam CD is again simply supported

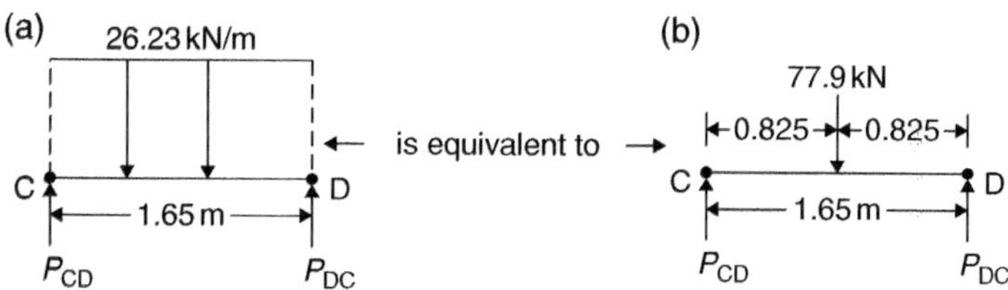

Figure 8.95

Sum of moments about D: $\overset{+}{\curvearrowleft}\Sigma M_D = 1.65 P_{CD} - 0.825 \times 77.9 = 0$

$$1.65 P_{CD} = 64.27$$

Therefore the force in strut C contributed by the loading on span CD is:

$$P_{CD} = \frac{64.27}{1.65} = 39\,\text{kN}$$

Summing the vertical forces to get P_{DC}:

$$+\uparrow \Sigma V = \overset{\uparrow}{P_{CD}} + \overset{\uparrow}{P_{DC}} - \overset{\downarrow}{77.9} = 0$$

$$39 + P_{DC} - 77.9 = 0 \quad \therefore \quad P_{DC} = 77.9 - 39 = 38.9\,\text{kN}$$

Beam DE is simply supported

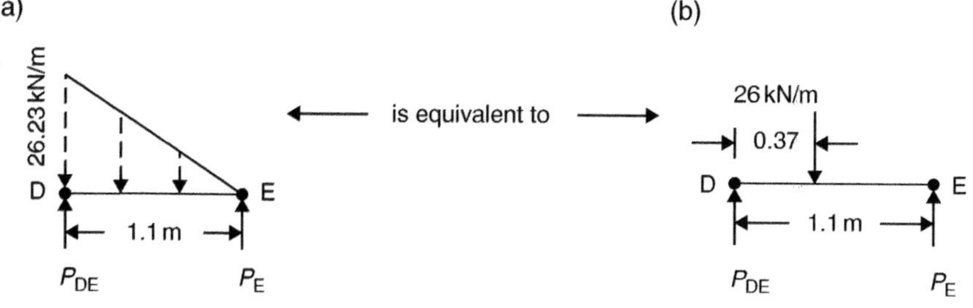

Figure 8.96

$$\circlearrowleft + \Sigma M_D = 26 \times 0.37 - 1.1P_E = 0 \qquad \therefore \quad P_E = \frac{9.62}{1.1} = 8.7\,\text{kN}$$

$$+\uparrow \Sigma V = P_{DE} + P_E - 26 = 0 \qquad \therefore \quad P_{DE} = 26 - 8.7 = 17.3\,\text{kN}$$

Summary of results:

$P_B = 70.8\,\text{kN}$
$P_C = P_{CB} + P_{CD} = 33.1 + 39 = 72.1\text{kN}$
$P_D = P_{DC} + P_{DE} = 38.9 + 17.3 = 56.2\,\text{kN}$
$P_E = 8.7\,\text{kN}$ (carried by the soil)

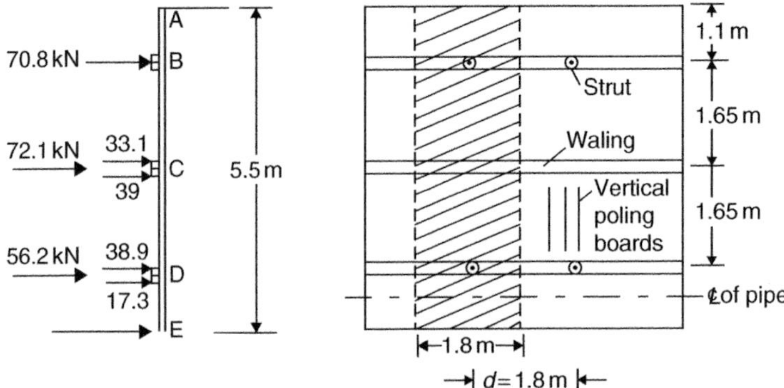

Figure 8.97

Notes:

1. A pressure diagram is not indicative of actual earth pressures, but is an aid for the determination of loads on the struts, as found by the site measurements. For this reason, only one side of the trench need to be drawn to solve problems.
2. The load on struts decreases with depth.
3. Failure of one strut transfers load to adjacent ones and could precipitate general collapse.
4. The timber or steel members of the bracing system must be designed as structural sections, subject to bending or compression.
5. It is advisable to check the estimated compressive force in a strut by in-situ measurements.
6. Check should be made as to the stability of base against heaving in clay or piping in water bearing sand.

8.18 Bentonite slurry support

This method is used as temporary support to the very narrow trenches. Thixotropic (see Chapter 6) bentonite slurry is poured into the trench as it is excavated. Being thixotropic, the slurry gels when undisturbed, but reverts to fluid if excavation is continued. The trench is then supported by the hydrostatic pressure of the slurry.

8.18.1 Trench in clay

The bentonite seals the vertical surfaces and prevents loss of pore water from the clay. The stability of the filled trench is analysed in terms of total stresses, as the clay is undrained and the trench is open in the short term only. As $\phi_u = 0$ for saturated clay, the factor of safety is derived in terms of c_u, assuming the inclination of the critical wedge at 45° to the wall.

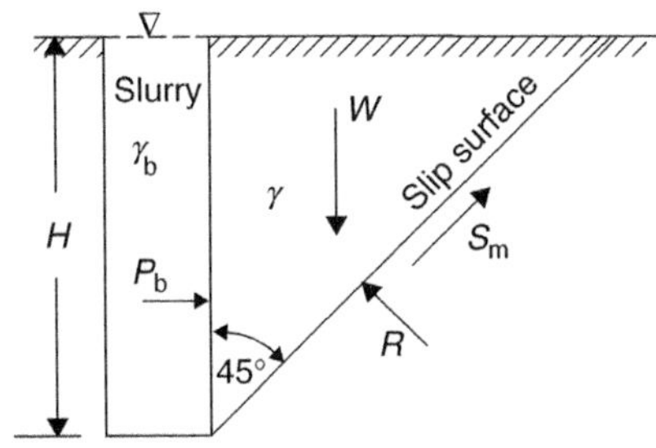

Figure 8.98

γ = unit weight of clay

γ_b = unit weight of slurry

L = length of slip surface

$L = \sqrt{H^2 + H^2} = \sqrt{2}H$

P_b = hydrostatic force exerted by the bentonite

$$P_b = \frac{\gamma_b H^2}{2}$$

W = weight of the wedge $\frac{\gamma H^2}{2}$

S_m = shearing force along the slip surface

S = available shear force along slip surface $= \sqrt{2}Hc_u$

Resolving all forces acting on the wedge along the slip surface and summing them: ignore reaction R

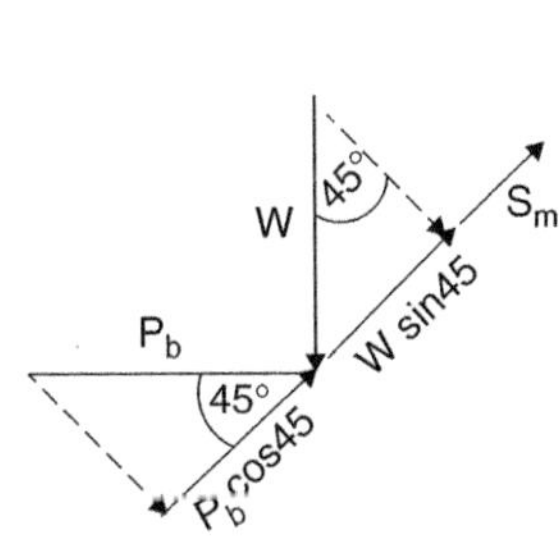

Figure 8.99

$$P_b \cos 45 + S_m - W \sin 45 = 0$$

$$S_m = W \sin 45 - P_b \cos 45$$

$$\text{But, } \sin 45 = \cos 45 = \frac{1}{\sqrt{2}}$$

$$\therefore \quad S_m = \frac{1}{\sqrt{2}}(W - P_b)$$

$$\text{or} \quad S_m = \frac{1}{\sqrt{2}}\left(\frac{\gamma H^2}{2} - \frac{\gamma_b H^2}{2}\right) = \frac{H^2}{2\sqrt{2}}(\gamma - \gamma_b)$$

The factor of safety may be defined as:

$$F_s = \frac{\text{Available shear force}}{\text{Induced shear force}} = \frac{S}{S_m} = \frac{\sqrt{2}Hc_u}{\frac{H^2}{2\sqrt{2}}(\gamma - \gamma_b)} \qquad \therefore \boxed{F_s = \frac{4c_u}{H(\gamma - \gamma_b)}} \tag{8.80}$$

Actual shearing stress on the slip surface is given by:

$$\tau = \frac{S_m}{L} = \frac{1}{\sqrt{2}H}\left[\frac{H^2}{2\sqrt{2}}(\gamma - \gamma_b)\right] = \frac{H}{4}(\gamma - \gamma_b)$$

Therefore, the alternative definition for F_s is:

$$F_s = \frac{\text{Available shear stress}}{\text{Actual shear stress}} = \frac{c_u}{\tau}$$

$$= \frac{c_u}{\frac{H}{4}(\gamma - \gamma_b)} = \frac{4c_u}{H(\gamma - \gamma_b)}$$

Example 8.17

A diaphragm wall 7 m deep has to be constructed in homogeneous clay, using bentonite mud of unit weight 11 kN/m³ to support the excavation. The clay has a shear strength of 85 kN/m² and a bulk unit weight of 19.6 kN/m³.

Evaluate the factor of safety against failure of the trench.

From the data: $H = 7\text{m}$

$\gamma = 19.6\,\text{kN/m}^3$

$\gamma_b = 11.0\,\text{kN/m}^3$

$c_u = 85\,\text{kN/m}^2$

$$F_s = \frac{4 \times 85}{7 \times (19.6 - 11)} = 5.6$$

8.18.2 Trench in sand

In this case, the existing ground water table as well as the height of the slurry column have some effect on the minimum unit weight of mud required. This is determined by means of the pressure distribution diagrams.

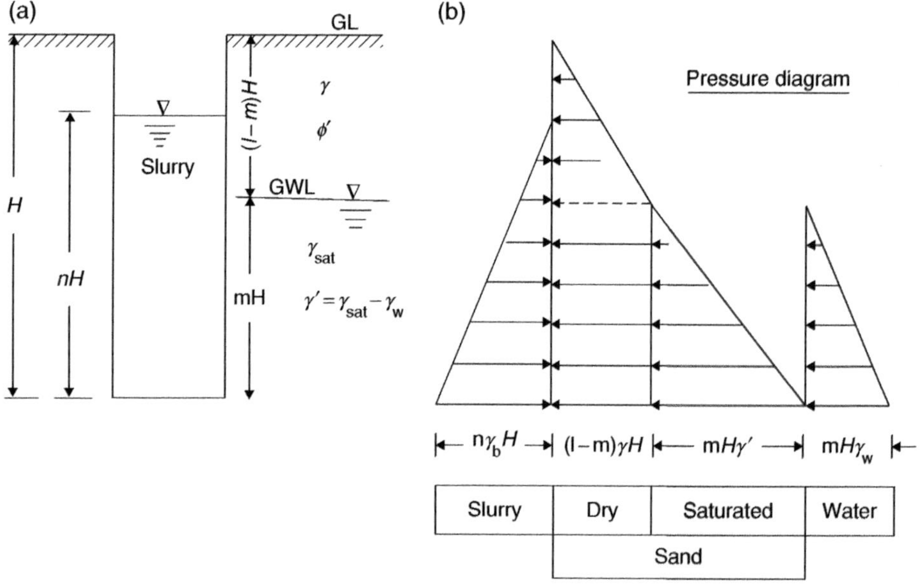

Figure 8.100

This is the general case, when the trench is only partially filled and the ground water level is mH distance above the base of the cut.

The forces acting are determined from the diagram

Slurry: $\overrightarrow{P_b} = \frac{1}{2}(nH \times n\gamma_b H) = \frac{\gamma_b}{2}(nH)^2 = n^2 \frac{\gamma_b H^2}{2}$

Dry sand: $\overleftarrow{P_1} = \frac{K_a}{2}[(1-m)H \times (1-m)H\gamma] = (1-m)^2 \frac{K_a \gamma H^2}{2}$

$$= (1 - 2m + m^2) \times \frac{K_a \gamma H^2}{2}$$

$$\text{where } K_a = \frac{1 - \sin\phi_m}{1 + \sin\phi_m} \text{ (mobilized)}$$

$$\overleftarrow{P_2} = (1-m)\gamma H \times mHK_a = (m - m^2)K_a \gamma H^2$$

Saturated Sand: $\overleftarrow{P_3} = \frac{K_a}{2}(mH \times mH\gamma') = m^2 \frac{K_a \gamma' H^2}{2}$

Ground Water: $\overleftarrow{P_4} = \frac{1}{2}(mH \times mH\gamma_w) = m^2 \frac{\gamma_w H^2}{2}$

Summing the forces for equilibrium:

$$\overset{+}{\rightarrow} \Sigma H = \overrightarrow{P_b} - \overleftarrow{P_1} - \overleftarrow{P_2} - \overleftarrow{P_3} - \overleftarrow{P_4} = 0 \quad \therefore \quad P_b = P_1 + P_2 + P_3 + P_4$$

$$n^2 \frac{\gamma_b H^2}{2} = (1 - 2m + m^2)\frac{K_a \gamma H^2}{2} + (m - m^2) K_a \gamma H^2 + m^2 \frac{K_a \gamma' H^2}{2} + m^2 \frac{\gamma_w H^2}{2}$$

Dividing both sides by $\frac{H^2}{2}$:

$$n^2 \gamma_b = (1 - 2m + m^2)K_a \gamma + (m - m^2)2K_a \gamma + m^2 K_a \gamma' + m^2 \gamma_w$$

Collecting: terms of $K_a \gamma$:

$$n^2 \gamma_b = (1 - 2m + m^2 + 2m - 2m^2)K_a \gamma + m^2 K_a \gamma' + m^2 \gamma_w$$

Substituting $\gamma' = \gamma_{sat} - \gamma_w$:

$$\begin{aligned} n^2 \gamma_b &= (1 - m^2)K_a \gamma + m^2 K_a(\gamma_{sat} - \gamma_w) + m^2 \gamma_w \\ &= K_a \gamma - m^2 K_a \gamma + m^2 K_a \gamma_{sat} - m^2 K_a \gamma_w + m^2 \gamma_w \\ &= K_a \gamma - m^2 [K_a \gamma - K_a \gamma_{sat} + K_a \gamma_w - \gamma_w] \\ &= K_a \gamma + m^2 (1 - K_a)\gamma_w + K_a(\gamma_{sat} - \gamma) \end{aligned}$$

Therefore, the minimum unit weight of the bentonite slurry is given by:

$$\boxed{\gamma_b = \frac{1}{n^2}\left[K_a \gamma + m^2 (1 - K_a)\gamma_w + K_a(\gamma_{sat} - \gamma)\right]} \tag{8.81}$$

If $\gamma_{sat}=\gamma$ then
$$\gamma_b = \frac{1}{n^2}\left[K_a\gamma + m^2(1-K_a)\gamma_w\right] \tag{8.82}$$

The factor of safety may be defined as:

$$F_s = \frac{\tan\phi'}{\tan\phi_m} \tag{8.83}$$

where ϕ' = effective angle of friction of sand

ϕ_m = mobilized friction angle obtained from (8.11), that is the mobilized value of K_a

$$\left.\begin{aligned} K_a &= \frac{1-\sin\phi_m}{1+\sin\phi_m} \\ K_a + K_a\sin\phi_m' &= 1-\sin\phi_m \\ (1+K_a)\sin\phi_m' &= 1-K_a \\ \sin\phi_m &= \frac{1-K_a}{1+K_a} \end{aligned}\right| \quad \therefore \phi_m = \sin^{-1}\left(\frac{1-K_a}{1+K_a}\right) \tag{8.84}$$

Expressing the mobilized value of the active coefficient from 8.81:

$$\begin{aligned} n^2\gamma_b &= K_a\gamma + m^2(1-K_a)\gamma_w + K_a(\gamma_{sat}-\gamma) \\ &= K_a\gamma + m^2\gamma_w - m^2K_a\gamma_w + K_a\gamma_{sat} - K_a\gamma \\ &= K_a(\gamma - m^2\gamma_w + \gamma_{sat} - \gamma) - m^2\gamma_w \end{aligned}$$

Therefore the mobilized
$$K_a = \frac{n^2\gamma_b - m^2\gamma_w}{\gamma_{sat} - m^2\gamma_w} \tag{8.85}$$

Assuming that γ_b and γ are known and that the slurry is up to ground level, that is $n=1$ then the coefficient becomes:

$$K_a = \frac{\gamma_b - m^2\gamma_w}{\gamma_{sat} - m^2\gamma_w} \tag{8.86}$$

There are two more cases to consider:

Case 1: Water table is at the surface

$$m=1 \quad \therefore \quad K_a = \frac{\gamma_b - \gamma_w}{\gamma_{sat} - \gamma_w} \tag{8.87}$$

Case 2: Water table at or below the base of the cut:

$$m=0 \quad \therefore \quad K_a = \frac{\gamma_b}{\gamma} \tag{8.88}$$

Example 8.18

Referring to the 7 m deep trench of example 8.17 assume that the soil is dense sand of $\phi'=30°$ and $\gamma=18.6\,\text{kN/m}^3$. The density of slurry is still $11\,\text{kN/m}^3$. The voids ratio and the specific gravity of the sand is 63% and 2.66 respective. Estimate the factor of safety for the following three cases, when the trench is completely filled with bentonite and the ground water table is at:

i. The surface
ii. 3 m below the surface
iii. 8 m below the surface

i. $m=1$ $\gamma_{sat}=\left(\frac{G_s+e}{1+e}\right)\gamma_w=\frac{2.7+0.63}{1.63}\times 9.81=20\,\text{kN/m}^3$

From (8.87): $K_a=\frac{11-9.81}{20-9.81}=0.117$

From (8.84): $\phi_m=\sin^{-1}\left(\frac{1-0.117}{1.117}\right)=\sin^{-1}\left(\frac{0.883}{1.117}\right)=52.2°$

From (8.83): $F_s=\frac{\tan 30}{\tan 52.2}=\frac{0.57735}{1.2892}=0.448\,\text{(failure)}$

ii. $mH=3$ $\quad\therefore\ m=\frac{3}{7}=0.4286$

From (8.86): $K_a=\frac{\gamma_b-m^2\gamma_w}{\gamma_{sat}-m^2\gamma_w}=\frac{11-0.184\times 9.81}{20-0.184\times 9.81}=0.505$

$$\phi_m=\sin^{-1}\left(\frac{1-0.505}{1.505}\right)=\sin^{-1}\left(\frac{0.495}{1.505}\right)=19.2°$$

$$F_s=\frac{\tan 30}{\tan 19.2}=\frac{0.57735}{0.3482}=1.66$$

iii. $m=0$ $\quad K_a=\frac{\gamma_b}{\gamma}=\frac{11}{18.6}=0.591$

$$\phi_m=\sin^{-1}\left(\frac{1-0.591}{1.591}\right)=\sin^{-1}\left(\frac{0.409}{1.591}\right)=14.9°$$

$$F_s=\frac{\tan 30}{\tan 14.9}=\frac{0.57735}{0.266}=2.17$$

Notes:

a) Safety decreases with increasing water level.
b) The factor of safety in clay is higher than in sand, under similar conditions.
c) If the required F_s is specified, then γ_b can be found. For instance, to increase the factor of safety to 2.1 in case (ii). Then the required ϕ_m is given by (8.83):

$$\tan\phi_m = \frac{\tan\phi'}{F_s} = \frac{\tan 30}{2.1} = 0.2749$$

$$\therefore \quad \phi_m = \tan^{-1}(0.2749) = 15.37°$$

$$K_a = \frac{1-\sin 15.37}{1+\sin 15.37} = \frac{0.7349}{1.26505} = 0.581$$

The required unit weight for the slurry is expressed from (8.86):

$$K_a\left(\gamma_{sat} - m^2\gamma_w\right) = \gamma_b - m^2\gamma_w$$

$$\therefore \gamma_b = K_a\left(\gamma_{sat} - m^2\gamma_w\right) + m^2\gamma_w$$

$$= 0.581 \times \left(20 - 0.4286^2 \times 9.81\right) + 0.4286^2 \times 9.81$$

$$= 0.581 \times (20 - 1.8) + 1.8 = 12.37 \text{ kN/m}^3$$

Note: A diaphragm wall is constructed within the slurry at the end of the excavation, the steel reinforcement is lowered through the mud and a trench is used to place concrete. During this process, the heavier materials displace the bentonite as the concrete fills the trench. After the concrete hardened and gained sufficient strength, the adjacent soil may be excavated to form the wall of a basement, for example.

Problem 8.1

The active thrust acting on a 9 m high wall, supporting sand and water, was determined in Example 8.3 (case 2) as 305 kN.

Also, the other forces acting on the wall were calculated in Example 8.8. These are:

Passive force: $P_p = 110.5\text{kN}$
Its line of action: $y_p = 0.67\text{m}$ from toe Q
Friction force: $F = 383\text{kN}$

Weight of wall

$W_1 = 216\text{ kN}$, $W_2 = 252\text{ kN}$, $W_3 = 144\text{ kN}$, $W = 612\text{ kN}$ acting at $x_1 = 3.50\text{ m}$, $x_2 = 2.67\text{ m}$, $x_3 = 2.00\text{ m}$ from the toe

Check the stability of the wall, shown in Figure 8.108, assuming $F_s = 2$ throughout.

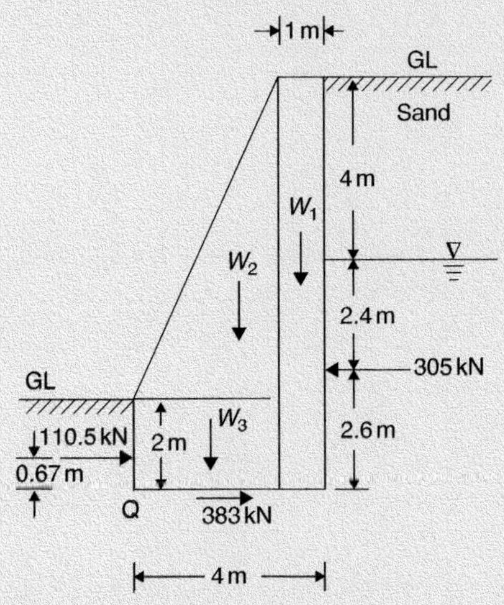

Soil characteristics for both active and passive cases

$\gamma = 17\text{ kN/m}^3$
$\gamma_{sat} = 19\text{ kN/m}^3$
$\phi = 32°$
$c = 0$

Bearing capacity of soil under the base $= 200\text{kN/m}^2$

Figure 8.101

Problem 8.2

A 9 m deep excavation, in dry sand, is to be supported by an anchored sheet pile wall. Calculate, by the Free-earth support method, the depth of penetration (z), if the working tension in the tie must not exceed 120 kN. Also, estimate the depth (h) and the approximate length (L) of the tie rod, when the height of the sunken anchor block is 2 m. Adopt 2 as the factor of safety.

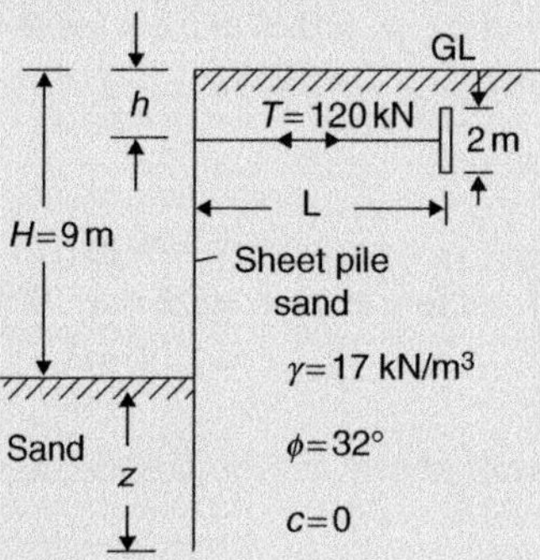

Figure 8.106

Problem 8.3

Derive the formula for the depth of tension crack, starting from the diagram in Figure 6.18, reproduced here.

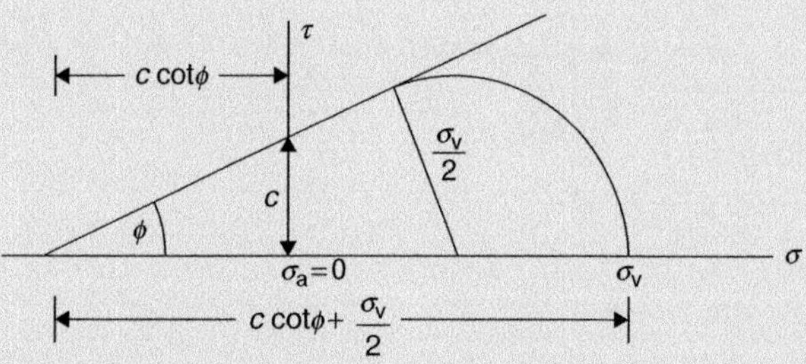

Figure 8.108

Problem 8.4

A 4 m high wall retains moist, coarse sand of $\phi=32°$ and $\gamma=17$ kN/m^3, as shown in Fig 8.109. Determine the active force acting on the wall by:

1. The Rankine theory
2. Graphical method
3. Trigonometry

Neglect wall friction. Ground water table is below the base.

Problem 8.5

Derive expressions for the passive force and the critical angle of the corresponding slip surface. These will be similar to those obtained in Problem 8.4. The main difference is due to the inclination of the reaction (R) to the slip surface, as the passive wedge tends to slide upwards. Neglect wall friction.

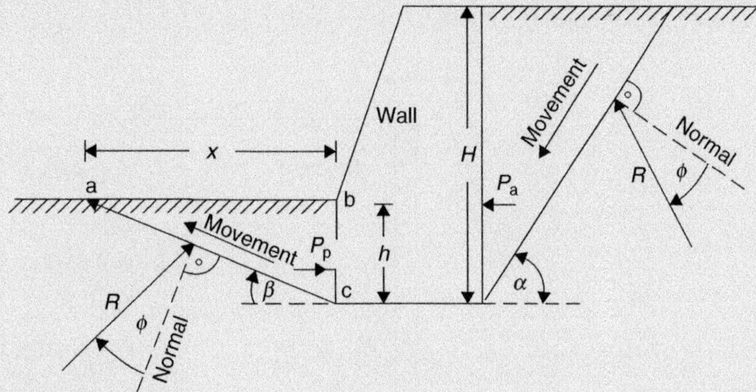

Figure 8.112

The reactions are inclined to resist movement.

The diagram below shows wedge abc and the forces acting on it, drawn at point a for convenience.

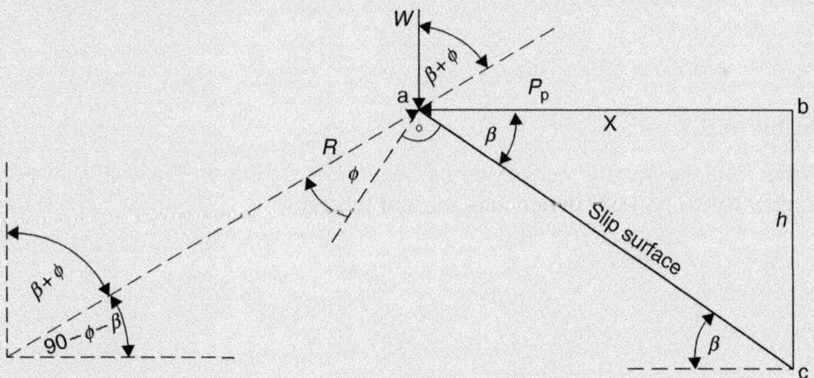

Figure 8.113

Problem 8.6

The smooth wall, shown in Fig 8.115, is supporting the moist sand of Problem 8.4. Determine:

a) The inclination of the active and passive slip surfaces.
b) The active and passive forces acting and compare the figures with those calculated in Examples 8.3 and 8.8.

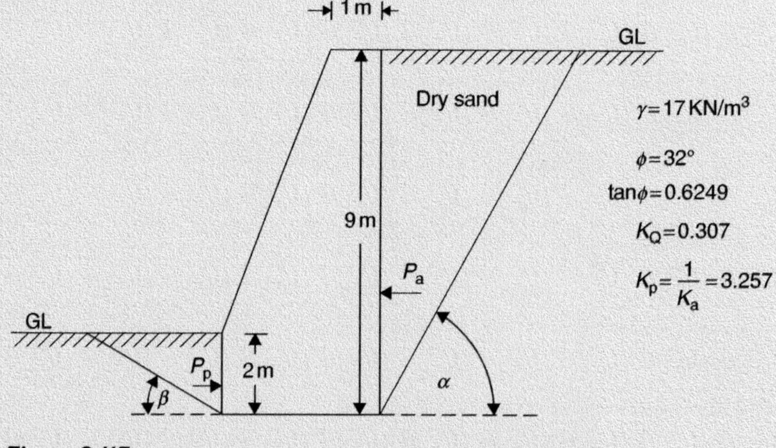

Figure 8.115

Problem 8.7

Derive formulae for the active and passive forces acting on the wall, shown in Graph 8.10, taking wall friction into account this time.

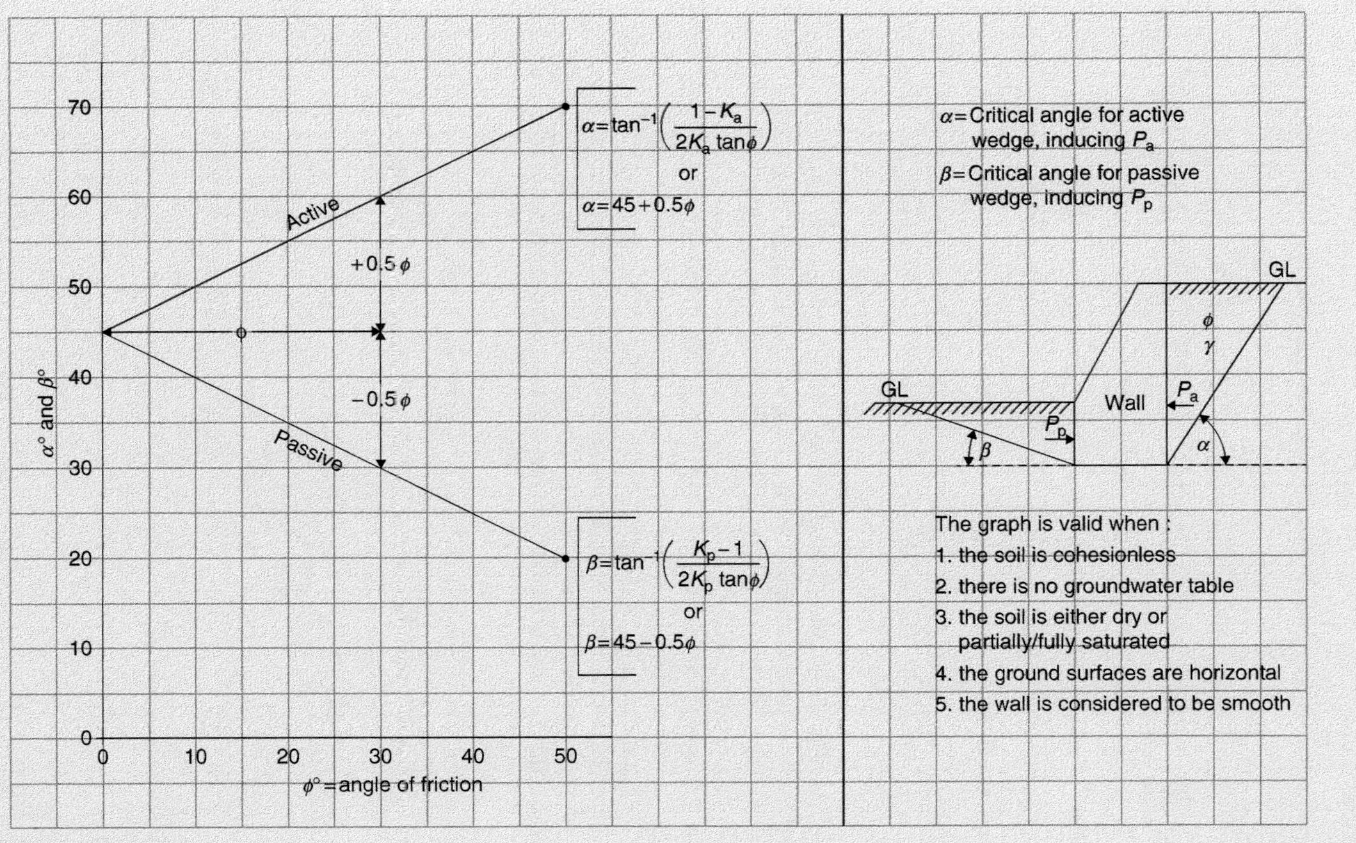

70
60
50
40
30
20
10
0
$\alpha°$ and $\beta°$
0
10
20
30
40
50
$\phi°$=angle of friction
Active
Passive
$+0.5\phi$
-0.5ϕ
$\alpha=\tan^{-1}\left(\frac{1-K_a}{2K_a \tan\phi}\right)$
or
$\alpha=45+0.5\phi$
$\beta=\tan^{-1}\left(\frac{K_p-1}{2K_p \tan\phi}\right)$
or
$\beta=45-0.5\phi$
α=Critical angle for active wedge, inducing P_a
β=Critical angle for passive wedge, inducing P_p
GL
GL
Wall
ϕ
γ
P_a
P_p
α
β
The graph is valid when :
1. the soil is cohesionless
2. there is no groundwater table
3. the soil is either dry or partially/fully saturated
4. the ground surfaces are horizontal
5. the wall is considered to be smooth

Graph 8.10

Problem 8.8

Suppose the retaining wall in Problem 8.6 has rough surface, with friction angle δ=17.6°.

Determine:

1. The maximum value of the active force and the angle of the relevant slip surface.
2. The minimum value of passive resistance and the relevant angle.

Problem 8.9

Determine the active force, acting on the 4 m high wall in Problem 8.4, by Culmann's method. The general procedure is described with reference to Figure 8.121 and Graph 8.13.

Step 1: Draw a trial surface

Step 2: Draw the baseline and the direction of P_a at ϕ=32° as shown.

Step 3: Calculate the weight (W) of the wedge and mark it along the baseline to any desired scale.

Step 4: Draw a line from the baseline at W, parallel to the direction of P_a (a – b), to meet the trial slip surface at point X. The distance between X and the baseline is the active force induced by the trial wedge.

Step 5: Repeat the process for a number of wedges and draw the Culmann line through each X point.

Step 6: Draw a tangent to the Cullmann line, parallel to the baseline to locate the maximum value of P_a and its slip surface.

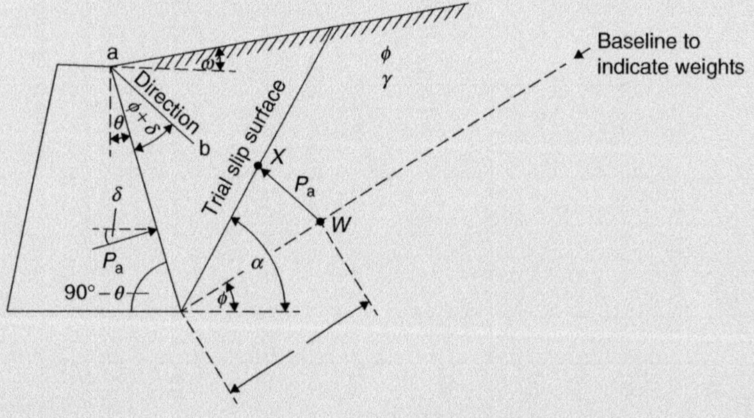

Figure 8.121

Problem 8.10

A reinforced concrete box section underpath is to be constructed below a main road. The surcharge on the section is expected to be 70 kN/m². The backfill is compact fine sand, having the following known characteristics:

Minimum voids ratio = 62%
Maximum voids ratio = 53%
Specific gravity = 2.65
Angle of friction = 30°

Determine the horizontal forces acting on the section shown, so that they can be taken into account during the design of its reinforcement. It is specified, that the:

a) Concrete surfaces should be smooth
b) Sand has to be compacted to a 'relative' density of 69%.

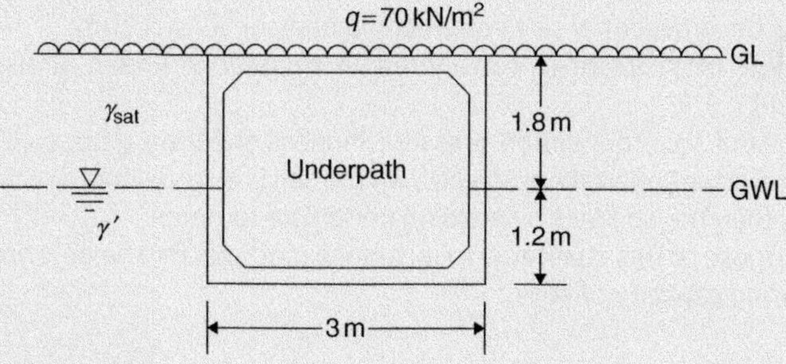

Figure 8.122

Problem 8.11

A retaining wall is to be designed with 3 m×2 m base area. It is specified, that the maximum and minimum pressure, below the base along the longitudinal axis, must not exceed 200 kN/m² and zero respectively.

Calculate the magnitude of the vertical force (R) acting on the base and its eccentricity (e) from the centre line, parallel to the shorter side.

Chapter 9
Bearing Capacity of Soils

When a building is constructed on top of soil layers, its structure could suffer extensive damage in two ways:

1. In clay soils, the excess pore pressure (Δu), induced by the excess total pressure ($\Delta\sigma$) could be so high, that the resulting consolidation of the soil would cause excessive settlement of the structure. Should various parts of building settle unevenly, the *differential settlement* could damage its structure.

   This type of problem is considered in Chapter 7 under 'Consolidation and Settlement'.
2. In all types of soil, foundation pressure induces shearing stresses.If the shearing resistance, often called *shear strength* of the soil is insufficient to resist the induced stresses, then the soil fails by yielding under the footing.

   The purpose of this chapter is to introduce methods for the determination of the load-bearing capacity of soils.

9.1 Terminology

Footing: The substructure of a building which transfers load to the ground

Foundation: This may mean 'footing' as well as the soil, upon which the base of the footing rests.

9.1.1 Foundation pressure (σ)

It is the total pressure of the structure, including the superimposed load on it, transmitted to the soil by the footing.

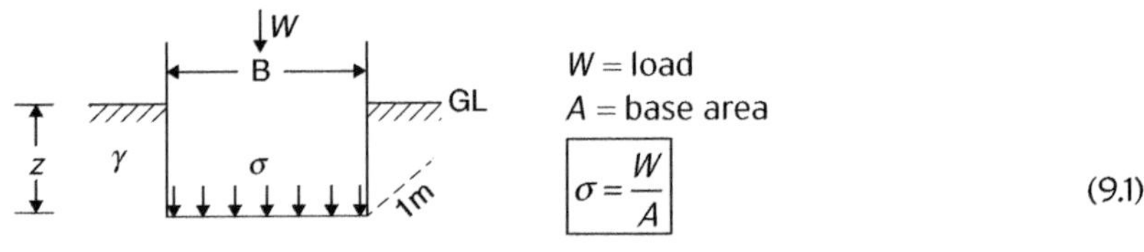

W = load

A = base area

$$\boxed{\sigma = \frac{W}{A}} \qquad (9.1)$$

Figure 9.1

Introduction to Soil Mechanics, First Edition. Béla Bodó and Colin Jones.

9.1.2 Net foundation pressure (σ_n)

It is the excess total pressure at base level, that is the safe foundation pressure, less the total overburden pressure (σ_0), that is the weight of the excavated soil.

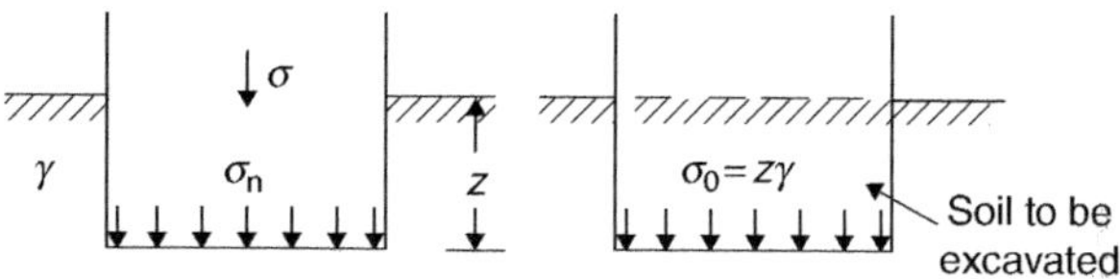

Figure 9.2

$$\boxed{\sigma_n = \sigma - \sigma_0} \quad (9.2)$$

9.1.3 Effective overburden pressure (σ_0')

It is the effective pressure at the proposed base level prior to excavation. It equals to the total overburden pressure (σ_0) minus any pore water pressure (u). (See also chapter 5).

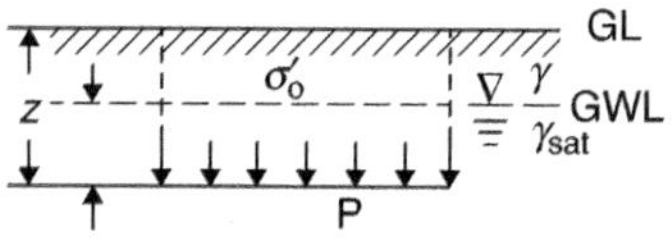

Figure 9.3

At point P:

$$\sigma_0 = (z-h)\gamma + h\gamma_{sat}$$

$$U_0 = h\gamma_w$$

$$\sigma_0' = \sigma_0 - u_0 \quad (5.1)$$

Note: The actual value of (σ_0') depends, of course on the stratification and on the position of the GWL relative to the proposed foundation level. It is, therefore, included in the bearing capacity formulae.

9.1.4 Ultimate bearing capacity (q_u)

It is that bearing pressure, which causes the soil to yield in general shear failure. Its value is calculated from one of the formulae based on the general equation, derived by Terzaghi et al, for the shear failure of shallow and deep foundations.

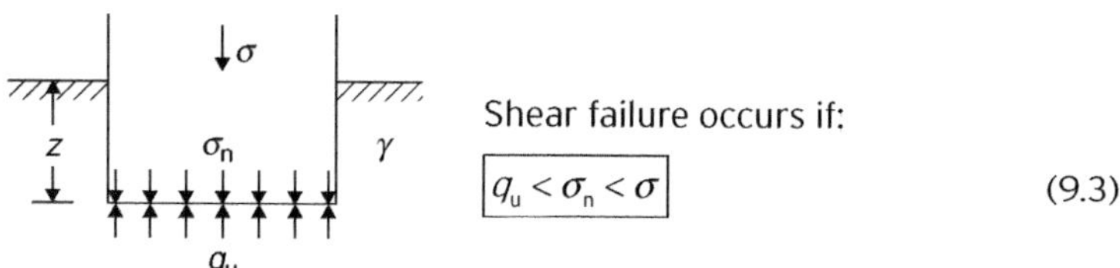

Shear failure occurs if:

$$\boxed{q_u < \sigma_n < \sigma} \quad (9.3)$$

Figure 9.4

9.1.5 Net ultimate bearing capacity (q_n)

It equals to the ultimate bearing capacity less the effective pressure of the excavated soil. Shear failure occurs if $\sigma_n > q_n$.

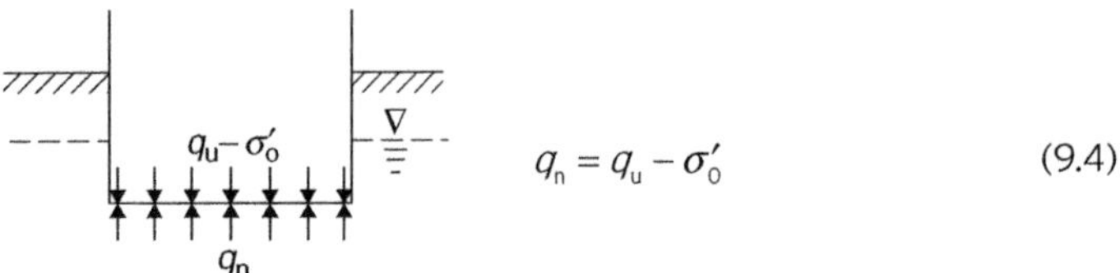

$$q_n = q_u - \sigma'_0 \qquad (9.4)$$

Figure 9.5

9.1.6 Safe net bearing capacity (q_{sn})

It is the net ultimate bearing capacity, reduced by a factor of safety of F_s

$$\boxed{q_{sn} = \frac{q_n}{F_s} = \frac{q_n - \sigma'_0}{F_s}} = \sigma_n \qquad (9.5)$$

9.1.7 Safe bearing capacity (q_s)

It is the safe net bearing capacity increased by the effective overburden pressure, thus providing added strength and stability against shear failure. The factor of safety is not applied to the overburden, as it is a constant quantity. The safe foundation pressure (σ) is equated to q_s.

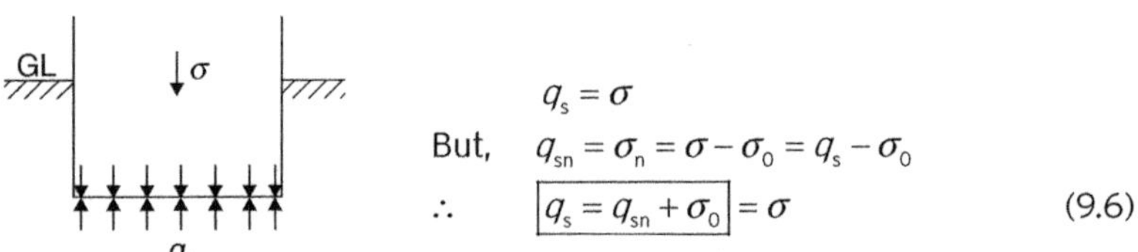

$$q_s = \sigma$$

But, $\quad q_{sn} = \sigma_n = \sigma - \sigma_0 = q_s - \sigma_0$

$$\therefore \quad \boxed{q_s = q_{sn} + \sigma_0} = \sigma \qquad (9.6)$$

Figure 9.6

The foundation pressure (σ) on the soil has to be less or equal to its safe bearing capacity.

$$\therefore \quad \sigma \leq q_s = \frac{q_n}{F_s} + \sigma_0 \quad \text{or} \quad \boxed{\sigma \leq \frac{q_u - \sigma'_0}{F_s} + \sigma_0} \qquad (9.7)$$

This formula is used normally in bearing capacity problems.

9.1.8 Allowable foundation pressure (σ_a)

It is the maximum foundation pressure that may be transmitted to the ground by the footing, taking into account the:

1. Safe bearing capacity of the soil and the adequacy of the chosen factor of safety (F_s).
2. Magnitude and rate of estimated settlement of various parts of the structure.
3. Ability of the structure to accommodate differential settlement.
4. Effect of the foundation pressure on adjacent structures and services.

Note: The consideration of differential settlement is especially important, when the structure is statically indeterminate. If one part of the structure settles more than the other, then the joints between its elements could be excessively overstressed. Take a rigid portal frame shown in Figure 9.7.

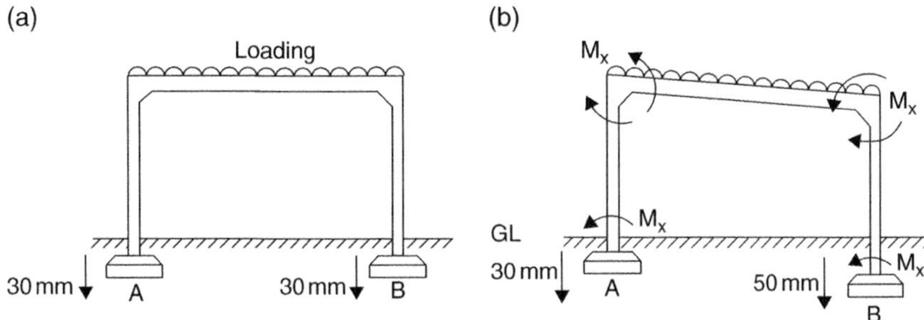

Figure 9.7

Suppose, the maximum settlement, predicted from a consolidation test result, was 30 mm for each footing.

In Figure 9.7(a), both footings settle equally, thus the stresses within the structure remain at their designed values.

In Figure 9.7(b), however, there is a differential, excess settlement of 20 mm at foundation B, which induces additional bending moment (M_x) in each rigid joint. Unless the frame is designed to carry the excess bending moment, it could exhibit structural distress or complete failure.

On the other hand, statically determinate structures, such as simply supported beams and slabs, can accommodate some differential settlement as shown in Figure 9.8.

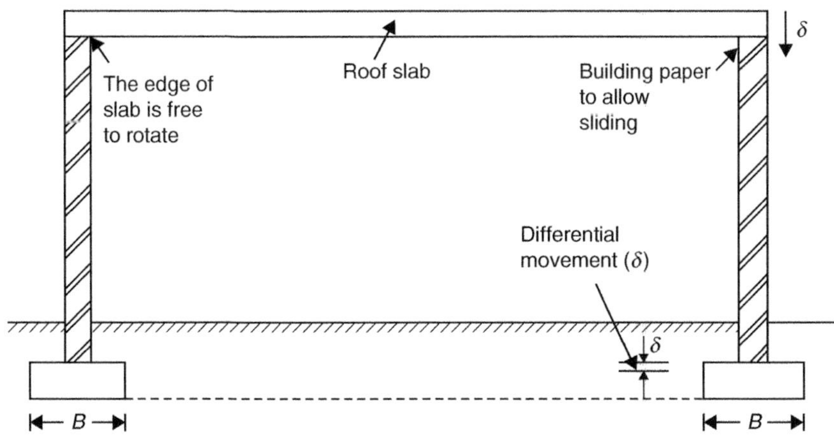

Figure 9.8

The rotation of the slab does not induce stresses at the supports as it is free to rotate there. As long as the differential settlement is not excessive, any damage would amount to an acceptable degree of cracking at the wall slab interface.

9.1.9 Presumed bearing values

In the absence of laboratory test results, the following bearing values may be assumed (BS 8004: 1986) in kN/m^2.

Table 9A

Dense gravel	>600	$B>1$ m Water table at least B m below base.
Medium dense gravel/sand and gravel	200–600	
Loose gravel	<200	
Compact sand	>300	
Medium dense sand	100–300	
Loose sand	<100	
Very stiff boulder clays/hard clays	300–600	Susceptible to long-term consolidation settlement.
Stiff clays	150–300	
Firm clays	75–150	
Soft clays and silts	<75	

Theory

Formulae (9.1–9.7) relate the bearing capacity of soil to the loading of the structure it supports, in general terms. In this chapter, the formulae to calculate the bearing capacity of the supporting soil will be introduced for shallow and deep foundations of various shapes.

A foundation is taken to be shallow if $B \geq z$.

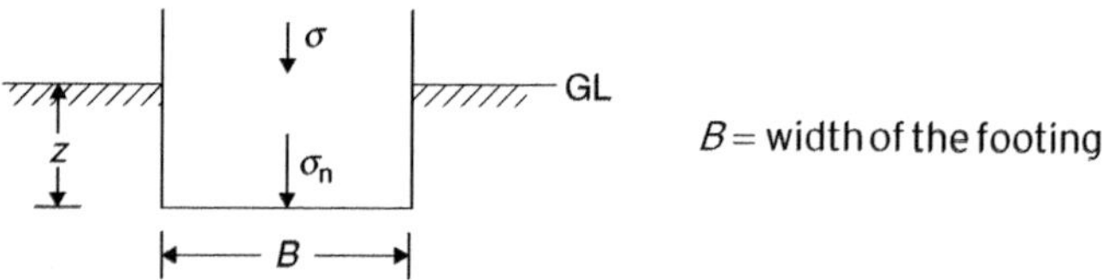

Figure 9.9

Formulae for the following four types of footing shapes, founded in cohesive or non-cohesive soils are considered:

a) Strip
b) Square
c) Circular
d) Rectangular

9.2 Shallow strip footing

The most general bearing capacity formula was given by Terzaghi, modified by Skempton, based upon the fact, that when soil in natural elastic state is loaded, it deforms. If the load is large enough to induce larger shearing stresses than the shear strength of the soil, then plastic deformation occurs. This load is the ultimate bearing capacity (q_u) of the particular soil. Plastic state is reached gradually in loose granular soils and soft clays, and suddenly in very compact, cohesive ones. At this juncture, the soil undergoes general shear failure, flowing from under the footing, sideways and upwards, as shown in Figure 9.10(b). The development of plastic state is shown by the load-settlement curves in Figure 9.10(a).

Net bearing pressure

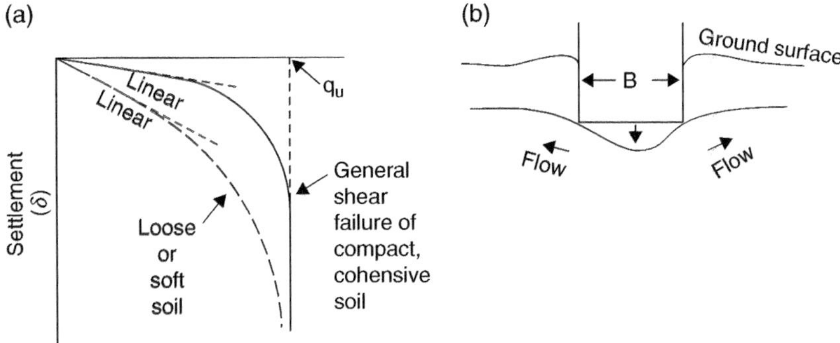

Figure 9.10

9.2.1 Terzaghi's equation for q_u

This equation was introduced by Terzaghi (1943), modifying the original solution of Prandtl (1920), assuming the curve of the failure surface as logarithmic spiral.

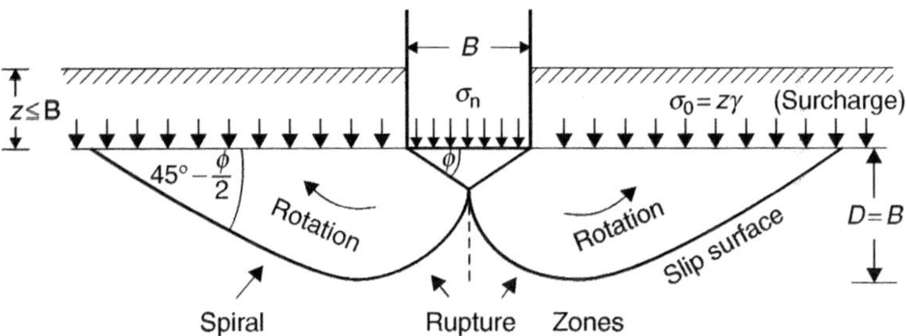

Figure 9.11

Depth (*D*) of the rupture zones is approximately equal to the width (*B*) of the footing. The ultimate bearing capacity of the soil is equal to the total shearing resistance available along the slip surfaces. It is assumed, that the surcharge is not contributing to the shearing resistance, being backfill normally. However, it resists the rotation of the soil mass, hence contributes to safety. The bearing capacity for this case is given by:

$$q_u = cN_c + \sigma_0' N_q + 0.5\gamma B N_\gamma \quad (9.8)$$

where

c = cohesion, often called 'apparent cohesion' of the soil
σ_0' = effective overburden pressure
N_c, N_q and N_γ are bearing capacity factors.

The term cN_c is contributed by the cohesion.
The term $\sigma_0' N_q$ is contributed by the surcharge.
The term $0.5\gamma B N_\gamma$ is contributed by the effect of width *B*.

The bearing capacity factors are given in Table 9.1, in terms of the angle of friction ϕ, assuming rough base.

Table 9.1

B.C. factor	$\phi°$ 0	5	10	15	20	25	30	35	40	45
N_c	5.7	7.2	9.9	13	17	25	37	58	95	172
N_q	1	1.5	2.7	4.5	7.5	13	22	42	81	173
N_γ	0	0	1	2.1	4.5	10	20	45	120	175

These figures are plotted on Chart 9.1.

The net bearing capacity can be expressed from (9.4) and (9.8):

$$\begin{aligned} q_n &= q_u - \sigma'_0 \\ &= cN_c + \sigma'_0 N_q + 0.5\gamma B N_\gamma - \sigma'_0 \end{aligned}$$

Hence,
$$\boxed{q_n = cN_c + \sigma'_0 (N_q - 1) + 0.5\gamma B N_\gamma} \tag{9.9}$$

Similarly, the safe bearing capacity can now be expressed from (9.6) and (9.9):

$$q_s = \frac{q_n}{F_s} + \sigma_0$$

$$\boxed{q_s = \frac{1}{F_s}\left[cN_c + \sigma'_0 (N_q - 1) + 0.5\gamma B N_\gamma\right] + \sigma_0} \tag{9.10}$$

The value of F_s is normally chosen between 2 and 3, for structures not sensitive to differential settlement.

The total safe foundation pressure (σ) is therefore given by:

$$\sigma \le q_s = \frac{1}{F_s}\left[cN_c + \sigma'_0 (N_q - 1) + 0.5\gamma B N_\gamma\right] + \sigma_0 \tag{9.11}$$

If there is restriction on the magnitude of settlement, then oedometer consolidation test has to be carried out, in order to ascertain the allowable foundation pressure and the corresponding consolidation.

Formula 9.11 is applicable to c-ϕ soils, that is where both friction and cohesion are present. It can easily be simplified for pure clay (c-soil) and pure sand (ϕ-soil).

For pure clay:

$$\begin{aligned} \phi &= 0° \\ N_c &= 5.7 \\ N_q &= 1 \\ N_\gamma &= 0 \end{aligned} \qquad \therefore \quad \boxed{q_s = \frac{5.7\,c}{F_s} + \sigma_0} \tag{9.12}$$

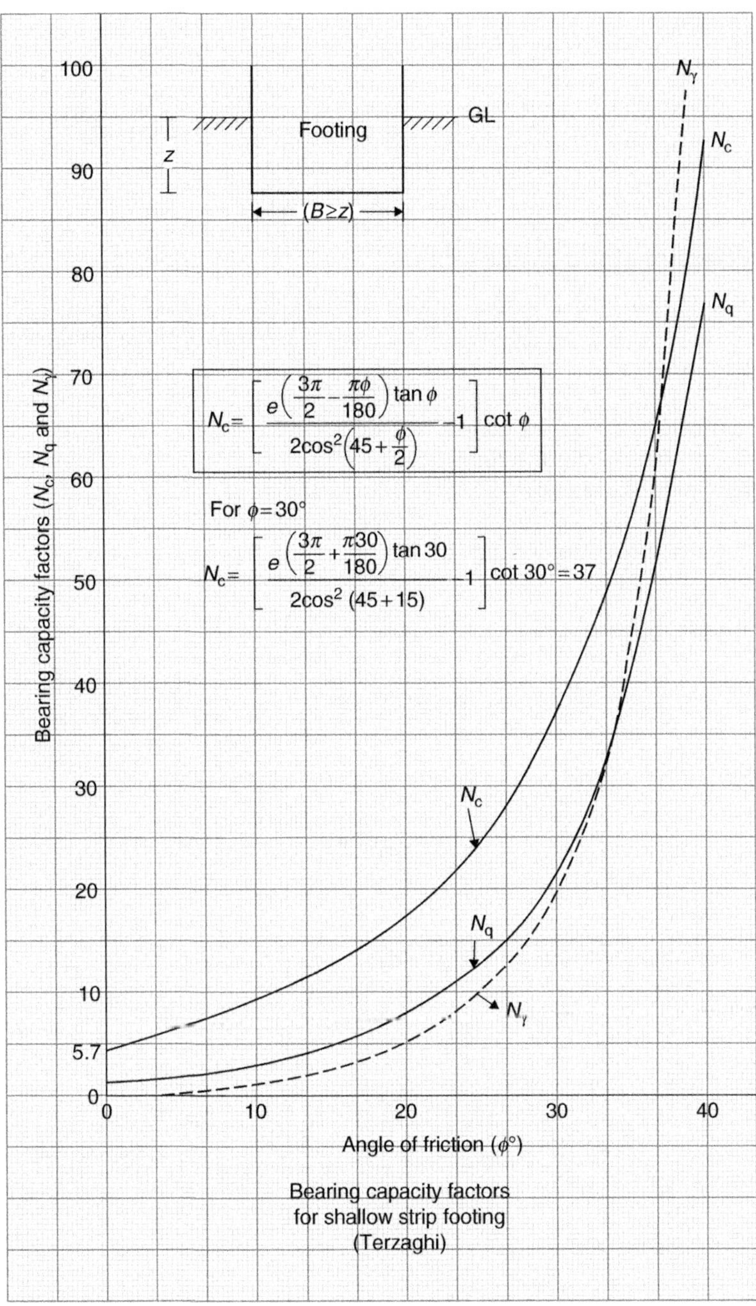

Footing
GL
z
(B≥z)
$N_c = \left[\dfrac{e^{\left(\frac{3\pi}{2} - \frac{\pi\phi}{180}\right)\tan\phi}}{2\cos^2\left(45 + \frac{\phi}{2}\right)} - 1 \right] \cot\phi$
For $\phi = 30°$
$N_c = \left[\dfrac{e^{\left(\frac{3\pi}{2} + \frac{\pi 30}{180}\right)\tan 30}}{2\cos^2(45 + 15)} - 1 \right] \cot 30° = 37$
N_γ
N_c
N_q
Bearing capacity factors (N_c, N_q and N_γ)
100
90
80
70
60
50
40
30
20
10
5.7
0
10
20
30
40
Angle of friction ($\phi°$)
Bearing capacity factors for shallow strip footing (Terzaghi)

Chart 9.1

For pure sand:

$$\phi > 0^\circ \quad N_c = 0 \quad N_q > 0 \quad N_\gamma > 0$$

$$\therefore \quad \boxed{q_s = \frac{1}{F_s}\left[\sigma_0'(N_q - 1) + 0.5\gamma B N_\gamma\right] + \sigma_0} \qquad (9.13)$$

9.2.2 Effect of static water table

The values of the effective pressure (σ_0') and the bulk density of the soil, hence the bearing capacity of the soil, depend on the ground water level (GWL). Three cases are depicted for uniform c-ϕ soil.

1. GWL is below the rupture zones
2. GWL is within the rupture zones
3. GWL is above base level.

1. GWL is below the rupture zone ($D>B$):

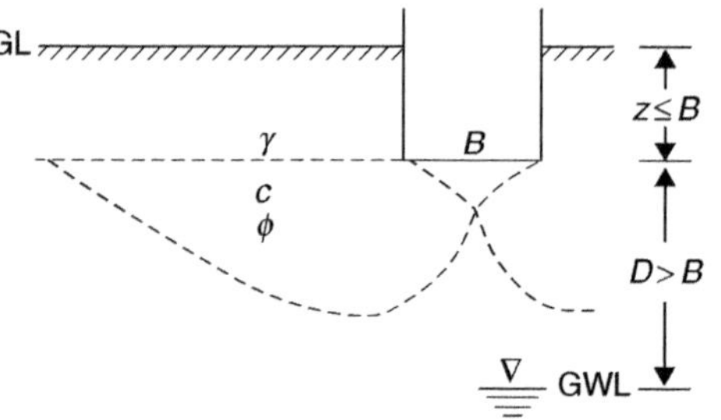

In this case, GWL has no effect on the bearing capacity. See section 5.8 on capillary movement however.

Figure 9.12

2. GWL is within the rupture zones ($D<B$):

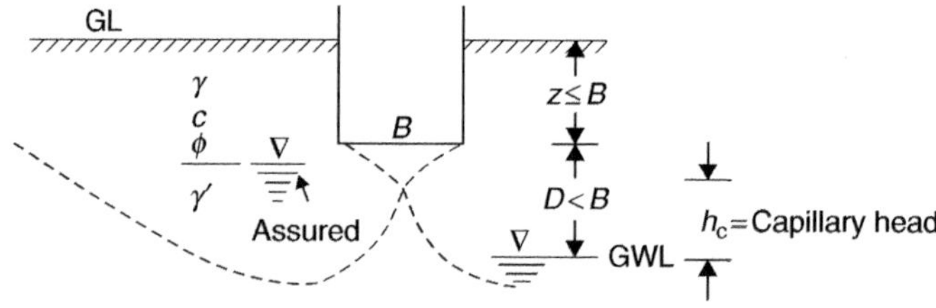

Figure 9.13

It may be assumed in this case that the soil below the footing is saturated because of capillary rise. For this reason, the submerged density (γ') is used in the third term of the bearing capacity formulae.

$$\therefore \quad \boxed{q_s = \frac{1}{F_s}\left[cN_c + \sigma_0'(N_q - 1) + 0.5\gamma' B N_\gamma\right] + \sigma_0}$$

where $\sigma'_0 = \sigma_0 = z\gamma$

3. GWL above base level ($a<z$):

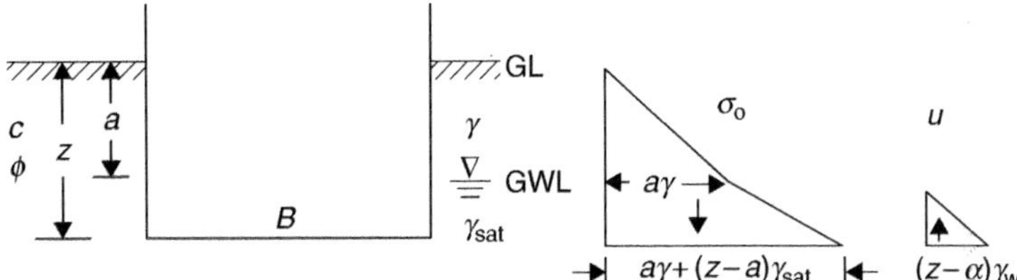

Figure 9.14

$$\sigma_0 = a\gamma + (z-a)\gamma_{sat}$$
$$u = (z-a)\gamma_w$$
$$\sigma_0' = z - u$$

$$\therefore \quad \sigma_0' = a\gamma + (z-a)\gamma'$$

Hence,

$$q_s = \frac{1}{F_s}\left[cN_c + \sigma_0'(N_q - 1) + 0.5\gamma BN_\gamma\right] + \sigma_0$$

Example 9.1

A 1.5 m wide strip footing is placed at a depth of 1.2 m, in uniform clay. The soil characteristics are:

$c = 50\,\text{kN/m}^2$
$\phi = 15°$
$\gamma = 17.7\,\text{kN/m}^3$
$\gamma_{sat} = 19.8\,\text{kN/m}^3$

Calculate the safe bearing capacity, when the water level is at:

1. 10 m below the foundation level
2. 0.9 m below the foundation level
3. 0.5 m below the ground level
4. 0.8 m above the ground level due to flooding

Apply a factor of safety of 3.

Obtain the bearing capacity factors from Chart 9.1

For $\phi = 15°$: $N_c = 13$
$N_q = 4.5$
$N_\gamma = 2.1$

Case 1

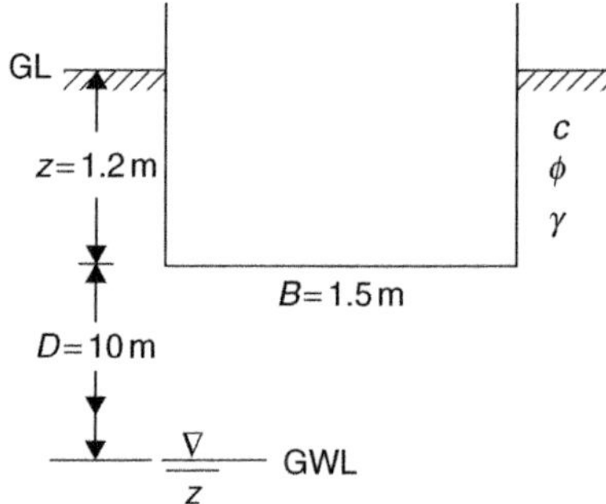

Figure 9.15

$D>2B$, therefore the bearing capacity is not affected by the position of the ground water table.

$$\sigma'_0 = \sigma_0 = 1.2 \times 17.7$$
$$= 21.2\,\text{kN/m}^2$$

$$q_s = \frac{1}{3} \times \left[50 \times 13 + 21.2 \times (4.5 - 1) + 0.5 \times 17.7 \times 1.5 \times 2.1\right] + 21.2$$
$$= \frac{1}{3} \times (650 + 74.2 + 27.88) + 21.2 = 271.9\,\text{kN/m}^2$$

Case 2

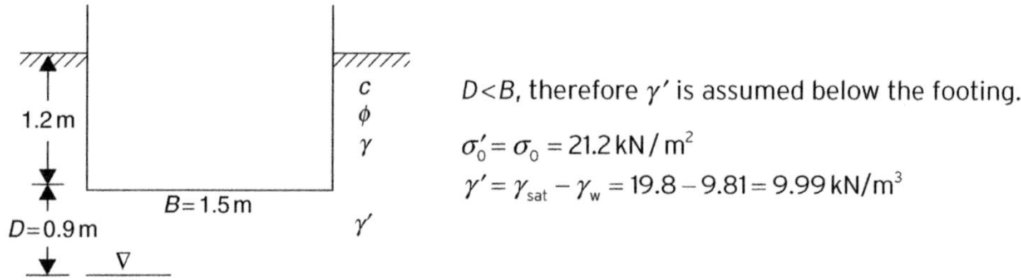

Figure 9.16

$D<B$, therefore γ' is assumed below the footing.

$$\sigma'_0 = \sigma_0 = 21.2\,\text{kN/m}^2$$
$$\gamma' = \gamma_{sat} - \gamma_w = 19.8 - 9.81 = 9.99\,\text{kN/m}^3$$

$$q_s = \frac{1}{3} \times (650 + 74.2 + 0.5 \times 9.99 \times 1.5 \times 2.1) + 21.2$$
$$= \frac{1}{3} \times (724.2 + 15.73) + 21.6 = \frac{739.93}{3} + 21.2 = 267.8\,\text{kN/m}^2$$

Case 3

(Neglecting capillary action arbitrarily)

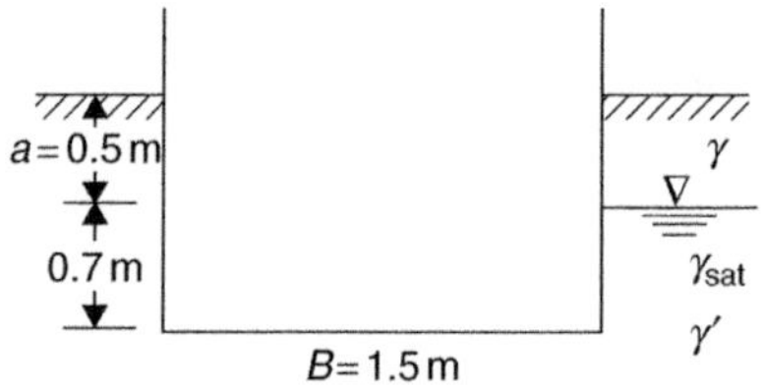

$$\sigma'_0 = 0.5 \times 17.7 + 0.7 \times 9.99$$
$$= 15.84\,\text{kN/m}^2$$
$$\sigma_0 = 0.5 \times 17.7 + 0.7 \times 19.8$$
$$= 22.7\,\text{kN/m}^2$$

Figure 9.17

$$q_s = \frac{1}{3} \times \left[650 + 15.84 \times (4.5 - 1) + 0.5 \times 9.99 \times 1.5 \times 2.1\right] + 22.7$$
$$= \frac{1}{3} \times (650 + 55.44 + 15.73) + 22.7 = 263\ \text{kN/m}^2$$

Case 4

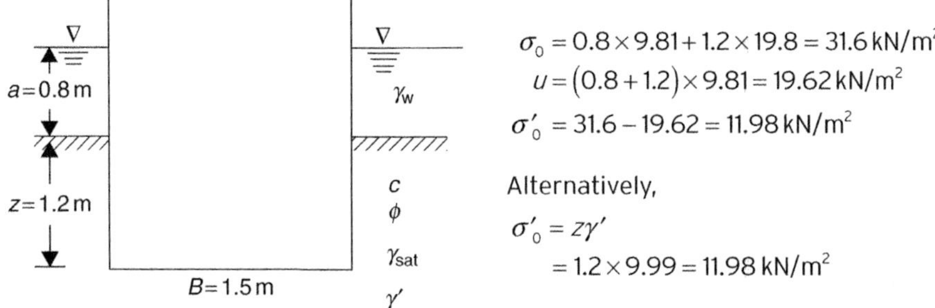

$$\sigma_0 = 0.8 \times 9.81 + 1.2 \times 19.8 = 31.6\ \text{kN/m}^2$$
$$u = (0.8 + 1.2) \times 9.81 = 19.62\ \text{kN/m}^2$$
$$\sigma'_0 = 31.6 - 19.62 = 11.98\ \text{kN/m}^2$$

Alternatively,

$$\sigma'_0 = z\gamma'$$
$$= 1.2 \times 9.99 = 11.98\ \text{kN/m}^2$$

Figure 9.18

$$q_s = \frac{1}{3} \times \left[650 + 11.98 \times (4.5 - 1) + 15.73\right] + 11.98$$
$$= \frac{1}{3} \times (650 + 41.97 + 15.73) + 11.98 = 247.9\ \text{kN/m}^2$$

Because of the reduction in the bearing capacity, due to the increase in water level, the safe bearing pressure is decreased from 271.9 kN/m^2 to 247.9 kN/m^2.

Example 9.2

Calculate the safe bearing pressure for the four cases in Example 9.1 for:

1. Pure clay
2. Pure sand

1. *Pure clay* $\quad c = 50\ \text{kN/m}^2$, $\phi = 0$

From (9.12): $q_s = \dfrac{5.7 \times 50}{3} + \sigma_0$

$$\boxed{q_s = 95 + \sigma_0}$$

Cases 1 and 2:	$\sigma_0 = 21.2\ \text{kN/m}^2$	$\therefore\ q_s = 116.2\ \text{kN/m}^2$
Case 3:	$\sigma_0 = 22.7\ \text{kN/m}^2$	$\therefore\ q_s = 117.7\ \text{kN/m}^2$
Case 4:	$\sigma_0 = 31.6\ \text{kN/m}^2$	$\therefore\ q_s = 126.6\ \text{kN/m}^2$

2. *Pure sand* $\quad \phi = 15°$, $c = 0$, $N_q = 4.5$, $N_\gamma = 2.1$

From (9.13):

$$q_s = \frac{1}{3} \times \left[\sigma'_0 \times (4.5 - 1) + 0.5\gamma \times 1.5 \times 2.1\right] + \sigma_0$$

$$\therefore \quad \boxed{q_s = 1.17\sigma'_0 + 0.53\gamma + \sigma_0}$$

Case 1:	$\sigma'_0 = 21.2\,\text{kN/m}^2$ $\sigma_0 = 21.2\,\text{kN/m}^2$ $\gamma = 17.7\,\text{kN/m}^3$	$q_s = 1.17\sigma'_0 + 0.53\gamma + \sigma_0$ $= 1.17 \times 21.2 + 0.53 \times 17.7 + 21.2$ $= 55.4\,\text{kN/m}^2$
Case 2:	$\sigma'_0 = 21.2\,\text{kN/m}^2$ $\sigma_0 = 21.2\,\text{kN/m}^2$ $\gamma' = 9.99\,\text{kN/m}^3$	$q_s = 1.17\sigma'_0 + 0.53\gamma' + \sigma_0$ $= 1.17 \times 21.2 + 0.53 \times 9.99 + 21.2$ $= 51.3\,\text{kN/m}^2$
Case 3:	$\sigma'_0 = 15.84\,\text{kN/m}^2$ $\sigma_0 = 22.7\,\text{kN/m}^2$ $\gamma' = 9.99\,\text{kN/m}^3$	$q_s = 1.17\sigma'_0 + 0.53\gamma' + \sigma_0$ $= 1.17 \times 15.84 + 0.53 \times 9.99 + 22.7$ $= 46.5\,\text{kN/m}^2$
Case 4:	$\sigma'_0 = 11.98\,\text{kN/m}^2$ $\sigma_0 = 31.6\,\text{kN/m}^2$ $\gamma' = 9.99\,\text{kN/m}^3$	$q_s = 1.17\sigma'_0 + 0.53\gamma' + \sigma_0$ $= 1.17 \times 11.98 + 0.53 \times 9.99 + 31.6$ $= 50.9\,\text{kN/m}^2$

The effect of the water table on different types of soil can be seen by tabulating the bearing pressures calculated in Examples 9.1 and 9.2.

Table 9.2

		$c-\phi$ soil	Clay ($\phi = 0$)	Sand ($c = 0$)
Case 1		271.9	116.2	55.4
Case 2	q_s	267.8	116.2	51.3
Case 3	kN/m²	263.0	117.7	46.5
Case 4		247.9	126.6	50.9

Note: Terzaghi's bearing capacity factors have been used extensively. They were determined after certain simplifying assumptions; hence their values may be regarded as reasonable estimates only. Further, the factors become unreasonable large beyond $\phi =$ 30° and therefore, should not really be applied above this value. Because of the uncertainties involved in the determination of the bearing capacity of soils, several theories have been advanced for the evaluation of factors N_c, N_q and N_γ One of these theories was evolved by Prandlt, Caquot and Reissner, deriving the following formulae for N_q and N_c:

$$N_q = \left(\frac{1+\sin\phi}{1-\sin\phi}\right)e^{\pi\tan\phi} \tag{9.14}$$

$$N_c = \left(N_q - 1\right)\cot\phi \tag{9.15}$$

The third factor N_γ may be calculated from Meyerhof's approximation:

$$\boxed{N_\gamma = (N_q - 1)\tan(1.4\phi)} \tag{9.16}$$

The factors calculated from these formulae are given in Table 9.2a and are plotted on Chart 9.2.

Table 9.2a

B.C. factors	$\phi°$									
	0	**5**	**10**	**15**	**20**	**25**	**30**	**35**	**40**	**45**
N_c	5.14	6.5	8.3	11	14.8	20.7	30.1	46.1	75.3	134
N_q	1.0	1.6	2.5	3.9	6.4	10.7	18.4	33.3	64.2	135
N_γ	0	0.07	0.37	1.11	2.9	6.8	16.6	37.2	93.7	263

Note: N_c and N_γ cannot be calculated at $\phi = 0$ as tan (0°) = cot (0°) = 0 there. They can, however, be approximated by choosing the value of ϕ near zero.

Let $\phi = 0.0001°$

Then $N_q = \dfrac{1+\sin(0.0001)}{1-\sin(0.0001)} \times e^{\pi \tan(0.0001)}$

$= \dfrac{1.74532925199 \times 10^{-6}}{0.999998254671} \times 1.00000548313$

$= 1.00000349067 \times 1.00000548313 = 1.00000897382$

$\therefore$ at $\phi = 0°$ $N_q \approx 1$

So, $N_c = (N_q - 1)\cot(0.0001) = 8.97382 \times 10^{-6} \times 572957.79513$

$= 5.14162012109 \approx 5.14$

and $N_\gamma = (N_q - 1)\cot(1.4 \times 0.0001)$

$= 8.97382 \times 10^{-6} \times 2.4434609528 \times 10^{-6} = 21.927178 \times 10^{-6}$

$\therefore$ at $\phi = 0°$ $N_c = 5.14$ and $N_\gamma = 0$

For pure clay, therefore, formula (9.12) is modified to:

$$\boxed{\sigma = \frac{5.14\,c}{F_s} + \sigma_0} \tag{9.17}$$

Note: The factors in Table 9.2a are smaller than Terzaghi's, giving correspondingly diminished values of safe bearing capacity.

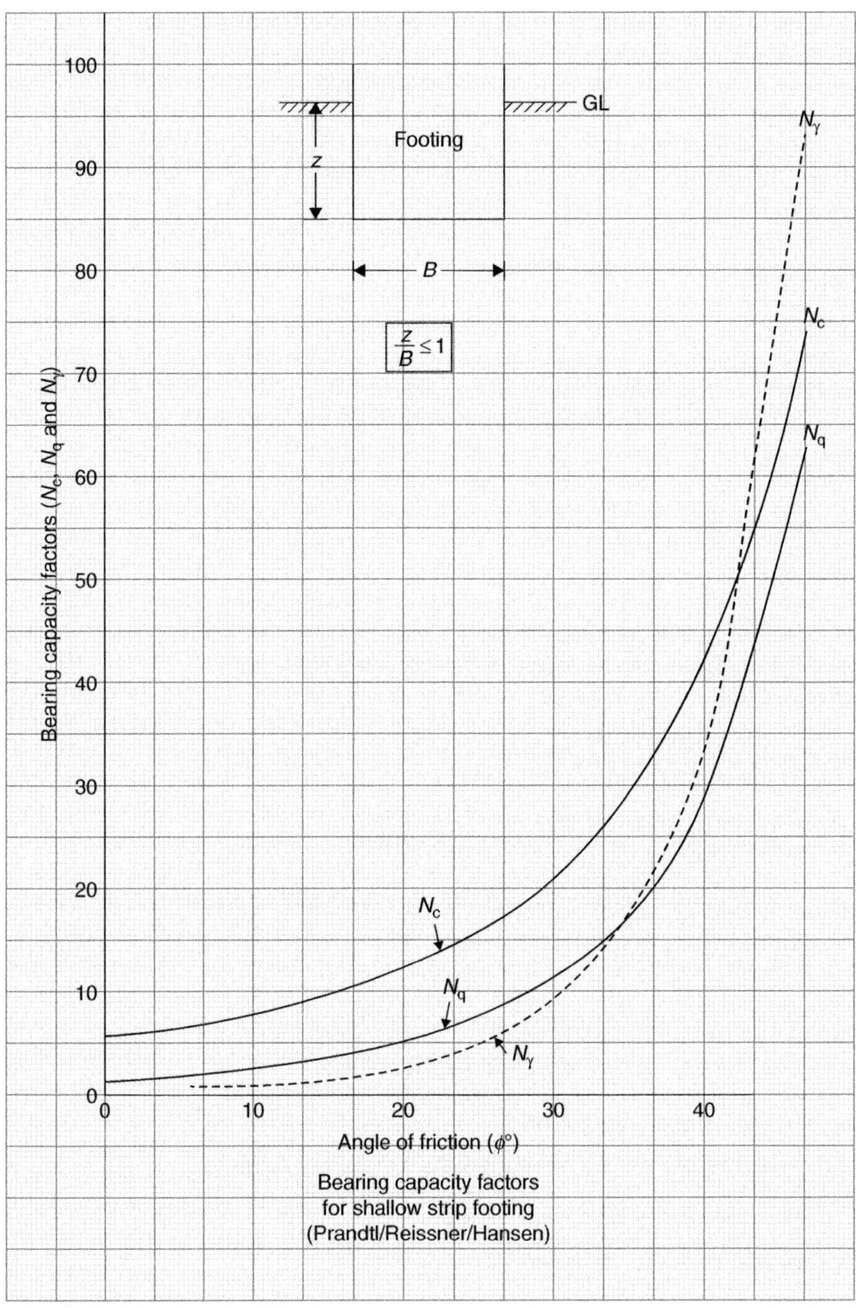

Footing
GL
z
B
$\frac{z}{B} \leq 1$
N_γ
N_c
N_q
Bearing capacity factors (N_c, N_q and N_γ)
0
10
20
30
40
50
60
70
80
90
100
Angle of friction ($\phi°$)
Bearing capacity factors
for shallow strip footing
(Prandtl/Reissner/Hansen)

Chart 9.2

9.3 Influence of footing shape

The magnitude of the safe foundation pressure depends on the bearing capacity and consolidation characteristics of the soil, as well as on the shape and size of the base area. Four shapes of the same area (A) are depicted in Figure 9.19, showing the plan view of the rupture zones around each.

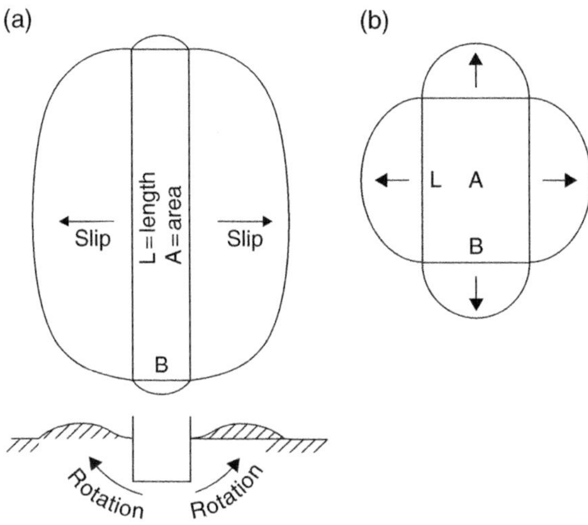

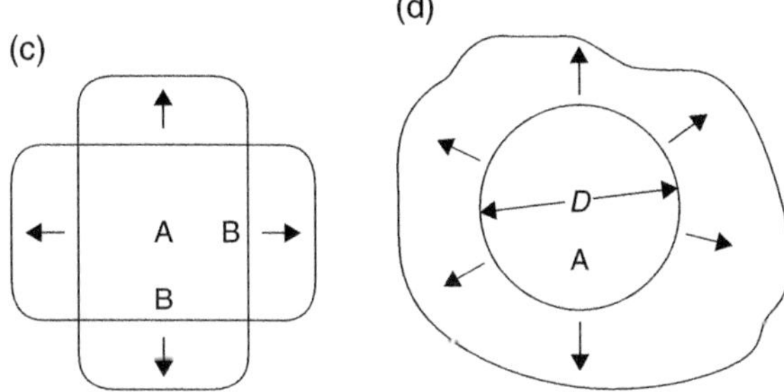

(a) *Strip footing*
This is a two-dimensional failure, as the soil at the two ends do not contribute to the shearing resistance. Failure occurs on one vertical plane, when the shear strength available under the two sliding masses is overcome.

(b) *Rectangular footing*
This is a three-dimensional failure, as the available shear strength has to be overcome on two vertical planes. The circumference for the same area is smaller than for strip footing.

(c) *Square footing*
This shape is slightly more efficient than a rectangle of the same area.

(d) *Circular footing*
The best shape for maximum bearing pressure, compared to strip footing of the same area.

Figure 9.19

Terzaghi modified his general equation for shallow square and circular footings and formula 9.8 becomes:

Square footing:
$$q_u = 1.3cN_c + \sigma'_0 N_q + 0.4\gamma BN_\gamma \qquad (9.18)$$

Circular footing:
$$q_u = 1.3cN_c + \sigma'_0 N_q + 0.3\gamma DN_\gamma \qquad (9.19)$$

where D = diameter of the circle

Note that in the modification there is no apparent distinction made between the load-bearing efficiency of the two shapes as predicted in Figure 9.19. See Example 9.3 however.

9.4 Shallow rectangular footing

The bearing capacity of a rectangular shape (Figure 9.19b) should, obviously, be between that of a strip and square base.

Skempton (1951), recommended the multiplication of N_c for strip footing by the factor $\left(1+0.2\frac{B}{L}\right)$. Formula 9.8 now becomes:

$$q_u = c\left(1+0.2\frac{B}{L}\right)N_c + \sigma'_0 N_q + 0.5\gamma B N_\gamma \qquad (9.20)$$

Hence, the total safe foundation pressure, given by 9.11 is changed to:

$$\boxed{q_s = \frac{1}{F_s^0}\left[c\left(1+0.2\frac{B}{L}\right)N_c + \sigma'_0\left(N_q - 1\right) + 0.5\gamma B N_\gamma\right] + \sigma_0} \qquad (9.21)$$

Example 9.3

Using the soil characteristics of Example 9.1, taking the safe bearing pressure for strip footing only from Case 1, calculate the dimensions of strip, rectangular, square and circular footings to support a vertical load of 1631 kN.

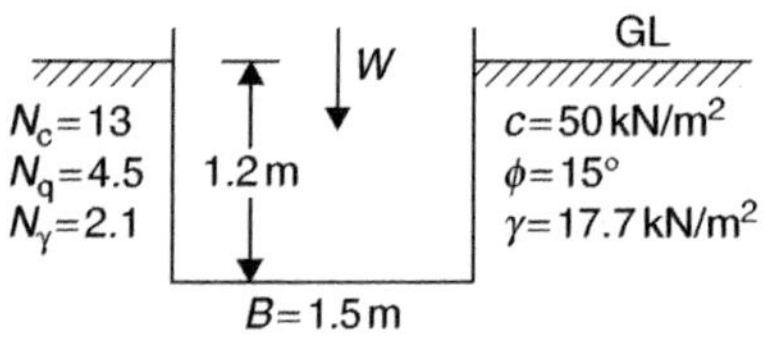

Ground water level is 10 m below GL.

$\sigma_0 = 21.2\,\text{kN/m}^2$

$F_s = 3$

Figure 9.20

a) *Strip footing*

From Case 1: q_s=271.9 kN/m

Vertical Load: W=1631 kN (including the weight of the base)

Base area required: $A = \frac{W}{q_s} = \frac{1631}{271.9} = 6\,\text{m}^2 (= 1.5\times 4)$

Therefore, each 4 m length of strip footing can carry 16.31 kN safely.

b) *Rectangular footing*

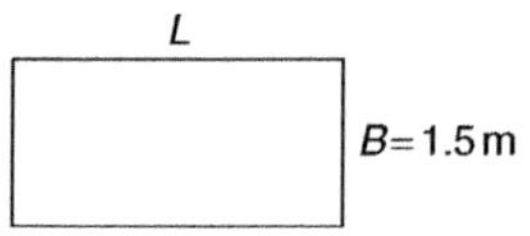

Calculate length (L) required to carry the same 1631 kN load.

Figure 9.21

Bearing pressure: $\sigma = \frac{1631}{1.5L} = \frac{1087}{L} = q_s$

But from (9.21):

$$q_s = \frac{1087}{L} = \frac{1}{3} \times \left[50 \times \left(1 + 0.2\frac{1.5}{L} \right) \times 13 + 21.2 \times 3.5 + 0.5 \times 17.7 \times 1.5 \times 2.1 \right] + 21.2$$

$$\therefore \quad \frac{1087}{L} = 216.7 + \frac{65}{L} + 33.43 + 21.2$$

$$= 271.3 + \frac{65}{L}$$

and $\frac{1022}{L} = 271.3 \qquad \therefore \quad L = \frac{1022}{271.3} = 3.77\,\text{m}$

Base area required: $A = 1.5 \times 3.77 = 5.66\,\text{m}^2$
This bearing area is smaller than required for strip footing.

(c) *Square footing*

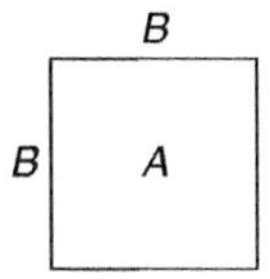

B
B A

Figure 9.22

Calculate the dimension (B) to carry 1631 kN.
$\sigma_0 = 21.2\,\text{kN/m}^2$

Foundation pressure: $\sigma = \frac{1631}{B^2}$

From (9.18): $q_u = 1.3 \times 50 \times 13 + 21.2 \times 4.5 + 0.4 \times 17.7 \times B \times 2.1$
$= 845 + 95.4 + 14.86\,B$

From (9.4): $q_n = q_u - \sigma_0 = 940.4 + 14.86\,B - 21.2$
$= 919.2 + 14.86\,B$

From (9.5): $q_{sn} = \frac{q_n}{F_s} = \frac{919.2 + 14.86\,B}{3} = 306.4 + 4.96B$

From (9.6): $q_s = q_{sn} + \sigma_0 = 306.4 + 4.96B + 21.2$
$= 327.6 + 4.96B$

From (9.7): $\sigma = \frac{1631}{B^2} = 327.6 + 4.96B \qquad (\sigma = q_s)$

$$1631 = 327.6B^2 + 4.96B^3$$

This is a cubic equation: $B^3 + 66.04B^2 - 328.8 = 0$

Solving $B = 2.2\,\text{m} \qquad \therefore$ Area $A = 2.2^2 = 4.84\,\text{m}^2$

Which is a smaller area than required for the rectangular shape.

(d) *Circular footing*

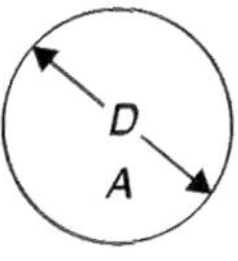

Figure 9.23

From (9.7) and (9.19), the diameter of the circle is found to be $D = 2.48\,\text{m}$, giving a bearing area:

$$A = \frac{D^2\pi}{4} = \frac{2.48^2\pi}{4} = 4.83\,\text{m}^2$$

The four results are tabulated for comparison.

Table 9.3

Shape of footing	Width B (m)	Length L (m)	Diameter D (m)	Area A (m²)	Pressure $\sigma = \frac{1631}{A} \text{kN/m}^2$
Strip	1.5	4	–	6	271.9
Rectangle	1.5	3.77	–	5.66	288.2
Square	2.2	2.2	–	4.84	337.0
Circle	–	–	2.48	4.83	337.6

These results show the variation of load-bearing efficiency of different shapes, as predicted in Figure 9.19.

9.4.1 Method of Fellenius

It applies to strip footing in purely cohesive soil. The footing fails by rotating about one edge.

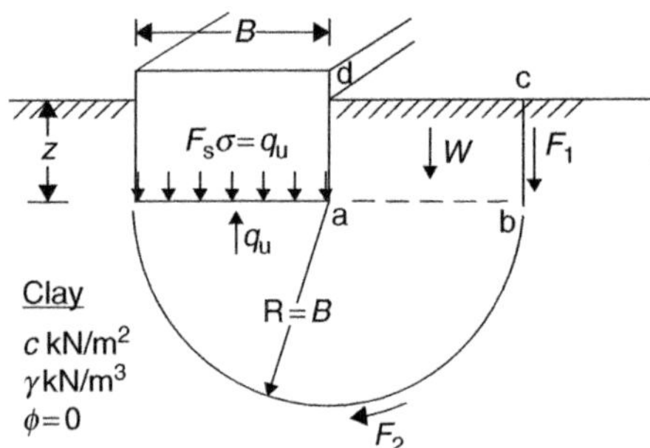

Let $F_s\sigma = q_u$

Disturbing force $= Bq_u$

Resisting cohesive force: $F = F_1 + F_2$

Where, $F_1 = cz$

$F_2 = c\pi B$

Resisting gravity force: $W = zB\gamma$

Figure 9.24

Overturning moment above corner 'a':

$$M_0 = q_u B \frac{B}{2} = \frac{B^2 q_u}{2}$$

Resisting moment about corner 'a'

$$\begin{aligned} M_R &= (F_1 + F_2)B + zB\gamma \frac{B}{2} \\ &= (cz + c\pi B)B + 0.5zB^2\gamma \\ &= czB + c\pi B^2 + 0.5zB^2\gamma \end{aligned}$$

For equilibrium: $M_0 = M_R$

$$\frac{B^2 q_u}{2} = czB + c\pi B^2 + 0.5zB^2\gamma$$

Expressing,
$$q_u = \frac{2}{B^2}\left(czB + c\pi B^2 + 0.5zB^2\gamma\right)$$
$$= c\left(\frac{2z}{B} + 2\pi\right) + z\gamma$$

Finally,
$$\boxed{q_u = 2c\left(\frac{z}{B} + \pi\right) + z\gamma} \tag{9.22}$$

Example 9.4

Figure 9.25 shows strip footing constructed in pure clay. Compare the values of q_u, calculated by formulae (9.22) and (9.8) when $\phi = 0$

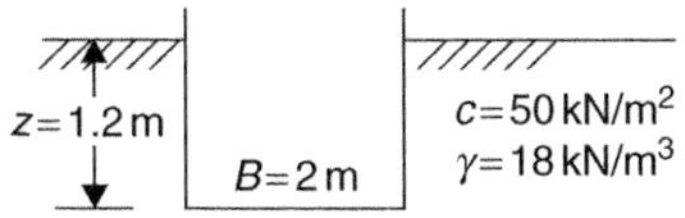

Figure 9.25

From (9.22): $q_u = 2 \times 50 \times \left(\frac{1.2}{2} + \pi\right) + 1.2 \times 18 = 396\,\text{kN/m}^2$

From (9.8): $N_c = 5.7$, $N_q = 1$, $N_\gamma = 0$ | $q_u = 5.7\,c + z\gamma$
$= 5.7 \times 50 + 1.2 \times 18 = 307\,\text{kN/m}^2$

Terzaghi's formula yields more conservative results than Fellenius'.

9.5 Deep foundations

Foundations may be categorized by their depth below ground level. There are three arbitrary general categories:

1. Shallow: $\frac{z}{B} \leq 1$
2. Moderately deep: $1 < \frac{z}{B} \leq 5$
3. Deep: $\frac{z}{B} > 5$

Deep foundations are to be discussed from Section 9.7 onwards.

9.5.1 Moderately deep foundations

Meyerhof modified Terzaghi's bearing capacity factors, taking into account the depth and shape of a footing (Chart 9.3).

Skempton introduced values for N_c for strip, square and circular footings in pure saturated clay (Chart 9.4). The bearing capacity formulae may be used in conjunction with these factors.

Example 9.5

A strip footing, shown in Figure 9.26 is placed 4 m below ground level. Determine the ultimate bearing capacity of the soil.

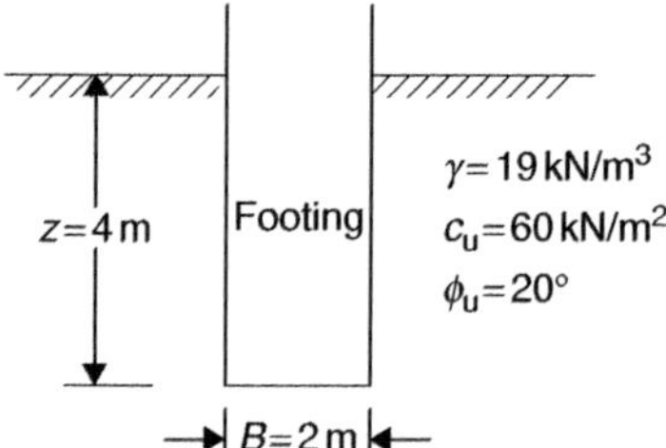

Figure 9.26

$$\frac{z}{B}=\frac{4}{2}=2>1$$

Therefore, the foundation is moderately deep.

From Chart 9.3 for $\phi_u = 20°$: $N_c = 56$

$N_q = 20$

$N_\gamma = 9$

Using (9.8): $q_u = cN_c + N_q + 0.5\gamma BN_\gamma$

$= 60\times 56 + 4\times 19\times 20 + 0.5\times 19\times 2\times 9 = 5051\,\text{kN/m}^2$

Example 9.6

Assume, that the soil in Example 9.5 is pure clay of $c_u = 60\,\text{kN/m}^2$.

Calculate the ultimate and safe bearing capacities of the foundation. Apply a factor of safety of 3.

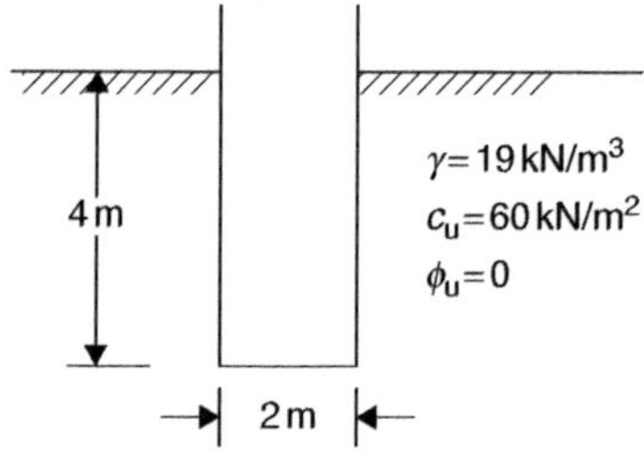

Figure 9.27

From Chart 9.3:

$\phi_u = 0$ | $N_q = 1$, $N_\gamma = 0$

From Chart 9.4:

$\frac{z}{B} = 2$, Shipfolling | $N_c = 7.1$

Using (9.8): $q_u = c_u N_c = 60 \times 7.1 = 426\,\text{kN/m}^2$

Using (9.10): $q_s = \frac{1}{F_s}(c_u N_c) + \sigma_0$

$$= \frac{1}{3}(426) + 4 \times 19 = 218\,\text{kN/m}^2$$

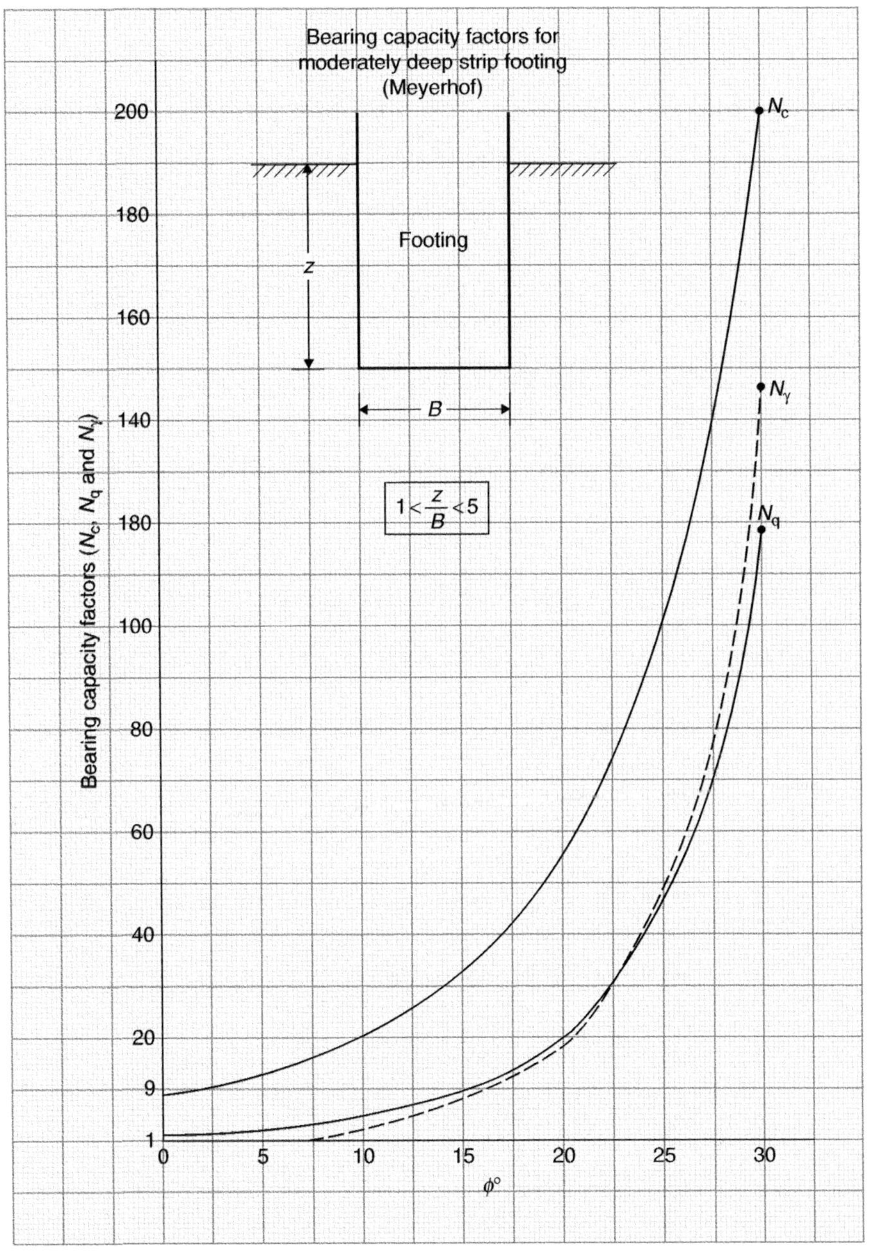

Chart 9.3

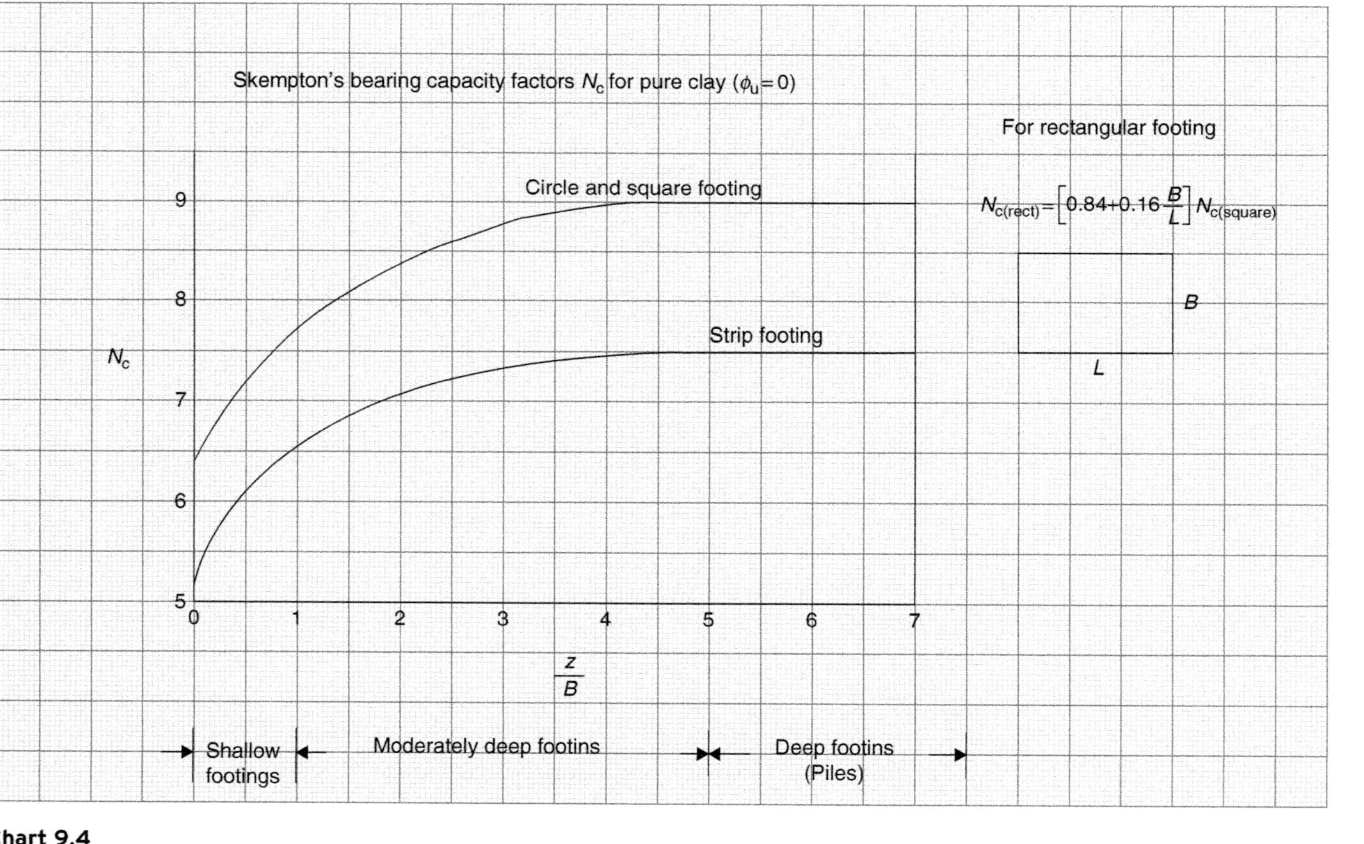

Skempton's bearing capacity factors N_c for pure clay ($\phi_u = 0$)
For rectangular footing
$N_{c(rect)} = \left[0.84 + 0.16\frac{B}{L}\right] N_{c(square)}$
B
L
Circle and square footing
Strip footing
N_c
9
8
7
6
5
0
1
2
3
4
5
6
7
$\frac{z}{B}$
Shallow footings
Moderately deep footins
Deep footins (Piles)

Chart 9.4

9.6 Standard penetration test (SPT)

This empirical in-situ test is carried out by means of a rotary drill. The borehole is drilled to the required depth and a split spoon sampler, attached to the drilling rods, is lowered to the bottom. It is then driven into the soil by hammer blows. The hammer, weighing 64 kg is dropped onto the top of the rods from a height of 760 mm. The sampler is driven into the soil through 450 mm, but the number of blows (N) is counted only over the last 300 mm.

Notes:

1. The test must start at the bottom of the casing, that is, in undisturbed ground.
2. If the test is carried out in sand or gravel, below ground water table, this water level must be maintained inside the casing, by pouring in sufficient amount, in order to prevent upward seepage and 'boiling'.
3. Correction has to be made to the value of N, when measured in fine sand under water. The corrected value is given by:

$$\boxed{N' = 15 + 0.5(N - 15)} \quad (9.23)$$

For $N > 15$ only

This takes into account, the excess pore water pressure induced during the test, which increases the resistance, hence the value of N.

Applications

The standard penetration results may be applied to both cohesive and cohesionless soils, correlating the number of blows with the:

a) Angle of shearing resistance ϕ (Chart 9.5)
b) Bearing capacity factors N_γ and N_q (Chart 9.5)
c) Allowable bearing pressure of foundations in sand (Chart 9.6)
d) Relative density of sand (Table 9.4)
e) Consistency/unconfined compressive strength of clay (Table 9.5).

Table 9.4

Number of blows/300 mm		Relative density of sand
N	$N'=15+0.5\ (N-15)$	
0–4	0–4	Very loose
4–10	4–10	Loose
10–30	10–22.5	Medium dense
30–50	22.5–32.5	Dense
>50	>32.50	Very dense

Table 9.5

Number of blows (N)	Unconfined compressive strength $2c_u$ (kN/m^2)	Consistency
<2	<25	Very soft
2–4	25–50	Soft
4–8	50–100	Medium stiff
8–15	100–200	Stiff
15–30	200–400	Very stiff
>30	>400	Hard

Note: Owing to the empirical nature of the SPT, its results are only approximations.

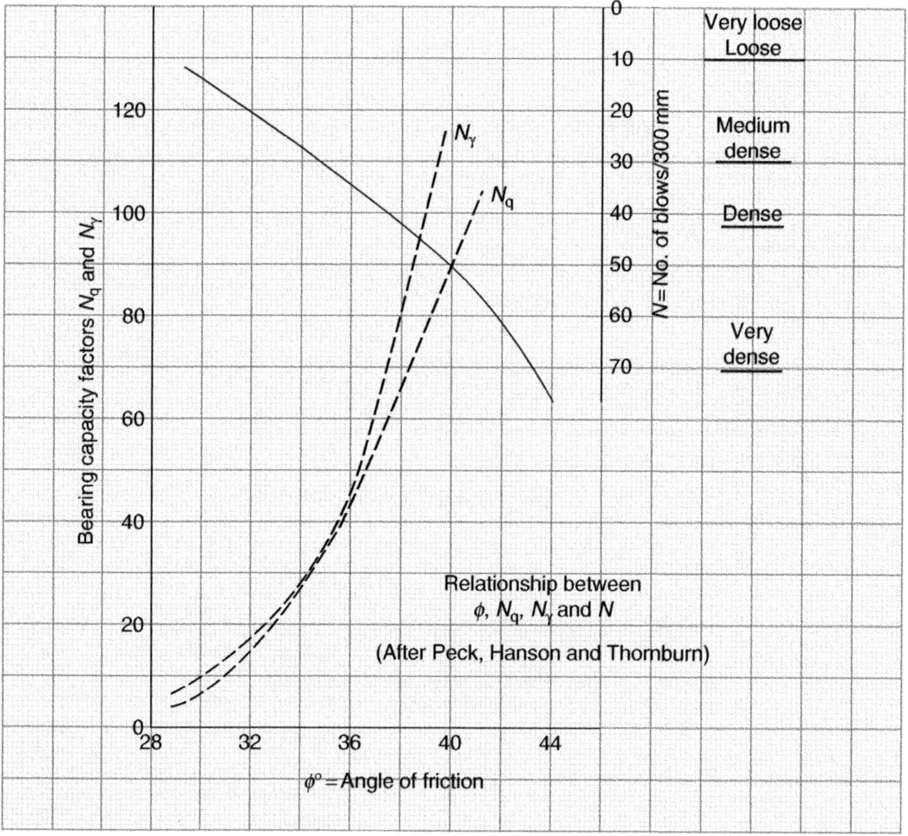

Chart 9.5

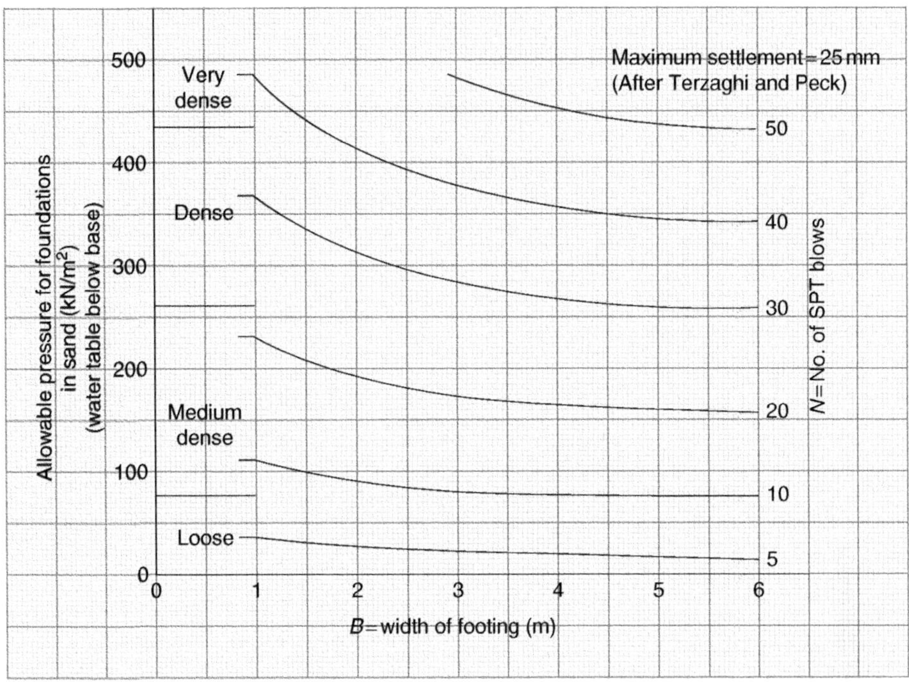

Chart 9.6

9.7 Pile foundations $\left(\frac{z}{B} > 5\right)$

Piles are structural elements, transmitting load from footings to the soil. The transmission occurs in three ways:

1. By friction or adhesion between the soil and the surface of the pile (Figure 9.28 (a))
2. By end bearing, where the pile acts as a column (Figure 9.28 (b))
3. By the combination of friction and end bearing (Figure 9.28 (c)).

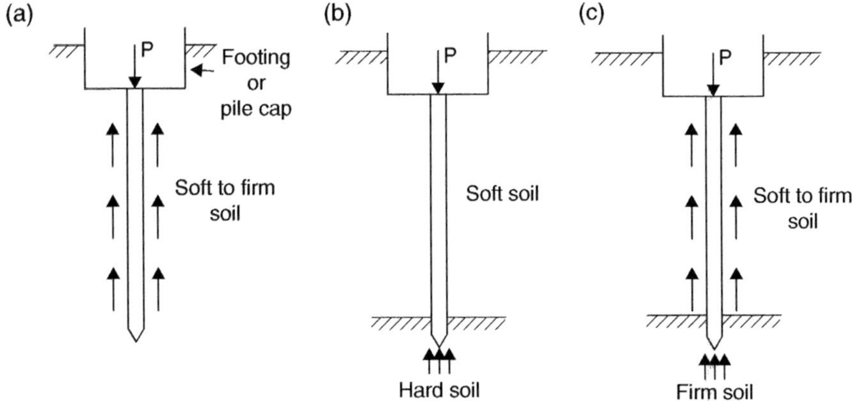

Figure 9.28

Piles are also used for purposes, other than the transference of vertical load to the ground. There are three special applications:

1. To increase the density of the surrounding soil be compaction, in order to stabilize it (Figure 9.29 (a))
2. To resist upward pull. The pile is in tension (Figure 9.29 (b))
3. To carry horizontal force, to support an anchorage for example (Figure 9.29 (c)).

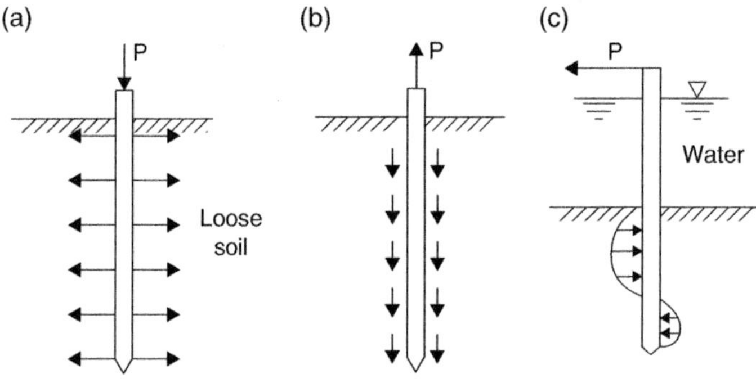

Figure 9.29

9.7.1 Types of pile

The method of installation of piles depends on the ground conditions as well as onother circumstances, such as the distance to nearby structures. The most frequently employed pile types are:

- Driven
- Jacked
- Screwed
- Bored
- Vibrated
- Jetted.

9.7.1.1 Driven piles

These are driven into the soil by a dropping hammer. The piles may be made of timber, precast, reinforced or prestressed concrete, or steel. Timber piles must remain under water to prevent decay.

Cast-in-place piles are formed by driving a steel or concrete tube into the ground and filling it with concrete. The tube can either be left permanently or removed gradually during the concrete pour.

It is advantageous to leave the tube, when the ground water contains sulphates or other chemicals detrimental to the concrete or steel reinforcement. The concrete must be well compacted for strength and impermeability.

Alternatively, precast concrete piles may be used in soil containing deleterious substances. Their maximum length, however, is limited to 20 m, because of increasing difficulty in driving them to larger depths.

One of the widely used driven, cast-in-place piling method is the Franki system. The process is illustrated in Figure 9.30 (a) to (e).

Step 1: A steel tube is positioned vertically and damp concrete or gravel rammed into its lower end to form a plug of about 1 m high (a).
Step 2: The plug is forced into the ground by the rammer. The tube follows the plug, due to the friction between the steel and concrete or gravel (b).
Step 3: As soon as the tube reaches the required bearing depth, it is fixed to the supporting frame and most of the plug is rammed out of the tube into the soil (c).
Step 4: Concrete is added and hammered out to form a bulb-shaped end, in order to increase the base area, hence the load carrying capacity of the pile (d).
Step 5: In most cases, a steel reinforcement cage is placed into the tube, which is gradually withdrawn as the concrete is poured and compacted by the drop hammer (e).

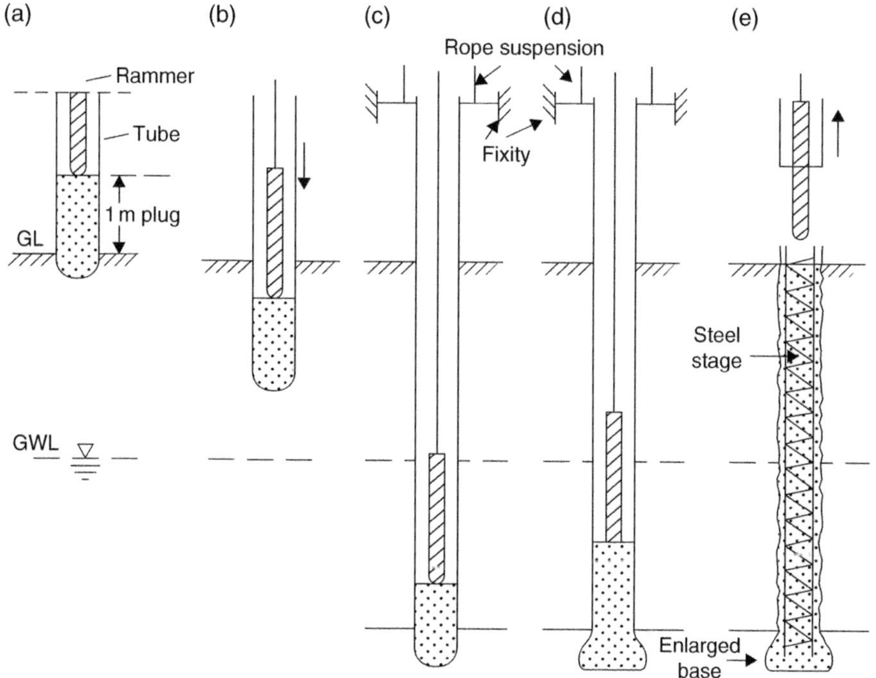

Figure 9.30

Notes:

1. The pile may be constructed below ground water level, as the rammed concrete plug is practically impervious.
2. The soil is compacted horizontally as well as around the enlarged bulb-shaped base, thereby increasing the shaft friction and the end resistance respectively.
3. When compared to other driven piles, the induced surface vibration is smaller, as the plug and the surrounding soil largely absorb the shocks.

9.7.1.2 Jacked piles

The piles are built up from short (750 mm) precast concrete sections, fitted together during the jacking process. The system is normally employed to underpin existing structures and transmit the structural load to some depth below the existing footing.

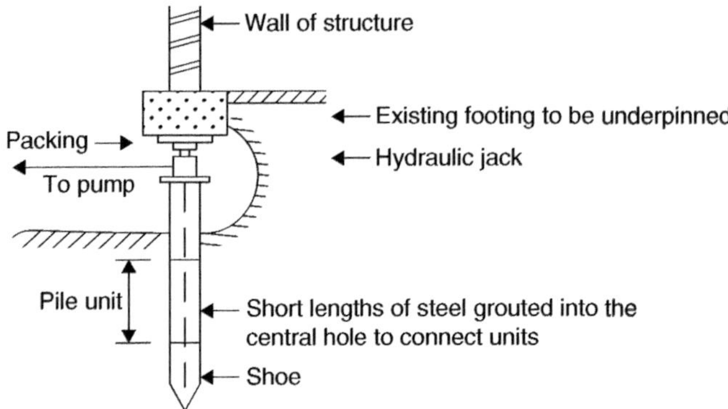

Figure 9.31

9.7.1.3 Screwed piles

These piles are normally used for structures built over soft alluvial soil, underlain by hard stratum. The lower end of the pile-tube is fitted with one or more helical blades. These are rotated, which forces the tube into the ground. When the base is at the desired level, the tube is filled with concrete.

9.7.1.4 Bored piles

Basically, the piles are constructed by filling drilled holes with concrete. They are, therefore, cast-in-place piles. The hole can be drilled with or without liners (tubes), depending on the soil. Also, the tubes can either be left or withdrawn, depending on the chemically aggressive nature of the ground water. The piles are assumed to be end-bearing only. For this reason, the base area is increased by under-reaming (Figure 9.32).

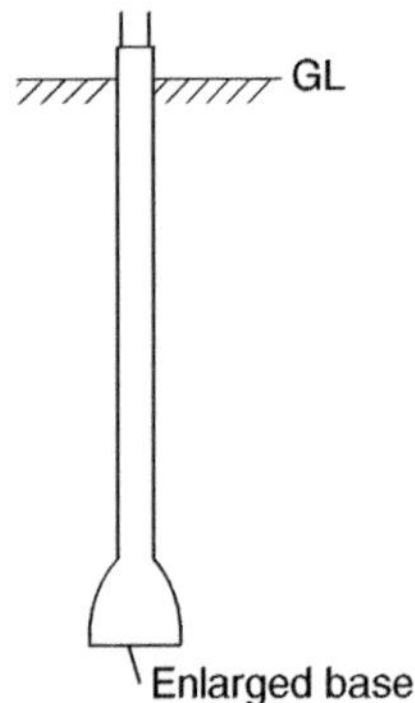

Figure 9.32

Advantages:

a) The surroundings are not subject to vibration and ground heave as with driven piles.
b) Soil profile can be verified and the length of the pile chosen during the boring process.
c) Pile diameter, hence load carrying capacity can be large, subject to the size of drill used.
d) Drilling can be carried out under low headroom, e.g. inside buildings.
e) The energy expended in drilling is much less than in pile-driving.

Boring without liner tube may be carried out in stiff cohesive soils above ground water level. The subsequently placed concrete should not be excessively wet as it would soften the soil, thus reducing its bearing strength.

Short friction piles, placed comparatively near to each other in cohesionless soil, should really be driven. It is easier, however, to drill long piles, when placed some distance apart and driving would be difficult. Plastic concrete is vibrated into the liner tube so, that its surface is always above any external ground water level. The disadvantage of tube withdrawal is, that the adjacent soil could become loose, decreasing the load bearing capacity.

Note: A bored pile requires more concrete than a driven one, because of the volume of the tube as well as the loosened surrounding soil.

There has to be a good reason to leave the tube in place after concreting. Some of these are:

a) To insulate the concrete from chemically aggressive ground water.
b) To minimise negative skin friction.
c) To prevent damage to the concrete caused by its arching (necking) in the tube, thus restricting extrusion.

9.7.1.5 Vibrated piles

These are applicable mainly to loose, cohesionless soils. The vibrator is positioned at the top of the tube, which can easily be pushed into as well as extracted from the ground whilst vibrated. This is due to the destruction of skin friction by the amplitude of vibration.

9.7.1.6 Jetted piles

The process is used to facilitate the installation of driven or screwed piles. The water is forced through a pipe located either inside or outside the pile, emerging at its tip. Jetting has to be stopped 1m above the final depth and the pile is driven normally to this level. Existing, adjacent jetted piles could be disturbed. It is advisable, therefore, to drive them a little deeper after the completion of piling.

The most suitable soil for this method is fine sand and silt. It is not used in clay as it can block the nozzles at the tip of the pile.

9.8 Some reasons for choosing piles

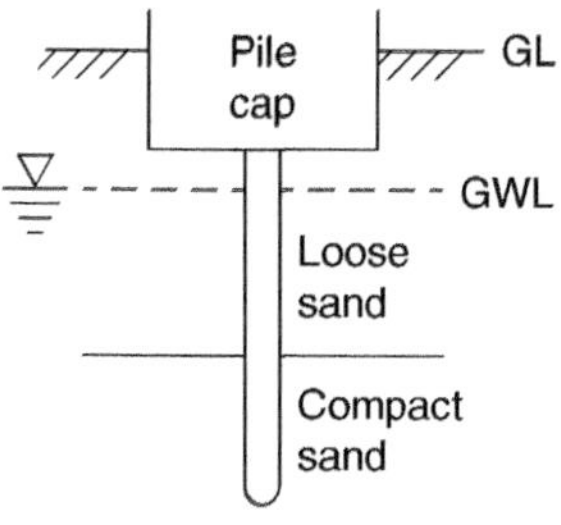

Figure 9.33a

a) Because of the high water table in the loose sand, difficulties in dewatering and piping failure can be expected during excavation. The friction pile should penetrate the compact sand deep enough.

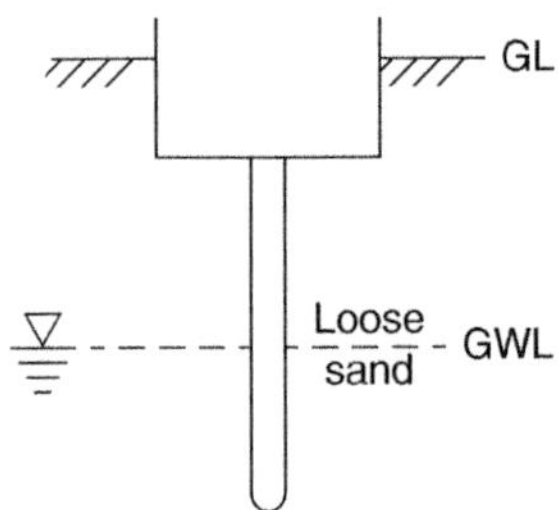

Figure 9.33b

b) Homogeneous, loose sand does not get stronger with depth. Moderately deep footing, approaching the GWL could prove expensive. Friction pile of sufficient length could suffice.

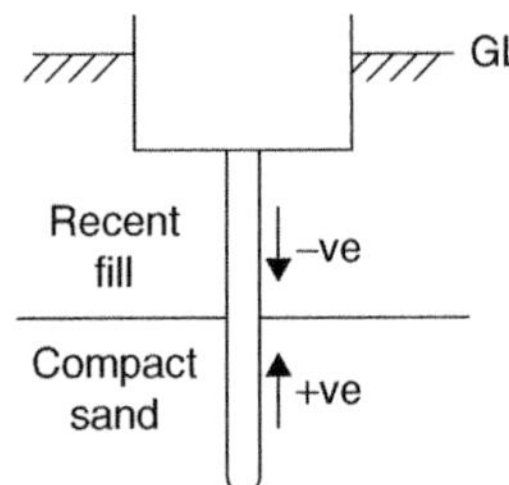

Figure 9.33c

c) The bearing capacity of recent fill layer is insufficient. Friction pile must penetrate the load-bearing compact sand. However, negative skin friction in the fill has to be taken into account.

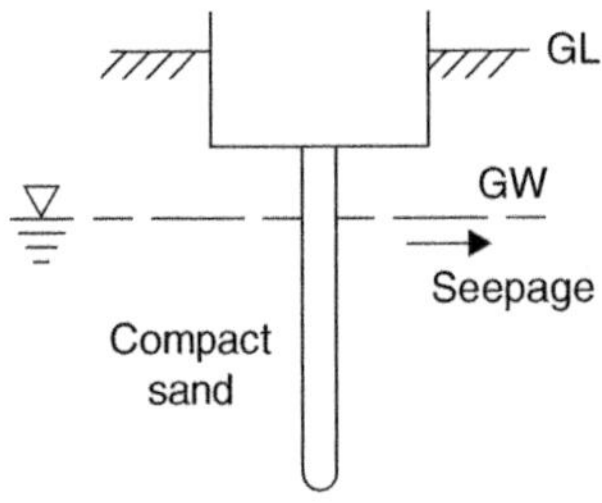
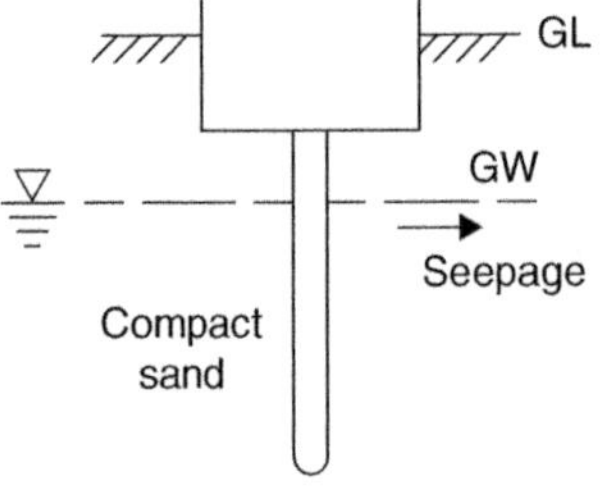

Figure 9.33d

d) Normally, shallow footing would be satisfactory in compact sand. However, seepage could cause washouts below it, hence friction piling is justified.

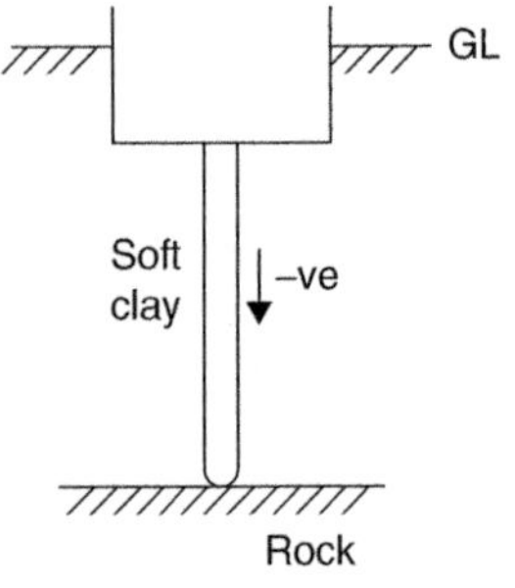

Figure 9.33e

e) The soft clay overlying hard material, such as rock, has insufficient bearing capacity end-bearing pile, seated on the rock, is required. Soft, sensitive clays can induce negative skin friction on the pile however.

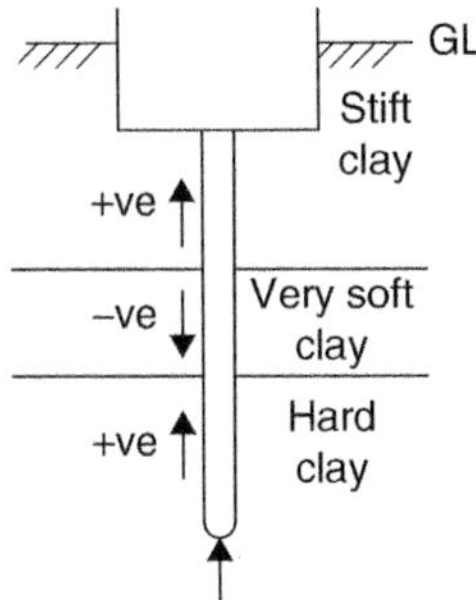

Figure 9.33f

f. The stiff clay has sufficient bearing capacity to support shallow footing, but the very soft clay cannot carry the transmitted (Boussinesq) load. Fiction/end-bearing pile may be chosen, taking the negative skin friction into account.

Also, piles are chosen in the following circumstances:

1. When the intensity of loading under parts of a large, heavy structure is different.
2. An existing building would be damaged by the construction of an adjacent shallow or moderately deep foundation. Bored piles are suitable in these circumstances.
3. A building to be constructed is not rigid enough and differential settlements could occur.

9.9 Some reasons for not choosing piles

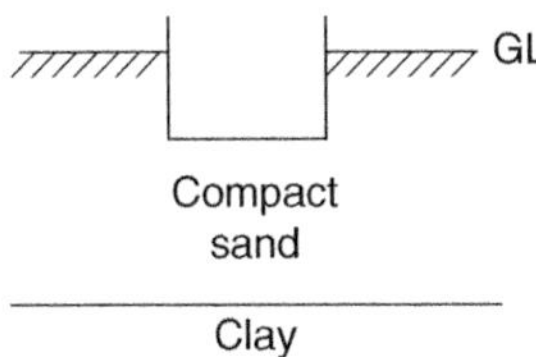

Figure 9.34

1. It is absolutely unnecessary to use piles, as the compact sand is strong enough to carry heavy shallow foundation load. Unless there is seepage of water, as shown in Figure 9.33d.

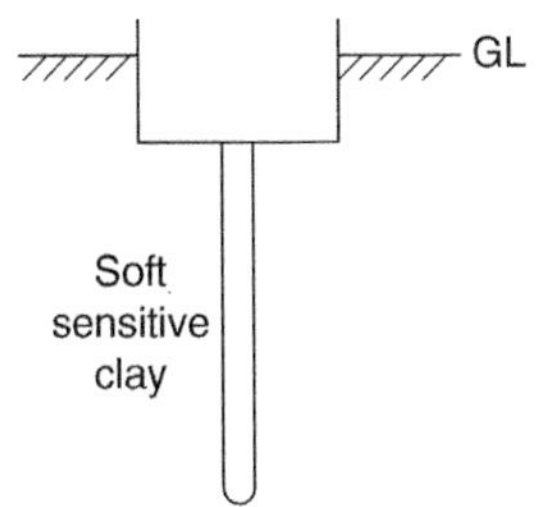

Figure 9.35

2. Driving piles into sensitive clay reduces its shear strength due to remoulding. The magnitude of strength loss depends on the degree of sensitivity.

9.10 Effects necessitating caution

(See Figures 9.36a–e).

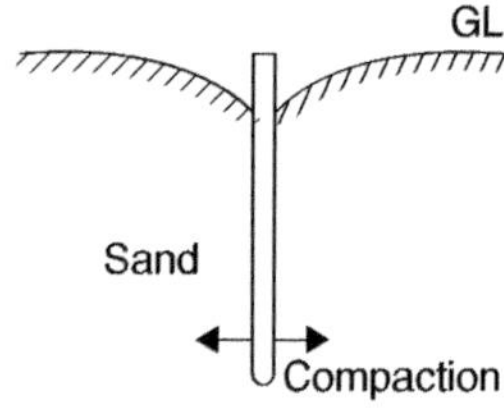

Figure 9.36a

a) When a pile is driven into granular soil, it caused ground surface lowering. More so with a group of piles. Horizontal compacting forces could affect adjacent structures.

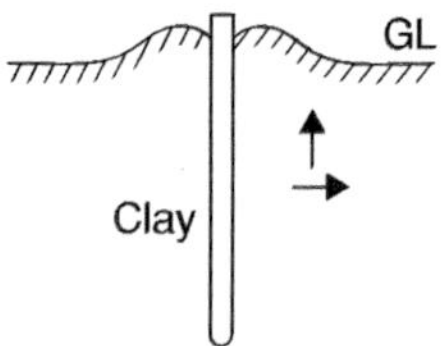

Figure 9.36b

b) In cohesive soils both horizontal and vertical movement (heave) would occur. This could induce negative skin friction. A group of piles could raise the entire surface in its vicinity.

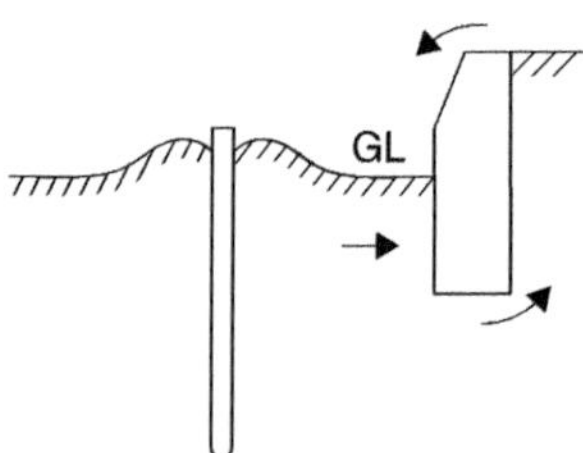

Figure 9.36c

c) The horizontal movement could exert undesirable forces on adjacent structures. Bored piles are more suitable in these situations.

Note: When a pile is forced into a cohesive soil, excess pore pressure is induced. As this pressure dissipates, the soil consolidates and the ground heave eventually disappears.

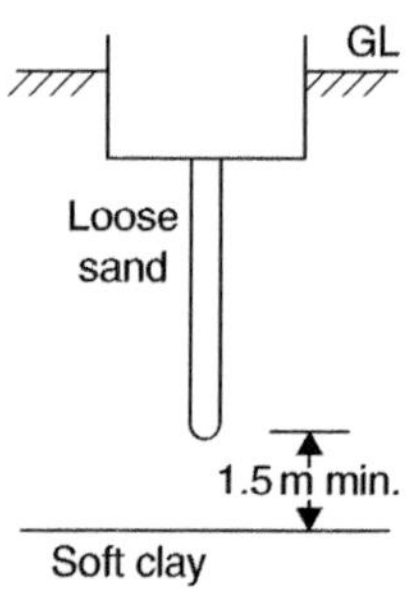

Figure 9.36d

d) Pile or pile group *must not* be extended down to the soft clay layer as it could become overstressed. The end of a pile should be at least 1.5 m above the clay.

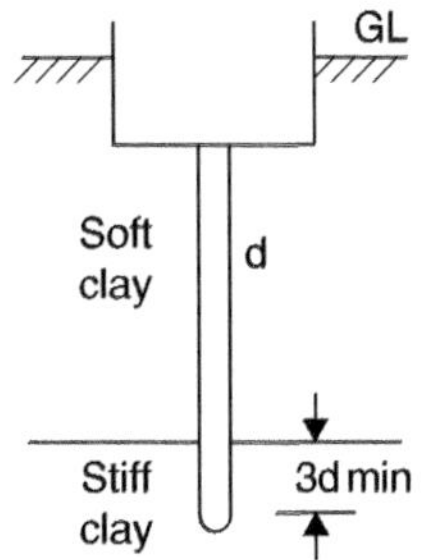

Figure 9.36e

e) The end of a pile must not be seated on the top of the stiff clay but penetrate it to a minimum depth of three times its diameter.

9.11 Negative skin friction

(See Figures 9.37a to e).

When a pile is loaded, it exerts downward drag on the soil. The skin friction, however, induces an upward force, which resists the load thus preventing relative movement between the pile and the soil. This is positive skin friction.

In certain circumstances, discussed below, the soil is displaced downwards, relative to the pile. The skin friction induces this time a downward force, thus increases the load on the pile. This is negative skin friction.

Piles must be designed to carry any additional load due to negative skin friction. Some possible causes are:

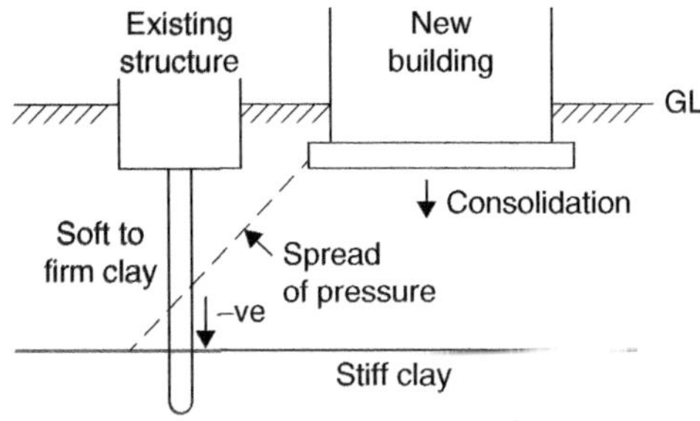

Figure 9.37a

a) The new building consolidates the clay, inducing negative skin friction on the pile surface.

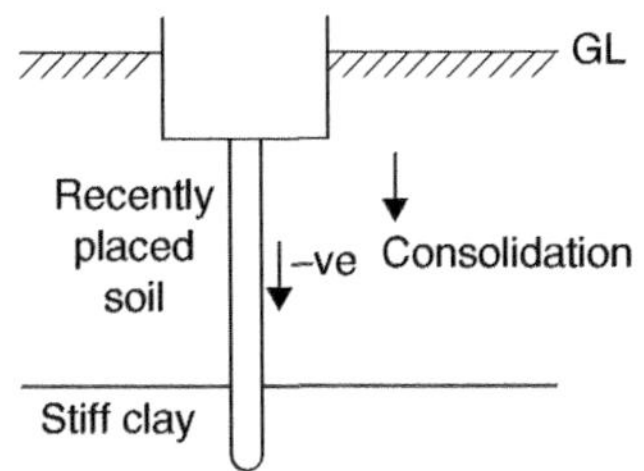

Figure 9.37b

b) The recently placed fill gradually consolidates under its own weight causing negative skin friction on the pile surface see also Figure 9.33 (c).

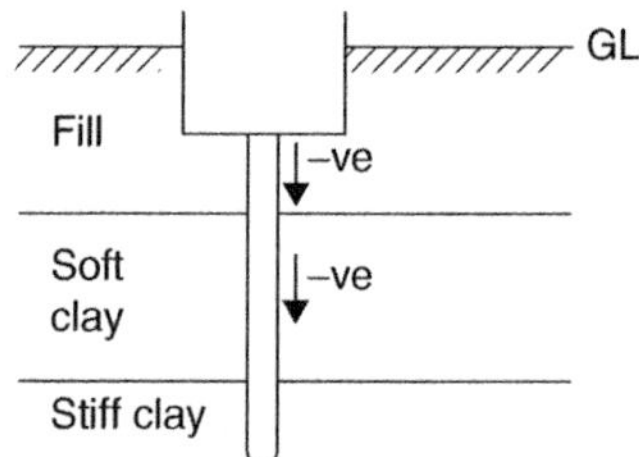

Figure 9.37c

c) The fill consolidates the soft clay. As both fill and soft clay move together, negative skin friction is induced along the pile in these layers.

Note: Negative skin friction can be high in granular fill. In extreme cases, piles could be forced away from the structure and driven further into the end-bearing layer. See also Figure 9.33 (f).

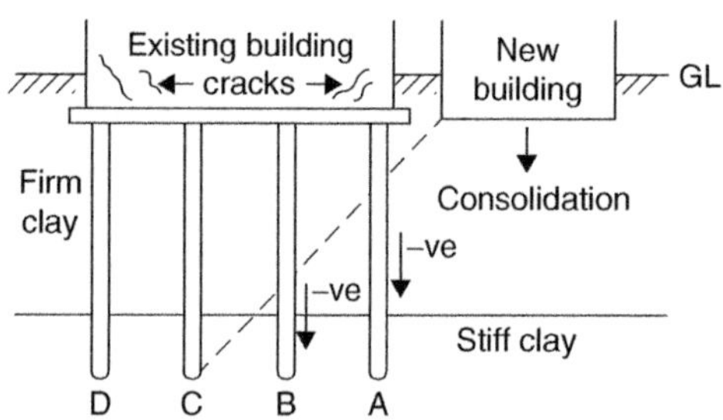

Figure 9.37d

d) Piles A and B are subjected to negative skin friction, whilst C and D are not. This could cause differential movement and cracks in the existing structure.

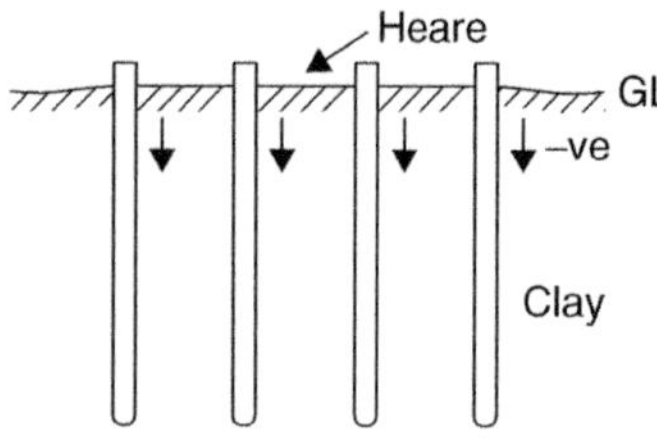

Figure 9.37e

e) When piles are grouped closely to each other in clay, excess pore pressure is induced and the ground heaves up. Reconsolidation occurs as the excess pressure dissipates, causing negative skin friction on the upper part of the piles.

Notes:

1. The effect of negative skin friction is greater on an end-bearing pile than on a friction one, as the former is 'fixed' and does not move with the consolidating soil, hence taking most of its weight.

   In contrast, a friction pile takes only some of the weight of the moving soil, because it moves with it to some extent. In this case, negative skin friction develops only, when the downward movement of the soil is greater than that of the pile.
2. Do not use raking piles subjected to this effect.

9.12 Stress distribution around piles

The distribution of stresses below shallow footings was discussed in Chapter 4. The formulae were derived on the basis of the Theory of Elasticity.

Stress distribution around piles has not been determined theoretically. Instead, the extent of compaction achieved by actual piles has been used as an indicator of the distribution.

In general, the region of pressure, significant enough to cause compaction, depends on the:

1. Type of soil
2. State of the soil
3. Width or diameter (d) of the pile
4. Length (*l*) of the pile
5. Method of placing.

It can be assumed, on the basis of site tests, that there is little compaction beyond the distance of four times the pile diameter (4*d*). The influence of a single pile, therefore, is limited to a small volume around its surface, as shown in Figure 9.38(a).

On the other hand, the pressure distribution of a closely placed pile group can extend to large depths. In effect, the group acts like a shallow raft foundation (Figure 9.38(b)).

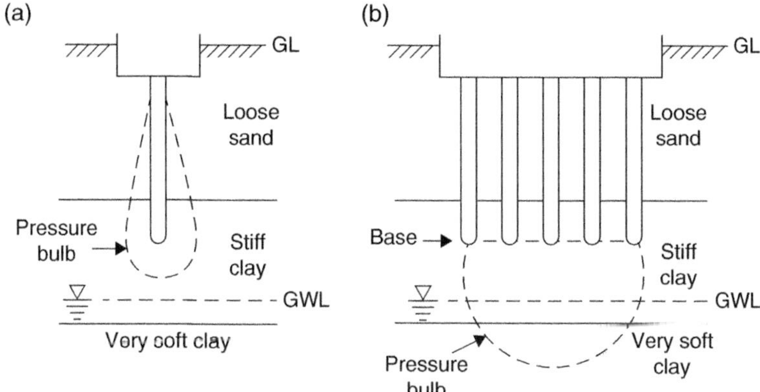

Figure 9.38

Note: The group of piles could overstress the very soft clay.

9.13 Load-carrying capacity of piles

The determination of the loading capacity of piles is complicated by the variability of soil characteristics as well as by the method of their construction. It may be estimated by:

1. Static, theoretical formulae
2. Dynamic, pile-driving formulae
3. In-situ, full-scale loading test.

In view of the empirical nature of the first two procedures, it is advisable to verify their results by in-situ loading test. As the last two procedures are normally carried out by specialist piling contractors on site, only the first method is to be discussed in this book.

9.13.1 Static formulae

The load on a pile, causing failure can be expressed in two ways:

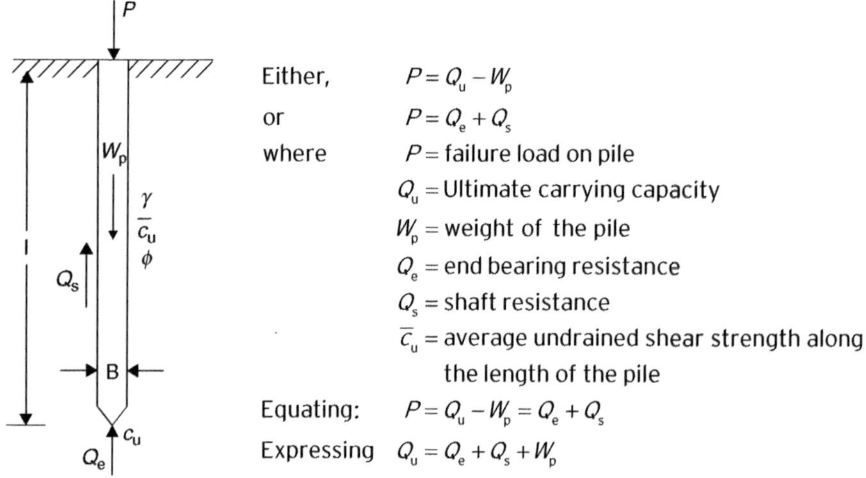

Either, $P = Q_u - W_p$

or $P = Q_e + Q_s$

where P = failure load on pile

Q_u = Ultimate carrying capacity

W_p = weight of the pile

Q_e = end bearing resistance

Q_s = shaft resistance

$\bar{c}_u$ = average undrained shear strength along the length of the pile

Equating: $P = Q_u - W_p = Q_e + Q_s$

Expressing $Q_u = Q_e + Q_s + W_p$

Figure 9.39

9.13.2 End-bearing resistance (Q_e)

It may be calculated from a modified form of the bearing capacity formula for shallow footings in uniform, homogeneous layer. Therefore,

$$q_n = cN_c + \sigma_0'\left(N_q - 1\right) + 0.5\gamma BN_\gamma \qquad (9.9)$$

But the width (B) or diameter (d) of a pile is very small, compared to its length (l), hence the third term is small enough to be neglected.

$$\therefore \quad q_n = c_u N_c + \sigma_0'\left(N_q - 1\right)$$

where σ_0' = effective overburden pressure at the base

c_u = undrained cohesion of soil at the base.

If A_e = end bearing area, then the base resistance is given by:

$$Q_e = A_e q_n = A_e\left[c_u N_c + \sigma_0'\left(N_q - 1\right)\right]$$
$$= A_e c_u N_c + A_e \sigma_0' N_q - A_e \sigma_0'$$

Weight of soil displaced by the pile = $W' = A_e\sigma'_0$

Therefore, $\boxed{Q_e = A_e(c_u N_c + \sigma_0' N_q - W')}$

Hence, $\boxed{Q_u = A_e(c_u N_c + \sigma_0' N_q) + Q_s + (W_p - W')}$

But, the term $(W_p - W')$ is usually small and negligible.

Therefore,
$$\boxed{Q_e = A_e\left(c_u N_c + \sigma_0' N_q\right)} \tag{9.24a}$$

And,
$$\boxed{Q_u = A_e\left(c_u N_c + \sigma_0' N_q\right) + Q_s} \tag{9.24b}$$

9.13.3 Shaft resistance (Q_s)

The nature of shaft resistance depends on whether the soil is cohesionless or cohesive. In cohesionless soil, the resistance is called 'skin friction' evaluated in terms of friction angle (ϕ). In cohesive soil, the resistance is called 'adhesion', evaluated in terms of average cohesion $\bar{c}_u$.

For a cohesionless soil, the shaft resistance is given by:

$$\boxed{q_s = K_s \overline{\sigma_0'} \tan\delta} \tag{9.25}$$

and
$$\boxed{Q_s = A_s q_s = A_s K_s \overline{\sigma_0'} \tan\delta} \tag{9.26}$$

where A_s = surface area of pile
K_s = average coefficient of earth pressure.
Its value depends on the relative density of the soil, which is normally found from Standard Penetration Test (see Table 9.6).
δ = angle of friction between the soil and the material of the pile.
$\overline{\sigma'}_0$ = Average effective overburden pressure, over the embedded length of pile.

Typical values, recommended by Broms (1966) are:

Table 9.6

Material of the pile	**$\delta°$**	**K_s Relative density: Loose**	**K_s Relative density: Dense**
Concrete	0.75ϕ	1.0	2.0
Steel	20	0.5	1.0
Wood	0.67ϕ	1.5	4.0

For a cohesive soil, the shaft resistance is given by:

$$\boxed{q_s = \alpha \bar{c}_u} \tag{9.27}$$

$$\boxed{Q_s = A_s q_s = \alpha \bar{c}_u A_s} \tag{9.28}$$

where: α = adhesion factor depending on the type of cohesive soil
$\bar{c}_u$ = Average undrained shear strength of undisturbed soil, in a layer, around the pile.

For bored piles and for design purposes, take $\alpha = 0.45$.

For driven piles $0.25 \le \alpha \le 1$ if $\bar{c}_u > 50\,\text{kN/m}^2$

$\alpha = 1$ if $\bar{c}_u < 50\,\text{kN/m}^2$

For negative friction α=0.2.

9.13.4 Ultimate carrying capacity of pile

In cohesionless soil: $$Q_u = A_e\sigma'_0 N_q + A_s K_s \overline{\sigma'_0} \tan\delta \quad (9.29)$$

Where, N_q is obtained from Chart 9.7 for various values of slenderness ratio l/B

In cohesive soil: $$Q_u = A_e c_u N_c + \alpha \overline{c}_u A_s \quad (9.30)$$

$$Q_u = Q_e + Q_s$$

where $N_c = 9$ as long as the pile penetrates the bearing layer to a depth not less than $5B$. (See Chart 9.4).

9.13.5 Allowable carrying capacity of piles (Q_a)

In order to limit the settlement of a pile to acceptable magnitude, separate factor of safety is applied to each component of Q_u.

$$\left.\begin{aligned} \text{For end bearing:} \quad & F_e = 13 \\ \text{For shaft resistance:} \quad & F_s = 1.5 \end{aligned}\right\} \text{Usual values}$$

Therefore, $$Q_a = \frac{Q_e}{F_e} + \frac{Q_s}{F_s} = \frac{Q_e}{3} + \frac{Q_s}{1.5} \quad (9.31)$$

Also, an overall factor of safety $F_0 = 2.5$ is applied to Q_u against shear failure.

$$Q_a = \frac{Q_u}{F_0} = \frac{Q_e + Q_s}{2.5} \quad (9.32)$$

The lesser of (9.31) or (9.32) is accepted as the allowable working load.

9.13.6 Negative skin friction (Q_f)

When a pile is penetrating layers, which consolidate for one reason or another, it is subjected to a downward drag (Q_f) or negative shaft friction resistance, as shown in Figure 9.40. This force has to be added to the design-load.

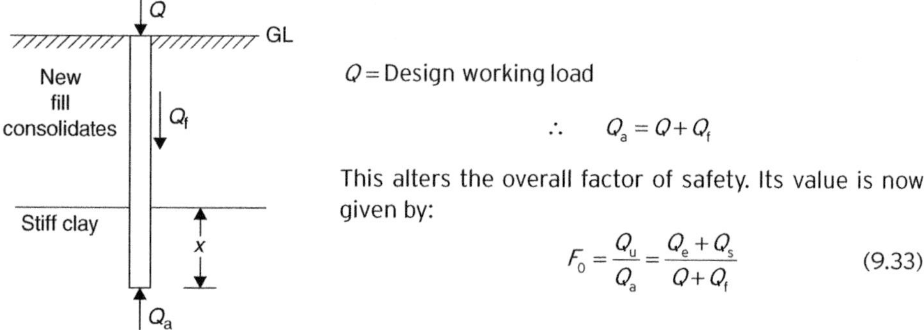

Q = Design working load

$$\therefore \quad Q_a = Q + Q_f$$

This alters the overall factor of safety. Its value is now given by:

$$F_0 = \frac{Q_u}{Q_a} = \frac{Q_e + Q_s}{Q + Q_f} \quad (9.33)$$

Figure 9.40

Notes:

1. When the soil profile consists of several layers of different strength, negative skin friction could occur in the soft, weak ones. It is important, therefore, to conduct detailed investigation into the nature of each stratum.
2. Negative skin friction can occur in cohesive soils after ground water lowering, and consequent shrinkage.
3. It is not advisable to place raking piles into soil, where negative skin friction could develop, as this would cause bending stresses in the pile and possible separation from the pile cap.

The value of Q_f may be estimated from the following empirical formulae for:

Cohesionless soil: $$q_f = K_s \overline{\sigma_0'} \tan\phi \quad (9.34)$$

$Q_f = A_s q_f$ $\quad \therefore$ $$Q_f = A_s K \overline{\sigma_0'} \tan\phi \quad (9.35)$$

Cohesive soil: $$q_f = 0.1 \overline{\sigma_0'} \quad (9.36)$$

$Q_f = A_s q_f$ $\quad \therefore$ $$Q_f = 0.1 \overline{\sigma_0'} A_s \quad (9.37)$$

Example 9.7

A 12 m long, 350 mm diameter, precast concrete pile was driven into the stiff clay layer, shown in Figure 9.41, through 5 m of recently deposited clayey ash fill. No ground water level was found in the borehole. Use the SPT results to verify, that the pile can carry at least 110 kN working load, adopting 2.5 as overall factor of safety.

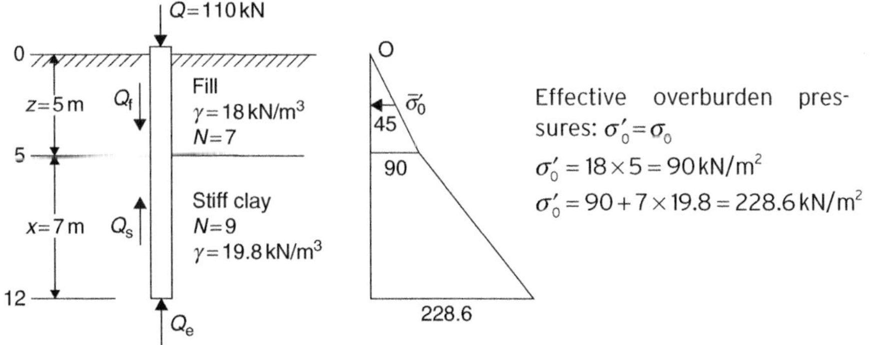

Figure 9.41

Recently deposited fill (0–5 m)
The layer will consolidate under its own weight, inducing negative skin friction, given approximately by formulae 9.37.

Surface area of pile in this layer:

$$A_s = \pi dz = \pi \times 0.35 \times 5 = 5.5\,\text{m}^2$$

Effective overburden pressure of 5 m depth

$$\sigma'_0 = 90\,\text{kN/m}^2$$

Therefore $$Q_f = 0.1\overline{\sigma'_0} A_s = 0.1 \times 45 \times 5.5 = 25\,\text{kN} \qquad (9.37)$$

Stiff clay layer ($\phi \simeq 0$)

From Table 9.5 for SPT $N=9$ $c_u=100\,\text{kN/m}^2$

In the absence of other information, the undrained shear strength is assumed to be uniform in the layer, hence $\bar{c}_u = c_u = 100\text{kN/m}^2$

End-bearing area: $A_e = \dfrac{d^2\pi}{4} = \dfrac{0.35^2\pi}{4}$

Surface area of pile: $A_s = \pi dx = \pi \times 0.35 \times 7 = 7.7\,\text{m}^2$

Adhesion factor: $\alpha = 0.45$

From (9.30): $$Q_e = A_e c_u N_c = \frac{0.35^2\pi}{4} \times 100 \times 9 = 65.6\,\text{kN}\uparrow$$

$$Q_s = A_s \alpha \bar{c}_u = 7.7 \times 0.45 \times 100 = 346.5\,\text{kN}\uparrow$$

Ultimate carrying capacity: $Q_u = Q_e + Q_s = 412\,\text{kN}\uparrow$

Allowable safe load carrying capacity:

From (9.31): $Q_a = \dfrac{Q_e}{3} + \dfrac{Q_s}{1.5} = \dfrac{65.6}{3} + \dfrac{346.5}{1.5} = 313\,\text{kN}$

From (9.32): $Q_a = \dfrac{Q_u}{2.5} = \dfrac{412}{2.5} = 165\text{kN} < 313\text{kN}$

Therefore, the safe load is: $Q_a = 165\,\text{kN}\uparrow$

But Q_a is reduced by: $Q_f = 25\,\text{kN}\downarrow$

Hence max safe working load is: $Q = 140\,\text{kN}\downarrow > 110\,\text{kN}$

As the maximum safe working load is larger than the proposed 110 kN, the pile is satisfactory. The overall safety factor is checked by formula (9.33):

$$F_0 = \frac{Q_u}{Q + Q_f} = \frac{412}{110 + 25} = 3 > 2.5 \text{ satisfactory}$$

Note: It has already been pointed out, that theoretically estimated values of safe loading must be considered as predictions only. As the actual carrying capacity and settlement depends on both the soil characteristics and the installation of a pile, the predictions can best be checked by loading tests. For detailed text on this subject, see Tomlinson (Ref: 21).

Example 9.8

Estimate the permissible working load on a 12 m long concrete pile, driven 5 m into stiff clay through recently deposited, well compacted sand and gravel layer and 4 m thick soft clay as shown in Figure 9.42. Ground water level is 4 m below the new surface. Pile diameter = 350 mm.

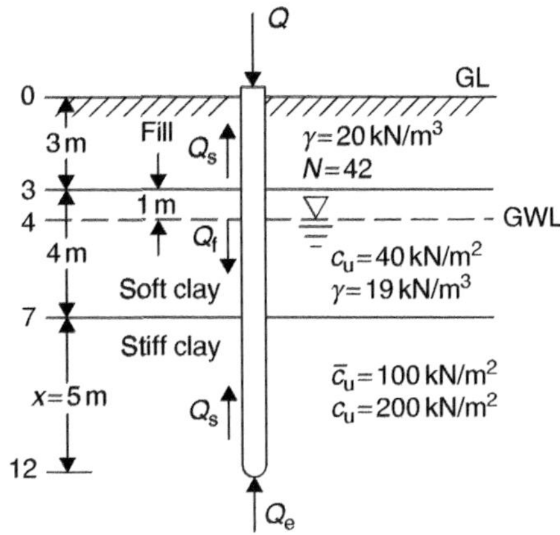

Note: The well compacted, dense fill layer does not consolidate, hence it contributes to the shaft resistance. However, its weight consolidates the soft clay stratum, which induces negative skin friction.

Figure 9.42

Recently deposited fill (0–3 m)

From Chart 9.5 for SPT $N=42$: $\phi=39°$

From Table 9.4 for concrete pile: $K_s=1$

$$\delta = 0.75\phi = 0.75 \times 39 = 29.3°$$

Surface area of pile in this layer: $A_s = \pi \times 0.35 \times 3 = 3.3\ \text{m}^2$

Average effective pressure: $\overline{\sigma'_0} = \dfrac{3\times 20}{2} = 30\ \text{kN/m}^2$

From (9.26): $Q_s = A_s K_s \overline{\sigma'_0} \tan\delta = 3.3 \times 1 \times 30 \times \tan 29.3 = 56\ \text{KN}$

Soft clay layer (3–7m)

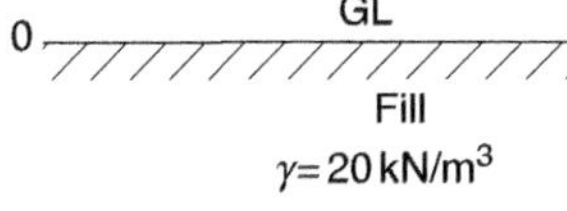

Effective overburden pressures:

$\sigma'_1 = 0\ \text{KN/m}^2$

$\sigma'_0 = 3 \times 20 = 60\ \text{KN/m}^2$

$\sigma'_1 = 30\ \text{KN/m}^2$

$\sigma'_1 = 60 + 4\times 19 - 3\times 9.81 = 106.6\ \text{kN/m}^2$

$\overline{\sigma'_0} = \dfrac{60 + 106.6}{2} = 83\ \text{kN/m}^2$

Surface area of pile in this layer: $A_s = \pi \times 0.35 \times 4 = 4.4\,\text{m}^2$

Average effective pressure: $\overline{\sigma'_0} = 83\,\text{kN/m}^2$

Negative friction from (9.37): $Q_f = 0.1 \times 83 \times 4.4 = 37\,\text{KN}\downarrow$

Stiff clay layer (7m+)

1. End bearing: Area: $A_e = \dfrac{0.35^2\pi}{4} = 0.096\,\text{m}^2$

   Shear strength: $c_u = 200\,\text{kN/m}^2$

   From (9.30) $Q_e = A_e c_u N_c$
   $= 0.096 \times 200 \times 9 = 173\,\text{kN}\uparrow$

2. Shaft Resistance: Area: $A_s = \pi \times 0.35x = \pi \times 0.35 \times 5 = 5.5\,\text{m}^2$

   $\alpha = 0.45$

   $\bar{c}_u = 100\,\text{kN/m}^2$

From (9.30): $Q_s = A_s \alpha \bar{c}_u = 5.5 \times 100 \times 0.45 = 248\,\text{KN}\uparrow$

Total shaft resistance: $Q_s = 56 + 248 = 304\,\text{kN}\uparrow$

Ultimate capacity: $Q_u = 304 + 173 = 477\,\text{kN}\uparrow$

Allowable capacity: $Q_a = \dfrac{Q_e}{3} + \dfrac{Q_s}{1.5} + = \dfrac{173}{3} + \dfrac{304}{1.5} = 259\,\text{kN}$

or $Q_a = \dfrac{Q_u}{2.5} = \dfrac{477}{2.5} = 191\,\text{kN}\uparrow$

The actual permissible working load is calculated from the lower figure:

$$Q = Q_a - Q_f$$
$$= 191 - 37 = 154\,\text{kN}\uparrow$$

Example 9.9

A bored concrete pile of 500 mm diameter is to carry a working load of 650 kN. Preliminary site investigation and laboratory tests revealed homogeneous clay to a depth 20 m underlain by sandstone. The undrained shear strength parameters and the ground water level are shown on Figure 9.43. Determine the required depth of penetration.

Try penetration length $x = 16\,\text{m}$

End bearing:
The value of undrained cohesion at 16 m depth is: $c_u = 374\,\text{kN/m}^2$

Table 9.7

Depth (m)	c_u (kN/m²)	ϕ_u^o	γ kN/m³
0	30	4	19
2	45	6	19.3
4	15	3	19.7
6	42	4	18.8
8	74	2	18.8
10	102	9	19.7
12	194	5	20.0
14	260	2	20.0
16	374	7	19.8
18	402	7	20.5
20	–	–	–

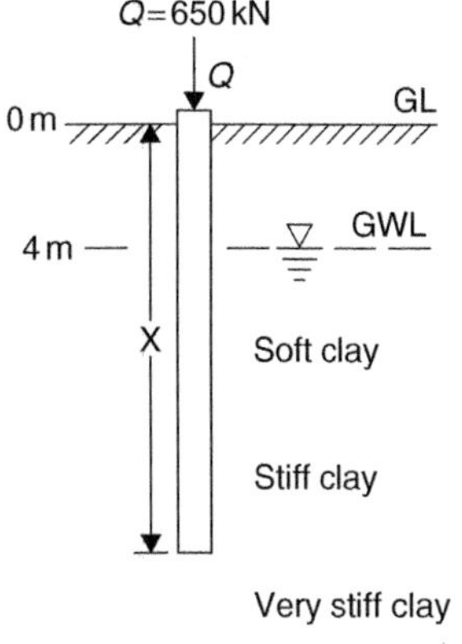

Figure 9.43

End bearing area: $A_e = \dfrac{0.5^2\pi}{4}$

From (9.30): $Q_e = 9A_e c_u = 9 \times \dfrac{0.5^2\pi}{4} \times 374 = 661\ \text{kN} \uparrow$

Shaft resistance:

The average shear strength $\bar{c}_u$ is evaluated in Table 9.8 and Q_s is obtained from (9.28) taking $\alpha = 0.45$.

Table 9.8

Interval (m)	$\bar{c}_u\left(\text{kN/m}^2\right)$
0–2	$\dfrac{30+45}{2} = 37.5$
2–4	$\dfrac{45+15}{2} = 30.0$
4–6	$\dfrac{15+42}{2} = 28.5$
6–8	$\dfrac{42+74}{2} = 58.0$
8–10	$\dfrac{74+102}{2} = 88.0$
10–12	$\dfrac{102+194}{2} = 148.0$
12–14	$\dfrac{194+260}{2} = 227.0$
14–16	$\dfrac{260+374}{2} = 317.0$
∴	$\bar{c}_u = \dfrac{934}{8} = 117.0$

Surface area of piles:

$A_s = \pi B x = \pi \times 0.5 \times 16 = 25\ \text{m}^2$

Shaft resistance:

$Q_s = \alpha_u \bar{c}_u A_s$

$= 0.45 \times 117 \times 25 = 1316\ \text{kN}$

Ultimate capacity:

$Q_u = 661 + 1316 = 1977\ \text{kN}$

Allowable loading:

$$Q_a = \frac{661}{3} + \frac{1316}{1.5} = 1098\ \text{kN}$$

or

$$Q_a = \frac{1977}{2.5} = 791\ \text{kN}$$

Accept $Q_a = 791\ \text{kN} \uparrow > Q = 650\ \text{kN}$

Therefore a 16 m long pile is satisfactory.

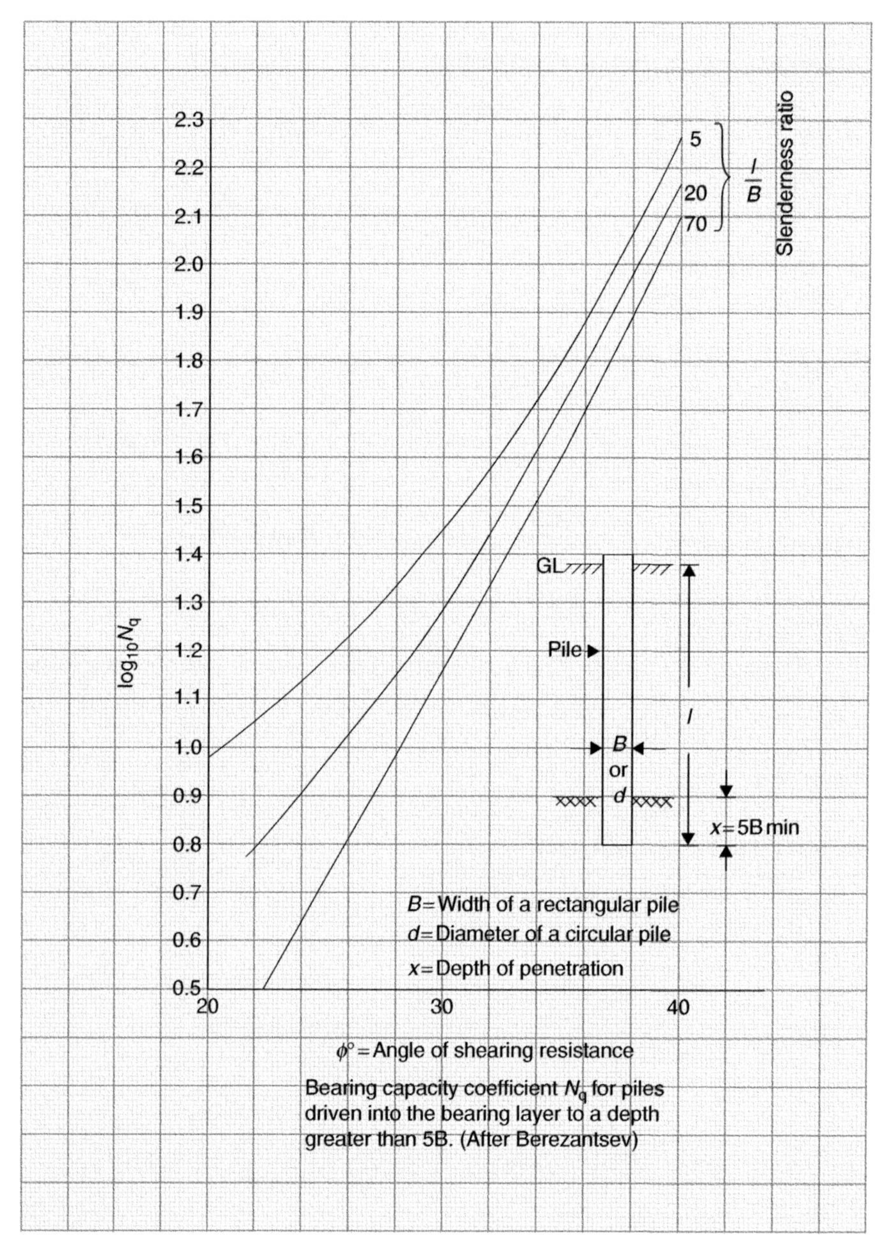

Chart 9.7

9.14 End bearing resistance and SPT

Meyerhof correlated q_e and q_s with the number of blows (*N*) obtained in standard penetration tests for sand, gravel and silt.

Table 9.9

Type of pile	Type of soil	q_e (kN/m²)	q_s (kN/m²)
Driven	Sand and gravel or sand	$\frac{40\,DN}{B} \leq 400\text{N}$	$2\,\overline{N}$
	Sandy silt or silt	$\frac{30\,DN}{B} \leq 300\text{N}$	
Bored	Gravel or sand	$\frac{14\,DN}{B}$	$0.67\,\overline{N}$
	Sandy silt or silt	$\frac{10\,DN}{B}$	

where $\overline{N}$ = average value of N over the embedded length

D = embedded length

q_e = end bearing resistance $= \frac{Q_e}{A_e}$

9.15 Influence of pile section on Q_u

It is implied in Figure 9.19 that circular footing could carry larger load than a rectangular one of the same base area. Therefore, for maximum end bearing resistance a circular pile section is marginally more economical.

Triangular pile section, however, has larger perimeter than circular or rectangular one of the same area, hence is supplies the largest shaft resistance. For the same reason, a square section is slightly more efficient than a circular one of the same cross-sectional area.

9.16 Group of piles

In most cases, not a single, but several piles are placed to support a structure. The settlement and load-transmitting characteristics of a pile group is different from those of an individual pile. In general, the allowable working load imposed on the group depends on the magnitude of its possible settlement. It is assumed, for the estimation of maximum settlement, that the pile group and the soil between the piles move together as a block, whose base may be regarded as a raft foundation. In general, the settlement depends on the:

1. Type of soil penetrated
2. Type of piles (friction or end-bearing)
3. Spacing of the piles.

Cohesionless soils

Driven piles compact the soil around them. The region of influence extends approximately to three times their diameter (Figure 9.44 (a)). The looser is the soil, the smaller is this distance.

Spacing of end-bearing piles:	$\rho \geq 2d$	(Figure 9.44 (b))
Spacing of friction piles:	$5d \leq \rho \leq 6d$	(Figure 9.44 (c))
Spacing of end bearing/friction piles:	$3d \leq \rho \leq 4d$	(Figure 9.44 (d))

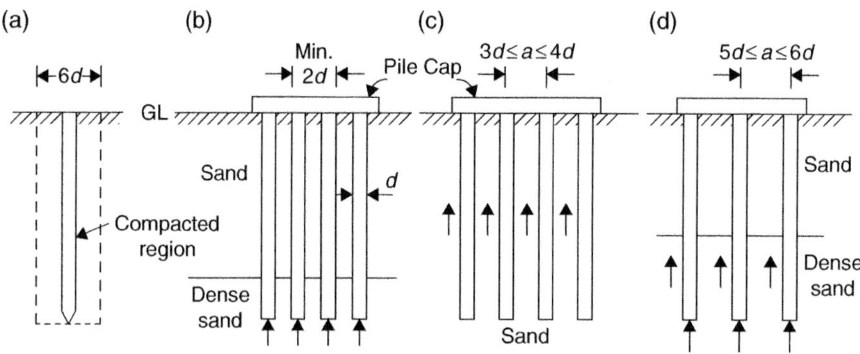

Figure 9.44

Cohesive soils

The consequences of driving piles into soft clay are considerable remoulding and heaving of the ground surface. In stiff clays, this upward movement could lift adjacent, previously placed piles, which should then be re-driven.

Block failure of a pile group

If the spacing of the piles is less than the diameter and the bearing capacity of the underlying cohesive soil is exceeded then the entire group could fail as a single block.

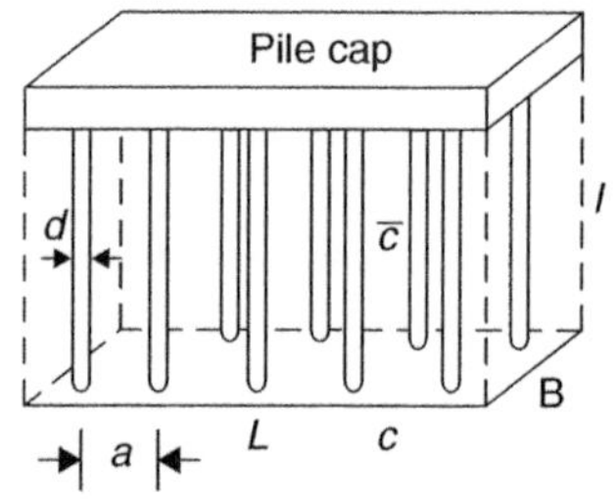

Figure 9.45

The soil enclosed by the piles moves with them as one mass. The ultimate bearing capacity of the group is given by:

$$Q_{ug} = c_u N_c A_b + A_p \bar{c}_u$$

where A_b = base area i.e. $A_b = BL$

A_p = perimeter area $A_p = 2(B+L)l$

N_c = Bearing capacity factor (usually $N_c = 9$)

Therefore,

$$\boxed{Q_{ug} = c_u N_c BL + 2(B+L)l \times \bar{c}_u} \tag{9.38}$$

Applying a factor of safety (F_s), the allowable carrying capacity of the group is:

$$\boxed{Q_{ag} = \frac{Q_{ug}}{F_s}} \tag{9.39}$$

The efficiency (η) of the group shows, that the ultimate carrying capacity of the group differs from the total capacity of n individual piles. It is expressed as:

$$\boxed{\eta = \frac{Q_{ug}}{nQ_u}} \tag{9.40}$$

Where, n = number of piles in the group
Q_u = ultimate carrying capacity of one pile.

Note: Block failure is unlikely to occur if the piles are placed not less than 3d apart.

Example 9.10

A structure weighing 2000 tonne is to be carried by a piled foundation. The soil profile is given in Figure 9.42 (Example 9.9). Check whether nine 16 m long piles, designed in that example, can carry this load safety. The piles are to be placed 3d apart, forming a 3.5 m square group. Assume a factor of safety of 3. Working load to be carried: $Q_g = 2000$ tonne $= 2000000$ kg

$$= 2 \times 10^6 \times 9.81 = 19.62 \times 10^6 \text{N}$$

$$= 19620 \text{ kN}$$

For a single pile, from example 9.9: Q_u=1977 kN
Average shear strength (Table 9.6): $\bar{c}_u = 117$ kN/m^2
Shear strength at the base: c_u=374 kN/m^2

For the group
$n = 9$

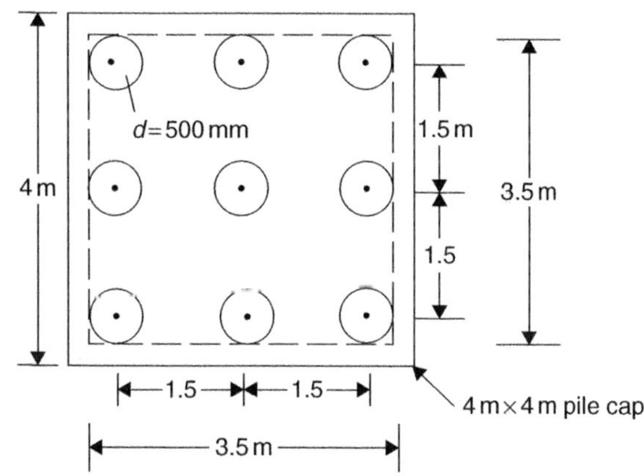

Figure 9.46

From (9.38): $Q_{ug} = 374 \times 9 \times 3.5^2 + 2 \times (3.5 + 3.5) \times 16 \times 117$

$$= 41234 + 26208 = 67442 \text{ kN}$$

From (9.39): $Q_{ag} = \dfrac{67442}{3} = 22481 \text{kN}$

Assuming that the reinforced concrete ($\gamma=24\,\text{kN/m}^2$) pile cap is 1 metre deep, the actual loading on the pile group is:

$$Q_g = 19620 + 1\times4\times4\times24 = 20004\,\text{kN}$$

From (9.40): $\eta = \dfrac{Q_{ug}}{nQ_u} = \dfrac{67442}{9\times1977} = 3.8$

Conclusions:

1. The piles are spaced at $3d = 1.5$ m apart, hence no block failure should occur
2. The allowable carrying capacity of the group is larger than $Q_g = 20004\,\text{kN}$
3. The ultimate capacity of the group is 3.8 times larger than that of nine piles summed.

The pile group is satisfactory.
Note: The unit weight of reinforced concrete is assumed to be $24\,\text{kN/m}^3$.

9.16.1 Eccentrically loaded pile group

In this case, the load carried by a pile depends on its distance from the centroid of the group. The load is calculated from formula (9.41), which is derived by the application of the following principle of statics.

"A force acting at point A can be transformed into the same force as well as a couple, acting at another point B". See also Appendix D.

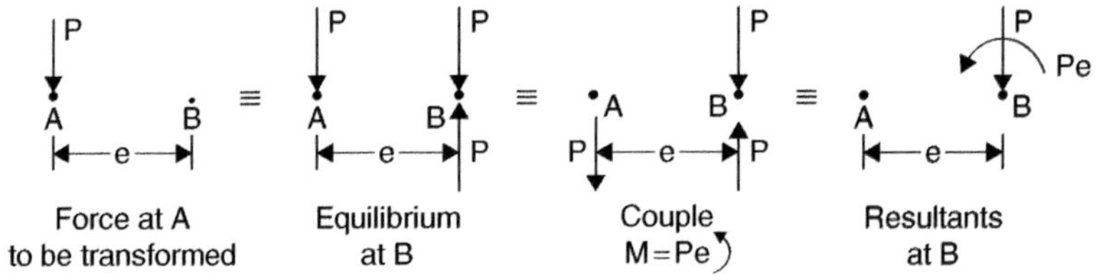

Figure 9.47

Derivation
Suppose, there are *n* piles in a group, made up of *r* rows. Each row contains n_r piles. The working load is acting eccentrically at distance e from the centreline of the group.

Step 1: Transfer force Q_g to the centreline and obtain the direct load *F*, carried by each pile. From Figure 9.48 (b) and (c): $F = \dfrac{Q_g}{n}$

Step 2: Apply the moment $Q_g e$ to the pile group, which tends to rotate about point c at the centreline of the pile cap. This induces additional load on each row of the group. The extra load R_r carried by the rth row is proportional to its distance from the centre of rotation. From Figure 9.48 (d), the proportions can be written as:

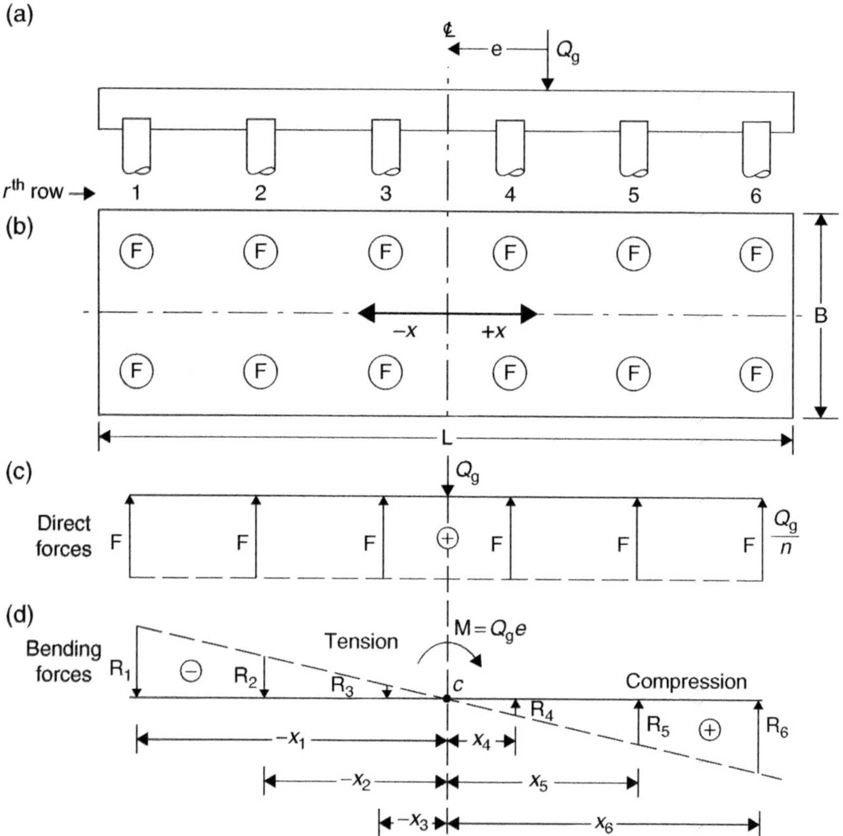

Figure 9.48

$$\frac{R_1}{x_1} = \frac{R_2}{x_2} = \frac{R_3}{x_3} = \frac{R_4}{x_4} = \text{........} = \frac{R_r}{x_r}$$

From which:

$$R_1 = R_r \frac{x_1}{x_r} \qquad R_2 = R_r \frac{x_2}{x_r}$$

$$R_3 = R_r \frac{x_3}{x_r} \qquad R_4 = R_r \frac{x_4}{x_r}$$

In general the nth pile:

$$R_n = R_r \frac{x_n}{x_r}$$

Also, for equilibrium of moments:

$$Q_g e = R_1 x_1 + R_2 x_2 + R_3 x_3 + R_4 x_4 + \text{......} + R_n x_n$$

Substituting,

$$Q_g e = R_r \frac{x_1^2}{x_r} + R \frac{x_2^2}{x_r} + R_r \frac{x_3^2}{x_r} + R_r \frac{x_4^2}{x_r} + \text{........} + R_r \frac{x_n^2}{x_r}$$

Rearranging,

$$Q_g e x_r = R_r \left(x_1^2 + x_2^2 + x_3^2 + x_4^2 + \text{.......} + x_n^2 \right)$$

Therefore, the load carried by any pile in the rth row is given by:

$$R_r = \frac{Q_g e\, x_r}{x_1^2 + x_2^2 + x_3^2 + x_4^2 + \ldots\ldots + x_n^2}$$

As there are n_r piles in a row, each carries an additional load of:

$$P_r = \frac{R_r}{n_r} = \frac{Q_g e\, x_r}{n_r \Sigma x_r^2}$$

Combining the direct and bending loading, the working load on any pile in a row is given by:

$$Q_r = F + P_r = \frac{Q_g}{n} + \frac{Q_g e\, x_r}{n_r \Sigma x_r^2}$$

Or

$$Q_r = Q_g\left(\frac{1}{n} + \frac{e\, x_r}{n_r \Sigma x_r^2}\right) \tag{9.41}$$

Where x_r is +ve on the compression side and −ve on the tension side.

Should Q_g be placed eccentrically, both in the x- and y-directions, then the expression is modified to:

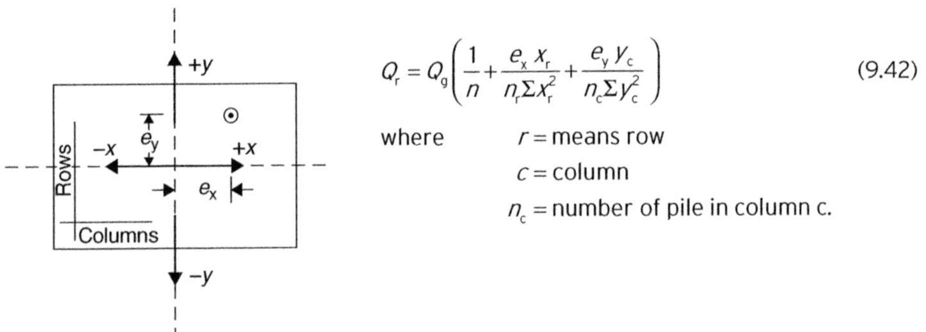

$$Q_r = Q_g\left(\frac{1}{n} + \frac{e_x\, x_r}{n_r \Sigma x_r^2} + \frac{e_y\, y_c}{n_c \Sigma y_c^2}\right) \tag{9.42}$$

where r = means row

c = column

n_c = number of pile in column c.

Figure 9.49

Example 9.11

A 15000 kN concrete structure is supported by a 25-pile group, arranged in square pattern. The 500 mm diameter piles are placed at 1.5 m, centre to centre. The structure is seated eccentrically to the centroid of the pile group. The eccentricity is 0.7 m. Calculate the load carried by each pile. Ignore the size of the pile cap.

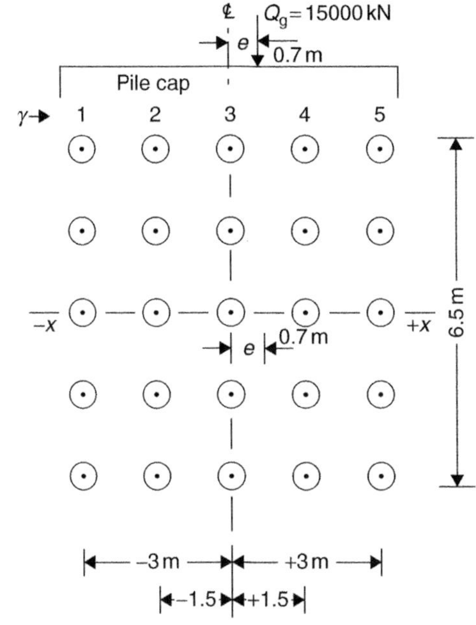

Given: $n = 25$

$n_r = 5$

$e = 0.7\,\text{m}$

$Q_g = 15000\,\text{kN}$

$$\Sigma x_r^2 = x_1^2 + x_2^2 + x_4^2 + x_5^2$$
$$= 9 + 2.25 + 2.25 + 9 = 22.5\,\text{m}$$

From (9.41):

$$Q_r = 15000\left(\frac{1}{25} + \frac{0.7\,x_r}{5 \times 22.5}\right)$$

$$\therefore \quad Q_r = 600 + 93.3\,x_r$$

Figure 9.50

Tabulating the calculations:

Table 9.10

Row r	x_r (m)	Load per pile (Q_r) in row r (kN)
1	−3	600 − 93.3×3 = 320
2	−1.5	600 − 93.3×1.5 = 460
3	0	600 − 0 = 600
4	1.5	600 + 93.3×1.5 = 740
5	3	600 + 93.3×3 = 880
		Load carried by one row = ΣQ_r = 3000

Check: Total load on group $= n_r \Sigma Q_r = 5 \times 3000 = 15000\,\text{kN}$

When it is not possible to eliminate the eccentricity, then its effect may be minimized by:

1. Altering the spacing of the piles or rows
2. Designing piles of varying length
3. Altering the size of the pile cap.

9.16.2 Settlement of pile groups

Whilst the settlement of single piles can best be determined by in-situ loading test, for pile groups the consolidation theory is applicable. Because the group behaves as a raft, its bearing pressure extends to some depth below it (see Figure 9.38). For this reason, the consolidation settlement of a group is usually much larger than the downward movement of a single pile. For the purpose of estimating the movement, it is normal practice to assume an equivalent raft, determined by a load spread of 1 in 4.

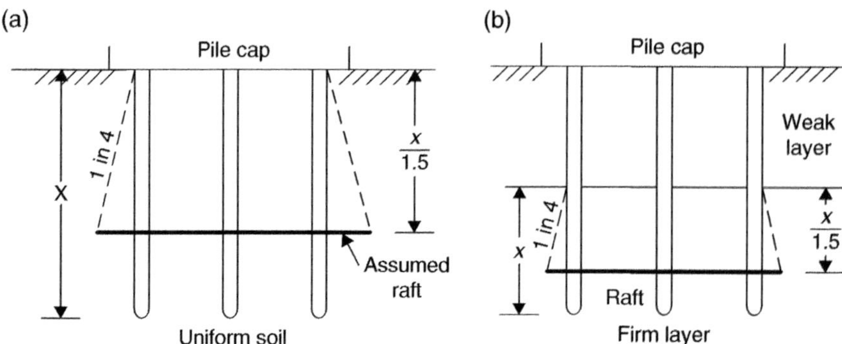

Figure 9.51

x = depth of penetration

In the knowledge of the consolidation characteristics of the soil, the approximate settlement of the group can be estimated as for shallow footings in Chapter 7.

9.16.3 Raking piles

When a structure imposes horizontal forces on a pile cap, only a small portion of this load can be carried by vertical piles, depending on the passive resistance mobilised along their lengths that is, on the shear strength of the adjacent soil.

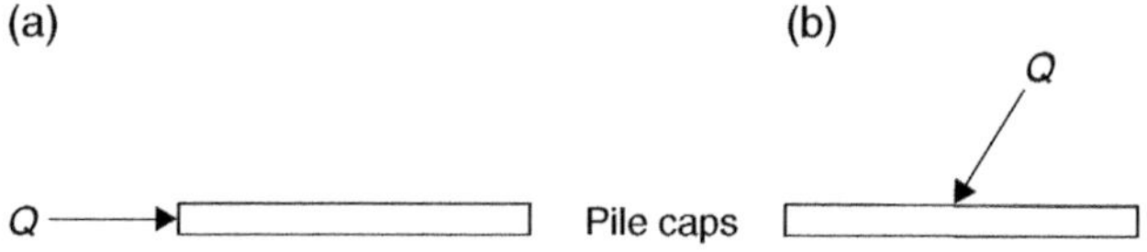

Figure 9.52

Raking or batter piles are used, as part of a group to carry the horizontal component of an imposed load. The force carried by each pile may be determined graphically, as long as there are only three non-parallel rows in the group. The problem becomes statically indeterminate for larger number of rows, hence beyond the scope of this book.

Typical problem:

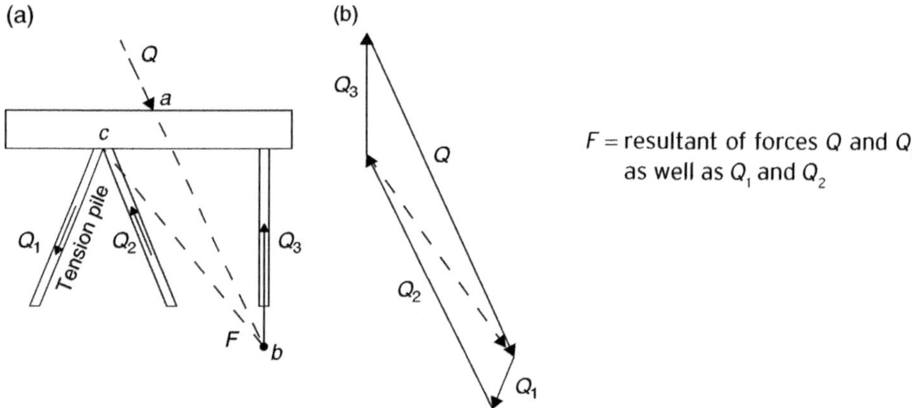

F = resultant of forces Q and Q_3 as well as Q_1 and Q_2

Figure 9.53

Note: If the tension force Q_1 is larger than the available shaft friction, then the pile-configuration has to be rearranged, so that point x is lowered and the resultant F is then located between Q_1 and Q_2 as in Figure 9.53(a). This can be done by either moving the load Q to the left or pile No. 3 to the right.

Example 9.12

Determine the forces in the piles, shown in Figure 9.54, subjected to an inclined load of 5000 kN.

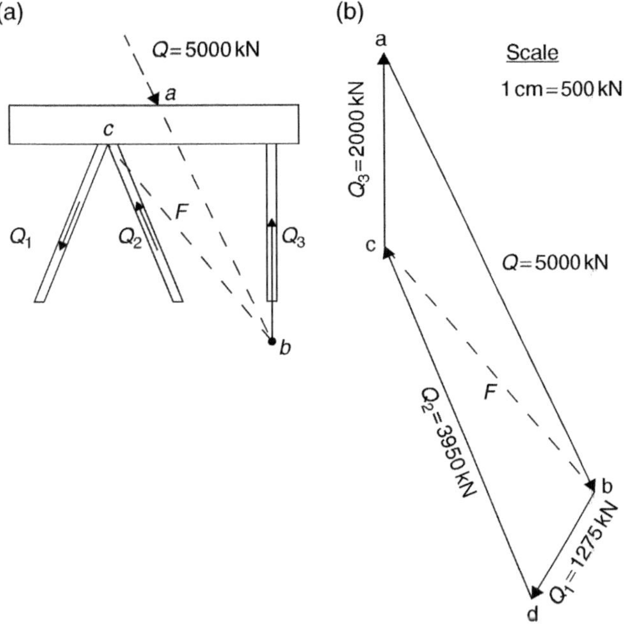

Figure 9.54

Step 1: Draw force Q (a–b) to scale
Step 2: Draw auxiliary force F (c–b)
Step 3: Draw force Q_3(a–c). This yields the magnitude of Q_3.
Step 4: Draw Q_1 from b and Q_2 from c, to intersect at point d, thus obtaining their magnitude
Step 5: Draw the arrows round the polygon for equilibrium.

Results: Q_1 = 1275 kN tension
Q_2 = 3950 kN compression
Q_3 = 2000 kN compression

Problem 9.1

Strip footing, 1.5 m wide and 1 m deep, is to be constructed in a 3 m thick soft clay layer, underlain by stiff boulder clay. There is no evidence of ground water table. The base transmits 200 kN/m^2 pressure to the soil.

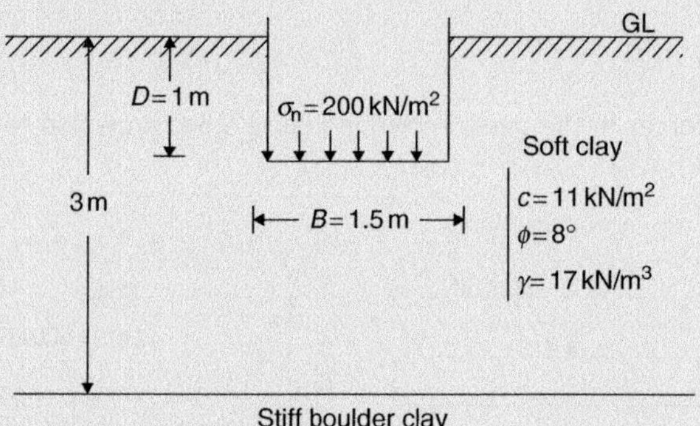

Figure 9.55

a) Check, whether the bearing capacity of the soft clay is sufficient enough to carry the footing.
b) It is proposed that, should the soft clay be unsatisfactory, then it is to be replaced by suitable, compacted material. The results of laboratory compaction and shear box tests on the imported material are:

Dry density	= 16.1 kN/m^3
Optimum water content	= 18%
Specific gravity	= 2.75
Cohesion	= 5 kN/m^2
Angle of friction	= 35°

Check the bearing capacity of the compacted layer:

1. In partially saturated state.
2. In fully saturated state after the area is flooded permanently to a depth of 0.5 m.

Apply Terzaghi's bearing capacity factors and a factor of safety F_s=3 throughout.

Problem 9.2

A 2 m × 4 m rectangular foundation, based 0.9 m below the surface, supports a column load of 1000 kN (including self-weight) and a moment of 500 kNm, as shown.

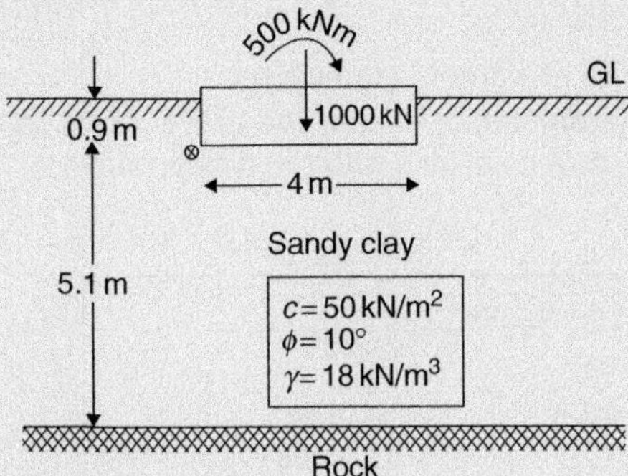

Figure 9.59

Estimate the bearing capacity of the clay, at factor of safety of 3, and state whether it is overstressed or not. Apply Terzaghi's bearing capacity factors.

Problem 9.3

Bored piles, 1 m diameter, are to be constructed in soft to stiff clay. Their total length must be 13.3 m. Figure 9.61 shows an average undrained strength of 174 kN/m^2 along the shaft and an undrained shear strength of 260 kN/m^2 at the base of the piles. Estimate the allowable carrying capacity, taking the adhesion factor as 0.45, of the:

a) straight pile
b) under-reamed pile, ignoring the adhesion above the tapered part, over a length twice the base diameter.

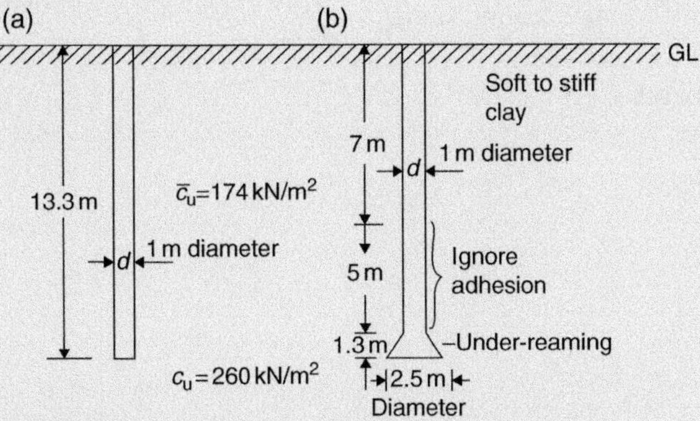

Figure 9.61

Problem 9.4

A 2 m wide strip footing is placed at 0.5 m below ground level in a sand layer, underlain by coarse gravel. The known characteristics of the sand are given in Figure 9.62. Determine the safe bearing pressure, taking the factor of safety as 3, if:

1. There is no evidence of water in either layer
2. The water in a piezometer, placed in the gravel, rises to 1.5 m above the water table, which now coincides with the ground surface.

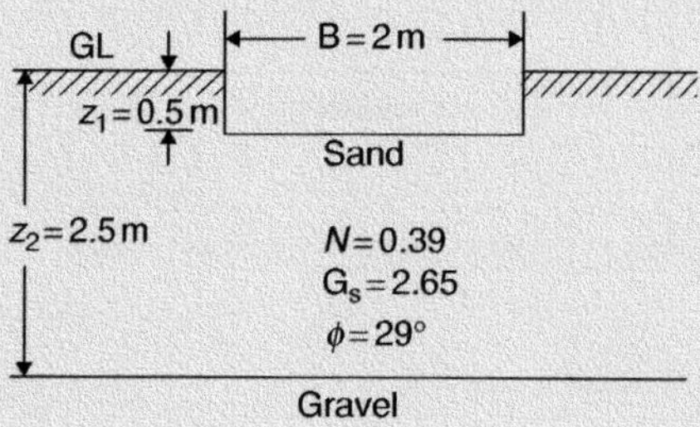

Figure 9.62

Problem 9.5

With reference to Figure 9.64, express bearing capacities q_u, q_n, q_{sn}, q_s and the net bearing pressure σ_n in terms of depth z, taking 3 as the factor of safety.

Given that $\sigma_n = 155\,\text{kN/m}^2$, determine the depth z and the value of the safe bearing pressure σ.

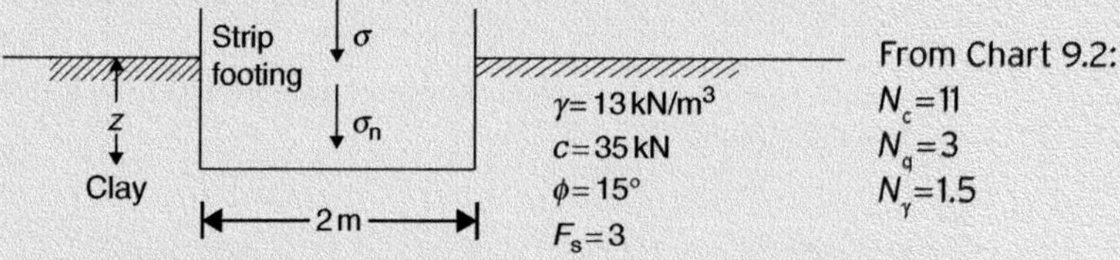

Figure 9.64

Problem 9.6

A 4m square footing, based at 2m depth, transmits 388kN/m² safe bearing pressure (self weight incl.) to the sand underlain by soft clay, as shown. The ground water table is 3m below the surface.

1. Determine the bearing strength of each layer, adopting 3 as factor of safety for the clay.
2. Compare the strength of each layer with the applied loading.

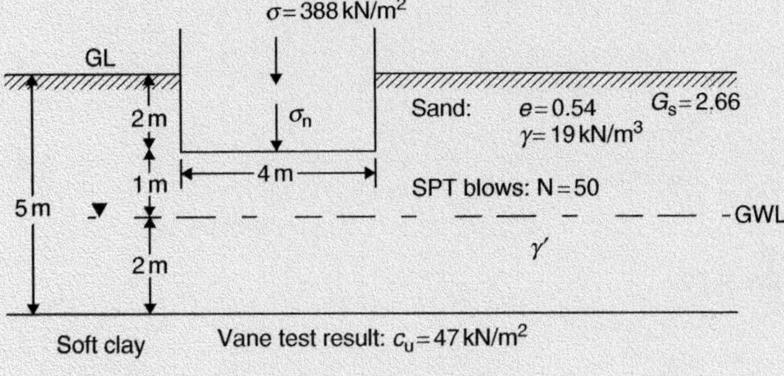

Figure 9.65

Problem 9.7

It is proposed, that a footing should be constructed within the clay, underlain by gravel, as shown in Figure 9.68. The gravel is under an artesian pressure of 26kN/m². Because of the frost-susceptibility of the clay, the depth of the base has to be 1m below the ground surface.

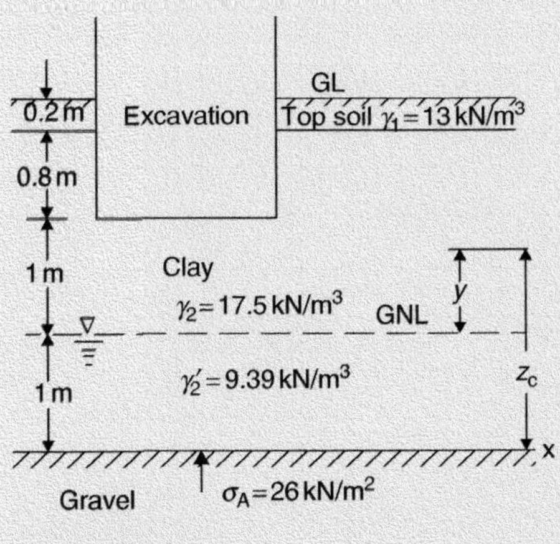

Figure 9.68

Check the feasibility of the scheme and recommend a suitable solution, if it is not acceptable.

Problem 9.8

A strip footing, 2 m wide, is based at a depth of 1 m in the stiff clay shown in Figure 9.70, transmits a net bearing pressure of 245 kN/m^2 to the soil. Determine whether the underlying medium clay is overstressed or not at $F_s = 3$.

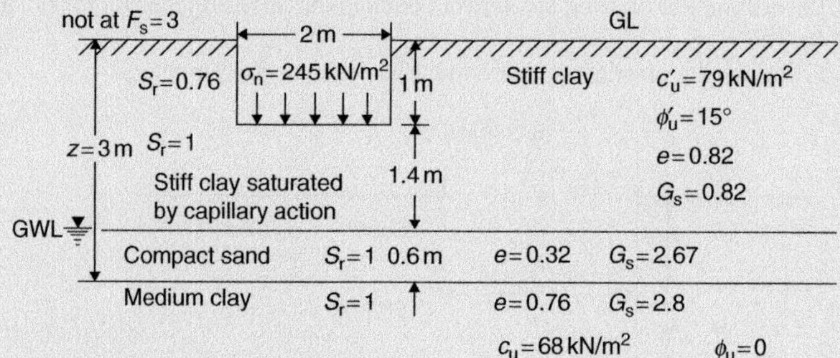

Figure 9.70

Chapter 10

Stability of Slopes

A frequently occurring engineering problem is the prevention of slope failure during and after the construction of earthworks, such as embankments, earth dams or road and railway cuttings, etc. Also, it is often necessary to stabilize natural, undisturbed slopes, before or after the onset of instability. In general, the aim is to provide sufficient margin of safety against shear failure, in any part of the soil mass.

10.1 Short-term and long-term stability

The method of analysis depends on whether the factor of safety required is for the short-term or the long-term stability of an earth structure.

1. *Short-term* stability calculations are relevant to safety at the end of construction (time $t=0$), when no appreciable dissipation of excess pore pressure is assumed. This condition is simulated in the undrained (QU) triaxial test, yielding the shear strength parameters c_u and ϕ_u. For saturated clays however, $\phi_u=0$, hence shear strength $\tau=c_u$.

   It was pointed out in Chapter 6, that the Mohr-envelope is expressible only in terms of total stress, in a QU-test on saturated clays, For this reason a short-term analysis is carried out in terms of total stresses, often referred to as $\phi_u=0$ analysis.
2. *Long-term* stability calculations are relevant to safety after the dissipation of excess pore pressure (time $t=\infty$), that is after the completion of consolidation. Failure is assumed to occur after this period. This condition is simulated in the consolidated, undrained (CU) triaxial test with pore pressure measurement, yielding the effective shear strength parameters c' and ϕ'. The long-term analysis is then carried out in terms of effective stresses.

   Note: When analysing the stability along a surface which has already slipped, then the residual strength (ϕ'_r) has to be used as the slip surface is in a remoulded state.

Introduction to Soil Mechanics, First Edition. Béla Bodó and Colin Jones.

10.2 Total stress analysis (cohesive soils)

10.2.1 Homogeneous, pure clay ($\phi_u=0$)

In general, a slope becomes unstable, when the effect of gravity forces, acting on a volume of soil, exceeds its shear strength along a slip surface. The usual gravity forces encountered in the total stress analysis are:

a) Weight of the sliding mass.
b) Surcharge or other surface loads.

Figure 10.1 depicts the simplest case of a rotational slip in homogeneous, saturated clay.

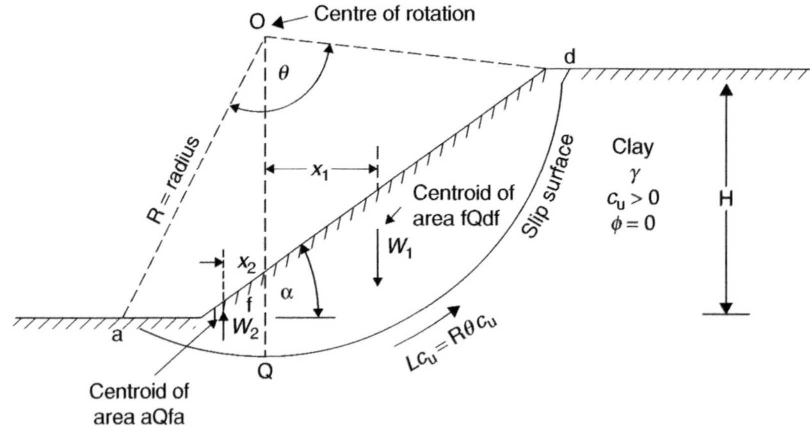

Figure 10.1

where γ=bulk unit weight of clay
W_1=weight of soil inducing slide
W_2=weight of soil resisting slide
L=Length of the slip surface a-Q-d
c_u=shear strength of clay
c_uL=force resisting shear along the slip surface
x_1 and x_2 are the moment-arms of W_1 and W_2 respectively from the centre of the circle.

Taking moment about the centre of rotation:

Disturbing moment: $\overset{+}{M_D} = W_1x_1 - W_2x_2 = \Sigma Wx$

Resisting moment: $\overset{-}{M_R} = c_uLR$

Factor of safety: $F_s = \dfrac{\text{Resisting moment}}{\text{Disturbing moment}}$

or $$F_s = \frac{M_R}{M_D} = \frac{c_uLR}{\Sigma W_x} \quad (10.1)$$

When $F_s=1$, then $M_D=M_R$ and the slope is said to be in a state of impending or incipient failure. The normally acceptable range for the factor of safety is: $1.25 \le F_s \le 1.5$

10.2.2 Increasing the value of F_s

Moments M_R and M_D can be varied in (10.1) by altering the values of W_1 and W_2. The factor of safety can be increased in several obvious ways by:

1. Decreasing M_D by benching (Figure 10.2 (a)).
2. Increasing M_R by loading the toe (Figure 10.2 (b)).
3. Decreasing M_D by flatter slope (Figure 10.2 (c)).
4. The combination of the above three methods.

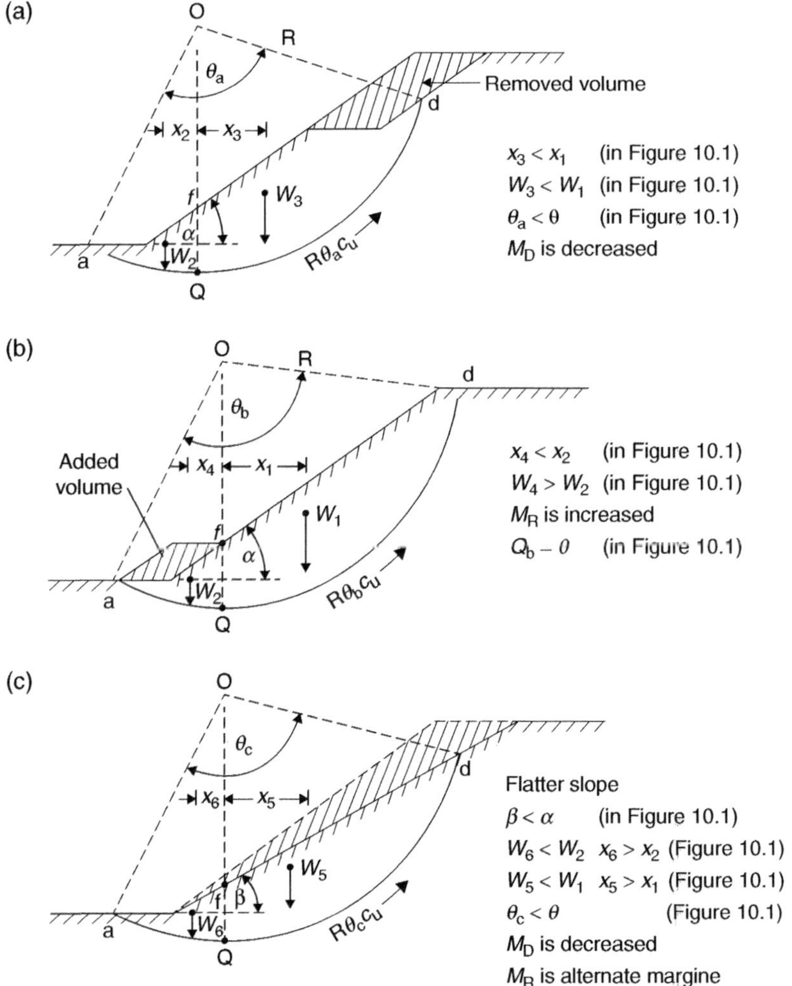

Figure 10.2

10.2.3 Minimum value of F_s

The aim at the design and analysis of slopes is to find that potential slip surface, for a given slope, which yields the smallest factor of safety. If this value is more than 1.25, no failure is expected. If less, then M_R has to be increased one way or another. The procedure is to analyse several trial circles and compare their factors of safety, until its smallest value is found.

10.2.4 Potential slip surface

The centres of the trial circles are normally located arbitrarily. This means lengthy search for the critical radius and its position. The procedure can easily be done by a computer nowadays, however, for homogeneous clay, the potential slip surface may be drawn approximately, considering that:

1. The failure surface should pass through the toe, if $\phi_u > 3°$ or $\alpha > 53°$, unless the base of the slope is of stronger material than the homogeneous clay. In that case the circle touches the base without intersecting it.

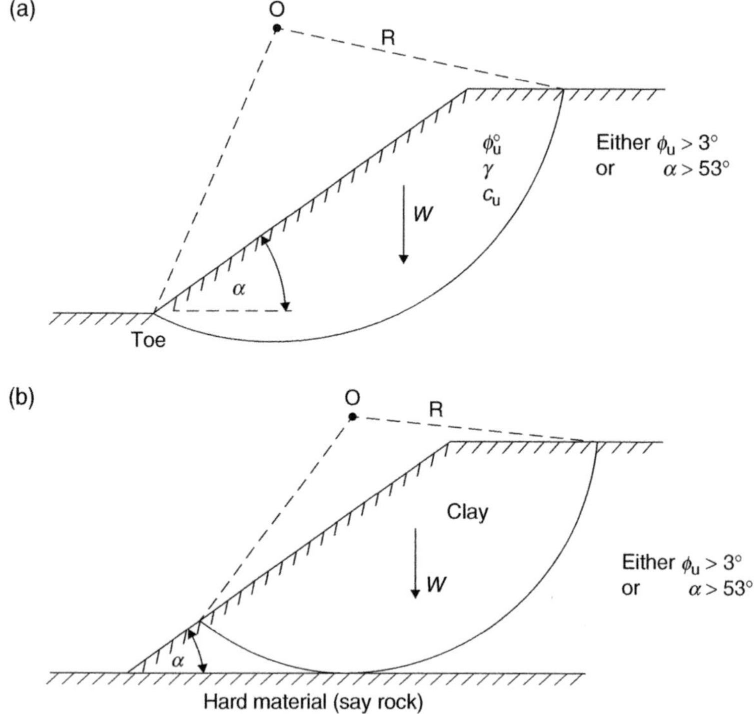

Figure 10.3

2. If $\phi_u > 3°$ or $\alpha \leq 53°$, then the critical slip surface intercepts the ground in front of the toe (Figure 10.4a) if the base of a slope is stronger material, then the

intersection could be in front, at the toe or on the slope, depending on the depth of the base (Figure 10.4b).

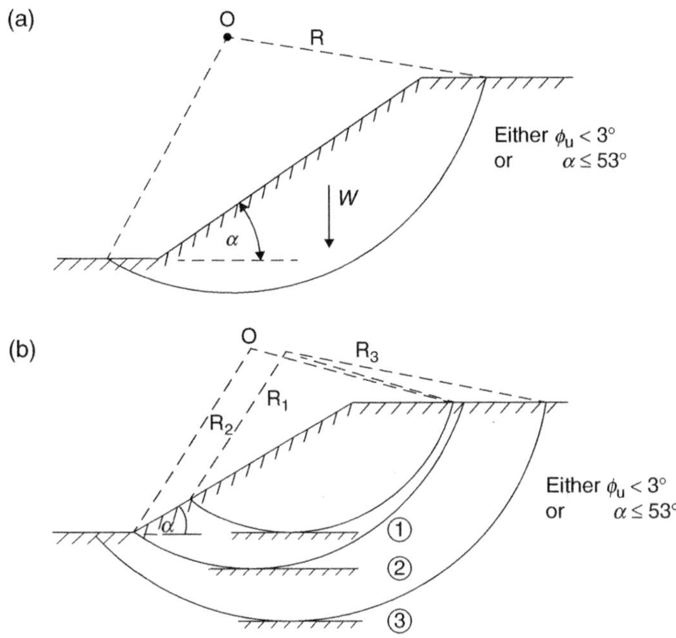

Figure 10.4

Notes:

a) These points should be considered as guidance only, applicable to uniform slopes in homogeneous clay, as the predicted circles are not always the most critical ones.
b) The possible centre of a critical circle can be located by means of Chart 10.1 (after Fellenius, for uniform, homogeneous clay ($\phi_u = 0$).

10.2.5 Determination of the factor of safety

There are several ways in which the disturbing and resting moments may be evaluated. Some of these are detailed in the following five examples in terms of total stress.

Example 10.1	Planimetric method	Saturated c-soil
Example 10.2	Method of slices	
Example 10.3	Radial procedure	
Example 10.4	Allowance for tension cracks	
Example 10.5	Analysis of partially saturated ($c-\phi$) soil	

Graph 10.1 shows a 6 m deep cutting slope, inclined at 1 in 2, and the slip circle of 13.5 m radius in homogeneous, fully saturated clay. Its properties are indicated on the graph.

This slope and slip surface will be analysed for various conditions in this chapter.

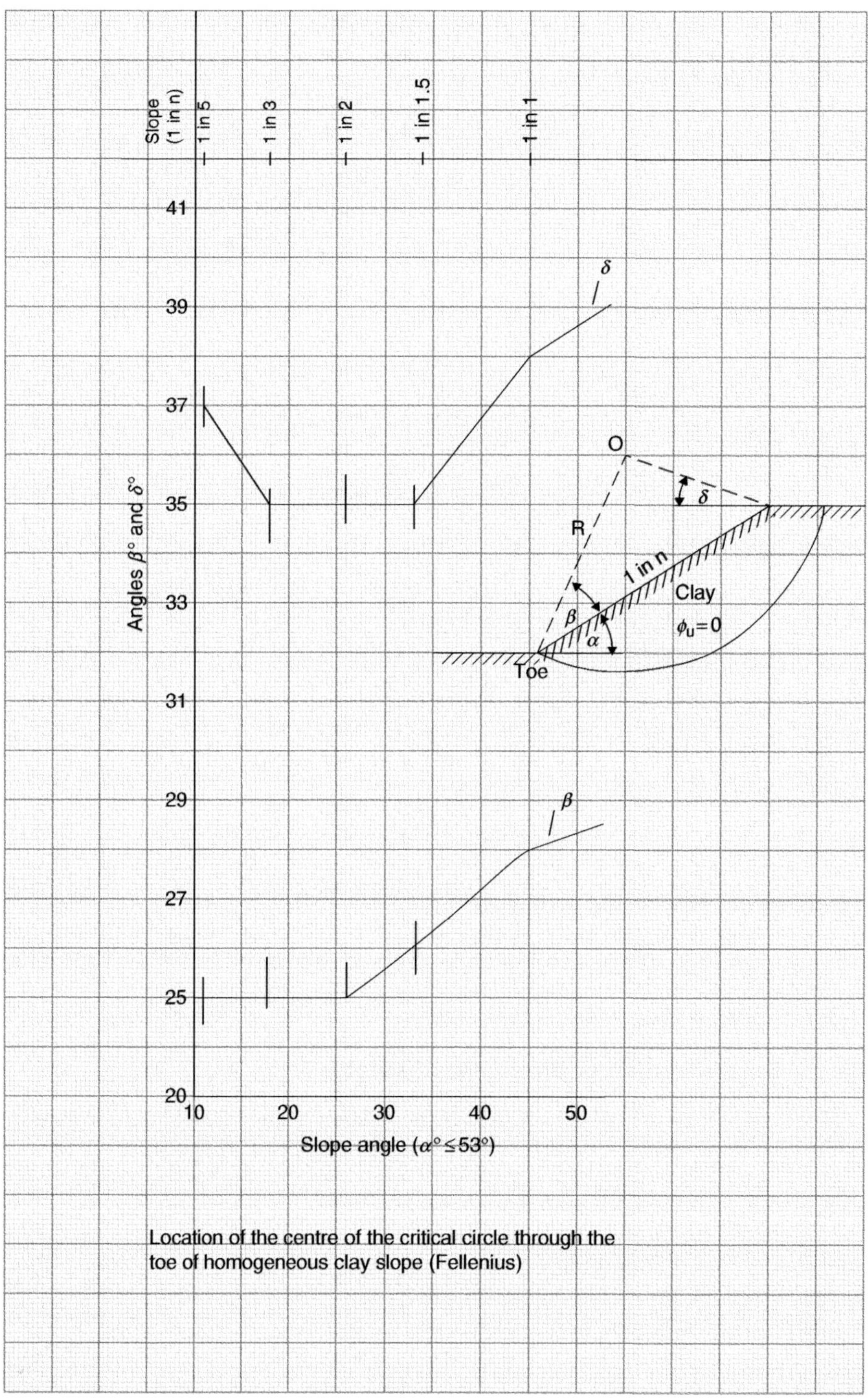
Slope
(1 in n)
1 in 5
1 in 3
1 in 2
1 in 1.5
1 in 1
41
39
37
35
33
31
29
27
25
20
δ
β
O
δ
R
1 in n
β
α
Clay
ϕ_u=0
Toe
Angles β° and δ°
10
20
30
40
50
Slope angle (α° ≤ 53°)
Location of the centre of the critical circle through the toe of homogeneous clay slope (Fellenius)

Chart 10.1

Example 10.1

(See Graph 10.1)

Step 1: Determine the cross-sectional area (abdeQa) of the potential slip by means of planimeter, Simpson's Rule or otherwise (see Appendix C).

In this example: $A \approx 82.7\,\text{m}^2$

Step 2: Calculate the weight of 1m long slope.

$$W = \gamma A = 19 \times 82.7 = 1571.3\,\text{kN}$$

Rounded to $W = 1571\,\text{kN}$

Step 3: Locate the position of the centroid ($\overline{x}$) of the cross section, relative to the centre of rotation (See also Appendix C). This can be done also by cutting a template of the section, out of cardboard or from other suitable material and bore at least three holes through it, at opposite extremities (x, y and z).

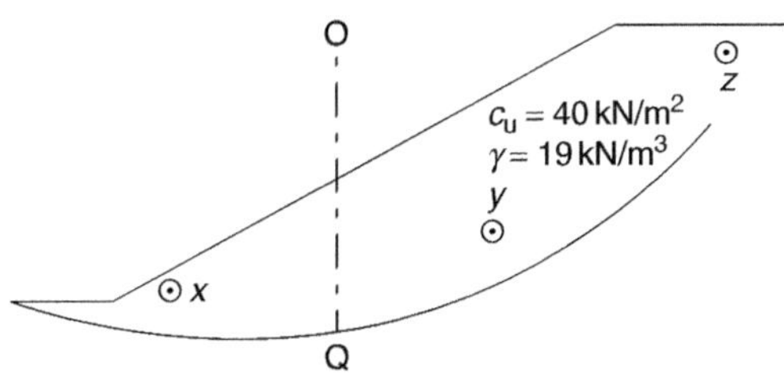

Figure 10.5

Step 4: Suspend the template at each point in succession and draw a vertical line from the pin-hole with the help of a spirit level or plumbob. The three lines intersect at the centroid.

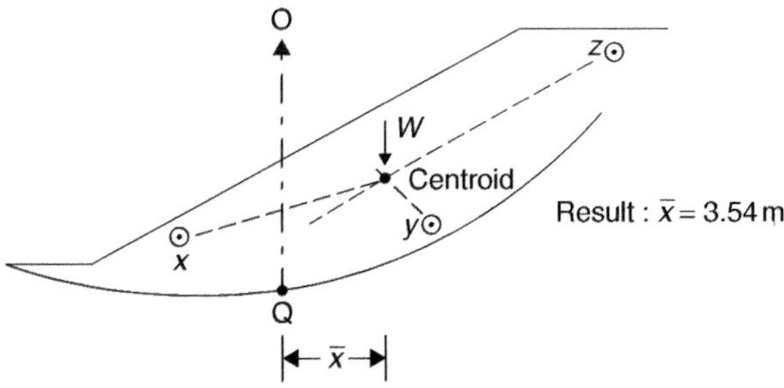

Figure 10.6

Note: This construction can be applied to any shape.

Step 5: Calculate the resisting shear force, due to the shear strength of the clay, along the potential slip surface.

Length of surface: $L = R\theta = 13.5 \times \frac{101 \times \pi}{180} = 23.8\ \text{m}$

Shear force: $S = c_u L = 40 \times 23.8 = 952\ \text{kN}$

Step 6: Calculate the moments about the centre of rotation and the factor of safety.

Disturbing : $M_D \times W\bar{x} = 1571 \times 3.54 = 5561\ \text{kNm}$

Resisting : $M_R = SR = 952 \times 13.5 = 12852\ \text{kNm}$

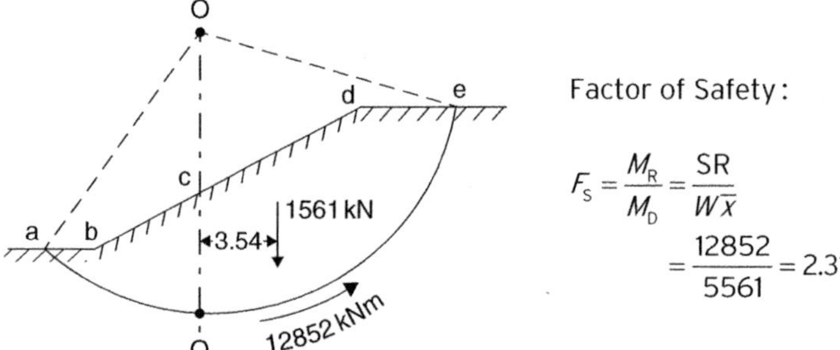

Factor of Safety:

$$F_S = \frac{M_R}{M_D} = \frac{SR}{W\bar{x}} = \frac{12852}{5561} = 2.31$$

Figure 10.7

Alternatively, find areas (abcQa) and (QcdeQ) with their centroids on separate templates, from which the forces (W_1 and W_2) and their position can be located as shown in Graph 10.1. The results are:

Areas			
	$A_1 = 62.8\ \text{m}^2$	$W_1 = 19 \times 62.7 = 1191.3$	$x_1 = 5.4\ \text{m}$
	$A_2 = 19.9\ \text{m}^2$	$W_2 = 19 \times 19.9 = \underline{378.1}$	$x_2 = 2.6\text{m}$
	$A = 82.7\ \text{m}^2$	$W = 1569.4\ \text{KN}$	

Disturbing moment: $M_D = Wx_1 - W_2x_2$
$= 1191.3 \times 5.4 - 378.1 \times 2.6$
$= 6433 - 983 = 5450\ \text{kNm}$

Resisting moment: $M_R = SR = 12852\ \text{kNm}$

Factor of Safety:

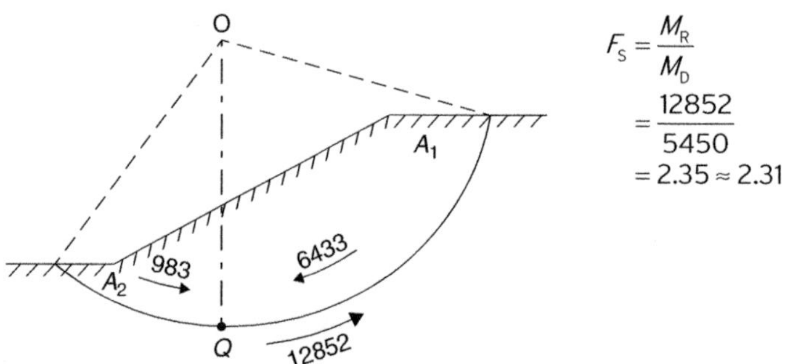

$$F_S = \frac{M_R}{M_D} = \frac{12852}{5450} = 2.35 \approx 2.31$$

Figure 10.8

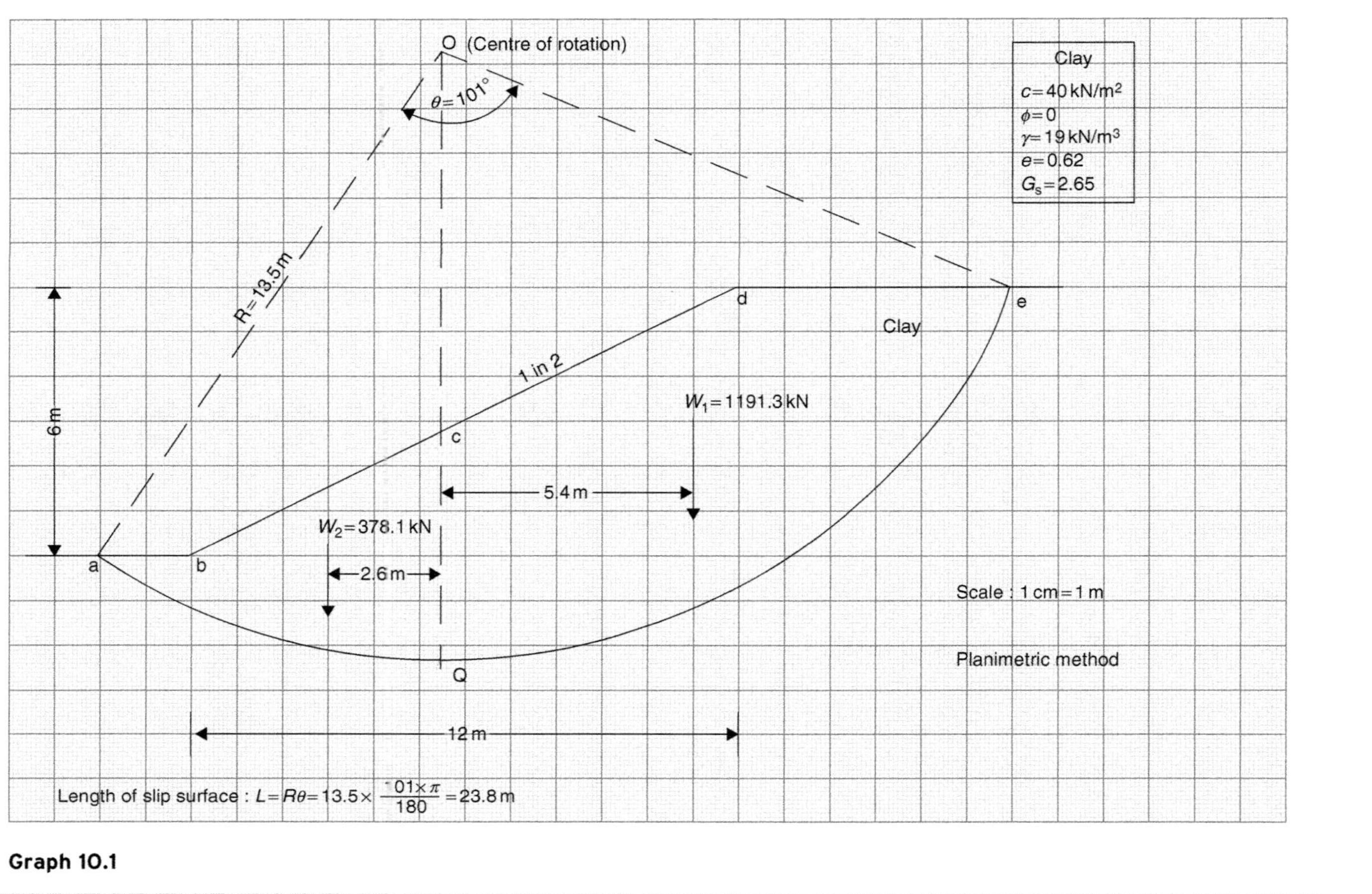
O (Centre of rotation)
θ=101°
R=13.5m
Clay
c=40 kN/m²
φ=0
γ=19 kN/m³
e=0.62
Gs=2.65
d
e
Clay
1 in 2
W1=1191.3 kN
6m
c
5.4 m
W2=378.1 kN
a
b
2.6m
Scale : 1 cm=1 m
Q
Planimetric method
12 m
Length of slip surface : L=Rθ=13.5× 101×π/180 =23.8 m

Graph 10.1

The determination of centroids of more complicated, irregular, shapes is somewhat tedious. It is much more easy to estimate the factor of safety by the Swedish Method of Slices. It is assumed in the theory, that the pressures (E_a and E_p) of soil acting on both sides of the slice cancel each other out, hence may be neglected.

Two procedures of the method are to be introduced by the next two examples.

Example 10.2

(See Graph 10.2)

Step 1: Divide the segmental slip area into a desired number of slices of equal width. The larger is the number, the more accurate is the answer. In this example, 20 slices of 1 m width are chosen for convenience. This way; Area of slice = height of slice.

Step 2: Measure the middle height (z) of each slice and calculate their weight per metre length of slope e.g.

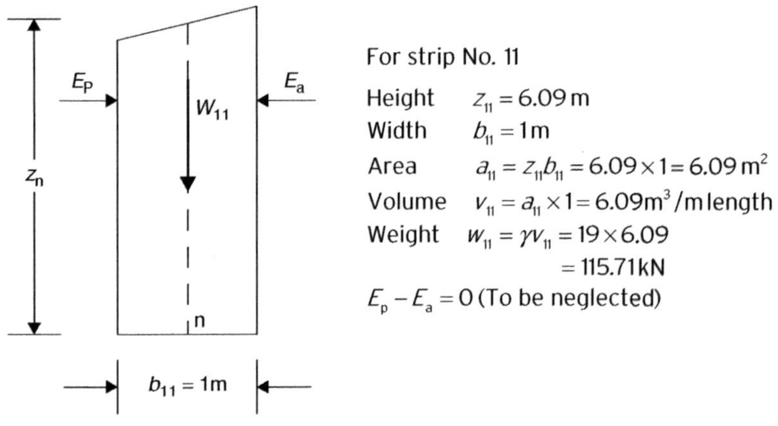

Figure 10.9

Step 3: Measure the moment arm (x) of each weight from the centre of the circle, taking ± signs into account.

Step 4: Calculate the disturbing moment M_D, by summing the positive and negative moments in Table 10.1.

$$M_D = \sum Wx$$

Step 5: Calculate the resisting moment, due to the shear strength of the clay.

Radius of the slip circle: $R = 13.5\ \text{m}$

Included angle:

$$\theta = 101^\circ = \frac{101 \times \pi}{180}\ \text{radian}$$
$$= 1.763\ \text{radian}$$

Table 10.1

No.	Height z (m)	Width b (m)	Weight W (kN)	Arm x (m)	$+Wx$ kNm	$-Wx$ kNm
1	0.14	1.0	2.66	−7.1		−18.89
2	0.70	1.0	13.30	−6.1		−81.13
3	1.41	1.0	26.79	−5.1		−136.63
4	2.27	1.0	43.13	−4.1		−176.83
5	3.05	1.0	57.95	−3.1		−179.65
6	3.75	1.0	64.03	−2.1		−134.46
7	4.37	1.0	83.03	−1.1		−91.33
8	4.91	1.0	93.29	−0.1		−9.33
9	5.38	1.0	102.22	0.9	92.00	
10	5.78	1.0	109.82	1.9	208.66	
11	6.09	1.0	115.71	2.9	335.56	
12	6.33	1.0	120.27	3.9	469.05	
13	6.49	1.0	123.31	4.9	604.22	
14	6.55	1.0	124.45	5.9	743.26	
15	6.26	1.0	118.94	6.9	820.69	
16	5.61	1.0	106.59	7.9	842.06	
17	4.81	1.0	91.39	8.9	813.37	
18	3.84	1.0	72.96	9.9	722.30	
19	2.62	1.0	49.78	10.9	542.60	
20	1.03	1.0	19.57	11.9	232.88	
	$\Sigma z = 81.39$		$\Sigma W = 1539.19$		+6426.65	−828.25
					$M_D = \Sigma Wx = 5598.4$ kNm	

Slice: No., Height, Width, Weight. Moment $= \pm Wx$: Arm, $+Wx$, $-Wx$.

Length of the slip surface: $\boxed{L = R\theta}$ (10.2)

$$= 13.5 \times 1.763 = 23.8\text{m}$$

Shear strength of clay: $\tau = c_u = 40\ \text{kN/m}^2$

Resting force provided by the shear strength:

$$\boxed{S = c_u L} \qquad (10.3)$$

$$= 40 \times 23.8 = 952\ \text{kN}$$

Therefore, the resisting (–ve) moment due to the shear strength is:

$$\begin{aligned} M_D &= SR \\ &= 952 \times 13.5 \\ &= 12852\ \text{kNm} \end{aligned} \qquad (10.4)$$

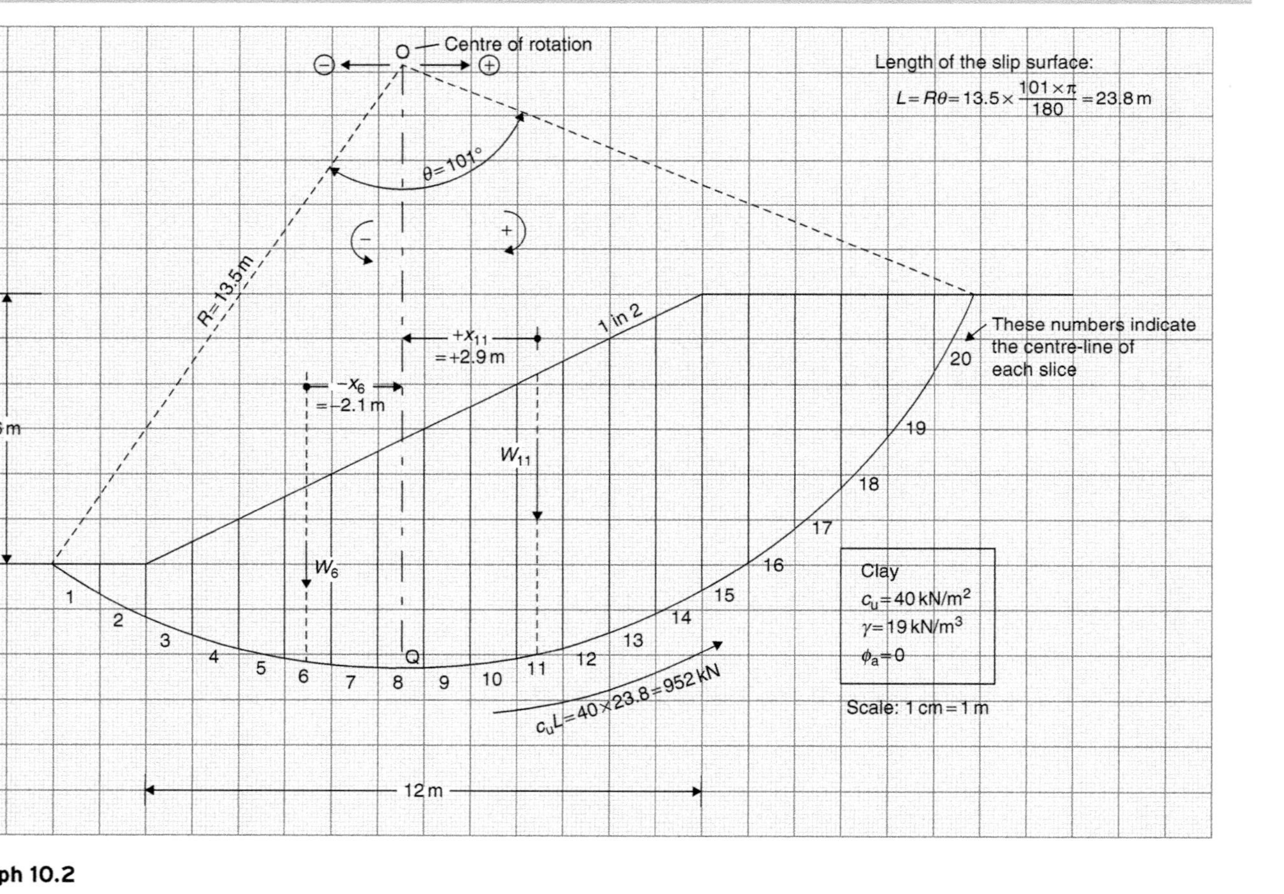
O — Centre of rotation
Length of the slip surface:
$L = R\theta = 13.5 \times \frac{101 \times \pi}{180} = 23.8$ m
$\theta = 101°$
R = 13.5 m
6 m
$+x_{11}$ = +2.9 m
$-x_6$ = −2.1 m
1 in 2
W_{11}
W_6
Q
These numbers indicate the centre-line of each slice
1 2 3 4 5 6 7 8 9 10 11 12 13 14 15 16 17 18 19 20
$c_u L = 40 \times 23.8 = 952$ kN
Clay
$c_u = 40$ kN/m^2
$\gamma = 19$ kN/m^3
$\phi_a = 0$
Scale: 1 cm = 1 m
12 m

Graph 10.2

Step 6: Calculate the factor of safety, by considering the moments acting on the soil mass as shown:

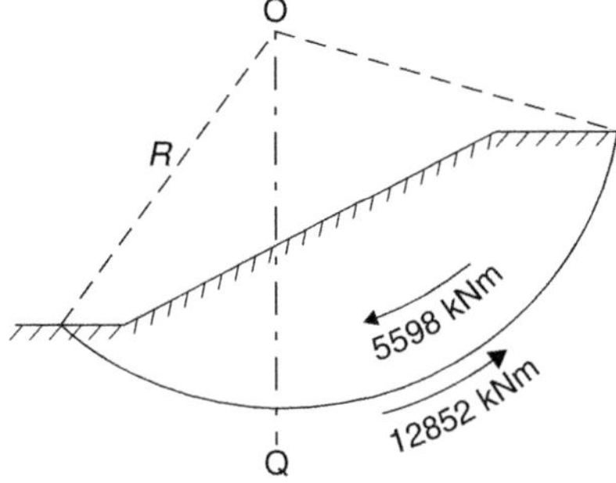

Disturbing moment from Table 10.1.

$M_D = \Sigma Wx = 5598\,\text{kNm}$

Resisting moment

$M_R = 12852\,\text{kNm}$

Figure 10.10

Hence the factor of safety: $F_s = \dfrac{M_R}{M_D} = \dfrac{12852}{5598} = 2.3$

This value is satisfactory from the safety point of view. However, assuming that this circle is the most critical one, F_s=2.3 may be reduced by steepening the slope, thus saving on expensive excavation in earthworks schemes for cuttings.

Example 10.3 Radial procedure

(See Graph 10.3).

In the previous example, M_D and M_R were determined by taking moments of weights about a vertical line through the centre of the circle. In this alternative method, the component of weight, tangential to the slip surface, is taken into account.

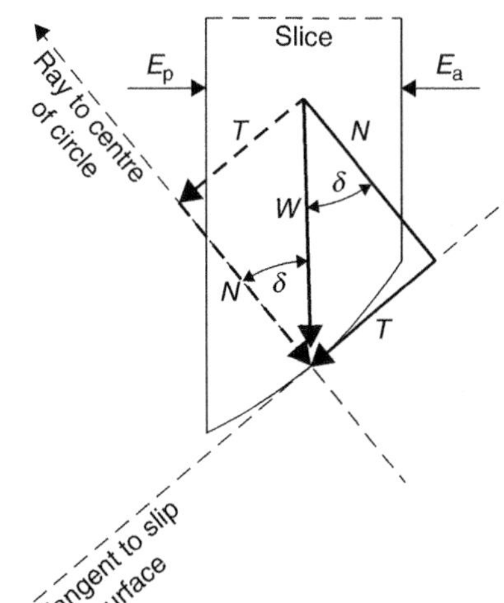

N=Radial component of W

T=Tangential component of W

$$\boxed{T = W\sin\delta} \qquad (10.5)$$

For slice 19, on Graph 10.3:

$$W_{19} = 49.78\,\text{kN}$$
$$\delta_{19} = 54^\circ$$

$$\begin{aligned} T_{19} &= W_{19}\sin\delta_{19} \\ &= 49.78\sin 54 \\ &= 40.27 \end{aligned}$$

Figure 10.11

For the slope given in Example 10.1, calculate the factor of safety by the radial method.

Step 1: Divide the segment into 20 slices and mark only the middle of each, on the slip surface.

Step 2: Draw a radial line through each mid-point and the centre of the circle.

Step 3: Calculate the weight (W) of each slice as in step 2 of Example 10.2.

Step 4: Measure the angle (δ) between the ray and the vertical line OQ

Step 5: Calculate the tangential component of W for each slice by formula (10.5): $T = W\sin\delta$

Step 6: Tabulate the results so far:

Table 10.2

Slice		Angle		Tangential force	
No.	Weight	δ	$\sin\delta$	$T = W\sin\delta$ (kN)	
	W (kN)	Degree		+ ←	− →
1	2.66	−31.8	−0.527		−1.40
2	13.30	−27	−0.454		−6.04
3	26.79	−22.7	−0.386		−10.37
4	43.13	−17.5	−0.301		−12.98
5	57.95	−13	−0.225		−13.04
6	64.03	−9	−0.156		−9.99
7	83.03	−4.5	−0.078		−6.48
8	93.29	−1	−0.017		−1.59
9	102.22	4	0.07	7.16	
10	109.82	8	0.139	15.26	
11	115.71	12.5	0.216	24.99	
12	120.27	17	0.292	35.12	
13	123.31	21.5	0.367	45.25	
14	124.45	26	0.438	54.51	
15	118.94	31	0.515	61.25	
16	106.59	35.8	0.585	62.36	
17	91.39	41.3	0.660	60.32	
18	72.96	47	0.731	53.33	
19	49.78	54	0.809	40.27	
20	19.57	62	0.883	17.28	
	Σ 1539.19			+477.10	−61.89
			Rounded to:	$\Sigma T = 415$ kN	

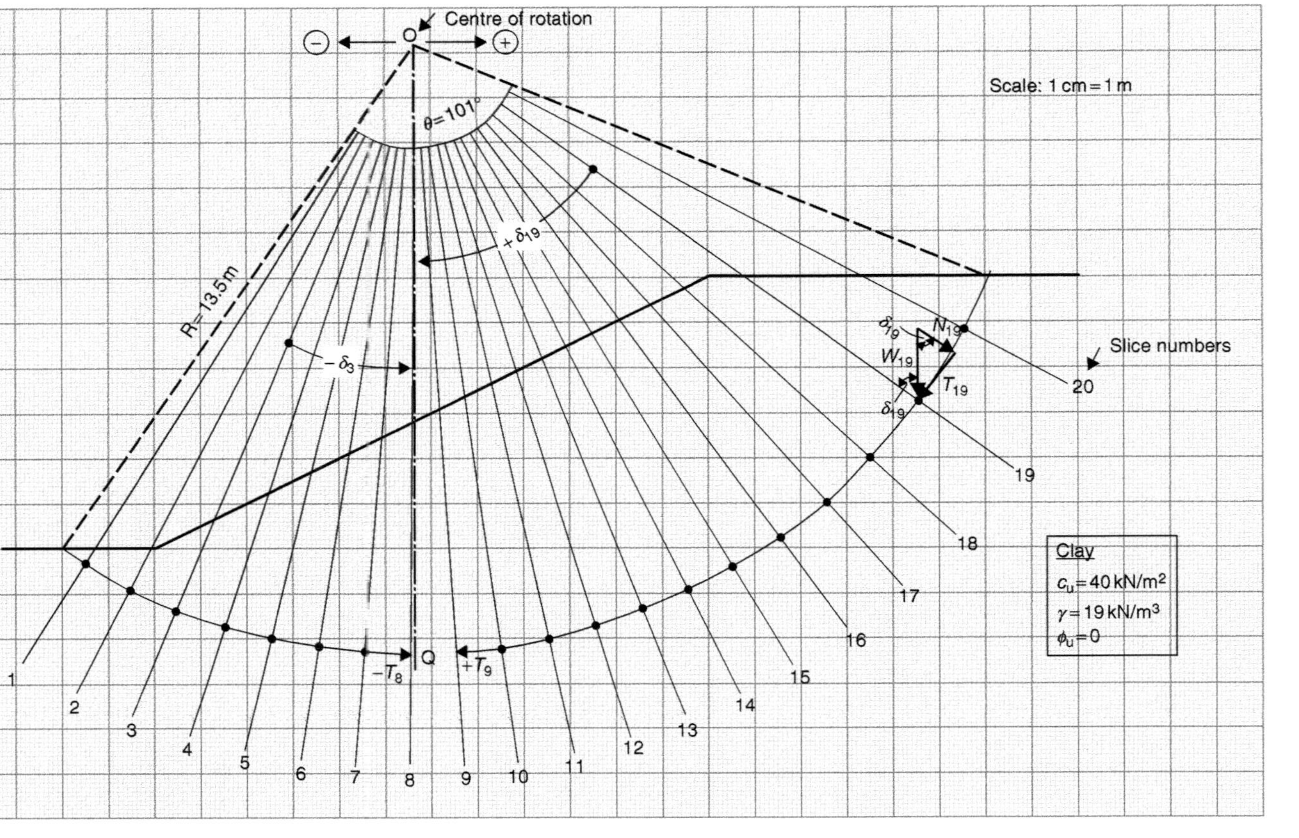
Centre of rotation
O
Q
Scale: 1 cm = 1 m
θ = 101°
R = 13.5 m
+δ19
−δ3
−T8
+T9
N19
W19
T19
δ19
Slice numbers
1
2
3
4
5
6
7
8
9
10
11
12
13
14
15
16
17
18
19
20
Clay
cu = 40 kN/m2
γ = 19 kN/m3
φu = 0

Graph 10.3

Step 7: Determine M_R and M_D by considering the forces acting along the slip surface as shown and taking moments about the centre of the circle.

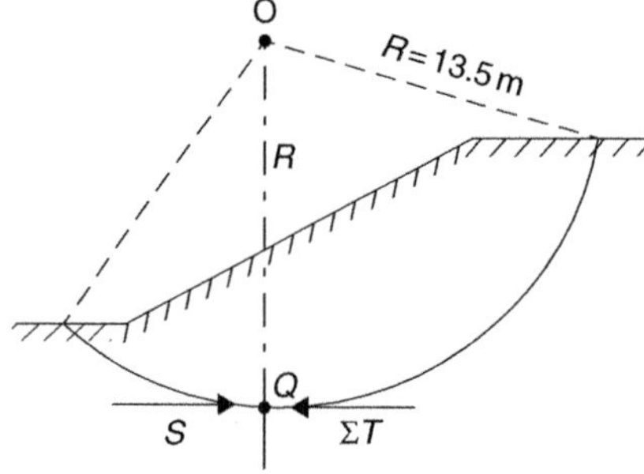

Net disturbing tangential force from Table 10.2: $\Sigma T = 415$ kN ←

Resisting force from Example 10.1: $S = c_u L = 952$ kN →

Figure 10.12

Disturbing moment: $M_D = R\Sigma T$
$= 13.5 \times 415.21 = 5605$ kNm ↻

Restoring moment: $M_R = RS$
$= 13.5 \times 952 = 12852$ kNm ↺

Factor of Safety: $$F_s = \frac{RS}{R\Sigma T} = \frac{S}{\Sigma T} \qquad (10.6)$$

$$= \frac{952}{415} = 2.3$$

Example 10.4 Tension crack

(See Graph 10.4).

It was shown in Chapter 8 that the depth of a tension crack in pure clay is given by:

$$z_0 = \frac{2c_u}{\gamma} \qquad (8.24)$$

If the crack is full of water, then the moment of the hydrostatic force P_w about the centre of rotation increases the overturning moment.

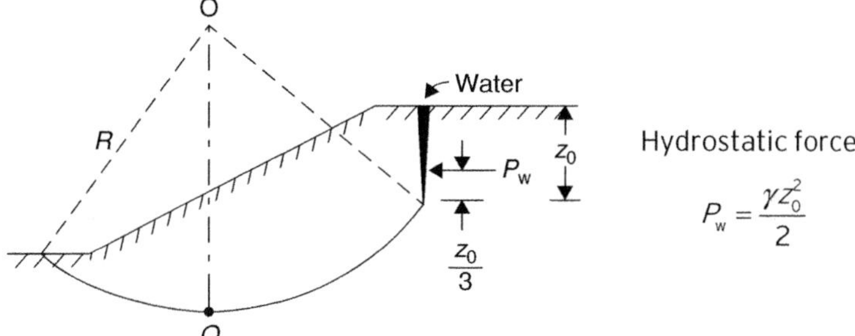

Figure 10.13

This is to investigate the effect of tension crack on the factor of safety, determined in Example 10.2.

Step 1: Calculate z_0 and P_w

For $\left| \begin{matrix} \gamma = 19\,\text{kN/m}^3 \\ c_u = 40\,\text{kN/m}^2 \end{matrix} \right|$ $\quad z_0 = \frac{2c_u}{\gamma} = \frac{2 \times 40}{19} = 4.21\,\text{m}$

Say $z_0 = 4.4\,\text{m}$

Then $P_w = \frac{\gamma z_0^2}{2} = \frac{19 \times 4.5^2}{2} = 192.4\,\text{kN}$

Line of action of $P_w = \frac{z_0}{3} = \frac{4.4}{3} = 1.47\text{m}$ from the bottom of the crack.

Step 2: It is seen from Graph 10.4, by comparing it with Graph 10.2, that the crack has separated slices 18, 19 and 20 from the circular segment, hence they do not contribute to the disturbing moment as calculated in Example 10.2. Determine the new value of M_D.

From Table 10.1, the sum of moments due to slices 18, 19 and 20 $= 722.3 + 542.8 + 232.88$
$= 1497.78\,\text{kNm}$

Remaining moment rounded to: $\Sigma Wx = 5598.4 - 1497.78$
$= 4101\,\text{kNm}$

From Graph 10.4, the disturbing moment due to the hydrostatic force $= 192.4 \times 7.16 = 1378$ kNm↻

And the final value of the total disturbing moment is:

$$M_D = 4101 + 1378 = 5479 \text{ kNm}$$↻

Therefore, the removal of three slices and the addition of hydrostatic force resulted in decreased disturbing moment, in this case.

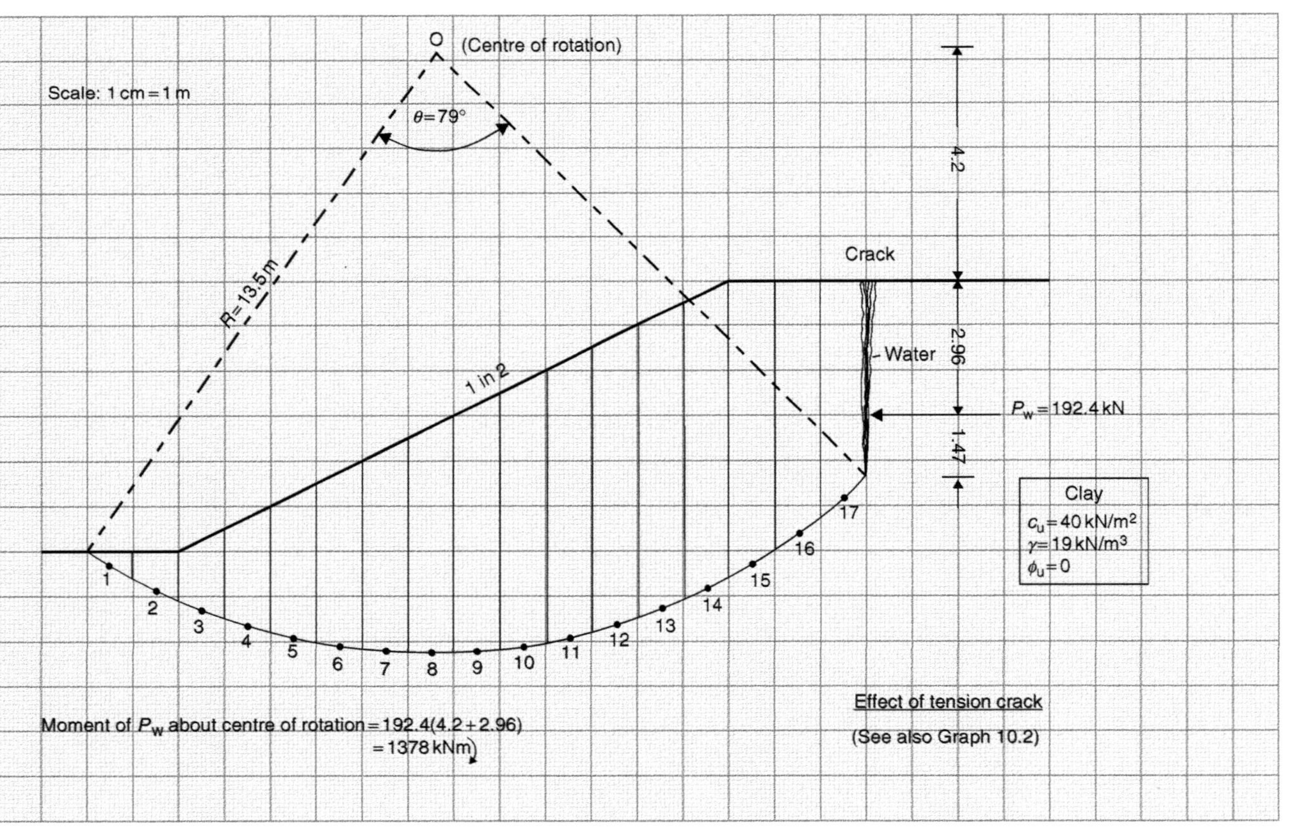
O (Centre of rotation)
Scale: 1 cm = 1 m
θ=79°
R = 13.5 m
1 in 2
Crack
Water
4.2
2.96
1.47
P_w = 192.4 kN
Clay
c_u = 40 kN/m²
γ = 19 kN/m³
φ_u = 0
1
2
3
4
5
6
7
8
9
10
11
12
13
14
15
16
17
Moment of P_w about centre of rotation = 192.4(4.2 + 2.96)
= 1378 kNm
Effect of tension crack
(See also Graph 10.2)

Graph 10.4

Step 3: Calculate the resistance contributed by the shorter slip surface and obtain the new value of M_R.

From Graph 10.4 $\theta = 79° = 1.379$ rad
Length of slip surface: $L = R\theta = 13.5 \times 1.379 = 18.62$ m
Resisting shear force: $S = c_u L = 40 \times 18.62$
$= 744.8$ kN
Resisting moment due to shear strength:

$$M_R = SR = 744.8 \times 13.5 = 10055 \text{ kNm}$$

Step 4: Calculate the factor of safety

$$F_s = \frac{M_R}{M_D} = \frac{10055}{5479} = 1.84$$

Therefore, the effect of tension crack is to reduce the factor of safety by 20% in this example.

10.2.6 Homogeneous $c-\phi$ soil (total stress analysis)

The radial procedure can easily be applied to partially saturated soils, that is where no pore pressure has to be considered. Because of the existence of friction angle ϕ, the normal component (N) of W has now to be taken into account, as well as the tangential one (T).

According to the mechanical theory of friction, the force (F) required to slide two bodies over each other is given by:

$$\boxed{F = W\tan\phi} \quad (10.7)$$

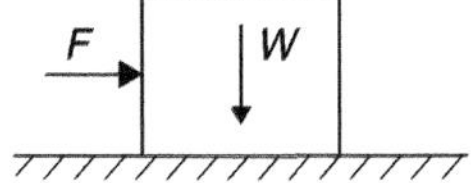

where W = normal force acting on the contact surfaces.
ϕ = angle of friction or angle of repose.
$\tan\phi$ = coefficient of friction

Figure 10.14

Conversely, when a body of weight W is placed on an incline, F is that friction force which prevents it from sliding. It is given by:

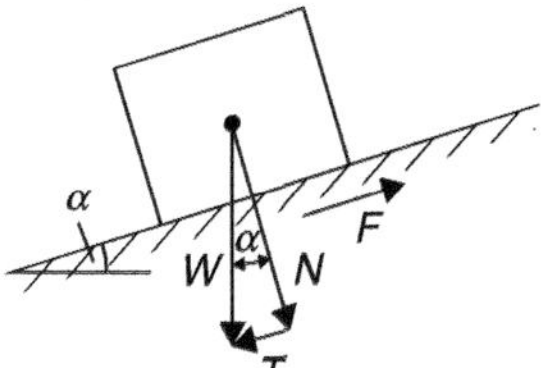

$$\boxed{F = N\tan\phi} \quad (10.8)$$

where N = component of W, normal to the contact surfaces.

$$\therefore \quad \boxed{N = W\cos\alpha} \quad (10.9)$$

Figure 10.15

Therefore, the friction force is: $$\boxed{F = W\cos\alpha\tan\phi} \quad (10.10)$$

This formula is applied to each slice separately, as shown in Figure 10.16, using the appropriate tangential angle δ at each midpoint, that is:

(10.10) becomes: $$\boxed{F = W\cos\delta\tan\phi} \quad (10.11)$$

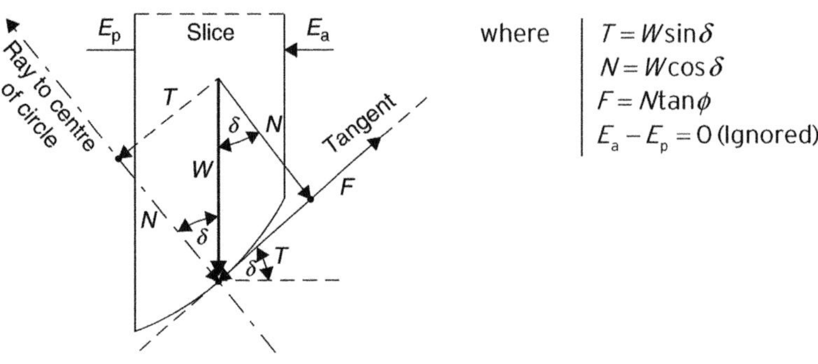

Figure 10.16

Example 10.5

Suppose the shear strength characteristics of the clay, in Example 10.1, are given as:

$$c_u = 40\,\text{kN/m}^3 \qquad \phi_u = 15° \qquad \therefore \tan\phi_u = 0.2679$$

Calculate the factor of safety for the same slope. Some results, extracted from Table 10.2, are tabulated:

Table 10.3

Slice		Angle	T=Wsinδ		Friction force (kN)
No.	W (kN)	δ	+ ←	– →	F=0.2679 Wcosδ
1	2.66	–31.8		–1.40	0.61
2	13.30	–27		–6.04	3.17
3	26.79	–22.7		–10.37	6.62
4	43.13	–17.5		–12.98	11.02
5	57.95	–13		–13.04	15.13
6	64.03	–9		–9.99	16.94
7	83.03	–4.5		–6.48	22.18
8	93.29	–1		–1.59	24.99
9	102.22	4	7.16		27.32
10	109.82	8	15.26		29.13
11	115.71	12.5	24.99		30.26
12	120.27	17	35.12		30.81
13	123.31	21.5	45.25		30.74
14	124.45	26	54.51		29.97
15	118.94	31	61.25		27.31
16	106.59	35.8	62.36		23.16
17	91.39	41.3	60.32		18.39
18	72.96	47	53.33		13.33
19	49.78	54	40.27		7.84
20	19.57	62	17.28		2.46
Σ	1539.19		+477.10	–61.89	371.38
		Rounded to:	ΣT=415 kN		ΣF=371 kN

The length of the slip surface has not changed, hence the resisting force contributed by the cohesion is still: $S = c_u L = 952$ kN

Resisting force contributed by ϕ: $\Sigma F = 371$ kN

Disturbing force: $\Sigma T = 415$ kN

Forces acting along slip surface are as shown.

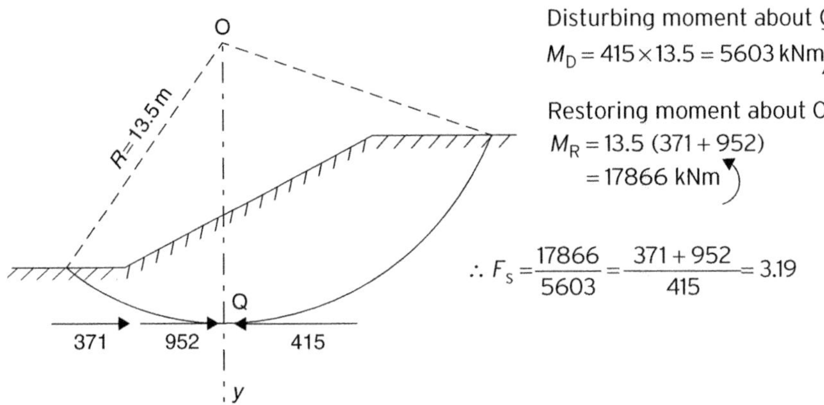

Disturbing moment about O

$M_D = 415 \times 13.5 = 5603$ kNm

Restoring moment about O

$M_R = 13.5\ (371 + 952)$

$= 17866$ kNm

$$\therefore F_s = \frac{17866}{5603} = \frac{371+952}{415} = 3.19$$

Figure 10.17

Therefore, the effect of $\phi_u = 15°$ is to increase the factor of safety by 29%, in this example.

Notes: a) It is not really necessary to calculate the disturbing and resisting moments as the radius is cancelled from:

$$F_s = \frac{M_R}{M_D} = \frac{R(S+\Sigma F)}{R\Sigma \pi} = \frac{952+371}{415} = 3.19$$

The formula, therefore, is transformed into a ratio of forces only.

$$\boxed{F_s = \frac{S + \Sigma F}{T} = \frac{cL + \tan\phi\ \Sigma W \cos\delta}{\Sigma W \sin\delta}} \qquad (10.12)$$

b) The depth of tension cracks for the $c-\phi$ soil is now estimated by:

$$z_0 = \frac{2c_u}{\gamma K_a}$$

$$\text{where} \quad K_a = \frac{1-\sin\phi_u}{1+\sin\phi_u}$$

10.2.7 Stratified slopes

The methods introduced in this chapter can easily be adapted to the analysis of multi-layered slopes. The general procedure is outlined below.

1. Planimeter method

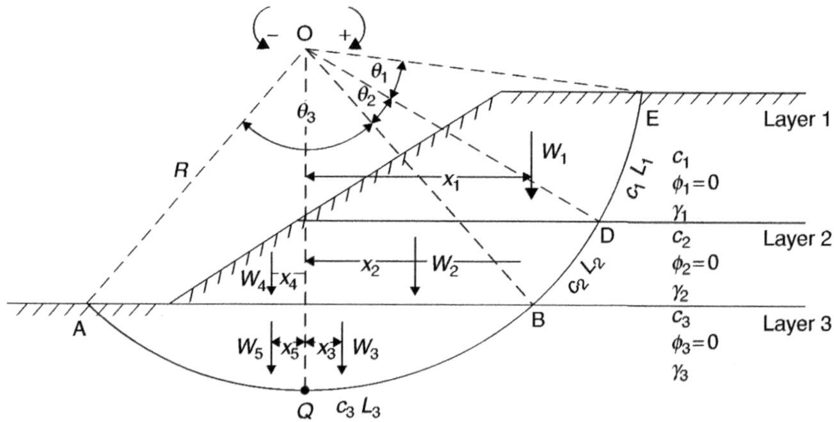

Figure 10.18

Determine the weights and moment-arms about the centre of rotation as described in Example 10.1.

Disturbing moment: $M_D = R\Sigma Wx$

$$= R\left[W_1x_1 + W_2x_2 + W_3x_3 - (W_4x_4 + W_5x_5)\right]$$

Calculate the length of arc in each layer:

$$L_1 = \overline{AB} = R\theta_1$$
$$L_2 = \overline{BD} = R\theta_2$$
$$L_3 = \overline{DE} = R\theta_3$$

Resisting shear force is the sum of resisting shear forces contributed by the layers.

$$S = \Sigma c_u L = c_1L_1 + c_2L_2 + c_3L_3$$

And resisting moment: $M_R = SR = R\Sigma c_u L$

Factor of safety: $F_s = \dfrac{M_R}{M_D} = \dfrac{R\Sigma c_u L}{R\Sigma W_x}$

Cancelling R

$$\therefore \boxed{F_s = \frac{\Sigma c_u L}{\Sigma W_x}} \qquad (10.13)$$

2. Method of slices

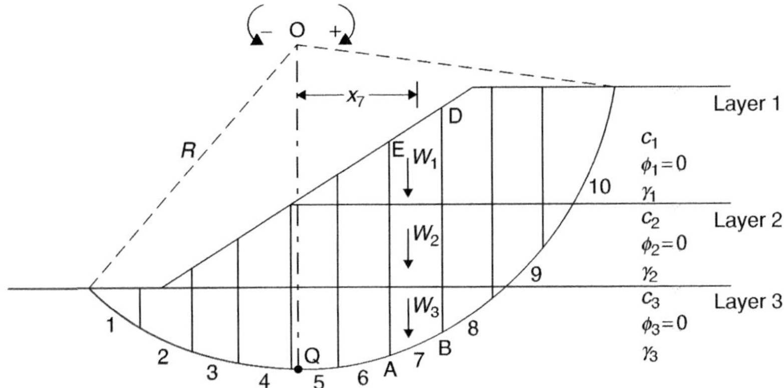

Figure 10.19

Determine the total weight of each slice, by summing the weight contributed by each layer e.g.

$$W_7 = W_1 + W_2 + W_3$$

The moments are then calculated as in Example 10.2.

Disturbing moment: $M_D = R\Sigma(Wx)$

Resting force due to cohesion: $S = \Sigma c_u L$ (as above)

Resisting moment: $M_R = R\Sigma c_u L$

Factor of Safety: $F_s = \dfrac{\Sigma c_u L}{\Sigma Wx}$

which has the same form as (10.13), but in this case ΣWx indicates the sum of moments of all slices about the centre rotation.

10.2.8 Slopes under water

There are many situations, when the slope is kept permanently under water, as in canals or reservoirs, without significant change in water level. In these circumstances there is no change in the pore pressure either, hence the stability of the slope may be evaluated in terms of total stresses. In general, higher water level means larger factor of safety. There are three cases to consider:

Example 10.4: Dry slope with tension cracks, as already analysed in Example 10.4.
Example 10.6: Totally submerged slope, disregarding tension cracks.
Example 10.7: Partially submerged slope. Water level below tension cracks.

1. Totally submergedslope

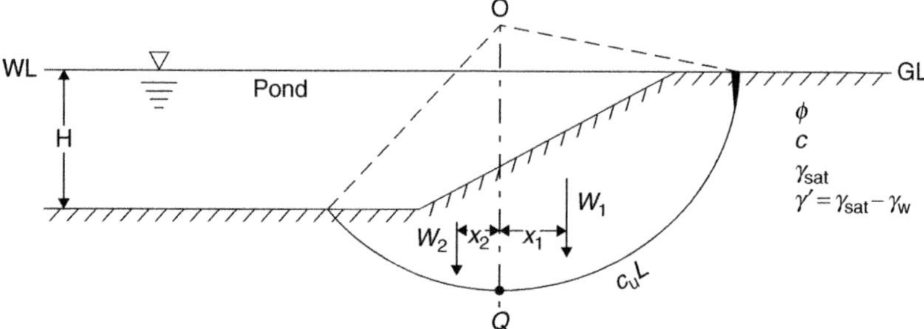

Figure 10.20

Notes: i. The horizontal pressure due to water in the tension crack is balanced by the pressure of water in the pond and has no effect on the factor of safety.

ii. It was established in Chapter 5, that the effective pressure at a depth z below water surface level is given by $\sigma' = z\gamma'$. It follows, therefore, that the weight of soil below this level is determined from $W = V\gamma'$, where V is the volume considered.

Example 10.6 (Total submergence)

With reference to Example 10.1 and Graph 10.1, estimate the factor of safety, if the slope is completely flooded.

Saturated unit weight:
$$\gamma_{sat} = \left(\frac{G_s + e}{1 + e}\right)\gamma_w$$
$$= \left(\frac{2.65 + 0.62}{1 + 0.62}\right) \times 9.81 = 19.8\,\text{kN/m}^3$$

Submerged unit weight:
$$\gamma' = \gamma_{sat} - \gamma_w = 19.8 - 9.81$$
$$= 9.99\,(\text{say } 10)\,\text{kN/m}^3$$

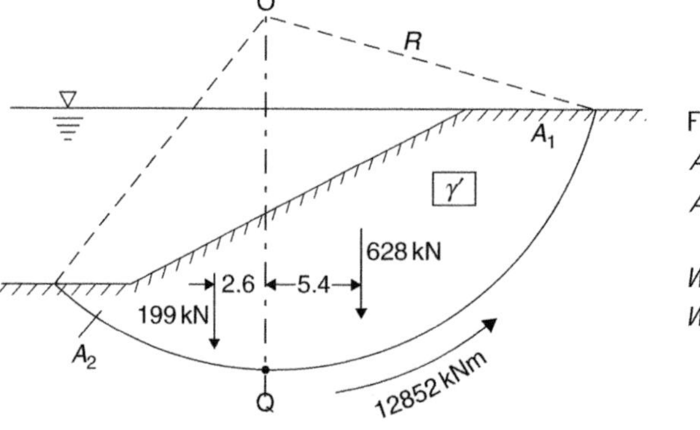

Figure 10.21

From Example 10.1:

$A_1 = 62.8\text{m}^2$ $x_1 = 5.4\text{m}$

$A_2 = 19.9\text{m}^2$ $x_2 = 2.6\text{m}$

$W_1 = 62.8 \times 10 = 628\text{kN}$

$W_2 = 19.9 \times 10 = 199\text{kN}$

Disturbing moment: $M_D = 628 \times 5.4 - 199 \times 2.6 = 2874\ \text{kN/m}$

Resisting moment: $M_R = 13.5 \times 952 = 12852\ \text{kNm}$

Factor of safety: $$F_s = \frac{12852}{2874} = 4.47$$

The problem may, of course, be solved by the other two methods.
Note, that the weight of water itself, within the soil as well as above the slope, is not taken into account, when γ' is used.

Example 10.7 (Partial submergence)

With reference to Example 10.4 and Graph 10.4, estimate the factor of safety if the water level in the pond is 4.5 m below crest level.

The solution differs from that in Example 10.6 in two respects:

1. The force P_w induced by water in the crack has to be applied
2. The weight of soil above the water level is estimated in terms of $\gamma = 19\ \text{kN/m}^3$.

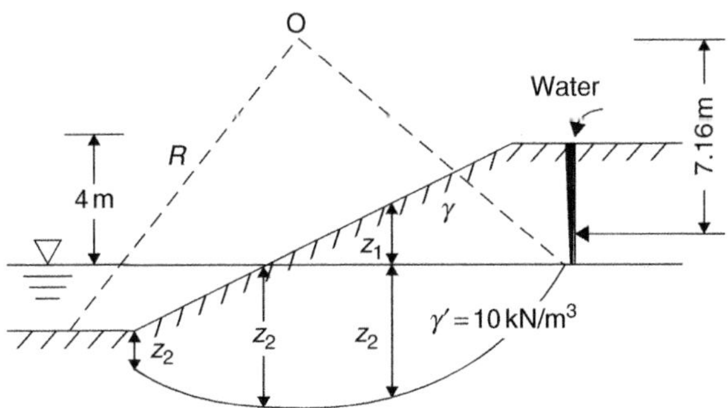

Figure 10.22

$$P_w = 192.4\ \text{kN (From Graph 10.4)}$$

$$W = \gamma z_1 + \gamma' z_2 = 19 z_1 + 10 z_2$$

The relevant calculations are tabulated below:

Table 10.4

Slice					Moment=±*Wx*	
	Height (m)		Weight	Arm	+*Wx*	−*Wx*
No.	z_1	z_2	*W* (kN)	*x* (m)	kNm	kNm
1		0.14	1.40	−7.1	–	−9.94
2		0.70	7.00	−6.1	–	−42.70
3		1.41	14.10	−5.1	–	−71.91
4		2.27	22.70	−4.1	–	−93.07
5		3.05	30.50	−3.1	–	−94.55
6	0.25	3.50	39.75	−2.1	–	−83.48
7	0.75	3.62	50.45	−1.1	–	−55.50
8	1.25	3.66	60.35	−0.1	–	−6.04
9	1.75	3.63	69.55	0.9	62.60	–
10	2.25	3.53	78.05	1.9	148.30	–
11	2.75	3.34	85.65	2.9	248.39	–
12	3.25	3.00	91.75	3.9	357.83	–
13	3.75	2.74	98.65	4.9	483.39	–
14	4.25	2.30	108.50	5.9	640.15	–
15	4.5	1.76	103.10	6.9	711.39	–
16	4.5	1.11	96.60	7.9	763.14	–
17	4.5	0.31	88.60	8.9	788.54	–
	Rounded to Σ		1047		+4204 kNm	−457 kNm
					$\Sigma Wx = 3747$ kNm	

Disturbing moment:

where $\Sigma Wx = 3747$ kNm is due to the moving soil mass

$7.16\,P_W = 7.16 \times 192.4 = 1378$ kNm is due to the water pressure in the tension crack.

Therefore, $M_D = 3747 + 1378 = 5125$ kNm

Resisting moment:

Resisting shear force has been determined in Example 10.4 as $S = 744.8$ kN;

And $M_R = SR = 744.8 \times 13.5 = 10055$ kNm

Factor of Safety:

$$F_s = \frac{10055}{5125} = 1.96$$

10.2.9 Taylor's stability numbers

Chart 10.2 facilitates the determination of minimum F_s for homogeneous slopes, in terms of total stresses. The method is based on the following assumptions:

a) For two slopes of the same shape, but of different size, the failure surfaces are in the same proportions.

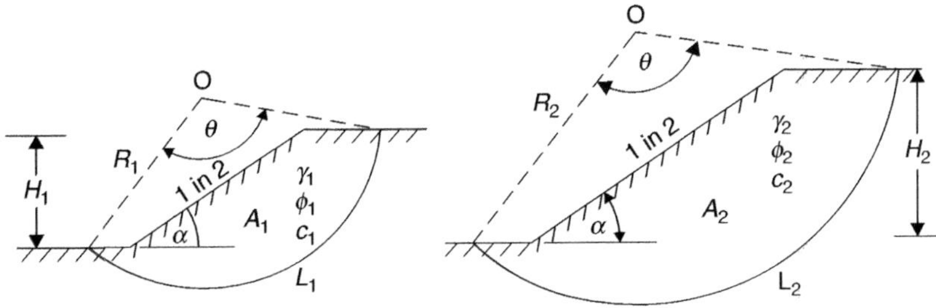

Figure 10.23

By geometric similarity: $\frac{R_1}{R_2} = \frac{H_1}{H_2}$

$L_1 = R_1\theta$

and $L_2 = R_2\theta$ $\quad \theta = \frac{R_1}{L_1} = \frac{R_2}{L_2}$

In general

$L \propto H$

or

$L = mH$

Similarly, the cross-sectional area: $A \propto H^2$

$A = nH^2$

where m and n are constants of proportionality.

b) Both slopes have the same angle of friction.

$$\phi_1 = \phi_2$$

c) Tension cracks may be ignored.

The factor of safety in respect of cohesion is given by: $F_c = \frac{\text{Resisting force}}{\text{Disturbing force}}$

$$= \frac{cL}{\gamma A} = \frac{c\,mH}{\gamma n H^2} = \frac{m}{n}\left(\frac{c}{\gamma H}\right)$$

The dimensionless stability number is expressed from this as: $\frac{n}{m} = \boxed{N_c = \frac{c}{\gamma H F_c}}$ (10.14)

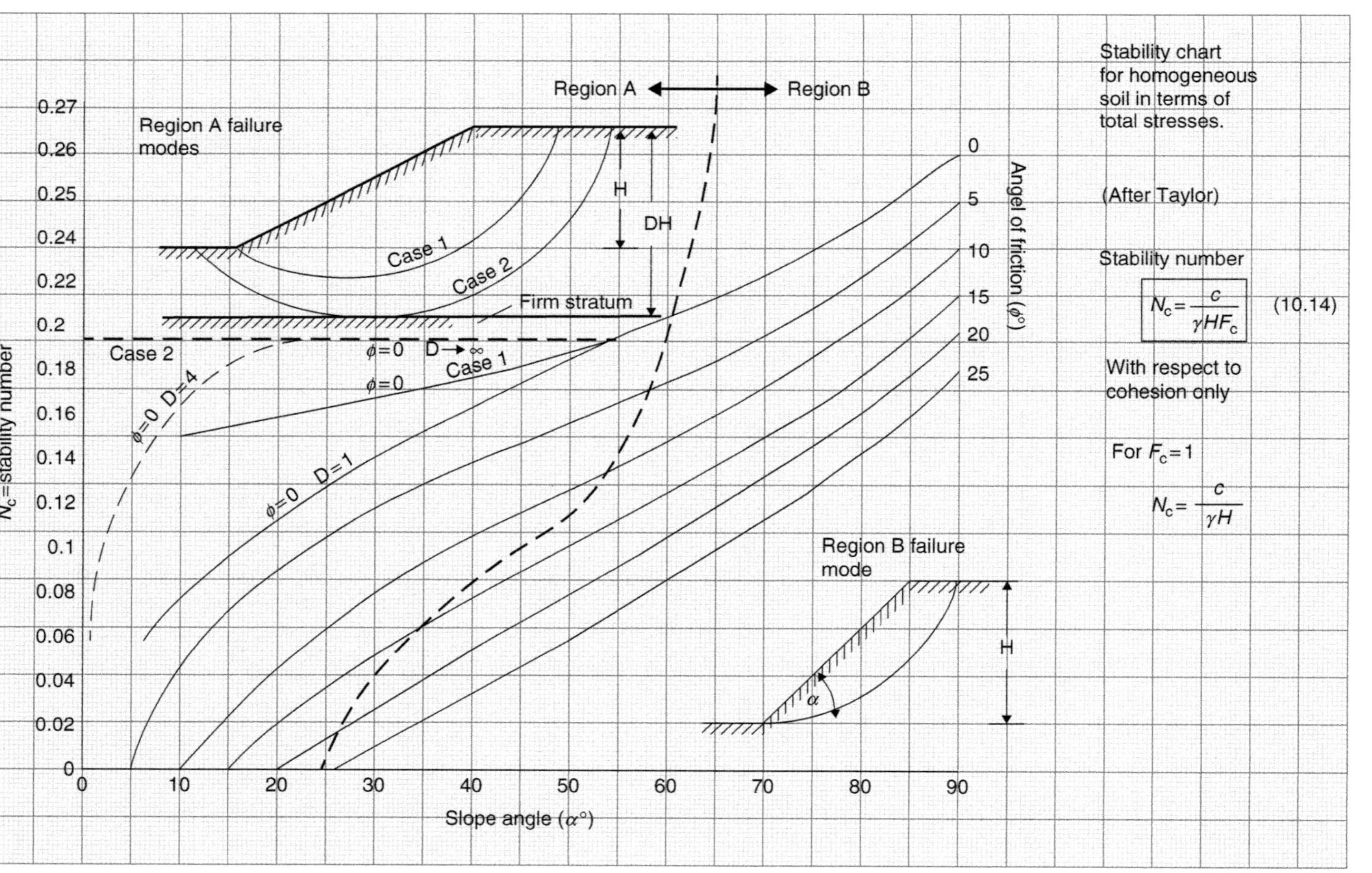
Stability chart
for homogeneous
soil in terms of
total stresses.
(After Taylor)
Stability number
$N_c = \frac{c}{\gamma H F_c}$ (10.14)
With respect to
cohesion only
For $F_c = 1$
$N_c = \frac{c}{\gamma H}$
Region A
Region B
Region A failure
modes
H
DH
Case 1
Case 2
Firm stratum
Case 2
$\phi = 0$ D → ∞
Case 1
$\phi = 0$
$\phi = 0$ D = 4
$\phi = 0$ D = 1
Angel of friction ($\phi°$)
0
5
10
15
20
25
Region B failure
mode
H
α
N_c=stability number
0.27
0.26
0.25
0.24
0.22
0.2
0.18
0.16
0.14
0.12
0.1
0.08
0.06
0.04
0.02
0
0
10
20
30
40
50
60
70
80
90
Slope angle ($\alpha°$)

Chart 10.2

Therefore, the factor of safety applies to cohesion only, as the stability number is the ratio of the cohesion developed to the total pressure of the base of the embankment.

Factor of safety in respect to c and ϕ

The shear strength in $c-\phi$ soils is made up of cohesion as well as friction. It is expressed by:

$$\tau = c + \sigma\tan\phi$$

The shear strength mobilized along the slip surface is given by:

$$\tau_m = c_m + \sigma\tan\phi_m$$

where c_m and ϕ_m are the available cohesion and angle of friction. The factor of safety may now be defined as:

$$F_s = \frac{\tau}{\tau_m} = \frac{c + \sigma\tan\phi}{c_m + \sigma\tan\phi_m}$$

Expressing the shear strength mobilized:

$$\tau_m = \frac{\tau}{F_s} = \frac{c}{F_c} + \frac{\sigma\tan\phi}{F_\phi}$$

where F_c = Factor of safety in respect to cohesion defined as:

$$\boxed{F_c = \frac{c}{c_m}} = \boxed{\frac{c}{\gamma H N_c}} \quad (10.15)$$

where c_m = amount of cohesion fully mobilized

$$\boxed{c_m = \gamma H N_c} \quad (10.16)$$

F_c = 1 means that cohesion is fully mobilized.

Also, F_ϕ = Factor of safety in respect to friction,

$$\boxed{F_\phi = \frac{\tan\phi}{\tan\phi_m}} \quad (8.61)$$

F_ϕ = 1 means that friction is fully mobilized.

The factor of safety in respect to strength (F_s) can be expressed in terms of F_c and F_ϕ. There are three cases:

1. If $\phi = 0$, then $\frac{\tau}{F_s} = \frac{c}{F_c}$ and $\boxed{F_s = F_c}$ (10.17)

2. If $F_c = F_\phi$, then $\frac{\tau}{F_s} = \frac{c}{F_c} + \frac{\sigma\tan\phi}{F_\phi}$

$$= \frac{c}{F_c} + \frac{\sigma\tan\phi}{F_c}$$

$\therefore$ $\boxed{F_s = F_c = F_\phi}$ (10.18)

It follows that

$$F_s = \frac{c}{c_m} = \frac{\tan\phi}{\tan\phi_m} \tag{10.19}$$

and

$$F_s = \frac{F_c + F_\phi}{2} \tag{10.20}$$

3. If $\phi>0$, then F_s is determined by successive approximation. In this, an arbitrary value of F_ϕ is adjusted successively, until $F_s=F_c=F_\phi$. Two procedures are introduced in the next example.

Example 10.8

A 6 m high cutting is to be made in normally consolidated clay. Figure 10.24 shows the dimensions of the slope and the shear strength parameters of the soil. Determine F_s for the slope. See also Example 10.5.

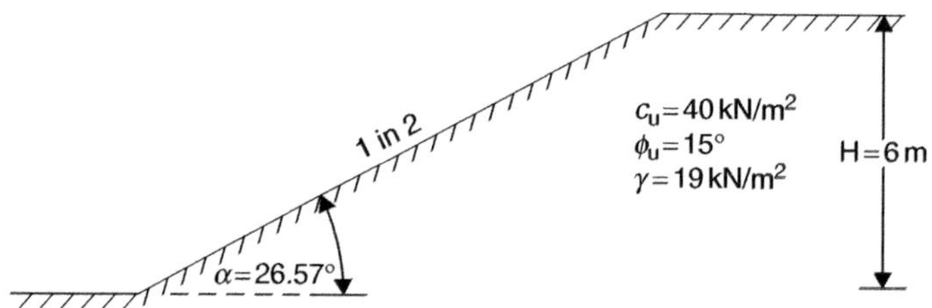

Figure 10.24

10.2.9.1 Semi-graphical method

Step 1: Assume the frictional resistance fully utilized, that is the mobilized friction angle ϕ_m equals to the total, available friction.

$$\phi_m = \phi = 15° \quad \therefore F_\phi = \frac{\tan\phi}{\tan\phi_m} = \frac{\tan 15}{\tan 15} = 1$$

Step 2: Using Chart 10.2, determine the stability number.

For $\phi_m = 15°$ and $\alpha = 26.57°$ | $N_c = 0.038$

Step 3: Calculate the factor of safety, in respect to cohesion from (10.14):

$$F_c = \frac{c}{N_c \gamma H} = \frac{40}{0.038 \times 19 \times 6} = 9.24$$

Note: $F_\phi=1$ and $F_c=9.24$, therefore $F_s \neq F_c \neq F_\phi$, hence repeat steps 1 to 3, for different values of ϕ_m.

Step 4: Tabulate the calculations for ϕ_m=0, 5, 10, 15, 20 and 25 degrees.

Table 10.5

ϕ_m ($\alpha = 26.57°$)	$\tan\phi_m$	N_c	$F_\phi = \dfrac{\tan 15}{\tan\phi_m}$	$F_c = \dfrac{40}{114N_c}$
0	0.0000	0.154	∞	2.28
5	0.0875	0.102	3.06	3.44
10	0.1763	0.064	1.52	5.48
15	0.2679	0.038	1.00	9.24
20	0.3640	0.016	0.74	21.93
25	0.4663	0.002	0.57	175.44

Step 5: Plot F_c against F_ϕ on Graph 10.5 and draw line $\overline{OA}$ at 45° to intersect the curve. The point of intersection at A locates the required $F_c = F_\phi = F_s = 3.25$. This value compares well with $F_s = 3.19$ in Example 10.5.

10.2.9.2 Method of successive approximation

Step 1: Assume $\phi_m = \phi = 15°$, that is at $F\phi = 1$
Obtain $N_c = 0.038$

Step 2: Calculate the amount of cohesion mobilized.

$$c_m = \gamma H N_c = 19 \times 6 \times 0.038 = 4.33\,\text{kN/m}^2$$

Determine:

$$F_c = \frac{c}{c_m} = \frac{40}{4.33} = 9.24$$

Note: $F_\phi < F_c$, therefore, $F_s \neq F_\phi \neq F_c$ and another, larger, value of F_ϕ has to be tried.

Step 3: The nth approximation is obtained by the following formulae:

From (6.61):

$$F_\phi = \frac{\tan\phi}{\tan\phi_m} = \frac{\phi}{\phi_m}$$

For $\phi < 30°$

$$\boxed{\phi_m = \frac{\phi}{F_\phi}} \qquad (10.21)$$

From (10.14) for the same slope $\dfrac{c}{\gamma H} = F_c N_c = \text{constant}$

So,

$$F_n N_n = F_c N_c$$

The nth approximation:

$$\boxed{F_n = \frac{F_c N_c}{N_n}} \qquad (10.22)$$

1st approximation (n=1)

Try $F_\phi = 1.5$	$\phi_m = \dfrac{15}{1.5} = 10°$	$F_1 = F_c = \dfrac{9.24 \times 0.038}{0.064}$
From Chart 10.2	$N_c = N_1 = 0.064$	$= 5.48$

$F_c > F_\phi$, therefore try larger value for F_ϕ.

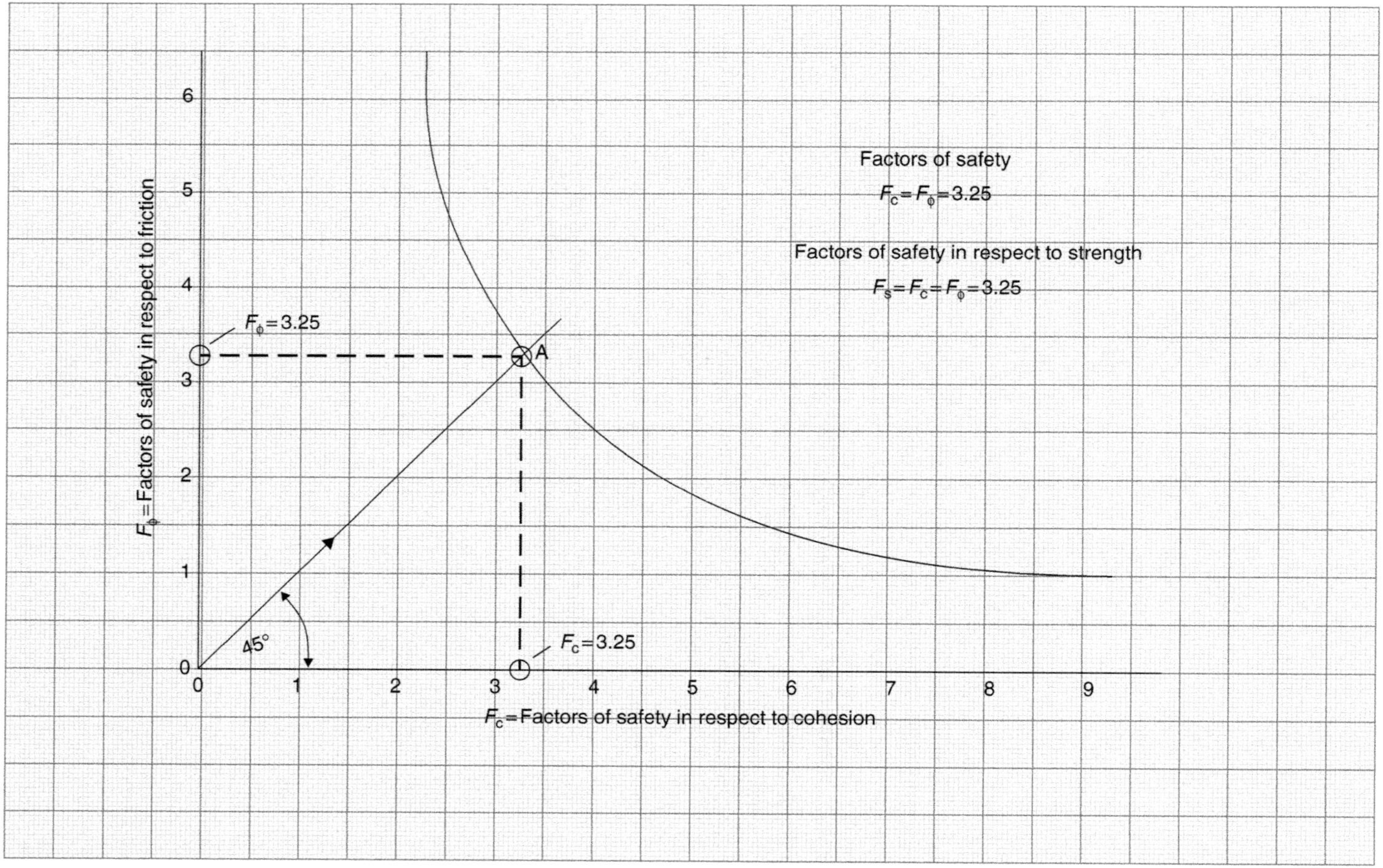

Factors of safety
$F_c = F_\phi = 3.25$
Factors of safety in respect to strength
$F_s = F_c = F_\phi = 3.25$
$F_\phi = 3.25$
A
45°
$F_c = 3.25$
0
1
2
3
4
5
6
7
8
9
F_c = Factors of safety in respect to cohesion
F_ϕ = Factors of safety in respect to friction

Graph 10.5

2nd Approximation (n=2)

Try $F_\phi = 2$ $\qquad \phi_m = \dfrac{15}{2} = 7.5°$

The stability factor is found by interpolation on Chart 10.2.

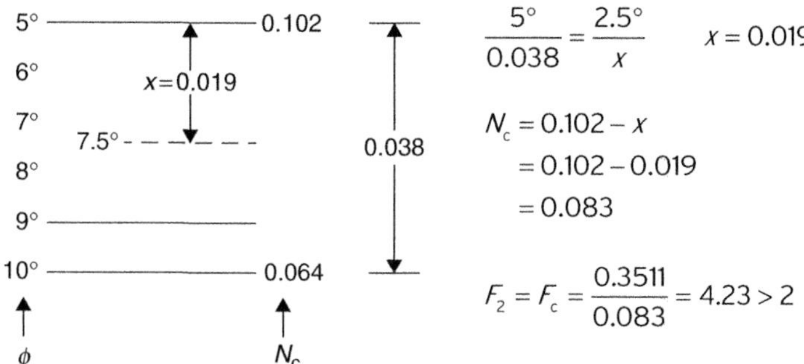

Figure 10.25

F_c is still larger than F_ϕ, therefore try F_ϕ=3 and tabulate all subsequent tries and calculations.

Table 10.6

Tries n	F_ϕ	$\phi^o_m = \dfrac{15}{F_\phi}$	N_c	$F_c = \dfrac{0.3511}{N_c}$
0	1	15	0.038	$9.24 > F_\phi$
1	1.5	10	0.064	$5.48 > F_\phi$
2	2	7.5	0.083	$4.23 > F_\phi$
3	3	5	0.102	$3.44 > F_\phi$
4	3.2	4.69	0.105	$3.34 > F_\phi$
5	3.3	4.54	0.107	$3.28 \approx F_\phi$

The 5th approximation yields $F_s \approx F_\phi \approx F_c = 3.28$.
Again, the result compares well with 3.19 and 3.25.

Example 10.9

Referring to Example 10.1 (Graph 10.1), calculate the factor of safety.

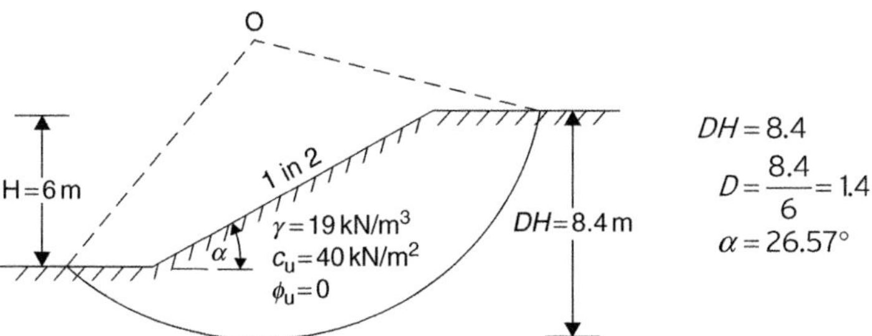

Figure 10.26

N_c is obtained by interpolation on Chart 10.2.

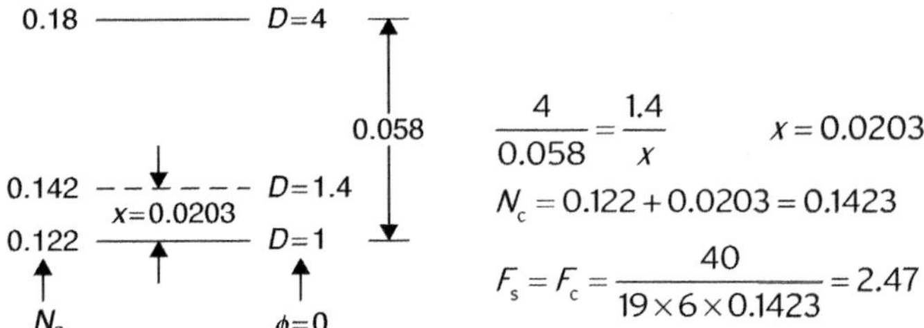

$$\frac{4}{0.058} = \frac{1.4}{x} \qquad x = 0.0203$$

$$N_c = 0.122 + 0.0203 = 0.1423$$

$$F_s = F_c = \frac{40}{19 \times 6 \times 0.1423} = 2.47$$

Figure 10.27

This value compares well with F_s=2.35 of Example 10.1.
Note: This method cannot take tension cracks into account.

Example 10.10

Suppose the slope in Example 10.9 is part of a reservoir. Estimate the factor of safety if:

1. The slope is flooded to its crest and compare it with the result of Example 10.6.
2. There is complete, sudden drawdown.

1. Flooded state: In this case the submerged density is applied because of buoyancy.

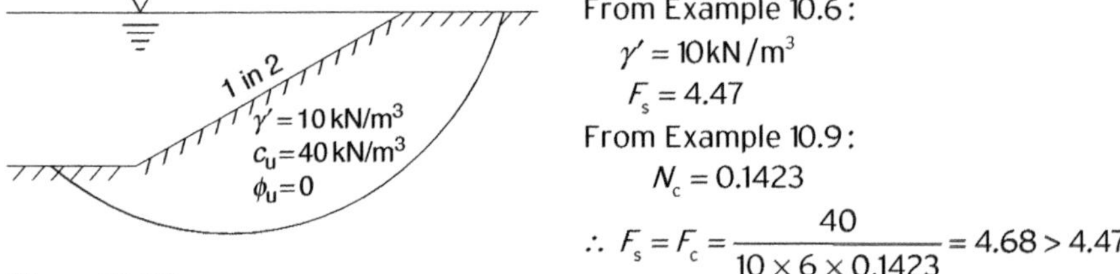

From Example 10.6:
$\gamma' = 10\text{kN/m}^3$
$F_s = 4.47$
From Example 10.9:
$N_c = 0.1423$

$$\therefore\ F_s = F_c = \frac{40}{10 \times 6 \times 0.1423} = 4.68 > 4.47$$

Figure 10.28

2. Sudden drawdown: in this case, buoyancy is absent, but the soil is still fully saturated. Therefore, the saturated density (γ_{sat}) is applicable.

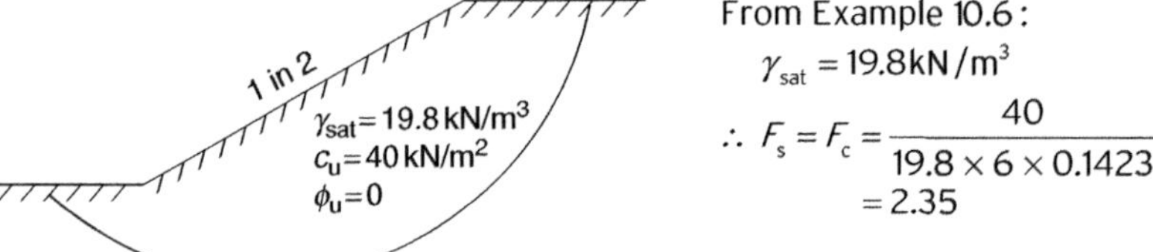

From Example 10.6:
$\gamma_{sat} = 19.8\text{kN/m}^3$

$$\therefore\ F_s = F_c = \frac{40}{19.8 \times 6 \times 0.1423} = 2.35$$

Figure 10.29

Example 10.11

The height (H) of the slope in Example 10.9 has to be increased. Estimate its maximum value, if the factor of safety specified for the slope is 1.5.

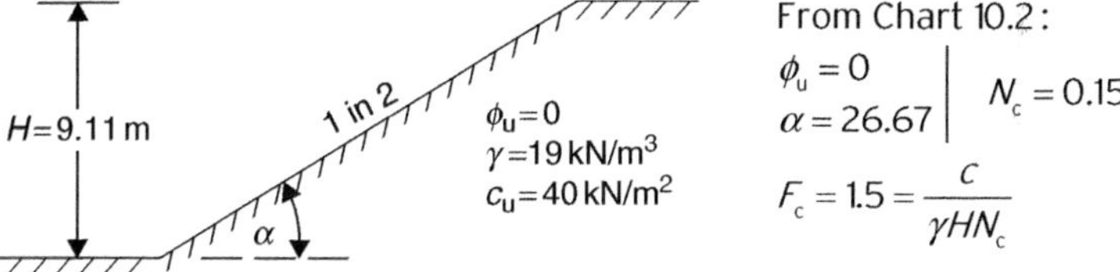

From Chart 10.2:

$\phi_u = 0$, $\alpha = 26.67$ | $N_c = 0.154$

$$F_c = 1.5 = \frac{c}{\gamma H N_c}$$

Figure 10.30

Expressing $H = \dfrac{c}{1.5\gamma N_c} = \dfrac{40}{1.5 \times 19 \times 0.154} = 9.11\,\text{m}$

10.3 Effective stress analysis (cohesive soils)

The total stress analysis is usually used to the estimation of short-term stability of slopes, in recently constructed embankments or new cuttings, in fully or partially saturated normally consolidated clays, where the only force inducing instability is weight and there is no change in pore water pressure.

The effective stress analysis is applicable to the long-term stability of slopes in normally as well as over consolidated clays, where ground water is present.

10.3.1 Method of slices (radial procedure)

In addition to the procedures applied in the total stress analysis, the pore pressure at the base of each slice has to be taken into account.

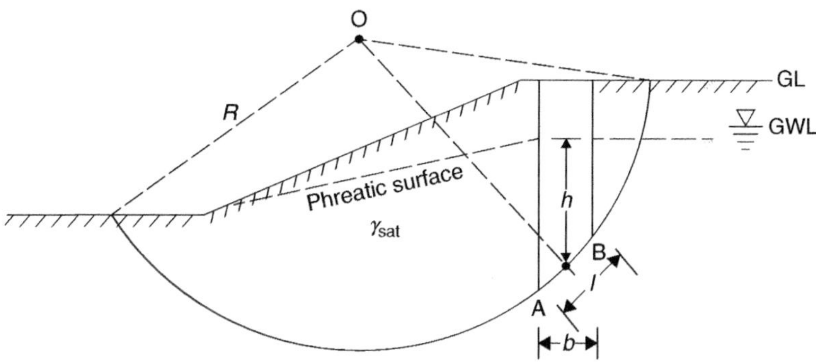

Figure 10.31

where b = width of slice

l = approximate length of arc $\widehat{AB}$ = straight line between points A and B.

h = height of water above the middle of the arc $\widehat{AB}$.

Forces on a strip

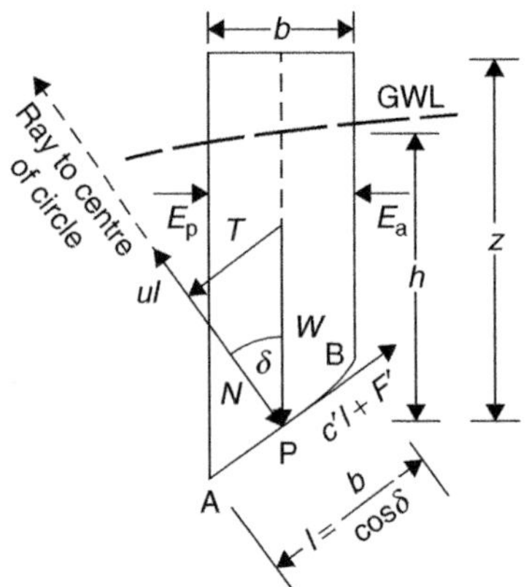

Figure 10.32

Lateral, opposing forces E_a and E_p are nearly equal, hence their effect is negligible.

Normal component of W:

$$N = W\cos\delta$$

Total pressure at P:

$$\sigma = \frac{N}{l} = \frac{W\cos\delta}{l}$$

Pore pressure at P:

$$u = h\gamma_w$$

Effective pressure at P: $\sigma' = \frac{N'}{l} = \frac{N}{l} - u$

Effective normal force on slip surface $\widehat{AB}$: $N' = N - ul$
$= W\cos\delta - ul$

Effective friction force: $F' = N'\tan\phi'$

Tangential force on $\widehat{AB}$: $T = W\sin\delta$

Resisting force due to cohesion: $S = \sum c'l$

Forces acting on the slip surface

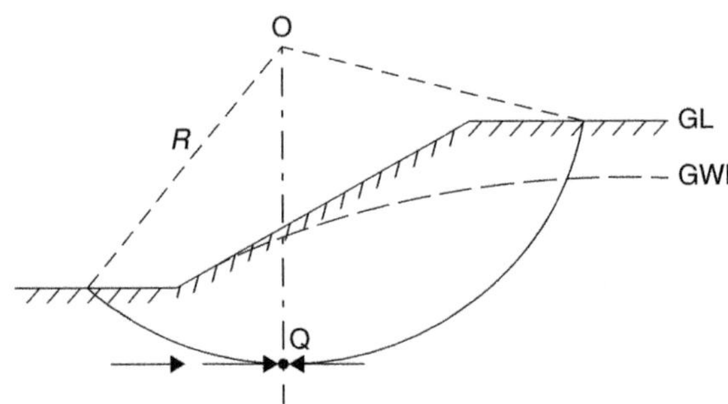

Figure 10.33

Disturbing moment:

$$M_D = R\Sigma T$$

Resisting moment:

$$M_R = R(S + \Sigma F')$$

Factor of safety: $$F_s = \frac{R(S+\Sigma F')}{R\Sigma T} = \boxed{\frac{S+\Sigma F'}{\Sigma T}} \qquad (10.23)$$

$$\boxed{F_s = \frac{S + (\Sigma N')\tan\phi'}{\Sigma T}} \tag{10.24}$$

$$= \frac{\Sigma c'l + \Sigma\left[(N - ul)\tan\phi'\right]}{\Sigma T}$$

Therefore,

$$\boxed{F_s = \frac{\Sigma\left[c'l + (W\cos\delta - ul)\tan\phi'\right]}{\Sigma T}} \tag{10.25}$$

For homogeneous clay: $\Sigma c'l = c'L$ and

$$\boxed{F_s = \frac{c'L + \Sigma(W\cos\delta - ul)\tan\phi'}{\Sigma T}} \tag{10.26}$$

Example 10.12

With reference to Example 10.5, assume that the shear strength parameters, in terms of effective stress, are:

$$c' = 40\,\text{kN/m}^2$$
$$\phi' = 15°$$
$$\gamma_{sat} = 19.8\,\text{kN/m}^3$$

The slope is assumed to be saturated above the phreatic surface, due to capillary action. Calculate the factor of safety.

Step 1: Measure the height (h) of water level above the slip surface in each slice on Graph 10.6.

Step 2: Calculate the weight of each slice and the pore pressure of the base from:

$$W = zb\gamma_{sat} = 19.8z$$
$$u = h\gamma_w = 9.81\,h$$

Note that $l = \dfrac{b}{\cos\delta} = \dfrac{1}{\cos\delta}$ in this example! Since $b = 1\,\text{m}$.

Step 3: For each slice, measure angle δ and calculate:

$$\pm T = W\sin\delta$$
$$N = W\cos\delta$$
$$N' = N - ul = N - \frac{ub}{\cos\delta} = N - \frac{u}{\cos\delta}$$
$$F' = N'\tan\phi' = 0.2679N'$$

Step 4: Tabulate the results so far:

Table 10.7

Slice						Components of W				Friction
	z	h	W	u	δ	$+T$	$-T$	N	N'	F'
No.	m	m	kN	kN/m²	deg.	kN	kN	kN	kN	kN
1	0.14	0.14	2.77	1.37	–31.8		–1.46	2.35	0.74	0.20
2	0.70	0.70	13.86	6.87	–27.0		–6.29	12.34	4.63	1.24
3	1.41	1.41	27.92	13.83	–22.7		–10.77	25.76	10.77	2.88
4	2.27	2.27	44.95	22.27	–17.5		–13.52	42.87	19.52	5.23
5	3.05	3.05	60.39	29.92	–13.0		–13.58	58.84	28.13	7.54
6	3.75	3.72	74.25	36.49	–9.0		–11.62	73.34	36.40	9.75
7	4.37	4.30	86.53	42.18	–4.5		–6.79	86.26	43.95	11.77
8	4.91	4.70	97.22	46.11	–1.0		–1.70	97.21	51.09	13.69
9	5.38	4.90	106.52	48.07	4.0	7.43		106.26	58.07	15.56
10	5.78	5.05	114.44	49.54	8.0	15.93		113.33	61.81	16.56
11	6.09	5.05	120.58	49.54	12.5	26.10		117.72	66.98	17.94
12	6.33	4.95	125.33	48.56	17.0	36.64		119.85	69.07	18.50
13	6.49	4.80	128.50	47.01	21.5	47.10		119.56	69.03	18.49
14	6.55	4.50	129.69	44.15	26.0	56.85		116.56	67.44	18.07
15	6.26	4.10	123.95	40.22	31.0	63.84		106.25	59.33	15.89
16	5.61	3.55	111.08	34.83	35.8	65.40		90.09	47.15	12.63
17	4.81	2.85	95.24	27.96	41.3	62.86		71.55	34.33	9.20
18	3.84	1.93	76.03	18.93	47.0	55.60		51.85	24.09	6.45
19	2.62	0.81	53.06	7.95	54.0	42.93		31.19	17.66	4.73
20	1.03	0	20.39	0	62.0	18.0		9.57	9.57	2.56
Rounded ΣW = 1613						499	–66			ΣF = 209
						ΣT = 433				

Step 5: Calculate the resisting force contributed by the cohesion: $c' = 40\,\text{kN/m}^2$:

$S = c'R = 40 \times 23.8 = \underrightarrow{952}\,\text{kN}$

Step 6: Determine the factor of safety from the forces, assumed to be concentrated at point Q.

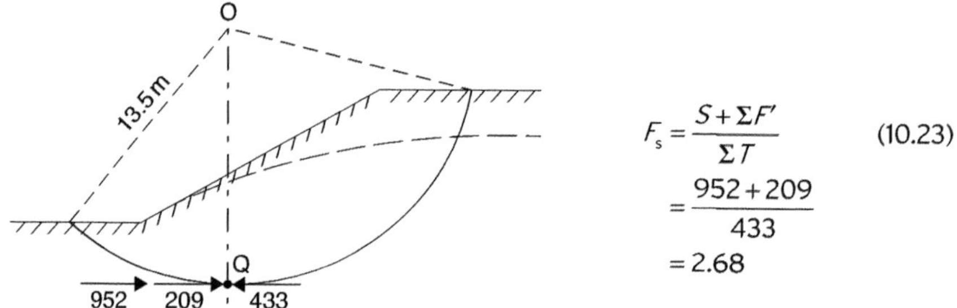

$$F_s = \frac{S + \Sigma F'}{\Sigma T} \quad (10.23)$$

$$= \frac{952 + 209}{433}$$

$$= 2.68$$

Figure 10.34

Any of the other formulae may be applied, if preferred.

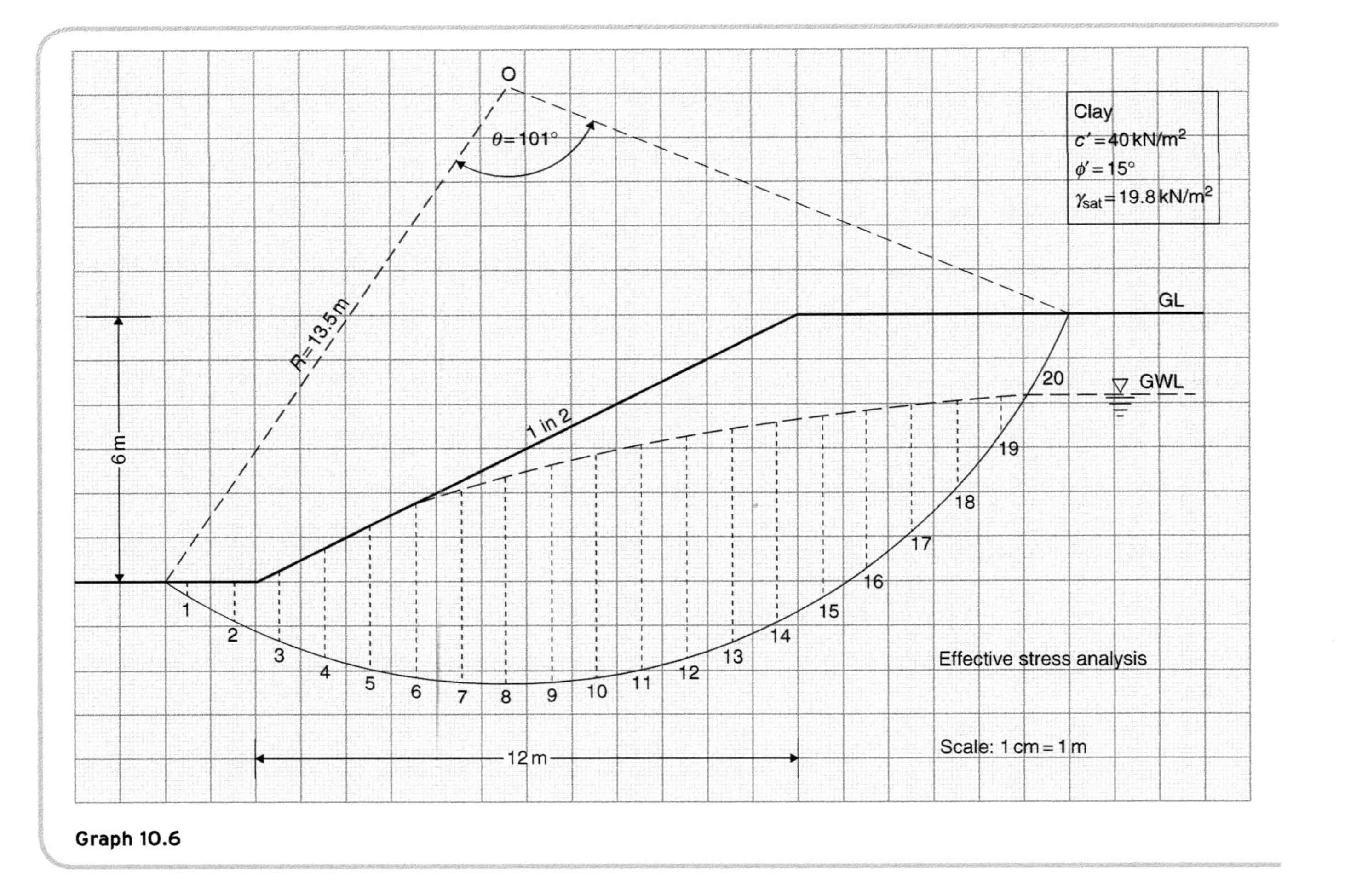
O
θ = 101°
R = 13.5 m
Clay
c′ = 40 kN/m²
φ′ = 15°
γsat = 19.8 kN/m²
GL
GWL
1 in 2
6 m
1
2
3
4
5
6
7
8
9
10
11
12
13
14
15
16
17
18
19
20
Effective stress analysis
Scale: 1 cm = 1 m
12 m

Graph 10.6

10.3.2 Bishop's conventional method

In this method the pore water pressure (u) is expressed as a ratio of the total stress

Pore pressure ratio: $$r_u = \frac{u}{\gamma z} \quad (10.27)$$

$$u = \gamma z r_u$$

But, the total stress at the base of a slice is:

$$s = \gamma z = \frac{W}{b} \qquad \text{hence,} \qquad u = \frac{W}{l} r_u$$

But, $l = b\cos\delta$ therefore, $u = \frac{W}{l\cos\delta} \times r_u$

and, $ul = Wr_u \sec\delta$

Substituting this into formula (10.26):

$$F_s = \frac{1}{\Sigma T} \Sigma\left[c'l + (W\cos\delta - W r_u \sec\delta)\tan\phi'\right]$$

Substituting $\Sigma T = W\sin\delta$, the factor of safety can be expressed as:

$$F_s = \frac{1}{\Sigma W\sin\delta} \Sigma\left[c'l + W\Sigma\left(\cos\delta - \frac{r_u}{\cos\delta}\right)\right]\tan\phi' \quad (10.28)$$

Example 10.13

Calculate F_s in Example 10.12 in terms of r_u.

Table 10.8

	Slice (from Table 10.7)						$\pm T$	F'
	z	h	W	u	δ	$r_u = \frac{u}{W}$	$W\sin\delta$	$W\left(\cos\delta - \frac{r_u}{\cos\delta}\right)\tan\phi'$
No.	m	m	kN	kN/m²	deg.	–	kN	kN
1	0.14	0.14	2.77	1.37	–31.8	0.494	–1.46	0.20
2	0.70	0.70	13.86	6.87	–27.0	0.496	–6.29	1.24
3	1.41	1.41	27.92	13.83	–22.7	0.495	–10.77	2.88
4	2.27	2.27	44.95	22.27	–17.5	0.495	–13.52	5.23
5	3.05	3.05	60.39	29.92	–13.0	0.495	–13.58	7.54
6	3.75	3.72	74.25	36.49	–9.0	0.491	–11.62	9.75
7	4.37	4.30	86.53	42.18	–4.5	0.487	–6.79	11.77
8	4.91	4.70	97.22	46.11	–1.0	0.474	–1.70	13.69
9	5.38	4.90	106.52	48.07	4.0	0.451	7.43	15.56
10	5.78	5.05	114.44	49.54	8.0	0.433	15.93	16.96
11	6.09	5.05	120.58	49.54	12.5	0.411	26.10	17.94
12	6.33	4.95	125.33	48.56	17.0	0.387	36.64	18.51
13	6.49	4.80	128.50	47.01	21.5	0.365	47.10	18.47
14	6.55	4.50	129.69	44.15	26.0	0.340	56.85	18.07
15	6.26	4.10	123.95	40.22	31.0	0.324	63.84	15.89

Table 10.8 (continued)

	Slice (from Table 10.7)						$\pm T$	F'
	z	h	W	u	δ	$r_u = \frac{u}{W}$	$W\sin\delta$	$W\left(\cos\delta - \frac{r_u}{\cos\delta}\right)\tan\phi'$
No.	m	m	kN	kN/m²	deg.	–	kN	kN
16	5.61	3.55	111.08	34.83	35.8	0.314	65.40	12.63
17	4.81	2.85	95.24	27.96	41.3	0.294	62.86	9.20
18	3.84	1.93	76.03	18.93	47.0	0.249	55.60	6.46
19	2.62	0.81	53.06	7.95	54.0	0.153	42.93	4.65
20	1.03	0	20.39	0	62.0	0	18.0	2.56
			1613				Σ 433	$\Sigma F'$=209.00

Again, $\Sigma T = 433\,\text{kN}$, $\Sigma F' = 209\,\text{kN}$, $S = 952\,\text{kN}$

$$F_s = \frac{952+209}{433} = 2.68 \text{ (as before)}$$

10.3.3 Bishop's rigorous iterative method

This procedure is somewhat tedious for hand computation, because it requires several iterations before the final value of F_s is found by trial and error. For this reason, problems are best solved by means of a computer. The derivation of the relevant formula is based on (10.23):

$$F_s = \frac{S + \Sigma F'}{\Sigma T} = \frac{\Sigma c'l + \Sigma(N'\tan\phi')}{\Sigma T}$$

Considering the forces acting on a slice, N' is expressed and substituted into this formula.

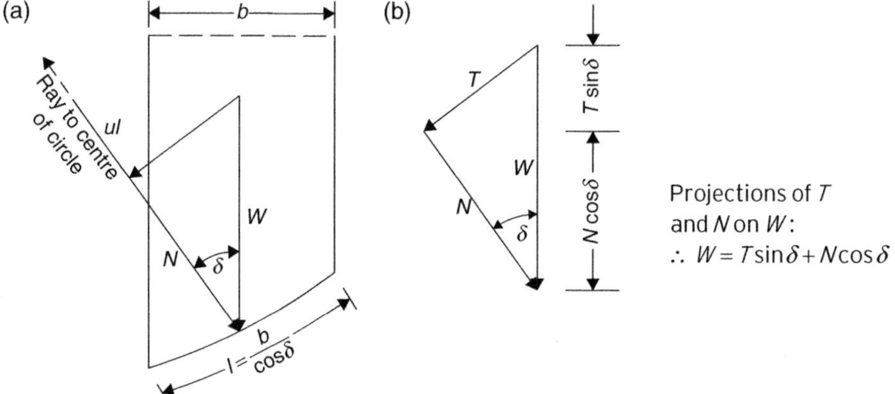

Figure 10.35

From (10.24) for one slice: $T = \dfrac{c'l + N'\tan\phi'}{F_s}$

Force due to pore pressure $= ul$

Effective normal force: $N' = N - ul \quad \therefore N = N' + ul$

Substituting into

$$
\begin{aligned}
W &= T\sin\delta + N\cos\delta \\
&= \frac{(c'l + N'\tan\phi')\sin\delta}{F_s} + (N' + ul)\cos\delta \\
&= N'\left(\frac{\tan\phi'\sin\delta}{F_s} + \cos\delta\right) + \frac{c'l\sin\delta}{F_s} + ul\cos\delta
\end{aligned}
$$

Expressing:

$$N' = \frac{W - \dfrac{c'l\sin\delta}{F_s} - ul\cos\delta}{\dfrac{\tan\phi'\sin\delta}{F_s} + \cos\delta}$$

But, $l = \dfrac{b}{\cos\delta}$ and $r_u = \dfrac{u}{\gamma z} \quad \therefore \quad u = r_u\gamma z$

also, $W = \gamma zb \quad \therefore \quad u = r_u\dfrac{W}{b} \quad \therefore \quad ul = r_u\dfrac{W}{\cos\delta}$

Substituting u and l into the expression for N':

$$N' = \frac{W - \dfrac{c'b\sin\delta}{F_s\cos\delta} - \left(\dfrac{r_u W}{\cos\delta}\right)\cos\delta}{\dfrac{\tan\phi'\sin\delta}{F_s} + \cos\delta}$$

Cancelling $\cos\delta$ and arranging:

$$N' = \frac{W(1 - r_u) - \dfrac{c'b}{F_s}\tan\delta}{\left(\dfrac{\tan\phi'\tan\delta}{F_s} + 1\right)\cos\delta}$$

Substituting N' and l into formula (10.25), thus summing the effective normal force of all elements.

$$F_s = \frac{1}{\Sigma T}\left[\frac{c'b}{\cos\delta} + \frac{W(1 - r_u)\tan\phi' - \dfrac{c'b}{F_s}\tan\delta\tan\phi'}{\left(\dfrac{\tan\phi'\tan\delta}{F_s} + 1\right)\cos\delta}\right]$$

This expression can be simplified further.

$$F_s = \frac{1}{\Sigma T}\Sigma\left[\frac{c'b\left(\dfrac{\tan\phi'\tan\delta}{F_s}+1\right)+W(1-r_u)\tan\phi-\dfrac{c'b}{F_s}\tan\phi'\tan\delta}{\left(\dfrac{\tan\phi'\tan\delta}{F_s}+1\right)\cos\delta}\right]$$

$$F_s = \frac{1}{\Sigma T}\left[\frac{\dfrac{c'b}{F_s}\tan\phi'\tan\delta+c'b+W(1-r_u)\tan\phi'-\dfrac{c'b}{F_s}\tan\phi'\tan\delta}{\left(\dfrac{\tan\phi'\tan\delta}{F_s}+1\right)\cos\delta}\right]$$

Cancelling $\frac{c'b}{F_s}\tan\phi'\tan\delta$, the final expression is obtained.

$$F_s = \frac{1}{\Sigma T}\Sigma\left[\frac{c'b+W(1-r_u)\tan\phi'}{\left(\dfrac{\tan\phi'\tan\delta}{F_s}+1\right)\cos\delta}\right] \tag{10.29}$$

Problems are solved by choosing a value for F_s on the left hand side and substituting it into the right hand one. If the two sides are unequal, then substitute the calculated F_s into the right side. Repeat the process, until the two figures are practically the same. Normally, the process converges rapidly to the solution, as seen in the next example.

Example 10.14

Re-evaluate the factor of safety in Example 10.13, by means of the rigorous method. Reproduce in Table 10.9, W, r_u and δ from Table 10.8 and define two auxiliary variables as:

$$K = c'b + W(1-r_u)\tan\phi' = 40 + W(1-r_u)\tan 15 = 40 + 0.2679W(1-r_u)$$

$$Q = \left(\frac{\tan\phi'\tan\delta}{F_s}+1\right)\cos\delta = \left(\frac{0.2679\tan\delta}{F_s}+1\right)\cos\delta$$

Also, $\Sigma T = 433$ kN remains the same.

Formula (10.29) now becomes: $F_s = \frac{1}{\Sigma T}\Sigma\left(\frac{K}{Q}\right) = \frac{1}{433}\Sigma\left(\frac{K}{Q}\right)$

Table 10.9

Slice No.	W kN	r_u	δ	K	ITERATION 1st try F_s=1.5 Q	K/Q	2nd try F_s=2.64 Q	K/Q	3rd try F_s=2.75 Q	K/Q
1	2.77	0.494	−31.8	40.38	0.944	42.77	0.903	44.69	0.901	44.80
2	13.86	0.496	−27.0	41.87	0.972	43.08	0.937	44.68	0.935	44.70
3	27.92	0.495	−22.7	43.78	0.991	44.17	0.962	45.52	0.960	45.59
4	44.95	0.495	−17.5	46.08	1.007	45.76	0.984	46.82	0.983	46.88
5	60.39	0.495	−13.0	48.17	1.002	47.36	0.987	48.31	0.996	48.35
6	74.25	0.491	−9	50.13	1.016	49.34	1.004	49.95	1.003	49.98
7	86.53	0.487	−4.5	51.89	1.011	51.33	1.004	51.64	1.005	51.65
8	97.22	0.474	−1.0	53.70	1.003	53.54	1.002	53.61	1.002	53.62
9	106.52	0.451	4.0	55.67	1.001	55.12	1.005	55.41	1.004	55.43
10	114.44	0.433	8.0	56.93	1.015	56.09	1.005	56.68	1.004	56.71
11	120.58	0.411	12.5	59.03	1.015	58.15	0.998	59.13	0.997	59.18
12	125.33	0.387	17.0	60.58	1.009	60.04	0.986	61.44	0.985	61.52
13	128.50	0.366	21.5	61.83	0.996	62.08	0.968	63.90	0.966	64.00
14	129.69	0.340	26.0	62.93	0.977	64.41	0.943	66.71	0.942	66.84
15	123.95	0.324	31.0	62.45	0.949	65.30	0.909	68.67	0.907	63.84
16	111.08	0.314	35.8	60.41	0.916	65.95	0.870	69.41	0.868	69.60
17	95.24	0.294	41.3	58.01	0.869	66.76	0.818	70.90	0.816	71.13
18	76.03	0.249	47.0	55.30	0.813	68.02	0.756	73.13	0.753	73.41
19	53.06	0.153	54.0	52.04	0.732	71.09	0.670	77.68	0.667	78.07
20	20.39	0	62.0	45.46	0.627	72.50	0.559	81.31	0.556	81.83
						Σ 1143.36		Σ 1189.59		Σ 1187.20
					$F_s = \frac{1143}{433} = 2.64 > 1.5$		$F_s = \frac{1190}{433} = 2.75 > 2.64$		$F_s = \frac{1187}{433} = 2.74 \approx 2.75$	
Therefore, the accepted result is: F_s=2.74.					Try F_s=2.64		Try F_s=2.75			

10.4 Stability of infinite slopes

Often, the failure surface is not curved, but parallel with the sloping ground surface at shallow depth. The general formula for the factor of safety is derived by assuming that:

1. The ground water level is parallel to the slope
2. Steady seepage occurs parallel to the slope
3. The forces (E_a and E_p) acting on the vertical sides of a slice are equal and opposite.
4. The failure surface is at depth z.

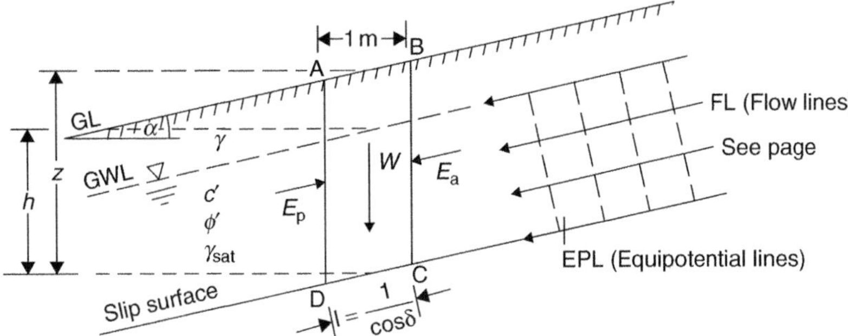

Figure 10.36

Free-body diagram of the slice ABCD shows the forces acting on the slip surface, neglecting E_a and E_p:

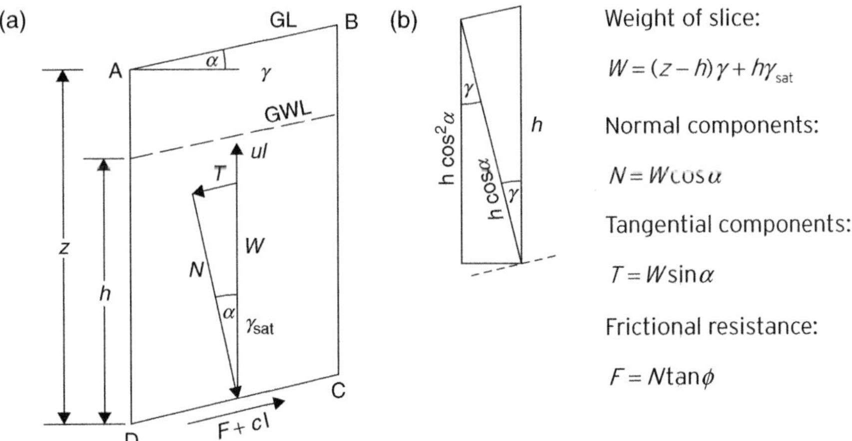

Weight of slice:

$$W = (z - h)\gamma + h\gamma_{sat}$$

Normal components:

$$N = W\cos\alpha$$

Tangential components:

$$T = W\sin\alpha$$

Frictional resistance:

$$F = N\tan\phi$$

Figure 10.37

Pore pressure: $$u = \gamma_w h\cos^2\alpha$$

Normal pressure: $$\sigma = \frac{N}{\frac{1}{\cos}} = N\cos\alpha = W\cos^2\alpha$$

Shear stress: $\tau = \dfrac{T}{\frac{1}{\cos}} = T\cos\alpha = W\sin\alpha\cos\alpha$

The factor of safety is given by:

$$\boxed{F_s = \frac{c' + \sigma'\tan\phi'}{c_m + \sigma'\tan\phi'}} = \frac{\tau}{\tau_m} \tag{10.30}$$

where c_m and ϕ_m are the mobilized shear stress parameters.

Effective stress:
$$\begin{aligned}\sigma' &= \sigma - u \\ &= W\cos^2\alpha - \gamma_w h\cos^2\alpha \\ &= [\gamma(z-h) + \gamma_{sat}h - \gamma_w h]\cos^2\alpha \\ &= [\gamma(z-h) + \gamma' h]\cos^2\alpha\end{aligned}$$

Shear stress mobilized:

$$\tau_m = c_m + \left[\gamma(z-h) + \gamma' h\right]\cos^2\alpha\tan\phi_m$$

From formula (8.61): $F_s = F_\phi = \dfrac{\tan\phi'}{\tan\phi_m}$ $\quad\therefore\quad$ $\tan\phi_m = \dfrac{\tan\phi'}{F_s}$

Also, from (10.15): $F_s = \dfrac{c'}{c_m}$ $\quad\therefore\quad$ $c_m = \dfrac{c'}{F_s}$

Remember! $F_s = F_\phi = F_c$

Substituting these into the expression for τ_m:

$$\tau_m = \frac{c'}{F_s} + \left[\gamma(z-h) + \gamma' h\right]\cos^2\alpha\frac{\tan\phi'}{F_s}$$

Equating, $\tau_m = \tau = W\sin\alpha\cos\alpha$

$$= \left[\gamma(z-h) + \gamma_{sat}h\right]\sin\alpha\cos\alpha$$

$$\frac{c'}{F_s} + \left[\gamma(z-h) + \gamma' h\right]\frac{\cos^2\alpha\tan\phi'}{F_s} = \left[\gamma(z-h) + \gamma_{sat}h\right]\sin\alpha\cos\alpha$$

$$\frac{1}{F_s}\left\{c' + \left[\gamma(z-h) + \gamma' h\right]\cos^2\alpha\tan\phi'\right\} = \left[\gamma(z-h) + \gamma_{sat}h\right]\sin\alpha\cos\alpha$$

Hence, the factor of safety is given by:

$$\boxed{F_s = \frac{c' + \left[\gamma(z-h) + \gamma' h\right]\cos^2\alpha\tan\phi'}{\left[\gamma(z-h) + \gamma_{sat}h\right]\sin\alpha\cos\alpha}} \tag{10.31}$$

This general formula is applicable to drained, infinite slopes, made of $c-\phi$ soil, where $h<z$.

There are five special cases to consider:

1. $h = z$ | Water table at GL.
   $c' - \phi'$ soil

$$F_s = \frac{c' + \gamma' z \cos^2\alpha \tan\phi'}{\gamma_{sat}\, z \sin\alpha \cos\alpha} \tag{10.32}$$

2. $h = z$ | Water table at GL.
   $c' = 0$ | cohesionless soil

$$F_s = \left[\frac{\gamma'}{\gamma_{sat}}\right] \times \frac{\tan\phi'}{\tan\alpha} \tag{10.33}$$

3. $h = 0$ | There is no water table.
   $c' - \phi'$ soil

$$F_s = \frac{c' + \gamma z \cos^2\alpha \tan\phi'}{\gamma z \sin\alpha \cos\alpha} \tag{10.34}$$

4. $h = 0$ | There is no water table.
   $c' = 0$ | cohesionless soil.

$$F_s = \frac{\gamma z \cos^2\alpha \tan\phi}{\gamma z \sin\alpha \cos\alpha} = \frac{\cos\alpha \tan\phi}{\sin\alpha}$$

Therefore,

$$F_s = \frac{\tan\phi'}{\tan\alpha} \tag{10.35}$$

Note that F_s is independent of depth when $c' = 0$.

5. $h = 0$ | There is no water table.
   $\phi = 0$ | Pure clay
   $c' = c_u$

$$F_s = \frac{c_u}{\gamma z \sin\alpha \cos\alpha} \tag{10.36}$$

Example 10.15

Figure 10.30 shows an infinite slope, inclined at an angle of 25° to the horizontal. The slope is underlain by solid rock. Calculate the factor of safety, when the water table is:

1. at the ground surface ($S_r = 1$)
2. at 2 m below the ground surface
3. non-existent ($S_r = 0.46$)

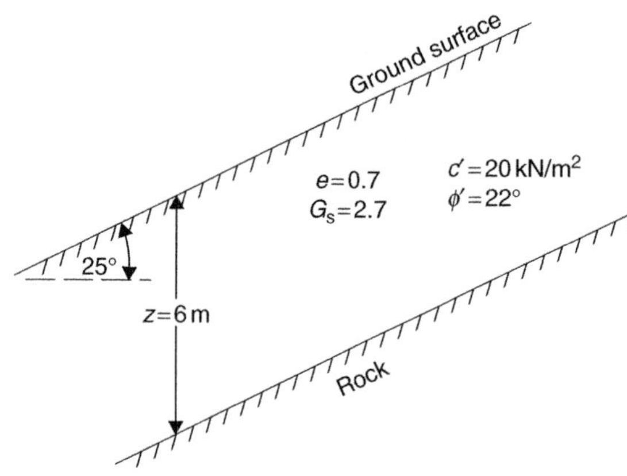

Figure 10.38

From (1.42):

$$\gamma_{sat} = \left(\frac{G_s + e}{1+e}\right)\gamma_w$$
$$= \frac{2.7+0.7}{1.7}\times 9.81$$
$$= 19.62\,\text{kN/m}^3$$

From (1.38):

$$\gamma = \left(\frac{G_s + S_r e}{1+e}\right)\gamma_w$$
$$= \frac{2.7+0.46\times 0.7}{1.7}\times 9.81$$
$$= 17.4\,\text{kN/m}^3$$

1. *Water at ground surface (S_r=1 and h=z=6 m)*

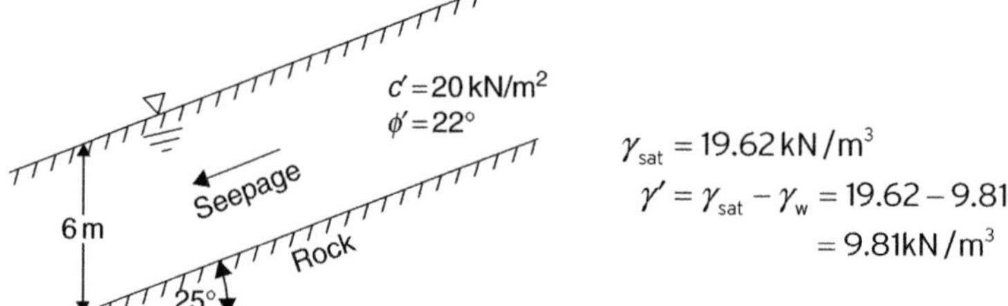

Figure 10.39

$$\gamma_{sat} = 19.62\,\text{kN/m}^3$$
$$\gamma' = \gamma_{sat} - \gamma_w = 19.62 - 9.81$$
$$= 9.81\,\text{kN/m}^3$$

From (10.32): $F_s = \dfrac{c' + \gamma' z\cos^2\alpha\tan\phi'}{\gamma_{sat} z\sin\alpha\cos\alpha} = \dfrac{20+9.81\times 6\times 0.821\times 0.404}{19.62\times 6\times 0.423\times 0.906}$

$= 0.88 < 1 \quad \therefore$ unsafe

2. *Water is at 2 m below the ground surface (h=4 m)*

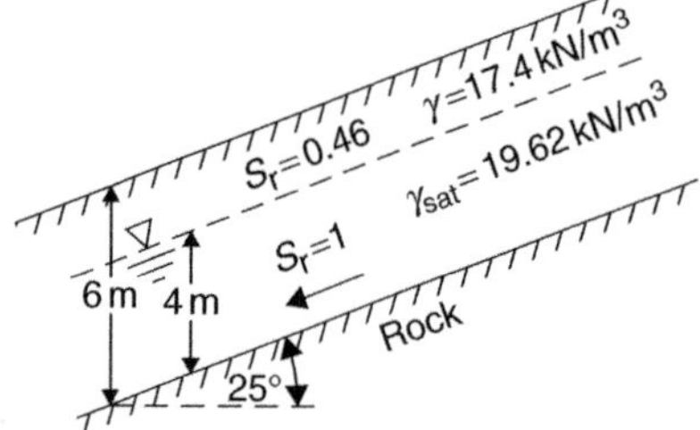

Figure 10.40

The soil is assumed to be partially saturated above the surface, ignoring capillary action.

From (10.31): $$F_s = \frac{c' + [\gamma(z-h) + \gamma' h]\cos^2\alpha \tan\phi'}{[\gamma(z-h) + \gamma_{sat} h]\sin\alpha\cos\alpha}$$

$$= \frac{20 + [17.4 \times (6-4) + 9.81 \times 4] \times 0.821 \times 0.404}{[17.4 \times (6-4) + 19.62 \times 4] \times 0.423 \times 0.906} = 1.02$$

3. *Wet soil ($S_r = 0.46$ and $h = 0$)*

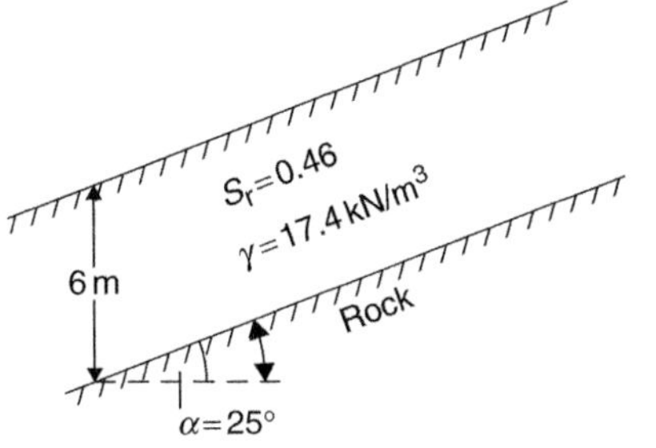

Figure 10.41

From (10.34):

$$F_s = \frac{c' + \gamma z \cos^2\alpha \tan\phi'}{\gamma z \sin\alpha\cos\alpha}$$

$$= \frac{20 + 17.4 \times 6 \times 0.821 \times 0.404}{17.4 \times 6 \times 0.423 \times 0.906}$$

$$= 1.37$$

Problem 10.1

A 4 m deep cutting is to be made in a 15 m thick pure clay layer, for a new road scheme. The inclination of the side slopes is to 1 in 2. The shear strength and unit weight of the clay is 35 kN/m² and 18 kN/m³ respectively.

Estimate, for the slip circle indicated in Figure 10.45, the factor of safety against rotational failure.

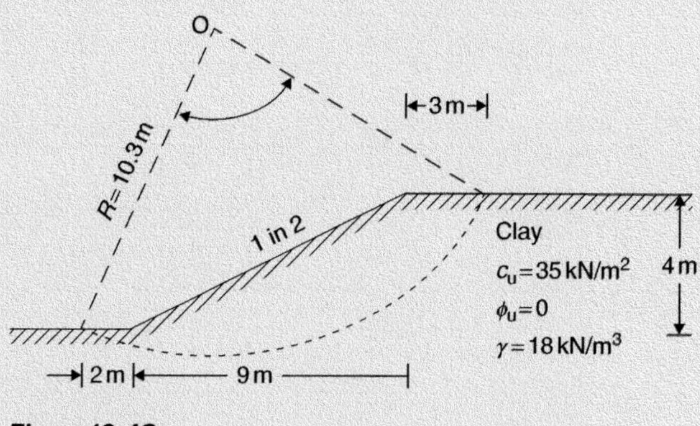

Figure 10.42

Problem 10.2

Figure 10.45 shows the section of a new canal to be built in a 20 m thick clay layer.

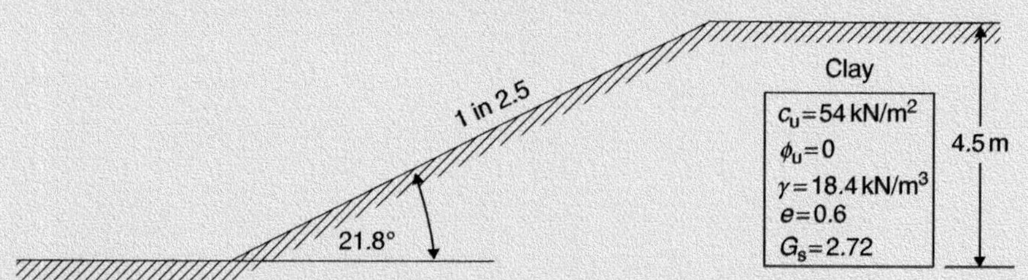

Figure 10.45

Estimate the factor of safety, just after:

a) Construction
b) It is flooded to the crest
c) Sudden drawdown

Apply Taylor's stability numbers.

Problem 10.3

A footpath is proposed to be constructed along a clay slope. In order to do this, the cut is to be supported by a gravity retaining wall. Its position and dimensions are shown below. There is no ground water table. Assume the stability of the wall as satisfactory. Estimate the factor of safety for the slip circle of 10.2 m radius, passing through points A and B on the slope:

a) In its original, undisturbed state
b) After the construction of the wall

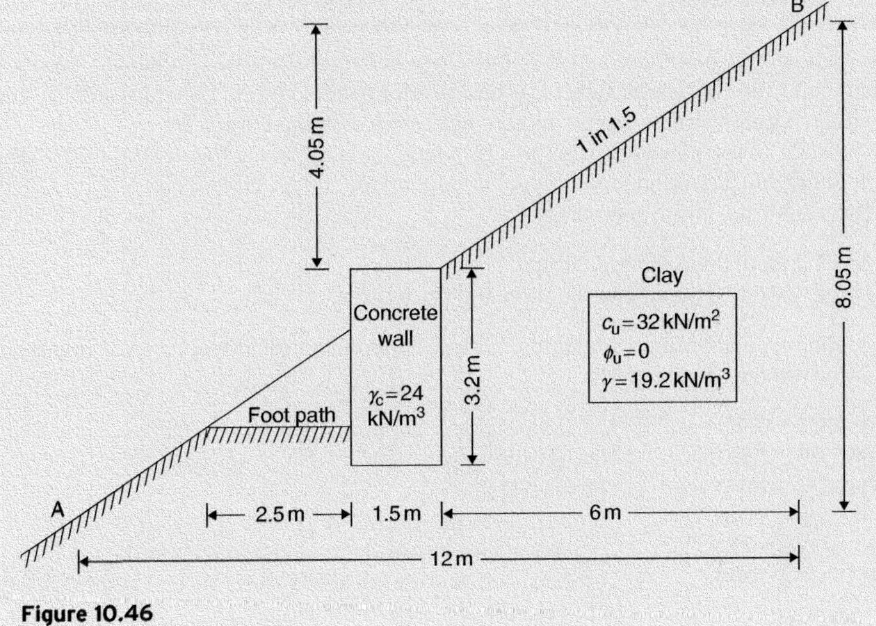

Figure 10.46

Chapter 11

Eurocode 7

11.1 Introduction

The basic purpose of Eurocode is to introduce common standards within the European community for consistent structural design. However, some of the standards of each member state will remain in use and are not covered by the Eurocodes.

The code concerned with structures is in nine parts, of which the seventh (EC7) deals with Geotechnical Design.

Eurocode 7 itself is in two parts:

Part 1 (EC7-1): Geotechnical design
Part 2 (EC7-2): Ground investigation and testing.

The following eight British Standards, relating to geotechnical aspects are not covered by EC7 and remain UK standards:

BS 1377-1: 1990:	Soil sample preparation
BS 1377-2: 1990:	Classification tests
BS 1377-3: 1990:	Chemical tests
BS 1377-4: 1990:	Compaction tests
BS 1377-5: 1990:	Compressibility, permeability and durability tests
BS 1377-6: 1990:	Consolidation and permeability tests in hydraulic cells and with pore pressure measurements
BS 1377-7: 1990:	Shear strength tests (total stress)
BS 1377-8: 1990:	Shear strength tests (effective stress)

11.2 Recommended units

For geotechnical calculations a slightly modified version of SI units are recommended.

Force	kN
Moment	kN m
Mass and weight density	kN/m^3
Stress pressure, strength and stiffness	kP$_a$
Coefficient of permeability	m/s
Coefficient of consolidation	m^2/s

Introduction to Soil Mechanics, First Edition. Béla Bodó and Colin Jones.

Note: $1\,kN/m^2 = 1\,kP_a$

The unit of pressure and stress is N/m^2 and its decimal multiple is kN/m^2 in the SI coherent system of units. In certain countries the name "Pascal" (Pa) was given to this unit in honour of Blaise Pascal, French mathematician and physicist.

Although the Eurocode recommends kP_a, this does not exclude the use of kN/m^2. However, it is prudent to choose the former, when a project has an international dimension.

11.3 Limit states

Eurocode sets out parameters used in a design so that the limit state considered is not exceeded.

There are five limit states, specified by EU7. These are:

1. EQU: Considering the equilibrium of a structure (e.g. overturning of a retaining wall).
2. GEO: Considering the failure of excessive deformation of the ground (e.g. slope stability or bearing strength of foundation soil).
3. STR: Considering the failure of a structure or its elements, due to soil pressure.
4. UPL: Considering the failure of a structure or ground, due to upward water pressure (e.g. dams or other water-retaining structures).
5. HYD: Considering hydraulic gradients in the ground causing piping and internal soil erosion (e.g. coffer dams, Earth dams etc): see Example 11.6.

Most geotechnical problems involve one or two of the first three limit states i.e. EQU, GEO and STR.

11.4 Design procedures

In general, a design applies partial factors for the determination of soil properties and 'actions' of forces and moments.

The recommended partial factors are given in Annex A, EC7.

Design approaches

These are defined in section 2 and Annex B of EC7. The general procedure is represented symbolically by:

$$\boxed{A\text{"+"}M\text{"+"}R} \tag{11.1}$$

[A] represents the partial factors applied for *actions* (e.g. earth pressure, weight of soil) or *effects of actions* (e.g. ground excavation).

[M] represents partial factors for soil or other material parameters (e.g. weight density, cohesion).

[R] represents partial factors for resistance to an action (e.g. passive force, friction etc).

["+"] implies 'to be combined with'.

There are three Design Approaches presented in section 2 and Annex B of EC7 for GEO and STR limit states:

Design approach 1: This itself has two methods or combinations of partial factors to ascertain that a structure is safe and that failure or large deformation is not going to occur. The combinations are given symbolically as:

$$\boxed{\begin{array}{ll}\text{Combination 1:} & \text{A1 "+" M1 "+" R1} \\ \text{Combination 2:} & \text{A2 "+" M2 "+" R1}\end{array}} \qquad \begin{array}{r}(11.2)\\(11.3)\end{array}$$

In these combinations, the partial factors are applied to actions (favourable or unfavourable) and to the soil parameters.

Design approach 2: (One combination)

Combination: $$\boxed{\text{A1 "+" M1 "+" R}_2} \qquad (11.4)$$

In this approach, the partial factors are applied to *actions* or *effects of actions* and ground resistances.

Design approach 3: (One combination)

Combination $$\boxed{(\text{A1 or A2}) \text{ "+" M2 "+" R3}} \qquad (11.5)$$

In this approach, the partial factors are applied to *actions*from the structure and to the soil parameters.

A1 is applied on structural actions
A2 is applied on geotechnical actions.

Design approach1 (for piles and anchorages)
There are two combinations:

$$\boxed{\begin{array}{ll}\text{Combination 1:} & \text{A1 "+" M1 "+" R1} \\ \text{Combination 2:} & \text{A2 "+" (M1 or M2) "+" R4}\end{array}} \qquad (11.6)$$

In Combination 1, the partial factors are applied to actions, and soil parameters. In Combination 2, the factors are applied to action, soil and ground strength parameters. Also, the factors of M1 are to determine the pile or anchor resistances and those of M2 to calculate the actions on piles due to unfavourable effects (transverse loads or negative skin friction).

Section 7.6.2.3(2) states that when the compressive resistance of a pile is determined from ground test results, a 'model factor' may be introduced as described in 2.4.1(9). The value of the model factor is given in The UK National Annex as 1.4.

11.5 Verification procedures

The results of a combination are verified by a formula or inequality given for each limit state by EC7.

1. Section 2.4.7.2(1)P requires that the following inequality should be satisfied for EQU limit state:

$$E_{dst;d} \leq E_{stb;d} + T_d \qquad (11.7)$$

where $E_{dist;d}$ = design value of the effect of a destabilising action, e.g. overturning moment on a retaining wall.
$E_{stb;d}$ = design value of the effect of a stabilising action, e.g. resisting moment.
T_d = design value of shearing resistance on the part of the structure in contact with the ground, e.g. rigid foundation on rock (usually very small).

2. Section 2.4.7.3.1 for STR and GEO limit states define the inequality

$$\boxed{E_d \leq R_d} \tag{11.8}$$

E_d = design value of the effect of actions, e.g. disturbing moment acting on a slope.
R_d = design value of the resistance to an action, e.g. resisting moment to sliding.

3. Section 2.4.7.4(1) P for uplift (UPL) gives the inequality:

$$\boxed{V_{dst;d} \leq G_{stb;d} + R_d} \tag{11.9}$$

where $V_{dst;d} = G_{dst;d} + Q_{dst;d}$

and $V_{dst;d}$ = Vertical destabilising action, e.g. upward force by water causing uplift of the ground.
$G_{stb;d}$ = Vertical stabilising action e.g. weight of soil resisting uplift.
R_d = Any additional resistance uplift, e.g. weight on top of the ground.
$G_{dst;d}$ = Destabilising, permanent actions.
$Q_{dst;d}$ = Destabilising, variable, permanent actions.

4. Section 2.4.7.5(1) P, for seepage (HYD) limit state defines two verifying inequalities:

$$\boxed{\begin{array}{l} U_{dst;d} \leq \sigma_{stb;d} \\ S_{dst;d} \leq G'_{stb;d} \end{array}} \tag{11.10}$$

Where, $U_{dst;d}$ = Destabilising pore pressure at the base.
$\sigma_{stb;d}$ = Stabilising pressure at the base.
$S_{dst;d}$ = Seepage force on the base.
$G'_{stb;d}$ = Submerged weight of soil column through which seepage occurs.

Note: The results of calculations are presented in terms of the inequalities, which are synonymous with F_s (Factor of safety).

For example, in Ex. 11.4
M_O = 727 kNm (Overturning moment)
M_R = 1599 kNm (Resisting moment)
$\therefore F_S = 2.19$

In terms of EC7:
E_d = 727 kNm (Disturbing moment)
R_d = 1599 kNm (Stabilising moment)
$\therefore E_d < R_d$

Conclusion: The stabilising moment is greater than the disturbing moment, therefore the overall stability, hence the GEO limit state requirement is satisfied.

11.6 Application of partial factors

In traditional British practice a 'Global factor of safety' has been applied (as divisor) to the final result, that is to the effect of an action, e.g. to the moment caused by the action of forces. EC7-1, however applies partial factors of safety to all soil parameters as well as to the actions during the design process.

Section 2.4.6.1 gives Formula 2.1a for the calculations of the design value of an action:

$$F_d = \gamma_F F_{rep}$$

where F_{rep} = representative value of an action
γ_F = partial factor for an action (multiplier)

Section 2.4.6.2 gives Formula 2.2 for the calculations of design values of soil parameters.

$$X_d = \frac{X_k}{\gamma_m}$$

where X_K = characteristics value of a soil property.
γ_m = partial factor for soil property (divisor)

Special case: For angle of shearing resistance:

$$\phi'_d = \tan^{-1}\left(\frac{\tan\phi'}{\gamma_m}\right)$$

The partial factors for various limit states are given various tables of Appendix A of EC7.

For EQU:	Tables A.1 and A.2
For STR and GEO:	Tables A.3 to A.14
For UPL:	Tables A.15 and A.16
For HYD:	Tables A.17

Example 11.1 Design of shallow footings (Eurocode 7, section B)

The ultimate limit state GEO is appropriate for a design as the failure or the excessive deformation of ground i.e. soil or rock is significant in providing resistance. Any accepted method of analysis may be applied.

Taking the soil characteristics of Example 9.3 for a short strip footing, covering an area of 6 m² at a global safety factor of $F_s = 3$. Check the value of F_s, by GEO limit state.

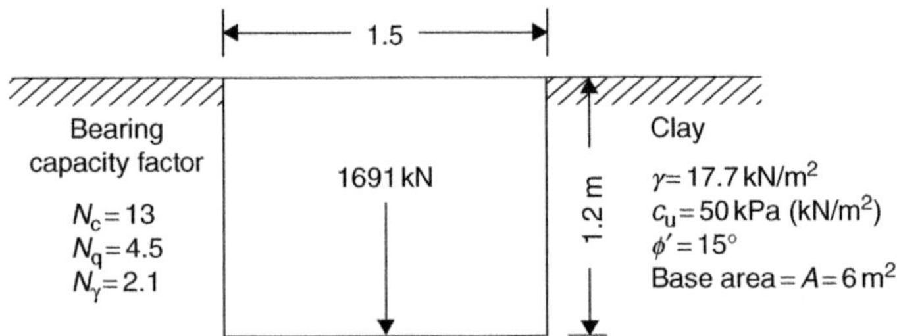

Figure 11.1

Ground water is 10 m below GL.
Combination 1: A1 "+" M1 "+" R1

A1 – Design actions

The partial factors for the effects of actions influencing stability either favourably or unfavourably e.g. disturbing force and resisting forces respectively. These factors are given in Table A3, Annex A of Eurocode 7.

Factor for unfavourable action of 1631 kN

$$\gamma_{\text{Gdist}} = 1.35$$

M2 – Material factors (Table A4)

$$\text{For } c_u: \quad \gamma_{cu} = 1.25$$
$$\text{For } \phi_u: \quad \gamma_Q = 1.25$$
$$\text{For } \gamma: \quad \text{For } \gamma = \gamma_\gamma = 1$$

*R*1 – Partial resistance factor (Table A5)

For bearing: $\gamma_{R;v} = 1$

Design values

$$\text{For } c_u : c_{u:d} = \frac{c_u}{\gamma_{cu}} = \frac{50}{1.25} = 40\,\text{kP}_a\left(\text{kN/m}^2\right)$$

$$\text{For } \phi_u : \phi_{ud} = \tan^{-1}\left(\frac{\tan 15^\circ}{1.25}\right) = 12.1^\circ$$

Design resistance

Bearing capacity factors taken from Chart 9.1.

For $\phi_{ud} = 12.1^\circ$ $\quad N_c = 11$

$N_q = 3.8$

$N_\gamma = 1.4$

Ultimate bearing capacity from Formula 9.8

$$\begin{aligned} q_u &= C_{ud}N_c + \sigma'_0\, Nq + 0.5\gamma_\phi \gamma B N_\gamma \\ &= 40 \times 11 + 1.2 \times 17.7 \times 3.8 + 0.5 \times 1 \times 17.7 \times 1.5 \times 1.4 \\ &= 440 + 80.7 + 18.6 = 539.3\,\text{kP}_a\left(\text{kN/m}^2\right) \end{aligned}$$

$$\text{Bearing resistance } R_b = \frac{qu}{\gamma_{R:v}} = \frac{539.3 \times 6}{1} = 3236\text{ kN}$$

Vertical force on foundation $W = 1.35 \times 1631 = 2202\text{ kN}$

$$\text{Factor of safety: } F_s = \frac{R_b}{W} = \frac{3236}{2202} = 1.45 > 1$$

Combination 2: A2 "+" M2 "+" R1

A2: For Action: $\gamma_G = 1$

M2: For c_u: $\gamma_{cu} = 1.25$

For ϕ_u: $\gamma_\phi = 1.25$

For γ: $\gamma_\gamma = 1$

R1: For bearing: $\gamma_{R:b} = 1$

As only γ_G has changed $W = \gamma_G = 1631\,\text{kN}$

And $F_s = \dfrac{3235.7}{1631} = 1.98 > 1$

Thus the GEO requirement is satisfied and the global factor of safety $F_s = 3$ is an overestimate.

Slope stability

The ultimate limit states GEO and STR are appropriate. Any of the accepted methods may be applied to estimate stability, using the partial factors of Eurocode 7.

Example 11.2

Figure 11.2 shows the details of the slope of Example 10.8.

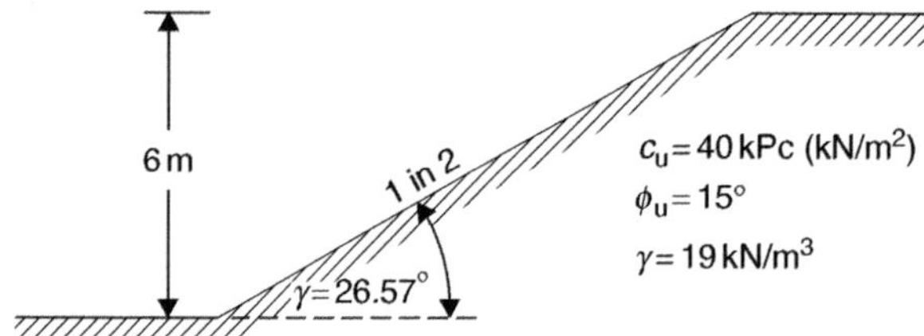

Figure 11.2

Check the global factor of $F_s = 3.28$, calculated in Example 10.8, adopting the radial procedure and results in Example 10.5 and Table 10.3.

Combination 1: A1 "+" M1 "+" R1

A1 - Design actions (Table A.3, Annex A, Eurocode 7)

Unfavourable (disturbing) action $\gamma_{Gdist} = 1.35$
Favourable (resisting) action $\gamma_{Gstb} = 1$

M1 - Material factors (Table A.4)

For ϕ_u $\gamma_\phi = 1$

For c_u $\gamma_c = 1$

For γ $\gamma_{\bar{a}} = 1$

R1 - Resistance factor (Table A14) $\gamma_{R;e} = 1$

Design parameters

As all partial factors for the soil are unity the design parameters are:

$$\phi_{d\phi} = \phi_u = 15^\circ$$
$$c_{dc} = c_u = 40\,\text{KPa}$$
$$\gamma_{dr} = \gamma = 19\,\text{kN/m}^3$$

Design calculations

In combination 1, all material factors are unity, hence the calculations in Example 10.5 are unaltered and listed below:

Resisting force contributed by cohesion: $S = \underrightarrow{952}$ kN

Resisting force contributed by friction: $\Sigma F = \underrightarrow{371}$ kN

Disturbing force contributed by weight (γ): $\Sigma T = \underrightarrow{415}$ kN

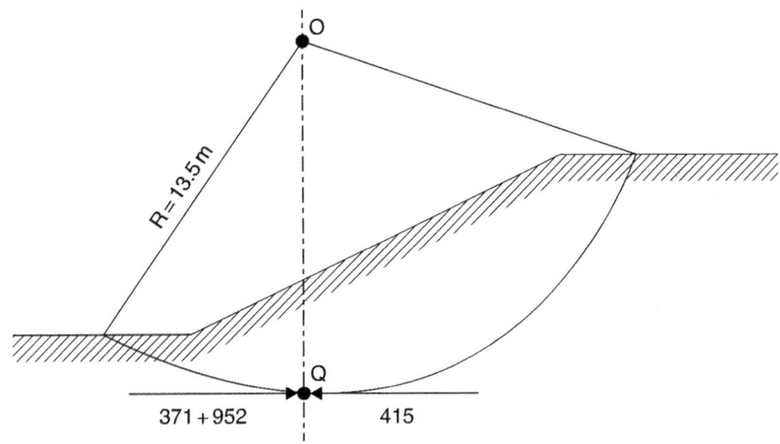

Figure 11.3

$$\text{Disturbing moment}: M_D = 415 \times 13.5 = 5603 \text{ kNm}$$

$$\text{Resisting moment}: M_R = (371 + 952) \times 13.5 = 17866 \text{ kNm}$$

Applying $\gamma_{Q:unf}$=1.35 to the unfavourable moment the factor of safety is:

$$F_s = \frac{17866}{1.35 \times 5603} = \frac{17866}{7564} = 2.36$$

The restoring (stabilising) moment is larger than the disturbing moment, therefore the stability of the slope is satisfactory.

Combination 2: A2 "+" M2 "+" R1

A2 - Design action (Table A.3)

Unfavourable action γ_{Gdist}=1
Favourable action γ_{Gstb}=1

M2 - Material factors (Table A.4)

For ϕ_u $\gamma_\phi = 1.25$
For c_u $\gamma_c = 1.25$
For γ $\gamma_\gamma = 1$

R2 – Resistance factor (Table A14)

$$\gamma_{R:e} = 1$$

Design parameters

For friction $\phi_{d\phi} = \tan^{-1}\left(\frac{\tan 15}{1.25}\right) = 12.1°$

For cohesion $C_{dc} = 40/1.25 = 32\,kP_a$

For weight $\gamma_{d\phi} = 19/1 = 19\,kN/m^3$

Design calculations

From Example 10.5, friction force $F = N\tan\phi_u$

$N = W\cos\delta$

$\therefore F = W\cos\delta \times \tan\phi_0$

$= W\cos\delta \times \tan 18.75 \quad \therefore \quad \boxed{F = 0.339W\cos\delta}$ (11.11)

Table 10.3 is reproduced below as Table 11.1, changing values of F according to formula 11.2.

From the table: Disturbing force: $\Sigma T = \underleftarrow{415}\,kN$

Friction force: $\Sigma F = \underrightarrow{469.75}\,kN$

Table 11.1

Slice		Angle	$T = W\sin\delta$		Friction force (kN)
No	***W* (kN)**	δ	+ ←	− →	$F = 0.339\ W\cos\delta$
1	2.66	−31.8		−1.40	0.77
2	13.30	−27		−6.04	4.02
3	26.79	−22.7		−10.37	8.38
4	43.13	−17.5		−12.98	13.94
5	57.95	−13		−13.04	19.14
6	64.03	−9		−9.99	21.43
7	83.03	−4.5		−6.48	28.06
8	93.29	−1		−1.59	31.62
9	102.22	4	7.16		34.56
10	109.82	8	15.26		36.87
11	115.71	12.5	24.99		38.29
12	120.27	17	35.12		39.00
13	123.31	21.5	45.25		38.89
14	124.45	26	54.51		37.92
15	118.94	31	61.25		34.38
16	106.59	35.8	62.36		29.30
17	91.39	41.3	60.32		23.27
18	72.96	47	53.33		16.87
19	49.78	54	40.27		9.91
20	19.57	62	17.28		3.11
	1539.19		+477.10	−61.89	469.73
			$\Sigma T = 415\,kN$		$\Sigma F = 469.73\,kN$

From Example 10.2 the length of the arc: $L = 23.8\,\text{m}$.

Therefore, the resisting force contributed by the cohesion is $S = C_d L = 32 \times 23.8 = 761.6\,\text{kN}$.

Restoring moment: $M_R = (469.73 + 761.6) \times 13.5 = 16623\,\text{kNm}$

Disturbing moment: $M_D = 415 \times 13.5 = 5603\,\text{kNm}$

Hence factor of safety: $F_s = \left(\dfrac{\gamma_{G:dist}}{\gamma_{G:stb}}\right) \times \dfrac{M_R}{M_D} = \dfrac{16623}{5603} = 2.96$

Thus the overall stability of the slope is satisfactory.

Conclusion: The average value of F_s is $= \dfrac{2.36 + 2.96}{2} = 2.66$

The minimum factor of safety is 2.36.

The stability moments in both combinations are larger than the disturbing moments, hence the limit state requirement is satisfied.

Example 11.3 Sheet pile in cohesionless soil

Figure 11.4 shows the anchored sheet pile was analysed in Example 8.10.

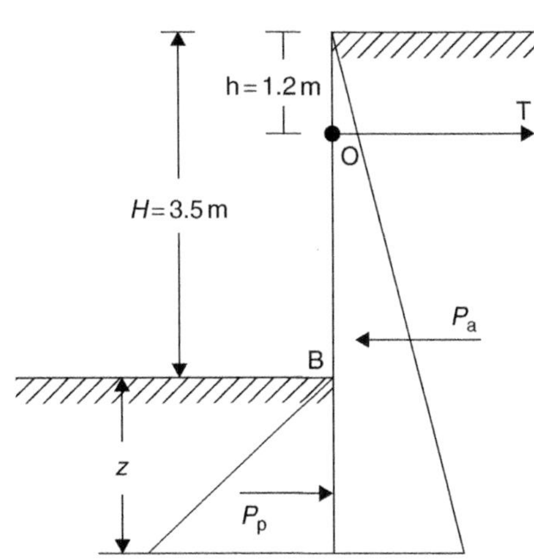

Soil paramaters:

$\phi = 34°$
$C = 0$
$\gamma = 19\,\text{kN/m}^3$

Check the value of Z by EC7, GEO Limit state.

Design approach 1

Combination 1: A1 "+" M1 "+" R1

Figure 11.4

A1 Design actions P_a and P_p are both permanent unfavourable ones, hence $\gamma_G = 1.35$ from Table A3

M1 $\gamma_\phi = \gamma_\phi = 1$ from Table A4

Design parameters: $\phi'_d = \tan^{-1}\left(\frac{\tan 34°}{1}\right) = 34°$

$$= \frac{\gamma}{1} = 19\,\text{kN/m}^3$$

Coefficients of earth pressure: $K_a = \frac{1-\sin 34}{1+\sin 34} = 0.283$

$$K_p = \frac{1}{K_a} = 3.537$$

Active force: $P_a = \frac{K_a \gamma_G \gamma (H+z)^2}{2} = \frac{0.283 \times 1.35 \times 19 \times (H+z)^2}{2}$

$$= 3.63\,(H+z)^2$$

Passive force: $P_p = \frac{K_a g_G g z^2}{2} = \frac{3.537 \times 1.35 \times 19 z^2}{2}$

$$= 45.4z^2$$

Moments of forces about tie rod T:

Active force: $\Sigma M_a = P_a\left[\frac{2}{3}(H+z) - h\right]$

$$= 3.63 \times (H+z)^2 \times \left[\frac{2}{3}(H+z) - h\right]$$

$$= 3.63 \times (12.25 + 7z + z^2) \times (1.53 + 0.666z)$$

$$= (44.5 + 25.41z + 3.63z^2) \times (1.53 + 0.666z)$$

$$= 68.1 + 29.64z + 38.88z + 16.9z^2 + 5.6z^2 + 2.42z^3$$

$$= 2.42z^3 + 22.5z^2 + 68.52z + 68.1$$

Passive force: $\Sigma M_p = P_p\left(\frac{2z}{3} + H - h\right) = 45.4z^2 \times (0.666z + 2.3) = 30.24z^3 + 104.42z^2$

Resultant moment: $\Sigma M_o = \Sigma M_a - \Sigma M_p = 0$

$$\Sigma M_O = 2.42z^3 + 22.5z^2 + 68.52z + 68.8 - 30.24z^3 - 104.42z^2 = 0$$

Collecting similar terms and changing signs

$$27.82z^3 + 81.92z^2 + 68.52z + 68.8 = 0$$

Simplifying: $\underbrace{z^3 + 3z^2}_{L} = \underbrace{2.46z + 2.47}_{R}$

Solving graphically:

Table 11.2

Z	0.6	0.8	1.0	1.2	1.4
L	1.3	2.4	4	6.1	8.6
R	3.95	4.43	4.93	5.42	5.91

The intersection point X on Graph 11.1 indicates that z=1.13 m for Approach 1.

Combination 2 A2 "+" M2 "+" R1

A2 For both P_a and P_p γ_G=1 from Table A3

M2 From Table A4 γ_ϕ=1.25 and γ_γ=1

Design parameters: $\phi'_d = \tan^{-1}\left(\dfrac{\tan 34^\circ}{1.25}\right) = 28.35^\circ$

$$\gamma_d = \gamma = 19\,\text{kN/m}^3$$

Coefficients of earth pressure: $K_a = \dfrac{1 - \sin 28.35^\circ}{1 + \sin 28.35^\circ} = 0.356$

$$K_p = \frac{1}{K_a} = 2.81$$

Active force: $P_a = \dfrac{K_a \gamma \gamma_G}{2}(H+z)^2 = \dfrac{0.356 \times 19 \times 1}{2}\left(H^2 + 2Hz + z^2\right)$

$$= 3.38 \times (12.25 + 7z + z^2) = 41.4 + 23.66z + 3.38z^2$$

Passive force: $P_p = \dfrac{K_p \gamma z^2 \gamma_G}{2} = \dfrac{2.81 \times 19 \times 1 \times z^2}{2} = 26.7$

Moments of P_a and P_p about tie rod:

$$\Sigma M_a = P_a\left[\frac{2}{3}(H+z) - h\right] = (41.1 + 23.66z + 3.38z^2) \times (1.53 + 0.666z)$$

$$= 63.34 + 27.57z + 36.2z + 15.76z^2 + 5.17z^2 + 2.25z^3$$

$$= 2.25z^3 + 20.93z^2 + 63.77z + 63.34$$

$$\Sigma M_p = P_p\left(\frac{2}{3}z + H - h\right) = 26.7z^2 \times \left(0.666z + 2.3\right) = 17.8z^3 + 61.41$$

$$\Sigma M_0 = \Sigma M_a - \Sigma M_p = 2.25z^3 + 20.93z^2 + 63.77z + 63.34 - 17.8z^3 - 61.4z^3 = 0$$

$$= -15.55z^3 - 40.47z^2 - 63.77z + 63.34 = 0$$

$$= -z^3 - 2.60z^2 + 4.1z + 4 = 0$$

Solving the cubic equation yields: $z = 1.58$ m

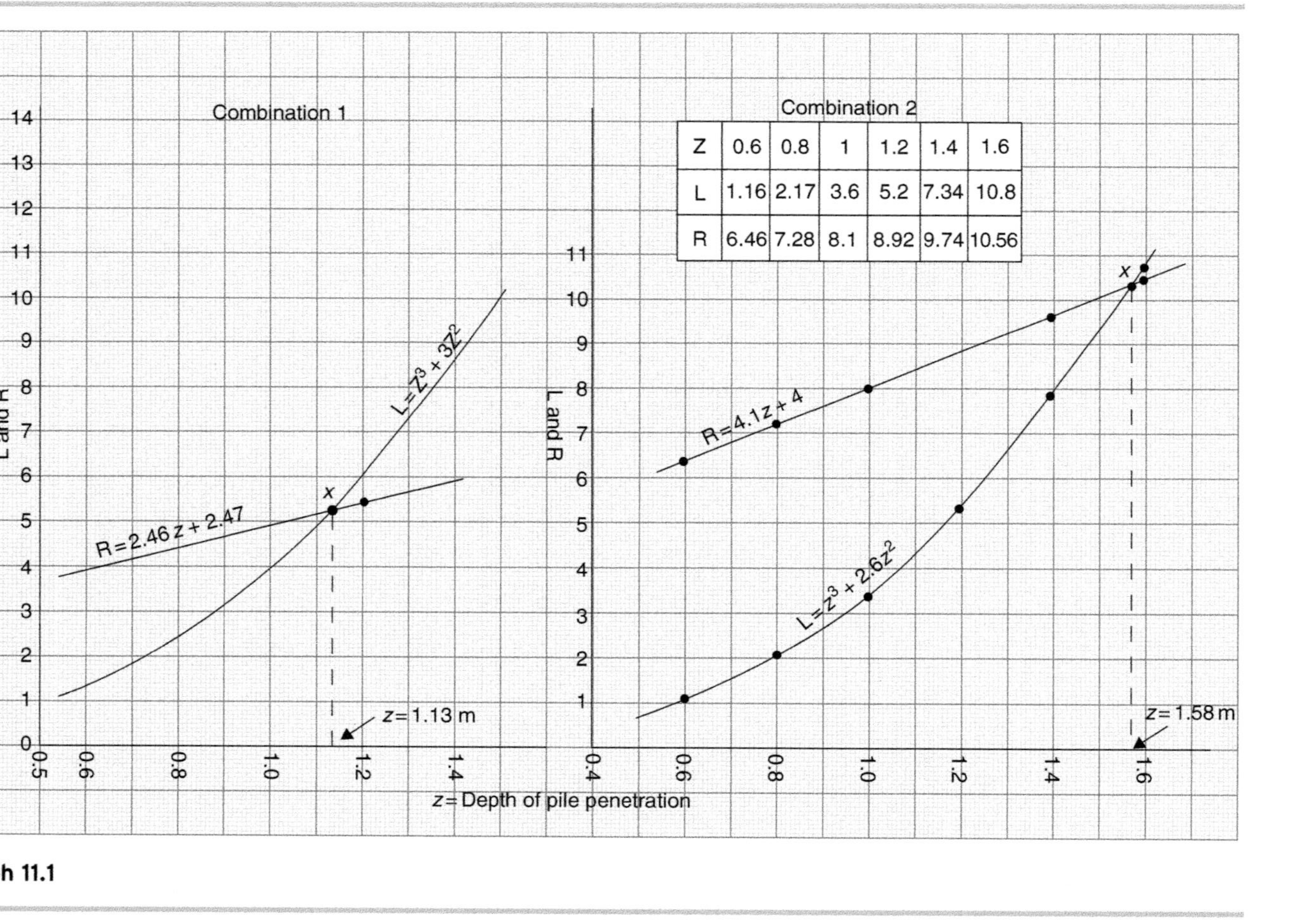

Z	0.6	0.8	1	1.2	1.4	1.6
L	1.16	2.17	3.6	5.2	7.34	10.8
R	6.46	7.28	8.1	8.92	9.74	10.56

Graph 11.1

Example 11.4 Gravity retaining wall

Figure 11.5 shows the gravity wall analysed in Examples 8.3, 8.4, 8.8 and Problem 8.1.

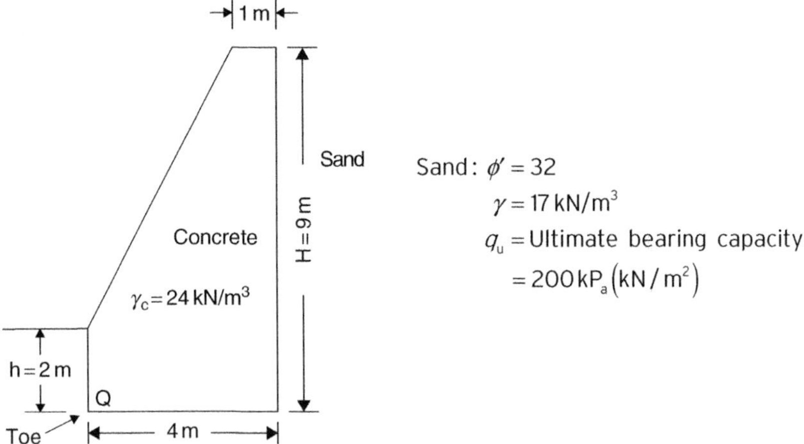

Figure 11.5

Check the design, using Eurocode 7 for

1. Overturning (EQU limit state)
2. Overturning when the passive resistance is removed
3. Sliding (GEO limit state)
4. Bearing failure (GEO limit state)

1. *Overturning*

Design approach 1: Combination 1 | A1"+" M1"+" R1 |

A1: Partial factors on actions (forces and moments from Table A.1, Annex A)
Permanent unfavourable $\gamma_{Gdist} = 1.1$
Permanent favourable $\gamma_{Gstb} = 0.9$

M1: Partial factors for soil parameters from Table A.2
Angle of shearing resistance: $\gamma'_f = 1.25$
Weight density: $\gamma_\gamma = 1$

$$\text{Design values: } \phi'_d = \tan^{-1}\left(\frac{\tan 32^\circ}{1.25}\right) = 26.6^\circ$$

It is stated in Section 9.5.1 (7) of Eurocode 7 that the design angle of friction (δ) between the soil and concrete is assumed to be equal to the critical state angle of shearing resistance ϕ'_d. The coefficient of active earth pressure is found in Figure C.1.1 (Annex C) for $\delta = \phi'_d$ or δ/ϕ'_d and $\phi' = 26.6^\circ$ as $K_a = 0.32$.

In Example 8.3 (case 1) the active force P_a was found to be 211.4 kN taking $K_a = 0.307$. Using Eurocode 7 value of $K_a = 0.32$.

$$P_a = \frac{K_a H^2}{2} = \frac{0.32 \times 17.9^2}{2} = 220.3\,\text{kN (unfavourable)}$$

Acting at $y_a = 3$ m from the base

The favourable passive force above the toe is calculated from:

$$K_p = \frac{1+\sin 26.6^\circ}{1-\sin 26.6^\circ} = \frac{1.4478}{0.5522} = 2.62 \downarrow$$

$$P_p = \frac{2.62 \times 17 \times 2^2}{2} = 89.1\,\text{kN}$$

Acting at $\frac{2h}{3} = 0.67$ m from the toe

In Example 8.8 the wall was divided into three areas and their weight as well as distances from the toe was calculated as shown in Figure 11.6.

Overturning moment $M_O = 3\gamma_{G,unf} \times 220.3$

$= 3 \times 1.1 \times 220.3 = 727$ kNm

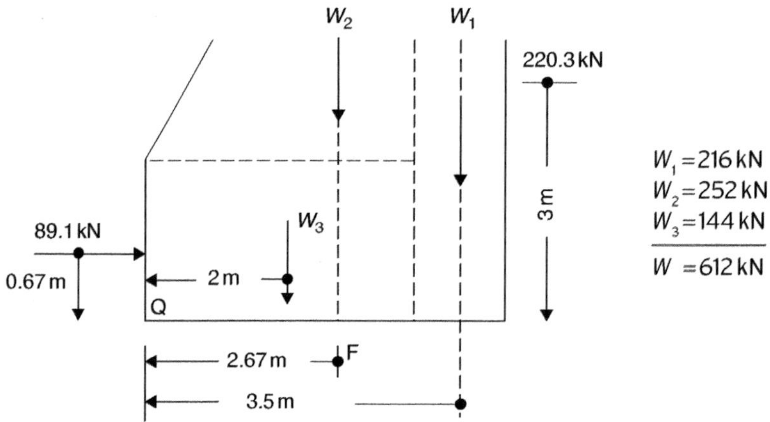

Figure 11.6

Resisting (favourable moment)

$$M_R = 0.9 \times (3.5 \times 216 + 2.67 \times 252 + 2 \times 144 + 0.67 \times 89.1) = 1599\,\text{kNm}$$

Factor of safety: $F_s = \frac{1599}{727} = 2.19$

As $M_R > M_O$ the EQU requirement is satisfactory.

2. Removal or passive resistance by mistake or carelessness. The favourable moment is decreased to $M_R = 0.9 \times (756 + 673 + 288) = 1545$ kN.

$$F_s = \frac{1545}{727} = 2.13$$

Thus in this case the wall is stable.

3. *Sliding* In the GEO limit states both combinations have to be applied.

Combination 1: A1 (Table A3) $\gamma_{G.unf} = 1.35$

$\gamma_{G.fav} = 1$

M1 (Table A4): For ϕ': $\gamma_f' = 1$

For γ: $\gamma_\gamma = 1$

Design value for ϕ': $\phi'_d = \tan-1\left(\frac{\tan 32°}{1}\right) = 32°$

Coefficient of friction between concrete and sand

$$\mu_1 = \tan 32° = 0.625$$

Friction force $F = W\tan\phi'_d = 612 \times 0.625 = 382.5\,\text{kN}$

Figure 11.7 shows the forces acting on the wall:

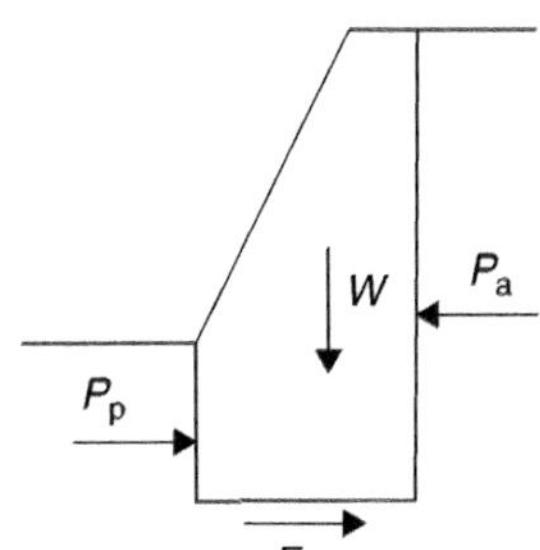

Force to cause sliding = P_a
Force resisting sliding = $F + P_p$
Factor of safety against sliding

$$F_s = \frac{1 \times (F + P_p)}{13.5\,P_a} = \frac{382.5 + 89.1}{1.35 \times 220.3} = 1.59$$

Figure 11.7

Combination 2: A2"+"M2"+"R1

Partial factors: A2 (Table A3) $\gamma_{Gdist} = \gamma_{Gstb} = 1$

M2 (Table A4) $\gamma'_\phi = 1.25 \quad \therefore \quad \phi'_d = 26.6°$

$\gamma_\gamma = 1$

$\mu_2 = \tan 26.6 = 0.5$ and $F = 612 \times 0.5 = 306\,\text{kN}$

Factor of safety $F_{s2} = \frac{306 + 89.1}{220.3} = 1.79$

Both combinations satisfy the GEO limit state requirements.

4. *Bearing* (Figure 8.62)

The calculations in Example 8.8 yield the following result:

Force $R = 612\,\text{kN}$ act at eccentricity: $e = 0.11\,\text{m}$. R is a permanent, unfavourable action hence, $\gamma_{Gdist} = 1.35$ from Table A3.

Check whether the maximum pressure on the foundation soil is larger than the bearing capacity of 200 kP_a (kN/m²).

Direct pressure: $f_d = \frac{\gamma_{Gdist}}{bd} = \frac{1.35 \times 612}{1 \times 4} = 206.6\,KP_a\ (\text{kN/m}^2)$

Section modulus: $z = \frac{bd^2}{6} = \frac{1.4^2}{6} = 2.67$

Bending pressure: $f_b = \pm\frac{1.35Re}{z} = \frac{1.35\times 612\times 0.11}{2.67} = 34\,\text{kP}_a\ (\text{kN/m}^2)$

Combined pressure: $f_{max} = 206.6 + 34 = 240.6\ \text{kP}_a\,(\text{kN/m}^2)$

As the maximum pressure is larger than the bearing capacity of the soil (206.6 >200), the GEO limit state requirement is not satisfied and the wall has to be redesigned and lower the value of f_{max}.

Example 11.5 Bored concrete pile

The bored concrete pileof Example 9.9 in soft to stiff clay was designed to be 16 m long as shown in Figure 11.8. The undrained cohesion of the clay varies with depth (see Table 9.5). Its average value if given in Table 9.6 as $\bar{c}_u = 117\,\text{kN/m}^2$ (117 kPa) and its value at 16 m depth as $\bar{c}_u = 317\,\text{kN/m}^2$ (317 kPa). Check the calculated length of the pile using EC7 procedure.

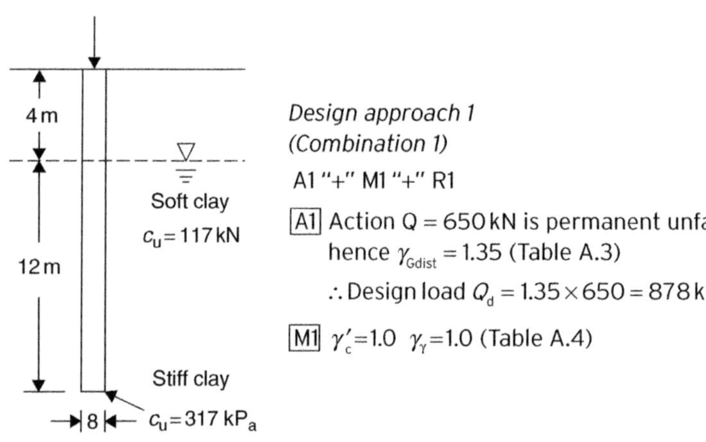

Figure 11.8

Design values of cohesion: $\bar{c}_{u:d} = 117\,\text{kP}_a c_u = 374\,\text{kP}_a$

R1 Base resistance: $\gamma_b = 1.25$ (Table A.7)
Shaft resistance: $\gamma_s = 1.0$
Model factor: $\gamma_m = 1.4$ (National Annex)

Corrected design partial factors
Base resistance: $\gamma_{b:d} = 1.4\times 1.25 = 1.75$
Shaft resistance: $\gamma_{s:d} = 1.4\times 1.0 = 1.4$

End bearing area: $A_e = \frac{0.5^2}{4} = 0.196\,\text{m}^2$

End bearing resistance: $Q_e = \dfrac{9 \times 0.196 \times 374}{1.75} = 377\,\text{kN}$

Shaft resistance: $Q_s = \dfrac{\alpha \bar{c}_u A_s}{1.4} = \dfrac{0.45 \times 117 \times \pi \times 0.5 \times 16}{1.4} = 945\,\text{kN}$

$$\therefore F_1 = \frac{Q_e + Q_s}{Q_d} = \frac{377 + 945}{878} = 1.5$$

Combination 2 A2 "+" (M1 OR M2) "+" R4

A2	For $Q = 650\,\text{kN}$:	$\gamma_G = 1$	(Table A.3)
M1	For cohesion:	$\gamma'_c = 1$	(Table A.4)
R4	For base:	$\gamma_b = 1.6$	(Table A.7)
	For shaft:	$\gamma_s = 1.3$	
	Model factor:	$\gamma_m = 1.4$	

Corrected design partial factors:

$$\gamma_{b:d} = 1.4 \times 1.6 = 2.24$$
$$\gamma_{s:d} = 1.4 \times 1.3 = 1.82$$

Using M1 $\bar{c}_{u:d} = 1 \times 117 = 117\,\text{kP}_a$

$$c_{u:d} = 1 \times 374 = 374\,\text{kP}_a$$

End bearing: $Q_e = \dfrac{9 \times 0.196 \times 374}{2.24} = 295\,\text{kP}_a$

Shaft resistance: $Q_s = \dfrac{0.45 \times 117 \times \pi \times 0.5 \times 16}{1.82} = 727\,\text{kP}$

$$\therefore F_2 = \frac{Q_e + Q_s}{Q_d} = \frac{1023}{878} = \underline{\underline{1.17}}$$

Conclusion: The 16m long, 0.5m diameter pile can carry the 650kN load. Combination 2 is critical as $F_2 < F_1$.

Example 11.6 Heave and seepage

Figure 11.9 shows the outline of one side of a sheet pile cofferdam. The depth of pile penetration was determined to be $z = 4.61$m in Example 8.14. As this did not take into account possible seepage under the piles, the problem was recalculated in Example 8.15 and the new depth was found to be $z = 5.5$m. The known details are shown below.

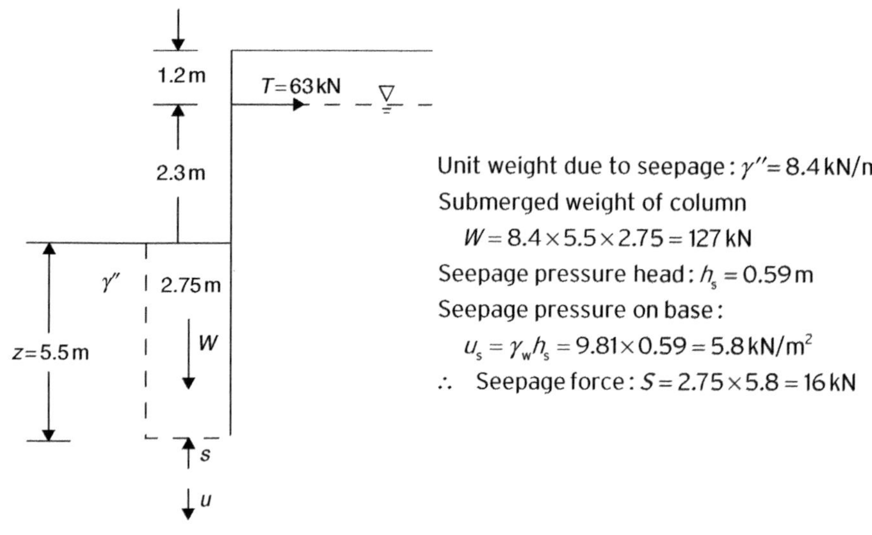

Figure 11.9

Check the stability of the cohesionless soil against uplift and internal erosion by EC7 methods.

Uplift (buoyancy)
Verification of (UPL) uplift limit state by inequality:

$$\boxed{\begin{array}{l} V_{dst;d} \leq G_{stb;d} + R_d \\ V_{dst;s} = G_{dst;d} + Q_{dst;d} \end{array}} \qquad (11.9)$$

In this example $R_d = Q_{dst;d} = 0$, thus the design inequality written in terms of Example 8.15 as:

$$V_{dst;d} = S_d \text{ and } G_{stb;d} = W_d \text{ or } S_d \leq W_d$$

The recommended values of partial factors for UPL are obtained from Tables A.15 and A.16:

For permanent unfavourable action S: $\gamma_{G:dst} = 1$
For permanent favourable action W: $\gamma_{G;stb} = 0.9$

$$\text{Design values: } \left.\begin{array}{l} S_d = 1 \times S = 16\,\text{kN} \\ W_d = 0.9W = 0.9 \times 127 = 114\,\text{kN} \end{array}\right\} \quad F_s = \frac{114}{16} = 7.1$$

In terms of EC7: $V_{sdt;d} < G_{stb;d}$ hence the requirement of UPL limit state is satisfied.

Seepage
Partial factors found in Table A.17

For permanent, unfavourable actions $\gamma_{G:dst} = 1.35$
For permanent favourable actions $\gamma_{G;stb} = 0.9$

$$\therefore \; U_{dst;d} = 1.35 \times u_s = 1.35 \times 5.8 = 7.83\,\text{kN/m}^2\ (\text{KP}_a)$$

$$\sigma_{stb;d} = 0.9\frac{W}{2.75} = \frac{0.9 \times 127}{2.5} = 45.7\,\text{kN/m}^2$$

$S_{dst;d} = 1.35\,S = 1.35 \times 16 = 21.6$ kN (see page force)

$G_{stb;d} = 0.9W = 0.9 \times 127 = 114.3$ kN (submerged weight of column)

$U_{dst;d} < \sigma_{stb;d}$ or $7.83 < 45.7$

$S_{dst;d} < G'_{stb;d}$ or $21.6 < 114.5$

Hence the requirements of HYD limit state are satisfied.

Sudden drawdown in a reservoir

Example 11.7

With reference to Example 10.1 and 10.6, Figure 11.10 shows the soil parameters and some of the calculated values of the problem.

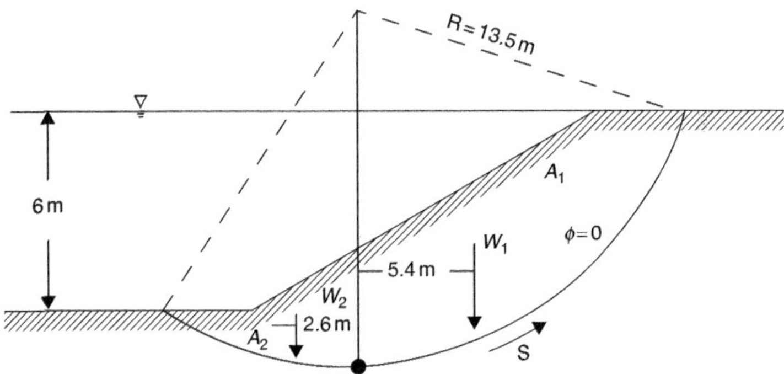

Figure 11.10

Crossectional areas

$A_1 = 62.8$ m²

$A_2 = 19.9$ m²

Soil: Clay of undrained shear strength: $c_u = 40\,KP_a$

and weight density: $\gamma = 19$ kN/m³

Saturated unit weight: $\gamma_{sat} = 19.8$ kN/m³

Submerged unit weight: $\gamma' = 10$ kN/m³

Length of slip surface: $L = 23.8$ m

Check the stability of the slope by EC7 when:

1. The reservoir is full
2. The reservoir is suddenly emptied

1. When the reservoir is full, the buoyant (submerged) unit weight of the clay is applied $\gamma' = 10\,\text{kN/m}^3$

   Weight of section A_1: $W_1 = 62 \times 8.10 = 628\,\text{KN}$
   Weight of section A_2: $W_2 = 19 \times 9.10 = 199\,\text{KN}$

   Combination 1: A1"+"M1"+"R1

   A1 Perm, unfavourable action: $\gamma_{G:unf} = 1.35$
   Perm, favourable action: $\gamma_{G:fav} = 1.0$

   M1 For $c_u = \gamma_{cu} = 1.0$ } Table A4
   For $\gamma' = \gamma_{\gamma'} = 1.0$

   R1 For the shear resistance against votation along the slip surface: $\gamma_{R:h} = 1.0$ (Table A.5)

   Disturbing moment: $M_D = 1.35\,(628 \times 5.4 - 199 \times 2.6)$
   $= 3880\,\text{kNm}$

   Shear force: $S = c_u L = 40 \times 23.8 = 952\,\text{kN}$
   Resisting moment: $M_R = RS = 13.5 \times 952 = 12852\,\text{kNm}$

   $\therefore$ Factor of safety $F_s = \dfrac{12852}{3880} = 3.3$

   Or in terms of EC7 (section 2.4.7.3.1 (1)P):

   Design effects: $E_d = 3880\,\text{kNm}$
   Design resistance: $R_d = 12852\,\text{kNm}$

   $E_d < R_d$ hence the GEO limit state requirement is satisfied for combination 1.

   Combination 2: A2"+"M2"+"R1

   A2 $\gamma_{G:dst} = \gamma_{G:fav} = 1.0$ (Table A.3)
   M2 $\gamma_{cu} = 1.4 = \gamma_\gamma = 1.0$ (Table A.5): $c_{u:d} = \dfrac{40}{1.4} = 28.6\,\text{KPa}$
   R1 $\gamma_{R:h} = 1.0$

   Disturbing $M_D = 1 \times (628 \times 5.4 - 199 \times 2.6) = 2874\,\text{KNm}$ ↘
   Shear force: $S = 28.6 \times 23.8 = 681\,\text{KN}$
   Resisting $M_R = R_S = 1 \times (13.5 \times 681) = 9189\,\text{kNm}$

   Therefore $F_s = \dfrac{9189}{2874} = 3.2 \quad \left(\dfrac{R_d}{E_d}\right)$

   Or $E_d < R_d$ hence the limit state requirement is satisfied for combination 2.

2. *Sudden drawdown*
   When the reservoir is suddenly emptied, the clay remains saturated as buoyancy has no effect, hence the saturated weight density $\gamma_{sat} = 19.8\,\text{kN/m}^3$ has to be applied.

As the partial factor for weight density is γ_γ in both combinations:

$$W_1 = 62.8 \times 19.8 = 1243\,\text{kN}$$
$$W_2 = 19.9 \times 19.8 = 394\,\text{kN}$$

Combination 1

A1	$\gamma_{G:dst}=1.35$	$\gamma_{G:stb}=1.0$	
M1	$\gamma_{cu}=1.0$	$\gamma_\gamma=1.0$	(As before)
R1	$\gamma_{R:h}=1.0$		

Disturbing: $M_D=1.35\times(1243\times54-394\times2.6)=7679$ kNm

Resisting $M_R=12852\,\text{kNm}$ ↶ (no change)

Factor of safety: $F_s = \dfrac{12852}{7679} = 1.7$

or $E_d=7679<R_d=12852$

Combination 2 A2"+"M2"+"R1

As all factors are the same as before $c_{u:d}=28.6\,\text{KP}_a$

Also, $M_D=1\times(1243\times5.4-394\times2.6)=5688\,\text{kNm}$

$M_R = 1\times9189 = 9189\,\text{kNm}$

$$F_s = \frac{9189}{5688} = 1.6$$

Or $E_d=5688<R_d=9189$ hence the limit state requirement is satisfied.

Appendix A
Mass and Weight

The terminology of the SI system of units does not specifically include the word 'weight'. However, the weight of materials are usually found to be tabulated either in terms of mass density (kg/m^3) or in terms of gravitational force (weight) density (kg/m^3). The clarification of these terms is the purpose of this appendix.

Mass (M)

It may be described as a quantity of matter, which occupies space. The location in space does not alter the quantity, be it on the earth's surface or in an orbiting satellite. The mass of a body, therefore is the same anywhere in the universe.

Weight (W)

It is the gravitational force in accordance with Newton's Second Law of motion; that is:

$$\text{Force} = \text{mass} \times \text{acceleration}$$

or $$W = Mg$$

Where g=gravitational acceleration (not to be confused with the unit 'gram'). Its value varies depending on the distance from the centre of the earth.

Table A1 Approximate values

Height above the poles (km)	g (m/s^2)
0 (sea level)	9.840
10	9.809
100	7.350
400,000	0.00242
Average value on earth	9.81

Introduction to Soil Mechanics, First Edition. Béla Bodó and Colin Jones.

When a body of mass M falls freely towards the earth at acceleration g (m/s^2), then there must be a force acting on it to produce this acceleration. Diagrammatically:

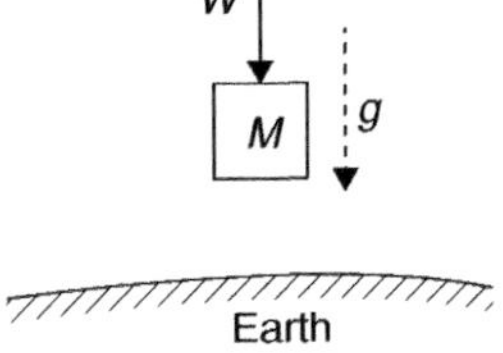

This gravitational force is called 'Weight' (W).

Figure A1

Units of *M* and *W*

The S.I. (System International d'Unites = International system of units) unit of mass is the gram or kilogram. In order to avoid confusing g (the acceleration) with gram, the kg unit is to be adopted in this appendix.

The S.I. unit of force is the Newton (N). It may be defined as the force which accelerates a mass of 1 kg at 1 m/s^2. Diagrammatically:

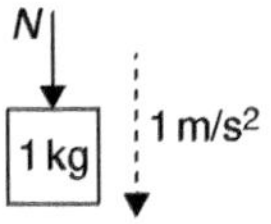

$$\therefore \quad \mathrm{N} = 1\left(\frac{\mathrm{kg\,m}}{\mathrm{s}^2}\right) \quad \therefore \quad \frac{\mathrm{m}}{\mathrm{s}^2} = \frac{\mathrm{N}}{\mathrm{kg}}$$

Figure A2

One Newton is called absolute unit of force, because its value does not depend on its location in space. Therefore, it can be used as the unit of force absolutely anywhere in the universe.

In order to know the weight of 1 kg at a particular place, it has to be multiplied by the value of the gravitational acceleration of that locality. Diagrammatically:

$$W = g \quad \mathrm{N/kg}\left(\text{or } \mathrm{m/s^2}\right)$$

Figure A3

For a mass M kg:

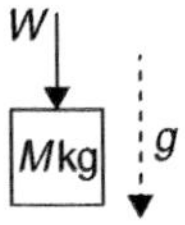

$$W = Mg \quad \text{Newton}$$

Figure A4

Table A.2 contains the weight of $1\,m^3$ of concrete at the altitudes listed in Table A.1, assuming its average mass as $M = 2400\ kg/m^3$

Table A.2

Altitude (km)	Mass (kg)	*g* (N/kg)	*W* (N)	*W* (kN)
0	2400	9.840	23616	23.62
10	2400	9.809	23542	23.54
100	2400	7.350	17640	17.64
400000	2400	0.00242	5.808	0.00581
Average value on Earth	2400	9.81	23544	23.54

There is insignificant variation in g over the Earth's surface, hence the use of the average value $g = 9.81\ m/s^2$ is satisfactory. From design and construction point of view, the gravitational force is of interest, being the load (weight) acting on structures.

The use of $g = 10\ m/s^2$ is acceptable in order to simplify engineering calculations. This approximation yields larger design-loads, hence is on the safe side.

Example A1

A 5 m long universal column section of mass 634 kg/m is supporting a load of 200 kN. Calculate the total load (P) in the column's base.

Weight if the column: $W = 5.634 \times 9.81 = 31098\ N \approx 31\ kN$

Total load on base: $P = 200 + 31 = 231\ KN$

Question: Why is it necessary to tabulate the mass density (kg/m^3) of materials?
Reason: Apart from the variation of g from place to place, the main reason is the method of measurement.

The mass of materials is normally 'weighed' by balancing it against another mass. As both masses are accelerated equally, at the same locality, they balance each other irrespectively of gravity as long as $M_1 = M_2$.

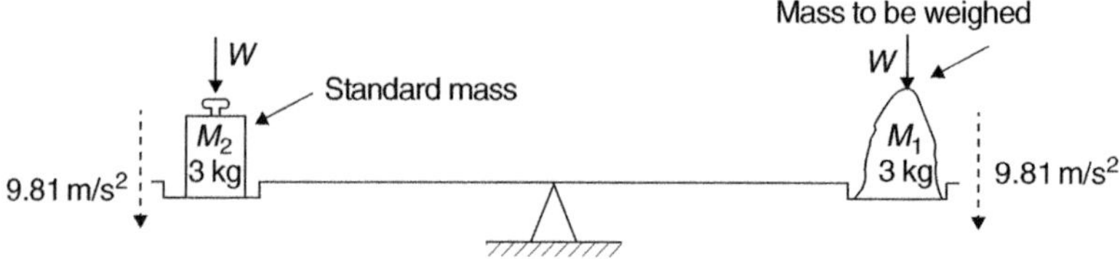

Figure A5

It is obvious from the figure, that the gravitational acceleration acts on both masses equally, hence the weight force is:

$$W = 9.81\,M_1 = 9.81\,M_2$$
$$\therefore M_1 = M_2$$

So, by weighing a body this way, we measure only its mass and not its weight.

Measurement of weight (*W*)

As the extension of a spring is independent of gravity, *W* can be measured directly by suspending the mass at the end of a spring, having stiffness *q* (N/mm).

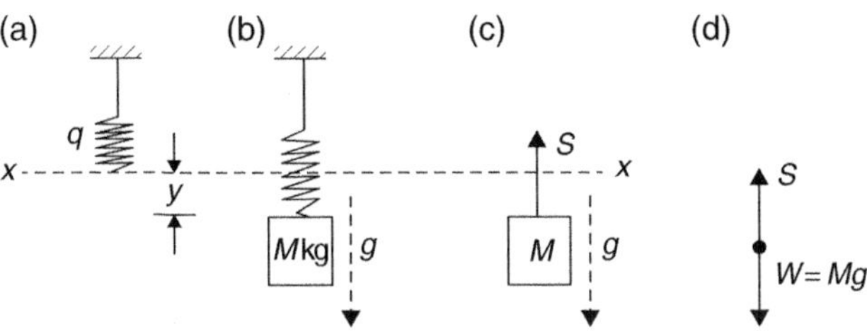

Figure A6

Fig. (a): Unloaded spring
Fig. (b): Loaded spring stretches *y* mm
Fig. (c): Force in spring: $S = qy$
Fig. (d): Weight equals to spring force $S = W = Mg$
or $qy = W = Mg$

Therefore, $\boxed{W = qy}$ and $\boxed{M = \frac{qy}{g}}$

It is physically inconvenient to arrange the measurement of either weight or mass by this method, compared to the balancing of masses. This is why mass density is tabulated.

Example A2

A large stone is attached to a spring of $q = 25$ N/mm and the extension measured as $y = 20$ mm.

Determine its weight and mass:

Weight: $W = qy = 25 \times 20 = 500\,\text{N} = 0.5\,\text{kN}$

Mass: $M = \frac{W}{g} = \frac{500}{9.81} = 51\ \text{kg}$

Appendix B

Units, Conversion Factors and Unity Brackets

The S.I. system of units has been in use for some years now. Inspite of this, it is often necessary to refer back to the imperial unit system when dealing with old drawings, books or for some other purpose. For this reason, units relevant to soil mechanics, and their conversion from one system to the other are listed in this Appendix. Also, the unity bracket method of conversion, initiated by A.C. Walshaw, is introduced.

Basic units

The S.I. and the imperial systems are based on the following three units:

Table B1 (a)

	S.I.	Imperial
Length (*L*)	Metre (m)	Foot (ft)
Mass (*M*)	kilogramme or kilogram (kg)	Pound (lb)
Time (*T*)	Seconds (s)	Seconds (s)

Note: The unit of force (*F*), in general, is F Newton.

The unit of gravitational force is $W=Mg$ Newton. The imperial equivalent is the pound force (lbf) at $g=32.2$ ft/s^2.

Table B1 (b)

Force (*F*)	Newton (N)	Pound force (lbf)
Weight (*W*)	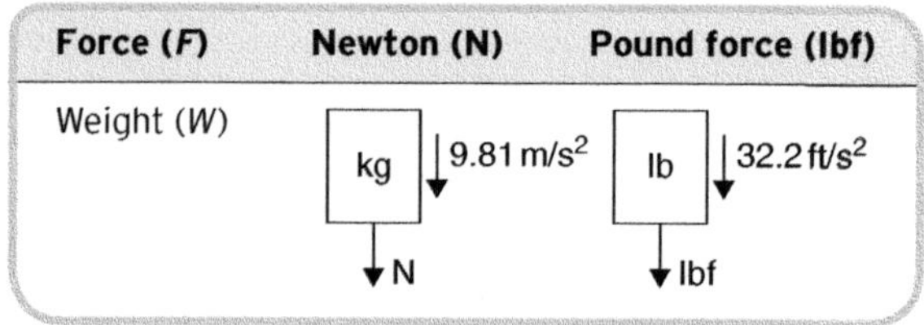	

Introduction to Soil Mechanics, First Edition. Béla Bodó and Colin Jones.

Multiplier prefixes

The basic units may be made larger or smaller by multiplying them by ± powers of ten. The standard multipliers, their symbols and names are:

Table B.1 (c)

Name	Symbol	Multiplier	Conversion factors
tera	T	10^{12}	} Not normally used in soil mechanics
giga	G	10^{9}	
mega	M	10^{6}	Mega gram: 1 Mg $=10^6$ g
kilo	k	10^{3}	Kilometre: 1 km $=10^3$ m
hecto	h	10^{2}	Hectometre: 1 hm $=10^2$ m
deca	da	10	Decametre: 1 dam $=10$ m
deci	d	10^{-1}	Decimetre: 1 dm $=10^{-1}$ m
centi	c	10^{-2}	Centimetre: 1 cm $=10^{-2}$ m
milli	m	10^{-3}	Millimetre: 1 mm $=10^{-3}$ m
micro	μ	10^{-6}	Micrometre: 1 μm $=10^{-6}$ m
nano	n	10^{-9}	} Not normally used in soil mechanics
pico	p	10^{-12}	
femto	f	10^{-15}	
atto	a	10^{-18}	

Note: M in this table, must not be confused with mass!

Unity bracket

Conversion factors may be transformed into unity by transferring the quantity from one side of the equality sign to the other and enclose it in a square bracket, as illustrated by the following simple example:

From Table B.1: Conversion factor: 1 cm = 0.01 m. The unity bracket may be written in two ways, as the reciprocal of unity is still unity.

Either $$1=\left[\frac{0.01\ \text{m}}{\text{cm}}\right] \quad \text{or} \quad 1=\left[\frac{\text{cm}}{0.01\ \text{m}}\right]$$

The expressions in brackets are therefore equal to one. The significance of this will be explained later.

Application of the unity brackets

The procedure is based on the two simple facts:

1. Unity multiplied by unity remains unity.
2. Unity raised to any power remains unity.

Because each bracket equals unity, they may be multiplied together or exponentiated and the final result is equated to unity. New conversion factors can be formulated in this way.

Table B.2 Conversion Factors

Quantity	Unit System				Conversion factors
	S.I.		Imperial		
Length	Metre	m	Foot	ft	1 m = 3.281 ft
					1 ft = 0.3048 m
			Inch	in	1 ft = 12 in
	Millimetre	mm			1 in = 25.4 mm
	Kilometre	km	Mile (statute)	mile	1 mile = 5280 ft
					1 mile = 1.609 km
			Yard	yd	1 yd = 3 ft
					1 yd = 0.9144 m
Area	Area = 10 m^2	are			1 are = 100 m^2
	Hectare	ha			1 ha = 10^4 m^2
	ha 100 m		Acre	acre	1 Acre = 4840 yd^2
	100 m				1 Acre = 0.4047 ha
	sq. kmetre	km^2	sq. mile	$mile^2$	1 $mile^2$ = 2.59 km^2
	sq. metre	m^2	sq. feet	ft^2	1 m^2 = 10.764 ft^2
	sq. cm	cm^2	sq. inch	in^2	1 cm^2 = 0.155 in^2
	sq. mm	mm^2			1 in^2 = 645.2 mm^2
Volume and capacity	Litre	*l*	UK gallon	gal	1 gal = 4.546 *l*
	l 10 cm 10 cm 10 cm		US gallon		1 gal = 3.785 *l*
					1 *l* = 1000 cm^3
			cu. feet	ft^3	1 ft^3 = 28.32 *l*
	Millilitre	ml			ml = 1 cm^3
	ml 1 cm 1 cm 1 cm		cu. inch	in^3	1 in^3 = 16.39 ml
		m^3		ft^3	1 m^3 = 35.32 ft^3
					1 m^3 = 1000 *l*
				yd^3	1 yd^3 = 0.7646 m^3
		ml	Fluid ounce	fl. oz	1 ml = 0.0352 fl. oz
		l	Pint	Pint	1 *l* = 1.76 pint

(continued)

Table B.2 (continued)

Quantity	S.I.		Imperial		Conversion factors
	Unit System				
Mass	gramme	g	Ounce	oz	1 oz = 28.349 g
	kilogramme	kg	Pound	lb	1 lb = 0.4536 kg
	tonne (1. Mg)	tonne	ton	ton	1 tonne = 0.9842 ton
					1 tonne = 1000 kg
			Hundredweight	cwt	1 cwt = 50.802 kg
			Stone	Stone	1 stone = 14 lb
		kg		Stone	1 stone = 6.35 kg
Force	Newton	N	Pound force	lbf	1 lbf = 4.45 N
			Ton force	tonf	tonf = 2240 lbf
	kilonewton	kN			1 kN = 1000 N
				tonf	1 tonf = 9.968 kN
Pressure and stress	Newton/m^2	N/m^2	Poundforce / ft^2	lbf / ft^2	lbf / ft^2 = 47.88 N/m^2
	pascal	Pa			1 Pa = 1 N/m^2
	bar	b			1 b = 10^5 N/m^2
	millibar	mb			1 mb = 10^2 N/m^2
	kilonewton/m^2	kN/m^2	Tonforce / ft^2	tonf / ft^2	1 tonf / ft^2 = 107.25 kN/m^2
	Newton/mm^2	N/mm^2	Poundforce / in^2	lbf / in^2	N/mm^2 = 145.14 lbf / in^2
		N/mm^2	Tonforce / in^2	tonf /in^2	tonf / in^2 = 15.44 N/mm^2
	meganewton	MN			1 MN = 10^6 N
		mm of water			1 mm H_2O = 9.81 N/m^2
		mm of mercury			1 mm Hg = 133.3 N/m^2
		Average atmospheric pressure at sea level			1 atm = 101.33 Pa; 1 atm = 14.7 lbf / in^2
Density (P)	kilogram / m^3	kg/m^3	Pound /ft^3	lb / ft^3	1 lb / ft^3 = 16.02 kg/m^3
	kilogram / litre	kg / l	Pound /in^3	lb / in^3	1 lb / in^3 = 27.68 Kg / l
	gram / m^3	g/m^3			1 lb / in^3 = 10^6 × 27.68 g/m^3
Unit Weight (γ)	kilonewton / m^3	kN/m^3	Poundforce / ft^3	lb / ft^3	1 kN/m^3 = 6.363 lb / ft^3
			Poundforce / in^3	lb / in^3	1 kN/m^3 = 3.683 lb / m^3

For water: $\gamma_w = 9.81\,kN/m^3 = 62.44\,lb_f/ft^3$

$$\rho_w = 1\,kg/m^3$$

Example B1

Prove that:

a) $1\,\text{tonf} = 9.968\,\text{kN}$

b) $1\,\text{N}/\text{mm}^2 = 145.14\,\text{lbf}/\text{in}^2$

c) $\gamma_w = 9.81\,\text{kN}/\text{m}^3 = 62.44\,\text{lbf}/\text{ft}^3$

a) Step 1

Start from another known conversion factor containing tonf. Choose say: 1 tonf=2240 lbf to get the bracket:

$$1 = \left[\frac{2240\,\text{lbf}}{\text{tonf}}\right]$$

Another factor had to be found, which contains 'lbf' in order to eliminate it from the first bracket. Choose, say:

$$1\,\text{lbf} = 4.45\,\text{N}$$

Hence,

$$1 = \left[\frac{4.45\,\text{N}}{\text{lbf}}\right]$$

Step 2

Multiply the brackets together and cancel lbf:

$$1 = \left[\frac{2240\,\text{lbf}}{\text{tonf}}\right]\left[\frac{4.45\,\text{N}}{\text{lbf}}\right] = \frac{2240 \times 4.45\,\text{N}}{\text{tonf}}$$

But, the result should be in terms of kN, so N has to be eliminated by:

$$1\,\text{kN} = 1000\,\text{N}$$

$$\text{or } 1 - \left[\frac{\text{kN}}{1000\,\text{N}}\right]$$

Step 3

Multiply the brackets together and cancel N.

$$1 = \left[\frac{2240 \times 4.45\,\text{N}}{\text{tonf}}\right]\left[\frac{\text{kN}}{1000\,\text{N}}\right] = \left[\frac{2240 \times 4.45\,\text{kN}}{1000\,\text{tonf}}\right]$$

$$= \left[\frac{9.968\,\text{kN}}{\text{tonf}}\right]$$

Converting it to factor: tonf = 9.968 kN ∴ true

Alternatively, any other initial choice may be made, say,

$$1\,\text{tonf}/\text{ft}^2 = 107.25\,\text{kN}/\text{m}^2$$

or

$$1 = \left[\frac{107.25\,\text{kN} \times \text{ft}^2}{\text{tonf} \times \text{in}^2}\right]$$

Now, m^2 and ft^2 have to be eliminated.

Choose, $1\text{ m} = 3.281\text{ ft}$

And $1 = \left[\dfrac{\text{m}}{3.281\text{ft}}\right]$

Squaring the bracket $1 = \left[\dfrac{\text{m}}{3.281\text{ft}}\right]^2 = \left[\dfrac{\text{m}^2}{10.765\text{ft}^2}\right]$

Multiply the brackets and cancel m^2 and ft^2,

$$1 = \left[\frac{107.25\,\text{kN}\,\cancel{\text{ft}^2}}{\text{tonf} \times \cancel{\text{m}^2}}\right] \times \left[\frac{\cancel{\text{m}^2}}{10.765\,\cancel{\text{ft}^2}}\right] = \left[\frac{9.963\,\text{kN}}{\text{tonf}}\right]$$

b) Choose: $1\text{lbf} = 4.45\text{N} \quad \therefore \quad 1 = \left[\dfrac{4.45\text{N}}{\text{lbf}}\right]$

Also, $1\text{in} = 25.4\text{mm} \quad \therefore \quad 1 = \left[\dfrac{25.4\text{mm}}{\text{in}}\right]$

Square the second bracket to get in^2 and mm^2

$$1 = \left[\frac{25.4\text{mm}}{\text{in}}\right]^2$$

Multiply the brackets: $1 = \left[\dfrac{4.45\text{N}}{\text{lbf}}\right] \times \left[\dfrac{\text{in}}{25.4\text{mm}}\right]^2$

$$1 = \left[\frac{4.45\text{N}\,\text{in}^2}{645.16\,\text{mm}^2 \times \text{lbf}}\right] = \left[\frac{\text{N}\,\text{in}^2}{145.\ \text{mm}^2 \times \text{lbf}}\right]$$

From which, $1\dfrac{\text{N}}{\text{mm}^2} = 145\dfrac{\text{lbf}}{\text{in}^2}$

NB: Any discrepancies is due to cumulative arithmetic errors.

c) $\gamma_w = 9.81\dfrac{\text{kN}}{\text{m}^3} = 9.81\left[\dfrac{\text{kN}}{\text{m}^3}\right]$

Choose: $1\text{lbf} = 4.45\text{N} \quad 1 = \left[\dfrac{4.45\text{N}}{\text{lbf}}\right]$

And: $1\text{kN} = 1000\text{N} \quad 1 = \left[\dfrac{1000\text{N}}{\text{kN}}\right]$

And: $1\text{m} = 3.281\text{ft} \quad 1 = \left[\dfrac{3.281\text{ft}}{\text{m}}\right]$

Multiply the brackets and eliminate kN, m^3 and N

$$\gamma_w = 9.81\left[\frac{\text{kN}}{\text{m}^3}\right] \times \left[\frac{1000\text{N}}{\text{kN}}\right] \times \left[\frac{\text{lb}_f}{4.45\text{N}}\right] \times \left[\frac{\text{m}}{3.281\text{ ft}}\right]^3$$

$$= 9.81\left[\frac{1000 \times \text{lbf}}{4.45\text{m}^3}\right] \times \left[\frac{\text{m}^3}{35.32\,\text{ft}^3}\right]$$

$$= 9.81\left[\frac{6.3624\text{ lbf}}{\text{ft}^3}\right] = 62.41\text{lbf/ft}^3 \qquad \text{True}$$

Note: Any other conversion factor not present in Tables B.1 and B.2 may be derived in this manner.

Appendix C
Simpson's Rule

This rule is referred to in connection with

a) Slope stability, where the area between the ground surface and the trial slip circle has to be determined.
b) Consolidation, where the area of the pore pressure isochrones has to be found.

The rule is applicable to any area (A), which is divided into EVEN numbers of strips of equal width.

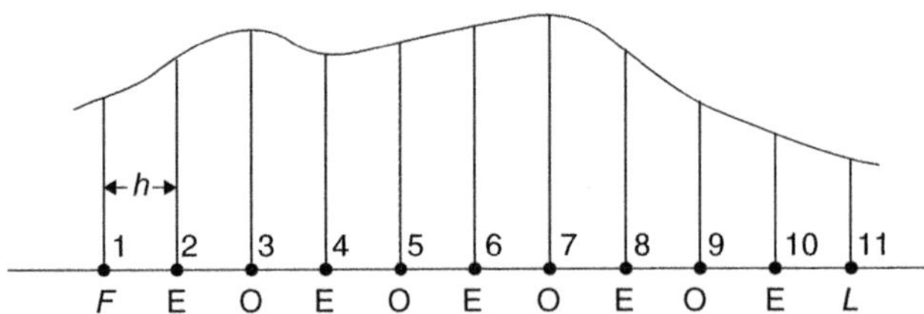

Figure C1

h = width of each strip
F = first ordinate
L = last ordinate
E = even ordinate
O = odd ordinate

The rule:

$$A = \frac{h}{3}(F + L + 2\Sigma\text{O} + 4\Sigma\text{E}) \tag{C1}$$

$2\,\Sigma$O = twice the sum of the odd ordinates
$4\,\Sigma$E = four times the sum of the even ordinates

Accuracy can be increased by choosing a large number of strips.

Introduction to Soil Mechanics, First Edition. Béla Bodó and Colin Jones.

Example C1

Evaluate the area under the $T_v = 0.3$ isochrone on Graph 7.3
Only the upper portion is reproduced here, as the figure is symmetrical.

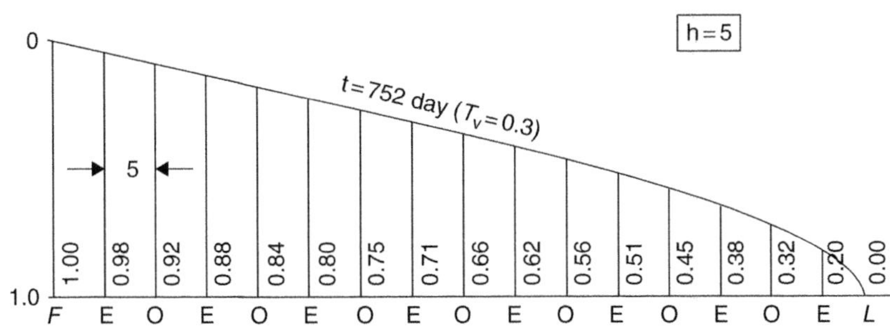

Figure C2

$F + L = 1 + 0 = 1$

$\sum O = 0.92 + 0.84 + 0.75 + 0.66 + 0.56 + 0.45 + 0.32 \quad = 4.5$

$\sum E = 0.98 + 0.88 + 0.80 + 0.71 + 0.62 + 0.51 + 0.38 + 0.20 = 5.08$

$2\sum O = 9$

$4\sum E = 20.32$

$$\frac{A}{2} = \frac{5}{3}(1 + 9 + 20.32) = 51$$

Therefore, area of the isochrone: $A = 2.51 = 102\,\text{cm}^2$
The same area measured by planimeter $= 100\,\text{cm}^2$

Mean ordinate rule

This rule provides a convenient procedure, applicable to the method of slices, for the evaluation of the:

1. Cross-section area (A) of the slope examined for stability
2. Position of its centroid ($\overline{x}$)
3. Disturbing moment (M_D).

As in the method of slices, the mid-ordinate of each slice and its distance from the vertical line, drawn through centre of rotation (see Graph 10.2), is to be used in order to determine the three unknowns.

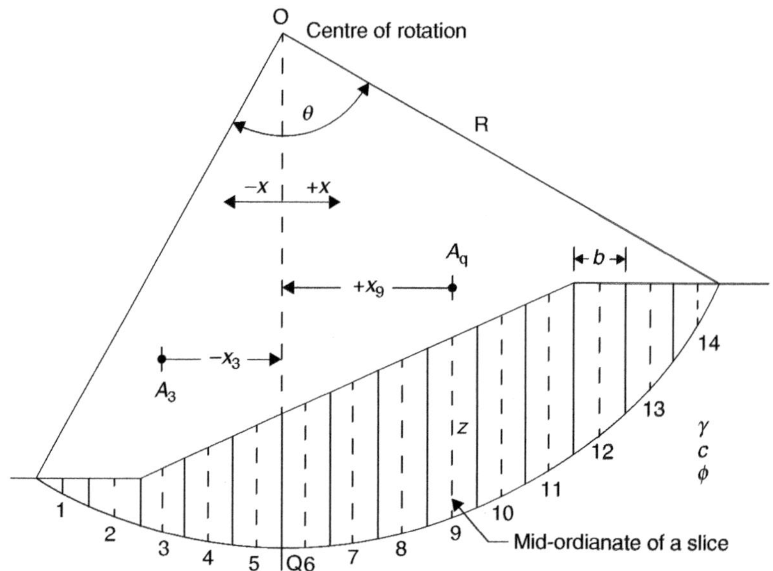

Figure C3

Procedure:

Step 1: Divide the area into strips of equal width (b) and measure the length (z) of each mid-ordinate. This gives:

The area of a slice: $A_n = bz_n$

and the total area: $\boxed{A = b\sum z_n}$ (C2)

Step 2: Sum the moment of each area about the centre of rotation. Clockwise moments are considered positive and anticlockwise ones negative.

$$\sum(-M) = A_1x_1 + A_2x_2 + A_3x_3 + A_4x_4 + A_5x_5 + \ldots.$$
$$\sum(+M) = A_6x_6 + A_7x_7 + A_8x_8 + A_9x_9 + A_{10}x_{10} + A_{11}x_{11} + A_{12}x_{12} + A_{13}x_{13} + A_{14}x_{14} + \ldots.$$

Step 3: The moment of the total area about the centre of rotation: $M = A\bar{x}$. This equals to the sum of the $\pm$moments. Therefore,

$$M = \sum(+M) + \sum(-M)$$

or $A\bar{x} = \sum(+M) + \sum(-M)$

Hence, $$\boxed{\bar{x} = \frac{\sum(+M) + \sum(-M)}{A}} \quad \text{(C3)}$$

Note: that the procedure is greatly simplified if the width of each slice is chosen to be unity. Then the total area equals to the sum of the mean ordinates, that is:

$$\boxed{A = \sum z_n} \qquad \text{(C4)}$$

Taking the unit weight (γ) into account, the total weight of 1 m long slope is given by:

$$W = \gamma A \quad \text{(or)} \quad \boxed{W = \gamma b \sum z} \qquad \text{(C5)}$$

Alternatively, the disturbing moment due to the weight of soil right of the centre of rotation is:

$$M_d = +Wx_n = \gamma \sum (+M)$$

Similarly, the resisting moment due to the weight of soil left of the centre of rotation is:

$$M_r = -Wx_n = \gamma \sum (-M)$$

Resultant disturbing moment: $\boxed{M_D = M_d - M_r}$ (C6)

Centroid: $$\bar{x} = \frac{M_D}{W} = \frac{\gamma(+M) + \gamma(-M)}{\gamma A}$$

Hence, $$\boxed{\bar{x} = \frac{M_D}{W} = \frac{\sum(+M) + \sum(-M)}{A}}$$ As before

Example C2

Using the results of Example 10.1 calculate $\bar{x}$.

From Table 10.1:

$$A = \sum z = 81.39\,\text{m}^2$$
$$+Wx = 6426.65\,\text{kNm}$$
$$-Wx = 828.25\,\text{kNm}$$

But $$\sum(+M) = \frac{+Wx}{\gamma} = \frac{6426.65}{19} = 338.2\,\text{m}^2\,\text{m}$$

And $$\sum(-M) = \frac{-Wx}{\gamma} = \frac{828.25}{19} = -43.6\,\text{m}^2\,\text{m}$$

Therefore, from (C3): $$\bar{x} = \frac{338.2 - 43.6}{81.39} = 3.6\,\text{m} \approx 3.54\,\text{m}$$

Evaluation of Angle θ

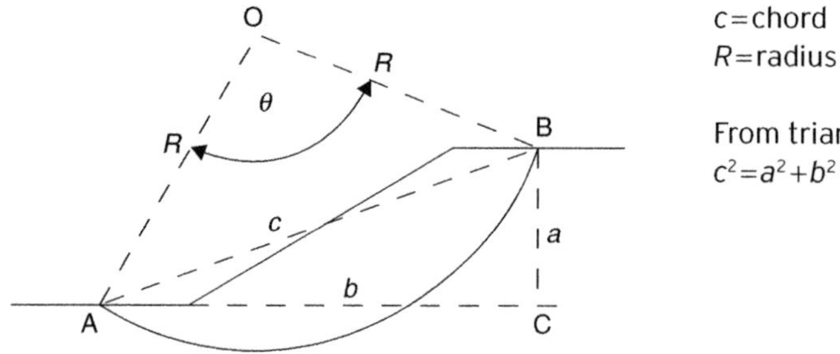

c=chord
R=radius

From triangle ABC:
$c^2=a^2+b^2$

Figure C4

By the consine rule:

$$c^2 = R^2 + R^2 - 2R \times R \times \cos\theta$$
$$c^2 = 2R^2 - 2R^2 \cos\theta$$
$$c^2 = 2R^2(1 - \cos\theta)$$
$$\cos\theta = 1 - \frac{c^2}{2R^2}$$

Therefore,

$$\boxed{\theta = \cos^{-1}\left(1 - \frac{c^2}{2R^2}\right) = \cos^{-1}\left(1 - \frac{a^2 + b^2}{2R^2}\right)} \tag{C7}$$

Appendix D

Resultant Force and Its Eccentricity

When a retaining wall is acted upon by a system of forces and moments, the system can be replaced by a single force (Resultant R) acting some distance away from the centroid of the base area. The resultant, therefore, acts eccentrically on the base of eccentricity e. The determination of R and e depends on the configuration of the loading system. Three typical problems are discussed below:

1. A given force W acting at distance x from the centroid (CG) of a rectangular base.

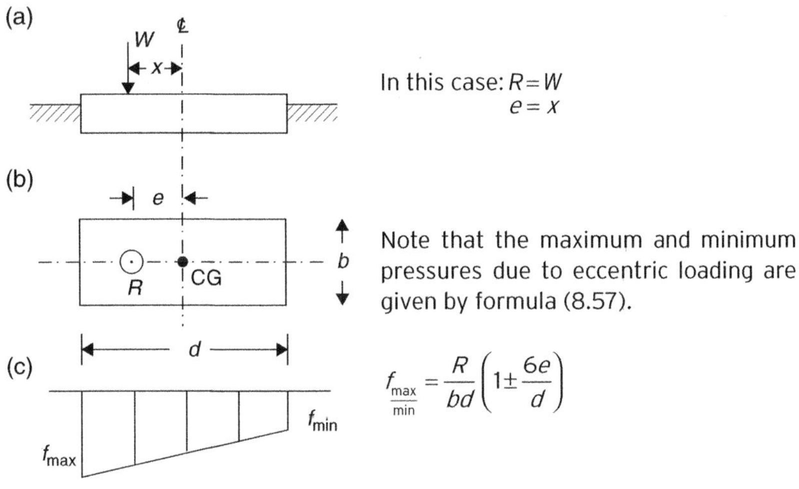

$$f_{\substack{max \\ min}} = \frac{R}{bd}\left(1 \pm \frac{6e}{d}\right)$$

Figure D1

2. A given force W acts at the centroid, as well as a moment (M).

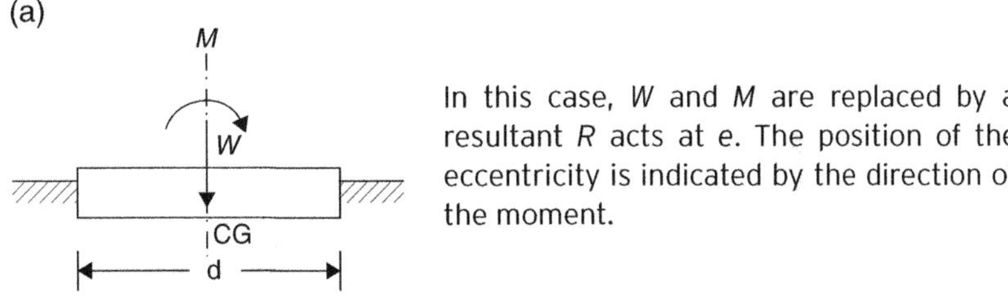

In this case, W and M are replaced by a resultant R acts at e. The position of the eccentricity is indicated by the direction of the moment.

Introduction to Soil Mechanics, First Edition. Béla Bodó and Colin Jones.

Step 1: Replace M by two forces, each equals to W, at distance e apart and place this pair of forces on diagram D2(b) as shown:

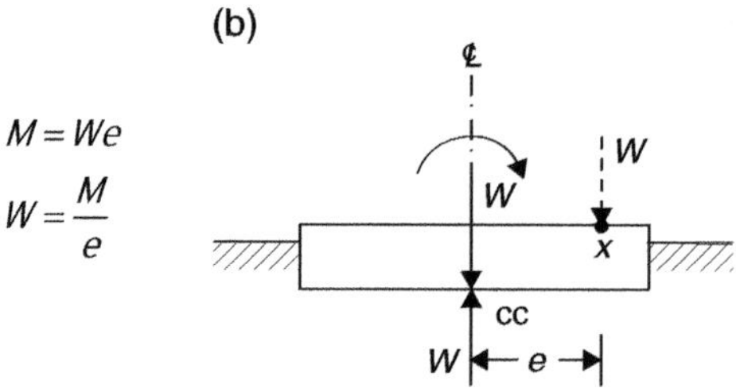

Step 2: The two forces at the centroid are equal and opposite, hence cancel each other. This leaves the resultant acting eccentrically at x.

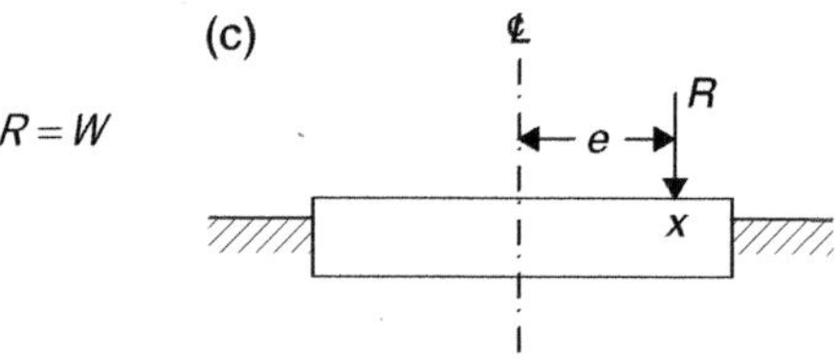

Figure D2 (continued)

Formula (8.57) can now be applied.

Example D1

Figure D3 shows a 1.2 m × 3 m raft foundation transmitting 600 kN central load (self-weight inclusive) and a moment of 400 kNm to the soil. Calculate the maximum and minimum pressures under the base.

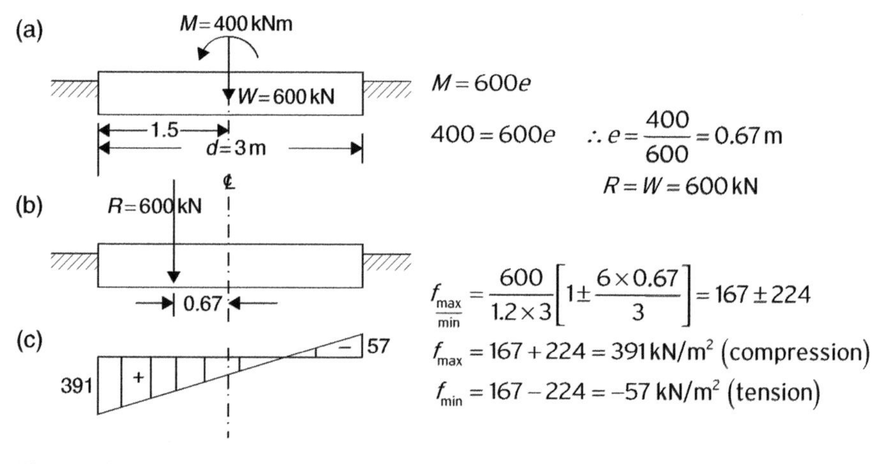

$$M = 600e$$

$$400 = 600e \quad \therefore e = \frac{400}{600} = 0.67\,\text{m}$$

$$R = W = 600\,\text{kN}$$

$$f_{\frac{max}{min}} = \frac{600}{1.2 \times 3}\left[1 \pm \frac{6 \times 0.67}{3}\right] = 167 \pm 224$$

$$f_{max} = 167 + 224 = 391\,\text{kN/m}^2\ \text{(compression)}$$

$$f_{min} = 167 - 224 = -57\,\text{kN/m}^2\ \text{(tension)}$$

Figure D3

When a footing or a retaining wall is subjected to several forces, then the resultant and its line action that is the eccentricity can be calculated as in Example 8.8 General case:

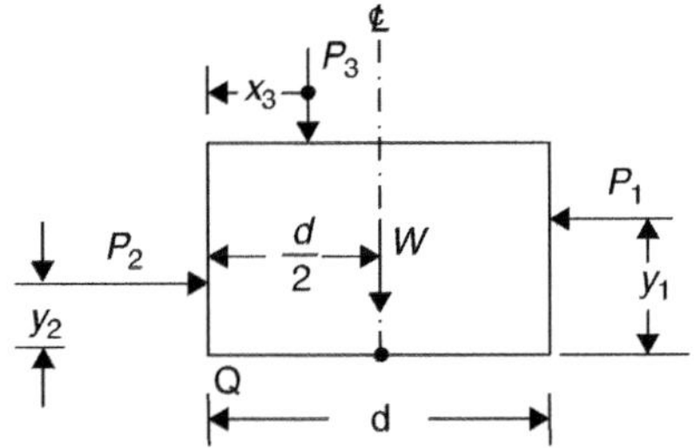

R = sum of vertical forces in this diagram:

$$R = W + P_3$$

Figure D4

For eccentricity, sum the moments of the forces shown about a point, say Q:

$$\overset{+}{\curvearrowleft}\Sigma M_Q = W\frac{d}{2} + P_2 y_2 + P_3 x_3 - P_1 y_1$$

This has to be equal to the moment of R about the same point Q, for equilibrium.

$$\overset{+}{\curvearrowleft}\Sigma M_Q = Rx$$

Equating, $$Rx = \frac{Wd}{2} + P_2 y_2 + P_3 x_3 - P_1 y_1$$

$$x = \frac{\frac{Wd}{2} + P_2 y_2 + P_3 x_3 - P_1 y_1}{W + P_3}$$

From which, $$e = \frac{d}{2} - x$$

The simplified diagram can now be drawn and formula (8.57) applied.

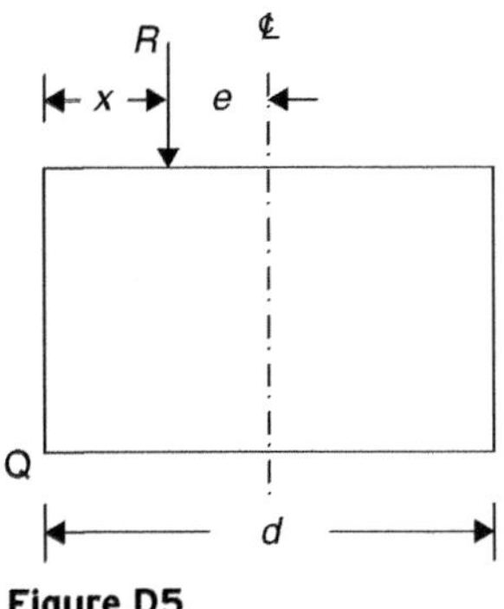

Note: The procedures are applicable to any combination of forces and moments. (see texts on statics).

Figure D5

Appendix E
References

Atkinson, G.H. and Bransby, P.L. (1978) The mechanics of soil - an introduction to critical state soil mechanics. McGraw-Hill.

Barnes, G.E. (1995) Soil Mechamics, Principles and Practice. Macmillan Press Ltd.

Bishop, A.W. and Henkel, D.J. (1974) The Measurement of Soil Properties in the Triaxial Test (Second Edition). Edward Arnold, London.

Capper, P.L., Cassie, W.F. and Geddes, J.D. (1974) Problems in Engineering Soils. E & F.N. Spon.

Cedergren, H.R. (1977) Seepage and Flow Nets (Second Edition). John Wiley & Sons, Inc., New York.

Craig, R.F. (2004) Soil Mechanics (Fifth Edition). Chapman & Hall, London.

Das, B.M. (1985) Principles of Geotechnical Engineering (Second Edition). PWS-Kent Publishers.

Das, B.M. (1985) Advanced Soil Mechanics. McGraw-Hill, London.

Jumikis, A.R. (1962) Soil Mechanics. D. Van Norstad Co. Inc., New York.

Harr, M.E. (1966) Foundations of Theoretical Soil Mechanics. McGraw-Hill, New York.

Head, K.H. (1982) Soil Laboratory Testing Vol. 1, 2 and 3. ELE International Ltd. Pentech Press Ltd., Plymouth.

Kezdi, A. (1958) Talajmechanika 1 & 2 (Second Edition). Tankonyvkiado, Budapest.

Kezdi, A. (1964) Talajmechanika Praktikum (Second Edition). Tankonyvkiado, Budapest.

Lambe, T.W. and Whitman, R.V. (1969) Soil Mechanics. John Wiley & Sons, Inc., New York.

Rosenak, S. (1966) Soil Mechanics. B.T. Batsford Ltd, London.

Resnick, R. and Halliday, D. (1966) Physics (Parts I and II). John Wiley & Sons, Inc.

Schofield, A. (2005) Disturbed soil properties and geotechnical design. Thomas Telford Publishing.

Schofield, A.N. (1963) Wroth CP. Critical state soil mechanics. McGraw-Hill.

Scott, C.R. (1969) An Introduction to Soil Mechanics and foundations. Applied Science Publishers, London.

Smith, G.N. (1990) Elements of Soil Mechanics (Sixth Edition). BSP Professional Books, London.

Smith, I. (2006) Elements of Soil Mechanics (Eighth Edition). Blackwell Publishing.

Smith, M.J. (1996) Examination Subject for Civil Engineers, Soil Mechanics. MacDonald & Evans Ltd, London.

Szechy, K. (1963). Alapozas I & II Muszaki Konyvkiado, Budapest.

Sutton, B.H.C. (1986) Solving Problems in Soil Mechanics. Pitman Publishers Ltd, London.

Terzaghi, K. and Peck, R.B. (1976) Soil Mechanics in Engineering Practice. Wiley, New York.

Tomlinson, M.J. (1986) Foundation Design and Construction (Fifth Edition). Longman Scientific and Technical.

U.S. Waterways Experimental Station (1949) Report No. 1., Tech. Memo. No. 3-286. Vicksburg, Mississippi, USA.

Introduction to Soil Mechanics, First Edition. Béla Bodó and Colin Jones.

Vickers, B. (1978) Laboratory work in Civil Engineering Soil Mechanics. Crosby Lockwood Staples/ Granada Publishing, London

Whitlow, R. (1992) Basic Soil Mechanics (Second Edition). Longman Group UK Ltd, Harlow.

Wilun, Z. and Starzewski, K. (1975) Soil Mechanics in Foundation Engineering I & II. Surrey University Press/International Textbook Ltd, London.

Index

Introduction to Soil Mechanics, First Edition. Béla Bodó and Colin Jones.